BIOLOGY &

its relation to mankind

5th Edition

A. M. WINCHESTER
University of Northern Colorado

D. VAN NOSTRAND COMPANY
New York/Cincinnati/Toronto/London/Melbourne

D. Van Nostrand Company Regional Offices:
New York   Cincinnati   Millbrae

D. Van Nostrand Company International Offices:
London   Toronto   Melbourne

Published by D. Van Nostrand Company
450 West 33rd Street, New York, N.Y. 10001

Published simultaneously in Canada by
Van Nostrand Reinhold Ltd.

10 9 8 7 6 5 4 3 2

Preface

The Fifth Edition of *Biology and Its Relation to Mankind* retains the straightforward treatment of basic biological principles that has characterized previous editions while incorporating the results of recent and rapidly expanding biological research on topics of special interest to the student: high-energy radiation, human disease, drug use and abuse, human sexuality, and ecological problems and their solutions.

The book is designed for a full-year course in biology, but it can easily be adapted to shorter courses by the omission of selected chapters or portions of chapters. Many photographic illustrations and drawings are included in each chapter as an aid to understanding the text. Questions are provided at the end of each chapter to allow the student to assess his comprehension of the chapter and to stimulate further thought. Suggestions for further reading are also provided at the end of each chapter for the student who wishes to explore a topic more fully.

The Fifth Edition has an entirely new section on drug use and abuse in Chapter 29. In Chapter 36, a section on the prenatal detection of genetic defects has been added. Chapter 35, Human Sexuality, is a new chapter dealing with sexual identity, forms of sexual expression, and birth-control methods. Another new chapter (Chapter 39) reviews ecological problems and the consequences of overpopulation.

Additional new material in the Fifth Edition deals with the effects of high-energy radiation (Chapter 4), the role of viruses in disease

(Chapter 14), practical applications of plant hormone research (Chapter 22), hereditary blood disorders (Chapter 30), the role of protein in nutrition (Chapter 31), the hormonal regulation of the menstrual cycle (Chapter 33), venereal disease (Chapter 34), sex determination (Chapter 36), and natural selection mechanisms (Chapter 40).

The basic biological principles which must be understood if new directions in biology are to be appreciated have not been neglected. Fundamental topics such as cell structure and function, cell duplication, the cell's genetic machinery, the kingdoms of living organisms, and basic vertebrate physiology recieve their full share of attention.

The author is indebted to those biology instructors and their students who have made comments on the previous editions. The author expresses his appreciation for the many constructive suggestions, a good number of which have been incorporated into this edition.

Contents

v

1

The Nature of Living Matter

The living things which inhabit the earth exist in great variety. The range of size in living organisms is enormous. There are tiny micro-organisms which can be seen only with the aid of a high-powered microscope and there are huge whales and elephants. The range in shape is just as striking. There are filaments of algae which are several feet long but so slender that they are barely visible, and puffer fish which can inflate themselves into a ball. Habitat can be just as variable. We have become aware of the impor-tance of habitat to the survival of living orga-nisms in recent years because the deterioration of habitat, which has resulted from pollution, the energy crisis, and overpopulation, has had adverse effects on many species. Life is abundant on the earth. No matter where you go you will find some form of life existing there. Living organisms are found in the cold of the Antarctic, the heat of the tropical deserts, and in the high pressures of the ocean depths.

Biology is the field of science dedicated to the study of these varied forms of life. No matter how varied living organisms may be, however, everything which is alive shares certain charac-teristics, which are not found in nonliving things, with all other forms of life. The distinctions be-tween living and nonliving things might appear easy to make. A chipmunk on a boulder is defi-nitely alive. He is active, he eats peanuts, and he may jump away quickly if you get too close. The boulder exhibits none of these properties. There are times, however, when confusion may arise.

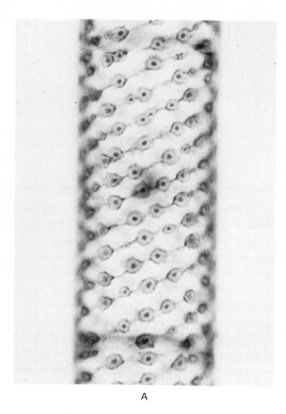

A

B

FIG. 1.1 Life exists in a great variety of shapes and sizes. The green alga *Spirogyra* is a long filament which is so small that it can be seen clearly only through a microscope. The elephant has a typical mammalian shape and weighs several tons. Both share many characteristics of living organisms.

Barnacles on a pier at the beach sometimes appear to be little pieces of rock, but barnacles are alive. The dried-up remains of the resurrection fern may appear to be dead, but after a good rain this apparently lifeless plant will open up and begin to grow. If a man stops breathing and his heartbeat ceases, the life processes can sometimes be restored by adrenalin injections, heart massage, or artificial respiration. A handful of dirt from the bottom of a dried pond appears lifeless, but if a bit of this dirt is placed in sterile water, a large variety of active organisms will appear in the water within a few days.

There are two quite different categories of nonliving matter, but the English language has no words to distinguish them. First, there are things which have never been alive, such as the boulder, a piece of turquoise, or a clay pot. Then there are things which have been alive but which are no longer living. A dog struck by a car on the highway, an Egyptian mummy, or a preserved frog in a laboratory all once exhibited the properties which characterize life. These "once lived" things will eventually deteriorate, and their material becomes indistinguishable from the "never lived" category.

THE NATURE OF LIVING MATTER

Living matter shows certain differences from nonliving matter, and a tabulation of these may serve as a basis of judgment in cases of doubt.

Made of Protoplasm

Everything which is alive contains a distinctive substance known as **protoplasm.** Protoplasm is viscid and it may contain many suspended solids and fibers. Protoplasm has the same basic nature whether it is found in the cells of tree leaves, in a bacterium, or in a cell from your skin. Protoplasm is made of a combination of proteins, carbohydrates, lipids (fatty materials), water, and nucleic acids (the material of which genes and ribosomes are made). We do not find this combination naturally in the "never lived" category. These substances are present in the "recently died" category, but they do not function and they soon break down. A scientist can assemble these compounds in the laboratory, but he can not create life. The com-

FIG. 1.2 A chipmunk on a boulder is definitely alive and the boulder is nonliving, but it is not always so easy to distinguish between living and nonliving matter.

ponents of protoplasm must be put together in a certain way to create life, and under present conditions of the earth, this can be done only by other living things. Some investigators have been able to synthesize some of the elemental substances which are found in living protoplasm,

A

B

FIG. 1.3 The dried-up plant on the left appears to have lost all characteristics of life, but after a good rain it becomes fresh and green. This plant is called the resurrection fern.

and some of these substances have been placed in living cells where they function normally, but this is still far from the creation of life.

While the protoplasm of all living matter is composed of the same basic components, these components can be assembled in many different ways. Even within the same species there are differences in the nature of the protoplasm taken from different individuals. The protoplasm of your cells has a unique assemblage of components which is not duplicated in anyone else, unless you have an identical twin. Identical twins start life as one cell, the fertilized egg, which splits to form two different individuals sometime during early embryonic development. The fertilized egg contains the genes, the master plans, which direct the production of protoplasm, and these genes are passed to both embryos.

Humans, as well as other vertebrate species, have the ability to recognize their own proto-plasm and to reject protoplasm of other organisms. Transplantation of organs is possible only when this rejection mechanism is depressed. A person can accept a kidney from an identical twin without a depression of the rejection mechanism. The recipient's body recognizes the transplanted protoplasm as being the same as his own. An attempted skin graft is used as a test to determine if a pair of twins are identical. A skin graft from one twin will be accepted by the other if they are identical twins, but the skin graft will be rejected if they are fraternal twins which developed from two different fertilized eggs.

Has Cellular Organization

Living matter is generally organized into small units known as cells. Each cell has certain components and is a life unit. The cellular nature

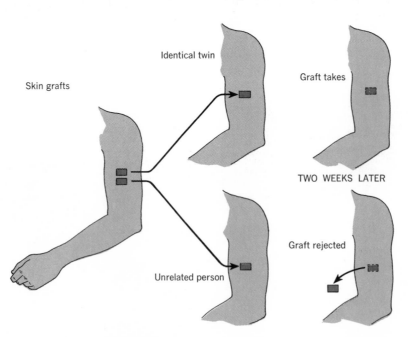

FIG. 1.4 Human skin grafting demonstrates the differences in protoplasm among different members of the same species. Donor and recipient must be identical twins to permit successful skin grafting from one to the other.

FIG. 1.5 A rapid homeostatic response is shown by the leaves of the sensitive plant. When the tip of the stem is touched, the leaves fold up instantly. This reaction protects the leaves from grazing animals which are not attracted to the folded leaves.

of protoplasm confers a great advantage on an organism. Many cells of an organism may be destroyed and the remaining cells will continue to function. In large multicellular forms of life, cells assume many different shapes and functions and work together to contribute to the maintenance of the entire body, but many of these cells can exist separately. Some body cells can be removed and placed in a culture medium where they will continue to grow and divide independently to form a **tissue culture.** A tissue culture grown from cells taken from a woman in 1944 is still alive, even though the woman died long ago. The tissue culture has continued to grow and divide, and it seems that a culture can be maintained almost indefinitely as long as it receives proper nourishment and is kept free of harmful bacteria. Frequently some of the tissue mass must be discarded, of course, or it would fill the entire laboratory. There are some science fiction stories about tissue cultures which have escaped from laboratories and spread over the entire countryside, engulfing all living things in their pathway. This is not likely to happen in reality because tissue culture must be maintained within narrow limits of temperature, nutrients, and sterility. In Chapter 6 we shall learn more about these units of life.

Exhibits Growth

All living things can become larger at some time during their life through the process of growth. Food is utilized to produce more protoplasm. A kitten grows into a cat, a pig into a hog, and a tiny embryo in a seed can grow into a giant tree. The potential for growth is enormous. Most forms of life begin as a microscopic bit of protoplasm which may grow into an organism as large as a whale.

No form of nonliving matter grows in the same way as biological organisms. Some nonliving things can enlarge by **accretion**: the addition of materials on the outside without causing a change in those materials. The beautiful terraces in Yellowstone Park have been built up by the addition of crystals of calcium carbonate which come from the hot water which boils up from the underground springs. The stalactites and the stalagmites in underground caves are formed by crystallization of minerals which come from water dripping from the roofs of the caves. Accretion is a form of growth entirely different from biological growth in which food is converted into protoplasm.

Expends Energy from Food

Life cannot be maintained without the intake and release of energy. Most of the food you eat is used for energy, and relatively little is used for maintenance and growth of tissue. In a young child more food is used for growth, but most of it is still used for energy. This is why the food shortage is one of the most urgent problems facing the world today. In most areas of the world, plant products are needed for food and they cannot be spared for animal fodder. If plant products were fed to animals, the animals would use most of the plant nutrients to supply themselves with energy and comparatively little of the plants products would be incorporated into the growth of flesh which would become food for people. People must therefore eat primarily plant foods, which may sometimes be deficient in some of the essential amino acids which are necessary for the best health and growth.

The amount of food used for energy varies greatly. Plants, as a whole, use much less energy than animals because plants are less active and need less energy to produce heat. Warm-blooded animals use an enormous amount of energy to maintain a high body temperature, especially when the surroundings are cool. Cold-blooded animals need less food because their body temperature fluctuates with the temperature of the surroundings. A highly active animal uses more energy than a sedentary animal. A lumberjack cutting trees may need to eat three times as much as a businessman who sits at his desk all day. The businessman may eat as much as the lumberjack, but the unneeded food is transformed into fat. Cattle being fattened for the market in feed lots are kept in small pens so they will not waste much energy in activity. A plant seed contains an embryo which expends very little energy. The plant embryo can stay alive for years within its outer protective coating. Mold spores, which are dry little bodies, can remain alive for years because the spores use very little energy, but they can start growth when nutrients become available.

Shows Homeostatic Response to Environment

Living things adjust to their environment. Each living organism functions best within a rather narrow range of temperature, moisture, atmospheric pressure, and other conditions of its environment. The environment surrounding the organism can vary greatly, however, and living things must have the ability to adjust to changing environmental conditions to maintain constant conditions in their internal environment. This ability to adjust to environmental change is known as a **homeostatic response.** The following examples illustrate this concept. You can go to the ski slopes or to a tropical beach, but the internal temperature of your body does not vary more than a fraction of a degree. Many adjustments take place constantly to maintain your internal temperature at the level best suited for cellular function. Your brain must have a constant supply of blood at a certain pressure. If your blood pressure drops, you will black out. The pull of gravity on blood going to your head varies greatly. You may be upright, lying down, or even upside down, yet homeostatic adjustments keep the pressure constant.

FIG. 1.7 The pupils of the eyes of this owl do not appear to be synchronized. When a light was shined into the eye on the left, the iris contracted making the pupil smaller. This reaction is typical of the rapid homeostatic responses of living organisms.

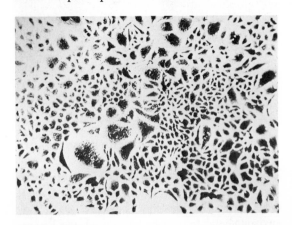

FIG. 1.6 Human tissue culture cells. These cells belong to the HeLa strain, which was developed in 1951 from tissue taken from a cervical cancer belonging to Henrietta Lacks. Although she died eight months later, the tissue culture is maintained in laboratories throughout the world. The cells shown here have been exposed to X rays. The very large cells have lost their ability to divide and will continue to grow until they die. (Courtesy of T. H. Puck, University of Colorado Medical School.)

Plant adjustments are not so readily apparent, but they occur. Stems grow up and roots grow down regardless of how seeds are planted. An evergreen tree responds to the coming of cold weather by increasing the sugar content of the cells in the leaves and stems as a sort of natural antifreeze. Leaves of a tree grow in such a way that each leaf will receive the best exposure to the light. Some plants respond quickly to environmental changes. The sensitive plant folds its leaves immediately when one is touched; the Venus's fly-trap closes its leaves quickly over insects which happen to alight on them.

The term **irritability** is used for rapid and short-term homeostatic responses. In animals these responses are sometimes called **reflexes.**

A

B

FIG. 1.8 Geranium leaves turn toward light coming from one side. This is a homeostatic response which gives the leaves better exposure to light. The top photo-graph shows the plant as it appears from the direction of the light. The bottom photograph shows the same plant from the side.

For example, you touch a hot stove and jerk your finger away immediately. Slower, longer-lasting responses are known as **individual adaptations.** If you have a fair skin and are exposed regularly to sunlight, you develop heavier deposits of melanin in your skin and these deposits form a filter which protects the sensitive lower layers of the skin against the sun's rays. The development of melanin deposits happens slowly, as anyone who has tried to get a tan in one day will testify. Once the tan has been built up, however, it remains long after the exposure to the sun has stopped. Fur-bearing animals develop heavier coats in cold winters, and these coats remain, to some extent,

all year, so the best furs come from the northern regions in winter.

As a rule, homeostatic responses are beneficial to the organism; they protect it or make it more efficient. Sometimes, however, when new environmental forces are introduced, the homeostatic responses may be harmful to the organism. Some moths respond to bright objects by flying toward them. This is an instinctive homeostatic response which leads them to bright flowers where they can suck the nectar. When man creates bright artificial lights, however, the moths make their instinctive response to their detriment.

Many homeostatic responses are related to

the maintenance of internal cell environment. Enzymes must be produced at the proper time and in the proper quantity to maintain the normal state of the cell. The intake and excretion of inorganic minerals by the cell must be kept within narrow tolerances. When one of the many homeostatic responses fails, serious impairment of function and death of the cell may be the result.

Shows Species Adaptation

Living things incorporate certain adaptations into their genetic systems, and so the adaptation is transmitted to their descendants. An entire species may thus show adaptations without each organism showing individual adaptations. A person may adapt to considerable exposure to sunlight by tanning (an individual adaptation), but some races of people, descended from people who lived in regions of the world where sunlight exposure is intense, are born with heavy protective melanin deposits (species adaptation). Species adaptations develop through selective survival of the fittest. Not all members of the same species are alike genetically. Some will have genetic attributes which make them better adapted to a particular environment than others. Those members of a species who possess these favorable genetic attributes are more likely to live to reproductive age and to reproduce themselves. After countless generations of selection, the entire species, or a race within the species, will show genetic characteristics which are most adaptive to the environment in which the species lives.

Individual adaptations are homeostatic adjustments which benefit the individual but which are not passed on to descendants. Each generation must make its own individual adaptations. **Species adaptations,** on the other hand, involve the alteration of the genetic makeup of an entire group so that individuals are born already adapted to some aspect of their environment. Natural selection is the method by which species adaptation is achieved.

Reproduces

No complex living organism exists forever. Death can come any time during the life period through the action of predators, parasites, accidents, lack of food, and so forth. Even though an organism escapes these destructive forces, however, it cannot go on living indefinitely. There comes a time when the internal forces which maintain life cannot keep up with the forces of deterioration, and there is old age and death. Hence, all organisms must have some method to reproduce themselves so that they can leave descendants which will continue the species. In some of the smaller forms of life, reproduction consists of a simple splitting. These simple organisms grow, split in two, grow some more, split again, and so on. Many plants and simpler animals produce buds or other outgrowths which can develop into entire new organisms. The great majority of living things, however, produce small reproductive cells which unite to form a single cell which grows into an entire new organism with the

FIG. 1.9 This whippoorwill is not easy to see because it blends in well with the leaves around it. The whippoorwill's coloration is a species homeostatic adaptation produced by natural selection.

characteristics of the species. This process is called **sexual reproduction.** Sexual reproduction has an important biological value beyond that of producing offspring. It allows a blending of hereditary characteristics from two different organisms, and this blending produces variety in the offspring which would not be possible with **asexual reproduction.** Genetic variety is the basis for species adaptation. Many forms of life have both asexual and sexual forms of reproduction.

Changes Through Mutation

In all forms of reproduction, the genes are passed from the parents to the offspring. You might think, therefore, that the offspring would be limited to the expression of the hereditary traits carried in the genes of their parents. This is generally true, but occasionally a gene will mutate or undergo a change. The mutated gene produces a different expression of a characteristic. Breeders of domestic animals know that a new trait will occasionally appear in their stock. Farmers know that, of the many seeds they sow each year, one will occasionally produce a plant

which is different from the others. These new types—**mutations**—can arise as a result of slight alterations of a gene.

Mutations provide an important source of variety which makes species adaptation possible. Sexual reproduction produces recombinations of existing genes and the recombinations provide variety, but the variety would be limited if it were not for the occurrence of mutations which produce new genes.

The characteristics of living matter are quite well defined and these descriptions should prevent confusion when you attempt to decide if something is living or not. As is true with all of man's attempts to set up organized categories of natural things, however, there are cases which do not fit exactly. The viruses, for instance, are very tiny entities which have some of the characteristics of living organisms but not others. Some scientists say that viruses are living organisms which have not developed every trait which characterizes other forms of life. Other scientists say that viruses are nonliving but have acquired some of the traits which we associate with life. We shall learn more about these ubiquitous entities in Chapter 14.

A

B

FIG. 1.10 This fruit fly expresses two gene mutations. It has white eyes (A) and miniature wings (B) rather than the wild type red eyes and long wings. This fruit fly has poor vision and it cannot fly, so these mutations are both harmful.

FALSE IDEAS ABOUT LIVING MATTER

Before we close our discussion of the nature of living matter, it might be well to clear up some confusing points. An old idea is that "life" is a mysterious force, and that when this vital force departs, the organism is no longer living. This concept was developed when many of the phenomena of nature were explained by supernatural means. When a person became sick, it was thought that devils had invaded his body and that a cure could be worked by driving the devils out. Knowledge, however, is the enemy of superstition, and when it was discovered that invading bacteria rather than devils caused the sickness, this old idea received a setback. Even today the idea of possession by devils is not dead and exorcists are engaged to drive them out. As we have learned more about living matter, it has become apparent that there are natural explanations for the life processes and that they follow the same laws which govern reactions in nonliving matter. Concepts die slowly, however, and there were many years of arguments between the vitalists, who believed in some mysterious vital force of life, and the mechanists, who felt that there was ample evidence for natural mechanistic explanations for life. Today practically all trained biologists recognize the fact that there is no peculiar spark of life which distinguishes living matter from nonliving matter.

SUBDIVISIONS OF BIOLOGY

Many colleges and universities of the last century had courses in natural history which were primarily surveys of the different kinds of plants and animals on the earth. As biological knowledge increased, courses in botany and zoology appeared. With continuing expansion of knowledge, further fragmentation of the field occurred until we now find as many as a half-dozen or more departments in some universities, all dealing with subject matter related to living things. In a way, such subdivision is unfortunate because a person can become a specialist in one branch without familiarity with the entire field of biology. Confusion arises when microbiologists use one term for a process and physiologists use another term for the same thing. With the great explosion of biological knowledge in this century, however, it is not possible for anyone to develop extensive knowledge of the entire field, so the subdivisions are necessary although some acquaintance with the entire field is valuable.

Subdivisions According to Subject Matter

1. **Morphology.** A subdivision of biology dealing with structure of living organisms. It is further broken down into:
   a. **Anatomy.** The study of gross structure (that which can be seen with the naked eye).
   b. **Histology.** The study of fine (microscopic) structure. This subdivision of biology is concerned primarily with the study of tissues from living organisms.
   c. **Cytology.** The study of the detailed structure of individual cells.
2. **Physiology.** The study of the functioning of living organisms. How food is digested or how a plant absorbs water and minerals from the soil are examples of the types of studies done in physiology. Much biological research today is concentrated on cellular physiology.
3. **Taxonomy.** The study of the classification of living things.
4. **Genetics.** The study of inheritance. This subdivision studies the way in which genes are transmitted from one generation to the

next and how genes determine the characteristics of each individual.

5. **Embryology.** The study of the development of an organism from its embryonic stages until it reaches adulthood. Since most forms of life begin with a single cell, embryology traces the divisions and differentiations of succeeding cells and studies how these cells develop into the adult form. Embryology is closely related to genetics because genes determine the changes in cells which lead to the formation of an adult organism.

6. **Ecology.** In the past ecology was a branch of biology known primarily to biologists, but today it is a word known to practically everyone because we can see its importance. Ecology is a study of organisms in relation to their environment. Since the environment has undergone great alterations as the result of man's meddling with it, some living organisms are under threat of extinction in many places. Man, himself, is beginning to feel the impact of changes in the environment. We shall spend considerable time on this subject in the latter part of the book.

7. **Evolution.** A study of the changes in species which take place as the species adapt to their surroundings. Evolution involves a study of the changes which have taken place in the past, the changes which are now taking place, and a prediction of possible changes in the future. Evolutionary studies extend back into the possible origins of living things.

8. **Biochemistry.** A study of the chemical nature of living matter and the chemical reactions associated with life.

9. **Biophysics.** A study of the principles of physics as applied to reactions in living organisms.

10. **Molecular biology.** A study of the molecular organization of living organsms. This subdivision of biology involves elements of chemistry, physics, physiology, and genetics.

It is closely related to the study of evolution since records of past changes in forms of life show most clearly in fossil records.

11. **Paleontology.** The study of life as it existed in the past. Paleontology depends primarily on fossil records of prehistoric life. It is closely related to the study of evolution since records of past changes in forms of life show most clearly in fossil records.

Subdivisions According to Organisms Studied

1. **Botany.** The study of plants.
2. **Zoology.** The study of animals.
3. **Algology.** The study of algae.
4. **Mycology.** The study of fungi.
5. **Bryology.** The study of mosses.
6. **Bacteriology.** The study of bacteria.
7. **Virology.** The study of viruses.
8. **Microbiology.** The study of microscopic organisms.
9. **Protozoology.** The study of one-celled animals.
10. **Helminthology.** The study of worms.
11. **Parasitology.** The study of parasitic forms of life.
12. **Entomology.** The study of insects.
13. **Ichthyology.** The study of fishes.
14. **Herpetology.** The study of amphibians and reptiles.
15. **Ornithology.** The study of birds.
16. **Mammalogy.** The study of mammals.

Many of these subdivisions overlap. Botany is a large subdivision that includes a number of the others, each of which deals with a more restricted group of organisms. Parasitology includes studies of both plants and animals, and microbiology includes studies of organisms from several of the other groups.

A biologist is likely to specialize in one or more of these subdivisions. While he needs a background of general information in all areas, the amount of information in biology is so great that he cannot hope to be an authority in all branches. To be sure, an ecologist studying chipmunks in a particular region will need to know some taxonomy in order to classify them, some anatomy to understand their structure, some physiology to understand the functioning of their bodies, and some parasitology to understand how they become infected with the parasites they carry. Still, his primary concentration of study will be in ecology.

In this book we shall offer some insight into each of the subdivisions of biology, but as you can see from the number of subdivisions, this insight must necessarily be brief. Yet the subjects are of such intrinsic interest that even a little information will stimulate a desire to learn more.

REVIEW QUESTIONS AND PROBLEMS

At the end of each chapter will be some questions which will help you to test yourself to determine how well you have learned the material in the chapter. These questions will include some which will require you to think out the answers using the subject matter in the chapter.

1. Protoplasm is made of certain basic parts, yet when these parts are mixed together in a test tube we do not have living matter. Explain why.
2. Both living and nonliving matter can increase in size. Explain the difference between the two types of growth.
3. A man once thought he had a great idea for raising minks without any cost for food. He would raise rats which he would feed to the minks and then, after he had killed and skinned the minks, he would feed the mink carcasses to the rats. Thus he would never have to buy food for either minks or rats. What important characteristic of life did he fail to take into account in this scheme?
4. Describe one of the many homeostatic adjustments which your body makes to changing conditions around you. (Do not select one which is similar to one mentioned in this chapter.)
5. Describe some homeostatic adjustment which may, at times, prove unfavorable to the organism because environmental conditions may have changed. (Do not use the example of the moth mentioned in this chapter.)
6. Distinguish between short-term and long-term homeostatic responses and illustrate your answer with an example of each in some organisms other than human beings. (Do not use examples described in this chapter.)
7. Many people confuse adaptation of an individual with adaptation of the species. Explain the difference between the two.
8. Since most mutations are harmful, would a species be better off if its genes never mutated? Explain your answer.
9. Before reading this chapter, or before you studied biology in high school, what criteria did you use to decide whether a thing was living or not. How did your criteria compare with the criteria given in this chapter?

10. Into which of the subdivisions of biology according to subject matter would each of the following studies fall?
     a. Nesting habits of the quail.
     b. Inheritance of eye color in man.
     c. Ancestry of modern horses, as determined from fossil remains of horses which have lived in the past.
     d. Effect of widespread use of DDT on the eggs of wild birds.
     e. Development of a plant within a seed.
     f. Structure of the human kidney as seen by the naked eye.
     g. Effect of water pollution on the growth of plants in a river.
     h. Appearance of small pieces of a root under the microscope.

FURTHER READING

Bonner, J. T. 1962. *Ideas of Biology*. New York: Harper and Row.
Moment, G. 1962. *Frontiers of Modern Biology*. Boston: Houghton Mifflin.
Moore, J. A. 1965. *Ideas in Modern Biology*. New York: Doubleday.
Waddington, D. H. 1962. *The Nature of Life*. New York: Atheneum.

2

Expansion of Biological Knowledge

Biology is not a static body of knowledge. Biology is constantly expanding and changing as research uncovers new information. In fact, discoveries have been coming so thick and fast during the past few years that the expansion of biological knowledge is in the nature of an explosion. We all benefit from the application of these discoveries, and we have gained a greater understanding of the nature of life. But we can better appreciate the discoveries if we know something about the ways in which biological research is conducted, and so the purpose of this chapter is to describe some of the methods used, as well as the progress of typical research investigations.

THE METHODS OF BIOLOGICAL RESEARCH

Similarity to Criminal Investigations

Biology is one of the sciences, and a procedure known as the scientific method is typically used to uncover new information. While there are no hard and fast rules which must be followed in this method of inquiry, an orderly progression of procedures which tends to follow a certain pattern is used. Different scientists and different problems may dictate variations.

Albert Einstein, the great mathematical physicist, pointed out the similarity of criminal investigations to scientific research. A detective is called upon to solve a murder. He collects all the evidence he can find. He analyzes the facts; he considers other crimes of a similar nature and how they were solved. He uses laboratory techniques to analyze every bit of material which might give a clue. His brain sorts out all this information and he may arrive at a prime suspect. He seeks more information to support or disprove his suspicion. He may have to switch to another suspect. In time, he may arrive at the correct identity of the criminal.

Such criminal investigations are fascinating to follow, as shown by the popularity of detective stories on television and in books and films. Biological investigations can have the same fascination. This is what holds biologists to long hours of exacting work in the hope of uncovering facts. Many frustrations and delays may occur, but there is great satisfaction when a correct answer is obtained. Students in the biology laboratory usually show more interest when they are called upon to conduct their own investigations and to draw their own conclusions.

Selection of a Problem

As a biologist develops knowledge in his field, he finds that there are many gaps, or things that are not yet known. One particular gap may interest him greatly. He reads all he can find on the subject, he attends meetings and hears reports on the subject, he talks to others who have worked in the field, he does preliminary investigations which give him ideas on the techniques which may be used. His brain is a marvelous computer which sorts out all the accumulated information, and then, perhaps as he awakes in the morning, perhaps during a golf game or on a fishing trip, he has an idea of a plausible solution to the problem. Such an educated guess, backed up by all the information he has accumulated, is known as a **hypothesis.**

Beginners have a tendency to choose a problem which is much too extensive. The simplest problems have a way of becoming complex as investigation proceeds. An eager but untrained person which may wish to tackle a problem so complex that it would require a lifetime of extensive work for any possibility of solution. An adviser to a high school student who wishes to do some research must warn him of this tendency. Perhaps the student has read that high-energy radiation causes mutations, and he wants to investigate the production of mutations in mice. He may think that he can simply expose some mice to such radiation and then find all kinds of strange traits in the offspring. Such a problem is far beyond his abilities, however. First of all, there are different kinds of high-energy radiation, and he would have to restrict himself to one of these. Then, he would find that he must raise enormous numbers of mice to hope for any statistically significant results. He would have to include nonirradiated controls as well as irradiated mice. Even if he did all this, he would probably not find much in the first generation because most mutations are recessive and do not show in the first generation. The study would have to extend over many generations with special breeding programs designed to detect any mutations which had been induced. Various physiological tests would have to be made on the mice to detect mutations which gave no visible physical indications. The student would quickly become discouraged and give up if he started such a great problem. If he were content with a much narrower problem, however, he would have a good chance of success. For instance, he could study the effect of X rays on the embryos in pregnant female mice. This would be a challenging problem, but within the realm of possible solution in a reasonable time.

Experimentation

Once a problem has been selected, a plan of procedure must be devised in which possible experiments are included. To determine the effect of vitamin A deficiency on rats, for instance, a number of rats, preferably all of the same sex and genetic background, would be divided randomly into two groups. These two groups would be treated in exactly the same way except for one variable, the amount of vitamin A in their diets. One group would be the experimental group and would have a diet deficient in this vitamin. The other group would serve as a control and would receive the same diet, except that it would contain the vitamin. Without a control, no definite conclusions can be drawn from the results. All the rats fed a vitamin A–deficient diet might show eye defects, but this would not prove that the vitamin deficiency caused the defects. The rats might have an infection that caused the defects, or they might be from a hereditary strain that had genes for such defects. Any one of many other factors could have been responsible. If the controls being raised beside the experimental group showed no defects, however, we would have reason to conclude that the vitamin A deficiency was responsible, provided that we had sufficient numbers to eliminate chance variations between the two groups.

Multiple controls are sometimes necessary in experiments where two or more variables are being studied at the same time. Such an experiment was conducted by the author on a possible chemical protection against damage from high-energy radiation. Other experiments had shown that 900 rads (R) of X rays applied to young rats would result in death of all within about ten days. The deaths seemed to result from chemical changes in the cells. Certain chemicals are known to neutralize the products of these changes and might serve as a means of radiation protection, provided they do not injure the organism. The chemical selected for the test was known as AET.

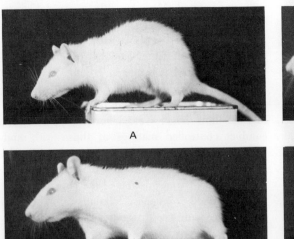

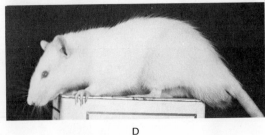

FIG. 2.1 Multiple controls are needed in experiments which deal with more than one variable. In this experiment which tests the value of a chemical (AET) as a protection against radiation damage, four different groups of rats were used. The rats shown here are typical representatives of the four groups. Each rat weighed about 150 grams at the beginning of the experiment. The rat in (A) at the top now weighs 123 grams. It received 900 R of radiation. The second rat (B) received an injection of AET and 900 R of radiation. It weighs 214 grams. The third rat (C) received AET and no radiation. It weighs 232 grams. The fourth rat (D) received nothing. It weighs 234 grams.

In this experiment, there were two variable factors, the radiation and the chemical, and so four combinations had to be tested. Male rats of the same strain, age, and approximate weight were selected and divided into four groups of ten each. All were fed the same food. One group served as controls and received no treatment. A second group received 900 R of X rays. A third group received an injection of AET. A fourth group received both X rays and AET. The first group were all alive and healthy after ten days. The second group all died. Two of the third group died, probably because of the toxic effects of

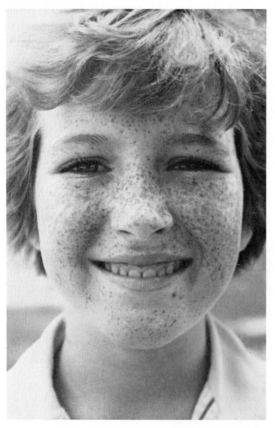

FIG. 2.2 Heredity may cause melanin in the facial skin to accumulate in spots called freckles. Determination of the method of inheritance of this trait is limited to observation of families in which it appears.

AET, and two of the fourth group died. These results show that AET does give protection if the animal can survive the injection of AET. It would not be practical for human beings to use it, however, because it is toxic to a degree. Other, less toxic compounds have been tried more recently and may eventually be cleared for use by people who must be exposed to levels of radiation considerably above what is considered safe. Men on long space missions who are exposed to considerable extra radiation in outer space would be one group who might use such a compound.

Some problems are encountered which do not lend themselves well to experimentation, and the biologist must limit his research to observation alone. In biology many investigations on human beings are so limited because we do not like the idea of experimenting on human beings. The Nazi experiments on people in concentration camps during World War II have been universally condemned. More recently a scandal arose when it came to light that a group of men in Alabama who had syphilis were not treated for the disease so that they could serve as controls and be compared to a treated group.

Hence, we are frequently limited to observation alone in studying human beings. If you want to learn how freckles are inherited, you do not conduct controlled matings of those with and without freckles. Instead, you find families in which some have freckles and make tabulations. The pattern of transmission of the trait then can be determined from these tabulations. When we are dealing with new drugs and methods of treating diseases, we do sometimes conduct experiments. After experiments on other animals show that no serious harm is likely to result, we try to find human volunteers who are willing to take a small risk so that the drug may be tried on people.

Great care must be taken in evaluating the possible danger of a product for use by people. Thorough investigations on lower animals must be done first, and then the product must be tried by volunteers. A tragic illustration of the results of too hasty approval of a drug occurred in 1961 and 1962 in Europe. A newly developed tranquilizing drug, known as **thalidomide,** had been tested on laboratory animals and seemed to have no ill effects. Even some people tried it with no apparent harm. Doctors began prescribing it for women in early pregnancy to help relieve the nausea which often comes at this time. Thousands of women took the pills for this purpose, but as their babies began to be born, a great tragedy unfolded. Many of the babies had only stumps for arms or legs, a condition known as **phocomelia.** Fortunately, a woman in the United States Food and Drug Administration felt that the tests had not been sufficient to warrant release of the drug. Many young men and women in this country today have normal arms and legs because of the vigilance of this one woman. The experiments on lower animals had not included pregnant females and therefore gave no indication that embryos would be affected. Some of the women who used the drug reported a temporary numbness in their arms and legs which soon went away. For the fetuses within their bodies, however, it was another story. They were growing rapidly, and anything which interferes with growth at this critical time can result in incomplete development.

Statistical Evaluations

When results of experiments and observations are quantitative, it is usually necessary to use mathematical methods to determine if they are significant. One of today's ecological problems is the effect of DDT on the development of bird eggs. There is some indication that this insecticide can cause defects in the shells so that the eggs do not hatch properly. The best way to settle the question is to conduct an experiment which can be controlled. The domestic fowl, better known as a chicken, is a bird which is easily raised and sudied. One group of chickens may be

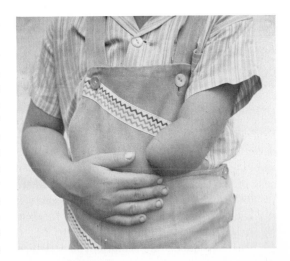

FIG. 2.3 Phocomelia, the absence of a portion of a limb, appeared in many babies born in Europe in 1961 and 1962. The mothers of these babies took thalidomide during early pregnancy. A more scientific investigation of the effects of the drug before approval for human use would have avoided such tragedies.

raised with no additions of DDT to the diet. Suppose we take 100 eggs from these and incubate them properly and find that 90 hatch. Eggs from chickens of the same breed which have had small quantities of DDT added to the diet are then tested. Suppose we find that 85 eggs out of 100 incubated hatch. Is this a significant difference from the controls, or could this merely be a chance variation which might have occurred anyway? Suppose only 75 eggs hatched, or suppose only 60 hatched. Where do we draw the line between significant deviations and insignificant or possible chance variations? Mathematical methods are available for such determination and would show that a hatching of 85 eggs in the experimental group, as compared to 90 in the controls, would not be significant. In fact, deviations of this magnitude would be found in almost half of the trials should the experiment be repeated, even though the DDT had no effect. Such deviation, therefore, could be assigned to possible chance variation and could not be used as evi-

TABLE 2.1.   Effects of Dried Pituitary Gland on Weight Gain in Rats

Amount of gland injected, in mg	0	4	8	12	16	20
Weight after two weeks, in g	213	224	235	247	254	262

dence that DDT caused harm. If only 75 eggs hatched, however, the analysis would show that deviations of this magnitude from the controls would occur by chance in slightly less than 5 percent of the trials. If only 60 eggs hatched, the results would be highly significant because a deviation this great would be expected to occur by chance in less than 1 percent of the trials.

Large numbers are important in statistical studies, especially when the effects of an agent being studied are not great. Had we used 1000 eggs instead of 100, for instance, we could have detected a significant effect even if the proportion of eggs hatching remained the same. If controls showed 900 eggs hatching and the experimental group showed 850 out of 1000, this would be a significant deviation. A deviation this great would be expected by chance in only about 2 percent of the trials. Generally we feel that when chance could account for the deviation in less than 5 percent of the cases, the results are significant. Of course, the smaller this percentage, the more significant the results are.

Sometimes statistical studies are used to determine if there is a **correlated variation** between two variables. For instance, we would find a positive correlation between human height and weight. Tall people tend to weigh more than short people. Of course, there are variations; there are many thin tall people who weigh less than obese short people, but on the average, height is positively correlated with weight.

As an example, suppose you wanted to determine if feeding dried pituitary gland to rats resulted in weight gain proportionate to the amount of the gland administered. Table 2.1 gives the results which might be obtained. The results show that weight gain is in proportion to the amount of the gland given. There is an average weight gain of 9.4 grams (g) for each 4 milligrams (mg) of the gland administered. There is the expected chance fluctuation, but there is no doubt of the effectiveness of the gland on stimulating weight gain. Had all the rats shown about the same gain in weight, allowing for chance variations, then we could conclude that there was no correlation.

Correlation may sometimes be negative. When one variable goes up, the other variable may go down. X rays are known to have a retarding effect on growth, as is demonstrated in the results of an experiment tabulated in Table 2.2. As the amount of radiation given increases, the weight gain of the rats decreases in an inverse proportion. Mathematical techniques make it possible to determine the percentage of correlation so that accurate conclusions may de drawn.

Conclusions

After all the evidence is in, the investigator must draw some conclusions, but he must be careful. All evidence may point to a certain conclu-

TABLE 2.2.   Effect of X rays on Weight Gain in Rats

Amount of X rays given, in R	0	100	200	300	400	500
Weight after two weeks, in g	218	209	193	175	159	124

sion, but as long as any reasonable doubt remains, the results must be considered tentative. Hasty conclusions can lead to false assumptions.

T. S. Painter, at the University of Texas in the 1920's, made extensive studies to try to determine the exact human chromosome number. He carefully examined thin sections of the seminiferous tubules from the testes. He saw chromosomes in cells which were in mitosis, but they were so small that he had difficulty in counting them. After much study he felt that the evidence showed that there were 48 and proposed this as the number, but because there was a chance for error, he made his proposal tentative. This number was accepted widely and for many years was given as the human chromosome number in textbooks and other publications. Then in 1956 a new method of studying cells was devised. It was known as the squash technique. Growing cells were placed on a microscope slide and squashed flat. This made it possible to see all the chromosomes in a cell clearly spread. J. H. Tjio applied this technique to cells taken from aborted early embryos and found only 46 chromosomes. This shows how improved techniques can alter previous conceptions and develop new knowledge. Today we use Tjio's technique on cells from human tissue culture, and any person can see what his chromosomes look like by contributing a few cells from his blood, skin, or other body parts. The cell can be grown in tissue culture and squashed for study.

The investigator must consider all possible variations before drawing conclusions. For example, statistical studies show that there is a greater incidence of stomach cancer among the people of Japan than among Americans. This might lead us to assume that the Japanese inherit a greater susceptibility to cancer, but investigation produces a better explanation. It has been found that Japanese who come to this country and adopt American dietary habits have no greater incidence of stomach cancer than do other American ethnic groups. In Japan the typical diet includes

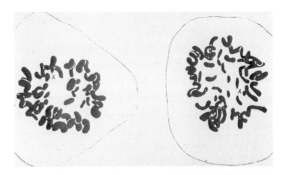

FIG. 2.4 Drawings of human chromosomes made by T. S. Painter in 1927. Using the techniques available at that time, he concluded that there were 48 chromosomes. Improved techniques provided additional evidence to prove this number to be slightly wrong.

smoked fish, and one of the substances in the smoke used to prepare this fish is **carcinogenic** (cancer causing). In this country smoked fish is not eaten frequently by most people. However, some people who live near Lake Superior in Minnesota do eat large quantities of smoked fish, and they also have a higher incidence of stomach cancer than other Americans. It seems, therefore, that dietary habits, rather than genetic back-

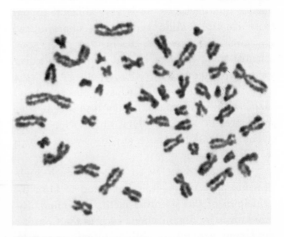

FIG. 2.5 Photograph of human chromosomes prepared by the smear technique by J. H. Tjio in 1956. This new technique led to the determination that the human chromosome number is 46. (Courtesy of Dr. J. H. Tjio.)

ground, are responsible for producing a suscepti-bility to stomach cancer.

Frequently it is difficult to draw any defi-nite conclusions. All the evidence may point to one explanation, but that final bit of proof is not forthcoming. In such cases the hypothesis may be called a **theory,** which is an assumption with a high probability of being correct. Theories can be valuable even though they may never be proved. For example, fossils and other evidence support theories about the nature of life on the early earth, but we cannot actually go back in time and verify these ideas. Sometimes the words *hypothesis* and *theory* are used interchangeably as synonyms, but in the strict sense a theory has more weight of evidence behind it.

THE SCIENTIFIC ATTITUDE

Evaluation of findings in biological investi-gations must be made on an objective basis, but human bias is one of the most difficult factors to eliminate in making such an evaluation. As hu-man beings we all have emotions which can in-fluence our decisions. It is very difficult to put these emotions aside and to use unprejudiced reasoning to arrive at conclusions. The good scientist must make a special effort to uncover facts which run counter to his hypothesis. He may hope that his research will bring him fame and monetary reward, and this may lead him to give much weight to those data which support his hypothesis and to ignore any discoveries which would disprove it.

In these days, when there is much high-pressure advertising, it is important to keep the principles of the scientific method in mind. Re-ports made by organizations such as the Consum-ers Union, which have nothing to gain or lose by the findings, are the most reliable. There are fed-eral agencies which monitor some advertising claims and which can require retractions when the claims are not founded on fact.

One of the problems facing mankind today, as we seek ways of obtaining energy from sources other than fossil fuels, is the danger that radio-active particles may be released by nuclear gen-erating plants operated by the Atomic Energy Commission. Many geneticists have warned that even low-level exposure to the by-products of atomic fission can cause harm to those exposed, as well as possible genetic damage which will affect future generations. On the other hand, scientists in the Atomic Energy Commission con-tend that their investigations show that such low-level radiation as may be spread from the gener-ating plants poses "no hazard to health." In one case we have a group of scientists who will not benefit from their findings. In another case, we have scientists whose very jobs depend upon find-ings which favor the continued operation of the plants. Which one of the two groups would be the most likely to give truly unbiased evaluations?

THE PRACTICALITY OF SCIENTIFIC RESEARCH

When an uninformed person is told of some scientific research project, he may frequently ask, "What good will it do to know about that?" The implication of his question is that no investigation is worthwhile unless the findings can have some direct practical use. The truth is that we cannot make practical applications until after the discov-eries have been made, and there is no way to predict in advance which discovery will and which will not have such an application. Even those discoveries which do not have practical uses are worthwhile because they contribute to man's knowledge about his universe.

Discovery of Penicillin

Penicillin was the first of the antibiotics to be discovered. Many of you would not now be alive had it not been for this discovery. At some time in your life it is likely that you were given an antibiotic which stopped an infection which otherwise would have been fatal. How was this valuable substance discovered? Did someone set out to find some substance which would cure many serious infections? Absolutely not. The discovery of penicillin was a byproduct of research on the growth of a soil bacterium. In the 1920's, **Sir Alexander Fleming** was growing soil bacteria in culture dishes. One day he noticed that one of his plates had been contaminated by a blue-green mold known as *Penicillium*. Other researchers had had such contaminations before, but they had simply discarded the cultures. Fleming, however, had a more analytical mind. He noted that no bacteria grew near the mold. He hypothesized that something from this mold had diffused into the culture medium and inhibited bacterial growth. This hypothesis was worth further investigation and proved to be true. An extract from the mold was found to inhibit bacterial growth in animal bodies without any harmful effects on the animals. Soon this extract was being used to cure human diseases, and today it is produced by the ton and is known as penicillin.

Determining Genetic Sex of Newborns

Of the thousands of babies born each year, there are some which have abnormalities of sex as a result of improper distribution of the sex chromosomes. Some of these abnormalities do not show at birth, and the parents do not realize that something is wrong until adolescence. Many hospitals now examine a bit of the amniotic membrane which has surrounded the fetus to detect abnormalities at birth. They look for darkly staining bodies lying against the nuclear membrane of the cells. These are the **Barr bodies.** They will be present in the cells of a normal girl but not in the cells of a normal boy. The distribution of Barr bodies was discovered by Murray

Barr at the University of Western Ontario. He did not set out to find a way to help detect abnormalities of sex in newborns. Instead, he was trying to determine the effect of fatigue on the nerve cells of cats. In the course of this experiment he noticed that cells from some cats had dark bodies while those from other cats did not. He found that only female cats had these bodies. The darkly staining bodies represent a coiled X chromosome. A female, who has two X chromosomes, has one X chromosome that coils up and one that remains uncoiled. The single X chromosome of the male is uncoiled. Any deviations in the sex chromosomes will show up as abnormalities in the number of Barr bodies. Thus, a practical application was the result of research which was not directed toward a practical end.

Male Birth Control Pills

Women sometimes complain that they are the ones who have to take the pills to prevent conception and wonder why there is not something for men to take. A male birth-control pill might be possible in the future as a result of research on an impractical problem on sterility in rats. Dr. Edward Peeple, one of the author's former students at the University of Northern Colorado, began a research project on why a particular gene

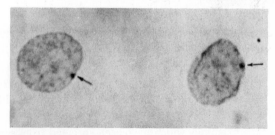

FIG. 2.6 Barr bodies. The arrow points to the dark spots next to the nuclear membrane. The presence of Barr bodies indicates that these cells were taken from a female. The significance of Barr bodies was discovered by Murray Barr. The cytoplasm and the cell membrane are not visible in this preparation.

caused sterility in rats. Biochemical assays showed that this gene caused the production of **cyclic AMP** (adenosine monophosphate) in the testes. The excess cyclic AMP inhibits cell division in the germinal epithelium, so the cells which produce sperm are in short supply. The discovery of the effect of cyclic AMP on male fertility led some scientists to think that sterility might be induced by administering extra cyclic AMP to rats that did not have the gene which causes excessive cyclic AMP production. Preliminary experiments have shown that sterility can be induced in this way in rats, and it is possible that, with further refinement of this technique, a wife of the future may remind her husband, "Dear, don't forget to take your cyclic AMP this morning." Thus, research on gene action in rats led to the discovery of a possible male birth-control technique.

Research Grants

Much research today is being funded by grants from governmental and private agencies, but there is always the danger that with money come controls and with controls the loss of freedom of research. Some of the granting agencies may not be familiar with the methods of science and may insist on some prospect of practical ap-

plication before they allocate money. One biologist related an interesting story about how he obtained a grant from a foundation. The administrator of the foundation listened to his plans and then asked: "What good is this to mankind: will it uncover a way to treat cancer, will it help prevent birth defects, or will it give us some valuable new source of food?" The biologist replied, "No, it is not designed to do any of these things, but what do you do when your watch stops running?" "Why, I take it to a watchmaker for repairs, of course, but what does this have to do with our conversation?" The biologist then said, "You take it to the watchmaker because you know he understands watches and what makes them run. He can thus figure out what is wrong with your watch and make the repairs. In my research I try to find out more about living cells, and the more we know about these the better we will be able to figure out the trouble when something goes wrong with the human body." The biologist got the grant.

Pure research leads to the discoveries which are the raw materials of applied research. The pure research investigator is trying to do only one thing—to extend man's knowledge into areas which are not understood. No one can foretell which discoveries will prove to be of great practical value and which will provide only additional theoretical knowledge.

REVIEW QUESTIONS AND PROBLEMS

1. Suppose you were planning a problem for research in the field of biological science. Outline a procedure which you would follow in selecting a problem and describe the pitfalls you would avoid.
2. Seedlings which have been sprouted from seeds in the dark are not green. Formulate a hypothesis to explain this and describe the procedure you would follow in testing this hypothesis.
3. Why should an investigator make a thorough survey of the work done by others on a problem before beginning research on it?
4. Describe a biological problem which could be solved by experimentation and one

which would have to depend upon observation alone. (Do not give any similar to those mentioned in this chapter).

5. Why are controls a vital part of scientific investigations?

6. Describe some very practical applications of discoveries which were not directed at practical ends. (Do not give any described in this chapter).

7. Suppose you read conflicting conclusions of two different investigations of the same problem. What criteria would you use in trying to decide which of the two is more likely to be correct?

8. Many people blame scientists for some of the horrible applications of scientific discoveries such as the atom bomb. These people say that we should restrict all research which might lead to such destructive applications. Would you agree with such restrictions? Explain your answer carefully.

9. In a certain country with a complete dictatorship, it was decreed that no money or time was to be wasted on impractical investigations and all research must have a practical end in view. In time this country fell far behind other countries where research workers were free to investigate all avenues regardless of possible practical results. Explain why this happened.

10. A manufacturing plant wants to expand its facilities in your area and representatives of the company report research studies which "prove" that there will be no significant pollution of the water or air as a result of this expansion. Some members of the Environmental Protection Agency, however, quote studies which show that plants of this nature do cause serious pollution of rivers and lakes. How would you evaluate the findings of these two? Give your reasoning.

FURTHER READING

Baker, J. J., and Allen, G. E. 1968. *Hypothesis, Prediction and Implication in Biology.* Reading, Mass.: Addison-Wesley.

Bertalanffy, L. 1960. *Problems of Life.* New York: Harper and Row.

Beveridge, W. I. B. 1960. *The Art of Scientific Investigation.* New York: Random House.

Bonner, J. T. 1962. *Ideas of Biology.* New York: Harper and Row.

Glass, B. 1965. *Science and Ethical Values.* Chapel Hill: U. of North Carolina Press.

Moment, G. 1962. *Frontiers of Modern Biology.* Boston: Houghton Mifflin.

Moore, J. A. 1965. *Ideas in Modern Biology.* New York: Doubleday.

Platt, J. P. 1962. *The Excitement of Science.* Boston: Houghton Mifflin.

3

The Chemical Basis of Life

All matter is composed of chemicals, and if we are to understand the nature and functioning of living matter, we must know something about its chemical makeup. Physical scientists have uncovered a wealth of information in recent years that makes possible an understanding of many chemical reactions within living systems that was undreamed of a few years ago. Much biological research today is on the chemical level of organization of living things. This chapter surveys some of the basic principles of chemistry with particular reference to their application to living matter.

THE STATES OF MATTER

Matter is anything that has weight and occupies space. It may exist as a gas, a liquid, or a solid, and it may change from one state to another as conditions of temperature or pressure change. Water, for example, is liquid when under the pressure found at sea level and within the temperature range of 0° to 100° Celsius (32° to 212° Fahrenheit). Below this range water exists as a solid (ice), and above this range it exists as a gas (steam). Regardless of the state in which it exists, water always has the same chemical composition, H_2O.

The state in which matter exists depends upon the speed of movement of the molecules which compose it. Molecules are in motion in all states of matter, but they may vary in their speed of motion and in their position with relation to other molecules. In a solid the molecules vibrate back and forth but remain in fixed relative positions. In a liquid they move faster and are free to move about one another. This gives the fluidity to a liquid. In a gas the molecules attain so high a speed that they move freely in wide separation from one another.

The movement of molecules is correlated with the energy of heat: the higher the temperature, the faster the movement. This explains why temperature changes are marked by changes in the states of matter. There is always some molecular movement, however, even in heavy solids. In theory this movement would stop at the temperature of absolute zero ($-273°$C or $-460°$F). This temperature has never yet been attained in the laboratory, although scientists have come within a fraction of a degree of it. It gives us a lower limit of temperature, but there seems to be no upper limit. There exist special ovens that produce temperatures of several thousand degrees Celsius. The center of a hydrogen bomb at the time of explosion reaches a temperature up in the millions of degrees.

All this movement of molecules may seem fantastic, but it can be demonstrated rather easily in the biology laboratory. If we suspend any very small particles—bacteria, fine soil, or the like—in water and view them under the high power of the microscope, they will be seen to move back and forth rapidly in different directions. Such movement is known as **Brownian movement.** Where does the energy involved in Brownian movements originate? It derives from the molecules of water which bombard the particles on all sides. The molecules are much too small to be seen, but they hit the suspended particles with great force and so can cause them to move. It is probably not a single molecular hit which causes visible movement in any direction but a brief imbalance in the number of hits from different sides. If more molecules hit a particle from one side than from another, it will be driven away from the side receiving the great-

est number of hits. We can often see Brownian movement of small particles inside living cells when viewed under the microscope.

DIVISIONS OF MATTER

All matter may be classified into elements, compounds, and mixtures. An **element** is a substance that cannot be broken down chemically into two or more different kinds of atoms. Iron, sulfur, oxygen, hydrogen, iodine, and calcium are typical elements. When any of these substances is broken down into the smallest divisible units which retain the characteristics of the element, we find that these are atoms of only one kind. A **compound** is a substance formed by the chemical union of two or more elements. Water, for instance, is a compound formed from two ele-

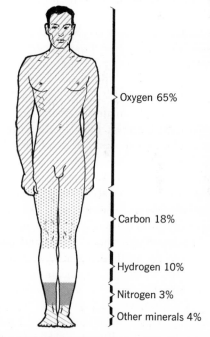

Oxygen 65%

Carbon 18%

Hydrogen 10%

Nitrogen 3%

Other minerals 4%

FIG. 3.1 The chemical elements which make up the human body are very common. It is not the nature of the elements but the way they are put together which produces the complexity of the body.

ments—hydrogen and oxygen—always combined in the ratio of two hydrogen atoms to one oxygen atom. By proper chemical techniques water can be broken down into these two component elements. If the proportion of the elements is changed, a different compound results. For instance, H_2O_2 is hydrogen peroxide, familiar as a bleach.

A **molecule** is the smallest unit of matter that retains all the properties of that matter. Water is formed of molecules composed of two atoms of hydrogen and one atom of oxygen. It is designated chemically as H_2O or HOH. All compounds are characterized by molecules, but not all molecules represent compounds; elements often occur in molecular form. For instance, oxygen as it exists in an oxygen tank in a hospital is an element, but it is composed of molecules. When oxygen exists in this pure state, the atoms always unite in pairs to form molecules of oxygen composed of two atoms. Hence, we represent the oxygen as O_2. Since there is only one kind of atom involved, however, we still refer to oxygen as an element. A molecular form is characteristic of most of the gaseous elements, for example, nitrogen (N_2) and chlorine (Cl_2), as well as oxygen. The majority of the liquid and solid elements, however, when in the pure state do not exhibit such a combination of their atoms into molecules.

For many years the number of different kinds of atoms was set at 92, and therefore the number of different elements was likewise 92. The atomic age has brought a widening of the physicist's knowledge about atoms, and one result has been the synthesizing of new atoms and elements. The number of elements now stands at about 105. Of these, 92 occur naturally and the rest are artificial products of the laboratory and reactor. This limited number of kinds of atoms can be combined, however, into an almost limitless number of different compounds. About 500,000 compounds have been identified and catalogued, but many more exist or can be pro-

duced. Water is one of the simpler compounds. Common table sugar (sucrose) is somewhat more complex; each of its molecules is composed of 12 atoms of carbon, 22 atoms of hydrogen, and 11 atoms of oxygen $(C_{12}H_{22}O_{11})$. Some complex compounds contain hundreds and even thousands of atoms in each molecule.

Mixtures are combinations of different elements or compounds in which there is little if any chemical union of the combined parts. Furthermore, substances occur in mixtures in different proportions, but compounds always have the same proportions of elements in their makeup. For example, if we mix water and alcohol, we get a homogeneous fluid, but not a compound. We can make water-alcohol mixtures of many different proportions. This is not true of compounds. In sucrose (table sugar) the proportions of carbon, hydrogen, and oxygen must always be the same. Since mixtures can be composed of different proportions of parts, it is evident that the chemical and physical properties of a mixture will vary with the variations in its component parts. A compound, on the other hand, having a fixed proportion of parts, will exhibit fixed chemical and physical properties.

The air around you is a familiar example of a mixture. Air is a mixture of nitrogen (an element), oxygen (an element), carbon dioxide (a compound), and an amount of vaporized water (a compound) that will vary in quantity depending upon the humidity. There will also be very small amounts of other gases, perhaps some smoke, and a few other assorted things in air. When you exhale air, its proportions of oxygen, carbon dioxide, and water vapor will change, but it will still be air.

ORGANIC AND INORGANIC COMPOUNDS

One hundred fifty years ago an organic compound would have been defined as one produced only in living systems, but with the development of organic chemistry, it became possible to synthesize organic compounds from inorganic substances. Today synthetic organic compounds include such important and diverse materials as drugs, explosives, hormones, synthetic fibers, vitamins, and plastics. In general, carbon-containing compounds are organic compounds, but a few simple compounds that contain carbon are inorganic. These consist of small molecules, for example, CO_2 (carbon dioxide) and Na_2CO_3 and other carbonates. Organic molecules have carbon in rings or chains and generally are large molecules; some organic compounds are composed of molecules containing thousands of atoms.

Living matter appears to be so different from nonliving matter that you might think living things would be made of elements which are quite unique and distinctive. A chemical analysis of any form of living matter, however, shows it to be made of some of the elements which are most common in nonliving matter. By weight 96 percent of the human body is composed of the very common elements oxygen, carbon, hydrogen, and nitrogen. The other 4 percent consists of elements that are not especially rare; it is the organization, rather than the composition, of the elements in living matter which makes it unique.

THE NATURE OF ATOMS

Atoms of different kinds have very different chemical properties, yet all atoms are composed of the same subatomic particles. Every atom has a heavy central **nucleus** of positive electrical charge. At varying distances from the nucleus, one or more much lighter, negatively charged electrons orbit rapidly around it. Consider the structure of a simple atom, that of carbon, as an example. Subatomic particles of two kinds are found in the nucleus; there are six **protons,** each of which has a positive electrical charge,

and six **neutrons,** each of which has about the same size and weight as a proton but carries no charge. Around this nucleus six **electrons** are orbiting, each with a negative charge. Two electrons are obiting in an inner orbit with a lower energy level, and four electrons are in an outer orbit at a higher energy level. The electrons are very light in comparison with protons and neutrons, weighing only about 1/1800 as much.

The structure of an atom is sometimes compared to the structure of our solar system, with the large sun equivalent to the nucleus and the smaller planets equivalent to the electrons. The gravity of the sun pulls the planets toward the sun, but this force is opposed by the momentum of the moving planets, which tend, like all moving bodies, to continue moving in a straight line. There exists a balance between these two tendencies, so the planets follow curved paths, their orbits around the sun. The orbit of a planet is not a perfect circle but is somewhat elliptical. Likewise, some electron orbits are elliptical, bringing these particles at times closer to the nucleus than at other times.

In this space age, there is a new illustration

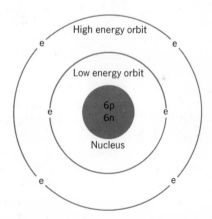

FIG. 3.2 Structure of a carbon atom in Bohr's theory. The central nucleus has 6 protons and 6 neutrons, and 6 electrons revolve around the nucleus at two energy levels. We now know that the electron orbits are not as regular as Bohr thought, but the basic structure is the same.

of the energy levels of electrons—the behavior of space ships. The thrust behind a space ship on takeoff is sufficient to establish a speed which just balances the gravitational pull of the earth, thus establishing an orbit at a certain average distance above the earth. (The orbit is always somewhat elliptical.) Rockets firing while the space ship is in orbit add additional thrust (energy) and raise the space ship to a higher energy level. It then orbits at a greater height from the earth.

On a recent space mission the original orbit was at an average height of about 250 miles, but this was raised to a height of over 800 miles by firing more rockets and increasing the speed of the vehicle. As more and more energy is applied, the speed becomes so great that the space ship escapes from the gravitational pull of the earth and goes out into space. This is the way we launch vehicles to the moon and the planets. A space ship can be brought into a lower orbit by firing retrorockets at the front of the vehicle.

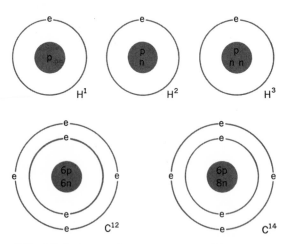

FIG. 3.3 Isotopes of common atoms. The top row shows the common form of hydrogen (H^1) and two isotopes of hydrogen which contain one and two neutrons in the nucleus. The addition of neutrons increases the atom's weight but does not change its chemical properties. The bottom row shows ordinary carbon-12 and the radioactive isotope carbon-14 which has two extra neutrons.

This slows the ship (lowers the energy level). When the speed is reduced below a certain level, the space ship is drawn back into the earth's atmosphere by gravity and thus can be returned to the earth.

It is possible for space ships to establish orbits out as far as the region where the energy expended is so great that the ships escape the gravitational pull of the earth and go into outer space. The orbits of electrons, however, must be at discrete energy levels which are known as **shells.** These range from shell number 1, which lies closest to the nucleus, up to number 7, which is at the highest energy level and lies farthest from the nucleus.

The **quantum theory** of energy helps explain why electrons must be at one of the seven particular energy levels. According to this theory, energy is discontinuous; it does not flow in a steady stream but instead consists of distinct units, or quanta, each **quantum** representing the smallest possible unit of energy. A quantum of light energy, for instance, is known as a **photon,** and a beam of light is regarded as a succession of photons. The shells in which the electrons of an atom orbit can be explained by the fact that a difference of only one quantum of energy exists between the different orbits, or shells. Since there is no unit of energy smaller than a quantum, there are no intermediate orbits. The electrons in shell number 3, for instance, have an energy of one quantum more than electrons in shell number 2, and one quantum less than in shell number 4.

In another respect orbiting electrons may remind us of space ships. Electron orbits are often not exactly circular, and the orbits of space ships around the earth are rarely circular. Instead, they are elliptical; at one point the ship may be 180 miles above the surface of the earth and at another point it may be 260 miles above the surface. Electron orbits may also be elliptical. Four types of orbits are recognized: circular, slightly elliptical, moderately elliptical, and extremely elliptical. All electrons in one shell have the same average energy, but there will be energy fluctuations at different points on the ellipse. Since there are four kinds of electron orbits, there are four possible **subshells.** The electrons in the innermost shell always have circular orbits only. An electron in the second shell may have a circular or a mildly elliptical orbit; an electron in the third shell may have one or another of three kinds of orbits; and in the fourth through seventh shells an electron may move in any of the four kinds of orbits or subshells.

An electron will change its orbit when energy is applied or taken away. When an orbiting electron gives off a quantum of energy, it will drop to the next lower shell. The energy may be given off in the form of a photon of light or X rays. If an electron absorbs a quantum of energy, it will move to the next outer shell of the atom. If the electron is already in the outermost shell and absorbs a quantum of energy it will be ejected from the atom, just as a space ship escapes into outer space when it adds energy while it is already at the outer limits of earth orbits.

A photon of light energy falling upon chlorophyll in green plants causes just such an ejection of an electron. The ejected electron gives up a quantum of energy in the food manufacturing process and then, once more at its original energy level, it returns to the chlorophyll molecule. More information about this is brought out in Chapter 10.

Carbon, with its six protons and six neutrons, has an atomic number of 6, the number of protons in the nucleus, and an atomic mass of 12, the total number of protons and neutrons in the nucleus. These numbers may be indicated as $_6C^{12}$. The lightest of the atoms is hydrogen, $_1H^1$, which has only one proton in its nucleus and no neutrons. It is the only atom without neutrons. Neutrons seem to be necessary in atoms with more than one proton. Even though like electrical charges repel one another, the nucleus of atoms can contain more than one positively charged proton if neutrons are also present. The exact stabilizing function of the neutrons is not completely understood, but we know that if a specific atom has too few neutrons or too many neutrons the atom may be unstable and will undergo rearrangement, with a consequent release of energy, until it attains a stable configuration.

Table 3.1 shows common chemical elements of particular importance in biology, along with some information about the atoms which form these elements. From this table you can see that the number of protons is always the same as the number of electrons. Thus the atom has electrical neutrality. Note also that the number of neutrons increases in comparison with the number of protons as the atoms get heavier. Uranium is the heaviest of the elements that occur naturally; it has 92 protons and 146 neutrons, written $_{92}U^{238}$. In the lighter atoms, such as oxygen and carbon, protons and neutrons are equal in number or nearly so.

Most of the volume of an atom is empty space, as is true of our solar system. The central nucleus forms a relatively large central core around which orbit the very small electrons. If we could enlarge an atom so that the nucleus was about the size of a grape, the electrons would be tiny specks extending out to a distance of 100 feet or more. Another correlation will emphasize the relative emptiness of the atom. If all the atoms that make up the Empire State Building in New York City were to be stripped of their electrons and if their nuclei were then compressed into one solid mass, that mass would be no more than a cubic inch in size; yet it would have the same weight that the building has as it now stands. Elements vary in their weight according to the number of their subatomic particles. A cubic inch of lead is much heavier than a cubic inch of magnesium. The atomic weight of lead is 207, while that of magnesium is only 24. Still, there is much empty space in the atoms of lead.

TABLE 3.1. Subatomic Nature of Some Common Chemical Elements

Element and Symbol	Atomic Number	Neutrons in Nucleus	Atomic Mass: Protons + Neutrons	Electron Shells (Orbits)	Electrons in Outer Shell
Hydrogen, H	1	0	1	1	1
Helium, He	2	2	4	1	2
Carbon, C	6	6	12	2	4
Nitrogen, N	7	7	14	2	5
Oxygen, O	8	8	16	2	6
Sodium, Na	11	13	24	3	1
Magnesium, Mg	12	13	25	3	2
Aluminum, Al	13	14	26	3	3
Phosphorus, P	15	16	31	3	5
Sulfur, S	16	16	32	3	6
Potassium, K	19	20	39	4	1
Calcium, Ca	20	20	40	4	2
Manganese, Mn	25	30	55	4	6
Iron, Fe	26	30	56	4	2
Cobalt, Co	27	32	59	4	2
Copper, Cu	29	34	63	4	1
Zinc, Zn	30	34	64	4	2
Strontium, Sr	38	50	88	5	2
Iodine, I	53	74	127	5	7
Gold, Au	79	118	197	6	1
Lead, Pb	82	126	208	6	4
Uranium, U	92	146	238	7	2

ISOTOPES OF ATOMS

If you weigh a certain quantity of an element and divide that weight by the number of atoms in the quantity of the element, you will find that the average weight per atom will deviate slightly from the weight calculated by the atomic mass. This is because some atoms in your sample will have an atomic mass number which is different from that of the great majority of the atoms. For instance, out of the many millions of carbon atoms in a small bit of soot scraped from inside a chimney, there will be a few atoms that have a mass number of 14 rather than the usual 12. These are heavy carbon atoms that have two extra neutrons each. The chemical properties of an atom are determined by the number of protons and electrons in it, so this heavy carbon still has the same chemical properties as the more common carbon, but it is heavier. We can represent it as $_6C^{14}$, or more commonly just C^{14} or carbon-14. Most oxygen atoms have a weight of 16, but some have two extra neutrons and form heavy oxygen, oxygen-18. There is also a form of heavy hydrogen with an atomic mass of 2. This hydrogen-2, or deuterium, has a neutron as well as a proton in its nucleus. There is even a hydrogen-3, or tritium, which has two neutrons in the nucleus.

We use the word **isotope** for atoms which have a number of neutrons different from that found in the most common form of the atom. Carbon-14 is an isotope of the more common carbon-12. The **atomic weight** of an element is the average weight of all atoms found in a sizable sample of the element as it occurs in its natural state. The most common oxygen atom has an atomic mass of 16, but the element oxygen, as it exists in the air, has an atomic weight of 15.9994. Some oxygen isotopes have an atomic weight of less than 16 and some have a weight of more than 16. In a large sample, where these are averaged together with the great majority of atoms that have an atomic weight of just 16, the result is an average which is slightly below 16. The commonest carbon atom has an atomic mass of 12, but the atomic weight of carbon is set at 12.01115.

Isotopes are valuable to biologists because they behave chemically about the same as the more common atoms, but they can be recognized by special techniques. Thus, they can serve as **tracers** of the movements of elements in living systems. For instance, the use of heavy oxygen has given us much information on the role played by oxygen in the manufacture of food by plants.

Because some isotopes have a number of neutrons which differs from the number in the more common atoms, they are unstable. An unstable isotope is liable at any time to revert to a more stable form, and as it does so it gives off energy. This energy may be in the form of high-energy radiation which can be detected by special detecting instruments. **Radioactive isotopes** in living systems can be traced with such instruments and by other techniques which are described in Chapter 4.

THE BONDING OF ATOMS

The bonding of atoms into molecules is dependent primarily upon the number of electrons in their outer shells, or energy levels. The maximum number of electron shells is seven. The inner, 1-shell can hold no more than two electrons, the 2-shell can hold no more than eight, the 3-shell can hold up to eighteen, and so on with a continuing increase in capacity as the

shells reach higher energy levels. No matter what its capacity, however, the outer shell normally has no more than eight electrons. Atoms tend to react with other atoms and become bonded together in such a way that their orbiting electrons achieve the most stable pattern possible, namely, an outer shell with a configuration of eight electrons (if they have more than one shell). Helium achieves stability with two electrons in its outer shell.

The number of electrons in the outer shell of an atom determines its **bonding capacity.** Hydrogen has an outer shell of only one electron but a capacity for two, so it has a bonding capacity of 1. Carbon has an outer shell of four electrons but a capacity of eight in the outer shell. Hence, it has a bonding capacity of 4.

One type of bonding, known as **covalent bonding,** involves a kind of **electron sharing.** By sharing the electrons in their outer orbits, two or more atoms can achieve a stable configuration. The formation of water from one atom of oxygen and two atoms of hydrogen is a good example. Oxygen has six electrons in its outer orbit, which has a capacity of 8. It needs two more to achieve stability. Hydrogen has one atom in its outer shell with a capacity of 2, so it needs one more to achieve stability. When a molecule of water is formed, each hydrogen atom shares its one electron with the oxygen atom, and the oxygen atom shares one of its outer electrons

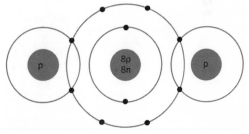

FIG. 3.4 Covalent bonding of two atoms of hydrogen with one atom of oxygen to form water. By sharing the electrons in their outer electron shells, the three atoms achieve a stable configuration.

with each of the hydrogen atoms. These shared electrons may orbit any atom in the molecule.

Carbon, with its bonding capacity of 4, may unite with four hydrogen atoms to form **methane gas.** The carbon and each of the four hydrogen atoms thus achieve stability. Nitrogen has seven orbital electrons, two in the inner shell and five in the outer shell. It has room for three more electrons in its outer shell, so its bonding capacity is 3. When combined with three atoms of hydrogen there is a stable configuration, and **ammonia,** NH_3 is formed.

We can represent these bondings as follows:

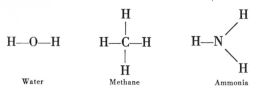

Some atoms have outer shells which are already complete, and they usually do not react with other atoms to form molecules. Helium has

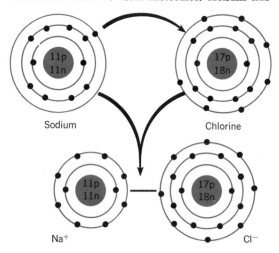

FIG. 3.5 Ionic bonding is achieved by electron transfer. The outer shells of sodium and chlorine are filled when an electron from the outer shell of sodium is transferred to the outer shell of chlorine. Both atoms then become ions. These ions are held together because chlorine has a negative charge and sodium has a positive charge. The compound formed by this bond is sodium chloride, table salt.

two electrons in its only shell, which is therefore full. Neon, argon, krypton, xenon, and radon are elements which have complete outer shells of eight electrons, and they constitute the **inert,** or nonreactive, gases.

A second type of bonding is by **electron transfer,** in which electrons are transferred from some atoms to other atoms. This is known as **ionic bonding** and can be illustrated by the formation of table salt. Sodium has two electrons in its 1-shell, eight in its 2-shell, and only one in a highly incomplete outer 3-shell. If a sodium atom could give up its single outer electron, the second shell would then become its outer shell and would be complete with eight electrons. Chlorine has three shells also, but there are seven electrons in its outer 3-shell. If it could pick up an electron it would become stable. Thus, it seems logical that sodium and chlorine could complement each other by an electron transfer. This frequently happens and the compound **sodium chloride,** NaCl, is formed. Since the nucleus of the atom is not disturbed by this reaction, the sodium is left with an extra overall positive charge; it now has eleven protons but only ten orbital electrons. The chlorine, on the other hand, has gained an electron, so it has an overall negative charge. We say that the two atoms have become **ions;** the sodium atom has become a positive sodium ion, Na^+, and the chlorine atom has become a negative chlorine ion, Cl^-. Since positively charged particles attract negatively charged particles, the two ions are held together and form a molecule, an **ionic compound.** It should be said that the ionic bonding is not quite complete; occasionally the transferred electron goes back and orbits the sodium ion, so there is some covalent bonding in the salt molecule.

When ionic compounds are dissolved in water, some of the molecules may become dissociated, and free ions will be liberated in the water. Sodium chloride molecules break into sodium and chlorine ions. When the water evaporates, the ions are again attracted to one another

BONDING CAPACITY

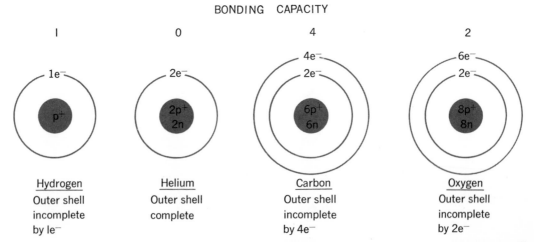

FIG. 3.6 The bonding capacity of several atoms. The capacity is determined by the number of electrons in the outer shell.

and the molecule is restored. Even water dissociates to some extent so that a small fraction of the particles which make up water are hydrogen ions, H^+, and hydroxide ions, OH^-. In this case, hydrogen gives up its single electron to the hydroxide combination.

What determines which atoms will give up electrons and which will receive them in ionic bonding? In general, the atoms with fewer than four electrons in the outer shells give up electrons, while those with more than four receive them. The atoms of hydrogen, sodium, potassium, calcium, and iron have three or fewer electrons in their outer shells, so these are **electron donors.** The atoms of oxygen, chlorine, sulfur, and iodine need one or two electrons to complete their outer shell configurations, so they are **electron acceptors.**

A third type of bonding, **hydrogen bonding,** is of great importance in biology. It is involved in holding the chains of proteins and the ladders of nucleic acids together. Hydrogen bonds are formed when hydrogen is found between atoms of nitrogen, oxygen, or fluorine. These three atoms have five, six, and seven electrons in their outer shells. When hydrogen shares an electron with nitrogen, the electron orbits around the nitrogen nucleus most of the time, leaving the "bare proton" with its positive charge on the surface of the molecule. When this proton comes near an oxygen atom, there is a static attraction between the positive hydrogen nucleus and the negative outer electrons of the oxygen. This attraction is not strong enough for the hydrogen to capture one of the oxygen electrons, but the oxygen is held to the hydrogen nucleus. The hydrogen bond is very weak, compared with bonds of the other two types, and is easily broken, since it is only as strong as the electrostatic attraction. Still, life would be impossible without hydrogen bonding.

ACIDS, BASES, AND SALTS

Some compounds, when mixed with water, dissociate in such a way as to yield hydrogen ions without corresponding hydroxide ions. These compounds are known as **acids.** You are familiar with many acids; vinegar contains acetic acid,

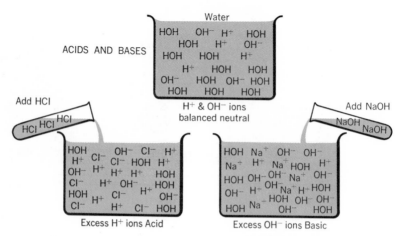

FIG. 3.7 When a neutral medium such as water is mixed with a compound which ionizes to release free hydrogen ions, an acid is produced. The addition of a compound which ionizes to produce hydroxide ions to water produces a base.

sour milk contains lactic acid, and lemon juice contains citric acid. They all have a sour taste. The human stomach produces a common acid, hydrochloric acid, in low concentrations.

Other compounds dissociate and release hydroxide ions without corresponding hydrogen ions. These are **bases** or **alkalies.** They commonly have a bitter taste, and their water solutions feel somewhat slimy.

Still other compounds ionize in water and produce ions which are neither hydrogen nor hydroxide ions. These are **salts.**

Whenever acids and bases are mixed together, the free ions of opposite charges are brought together and a salt is formed. People sometimes take an alkaline substance to relieve a "sour stomach." We can illustrate this reaction by mixing hydrochloric acid and sodium hydroxide. This yields table salt and water. Such a mixing is known as a **chemical reaction** and may be illustrated as follows:

$$HCl + NaOH \rightarrow NaCl + H_2O$$

Some acids are much weaker than others of the same concentration. For instance, common vinegar contains about 3 percent acetic acid, but it does not burn the month when used in foods. Neither does it react violently with aluminum cooking vessels when used in cooking. A 3 percent solution of hydrochloric or sulfuric acid, on the other hand, would severely injure the mouth and literally eat up any aluminum containers into which it was placed. What is the difference? Hydrochloric acid is almost completely ionized in water. On the other hand, the molecules of acetic acid are less than 1 percent ionized. Thus, there are many more free hydrogen ions in a 3 percent hydrochloric acid than in a 3 percent acetic acid, and it is the free hydrogen ion that gives acid its reaction properties.

In studies of living reactions it is often important to know the degree of acidity or alkalinity of a solution. This can be determined by electrical instruments which measure the amount of free hydrogen or hydroxide ions, or it can be determined more easily, but less accurately, by adding color indicators to the solution or using colored papers containing such indicators. Litmus, for example, is an indicator that turns red when an excess of hydrogen ions is present; it is lavender which the solution is neutral and blue when basic. Other indicators measure the degree of acidity or alkalinity.

We can express the degree of ionization of an acid or base in objective terms by means of **hydrogen ion concentration, or ph.** Pure water is ionized to the extent that it has 0.0000001 gram of hydrogen ions per liter. This can be expressed more conveniently as 10^{-7} gram of hydrogen ions. We say that the water has a pH of 7, the negative logarithm (log) of the hydrogen ion concentration.

Note that there are seven numbers after the decimal point. If there were only six numbers, 0.000001 gram of hydrogen ions per liter, the solution would be acidic, with a pH of 6 (10^{-6}). It would have a hydrogen ion concentration ten times that of plain water. A solution with a pH of 5 would have ten times the amount of hydrogen ions present in a solution with a pH of 6. Thus, there is a tenfold increase in the amount of hydrogen ions with each descending point of the pH scale. Each ascending point, on the other hand, represents a tenfold gain in hydroxide ions. The two kinds of ions are equal in amount at a pH of 7; 0 to 7 is the acidic pH range, and 7 to 14 is the alkaline range.

Living matter has a rather narrow tolerance of variation in the pH of its surroundings. Human blood normally has a pH of about 7.3, but can vary slightly above or below this point. If the blood pH ever falls as low as 7.0 or goes as high as 7.8, however, death will result. Hence, we have a tolerance of less than one point on the pH scale in our blood. Plants are likewise sensitive to variation in the pH of the water in the

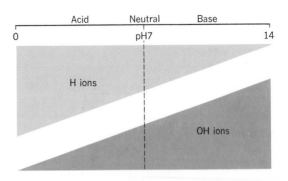

FIG. 3.8 The pH of a solution varies with the number of free hydrogen and hydroxide ions. At pH 7 the numbers of free hydrogen and hydroxide ions are equal and the solution is neutral. An increase in hydrogen ions is accompanied by a corresponding decrease in hydroxide ions and vice versa.

soil. There is an optimum point at which they grow best, and a minimum point and a maximum point compatible with growth. Plants grow very poorly when the pH approaches either extreme, and outside this range they will not grow at all. Farmers frequently add chemicals such as lime to the soil to alter the pH value when they find that it is not at the optimum for the particular plants they plan to grow on the soil.

KINDS OF MIXTURES

There are nine possible kinds of mixtures—gas in gas, gas in liquid, gas in solid, liquid in gas, liquid in liquid, liquid in solid, solid in gas, solid in liquid, and solid in solid. Since water is

TABLE 3.2. Equations of Ionization Reactions of Some Common Acids

Acid Formed	Equation	Name of Negative ion
Acetic	$CH_3COOH \rightleftharpoons H^+ + CH_3COO^-$	Acetate ion
Carbonic	$H_2CO_3 \rightleftharpoons H^+ + HCO^-$	Bicarbonate ion
Hydrochloric	$HCl \rightleftharpoons H^+ + Cl^-$	Chloride ion
Phosphoric	$H_3PO_4 \rightleftharpoons H^+ + H_2PO_4^-$	Diphosphate ion

TABLE 3.3.   Hydrogen Ion Concentration at Different pH Values

Grams Hydrogen Ions per Liter	pH	Examples
$10^0 = 1.0$	0	
$10^{-1} = .1$	1	
$10^{-2} = .01$	2	
$10^{-3} = .001$	3	Lemon juice 2.3
$10^{-4} = .0001$	4	acid
$10^{-5} = .00001$	5	
$10^{-6} = .000001$	6	
$10^{-7} = .0000001$	7	neutral Fresh cow's milk 6.6
$10^{-8} = .00000001$	8	Human blood 7.3
$10^{-9} = .000000001$	9	Sodium bicarbonate tenth normal 8.4
$10^{-10} = .0000000001$	10	
$10^{-11} = .00000000001$	11	basic
$10^{-12} = .000000000001$	12	
$10^{-13} = .0000000000001$	13	Lime water (saturated solution CaOH) 12.3
$10^{-14} = .00000000000001$	14	

The amount of hydroxide ions (OH$^-$) is in inverse proportion to the amount of hydrogen ions (H$^+$). At a pH of 7 the amounts are equal; a tenfold increase of one and decrease of the other is reflected at each point on the pH scale.

universally present in living matter and since water is a liquid, we shall learn more about the kinds of mixtures which substances can form with a liquid, using water as a specific example.

Solution

When mixed with water, some substances break down to form individual molecules, ions, or even atoms. Such a substance is said to be dissolved in the water, the water being known as the **solvent** and the dissolved substance as the **solute.** A mixture of this type is known as a **solution,** and the dissolved substance may be a solid, a liquid, or a gas. Sugar, a solid, dissolves in water to yield a clear solution. Alcohol is a liquid which dissolves in water, and oxygen is a gas which dissolves in water. (Were it not for dissolved oxygen, fish and other water animals could not live.) In every case, the physical and possibly the chemical properties of the water are affected by the dissolved substance. Water's boiling and freezing points may be altered, as well as its ability to transmit electric currents, and so on. In a true solution, the dissolved particles do not settle out or rise to the surface on standing. When water containing solids in solution is evaporated, the solids will form in characteristic crystals in most cases. We call such substances **crystalloids.**

Suspension

If we place some powdered clay in water and shake it thoroughly, the water becomes cloudy and the particles will be suspended in it. The clay particles have not gone into solution, however, for each particle is composed of many molecules and these will settle to the bottom if this mixture is allowed to stand for a time. The particles can be seen under the microscope and appear as rather large bodies. This type of mixture is known as a **suspension.** Water in such a suspension has the same boiling and freezing points, as well as the other physical and chemical properties, that are found in pure water.

Emulsion

If we put a little olive oil in water and shake it vigorously, the oil will break up into many tiny particles which become evenly distributed in the water. The water will appear milky. Each of these tiny droplets of oil will consist of many molecules, because olive oil will not go into solution in water. Such a mixture of a liquid in a liquid is known as an **emulsion.** If we allow the mixture to stand for a time, the oil droplets will rise to the surface and soon the oil will be separated from the water.

It is possible to prevent the separation of the two parts of an emulsion by the addition of an emulsifier. An **emulsifier** is a substance in

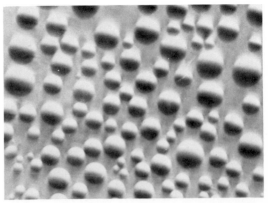

FIG. 3.9 Microscopic appearance of an emulsion of oil and water. The oil does not mix evenly in the water but accumulates in small droplets. To the naked eye, the mixture appears milky.

which the ends of the molecules are different— one end is soluble in water and the other in oil. Thus, when an emulsifier is added to a mixture of oil and water it tends to bind the two together and prevent separation. Ordinary soap or any detergent is an example of an emulsifier. In soap (sodium stearate), the sodium end of the molecule is soluble in water and the stearate end is soluble in oil. Thus soap can remove oil from your hands, whereas plain water cannot. The bile produced by the liver acts as an emulsifier of the fats which we eat. When there is a deficiency of bile, there cannot be a proper digestion of fats.

Colloidal Dispersion

A **collodial dispersion** is composed of particles that are larger than the tiny units of matter in a solution but smaller than the relatively large particles of matter in a suspension or an emulsion. The size of collodial particles (**colloids**) is usually stated to be between 0.001 and 0.1 micron in their greatest dimension. (A micron is a thousandth of a millimeter.) This is the size that is just below the limits of the ordinary

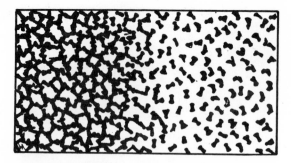

FIG. 3.10 The shift from a sol to a gel state in a colloidal dispersion. At left the solid (dark area) is discontinuous while the water (white area) is continuous (sol state). On the right the solid becomes the continuous phase while the water becomes discontinuous and the mixture becomes a solid (gel state).

optical microscope, and thus colloidal dispersions appear homogeneous under such microscopes. Molecules as a whole, on the other hand, lie below the smaller limits of size of colloids. In colloids particles will not settle out or rise to the top upon standing. The particles are so small that they are kept evenly dispersed by the motion of the water molecules around them. In this respect collodial dispersions are like solutions. But, unlike solutions, colloids do not cause any great change in the chemical properties of the water in which they are dispersed.

The white of an egg is a good example of a colloidal dispersion. The albumin is present as finely divided particles dispersed in water. This type of collodial dispersion is sometimes called a colloidal suspension but should not be confused with the true suspension just described. Homogenized milk is another example of a colloidal dispersion. In fresh cow's milk the oil or fat is dispersed through the milk in large droplets and forms an ordinary emulsion, evidenced by the fact that the cream will rise to the top of the milk upon standing. In homogenized milk, the fat droplets are broken into smaller sizes that fall within the range of

colloids. Thus, no cream will rise to the top of homogenized milk no matter how long it stands. Such a colloidal dispersion may sometimes be called a colloidal emulsion, but we should distinguish this from the true emulsions. Except for a very few cases, solid colloidal particles do not form crystals when the water has evaporated.

Sols and Gels

Colloidal dispersions can change from a liquid to a solid or semisolid state and back again. If we mix gelatin with hot water, the particles of gelatin form a colloidal dispersion and a liquid results. This is the **sol state.** The water forms a **continuous phase** and the particles of gelatin are in a **discontinuous phase.** As the mixture cools, however, the gelatin particles come together and form the continuous phase while the water is in the discontinuous phase, and the mixture becomes solid; it is then in the **gel state.** We can reheat the mixture and restore the sol state. The white of an egg is in a liquid state when it comes from the shell; but sufficient heat will cause the colloidal particles to fuse together, and the mixture goes into the gel state. We use gelatin, cornstarch, egg white, and flour to form colloidal systems in the gel state to make many of our everyday foods more attractive and tasty.

Changes in the viscosity of protoplasm representing slight changes in the sol-gel relationship account for many of the activities within a cell. When certain cells are viewed under the microscope, we can see an active movement of the protoplasm as it flows around in the cell. This is known as protoplasmic streaming and is brought about by changes in viscosity of protoplasm in different parts of the cell.

REVIEW QUESTIONS AND PROBLEMS

1. Chemical reactions take place more rapidly at high temperatures than at low temperatures. In view of what you have learned about the movement of molecules, why, do you think, is this true?

2. How may the movement of molecules be demonstrated visually even though the molecules themselves cannot be seen? What is this movement called?

3. The number of different kinds of elements is the same as the number of different kinds of atoms? Explain why this is true.

4. The oxygen of the air exists as a compound, yet oxygen is an element. How is it possible for an element to be a compound?

5. How do mixtures differ from compounds?

6. Explain the similarity and differences in the sun's solar system and the structure of an atom.

7. Describe the nature of each of the three basic subatomic particles which make up atoms.

8. Hydrogen has one proton and one electron; helium has two protons and two electrons; yet helium is four times as heavy as hydrogen. How can you explain this?

9. Describe the energy levels of electrons according to the quantum theory. How can an electron change from one energy level to another?

10. The common oxygen atom has eight protons, eight neutrons, and eight electrons. An isotope of oxygen weighs one-eighth more. How many protons, neutrons and electrons would this isotope have?

11. The great majority of the atoms found in a typical sample of a certain element have 27 protons and 27 electrons each with an atomic mass number of 59. How many neutrons would each such atom have in its nucleus? What is the atomic number of the element?

12. An iron skillet is heavier than an aluminum skillet of the same size. What differences would you expect to find in the atoms of these two elements?

13. Of what value are isotopes of atoms in biological research?

14. Helium is an element which does not form compounds with other elements, but carbon forms a great variety of compounds. Explain.

15. Describe the three methods of bonding of atoms.

16. How does an ion differ from an atom?

17. Vinegar is often called an acid. What does this tell you about the nature of the hydrogen ions and hydroxide ions in vinegar?

18. A liquid with a pH of 5 is ten times as acid as one with a pH of 6. Explain what is meant by this.

19. If you put some grains of an unknown substance in water and stir it up, how can you tell if the substance has gone into solution?

20. How does a suspension differ from a colloidal dispersion?

21. A farmer finds that his corn plants grow better if he adds lime, that is, calcium

hydroxide (CaOH), to his soil. From this information what can you say about the pH of the soil and pH optimum for corn plants?

22.   Explain the changes which occur when gelatin is stirred in hot water and the water is allowed to cool.

FURTHER READING

Borek, E. 1961. *The Atoms within Us*. New York: Columbia University Press.

Conn, E. E., and Stumpf, P. K. 1972. *Outlines of Biochemistry*. New York: John Wiley and Sons.

Haynes, R. H., and Hanawalt, P. C. 1968. *Molecular Bases of Life*. San Francisco: W. H. Freeman.

Lehninger, A. L. 1973. *Short Course in Biochemistry*. New York: Worth Publishers.

McElroy, W. D. 1971. *Cell Physiology and Biochemistry*. Englewood Cliffs, N.J.: Prentice-Hall.

Steiner, R. F., and Edelhoch, H. 1965. *Molecules and Life*. New York: Van Nostrand Reinhold.

4

Radiation and the Living Organism

In recent times we have become greatly concerned over the effects of high-energy radiation on living organisms. With the development of nuclear weapons and nuclear generating plants, the expanding use of X rays and radioactive isotopes in medical diagnosis and treatment, and the use of radiation in industry, man is being exposed to high-energy radiation more frequently than ever before. We know that large amounts of radiation can cause injury and death of living tissue. We also know that genetic alterations can be induced by radiation and that these changes in genes and chromosomes can be passed on to future generations. It benefits us all, therefore, to know something about the nature of radiation and how it affects man and other forms of life.

THE SOURCES OF HIGH-ENERGY RADIATION

Radiation is energy which travels in waves or particles. Visible light, heat, ultraviolet rays, radio waves, and TV waves are all radiation energies. One form of radiation energy which we have heard about, but with which we may not too familiar, is **high-energy radiation,** also called **ionizing radiation** because it creates ions as it passes through matter. We can neither see, hear, feel, smell, or taste this type of radiation, yet it can have powerful effects on our bodies. In common usage the word **radiation** is used to refer to this high-energy form, and we shall use it in that sense.

Einstein's Equation

In 1919 Albert Einstein reported the development of his famous equation $e = mc^2$ before a scientific meeting in London. The London *Times* reviewed his presentation and closed the review with the statement: "While this is an important theoretical presentation, it is agreed that it will make little difference to the practical world."

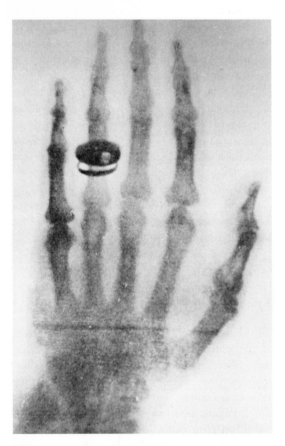

FIG. 4.1 The first X ray photograph. This X ray was taken by Konrad Roentgen in 1895. It shows bones inside the hand of a living person.

This may have been the greatest understatement of the century, since this simple equation opened the way for the release of energy from the atom with all its potential for destructive, as well as constructive, uses. This incident illustrates the point that it is not possible to predict in advance the ultimate application of scientific discoveries.

Discovery of X Rays

Man has always been exposed to low-level radiation, which comes from natural sources such as cosmic rays and radioactive materials in the earth. Since radiation is not perceived by any of the senses, however, he never knew it. The first demonstration of this form of energy came in 1895 when Wilhelm Konrad Roentgen, a professor at the Julius-Maximilian University in Wurzburg, Germany, devised a vacuum tube through which he passed a spark of high-voltage electricity. In the dim light of a late afternoon, Roentgen noticed that certain mineral samples would light up with an eerie, blue-green glow when the spark was passing through the tube. He reasoned that some sort of invisible rays must be passing from the tube to cause the minerals to glow. He called these **X rays**, X standing for the "unknown."

Roentgen was also experimenting with photography, which was just coming into its own at this time, and while he was experimenting with X rays, he happened to have photographic plates in their lightproof holders lying on the table. When he developed the plates after the experiment, they all turned black, as if they had been exposed to light. One plate particularly interested him. On it he had placed some of his keys, and the part of the plate underneath the keys was clear; this part of the plate had not been exposed. He reasoned that the X rays must have penetrated the plates and exposed them, but that the metal of his keys was so dense that the keys absorbed the X rays and did not allow them to reach that part of the plate. Then he had a brilliant idea. He decided that it might be possible to make pictures of dense organs inside the human body. To test this hypothesis, he asked his wife to place her hand on one of the plate holders while he turned on the electricity in the tube. You can imagine his great excitement while he watched the plate develop in the red light of his darkroom and saw the bones of her hand come out clearly, while the softer flesh was barely visible.

When the discovery of this use of X rays was reported in the spring of 1896, the scientific world became excited about the possible applications of this technique. Physicians especially saw the potential value of the use of X-ray pictures to detect abnormalities within the human body without surgery. X-ray machines were soon being widely used for this purpose. It became evident that the discovery was a mixed blessing. Many physicians who used the X-ray technique began to develop cancers on their hands, and some had to have their hands amputated as the tissues deteriorated. In some the damage extended to their whole bodies and they died.

In America, Thomas Edison began making X-ray tubes and devised a screen of fluorescent minerals which would glow when struck by X rays. This made it possible to see an X-ray image without having to develop film and to see the activity of organs in the body. In 1896 Edison put on an exhibit of this great invention in New York City. People stood before the machine to make their internal organs visible on the screen. One of Edison's assistants repeatedly exposed himself to the X-ray machine to show that the X rays were harmless. The assistant died of overexposure to X rays a few months after the exhibition. His death showed that X rays have a cumulative damaging effect. A single exposure seems to be harmless, but if exposures are continued over a period of time, the effects of the X rays build up and cause severe injury to body tissues.

Radioactive Elements

Within a few months after the discovery of X rays the French scientist Henri Becquerel found that similar radiation was given off by certain uranium salts. He wrapped some of these salts in black paper and placed them on a photographic plate in a darkroom. Later, when he developed the plate, he found a darkened outline of each crystal. From this he concluded that the crystals were giving off radiation like the X rays which caused exposure of photographic film. This was the first demonstration of the principle of **autoradiography**—pictures made by means of radiant energy given off by the subject being photographed.

Two other French scientists, Pierre Curie and his wife Marie, became interested in rays from uranium ore and began to try to isolate the substance in the ore which was emitting them. Madame Curie coined the word **radioactive** to describe minerals with this property. Their research led them to the discovery and isolation of radium, which later proved so valuable in the treatment of cancer. It was soon found that radioactive substances, like X rays, could cause damage to living tissue. Becquerel borrowed a small glass vial of radium from the Curies and carried it in his waistcoat pocket for several days. Shortly thereafter he developed a "burn" on his skin which had been near the radium.

VARIETIES OF HIGH-ENERGY RADIATION

Many radioactive isotopes have been discovered since the time of Becquerel and the Curies. Research on these isotopes showed that the type of radiation given off often differed from one isotope to another, and two main categories were recognized. **Electromagnetic radiation** is in the form of short-wave, high-energy radiation similar to X rays. **Particulate radiation** is in the form of actual subatomic particles speeding away from the atom with high energy. Let us survey the different forms of radiation to which man and other living things may be exposed.

Alpha Particles

Alpha particles consist of two protons and two neutrons. They are by far the largest of the forms of particulate radiation. Each particle carries two positive charges. With such a charge, they are deflected by any negatively charged ions that may lie near their pathway. As a result alpha particles are not highly penetrating; the outer layer of the human skin will deflect and absorb practically all the alpha radiation to which a person is exposed. Internal body organs may be exposed, however, if some radioactive isotopes are inhaled or swallowed and the alpha particles are emitted inside the body.

Beta Particles

Beta particles are single electrons emitted from radioactive isotopes at a high energy. They are very small in comparison with alpha particles, since an electron weighs only about 1/1800 as much as a proton or a neutron. Beta particles, being electrons, have a negative charge, so they will be deflected by positive charges in the matter through which they pass. They vary greatly in their penetrating power because they are emitted from the atom with varying degrees of energy. Even those with highest energy, however, are not very penetrating because of their charge, yet as a whole they are more penetrating than alpha particles. Like alpha particles, they can cause internal damage to a person's body if inhaled or swallowed.

Neutrons

Individual neutrons are emitted by some radioactive isotopes. A neutron is extremely penetrating because it has no electrical charge and is neither deflected nor slowed by passage near charged particles. Instead, it tends to move in a straight path until it collides with the nucleus of an atom. Since atoms consist largely of empty space, a neutron may travel a considerable distance before such a collision takes place. The cells inside an animal body receive just as much neutron radiation as is received by the skin on the outside of the body.

Neutrons are also dangerous for another reason; they can cause atoms to become radioactive. When the stable C^{12} is bombarded with neutrons, some of the carbon atoms will receive extra neutrons and C^{14} is created. The scientists in atomic energy laboratories create radioactive isotopes by neutron bombardment, many of which are used in biological research. The neutrons released in the fission of U^{235} when an atom bomb or a hydrogen bomb is exploded form many radioactive isotopes from the elements in the matter near the explosion, and these become a part of the radioactive fallout from such bombs.

Protons

Individual protons may be given off from isotopes, but these are much less common than the three forms of particulate radiation just described. They range between alpha and beta particles in penetrating power.

Gamma Rays

Gamma rays are short-wavelength, high-energy rays and thus represent electromagnetic radiation. They are given off as a form of energy release from some radioactive isotopes, or they may be generated from modern high-voltage X-ray tubes. Gamma rays are highly penetrating, but they gradually lose energy as they pass through matter, in proportion to the density of the matter. This characteristic makes X-ray pictures possible. The less dense parts of the body allow considerable penetration of the rays, and film behind the softer body tissues will thus be exposed. Bones and other denser body parts, on the other hand, absorb more of the rays, and the film behind them receives less exposure.

Cosmic Radiation

Cosmic radiation is a type of radiation constantly bombarding us from origins somewhere in outer space. Primary cosmic radiation reaches the earth in the form of stripped nuclei of elements such as carbon, nitrogen, and oxygen. Few of these nuclei ever penetrate the atmosphere, however, because our atmosphere acts as a shield, as effective at sea level as three feet of lead. When an invading particle collides with the nucleus of an atom of the atmosphere, however, it gives rise to a shower of both particulate and electromagnetic radiation known as secondary cosmic radiation. This reaches us at a very high energy level and is very penetrating. Even on the lower floors of buildings several stories tall this cosmic radiation can be detected. Only by going into a cave deep underground can one escape it. Fortunately, the amount of such

FIG. 4.2 Although human beings cannot detect high-energy radiation with any of their senses, some lower animals can. The animals shown here are planaria. In the top photograph the planaria have crawled under a heavy lead shield after being exposed to cobalt-60 radiation. In the middle picture, the shield has been removed. In the bottom picture, the planaria have become randomly distributed after the radiation source was removed.

radiation is very low; otherwise life on the land surfaces of the earth would be impossible.

MEASURING RADIATION

High-energy radiation cannot be detected by any of man's senses. A piece of radioactive cobalt (Co^{60}) the size of a pea could be attached to a chair and any person sitting in that chair for even a few minutes would receive a lethal dose of radiation without ever knowing it. Recent research by one of the author's students, David Hunsaker, has shown that some species of flatworms, **Planaria,** can detect high-energy radiation and respond to it. A group of these worms, placed near a Co^{60} source exposing them to gamma radiation of an intensity of about 50 R per hour, would crawl under a lead shield which absorbed about 75 percent of the radiation. If the two eyespots on the head of the worm were removed, the worms no longer responded. This showed that the eye spots were the areas of sensitivity. There is no evidence, however, that man can detect high-energy radiation. Hence, we must depend upon instruments which are sensitive to such radiation.

In the early days of X-ray use, it was customary to use the voltage of the tube and the distance to the subject as measurements of radiation. This measurement was not accurate because voltage can fluctuate and other conditions can cause considerable variation in the output of the tube. Then the **Geiger-Muller tube** was invented. It contains a gas which is ionized by radiation. The ionization causes pulses which can be detected by a counter. The first unit of X-ray measure, the **roentgen,** was based upon this kind of gas ionization. The roentgen is said to equal one electrostatic unit of charge in one milliliter of air at a standard temperature and pressure. As the study of particulate radiation progressed and revealed that ionization in a gas is not equivalent to the effects on matter, other units were introduced. Today we more commonly use the **rad** (radiation absorbed dose), which is represented by a capital **R.** The rad represents the amount of energy absorbed per unit of matter. In studies of the effects of radiation on man, we also sometimes use the **rem** (radiation equivalent man).

BIOLOGICAL USES OF RADIOACTIVE ISOTOPES

As Tracers in Living Systems

Radioactive isotopes behave chemically much like the stable elements. This means that most radioactive isotopes of stable elements can be used as **tracers** to determine the pathway of specific elements in living organisms. We can give a rat a small amount of radioactive iodine and determine how his body uses iodine by measuring the amount of radioactivity which appears in the rat's body organs. In this case the thyroid gland will show the greatest radioactivity because it is the part of the body which uses the most iodine.

The same tracer technique has medical value. Sometimes when there is cancer of the thyroid gland, parts of the cancerous tissue break off and lodge in other parts of the body. By giving the patient a very small quantity of the radioactive iodine-131 we can later check the areas of greatest activity and thus locate islands of migrated, cancerous thyroid tissue. This can then be removed surgically.

In plants the tracer technique is equally valuable. If we wish to learn something about the uptake of phosphorus by a plant, we can add a measured quantity of radioactive phosphorus (P^{32}) in the form of a phosphorus compound in solution to the soil at the base of the plant. Then with measuring instruments we can determine how rapidly the phosphorus is taken up, which parts of the plant receive it,

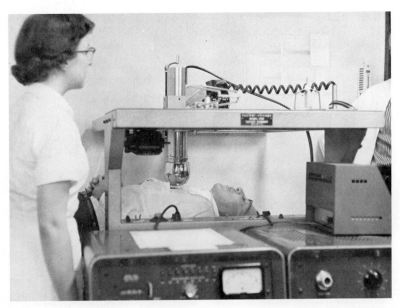

FIG. 4.3 Detection of thyroid cancer by using radioactive iodine-131. This man was given a small quantity of iodine-131. A radiation detecting tube is then passed over his body. This test has detected some thyroid cancer tissue which has migrated to his chest.

and how much remains in the soil. Such studies can have value in determining the best time for application of fertilizers on crops. By placing a plant in a sealed jar and adding carbon-14 in the form of carbon dioxide, we can learn much about how carbon enters into food manufacture and the various intermediate steps involved.

Radioactive tracers also have value in ecological studies. By feeding small water animals food containing tracers, we can learn how far they migrate, what other animals use them for food, how long they live, and other important information—all thanks to the telltale radiation.

Autoradiography

Since the radiation from isotopes affects photographic film, we can use the principle first discovered by Becquerel and make autoradiographs which are valuable in showing the uptake of minerals in various body parts. Fig. 4.4 shows how the uptake of four elements was demonstrated in young frogs. Each was injected with a small quantity of a compound containing a radioactive element. Twenty-four hours were allowed for absorption, and then the frogs were killed and placed on photographic film in a darkroom. After another twenty-four hours the films were developed. The pictures show that phosphorus goes primarily to the bones, iodine to the thyroid glands and to the skin to some extent, and sodium to the softer body parts; zinc remains rather localized where it was injected. Zinc is one of the trace elements of the body, an element needed only in very small quantities. These four autoradiographs tell us more about the absorption and distribution of minerals in an animal body than could be discovered by many years of careful chemical analysis of the various body parts.

Plants also serve for studies by autoradiography. Fig. 4.5 shows how the rapidity of uptake of phosphorus from the soil by a coleus

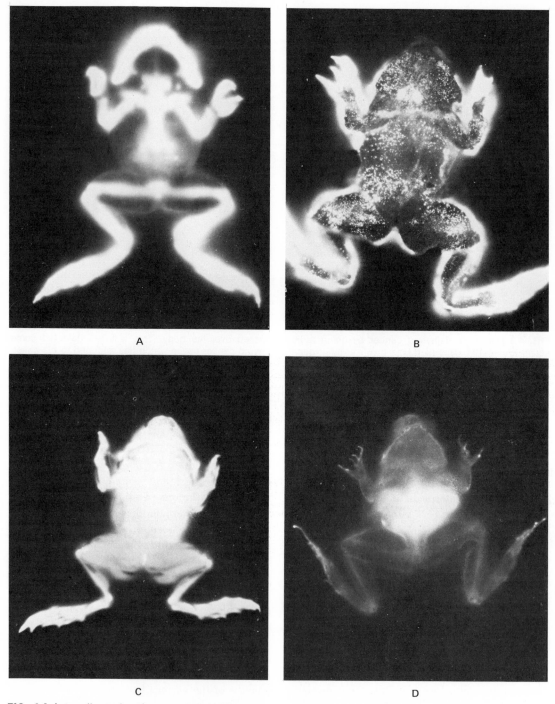

FIG. 4.4 Autoradiographs of young frogs which have been injected with four radioactive isotopes. These radioactive isotopes, or tracers, make it possible to see which areas of the body take up different ele-ments. The isotopes used were: phosphorus-32, upper left; iodine-131, upper right; sodium-22, lower left, and zinc-65, lower right.

plant is determined. A leaf removed fifteen minutes after the isotope was placed around the roots shows some radioactivity in the petiole, which attaches the leaf to the stem, but very little radioactivity in the blade of the leaf. A leaf removed after thirty minutes shows the radioactivity, and hence the phosphorus, already spread to the major vein of the leaf and beginning to spread out into the smaller veins. This spread continues until all parts of the leaf show the radioactivity after two hours.

It is even possible to trace the utilization of minerals within cells by means of microautoradiography. Very thin photographic emulsions can be fitted over microscopic preparations previously exposed to radioactive elements. After time has been allowed for exposure of the film, the entire slide can be immersed in developing solutions and dark spots will appear on the photographic emulsion where it has been exposed. These spots represent grains of silver which develop in the emulsion where the radiation reached it. By this method much information has been obtained on the functioning of cell parts. Some of this will be discussed as we continue our study (see Fig. 9.3).

Dating by Means of Isotopes

As understanding of radioactive isotopes advanced, a way opened for determining the dates of forms of life which lived on the earth in the past. The decay of radioactive isotopes takes place at a precise rate. This rate is expressed in terms of **half-life.** We know, for example, that any given mass of carbon-14 will lose half its radioactivity in about 5600 years. Let us say that you assemble a very small quantity of this isotope containing only 1,000,000,000 atoms. After 5600 years there will be only 500,000,000 carbon-14 atoms remaining; the rest will have decayed into stable nitrogen atoms. After another 5600 years you might expect that there would be none re-

maining, but instead there would be 250,000,000 atoms. In other words, the mass has again lost half of its radioactivity. When another such period of time has passed, the number will be reduced to 125,000,000 atoms and the mass will be only one-eighth as radioactive as it was at the beginning. The same principle holds true for other radioactive isotopes. The half-life varies greatly—ranging from fractions of a second up to millions of years—but for any specific isotope it is constant. Table 4.1 shows the half-lives for a number of isotopes.

Carbon-14 is one of the most valuable isotopes for dating. A certain very small proportion of the carbon in the carbon dioxide of the air has this radioactive form. Since carbon dioxide is used by plants to manufacture food, there will be an equal proportion of carbon-14 among the carbon atoms of all living things. Upon death, however, there is no more intake of food and the radioactive carbon already in the body will lose its radioactivity at the half-life rate. If the remains of an animal show one-eighth the carbon radioactivity that is found in living animals, we can assume that the animal died about 16,800 years ago (three half-lives).

TABLE 4.1. Some Radioactive Isotopes and Their Half-lives

Isotope	Radiation Emitted	Half-life
Nitrogen-16	Beta and gamma	7.4 seconds
Sulfur-37	Beta and gamma	5 minutes
Sodium-24	Beta and gamma	15 hours
Gold-198	Beta and gamma	2.7 days
Iodine-131	Beta and gamma	8 days
Phosphorus-32	Beta	14.5 days
Iron-59	Beta and gamma	45 days
Cobalt-60	Beta and gamma	5.2 years
Strontium-90	Beta	28 years
Radium-226	Alpha and gamma	1,620 years
Carbon-14	Beta	5,600 years
Chlorine-36	Beta	310,005 years
Potassium-40	Beta and gamma	300 million years
Uranium-235	Alpha, beta, gamma, and neutrons	710 million years

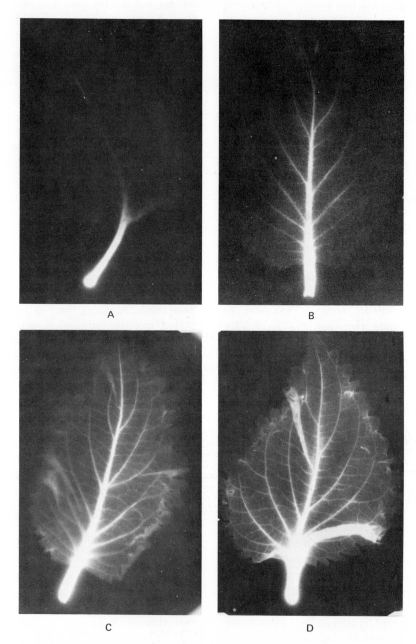

A

B

C

D

FIG. 4.5 Autoradiographs of leaves from a coleus plant showing the rate of uptake of phosphorus-32. The isotope was placed around the roots and leaves were removed from the plant at different times. The leaf shown in the upper left was removed after 15 minutes, the one at upper right after 30 minutes, the one at the lower left after 1 hour, and the one at lower right after 2 hours.

Many questions concerning the age of the remains of prehistoric man and other organisms have been answered by carbon-14 dating.

When we find remains which are more than about 50,000 years old, the carbon-14 dating cannot be used because the amount of radioactivity remaining is small and difficult to measure accurately. We can turn to other isotopes, however, which have longer half-lives. One method which has proved to be valuable is **potassium-argon dating.** Potassium-40 is a radioactive isotope. When the atoms of this isotope give off radiation, they become calcium-40 and argon-40. Argon-40 can be recognized by chemical analysis. Now suppose an animal is embedded in a lava flow which forms rock. There will be some potassium-40 in the rock but no argon-40, because argon is a gas that is given off as fast as it is formed. Within the rock, however, the argon-40 is trapped and builds up through the centuries. When we uncover the fossil, we can compare the amount of potassium-40 with the amount of argon-40 and determine the age of the rock and the animal embedded in

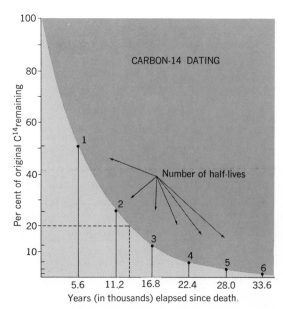

FIG. 4.6 The loss of radioactivity by carbon-14 proceeds at a definite half-life rate. Knowing the half-life of carbon-14 makes it possible to determine the age of fossils by measuring carbon-14 activity. If the present activity of carbon-14 in the fossil remains of an organism is 20 percent of the original activity, then the organism must have died about 13,250 years ago.

FIG. 4.7 The approximate age of the mummified remains of this Indian woman has been determined to be about 700 years by the amount of radiation given off by the carbon-14 in her bones. The mummy was found in a cave in New Mexico.

it. Since potassium-40 has a half-life of 300 million years, we can go far back into the prehistoric past with this method.

When we are dealing with the early formation of the earth in epochs whose remoteness as measured in years runs into the billions, there is still another radioactive element which can be utilized. This is uranium-238, ordinarily thought of as the stable form of uranium. It gives off some radiation and has a half-life of 4.5 billion years. It becomes an unusual form of lead when it decays, lead-206. When we find rocks containing uranium-238, we can measure the relative proportion of lead-206 and establish the approximate date when the rocks were formed. This method was used extensively in determining the dates of events during the formation of the earth, as discussed in Chapter 5.

RADIATION DAMAGE TO BIOLOGICAL SYSTEMS

Ionization

The damage done by radiation appears to be brought about largely by ionizations. Both electromagnetic and particulate high-energy radiations cause ionization of matter through which they pass. Within a living cell such ionization can produce chemicals that are highly reactive. Such ions may react with oxygen and produce compounds which in turn react with the chemical constituents of vital cell parts. This concept is supported by the fact that cells kept at a very low concentration of oxygen receive less damage from radiation than cells with normal oxygen concentration. Also, it is known that if the cells receive treatment with certain chemicals, such as AET (described in Chapter 1), they are less sensitive to radiation damage. This is because the chemicals combine with the highly

reactive compounds formed in the cell and prevent the compounds from reacting with the cell parts.

The different kinds of radiation cause ionization for different reasons. When an alpha particle passes near an atom, it may pull an orbital electron out of its orbit because of the attraction of the positive charges of the alpha particle for the negative charge of the electron. The electron lost from one atom may be captured by a nearby atom, and thus two ions are created, the first positive and the second negative. A beta particle will also produce ionization by causing electrons to be thrown off as a result of the mutual repulsion of negative charges. Neutrons do not have a charge and thus do not draw electrons out of orbit as they pass by. When a neutron strikes the nucleus of an atom, however, it causes the atom to become "excited"; some of the energy of the neutron is transferred to the atom. An "excited" atom is unstable and may give off excess energy by throwing off an electron. Gamma rays cause ionizations in a similar way. When the energy

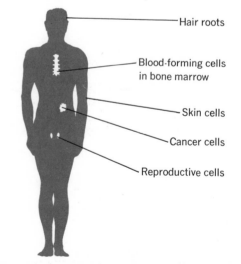

Hair roots

Blood-forming cells in bone marrow

Skin cells

Cancer cells

Reproductive cells

FIG. 4.8 Body tissues vary widely in their susceptibility to damage by radiation. This diagram shows the parts of the human body which are most readily damaged. These areas contain rapidly growing cells.

FIG. 4.9 Effect of high-energy radiation on barley seed. The controls received no radiation and show normal seedling growth. The experimental groups show increased retardation of growth as the amount of radiation received increases.

CONTROL 20,000 R 30,000 R 40,000 R 50,000 R

of the rays is absorbed by an atom, that atom becomes "excited" and throws off one of its electrons. We sometimes speak of **ion pairs** created by radiation, because whenever one ion is created, the lost particle joins another atom and creates a second ion. One rad of gamma radiation creates about two billion pairs of ions in matter, yet this represents ionization of only about one atom out of each ten billion in one gram of living tissue.

If a person's entire body is exposed to rather heavy radiation, certain parts of the body will be affected more severely than others. The skin will redden and develop many small lesions. The lining of the intestine will be damaged, and the person will become nauseated and suffer from diarrhea. As the days pass, there is a drop in the number of his blood cells, and so he becomes anemic, his blood does not clot properly, and he becomes less resistant to disease. Sterility may result because of the destruction of reproductive cells. Other body tissues, such as bone, muscle, nerve, and the tissues of the liver and kidneys, are less sensitive and are damaged only with very heavy doses of radiation. This differential is caused by the fact that the cells which are

growing and dividing rather rapidly are more sensitive to radiation. Fortunately, cancer cells come in this category, and cancer can be treated with high-energy radiation. The cancer cells can be destroyed with a dose which will not kill the surrounding tissue. Most of the cells of the nervous system, muscles, and bone are laid down in embryonic development and do not divide thereafter. Hence, these tissues are very resistant to radiation damage.

The damage done to cells by radiation exposures in the medium range seems to be due primarily to breakages and rearrangements of the chromosomes and their vital genes. If a cell does not divide after breakage and rearrangements of its chromosomes, it can continue with normal activities because all the genes are present and continue to function. If the cell divides, however, some of the broken segments of chromosomes may be lost or there may be an irregular distribution of the genes to the daughter cells. Both cells may die or behave abnormally as a result of the changes. This fact explains why the cells which are growing and dividing are the most sensitive to radiation damage. Sometimes radiation may bring about changes within grow-

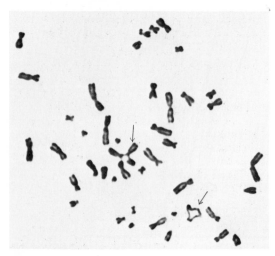

FIG. 4.10 Radiation damage to human chromosomes. These chromosomes come from tissue cultures of human cells which were exposed to 50 R of radiation. Arrows indicate two places where chromosome aberrations have occurred.

ing cells which deprive them of the ability to divide. The stimulus to divide, whatever it is, is destroyed or inhibited, and so the cells continue to grow until they become giant cells and eventually die because they become inefficient as a result of their large size.

Besides damage to chromosomes, radiation may bring about changes in the individual genes. Such a change is known as a **gene mutation.** Many mutations have no effect on the body cells in which they occur; a mutation for an abnormal type of eye would not be "expressed" at all if it occurred in a cell on your elbow. Even if it occurred in a cell of the eye of an adult person, the amount of tissue influenced by this mutant gene would be so small that no effect would be noticeable. If such a mutation took place in a reproductive cell, however, a child developing from that cell would express the abnormal eyes. Since most gene mutations are harmful, it is important that the reproductive cells be protected from undue exposure to radiation.

Cumulative Radiation Damage

Many of the early workers with X rays and other forms of radiation suffered severe damage to their bodies because they failed to realize the cumulative nature of the radiation. Over the course of years a person may accumulate sufficient radiation to bring about his death without ever suffering any of the symptoms of radiation sickness. To understand this we need to differentiate between recoverable and irrecoverable radiation damages. If a person receives as

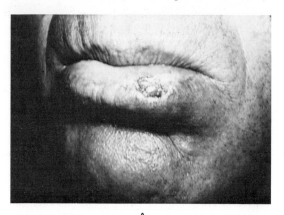

A

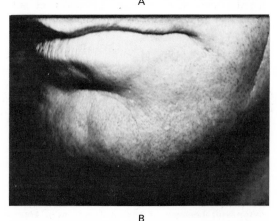

B

FIG. 4.11 Cancer destruction by radiation. The lip cancer was treated by careful application of a beam of cobalt-60 radiation (A). The same lip is free of the cancer three months later (B). (Courtesy of J. F. Mullins, Medical Branch of the University of Texas.)

much as 250 R of radiation, he will surely show symptoms—reddened and ulcerated skin, loss of hair, nausea, anemia, and other symptoms of radiation sickness. This damage, however, is recoverable. Many skin cells will have been killed, but some will have survived and they can repopulate the damaged area and eventually whole skin will be restored. The same is true of the blood-forming cells and other growing cells of the body. There may be temporarily sterility, but as surviving reproductive cells repopulate the gonads, fertility may return.

There is some lingering effect on the body, however, from which the person does not recover. If exposure to the 250 R occurs a little at a time over a span of years, the same effect will result, even though there is no radiation sickness. Whenever any part of the body accumulates a dosage beyond a certain level (about 6000 R), that part of the body dies. It may be only a hand that gets this much, as in the case of the early X-ray technicians, or it may be an entire body. Even though a person may never accumulate radiation reaching the lethal level, evidence indicates that there can be a shortening of life span in proportion to the amount of radiation received. This does not mean that we should not take full advantage of the medical value of X-ray diagnosis and treatment of human illnesses, but it does mean that we should avoid unnecessary exposures to radiation.

Variation in Sensitivity of Organisms

After the bomb blast over Hiroshima, Japan, many women had abortions and miscarriages. The embryos were killed by exposure to a dosage of radiation which did not cause serious damage to the women. Embryos are more sensitive to radiation damage because they contain

TABLE 4.2. Variation in Sensitivity of Rat Embryos to Radiation Damage*

Age of Embryo	Radiation Dosage to Produce 100 Percent Mortality
8 Days	200 R
9 Days	400 R
11 Days	600 R
6 Weeks (birth)	750 R
Mature adult	900 R

*Data from James Wilson.

rapidly growing and dividing cells in all their body parts. The degree of sensitivity varies at different embryonic stages, the younger embryos being the most sensitive because their cells are growing and dividing most rapidly. Table 4.2 shows how this varies in rat embryos.

There is also variation in the sensitivity of different species of organisms. A dosage of about 600 R to the whole body, when administered within a period of a day or so, will kill a human being, but it requires about 900 R to kill rats. An onion plant can withstand dosages up to about 25,000 R before it is killed, and many insects can withstand dosages twice that great. Some of the protozoa can survive dosages even greater than this. As a general rule, the less complex the organism the greater its resistance to radiation damage. Since complexity is correlated with the amount of DNA (and genes) within the cells, this factor might explain the difference in resistance, at least to some extent. It seems logical that with more DNA and greater chromosome area there would be a greater chance of damage to DNA and to chromosomes as a whole. Certainly the physiological state of the organism and the rapidity of cell growth and division at the time of radiation also play a part.

REVIEW QUESTIONS AND PROBLEMS

1. How and by whom were X rays discovered?

2. What important contribution was made by Becquerel to our modern techniques of the use of radioactive materials in biological research?

3. How does particulate radiation differ from electromagnetic radiation? Describe the different kinds of radiation which come under each category.

4. In atomic-energy plants great care is taken to shield workers from neutron radiation, but fewer precautions are observed when only beta radiation is being emitted. Why should this be?

5. Alpha particles are much heavier than beta particles. Explain why.

6. One study seemed to show a significantly greater number of embryonic abnormalities among children born to women living at very high altitudes than to those living at lower altitudes. Formulate a hypothesis to explain this difference after considering the topics studied in this chapter.

7. Describe an experiment which could use radioactive tracers in living organisms. Do not describe any of those used as illustrations in this chapter.

8. Iodine-131 is frequently put into the human body for purposes of medical diagnosis, yet carbon-14 is not used for such diagnosis. Consider the various properties of these isotopes and tell why you think this is so.

9. An ancient mummy is found in Egypt, and the date on the tomb indicates that it was placed there about 11,200 years ago. What proportion of carbon-14 would you expect to find in this mummy in comparison with the bodies of persons who have died in recent times? Explain how you arrived at the answer.

10. Assume that the radioactive carbon-14 in a gram of dry bone of animals which have died recently gives 240 counts per hour on a particular counter. The bones of a man are excavated and it is found that this bone gives only 72 counts per hour from its carbon-14. Use the half-life chart to determine the approximate age of these bones and explain how you arrive at your conclusions.

11. The potassium-argon "clock" is used rather than carbon-14 for establishing the age of some of the most ancient remains of human life. Why?

12. Why are rapidly growing and dividing cells more sensitive than other cells to radiation damage?

13. Explain how ionization in living matter is induced by a beta particle. By a gamma ray.

14. A mutation in most body cells may have no significance, but if the mutation is in a reproductive cell, it can be quite damaging. Explain.

15. Why are embryos of higher animals more sensitive to radiation damage than the adults of the same species?

16. What are possible explanations of variations in sensitivity to radiation damage by organisms of different species?

17.   What is meant by recoverable and irrecoverable radiation damage and how is this related to the cumulative effect of radiation?

FURTHER READING

Bleich, A. 1970. *The Story of X Rays*. New York: Dover Publications.
Clegg, E. J. 1968. *The Study of Man*. New York: American Elsevier.

5

The Beginning of Life on the Earth

On almost every part of the earth, a great variety of living things can be found. There was a time, however, when the earth was a hot, sterile ball on which life was impossible. Geological records show that the earth was once a great mass of molten lava. Sometime after it had cooled sufficiently, life appeared. How did this happen? One theory, which is an offshoot of our space age, claims that life was brought to earth in space ships from other planets. People who hold this theory point to diagrams of what they interpret to be space ships which were inscribed on rocks by ancient man. Another theory says that life in a primitive form may have reached the earth from another planet embedded in a meteorite which burst open when it hit the earth's surface. Most scientists feel, however, that it is more logical to assume that life developed here on earth from nonliving precursors. Concepts of the way in which the development of life happened differ. **Theistic explanations** say that life developed under the direction of a supernatural power, while **mechanistic theories** say that the development of life was the result of the operation of natural laws.

Of the many theistic views, we are most familiar with the biblical account of the origin of life which is found in Genesis. Some interpret the biblical account to mean that sometime in the relatively recent past each species was created in the same form in which it exists today. Discoveries which indicate that species have undergone change in the past and are now undergoing change have caused people who believe in theistic explanations to decide that one or several forms of life were first created by God but that God instilled in these forms the power to change and to adapt to their changing surrounding. Some theists believe that these species changes were divinely directed all along, while other feel that, once created, each species was allowed to develop according to natural laws without further divine direction.

Many other religions besides Judaism and Christianity have accounted for the origin of life. Ancient Hindu writings tell of the beginning when there was only one person, the Hindu equivalent of Adam. Like Adam, this man became lonely. He doubled in size and then split himself in two; one part became a man and the other, a woman. They mated and started the human species. In time the woman became ashamed of her relations with the man and tried to hide by turning herself into a cow. The man found her, however, and changed himself into a bull, and from their matings cattle were produced. Again she tried to hide by turning into a mare, but he became a stallion and horses were born. This happened again and again, and eventually all the species of the world were produced.

Supernatural explanations of the origin of life were developed at a time when many phenomena were not understood. As scientific techniques of inquiry were developed, however, it became apparent that some of the old supernatural accounts needed modification. When telescopes were invented, for example, beliefs about the nature of the earth and outer space changed. Biological studies provided information that challenged the idea that each species was specially created. Great controversies arose between those who felt that all living species had undergone some evolutionary development and those who felt that evolutionary development of living organisms was contrary to religious doctrine. Attempts, often successful, were made to suppress the spread of biological knowledge which supported evolutionary theory; laws were passed to prohibit such ideas from being discussed in the classrooms. In time, however, most people came to hold the view that the exact method of the development of life was a scientific, rather than religious, question. A person could be a devout adherent of a religion, or of no religion, and still accept scientific findings about the evolutionary nature of living organisms. Recently, however, there has been a resurgence of this conflict. Some well-organized groups are agitating for laws to force the inclusion of their theistic theories of creation in science textbooks.

This chapter examines some hypotheses about the beginnings of the earth and of life. The discussion is restricted to those hypotheses which have verifiable scientific evidence behind them because these hypotheses are the only ones which can be universally considered by anyone, regardless of their religious beliefs.

THE EARLY EARTH

According to the best available scientific estimates, the earth had its origin as a separate entity some $4\frac{1}{2}$ to 5 billion years ago. A hypothesis that seems in good accord with the evidence holds that before this time our entire solar system was a rapidly rotating ball of gas. This gas was composed of free atoms, with hydrogen atoms in greatest abundance. The friction of the atoms on one another generated great heat. Gravity pulled the atoms toward a center, and the great majority of them united to form the sun. Some small whorls of the gas, however, were thrown off in eddies and formed the planets. Still smaller whorls from the condensing planets formed the planet satellites, such as our moon. As the hot gaseous predecessor of the earth continued to whirl, the heavier atoms, such as iron, accumulated in the center. They are still there, and they are still very hot. Geologists estimate that the core of the earth is a white-hot ball of

FIG. 5.1 Diagram to illustrate how the solar system may have been formed. Condensing gases may have given rise to the sun and the planets.

iron twice the size of the moon. Atoms of medium weight, such as silicon and aluminum, formed a middle shell. The lightest elements, such as hydrogen, nitrogen, oxygen, and carbon, remained in the outer layers, and these four make up nearly all our atmosphere today. In the course of many millions of years there was a gradual cooling of the earth and the free atoms began to be bonded into molecules. A crust solidified on the outer surface of the earth, and beyond it there remained a gaseous atmosphere.

CHEMICAL PRECURSORS OF LIVING MATTER

Of the four kinds of atoms in the atmosphere of the early earth, hydrogen was the most reactive—that is, it combined with other atoms more readily than any of the other three. As a result, three compounds of hydrogen must have been formed. Hydrogen combined with oxygen and formed **water,** H_2O, the first water on earth. Hydrogen combined with nitrogen and formed ammonia, NH_3. It combined with carbon and formed **methane,** CH_4, a flammable gas that is obtained as a byproduct from oil wells today. On the planet Jupiter we have been able to detect these three compounds of hydrogen, but they exist as permanently frozen solids. Perhaps they were formed as on the earth, but Jupiter is so far from the sun that it cooled quickly, and these compounds were frozen before they could form other compounds. In addition, on the earth it is likely that carbon and oxygen united to form **carbon dioxide,** CO_2, and that hydrogen, carbon, and nitrogen united to form **hydrogen cyanide,** HCN. Also, the free hydrogen tended to unite into molecules, H_2. These simple compounds could have been the precursors from which living matter had its origin. Living matter is composed primarily of the four elements in these particular compounds.

The early earth was much too hot for water to exist as a liquid—it existed as steam in the atmosphere. The solid crust of the earth was frequently punctured by great eruptions of molten rock and heavy metals from the interior. There are still remnants of such eruptions on a comparatively small scale in volcanoes today. The conditions on Venus, which is much nearer the sun than the earth, today probably resemble those on the earth in these ancient times. The Venus-probe rocket showed the surface temperature of that planet to be about the melting point of lead, and over this hot surface lie great clouds of vaporized gases hundreds of miles thick. The early earth was also bombarded with intense high-energy radiation from the sun. The sun was much hotter then than now; there were tremendous explosions as its hydrogen atoms fused as in a monstrous hydrogen bomb. Cer-

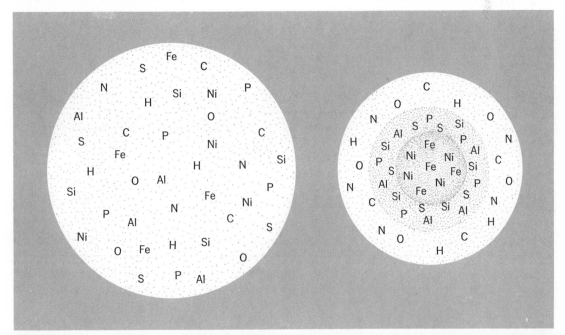

FIG. 5.2 Possible method of the formation of the earth and its atmosphere. At first all elements may have been distributed evenly, but as their whirling action slowed, the heavier elements such as iron and nickel gravitated toward the center. Elements of intermediate weight formed the earth's crust, while the lighter elements remained in gaseous form and formed the atmosphere.

tainly conditions were not yet right for the formation or existence of life.

As the earth continued to cool, the steam in the air began to condense and form great clouds covering the earth, and with continued cooling these clouds began to condense into raindrops. Then the downpour started. Rains such as man has never seen came down. For thousands of years the water fell, filling the depressions on the crust of the earth and forming oceans. As the water fell, it brought with it some of the methane, ammonia, hydrogen cyanide, carbon dioxide, and hydrogen, which remained in solution in the oceans. And as the water rushed over the land, it picked up many minerals, some of which reacted to form the salts of the oceans.

FORMATION OF EARLY ORGANIC COMPOUNDS

Carbon is an element that occupies a prime position in all forms of living matter. The emergence of carbon compounds, therefore, was an important step in the events leading up to the formation of living matter. The carbon compounds associated with living matter are com-

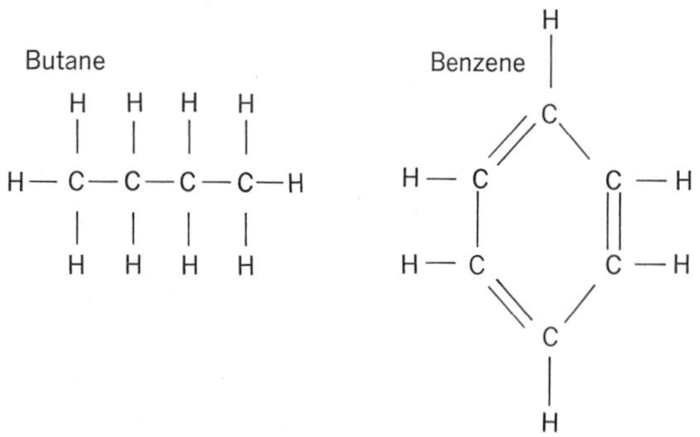

FIG. 5.3 A carbon chain and ring. The versatile bonding capacity of carbon is illustrated by the structural formulas of butane and benzene, typical organic compounds. Each carbon atom has four lines which represent bonds.

monly called **organic compounds,** discussed in Chapter 3. The carbon atoms of such compounds are in the form of chains or rings which can be joined together to form very large and complex molecules. **Inorganic compounds** either do not contain carbon, or the carbon is not in the form of chains or rings. The variety of organic compounds is much greater than that of inorganic compounds because of the size of the molecules and the versatile bonding capacity of carbon.

In 1828 the German chemist Friedrich Wöhler succeeded in transforming the inorganic compound ammonium cyanate into urea, an organic compound produced abundantly by living systems. Today, the synthesis of organic compounds is an everyday undertaking in chemical laboratories. Even such complex organic compounds as hormones, enzymes, and antibiotics are synthesized. There is also some natural synthesizing of organic compounds. In volcanic eruptions, for instance, inorganic carbonates are brought up, and in the presence of great heat they unite with hydrogen from the water vapor of the air to form simple organic compounds. This method of synthesis is probably not greatly unlike the origin of such compounds on the early earth.

The first simple organic compounds formed on the earth were not living, but they did furnish the building materials from which living matter could be constructed. Among these must have been the basic life-building materials listed below:

1. Sugars. Simple sugars could easily have been formed. Sugar molecules are carbon atoms arranged in a chain with hydrogen and with oxygen attached to the sides and ends. The hydrogen and oxygen atoms are present in the same proportion as they have in water, 2:1. Glucose is a common simple sugar, with the formula $C_6H_{12}O_6$.

2. Glycerol (glycerin). This compound also consists of carbon, hydrogen, and oxygen, but the hydrogen and oxygen atoms are not in the 2:1 proportion found in the sugars. Glycerol has a chain of three carbon atoms connected on one side with three oxygen atoms. The formula is $C_3H_8O_3$ (see Fig. 5.3).

3. Fatty Acids. There are several varieties of these, but all are also composed of carbon, hydrogen, and oxygen. The carbon atoms are in a chain of from two to twenty or more and the number of oxygen atoms is very low in proportion to the hydrogen. At one end of the chain two oxygen atoms are always found attached to the final carbon atom.

4. Amino acids. These include nitrogen as well as carbon, hydrogen, and oxygen in their molecules. A few also have sulfur. Twenty basic, different, naturally occurring amino acids have been identified, and there is considerable variation in their structure. The number of carbon atoms varies considerably, and they may occur in the form of either chains or rings, but they all include the **amino group,** NH_2.

5. Purines. These compounds have a double ring of carbon and nitrogen atoms. The presence of nitrogen in the rings distinguishes these from ring-shaped amino acids, which have only carbon in the rings.

6. Pyrimidines. These are similar to the purines, but have only one ring of carbon and nitrogen in each molecule.

The view that the first organic compounds were produced in the way described is speculation, but it is speculation based upon scientific facts. The feasibility of such generation was demonstrated dramatically in some recent experiments. The compounds of the ancient earth—water, methane gas, and ammonia gas—were mixed together in a flask, and a strong electric spark—to simulate lightning—was discharged through it for about a week. At the end of this period it was found that many sugars, amino acids, fatty acids, and other organic compounds had been generated. This experiment was first performed by Stanley Miller at the University of California.

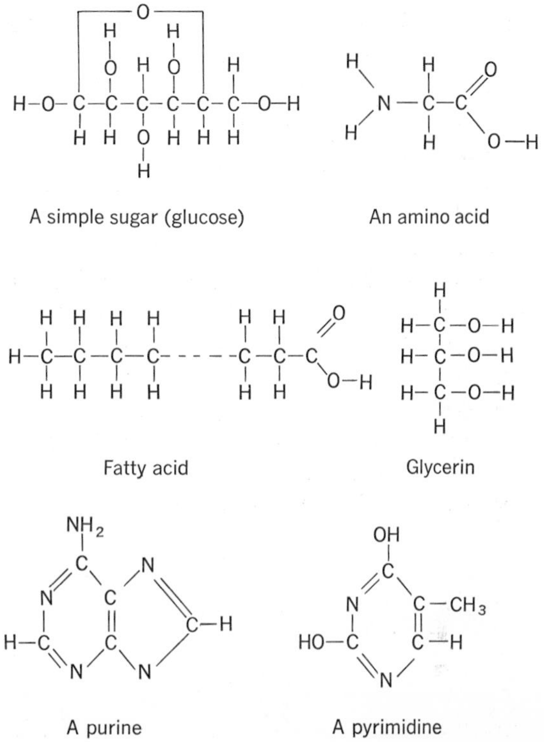

A simple sugar (glucose)   An amino acid

Fatty acid   Glycerin

A purine   A pyrimidine

FIG. 5.4 Some early organic compounds which may have been produced on the earth under conditions existing before life began. These compounds could have served as building blocks from which simple living organisms could have been formed.

THE EMERGENCE OF LIFE

Now the stage was set for the emergence of life. The oceans of the earth were gradually turned into a "broth" of mixed organic molecules. These molecules must have had frequent contacts and under the stimulation of the available energy, some must have united to form more complex molecules sometimes joining together into multimolecular aggregates. This must have occurred billions upon billions of times. In all these multiplied billions it is not only possible but almost inevitable that some of the multimolecular aggregates would be of such forms as to have characteristics of living matter. We are not speaking now of a complex form of life, such as a worm, or even of a one-celled organism, such as an amoeba. We are speaking of a complex molecule with self-replicating characteristics. Such molecules constitute the units of heredity, known as **genes,** which are found in all forms of life. It is only reasonable to assume that molecules possessing the properties of genes were the precursors of life. Many compounds resembling these large "genic" molecules must have been formed before the right combination appeared, but once molecules with this combination came

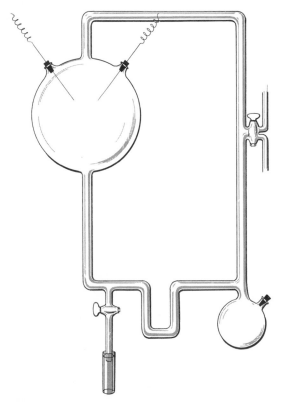

FIG. 5.5 An apparatus devised by Stanley Miller which produced organic compounds under conditions similar to those which were probably found on the young earth. Heat is applied to the flask at the right to keep the water in vapor form and an electric spark is discharged in the large flask to simulate the lightning on the young earth.

into being, the stage was set for the development of living forms.

A gene has the basic characteristic of life—the power of self-duplication—when it is in the proper environment. Today such an environment is to be found only within living cells, but the ancient oceans contained all the necessary ingredients found only in living cells today. Genes are made of combinations of purines, pyrimidines, simple sugar, and phosphate (PO_4). All these ingredients were present in the early oceans and would have permitted gene formation and continued multiplication.

The particular aggregation of elements that forms a gene is known as **deoxyribonucleic acid,** generally abbreviated to **DNA.** According to the Watson-Crick theory of gene structure, the DNA molecule is shaped like two pieces of string twisted around each other and held together by projections from each string joined by weak hydrogen bonds. Gene duplication is accomplished when these bonds break and the two halves unwind and separate. Then each half attracts to itself the missing chemical substances that were present in the other half; two complete genes are formed, each a perfect replica of the original.

That such a synthesis of simple organic compounds to form genes could have taken place was indicated by research done by the German biologist Gerhard Schramm at the Max Planck Institute for Virus Research in 1963. He combined some of the simple organic compounds which evidence indicates were present on the early earth before there was any life. These included simple sugars, purines, and pyrimidines. He added some inorganic compounds, including simple phosphorus compounds. This mixture was then subjected to moderate heat, pressure, and electric discharges such as were probably present on the early earth. When Schramm analyzed the contents of the flasks at the end of the experiment, he found a nucleic acid that showed the twisted, ladderlike structure characteristic of

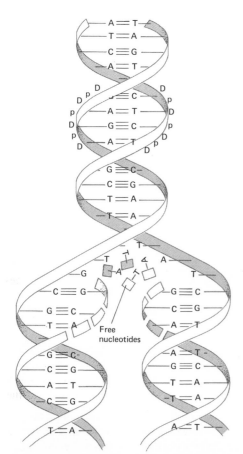

FIG. 5.6 Structure of DNA and its method of duplication according to the Watson-Crick theory.

in the oceans, they would have multiplied until gradually, after millions of years, they became very abundant. Some genes of this vast group may have clung together after duplication and formed gene aggregates. Perhaps a chain of genes attached end to end appeared. Then, on rare occasions, slight errors of duplication may have taken place so that some genes in the chain were different from those that produced them; such errors are known as **mutations.** We know that gene mutations occur today and that the rate of mutations is increased by high-energy radiation such as X rays or the radiation from radioactive isotopes. The early earth was bombarded with intense radiation from the sun. Water absorbs such radiation; in fact, life could not have existed otherwise, because this radiation in heavy doses will kill all forms of life. Some of the gene aggregates that floated near the surface of the oceans would have received enough radiation to

DNA. This was not a living gene—he had not created life—but he had shown that it was possible for nucleic acids to have been formed under conditions similar to those of the early earth. Some of these early nucleic acids could have had all the properties of genes, with the power of self-duplication. It is possible that some biologist in the future will be able to create a nucleic acid which can be instilled into a simple cell, such as a bacterium, and there carry on duplication. This would be a true creation of life, even though it would be life in its most elementary form.

Once self-duplicating nucleic acids appeared

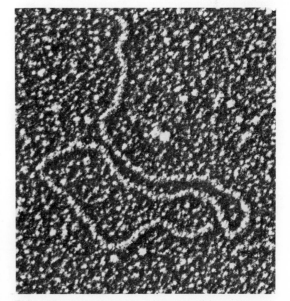

FIG. 5.7 Photograph of a gene aggregate. Since all forms of life have genes, something like this must have been among the precursors of living matter on the earth. (This gene aggregate is from the bacterium *Escherichia coli.*) (Courtesy of Jon Beckwith, Harvard Medical School.)

show a greatly increased mutation rate, but not enough to be destroyed. Gene mutation may have been much more frequent then than it is today.

At this point in the development of life, natural selection must have come into play, because the gene aggregates now contained different kinds of genes. As the gene aggregates became numerous, there must have been competition for survival, and those which were most successful in the environment survived while the others perished.

The next step toward the development of living forms which resemble those on earth was a big one. Modern viruses are gene aggregates surrounded by protein coats. We must assume that early gene aggregates developed the power to attract amino acids to themselves and to assemble these amino acids into polypeptide chains which make up the proteins. Once this was accomplished, there would have been forms of life similar to modern viruses. Modern viruses can duplicate themselves only within living cells, but we must realize that the primitive oceans contained materials resembling those found in living cells today.

THE DEVELOPMENT OF ENZYMES AND COMPLEX ORGANIC COMPOUNDS

The great advance from single gene to gene aggregates surrounded by protein coats must have been very slow in developing. Possibly millions of years passed, years in which untold millions of combinations of genes came about before one was successful in becoming established. The next advance, from a simple viruslike body to a cell, must likewise have been very slow; yet in the countless eons which passed on the primitive earth, sooner or later this step was taken. The development of living matter and living cells is a highly unlikely event, yet given sufficient time and trials, even a highly unlikely event may take place. With this in mind we can better comprehend the highly unlikely development of the living cell.

Before the gene aggregates could develop into anything like a cell, it was necessary that they produce **enzymes.** Enzymes play a very important role in cell physiology, and it is reasonable to assume that they appeared early. Enzymes are special proteins which have the power to stimulate reactions among other molecules. Thus enzymes fall into the class of substances known as **catalysts.** Chemists have long recognized the value of catalysts in inorganic reactions. For instance, if potassium chlorate ($KClO_3$) is heated to 368°C, very little reaction occurs, but if we add manganese dioxide (MnO_2) as a catalyst, there will be a rapid change of the potassium chlorate to potassium chloride and oxygen; yet the manganese dioxide remains unchanged. It triggers the reaction without being used up in the reaction. This is represented in the following equation:

$$2 \ KClO_3 \xrightarrow{\text{MnO}_2} 2 \ KCl + 3 \ O_2$$

As enzymes were formed, the gene aggregates became surrounded with material which we now know as cytoplasm, around which was a very thin membrane which permitted diffusion of dissolved particles both in and out. Three great classes of energy-rich compounds were formed as a result of these enzyme-stimulated reactions.

Carbohydrates

The simple sugars, **monosaccharides,** such as **glucose** and **fructose,** were combined into more complex carbohydrates. These include the double sugars, **disaccharides,** in which two molecules of simple sugar are combined. Common table sugar, **sucrose,** is a good example of this. It is formed as follows:

$$C_6H_{12}O_6 + C_6H_{12}O_6 \xrightarrow{\text{enzyme}} C_{12}H_{22}O_{11} + H_2O$$

Glucose   Fructose   Sucrose   Water

Glucose and fructose have the same chemical formula, but the atoms are arranged differently, so the two compounds show different chemical properties. Other disaccharides formed by the union of two monosaccharides are **maltose** (malt sugar), found in the sprouting seeds of cereal plants, and **lactose** (milk sugar), found in milk.

Polysaccharides are formed when there is a union of a large, even number of monosaccharides with the removal of hydrogen and oxygen in the form of water. The overall equation for the formation of **starch** under the catalytic influence of enzymes can be represented as

$$n(C_6H_{12}O_6) \xrightarrow{\text{enzyme}} n(C_6H_{10}O_5) + n(H_2O)$$

Glucose   Starch   Water

The letter n represens the number of glucose molecules. This number is not known exactly, but it appears to be as high as 2500. The simple sugars manufactured by plants are usually converted into starch, which is more easily stored than sugar. In animals the carbohydrate storage is usually in the form of **glycogen,** which has up to 1000 of the monosaccharide units in each molecule. **Cellulose** is a very large polysaccharide carbohydrate molecule formed by plant cells. It is one of the primary substances used in the construction of plant cell walls.

Carbohydrates are important components of the foods of almost every form of life. For most forms the energy from food is made available in an enzyme-mediated breakdown of carbohydrates through combination with oxygen (**oxidation**), with the release of the bond energy of the molecules. Glucose is the most common carbohydrate used as the energy fuel of cells. Some more complex carbohydrates are first broken down into simple sugars by enzymes. Most forms of life, however, do not have enzymes which can break down cellulose or chitin. Not even termites produce such enzymes, but certain protozoa that live in the intestine of the termites do produce them, and both termites and protozoa use the simple sugars that result from the breakdown of the cellulose of the wood ingested by the termites.

Lipids (Fatty Compounds)

These compounds are an important part of protoplasm. A lipid molecule, like the carbohydrate molecule, contains carbon, hydrogen, and oxygen, but unlike the carbohydrates, the lipids show a ratio of hydrogen to oxygen much greater than $2:1$. The most common of the lipids are the fats, which are formed by an enzyme-stimulated combination of three molecules of fatty acid with one molecule of glycerol:

$$3 \text{ fatty acid} + 1 \text{ glycerol} \xrightarrow{\text{enzyme}} 1 \text{ fat} + 3 H_2O$$

There are different kinds of fats, depending upon which fatty acids are combined in the molecules. Beef fat (tristearin) has the formula $C_{57}H_{110}O_6$. Note the low proportion of oxygen in the molecule.

Most animals store food in the form of fats. This is an efficient storage because fats are the most concentrated source of biological energy available. The low oxygen concentration of fats permits a greater degree of oxidation than is possible with carbohydrates. Not only do the fats supply more energy than carbohydrates; they also supply more water when they are oxidized. Thus, a bear heavy with stored fat can go into hibernation and have not only enough energy to go through the winter but also enough water to supply his needs. The hump on the back of a camel is another case of fatty issue supplying both energy fuel and water to a living organism.

Proteins

Some proteins known as **structural proteins** make up about 15 percent of the protoplasm of the cell. It is these structural proteins which must be duplicated if there is to be any growth of living matter. **Functional proteins,** primarily the enzymes, are also necessary for the many chemical reactions within living matter. Each cell must have thousands of such enzymes in order to carry out the complex reactions associated with living.

Proteins are made of amino acids. Although there are only twenty fundamental amino acids, they can be put together in a countless variety of ways in the large protein molecules. A single protein molecule is made of from several hundred to several thousand amino acids, and a change of even one of these amino acids can alter the nature of the entire protein. The amino acids are joined together in long chains, **polypeptide chains,** and there may be one, two, or four of these chains in a protein molecule. **Hemoglobin,** the oxygen-carrying component of human red blood cells, has a molecule consisting of two paired polypeptide chains (a total of four chains), each with almost 150 amino acids. These chains are not strung out but are folded and twisted together to form a somewhat compact molecule. This is true of most large protein molecules.

When proteins are eaten as food by animals, the digestive enzymes first break the polypeptide chains apart at certain bonds, leaving small chains of amino acids, the **peptides.** Further digestion breaks the peptides down into separate amino acids. These can pass into the cells, where cellular enzymes can put them together again into polypeptide chains with a sequence peculiar to the organism doing the assembling. We learned in Chapter 1 that the structural proteins of protoplasm are different in all organisms except where two or more organisms have identical genes, as is true of identical twins. Genes determine the sequence of amino acids which make the proteins, and different genes produce proteins with different sequences. The glucose produced by one plant will be exactly like the glucose produced by another plant and

it will exist in the same form in an animal body. Likewise, a specific type of fat in one animal will be exactly like the same fat in another animal.

THE EMERGENCE OF CELLULAR LIFE

Let us continue with our hypothetical account of the development of life on the earth. We would expect cellular life to have made its appearance as enzymes and complex organic compounds began to accumulate around the gene aggregates. Very simple cells would have formed at first, cells without a definite nucleus and without many of the parts found in most cells today. Many of the bacteria and some of the blue-green algae of today have cells which are similar to the earliest fossilized remains of living matter which have been found (about two billion years old).

These early cells had no problem of nutrition at first; the seas were rich in simple soluble organic compounds that could diffuse into the cells and, through the action of cell enzymes, could be broken down and re-formed into cell protoplasm. Thus, growth of the cells could take place. Continued growth, however, created a problem. As cells get larger, they soon become too large for efficient operation. The movement of materials from one part of the cell to another becomes too slow. Hence, some means of cell division had to develop which could separate the primitive cells into smaller units. This division had to be in the nature of a duplication, since each cell formed by the division must retain all the properties of the original cell. The duplication of genes must have taken place first, followed by the division of the rest of the cell, giving each daughter cell a full complement of genes.

Another new situation must have arisen as the cells became more numerous. If the cells had continued to use up the organic nutrients in the seas, the supply of food would have dwindled, creating the great danger that life so long in developing would perish, leaving the earth once again a sterile, lifeless planet. Such a catastrophe did not take place because some one or more of the cells developed enzymes that permitted them to obtain nourishment from the abundant store of inorganic chemicals in the oceans. This was the beginning of **autotrophic nutrition,** and through natural selection it became widespread. Some bacteria today can obtain nourishment in this way. The iron bacteria, for instance, can combine iron, oxygen, and water and obtain iron hydroxide, hydrogen, and energy. This energy can be incorporated into other inorganic materials and yield food for growth. Likewise, the sulfur bacteria can combine hydrogen sulfide and oxygen and obtain water, sulfur, and energy. Such a method of obtaining energy through chemical recombination with a release of a portion of the atomic energy is known as **chemosynthesis.** Although there was no free oxygen in the atmosphere at the time of the conjectured origin of life, it was possible for certain forms of chemosynthesis to take place.

DEVELOPMENT OF PHOTOSYNTHESIS AND CELL RESPIRATION

Even the inorganic compounds which would yield energy were limited in quantity, and eventually still another means of obtaining energy arose—**photosynthesis.** There was abundant energy available in the form of sunlight, and sometime before the nutrients in the oceans were exhausted, some organisms developed the power to utilize this energy in producing food from inorganic materials. This method of food production is known as photosynthesis, a word which means "manufacture by means of light." **Chlorophyll,** the green material so abundant in modern plant life, was developed as the catalyst which enabled the organism to combine water and carbon dioxide in the presence of light and obtain food. There was abundant carbon dioxide at this time because this gas was released by the organisms which utilized the nutrients of the oceans.

Photosynthesis made possible a much more efficient use of food than had previously been possible. Before the advent of this process there was no free oxygen on the earth and energy was extracted from the food by a process known as **fermentation (anaerobic respiration).** Fermentation, however, is a very inefficient means of obtaining energy from food. Only a small amount of the potential is extracted and it leaves poisonous waste products—alcohol, lactic acid, acetic acid, and others. Photosynthesis, however, releases oxygen as a byproduct, and as this gas became abundant in the atmosphere, a new method of extracting energy from food became possible. This was **aerobic respiration,** in which food is combined with oxygen to yield

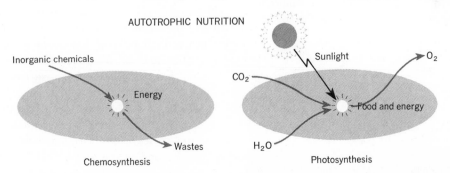

FIG. 5.8 Distinctions between chemosynthesis and photosynthesis, both forms of autotropic nutrition.

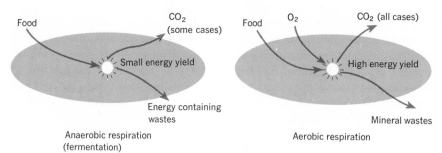

HETEROTROPHIC NUTRITION

FIG. 5.9 Distinctions between anaerobic and aerobic respiration, both forms of heterotrophic nutrition. A more efficient use of food for energy production is possible in aerobic respiration.

carbon dioxide, water, and energy. The efficiency of this process of energy release is about twenty times as great as that of the anaerobic process. Aerobic respiration has another advantage—it releases carbon dioxide and water as byproducts, and these are the raw materials which the photosynthetic organisms need to continue manufacturing food. Thus an interdependent relationship came into being betwen the **heterotrophic organisms,** which cannot manufacture their own food, and the **autotrophic organisms,** which manufacture food. Most of the hetrotrophic organisms of today use aerobic respiration.

IMPORTANCE OF ATMOSPHERIC OXYGEN

The entry of oxygen into the atmosphere was a major step in making possible the advanced forms of life that have developed on the earth. Not only did it make possible cell respiration, but it also brought about a great change in the composition of the atmosphere. It reacted with methane and transformed it into carbon dioxide and water:

$$CH_4 + 2\,O_2 \rightarrow CO_2 + 2\,H_2O$$

This provided an additional source of carbon dioxide for photosynthesis and changed the atmosphere into a form in which animal life could exist. Oxygen also reacted with ammonia and converted it to nitrogen and water:

$$4\,NH_3 + 3\,O_2 \rightarrow 2\,N_2 + 6\,H_2O$$

Oxygen in the air had another effect of great importance. It liberated living organisms from the oceans. Much of the sun's high-energy radiation, which is highly destructive to life, reached the earth, but water is a good absorber of such radiation. Thus, life could exist in the oceans but not on the land. As oxygen was liberated, however, some of it rose and formed a layer of ozone high in the atmosphere, and this absorbs most of the radiation from the sun. For the first time it became possible for living organisms to leave the oceans and take up life on the land and in the air.

At last we come to the point where the earth and atmosphere were somewhat as they

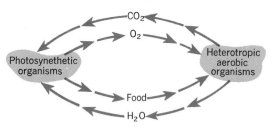

FIG. 5.10 Relationship between photosynthetic and heterotrophic nutrition.

are today. The planet had cooled to a tolerable temperature; the injurious radiation was effectively blanketed; the methane and ammonia in the air were replaced by nitrogen, oxygen, and carbon dioxide; photosynthesis had become the dominant form of energy capture; life was now free to develop into forms such as occur today. With these changes however, conditions were no longer conducive to the spontaneous generation of life. From this point on, all living things had to descend from living things already in existence.

LIFE ON OTHER PLANETS

Does life exist on any other planet? A first examination may make it seem very unlikely. Life, as we know it on earth, exists in a very narrow range of conditions. The chance that another planet could exist with just about the same conditions of temperature, light, and chemical composition as the earth is very small indeed. In our own solar system Mars seems to be the most likely candidate to support life as we know it because the conditions on Mars most closely resemble those on the earth. Even so, any form of life would have to be very hardy to withstand the environment on Mars. Attempts have been made to duplicate the Martian environment here on earth, and certain primitive organisms have been able to survive in these simulated environments. On the other hand, our Mars probe rocket, which took photographs at a very close range in 1974, showed no indication of any form of life.

The universe is so large, however, that life might exist in other solar systems. About 100 million galaxies lie within the range of our telescopes. In our own galaxy, the Milky Way, there are millions of "suns," each of which could have its system of planets. In this almost incomprehensible number of planets in the universe, the very unlikely set of conditions present on our earth would certainly be repeated many times over. It is therefore not only conceivable but very likely that the same set of developmental processes which led to our present forms of life has been repeated on some of these planets— not to result in human beings as we know them, perhaps, or the big-eared, bug-eyed little Martians of the comic strips, but in some form of intelligent creatures that could rival man in creative accomplishments.

REVIEW QUESTIONS AND PROBLEMS

1. All the early accounts of the origin of life on earth are based upon magical and supernatural happenings. Explain why this is true.
2. If some pine trees of today could be transplanted back to the early earth as soon as it had cooled sufficiently to tolerate life, would these trees survive? Explain your answer thoroughly.
3. If deer of today could be placed back on the early earth as soon as it had cooled, could they have survived? Explain your answer thoroughly.
4. How can you recognize sugars, glycerol, fatty acids, and amino acids by their chemical formulas?
5. How can you distinguish between a pyrimidine and a ring-form amino acid?
6. Describe Stanley Miller's experiment on the development of life on the earth and its significance.

7. Why is it logical to assume that the formation of genes was the first step in the development of life on the earth?

8. Describe the structure of a gene according to the Watson-Crick hypothesis.

9. Why were enzymes necessary before there could be any cellular forms of life on the earth.

10. Why are enzymes classed as catalysts?

11. Fats are the most efficient energy fuels of cells. Explain.

12. Distinguish between structural and functional proteins.

13. Explain how proteins are formed within the cells.

14. How does autotrophic nutrition differ from heterotrophic nutrition?

15. Why did photosynthesis become the dominant method of autotrophic nutrition rather than chemosynthesis?

16. Why did aerobic respiration become the dominant method of heterotrophic nutrition rather than fermentation (anaerobic respiration)?

17. In what way are autotrophic and heterotrophic organisms interdependent?

18. Why was the formation of atmospheric oxygen of such great importance in the development of life on the earth?

19. We say that spontaneous generation of life on the earth was possible at one time but that it is no longer possible today. Explain.

FURTHER READING

Dodson, E. O. 1960. *Evolution, Process and Product*. New York: Van Nostrand Reinhold.

Hamilton, T. 1967. *Process and Pattern in Evolution*. New York: Macmillan.

Oparin, A. I. 1957. *The Origin of Life on Earth*. New York: Academic Press.

Ponnamperuma, C. 1972. *The Origins of Life*. New York: E. P. Dutton.

Simpson, G. G. 1964. *This View of Life*. New York: Harcourt, Brace and World.

Stebbins, G. L. 1971. *Process of Organic Evolution*. Englewood Cliffs, N. J.: Prentice-Hall.

6

The Cellular Organization of Living Matter

Before microscopes were invented, people did not know that living matter had a **cellular organization,** because cells are generally too small to be seen with the naked eye, and each living organism was, therefore, considered to be a single unit. When man began to produce glass, however, he noticed that when the glass is curved, the light rays which pass through it are bent and objects seen through the glass are magnified. Convex lenses were used to make primitive microscopes which made visible things which had never before been seen. Water from a stagnant pond which appeared clear to the naked eye was found to be filled with many active organisms when seen under the microscope. Human semen was found to contain many tiny, tadpole-shaped bodies which we know as sperm. In 1665, when **Robert Hooke,** an Englishman, studied thin slices of cork under a microscope, he observed many small compartments. They reminded him of the cells of a honeycomb, and he called them "cells," the name we use today. Hooke did not really see complete cells; he saw only the dried cell walls of cork, a tissue which is found in the bark of a type of oak tree.

Microscopes continued to be improved, and more observations were made on cells from different organisms. In 1838 two German biologists, **Matthias Schleiden** and **Theodor Schwann,** proposed the **cell theory** which states that all forms of life have cellular organization. Schleiden had studied the embryos of plants and observed that the cells of the plant embryos had counterparts in mature plants. Schwann had studied the embryos of animals and had seen cells which were precursors of those which were present in mature animals. Continuing investigation has supported this theory about the organization of living matter, and today it is accepted as the definitive theory.

Great advances have been made in our understanding of cells during the past 30 years because there have been great improvements in microscopic technology and in physiological techniques. Since growth, reproduction, heredity, and the physiological reactions of living organisms all take place in cells, it is important that we understand the structure and functions of cells before we proceed with our study of biology.

METHODS OF STUDYING CELLS

The Compound Microscope

The **compound microscope** combines two or more lenses so that each adds to the microscope's total magnification. The early compound microscopes produced a distorted image because the curved lenses did not bring the entire field of vision into focus. **Spherical aber-**

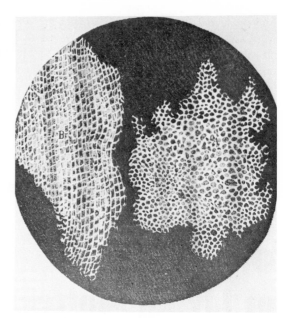

FIG. 6.1 Robert Hooke's drawings of "cells" in a piece of cork as he saw them under a primitive microscope in 1665. The compartments reminded him of the cells of a honeycomb and he therefore called them "cells," the term which we retain today.

rations of the lenses caused the periphery of the field to be fuzzy when the center was sharp and vice versa. **Chromatic aberrations** in the lenses were caused by the property of glass to bend

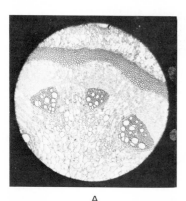
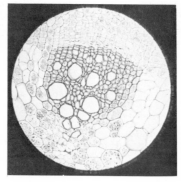
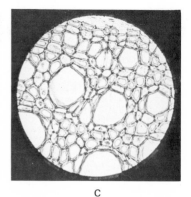

A   B   C

FIG. 6.2 Three degrees of magnification which can be achieved with a modern compound microscope. The low power view at left represents a magnification of about 100 diameters. The high power in the center is about 400-diameter magnification. The oil immersion at right is about 970 diameters. The object viewed in these pictures is a cross-section of a Dutchman's pipe *(Aristolochia)* plant stem.

different wavelengths of light to different degrees, so that images seen through glass tend to have rainbowlike colors at their borders. Today the chromatic aberrations of one kind of glass are corrected by the aberrations of another.

Most modern microscopes have condenser lenses under the stage to concentrate a bright beam of parallel light waves to provide illumination of the object. The amount of light which reaches an object is diffused as the object is magnified, and so a bright light is necessary if there is to be enough light for clear vision when the object is magnified.

The magnification of the compound mircoscope is limited to about 500 diameters. This means that a flea which is 2 millimeters (mm) long (about 1/12 inch) would appear to be 1000 mm long (almost 3½ feet) when it is magnified to this degree. Five-hundred diameters is a great level of magnification, but it is still not powerful enough to make visible all the fine details of cell structure. Lenses can be ground with greater curvatures, but they will not increase the magnification of the lens because of certain properties of light waves. Air and glass bend light waves to different degrees; the different **refraction indexes** of air and glass are one reason for the limitation of the magnification which may be obtained using a compound microscope. This problem was solved by immersing the lens in oil which has the same refractive index as glass. Oil may be applied between the condenser lens and the bottom of the microscope slide as well as between the slide and the objective lens of the microscope. With this **oil immersion** system, a magnification of 2000 diameters is possible. The nature of light makes magnification beyond 2000 diameters impossible.

Sectioning and Staining Techniques

Transmitted light was the original method of illumination, and it is still the most widely used. **Transmitted light** shines up through an object. This kind of light can be used, however, only for very small organisms or for very thin slices of larger organisms. A special instrument, the **microtome,** is used to make the thin sections which are required. Most material is too soft to be cut by this instrument directly and must first be embedded in a medium such as paraffin. It is also possible to freeze the material and then section it. Medical pathologists use this technique when they need quick results. During an operation, for example, a quick analysis of tissue is done to determine if tissue is cancerous so that the surgeon knows whether to remove it.

Contrast is another problem of microscopy. Most parts of protoplasm are of about the same density and details are difficult to see because the cell organelles do not look different from or contrast with one another. Various dyes are used to create contrast between cell organelles, which have different chemical properties and will absorb dyes to different degrees. For instance, chromosomes are rich in nucleic acid and therefore stain heavily with basic dyes. By using a number of different dyes on the same material, some beautiful and distinct images of cell organelles have been obtained.

Darkfield Illumination

Some things are too small to be seen clearly with transmitted light, but when they are illuminated from a circle around them, they glow brightly. The circle illumination produces a dark background which surrounds the objects. This is the same principle which illuminates dust particles in the air when they are struck by a beam of light in a darkened room; otherwise these particles would be invisible. Darkfield illumination has been used extensively for studying syphilis organisms. Syphilis organisms are shaped like very thin spirals and are practically invisible with transmitted light, but they show very brightly under darkfield illumination.

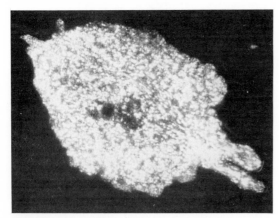

FIG. 6.3 An amoeba seen with darkfield illumination. Using surround lighting causes the object to glow brightly against a dark background.

Phase Contrast Illumination

During the 1950's another method of lighting was developed which utilized the visual contrast generated by the differences in refractive index of the parts of an object. When light hits a cell from an angle, some parts of the cell will bend the light more than other parts, and this difference in refractive index makes some parts darker than others. This technique is known as **phase contrast microscopy**; it is a very good technique for studying living material. There is always the danger that extensive treatment of cells with dyes and stains will alter the cell and create artifacts. The phase contrast microscope solves this problem.

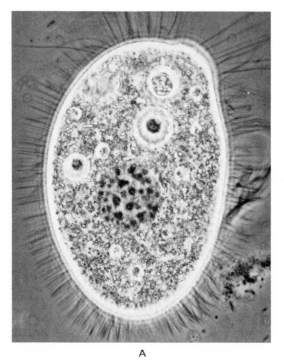

A

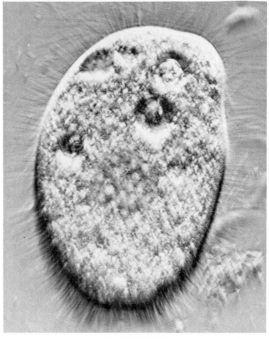

B

FIG. 6.4 Two methods of illumination, shown with a slide of a ciliated protozoan *Nassula ornata*. Phase contrast illumination at left brings out many details, including the multiple nature of the nucleus, which cannot be seen with transmitted light. With side lighting, at right, a different image which also shows important details is obtained.

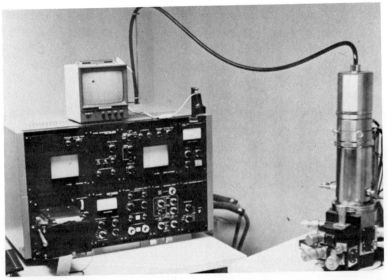

FIG. 6.5 An electron microscope. The scanning electron microscope bears little resemblance to the compound microscope. All this equipment is needed to produce and focus beams of electrons on the object to be photographed. A fluorescent screen at the top of the control panel enables the user to view the object being studied.

During the 1960's a variation of the phase contrast system of lighting, the Nomarski system, was developed. The Nomarski system produces a shadowed image, as does side lighting. The shadowing brings out details which cannot be seen otherwise, in the same way that the details of the mountains on the moon can best be seen when the sunlight strikes the mountains from the side.

Fluorescent Illumination

Certain parts of the cell glow brightly when stained with acridine dyes and viewed under a microscope in ultraviolet light. The method of fluorescent illumination has proved to be valuable in analyzing details of human chromosomes.

The Electron Microscope

To circumvent the limitations on magnification which are imposed by the wavelengths of light, scientists turned to the use of electrons as a source of energy. Electron beams travel with much shorter wavelengths than do light waves; therefore, using electron beams rather than light permits higher magnification. Glass lenses cannot be used to focus electron beams because these beams are not bent by glass. Electron beams can be spread, however, by magnetic fields. Electrons cannot be seen by the human eye, but they can change photographic film and they cause fluorescence of minerals on a screen. Your television tube is a good example of this principle. Electron beams generated in your set are spread into a pattern by a magnetic field, and they induce an image on the front of the tube which is coated with fluorescent minerals. The electron

FIG. 6.6 Electron photomicrograph of virus particles shadowed with chromium. By spraying colloidal particles of a metal from the side, an illusion of side lighting is achieved and this provides contrast. This is a virus which infects bacteria.

More recently a **scanning electron microscope** has been developed, and this microscope is unexcelled in bringing out details at the lower ranges of magnification. The object to be studied is first sprayed with colloidal metal; the electron beam is then directed on the object from the side and above. The electron beam bounces off the metal coating and is focused by magnetic fields. Great depth of field can be achieved with this method.

Units of Microscopic Measure

It is not practical to use units with many decimal places as units of measure for microscopic objects. To say that a part of a cell has a diameter of 0.003 mm or, even worse, 0.00012 inch, has little meaning. Smaller units are needed. One of the most common is the **micron,** written as the Greek letter μ, which is one-thousandth of a millimeter. By this unit, the cell part would have a diameter of 3 μ. Some prefer to use the term **micrometer,** a millionth of a meter, but most biologists still prefer to call this unit a micron. The optical microscope can achieve a resolution down to about 0.2 μ. This means that this type of microscope can distinguish two separate lines when they are separated by a distance of 0.2 μ or more. Violet light has a wavelength of about 0.4 μ, so the resolution of the optical microscope is about half the wave length of these shortest of light waves.

When the electron microscope greatly extended the power of magnification, units of measure had to be used which were even smaller than the micron. A unit known as the **millimicron,** mμ, which is one-thousandth of a micron, came into use. This may also be called a **nanometer,** a billionth of a meter. Sometimes biologists turn to a term used greatly by physicists for measuring light waves. This is the

microscope, which operates in a similar way, can provide a magnification power of over 100,000 diameters, and its use has revealed many details of cell structure.

The electron microscope has several drawbacks. It must operate in a vacuum because electrons do not respond properly to magnetic fields at atmospheric pressure; therefore, all objects which are to be viewed under the electron microscope have to be specially prepared. Contrast is difficult to obtain, and it is sometimes necessary to shadow the object with some metal to achieve contrast. A metal spray, directed at the object from the side, is used to give contrast and depth to the image.

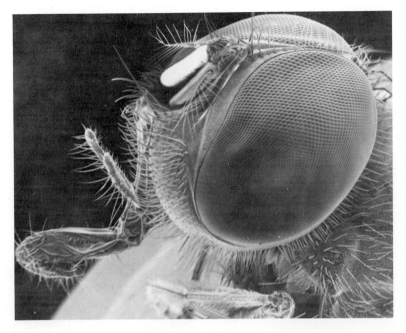

FIG. 6.7 The head of a house fly viewed under a scanning electron microscope. This microscope can bring out remarkable detail in opaque objects. The use of the scanning electron microscope has greatly extended our knowledge of the structure of small objects. (BioInformation Associates.)

angstrom unit, Å, which is one-tenth of a mμ, or one ten-thousandth of a μ.

The Size of Cells

Most cells are in a size range conveniently measured in microns. The smallest cell known is the **pleuropneumonia** organism, which is a ball-shaped form about 0.2 μ in diameter. This is the smallest a cell can be, since calculations of the number of cell parts necessary to sustain cellular life indicate that this much space is required in a cell. At the other extreme, birds' eggs, which contain a single nucleus, may be classified as cells. Actually, the yolk is the cell, and the albumen and shell are added on as the cell descends the oviduct before the egg is laid. The great majority of cells lie in the size range of 3 to 30 μ. This seems to be the size at which cells function best.

Separation of Cell Parts

Sometimes the parts of cells can best be studied when they are separated from one another. The separation of cells into their constituent parts is done by delicate centrifugation. First the cells are broken open in a blender. Then the opened cells are centrifuged. At a certain speed the nuclei are thrown to the bottom of the centrifuge tube because they are heavier than the cytoplasmic components. The nuclei are removed and the cytoplasmic material is then spun at a higher speed, which causes certain

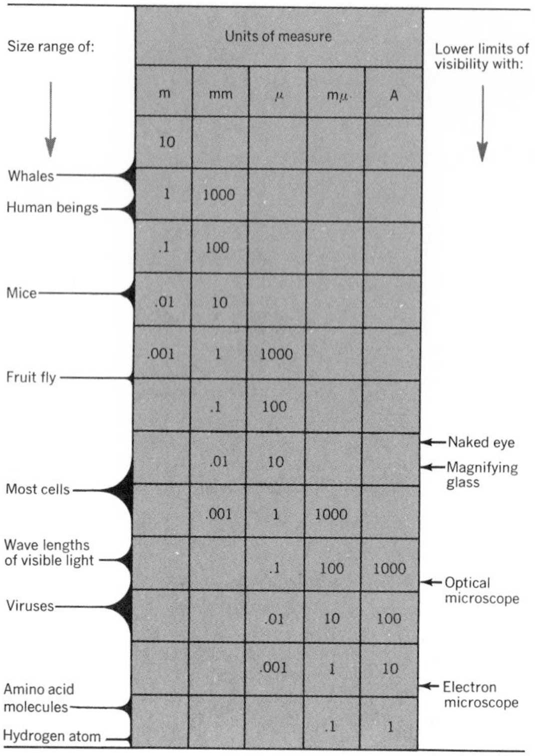

Size range of:	Units of measure					Lower limits of visibility with:
	m	mm	μ	mμ	A	
	10					
Whales — Human beings —	1	1000				
	.1	100				
Mice —	.01	10				
Fruit fly —	.001	1	1000			
		.1	100			
		.01	10			← Naked eye ← Magnifying glass
Most cells —		.001	1	1000		
Wave lengths of visible light —			.1	100	1000	← Optical microscope
Viruses —			.01	10	100	
			.001	1	10	← Electron microscope
Amino acid molecules — Hydrogen atom —				.1	1	

m-meter: mm-millimeter: μ-micron: mμ-millimicron: A-angstrom unit

FIG. 6.8 Units of measure used in biology. The last three columns are used for objects which lie below the visual range of the unaided human eye.

organelles to separate out. By putting the cellular material in fluids of different densities and running the centrifuge at different speeds, each of the cell parts can be separated from the others. Then chemical tests and other methods of analysis can be conducted on each part of the cell individually.

CELL ORGANIZATION

Prokaryotic and Eukaryotic Cells

Cells fall into two distinct categories. Some of the more primitive one-celled organisms are **prokaryotic** (also spelled **procaryotic**). Pro-

karyotic cells do not have a nucleus surrounded by a membrane, but they do have a central region where the genes are located. The genes consist of long strands of DNA which is not associated with protein to form true chromosomes. Prokaryotic cells do not undergo mitosis. Instead, the strands of DNA duplicate and the cell divides by fission. Prokaryotic cells also lack many of the organelles which are found in the cytoplasm of other cells. The bacteria and blue-green algae are prokaryotic cells.

All other cells are **eukaryotic** (also spelled **eucaryotic**). Eukaryotic cells have a definite nucleus surrounded by a membrance, chromosomes which contain both DNA and protein,

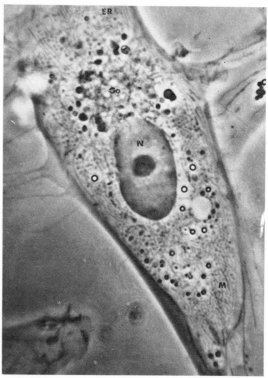

FIG. 6.9 Structure of a human nerve cell showing many of the parts of an animal cell. ER—endoplasmic reticulum; Go—Golgi apparatus; N—nucleus (note dark nucleolus within); M—mitochondria. (George E. Rose in *Journal of Cell Biology*.)

mitosis before cell division, and a cytoplasm containing many organelles which carry on specific cell functions. The following survey of cell organization will be concerned primarily with eukaryotic cells, since most cells fall into this category.

The Nucleus

Eukaryotic cells have two basic parts; the cytoplasm and the nucleus. The nucleus is the control center of the cell. From the nucleus come messages which direct the synthesis of protein in the cytoplasm of the cell. The **chromosomes** are found in the nucleus; they are most clearly seen when they are shortened and thickened into rods during mitosis, but usually they are extended into long, slender, threadlike bodies.

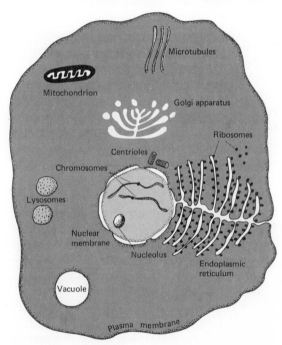

FIG. 6.10 Diagrammatic representation of the parts of an animal cell which can be brought out by a variety of techniques. The parts are not drawn in proportion to actual size.

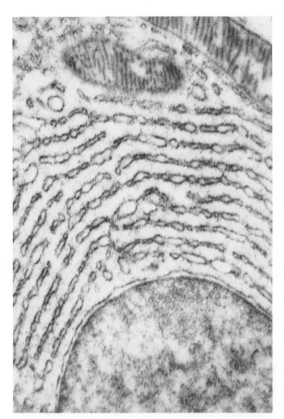

FIG. 6.11 Portion of a cell from the pancreas of a guinea pig showing the double nature of the nuclear membrane and the small spherical ribosomes located along the endoplasmic reticulum. Three mitochondria can be seen at the top. (Electron photomicrograph, courtesy of George E. Palade, Rockefeller University, New York.)

Chromosomes are so thin that they cannot be seen with the optical microscope, but there may be places along their length where they are compacted into coils which are visible. These visible coils are sometimes called **chromatic granules.** Closely associated with the chromosomes, there may be one or sometimes two or more spherical or ovoid bodies known as **nucleoli.** Differential staining shows the nucleoli to be rich in ribonucleic acid, RNA. Some evidence indicates that ribosomal RNA, which makes up part of the ribosomes in the cytoplasm, is produced in the

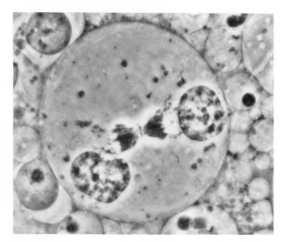

FIG. 6.12 In this grasshopper cell the two nuclei have already divided, but the mitochondria can be seen in the center, still in the process of replication.

nucleoli. The nucleoli are closely associated with the chromosomes and seem to be produced by them. The nucleoli disappear during mitosis and reappear after cell division.

A **nuclear membrane** surrounds the nucleus. The nuclear membrane is similar to the plasma membrane which surrounds the cytoplasm, but there is one important difference. The nuclear membrane seems to be perforated with many small pores which are not found in the plasma membrane. A closer examination of the nuclear membrane with the electron microscope shows that the pores are actually areas where the membrane becomes much thinner than the rest of the membrane. These areas are the sites at which the messages from the genes, which are part of the chromosomes, exit from the nucleus to the cytoplasm, but these areas do not allow a free flow of cytoplasm into the nucleus.

The Endoplasmic Reticulum and Ribosomes

Before the discovery of the electron microscope, the cytoplasm was thought to be a viscid liquid, somewhat like the albumen of an egg. A few cytoplasmic inclusions had been seen. Electron photomicrography, however, showed the cytoplasm to be interlaced with an intricate network of channels, the **endoplasmic reticulum,** which extend all the way from the nucleus to the outer plasma membrane. These channels are enclosed with membranes which are similar to the plasma membrane.

Many tiny, spherical bodies are attached to the endoplasmic reticulum membranes. These bodies are the **ribosomes,** which are centers of protein synthesis. Ribosomes are about 17 mμ in diameter, so they are visible only with the electron microscope. It is thought that proteins synthesized by the ribosomes pass directly into the channels of the endoplasmic reticulum and are conducted to the outside of the cell. This would explain how cells can secrete extracellular enzymes and other products which include proteins. Proteins do not easily pass through the plasma membrane of cells. Some of the channels may also allow materials from outside the cell to pass directly into the cell and in to the nucleus. Ribosomes are also found in the cytoplasm in a free form unassociated with the endoplasmic reticulum. These free ribosomes can aggregate into chains, known as **polyribosomes,** and synthesize proteins under the direction of RNA messages received from the genes in the nucleus. The ribosomes are made of a combination of protein and RNA. We shall learn how the ribosomes synthesize protein in Chapter 9.

Mitochondria

Mitochondria are rod-shaped or spherical bodies, typically about 0.5 μ in their longest dimension. They are visible under the optical microscope, but they usually are not visible in cells which have been stained. One stain, Janus green, will bring out the mitochondria, but they are best studied in living cells with the phase contrast microscope. Mitochondria have an outer membrane and a greatly involuted internal membrane (see Fig. 6.11). Mitochondria are the powerhouses of the cell; they contain enzymes which release most of the energy in food and form ATP, an energy fuel which can be used by all parts of the cell.

Some recent investigations indicate that the mitochondria might have originally been parasites which invaded the cell. Primitive cells can carry on only **anaerobic respiration,** which is the process of extracting energy from food without using oxygen. This process is very inefficient, however, and obtains only about one-twentieth of the energy which is in food. The mitochondria may have been cellular parasites which, because of their enzymes, could carry on aerobic respiration, a process which uses oxygen to extract energy from food. After invading the cells, the mitochondria may have set up a mutually beneficial association with the cells and become a permanent part of them. The mitochondria have their own DNA and a self-replicating system which is independent of mitosis of the cell. Mitochondria can often be seen dividing and segregating just before the cell splits in two. A final bit of evidence for this theory is that mitochondria can be removed from cells and, when placed in the proper culture medium, can grow and duplicate independent of all cell connections.

Plastids

Another group of self-replicating bodies found in the cytoplasm are the **plastids,** which are found only in the cells of plants and some of the more primitive plantlike organisms. Plastids have a kind of DNA which is not found in the nucleus, and they may also be descended from a parasitic organism which invaded the cell and established a permanent, mutually beneficial association with the host.

Chloroplasts are the most important plastids. They contain chlorophyll, a pigment, which takes part in the manufacture of food in the presence of light, a process known as photosynthesis. Each chloroplast is made up of many layers, known as **grana,** which are arranged like stacks of coins. **Leucoplasts** are plastids where the synthesis of starch, and storage of this polysaccharide, takes place. **Chromoplasts** contain yellow, orange, and red pigments. We see these pigments prominently in the fall, when the green chlorophyll fades away. Most unripe fruit is green because chlorophyll is present in the outer skin. As the fruit ripens, the chlorophyll decreases and the chromoplasts increase in number to give the ripened colors to the fruit.

The Golgi Apparatus

The **Golgi apparatus** lies near the nucleus and is made up of a series of three, four, or more curved membranes which are stacked like a pile of saucers. The Golgi apparatus functions in cell secretion. Cells which produce large quantities of secretions, such as cells in the liver, pancreas, and salivary glands, have a well-developed Golgi apparatus. The Golgi apparatus may serve as a storehouse for secretory products, to separate them from the rest of the cell. Vesicles may be seen coming off from the apparatus and these may carry the secretions from the apparatus to the outer membrane. The Golgi apparatus is not well defined in plant cells, but many botanists think it may function in the production of cellulose, which is found in the cell wall.

Lysosomes

Lysosomes are spherical or irregularly shaped bodies in the cytoplasm which contain enzymes which break down (**lyse**) cellular materials. Lysosomes isolate these enzymes from the rest of the cell to prevent the cell from digesting itself. Some biologists call lysosomes "suicide bags" because when they break open, the cell destroys itself. The deadly effect of some poisons may be caused by their ability to burst lysosomes and cause cell destruction. Lysosomes are also the culprits in cases of sunburn. Excessive exposure to the ultraviolet rays of sunlight will cause the lysosomes in the delicate lower layers of the skin to break and this, in turn, causes the destruction of many cells. Melanin, a pigment found in the outer layers of the skin, filters out most of the ultraviolet rays, but when there is insufficient melanin, sunburn occurs. When a cell is injured or dies, the lysosomes break and self-digestion begins. Beef is more tender after aging because autolytic enzymes contained in the lysosomes have had time to break down many of the tough fibers.

A sad fact of life is that we begin aging as soon as we stop growing. Aging is characterized by the destruction of tissue faster than it can be replaced. A possible explanation of aging is that, with the passage of time, there is an increased tendency for the lysosomes to rupture. Research on aging is concentrating on this hypothesis to try to find a way to inhibit the increased rate of lysosome rupture.

Vacuoles

Vacuoles are cell inclusions which contain nonliving materials. Vacuoles are bound by membranes which are similar to the plasma membrane, and, in fact, vacuole membranes may be derived from the plasma membrane in some cases. Vacuoles are quite large in mature plant cells, although they are small when the cells are young. The plant vacuoles generally contain water and dissolved minerals, along with some organic materials produced by the cell.

Some of the microorganisms and simpler animals engulf food. The cytoplasm of these organisms surrounds a food particle and closes around it. The food is then taken inside the cell, surrounded by a part of the plasma membrane. Lysosomes may then move to these food vacuoles and empty their contents into the vacuole so that the lysosomal enzymes can digest the food in the food vacuole. The digested food then diffuses out into the cytoplasm. In higher animals, fat molecules may be stored in fat vacuoles. Some human fatty tissue consists of large fat vacuoles surrounded by a very thin layer of cytoplasm and the nucleus. Some small freshwater organisms have contractile vacuoles which absorb and expel excess water. The protoplasm of the cells of these organisms contains a higher concentration of minerals than does the water outside. As a result, more water enters the cell, by osmosis, than is needed. The excess water collects in contractile vacuoles, which gradually swell and then suddenly contract and expel the water outside the cell.

Centrioles

Centrioles are small cylindrical bodies located near the nucleus, usually near the Golgi apparatus. Centrioles are common in animal cells but rare in plant cells. They contain nine groups of tubules with three tubules in each group. These tubules are spaced around the outside of the cylinder. Centrioles are about 150 mμ in diameter and about twice this length. Each centriole seems to be closed at one end and open at the other. There are two centrioles in each cell and they lie at right angles to each other except during mitosis. Immediately before mitosis occurs, the centrioles move apart, and as they move, they spin tiny microtubules of protein from their open ends. These microtubules form the **spindle figure** seen during mitosis. A centriole is found at either end of the spindle figure. At about the time the protoplasm of the cell divides, the centrioles duplicate so that each daughter cell has two.

The Plasma Membrane

We have already said much about this membrane which marks the outer boundary of the cytoplasm. The **plasma membrane** is a living membrane; through it pass all the materials which enter or leave the cell. The vital function of the plasma membrane is to regulate the passage of materials into and out of the cell. The plasma membrane is very thin, about 90 Å in width, so it can be seen only with the electron microscope. Under the electron microscope, the membrane appears to be composed of three layers, with protein molecules as the outside layers and lipid molecules as the inside layer. At various points along the membrane, there seem to be areas where the lipid layer fills the entire width of the membrane, somewhat like the soft filling of a sandwich which has oozed out at the edges of the bread. In the next chapter we shall learn how the plasma membrane functions in regulating passage of material into and out of the cell.

Cell Coverings

There are a few cells, such as red blood cells and amoebas, which have no covering outside the plasma membrane. The plasma membrane is so thin and delicate, however, that most cells have an additional covering. This covering is produced by the cell but is not a living part of the cell. In most plants it is a rigid **cell wall.** The cell wall keeps the cell in a particular shape and helps support the entire plant. The chief component of the cell wall is **cellulose,** a complex carbohydrate. Wood is primarily cellulose. The cell wall may also contain **lignin,** which adds rigidity to the wall. Adjacent cells share an intercellular structure, called the **middle lamella,** which contains a gummy substance known as **pectin.** Pectin is especially abundant in certain fruits and it causes jelly to become firm.

Animal cells do not have a cell wall, but most of them do have a thickened membrane around the outside of the cell which gives protection and a degree of rigidity to the cell. This thickened membrane is called the cell membrane by some biologists, but since this term is also used as a synonym for plasma membrane, confusion can result.

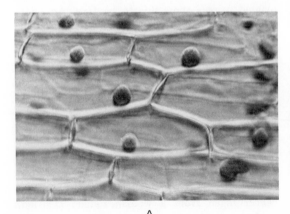
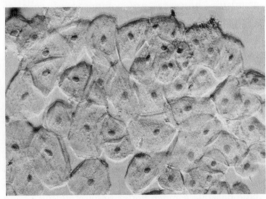

A   B

FIG. 6.13 Side lighting brings out the distinctions between typical plant cells (A) and animal cells (B). Cells from an onion (A) and from the human mouth (B).

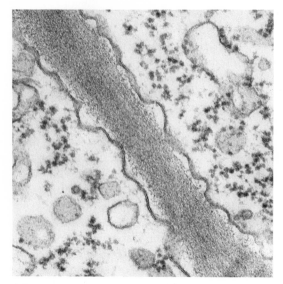

FIG. 6.14 The cell wall and adjacent plasma membranes can be seen in these portions of two adjacent cells of the African violet. The dark, dotlike ribosomes can also be seen. (Electron photomicrograph by K. E. Porter, Rockefeller University.)

CELLS AND TISSUES

The simplest forms of organisms have only one cell. Other organisms are aggregates of cells which are all almost alike. Most multicellular organisms, however, are composed of cells of different kinds which perform specific functions. Cells of the same kind, which are found in a group together with the intercellular substances produced by them, are known as **tissues.** Similar tissues tend to serve similar functions.

Tissue Culture

The discovery of a method for culturing cells taken from a multicellular organism has made possible a much better understanding of the different kinds of cells and how they function. A few cells from skin or blood can be placed on a sterile nutrient medium where they will grow and divide. Experiments can be conducted on these cells which could not be conducted on cells which are a part of a living organism. Our knowledge of effects of radiation on human tissues has been greatly expanded by tissue culture experiments. Cells from human tissue cultures can be exposed to doses of radiation which could never be used on a person. The effect of other possible harmful treatments can also be tried on cells from tissue cultures before these treatments are used on people. Human chromosomes are usually studied by tissue culture methods.

Tissue culture techniques were first used in 1912 by **Alexis Carrel.** He took bits of tissue from the heart of an embryonic chicken and placed them in a sterile solution which contained oxygen and all the nutrients which are needed to support life in this type of tissue. The tissue was kept alive and growing for many years, a period much longer than the normal life span of a chicken. The cells in the tissue divided about once a day, doubling their volume every day. From time to time much of the tissue had to be removed to keep its size within bounds.

Studies show that death does not come inevitably to individual cells. Death is a characteristic of biological entities which is inevitable for organisms but not for individual cells. Transplanted organs show this to be true. An ovary from an aging female mouse can be transplanted into a younger female mouse and the ovary from the older mouse will continue to live and function. As the younger mouse grows older, the same ovary can be transplanted into another young female, and so on. It would seem that a deficiency in the necessary nutrients or some factor produced by an aging body causes old age and death in animals.

We have learned much about cancer from tissue cultures. A tissue culture made from a cancer taken from a woman named Henrietta Lacks over thirty years ago is still being used in many laboratories today, even though Henrietta Lacks died many years ago.

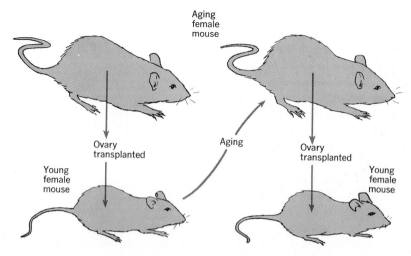

Aging
female
mouse

Ovary
transplanted

Aging

Ovary
transplanted

Young
female
mouse

Young
female
mouse

FIG. 6.15 The ovary of a mouse can continue to live and function through many generations if it is transplanted to young mice as each carrier grows old. This experiment shows that tissues do not have a limited life even though whole organisms do.

Viruses are also grown in tissue cultures. Viruses will multiply only within living cells of the proper type; they cannot be cultured on media like bacteria. Without the technique of the tissue culture, we would never have developed a vaccine for polio. It was found that the polio virus would grow in cells from the kidney of the rhesus monkey. This made it possible for virologists to experiment with different strains of the polio virus and come up with the vaccine which has reduced the incidence of polio to a very low level in the United States.

REVIEW QUESTIONS AND PROBLEMS

1. Why is an understanding of the functioning of the human body dependent upon an understanding of the nature of cells and how they work?
2. How do modern microscopes correct for the aberrations which are a natural part of curved glass lenses?
3. What factor sets a limit on the degree of effective magnification which can be achieved by the optical microscope?
4. What are the disadvantages of the electron microscope in contrast to the optical microscope?
5. Suppose an object is 0.1 mm in diameter. What is its diameter in microns, millimicrons, and in angstrom units?
6. Ribosomes were unknown before the development of the electron microscope. Explain why.
7. How does a prokaryotic cell differ from a eukaryotic cell?

8. Describe the evidence which supports the hypothesis that mitochondria became a part of eukaryotic cells through parasitic invasion.

9. What general difference exists between most plants and most animals which would make the Golgi apparatus less highly developed in the plant cells.

10. When a warm-blooded animal dies, its body stiffens in rigor mortis because of protoplasm coagulation. Within a few days, however, its body becomes limp. Give a possible explanation in light of the facts presented in this chapter.

11. Large animals generally have some sort of supporting skeleton, but a skeleton is not found in larger plants. Explain.

12. Give a possible reason why mature plant cells have larger vacuoles than mature animal cells.

13. Genetic studies show that certain characteristics of the chloroplasts are passed through the female line only, not from the plant which furnished the pollen grain. Give a possible explanation.

14. The leaves of trees which have been green all summer suddenly become brilliantly colored in the fall. Explain what happens.

15. Without tissue cultures much of our knowledge about human cells would never have been discovered. Explain why.

16. Formulate a biological problem which might be solved by tissue culture techniques and tell how you would solve the problem.

FURTHER READING

Buvat. R. 1969. *Plant Cells, an Introduction to Protoplasm.* New York: McGraw-Hill.

DeRobertis, E. D. P. 1970. *General Cytology.* Philadelphia: W. B. Saunders.

Faucett, D. W. 1966. *The Cell. Its Organelles and Inclusions.* Philadelphia: W. B. Saunders.

Freeman, J. A. 1964. *Cell Fine Structure.* New York: McGraw-Hill.

Ledbetter, M. C., and Porter, K. 1970. *Introduction to Fine Structure of Plant Cells.* New York: Springer-Verlag.

Loewy, A. G., and Siekevitz, P. 1970. *Cell Structure and Function.* 2nd ed. New York: Holt, Rinehart, and Winston.

Novikoff, A. B., and Holtzman, E. 1970. *Cells and Organelles.* New York: Holt, Rinehart and Winston.

Wilson, G. B., and Morrison, J. H. 1966. *Cytology.* 2nd ed. New York: Van Nostrand Reinhold.

7

Movement of Materials into and out of Cells

The life of a cell is dependent upon the proper movement of materials through its outer covering. Food and other vital dissolved substances must come into the cell, while metabolic wastes and other substances present in excess within the cell must be conveyed outside. The plasma membrane is the outer covering of the protoplasm and serves an important regulatory function in this regard. The movement of materials into and out of cells accords with the physical principles of the movement of dissolved substances in any system, living or nonliving.

DIFFUSION

Drop a crystal of soluble dye into a beaker of water. The color of the dye will gradually spread until all the water in the beaker shows some color. If left undisturbed, the water will eventually become uniformly colored throughout. The dispersion of the dye in the water is a typical case of **diffusion.** When the crystal of dye is dropped in the water it begins to dissolve. The outer molecules come off the solid into the surrounding water. Thus, they are more concentrated in the water immediately around the crystal than they are in the water further removed. According to the principle of diffusion, dissolved particles tend to move from regions where they are more concentrated to regions where they are less concentrated. Thus, there is an outward movement of the molecules of the

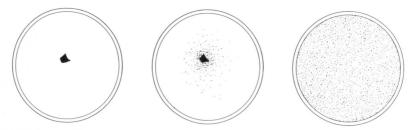

FIG. 7.1 Diffusion of a solid in a liquid. When a crystal of soluble dye is placed in water some of the outer molecules begin to break away and move outward toward an area in which a lesser concentration of these molecules is found. Given sufficient time, the crystal completely dissolves and its molecules are evenly distributed throughout the water.

dye until they eventually become evenly distributed in the beaker.

The movement of molecules or ions in diffusion is an expression of energy, as is true of all types of movement. In this case it is the energy of heat which is responsible for the movement. All the matter on the earth, being at some temperature higher than absolute zero ($-273°C$), exhibits heat energy, that is, the particles are in motion. Accordingly, there is always movement of molecules or ions in any solution. The amount of energy and the speed of movement of the particles is related to the **temperature** of the solution. If the particles of a solute are evenly

FIG. 7.2 The ice fish, *Chaenocephalus aceratus*, from the Antarctic. Sometimes called the white-blooded fish, this animal has no hemoglobin in its blood. Its blood plasma carries sufficient oxygen without hemoglobin because the very cold waters in which it lives hold high concentrations of oxygen. (Courtesy of Paul Richard, University of Northern Colorado.)

distributed throughout a solution and extra heat is applied to one part of the solution (and convection currents are prevented), the particles of the solute which are warmest will have greater free energy than those which are cooler. Diffusion of particles will be more rapid from the hot to the cool region than from cool to hot regions. As a result, the cool region will become more concentrated than the hot region. Thus, temperature as well as concentration affects the direction of diffusion.

The effect of temperature on the amount of **free energy** of dissolved particles is demonstrated in the oceans. It may come as a surprise to many people to learn that the greatest concentration of marine animal life is found in the cold waters of the northern and southern oceans, rather than in the warm waters of the tropics. Oxygen concentration is frequently the limiting factor for such forms of life. In warm tropical waters the oxygen concentration is relatively low because oxygen molecules have greater free energy and tend to pass readily out of the water into the air above. In cold ocean waters, however, the free energy of the oxygen molecules is lower, and the concentration of oxygen in the water is consequently higher.

People who raise tropical fish know that the water in the aquarium must be kept warm to correspond to the warm waters of their natural environment. This creates a problem with respect to the maintenance of the proper oxygen concentration of the water. If many fish are placed in the aquarium, there will not be sufficient oxygen to support them and they will begin to die. To get around this difficulty, a constant stream of air bubbles may be forced through the water. Oxygen from the bubbles dissolves in the water and keeps the oxygen concentration high in spite of some crowding. If you raise fish like goldfish, which can live in water at a cooler temperature, the aquarium can be maintained without bubbles even though it is relatively crowded. The cooler water can hold a higher concentration of the oxygen absorbed from the surface. Some fish may die, however, if you allow the water to get warm.

There is one species of fish living in the very cold waters of the antartic which has no hemoglobin (the oxygen carrier) in its blood. The oxygen concentration of the water is so high that the liquid plasma of the blood can absorb and transport all the oxygen needed by the fish.

Other factors also affect the free energy of particles in solution. **Concentration** is one of these; the higher the concentration, the greater the amount of free energy. In an area with many moving particles, it is logical that there will be a greater amount of free energy than in an area in which there are few or no moving dissolved particles. Hence, in the case of a dissolving crystal of dye, the free energy of the dissolved particles will be greatest immediately around the crystal where the concentration is greatest. Diffusion would, therefore, proceed from this area of greater free energy to the areas of lesser free energy farther away from the crystal.

Pressure is another factor affecting free energy; as pressure is increased there is a corresponding increase in the free energy of the particles. To illustrate, place some sugar solution in a porous membrane, one which will allow sugar molecules to pass through freely, and tie this membrane together tightly to make a bag. Now put this bag into a beaker containing a sugar solution of equal concentration and temperature. The sugar molecules will move back and forth through the membrane with equal speed in both directions. But if you apply pressure to the bag, there will be an increase in the free energy of the sugar molecules in the bag and they will move out more rapidly than they move in. Soon the concentration will be greater in the beaker than in the bag.

To sum up these observations on diffusion, we can say that diffusion is the movement of molecules or ions from an area where they have a greater free energy to an area where they have a lesser free energy. To put it more succinctly, diffusion is the movement of molecules or ions downward along the **free energy gradient.** This movement is most commonly downward along the concentration gradient, since concentration is the factor most usually affecting the free energy level of molecules or ions in solution.

In biology we are primarily concerned with the diffusion of particles in solution in water, but there can also be diffusion of particles in gases. If some ether is poured in a flat dish in one corner of a room, the odor of ether can soon be detected at the opposite end of the room. The ether molecules coming off the liquid have great free energy and so they travel rapidly across the room.

Remember that the molecules of the solvent (water or another liquid) are also moving in accordance with the free energy concept, just as the particles of the solute (sugar, dye, or the like) are. Such movement of the molecules of the solvent furthers the general dispersion of the solute.

A membrane does not prevent the diffusion of the particles of a dissolved substance if the membrane is permeable to the particles. A sheet of rubber stretched between two solutions of different concentrations would prevent diffusion because the rubber molecules fit together so tightly that they will not permit any dissolved particles to pass through. There are many mem-

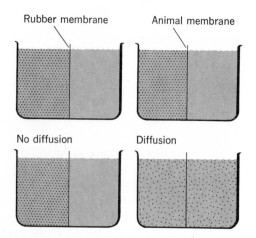

Rubber membrane   Animal membrane

No diffusion   Diffusion

FIG. 7.3 Diffusion through a membrane. No diffusion occurs when the membrane is impermeable to dissolved particles. Diffusion occurs when the membrane permits the passage of the particles.

branes, however, which contain openings between the molecules that are large enough for dissolved particles to pass through. Diffusion can take place through such permeable membranes.

Since the dispersed particles in a mixture vary greatly in size, it stands to reason that not all such particles can pass through the openings between the molecules of membranes. A membrane permeable to the small dissolved particles of a true solution will not ordinarily be permeable to the larger particles of colloids, emulsions, and suspensions. Such membranes are called **differentially permeable membranes.**

Suppose we have a room divided down the center from floor to ceiling with a net like a tennis net. On one side of the net suppose we have a general assortment of fruit—grapes, limes, grapefruit, cantaloupes. We put several children on each side of the net and tell them to start throwing this fruit at the net. When the grapes and limes are thrown, some of these will go through the openings in the net, but the grapefruit and cantaloupes will be stopped by the net and will remain on the same side. As the grapes and limes go through, the children on the other

side will begin throwing them back. Allow this to go on for some time (assuming that the fruit is not crushed in the process) and we will find that the limes and grapes are evenly distributed on the two sides of the net, but the grapefruit and cantaloupes remain on the side where they were at the beginning. Now think of the grapes and limes as small molecules of a dissolved substance and the grapefruit and cantaloupes as colloidal particles, and perhaps you can appreciate how a differentially permeable membrane allows even diffusion of dissolved particles but holds back larger particles.

OSMOSIS

Osmosis is a term which is well known but frequently misunderstood. The process can be explained logically on the basis of the free energy

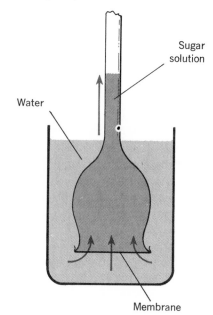

Sugar solution

Water

Membrane

FIG. 7.4 Apparatus to demonstrate osmosis. Water passes through the membrane to the region of lower water concentration in the tube. The movement of water causes the sugar solution to rise in the tube.

concept of the movement of molecules. Consider a demonstration such as the one commonly performed in a biology laboratory. A 5 percent solution of sucrose in water is put into a bag made of a membrane permeable to water but not readily permeable to sucrose molecules. The membrane can be prepared by spreading a thin film of liquid collodion over the inside of a beaker and allowing it to dry, or some natural membrane such as the bladder from a pig or the skin from the leg of a frog can be used. The membrane bag containing the sugar solution is tied tightly around a glass tube and suspended in a jar of water, as shown in Fig. 7.4. Within a few minutes the volume of the solution will begin increasing and forcing the solution up in the open tube. Water can pass through the membrane in both directions, but it moves into the solution faster than it moves out. This more rapid movement of water through a differentially permeable membrane toward an area of greater concentration of a dissolved substance is what is known as osmosis. To find an explanation for osmosis, we must determine why the water moves more rapidly in one direction than the other.

Various hypotheses have been proposed in the past to explain osmosis, but not until the development of the free energy concept has any hypothesis been very satisfactory. When sucrose is dissolved in water, there is a decrease in the free energy of the water molecules. Their movements are slowed down because there has been a decrease in the concentration of water molecules; the added sugar molecules have occupied some of the space formerly occupied by water molecules. This is expressed in the physical nature of the mixture; the sugar solution becomes more viscid or syrupy. Now it is easy to understand why water moves into the solution within the membrane faster than it moves out. The water molecules have a higher free energy on the outside and they are obeying the laws of diffusion by moving to an area of lower free energy on the inside of the membrane.

Some sucrose molecules will also diffuse out of the membrane because the membrane is not completely impenetrable to them and they have a greater free energy within the solution, where they are more concentrated, than outside. The membrane is differentially permeable, however, with pores so small that the water molecules pass through much more readily than the much larger molecules of sucrose. Hence, water passes into the membrane much faster than the sugar passes out.

This movement of water into the membrane exerts a pressure, known as **osmotic pressure,** within the membrane, and this causes the solution to rise up the tube for a considerable distance. If the membrane is completely closed, tied into a tight bag without a protruding tube, the pressure builds up within the membrane. Whenever the pressure within the membrane reaches a point where it balances the osmotic pressure, osmosis will stop. This will occur in the membrane with the tube because of the weight of the column of water, as well as within the closed membrane. An equilibrium is reached where the **hydrostatic pressure** within the membrane balances the osmotic pressure. The free energy of the water molecules becomes the same on both sides of the membrane even though there is still a difference in concentration. The water molecules continue to pass through the membrane, but the passage is equal in both directions.

Another factor is also involved. The incoming water dilutes the solution within the membrane and there is, therefore, a corresponding increase in the free energy of the water molecules within the membrane. This results in a gradual slowing of the rate of osmosis even though there is no alteration in pressure on the two sides of the membrane.

We can test this **thermodynamic concept** of osmosis by introducing the third factor which is known to influence the free energy of molecules, namely, temperature variation. A warm rod can be inserted inside the solution within the membrane and this will raise the temperature and increase the free energy of the water molecules. If the water outside is kept cold, there will be a faster movement of the water molecules from the inside of the membrane to the outside than in the reverse direction. Thus, osmosis takes place upward along the concentration gradient but downward along the free energy gradient. Concentration alone does not determine the movement of water in osmosis. Any of the factors which affect the free energy of the water molecules can influence the direction of osmosis.

In living systems osmosis is of great importance for the maintenance of life. A human red blood cell is surrounded by a thin, delicate plasma membrane, as is any cell, but the red blood cell lacks a tougher outer membrane around the plasma membrane. As a result, the red cell is very delicate. When a human red blood cell is placed in distilled water, osmosis causes water to enter the cell faster than it will come out. The concentration of dissolved substances within the cell is, of course, greater than in the water, which contains no dissolved sub-

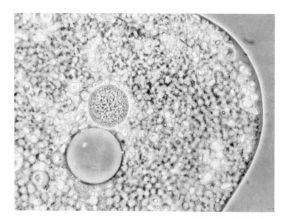

FIG. 7.5 A portion of an amoeba showing the clear contractile vacuole just below the nucleus. The contractile vacuole accumulates excess water which is taken into the cell by osmosis. The vacuole then contracts and expels the water to the outside.

stances. Since the red blood cell is a closed system, pressure begins to build up within it. This pressure soon becomes so great that the plasma membrane bursts and the cell is destroyed. Most other cells of the human body have a tough outer membrane around the plasma membrane. These cells can sustain sufficient pressure to equalize the movement of water into and out of the cell without bursting.

When cells are placed in water, or in water with dissolved particles of a lesser concentration than within the cells, we say they are in a **hypotonic medium.** Human red blood cells have a concentration of dissolved particles equal to about 0.87 percent. If these cells are placed in a salt solution of this same concentration, they do not burst; the water molecules move equally in both directions. They are in an **istonic medium.** Any salt solution with a concentration greater than this would be a **hypertonic medium.** Red blood cells placed in 1 percent salt solution will lose water and shrivel. Sometimes physicians will test the strength of red blood cells by placing them in various concentrations of hypotonic salt solutions. In certain diseases the plasma membranes become more fragile and burst more easily than those of normal cells. Normal red blood cells can remain intact when the outside solution is as low as 0.5 percent. In certain diseases the cells may begin to burst when the concentration is as high as 0.65 percent. This well illustrates the fact that osmotic pressure is directly related to the difference between the concentration of dissolved particles on the two sides of a membrane.

A garden plant will be killed if too much fertilizer is placed around the roots. As the fertilizer goes into solution it will form a solution which is hypertonic to the solution within the root hairs, which normally absorb water through osmosis. In this case, however, the process is reversed and water is drawn out of the root hairs. These plants are also constantly losing water through their leaves and will soon shrivel and die unless sufficient water is placed around the roots to lower the fertilizer concentration until the solution is hypotonic to the protoplasm within the cells of the roots.

In some areas of the world the concentration of minerals in the soil is very high. In dry years no plants can be grown on this soil because the soil moisture is hypertonic to the plant cells. In wet years the extra water dilutes the minerals so that the concentration is hypotonic to the plant cells and crops can be grown. This is also an important factor in hot, arid regions where irrigation of land is widespread. The water used for irrigation contains some dissolved minerals. Much of this water evaporates, concentrating the minerals in the soil. After many years of this concentration the soil may have to be abandoned because the mineral concentration becomes hypertonic to the plants.

Because of osmosis, most drugs that are prepared for injection into the human body are made isotonic before use. This can be done by adding just enough sodium chloride (table salt) to bring the total concentration to about 0.87 percent. As a result, there will be no swelling or bursting of cells when the injection is given. Nasal sprays and drops are also in an isotonic solution, which prevents swelling or shrinking of the tissues of the nasal cavity. We have all experienced the unpleasant sensation resulting from the accidental drawing of water into the nose while swimming in fresh water. It is also unpleasant to draw salt water into the nose when swimming in the ocean. In the first case there is a swelling of the membranes because fresh water is hypotonic to the cells; in the second case there is shrinkage because sea water, with a salt concentration of about 3.5 percent, is hypertonic to the cells. An isotonic solution can be drawn into the nose without discomfort.

Physicians often advise patients to soak a sprained ankle in hot water containing Epsom salts in a rather high concentration. The salts in the water prevent undesirable swelling of the cells of the skin and help to reduce a swollen area by removal of water through osmosis. The laxative effect of Epsom salts also is due to osmosis. When a hypertonic concentration of Epsom salts is present within the intestine, water is drawn by osmosis from the cells lining the intestine, and this keeps the contents of the intestine in a highly liquid state.

Small unicellular fresh-water animals like *Amoeba* and *Paramecium* have a problem because of osmosis. Their protoplasm has a higher concentration of dissolved particles than the water in which they live, so water is constantly coming in by osmosis. They have contractile vacuoles which expel the excess water and thus prevent the pressure within the cell from becoming too high. These vacuoles can be seen slowly filling and then suddenly contracting as they squirt the water to the outside. The protozoans that live in salt water do not have these contractile vacuoles; their protoplasm is isotonic to the water outside the cells.

Plant cells living in fresh water have hard cell walls, and they develop hydrostatic pressure which balances the osmotic pressure. Plant cells of this nature thus develop turgor, or they

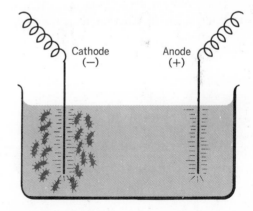

FIG. 7.6 Paramecia tend to gather around the cathode, or negative pole, when a weak electric current is induced in water. This experiment indicates that paramecia have an overall positive charge on the outer cell surface.

become **turgid** when in fresh water. If placed in a hypertonic solution they become limp as the protoplasm shrinks away from the cell wall. We say that they are **plasmolyzed,** or they exhibit **plasmolysis.** Potato slices placed in water become turgid and are crisp and firm. When placed in 5 percent salt solution, however, they are plasmolyzed and turn limp. It is even possible to tie shoestring potatoes into bow knots when they are plasmolyzed.

In the human body osmosis against the concentration gradient frequently occurs. The heartbeat generates blood pressure, which causes both the water and the dissolved materials to have a greater free energy than is found in the fluids outside the blood vessels. When the blood reaches the small capillaries, the water and dissolved materials pass out, forming the **tissue fluid** which bathes the cells and makes possible an efficient absorption of food and excretion of waste products by the cells. The fluid then moves into larger spaces, where it is known as **lymph,** and re-enters the blood stream through lymph vessels.

The pressure within the capillaries of the kidneys is even higher than in other body organs because the small vessels leading from the arteries are larger in diameter than the vessels collecting the blood from the capillaries. This forces water and dissolved material out of the blood because of the very high free energy of the molecules within the capillaries. Later there is a selective reabsorption of the water and dissolved materials needed by the body.

SELECTIVE NATURE OF LIVING MEMBRANES

Research in cell physiology has shown that the living plasma membrane can exert a selectivity on the passage of materials, and there can be a movement against the concentration (free energy) gradient. A root hair, which is an extension of a single cell, will absorb calcium from the water in the soil, even though there is more calcium inside the cell than there is outside. Some marine algae (kelp) will have a concentration of iodine within the cells up to a million times greater than that found in the sea water in which they live. An animal cell can continue to take in glucose even though there is more glucose inside than there is outside. Potassium ions can enter a cell that has a much higher concentration of these ions inside the cell than outside. Sodium ions, on the other hand, are usually to be found in much greater concentration outside the cell than inside. Such a movement against the free energy gradient requires energy from cell metabolism and is known as **active transport.** If a cell of an aerobic organism is deprived of oxygen, the energy-releasing mechanism is impaired and the cell loses its power of active transport. Various protoplasmic poisons which interfere with cell metabolism have the same effect.

How does a cell carry on active transport? Much research is being directed at this problem, but at present we have only partial and theoretical answers. Studies on human muscle cells show that the potassium ions inside the cell begin to diffuse outward according to the

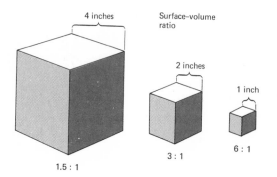

FIG. 7.7 The size of an object is related to its surface-volume ratio. These cubes are the same shape, but as they get smaller they have relatively more exposed surface in proportion to volume.

law of diffusion but are immediately brought back in again. It is as though something took hold of them and brought them back in forcibly. Sodium ions, like potassium ions, are positive, yet they are forcibly carried back outside when they begin to enter the plasma membrane. Something in the cell acts much like a butler, carefully guarding a door to a home, who takes hold of unwanted gate-crashers as they enter the door and immediately and forcibly escorts them back outside, but who also seizes wanted guests by the arm as they start out the door and escorts them back inside. Thus, the concentration of unwanted guests is greater outside and the concentration of wanted guests is greater inside. If the butler goes to sleep on the job, the concentration tends to equalize. The active transport of potassium or sodium ions is known as the **potassium pump** or the **sodium pump,** and one theory holds that there are molecules which actually unite with these ions and carry them across the membrane, releasing them outside or inside as the case may be. This union and movement require energy. It is estimated that as much as one-half of the total energy expended by a cell is used for active transport.

Definite information is available about the active transport of glucose. While some glucose enters the cell by simple diffusion, it is known that the movement can continue after the concentration within the cell becomes greater than outside. At the plasma membrane, the glucose is combined with a phosphate group (PO_4); that is, the glucose becomes phosphorylated, and glucose phosphate is formed. Energy is required for this synthesis. The glucose phosphate is then carried across the plasma membrane. In the course of glucose metabolism within the cell, the phosphate group will eventually be split off and become available for the transport of more glucose across the plasma membrane.

In the human kidneys active transport is important in the concentration of waste materials in the urine and in the prevention of loss of useful products from the blood. Pressure forces water and dissolved substances from the blood into collecting tubules in the kidneys. Later there is an exposure of the tubules to small capillaries where the blood is not under high pressure, and a selective reabsorption takes place. The glucose level of the blood must be maintained within a rather narrow range. Glucose passes out into the kidney tubules but is selectively reabsorbed by active transport, so that normally there will be practically no glucose in the urine which passes from the body. If you eat a great excess of sugar, however, some of this (as glucose) will remain in the tubule fluid and will be excreted in the urine. Thus, through active transport the kidneys help regulate the concentration of dissolved substances in the blood.

The plasma membrane also exerts some selectivity according to the solubility of the dissolved molecules in fats. The relatively large molecules of fatty acids and glycerol pass through the membrane more readily than some smaller molecules which are not fat-soluble. This seems to be due to the high proportion of lipids (fatty compounds) in the plasma membrane. Fat-soluble molecules can be taken up by the lipid portion of the membrane and transported across. Some alcohols have very large molecules, yet they pass readily into the cell because they are highly soluble in fats. In fact, the effect of alcohol on the nerve cells of higher animals is thought to be due to the fact that the alcohol alters the permeability of the nerve cells as a result of its reaction with the lipid portion of the plasma membranes. The same is true of ether. When a high concentration of either of these encounters a cell, death of the cell results because of this action on the plasma membrane.

The electrolytic charge on the plasma membrane also affects the absorptive capacity of the cell. If you place protozoans such as *Paramecium* under the microscope and touch the water at either side with electrodes carrying a weak electric current, the protozoans will move toward the cathode (negative pole). This reaction is the basis of an interesting demonstration to an audience. The microscopic image can be projected on a screen, and upon command of the operator all the organisms turn and swim to the left. Then he reverses the current and commands them to swim to the right, which they do in apparent obedience. Then he can say, "At ease," and turn the current off, and they will swim about at random. These reactions indicate that the surface charge of the organisms is positive.

Some cells from higher animals, when suspended in water, will be attracted to the anode (positive pole), indicating that they have an outer negative charge. Such charges are probably on the protein molecules which form the outer layer of the plasma membrane. Amino acids, which make up the proteins, can have either positive or negative charges and the particular charge of the outer layer of the membrane will result from the combination of amino acids present. A charged surface membrane can have an effect on the absorption of ions by the cell. The interior of the plasma membrane may have an overall charge which differs from the charge on the outside. This difference in charge has been demonstrated in nerve cells, and the passage of nerve impulses is related to this difference, as we shall learn in Chapter 29.

ENDOCYTOSIS

Some material may enter the cell through **endocytosis,** an engulfing process. In this process part of the plasma membrane spreads around the material and engulfs it. This is the way that small organisms, such as *Amoeba*, take in food and water, and we now know that endocytosis occurs in many other cells. **Pinocytosis** refers to the intake of liquids through endocytosis. The word literally means "cell drinking." **Phagocytosis,** which means "cell eating," refers to the in-

take of solids. After the liquid or solid material is taken into the cell, the material is still surrounded by a portion of the plasma membrane. If the material is food, it is digested within the food vacuole which has been formed as the food was engulfed. Liquids pass out into the cytoplasm of the cell by osmosis, and digested solids, by diffusion. Because enzymes are proteins, they are molecules that are too large to diffuse into a cell through the plasma membrane, but enzymes may be taken into the cell along with liquids during pinocytosis. Tissue culture studies show that some enzymes enter human cells in this way.

SURFACE-VOLUME RATIO

The efficiency of a cell depends upon its size. A small cell has a greater surface area in proportion to its volume than a larger cell of the same shape. Since the amount of material which can enter or leave a cell within a given time depends upon the amount of exposed surface, cells would not function efficiently if they were large. A cube which measures 4 inches in each direction would have 64 cubic inches of volume and 96 square inches of surface. This is a surface-volume ratio of 1.5 to 1. A cube measuring 2 inches in each direction would be 8 cubic inches in volume, and have 24 square inches of surface, a ratio of 3 to 1. A 1-inch cube would have 1 cubic inch of volume and 6 square inches of surface, a ratio of 6 to 1. Thus, we see how the ratio increases as objects get smaller.

The proper functioning of your red blood cells depends upon the rapid absorption and release of dissolved oxygen. These cells are very small. If their diameter were doubled, the same cells would be only half as efficient. If the diameter were quadrupled, they could not carry enough oxygen to keep us alive, even though, in this enlarged condition, each would still have a diameter of only about 1/1000 inch.

When a portion of a human lung is viewed under the microscope, the tissue appears to be composed of many very tiny air sacs. These give a vast area of exposed surface, which allows ready diffusion of oxygen and carbon dioxide. If the lungs were hollow like a balloon, there would be so little surface exposed to the air that it would be impossible to obtain sufficient oxygen to support life.

The surface-volume ratio of the entire body is also very important in determining many of the physiological processes and life habits of animals, as will become evident during the progress of our study.

REVIEW QUESTIONS AND PROBLEMS

1. Sugar dissolves more rapidly in a cup of hot coffee than in a glass of iced coffee. Explain.
2. How is the concept of free energy related to diffusion?
3. What factors affect the total free energy of the particles of a substance dissolved in water?
4. Fishermen obtain a more productive catch in the cold northern waters than in the warm tropical waters. Explain.
5. Goldfish can be raised in relatively small bowls without much worry about the exposed surface of the water, but tropical fish must not be crowded, or air must be bubbled through the water. Explain.
6. A little fertilizer on a lawn will make it grow better, but a large amount may kill the grass. Explain.
7. Osmosis can generate a pressure which will cause a sugar solution to rise in a tube up to a certain height. Then the rise stops even though the concentration of sugar in the tube is still greater than in the water outside. Explain in terms of the free energy concept.
8. If human red blood cells are placed in water they will burst, but human skin cells will not burst in water. Explain.
9. Explain what is meant by a hypotonic medium, an isotonic medium, and a hypertonic medium.
10. How does a protozoan like *Paramecium*, living in fresh water, prevent a buildup of excessive osmotic pressure within the cell?
11. How can it be proved that energy is required in the active transport of sodium and potassium ions across the plasma membrane?
12. The glucose level of the human blood remains fairly constant, but glucose will only be found in the urine after an excessive intake of sugar. Explain the role of active transport in this process.
13. How does phosphorylation play a part in active transport?
14. Some alcohol molecules of rather high molecular weight pass through the plasma membrane more readily than some smaller molecules. Explain.
15. Explain the factor which results in an overall electric charge on the outer surface of the plasma membrane.

16.   Protein molecules are too large to diffuse across the plasma membrane, yet some such molecules do enter the cell. Explain.

17.   A bacterium may double its size in as short a time as 20 minutes, while an amoeba, which is much larger, will require about 12 hours for such a doubling of its size. Explain fully.

FURTHER READING

Geise, A. C. 1973. *Cell Physiology.* Philadelphia: W. B. Saunders.

Kennedy, D. 1965. *Living Cell: Readings from Scientific American Magazine.* San Francisco: W. H. Freeman.

Loewy, A. G., and Siekevitz, P. 1970. *Cell Structure and Function.* 2nd ed. New York: Holt, Rinehart and Winston.

McElroy, W. D. 1971. *Cell Physiology and Biochemistry.* 3rd ed. Englewood Cliffs, N. J.: Prentice-Hall.

Murphy, Q. R. 1965. *Metabolic Aspects of Transport across the Cell Membrane.* Madison: U. of Wisconsin Press.

8

Cell Duplication

Self-duplication is a unique characteristic of living organisms. The *Amoeba* engulfs food and uses much of it for energy, but some nutrients go to build protoplasm and the *Amoeba* increases in size. The efficiency of the *Amoeba* declines, however, when its size increases beyond a certain point. The *Amoeba* divides before this point is reached, and this division is not a simple splitting in two. Certain events take place within the cell before division occurs. The chromosomes duplicate themselves and the duplicates **segregate** so that the daughter cells which arise from cell division have the same number and kind of chromosomes as those in the original cell. **Mitosis** is the name of the process in which chromosome duplication and segregation occur. The actual splitting of the cell into two parts is called **cytokinesis.** Frequently, however, the word "mitosis" is used to indicate *both* the chromosome duplication and segregation and the cell division. Let us discuss the process of mitosis first.

MITOSIS

For purposes of study, **mitosis** is divided into stages or phases, but in actuality mitosis is a continuous process and each phase passes gradually into the next.

Interphase

Interphase is the period between mitoses. Cells are found in this state most of the time.

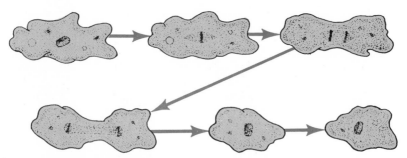

FIG. 8.1 Cell duplication is also reproduction in one-celled organisms. When an amoeba undergoes mitosis and cell division, two organisms are produced.

During interphase, cells are most active metabolically. The chromosomes appear as very long, threadlike bodies and the genes which are part of the chromosomes put out the messages which direct protein synthesis, which is carried out by the ribosomes in the cytoplasm. Biochemical analysis of the amount of DNA in the nucleus shows that DNA remains constant for a time after mitosis and then increases rapidly until it has doubled in quantity. This doubling marks the time of chromosome duplication. After this, there is no further DNA increase and the cell enters the first phase of mitosis.

Prophase

The first sign that a cell is entering **prophase** is that the chromosomes become visible under an optical microscope. The chromosomes appear quite long and slender at first, but they are shorter and thicker than during interphase. The shortening and thickening occurs as the chromosomes coil and fold. The shortening and thickening continues during early prophase until the chromosomes appear rod-shaped. In many cells, including human cells, it is possible to see that the chromosomes were duplicated in the interphase. Each prophase chromosome appears as a doublet, or **dyad,** made of two **chromatids,**

which are held together at a constricted region known as the **centromere.** The chromosomes are usually of different lengths and have different points of centromere location. The longest human chromosome is about five times the length of the shortest one, and the centromere attachment may vary from median all the way to subterminal.

As the chromosomes shorten, other changes take place in the cell. The nuclear membrane gradually breaks down until, by late prophase, the chromosomes lie free in the cell. If the cell has centrioles, the two centrioles separate and begin to move apart. As they do, they produce the fine microtubules of protein which form the **spindle figure.** Even though no centrioles are

FIG. 8.2 Prophase in an onion root cell. Side lighting shows that the chromosomes have already duplicated and have become much shorter than they were in interphase by coiling.

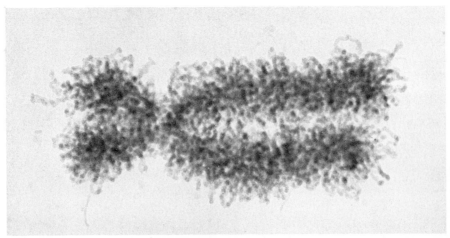

FIG. 8.3 A human chromosome in prophase. The electron microscope brings out the intricate folding and coiling of the very slender chromosome strands which are extended during interphase. [From DuPraw (1970). *DNA and Chromosomes.* Holt, Rinehart & Winston, N.Y.]

seen in some cells, the spindle still appears. Soon the spindle envelops the chromosomes and typically extends across the entire cell.

Metaphase

In **metaphase** the centromeres of the chromosomes are attracted to the equator of the spindle, which lies half way between the poles. Only the centromeres are attracted to the spindle. The rest of the chromosome extends out on either side of the equator. When a piece of a chromosome breaks off, the broken piece, which lacks a centromere, will not move on the spindle and will be lost when new daughter cells are formed. A spindle fiber becomes attached to the centromere on either side. The centromere has been single, even though the chromosome is made of two chromatids. At metaphase, however, the centromere duplicates and the spindle fibers seem to drag one duplicate toward each pole of the spindle.

Anaphase

As the centromeres move toward the poles, they drag the chromatids along with them. When the two chromatids, which make up the metaphase chromosome, are separated, each becomes known as a **chromosome.** It is at the **anaphase,** therefore, that the chromosome number is actually doubled. Anaphase lasts from the time of the separation of the chromatids until the chromosomes reach the poles of the spindle. During anaphase the chromosomes assume characteristic shapes, depending upon the location of the centromere. A medium centromere will produce a V-shaped chromosome. A terminal centromere produces a rod-shaped chromosome. If the centromere lies between the end and the middle, the chromosome may appear J-shaped.

Telophase

Once the chromosomes reach the poles, they form a tight aggregate and **telophase** begins.

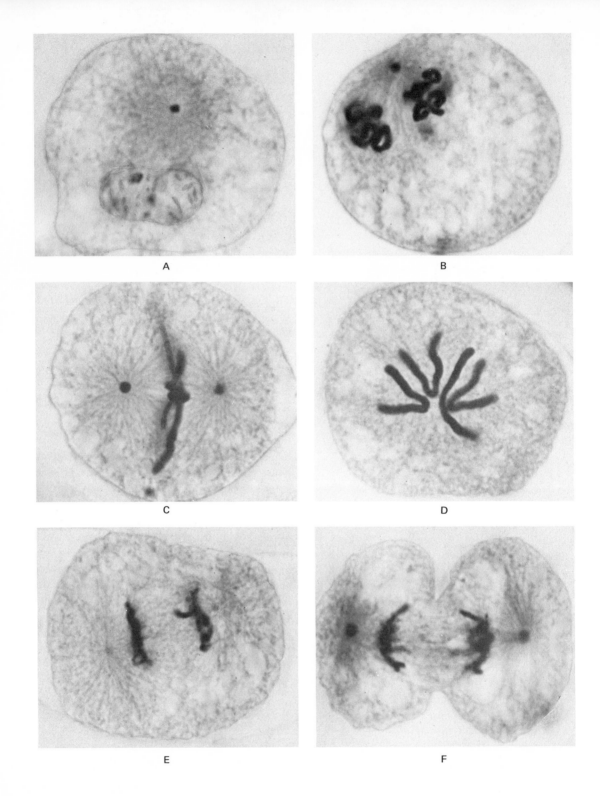

A

B

C

D

E

F

Telophase is an almost exact reversal of prophase. The chromosomes become longer and thinner by uncoiling and unfolding. The nuclear membrane reappears. The centrioles, if present, duplicate so that there are two beside each newly formed nucleus. Nucleoli begin to reappear. In many microorganisms and animal cells, a **cleavage furrow** begins to pinch the cell in two during the latter part of the telophase. The mitochondria undergo a duplication pattern of their own during telophase and just before the cell splits, the mitochondria separate, so that each daughter cell gets half. Plant cells, which have rigid cell walls, do not undergo an actual cleavage. Instead, a cell plate forms between the two new nuclei and the cell plate thickens to form a new cell wall.

Daughter Cells

When mitosis and cytokinesis are complete, two cells have been formed, each a duplicate of the original. Each cell will be only about half the size of the original, but through growth the daughter cells will soon become as large as the original cell. Each daughter cell has the same number and kind of genes and therefore has the same characteristics as the original. This can be demonstrated by an experiment with a fertilized salamander egg. After the egg has undergone mitosis and the first cleavage, the two daughter cells normally stick together and each cell will give rise to about half a salamander. If we pinch the two cells apart, however, each cell will make a complete salamander.

Identical human twins are the result of an accidental splitting of an embryo during early development. Identical twins begin as a single fertilized egg, but when the fertilized egg is split in two parts during early development, each half grows into a complete person with genes identical to those of its twin. Identical twins should be distinguished from fraternal twins, which arise as two separate fertilized eggs and have no more in common than brothers or sisters who are born at different times.

GENE AND CHROMOSOME DUPLICATION

A gene is the basic unit of heredity. Each gene consists of a sequence of **deoxyribonucleic acid (DNA) residues** which are shaped into a double-stranded helix. Each gene is thought to perform one particular function, such as providing the information necessary for the synthesis of a cellular enzyme. Genes are joined to one another to form long double-stranded helices of DNA. A chromosome is composed of a long sequence of genes (DNA) and some structural proteins. During the interphase period of mitosis, each gene duplicates itself, a process which we shall describe later in this chapter. All genes in a cell duplicate interphase, and the doubling of the chromosomes, which is visible during the prophase period of mitosis, is the result of gene duplication during interphase.

There have been many speculations about the method of gene duplication. One theory was that the genes served as templates upon which the duplicate genes were produced. Another theory stated that the genes split in half (the two strands of DNA separated) and that each half then at-

FIG. 8.4 Stages of mitosis in a fertilized egg of the parasitic worm, *Ascaris.* During interphase (A) the male and female nuclei come together. In prophase the chromosomes become visible (B). In metaphase the chromosomes line up along the equatorial plate of the spindle figure (C). The centrioles show clearly.

An end view of a cell in metaphase shows the four chromosomes are arranged in the same plane (D). In anaphase the choromosomes move toward the poles (E). In telophase the chromosomes reach the poles and the cell splits in two (F).

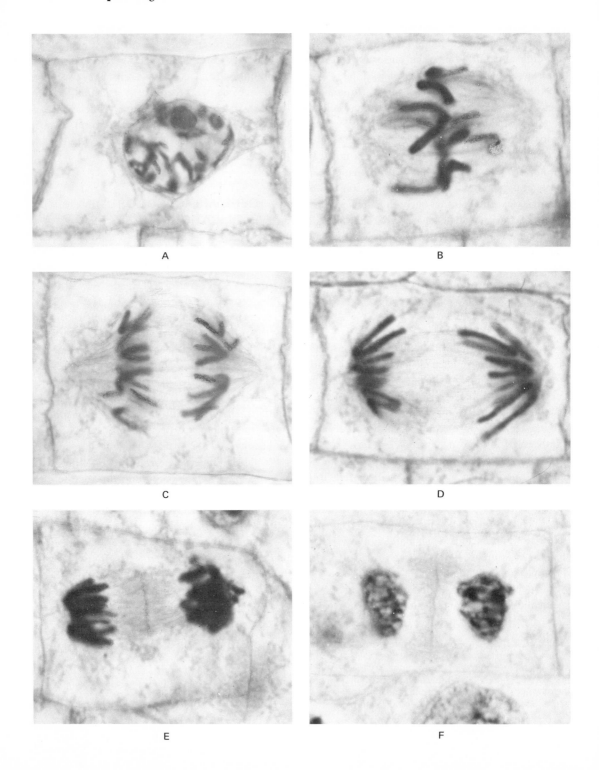

A

B

C

D

E

F

FIG. 8.5 Mitosis in cells of an onion root tip. In prophase (upper left) the beginning of the spindle figure can be seen. In metaphase the chromosomes are arranged in a line at the equator of the spindle. The center two photographs show early and late ana- phase when the chromosomes move toward the poles. The lower two photographs show early and late telophase and the formation of the cell plate which will separate the cell into two parts.

tracted to itself the parts needed to make up the lost half. This second theory has been proven to be correct. **Watson** and **Crick,** who discovered the structure of genes, also proved that genes duplicate themselves by this second method.

The Watson-Crick theory holds that a gene resembles a long, flexible ladder that is twisted into the shape of a double helix. The ladder is made of **deoxyribonucleic acid (DNA).** The outer supports of the ladder consist of mole- cules of a five-carbon sugar, **deoxyribose sugar,** held together by **phosphate bonds, PO_4.** The rungs of the ladder are made of nitrogenous bases, **purines** and **pyrimidines,** which were probably among the early organic compounds formed on the earth. Each rung consists of two of these compounds, one a purine and the other a pyrimidine. These are joined together by weak hydrogen bonds.

There are two purines, **adenine** and **gua- nine,** and two pyrimidines, **cytosine** and **thy- mine.** The purines are double-ring compounds and the pyrimidines are single-ring compounds, as discussed in Chapter 5. These compounds are of such a chemical nature that the hydrogen bonds are formed only between adenine and thy- mine and between cytosine and guanine. Thus, there are only two kinds of pairs as rungs on the DNA ladder, but there are four variations in the rungs, because the position of the rungs alters their effect on the gene. For instance, if one pair is adenine-thymine (AT), it will have an effect different from a pair which is thymine- adenine (TA). Still, with only four pairs in the

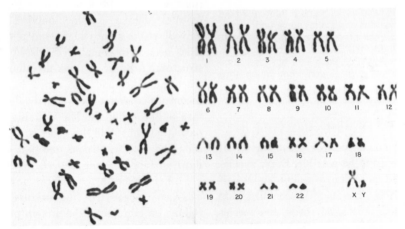

FIG. 8.6 A squash preparation of human chromo- somes in prophase (left) and a karyotype prepared by cutting up a photomicrograph of human chromo- somes and matching them in pairs according to length and centromere location (right). The two chromatids in each chromosome show clearly. This is a male cell because one pair of chromosomes, the X and the Y, are of unequal size. These chromosomes de- termine sex.

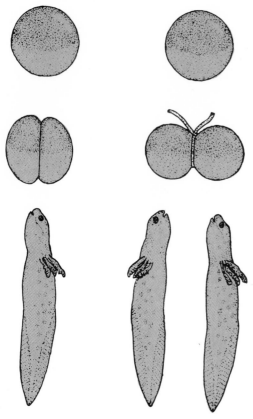

FIG. 8.7 Two salamanders produced from a single fertilized egg. When the cells are separated at the two-cell stage, each cell produces a complete animal. This experiment shows that duplication of the genes occurs during mitosis since each cell has the same characteristics as the original cell from which they both were produced.

gene, one might not expect the great variety of genes known to exist to be possible. Every species on the earth—and more than a million and a half have been identified—has its own peculiar group of genes, and each individual within the species can have a gene combination which is different from others in the species. The number of different kinds of genes is so great as to stagger the imagination. Such variety is possible because of the extreme length of each gene. There will be from several hundred to several thousand rungs in

the DNA ladder constituting one gene, and a loss, gain, or change of a rung can alter the operating effect of the gene. If one rung is taken out, turned around, and put back, the gene will operate differently. We can compare constructing a gene to making words of thousands of letters using a four-letter alphabet. Some have compared it to the Morse code, which, with only two symbols—the dot and dash—can form any word in any language. It has been estimated that if all the genes in just one human cell were removed and joined end to end they would make a string about five feet long but so thin that it could not be seen with the optical microscope.

The development of the Watson-Crick theory of gene structure opened the way for plausible and logical theories concerning the method of gene duplication. Remember that the base pairs which form the rungs of the DNA ladder are held together by hydrogen bonds, which are the weakest of molecular bonds. When the stimulus for gene duplication comes, the rungs break apart at these weak bonds. The reaction is somewhat like a zipper becoming unzipped. As the unzipping proceeds, however, the half-gene portions begin to attract to themselves free molecules of adenine, cytosine, guanine, and thymine, as well as the ribose sugar and phosphate molecules needed to complete the DNA double helix structure. Adenine attracts thymine, cytosine attracts guanine, and vice versa. In this way the reconstituted genes have exactly the same structure as the originals. The individual component parts that constitute the genes are present in the cell. They are formed as one part of the breakdown of food entering the cell.

Each gene in the nucleus now has become two genes, each like the original. Next there follows an increase in the protein portion of each chromosome under the direction of cell enzymes, and each chromosome becomes double—composed of two chromatids. The complex part of cell duplication has now been completed, although it will usually be about two hours before the events of mitosis are evident.

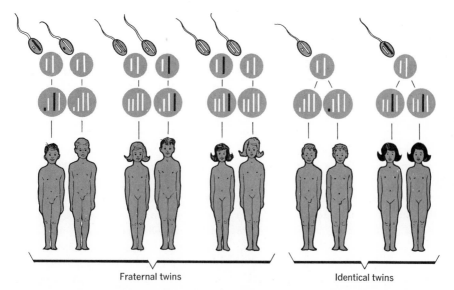

FIG. 8.8 Fraternal twins are formed by the union of two sperm and two eggs. They can differ in many characteristics, including sex, because they have different genetic backgrounds. Identical twins, however, come from a single fertilized egg. They have the same genes, so they are always of the same sex and show the same inherited characteristics.

INTERVALS BETWEEN MITOSES

After mitosis and cell division are complete, the daughter cells grow and in time undergo mitosis. The interval of time between the completion of one cell division and the initiation of mitosis in the daughter cells varies considerably in different cells and at different times. In the human embryo the cells grow and divide rapidly, some as often as once every twelve hours. As the body matures, this rate slows and some cells, such as those of the brain and nerves, may never divide again. These cells are said to be in the G_0, **growth rate zero** stage. Others, such as cells in the bone marrow which produce blood cells, continue rapid division throughout life. If the rate of growth of these cells slows down, you will die of anemia because you are constantly losing red blood cells. Other cells divide only when there is an injury which necessitates their replacement. Bone cells are of this type.

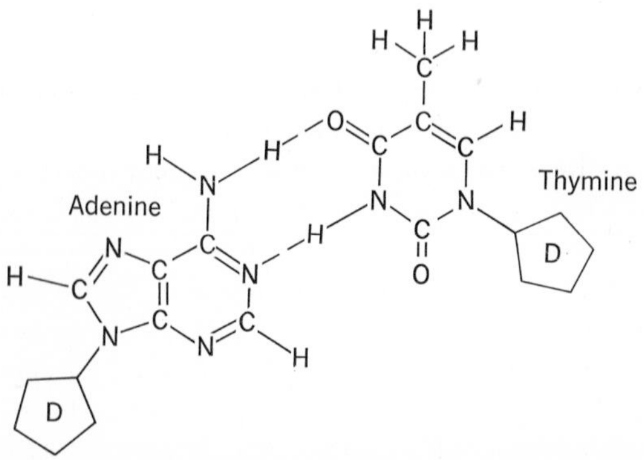

FIG. 8.9 Hydrogen bonding of adenine and thymine in a DNA molecule. The bases break apart at these bonds in DNA duplication.

Mitosis Regulators

If each cell underwent unrestricted growth and division, you would end up as an unorganized mass of cells of all kinds. There must be some factor which regulates cell growth, therefore. Your skin cells grow and divide at a rate just equal to the loss of these cells as they wear off. If you scrape some skin from your knuckle, the cells around the injured region start dividing and growing rapidly until the injury is repaired, at which time they return to their normal maintenance rate.

A loss of the ability to regulate cell growth results in a **cancer.** When cell growth is unregulated, some cells grow and divide rapidly, producing a malignant tumor which does not contribute to the body but instead becomes a parasite upon it. Some of the tumor cells may break away from the initial site of growth and migrate over the body where they start new growths. We can stop the cancerous growth by cutting out all parts of the body which contain any of the abnormal cells. Or we can treat the areas with high-energy radiation because it kills cells which are in a period of rapid growth and division more readily than cells which are slow-growing or which are in state G_0. If we could discover what regulates the rate of

mitosis, it might be possible to apply mitosis inhibitors to cancer cells which would stop their uncontrolled growth. An understanding of what happens during interphase has been helpful in the search for mitosis inhibitors.

Stages of Interphase

After mitosis and cell division are completed, the daughter cells enter a period of growth called the G_1, or the first growth stage. When conditions for growth are ideal in tissue cultures of human cells, for example, this stage lasts about eight hours. Then the cell enters the **S**, or synthesis, stage. During the S stage, the DNA replicates. The DNA does not replicate all at the same time, however, so this stage lasts about six hours. At the end of this stage, the chromosomes have duplicated themselves, but they adhere to each other rather tightly and do not appear to be doubled. Another growth stage, the G_2, follows the S stage. The G_2 lasts about four hours. It is followed by mitosis, which lasts about forty-five minutes.

When conditions of growth become less favorable, there is an extension of the G_1 stage but the length of the other stages remains about the same. Therefore, the initiation of the S stage starts the process which ends in mitosis and cell division and controls the time between mitoses. The factor which stimulates the beginning of DNA replication is also the mitosis-stimulating factor. There is some evidence that the cells at the site of a wound secrete a wound hormone which acts as a mitosis stimulant. A mitosis-stimulating substance is also produced by certain insects which lay their eggs in leaves and twigs of trees. The witch hazel gall wasp, for instance, has a sharp ovipositor which enables the female to lay eggs underneath the bark of a twig of the witch hazel tree. Along with the eggs, this insect deposits the mitosis-stimulating substance which causes the tree cells to begin rapid

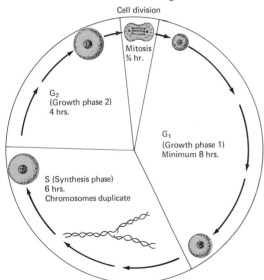

FIG. 8.10 The complete cell cycle. Cell DNA duplicates during interphase. A gap in time follows before the cell goes into mitosis. The length of the first growth stage can be variable, but the length of the other stages are fairly constant.

growth and division around the eggs. The plant cells form a gall within which the insect larvae develop. Coconut milk also contains a substance which stimulates mitosis in cells which normally would not undergo mitosis again. If mature tissues which have stopped growth and division are taken from the interior of a carrot and treated with coconut milk, the cells begin mitosis again. Human cells which are in the G_0 stage can also be stimulated to resume growth by an application of a mitosis stimulator. The white blood cells normally never undergo mitosis once they have been released into the blood stream. When these cells are treated with an extract of the kidney bean, however, they resume mitotic divisions and can be grown in tissue culture. White blood cells treated with kidney bean extract are used for human chromosome studies.

MEIOSIS

The Chromosome Number

All members of the same species have the same number of chromosomes in their body cells. A domestic chicken has 18, a dog has 22, a corn plant has 20, an onion has 16, a mouse has 40, and a man has 46. Since these numbers remain constant from generation to generation, a problem must be solved when **gametes,** or reproductive cells, are produced. If both male and female gametes had the full chromosome number of the species then during fertilization, when the male and female gametes join, a doubling of the chromosome number would occur. We know that this does not happen. There is no doubling of chromosome number during fertilization because during gamete formation a process known as **meiosis** occurs. Meiosis involves two cellular divisions and a reduction of the number of chromosomes to one-half. Then, when the gametes unite during sexual reproduction, the full number of chromosomes is restored. The number of chro-

mosomes in gametes is called the **haploid number,** or the **monoploid number.** The cells which are formed by a union of gametes, therefore, are **diploid.** The human diploid number of chromosomes is 46 and the haploid number is 23.

In animals meiosis occurs before the reproductive cells are produced and the gametes do not undergo mitosis after this. In plants meiosis takes place at one stage in the life cycle and the haploid cells produced by meiosis undergo mitotic division many times before gametes are produced. In a moss plant the major portion of the plant is haploid tissue produced from cells which have undergone meiosis. The amount of such tissue is reduced as plants become more complex. This will be considered in later chapters. Microorganisms maintain their chromosome number in various ways which also will be discussed in later chapters. For now, we will consider the production of human reproductive cells, or gametes, as an illustration of meiosis.

Human Spermatogenesis

The human testes are packed with many small, threadlike tubules, the **seminiferous tubules.** A microscopic examination of a cross section of a seminiferous tubule reveals that there is a **germinal epithelium** around the outside. When a cell from the germinal epithelium begins to develop into a sperm cell, it moves toward the inside of the tubule and is called a **primary spermatocyte.** At this stage it undergoes meiosis. In the prophase of the first meiotic division the chromosomes are found in pairs. Each chromosome is paired with a **homologous** chromosome which contains homologous genes. The pairing of homologous chromosomes is known as **synapsis.** In the first meiotic prophase, therefore, we see 23 pairs of chromosomes rather than the 46 separate chromosomes which we see in mitosis. In men, but not in women, one of the chromosome pairs is composed of a medium length chromosome and

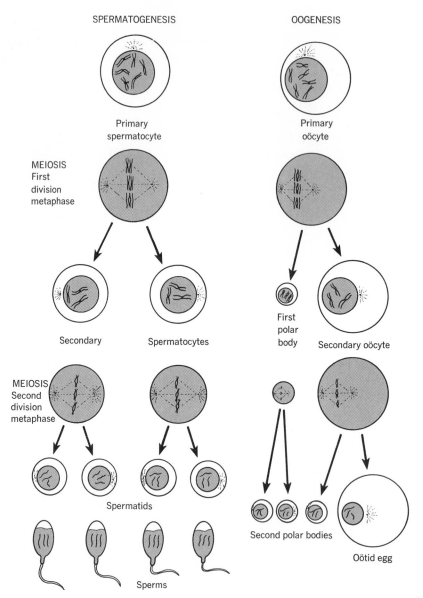

SPERMATOGENESIS

OOGENESIS

Primary
spermatocyte

Primary
oöcyte

MEIOSIS
First
division
metaphase

Secondary   Spermatocytes

First
polar
body   Secondary oöcyte

MEIOSIS
Second
division
metaphase

Spermatids

Second polar bodies

Oötid egg

Sperms

FIG. 8.11 Spermatogenesis and oogenesis. The similarities and differences in the formation of sperm and eggs are shown in these diagrams. Only six chromosomes are shown, although the human diploid number is 46.

a very short chromosome. These are the chromosomes which determine sex. In women, the sex chromosomes consist of two medium length chromosomes. Since each chromosome doubles during the interphase preceding meiosis, each chromosome pair is actually made up of four chromatids which form a **tetrad.** The tetrads line up at the spindle equator, and since each tetrad has two centromeres, there is no centromere duplication. The two centromeres move to opposite poles, pulling the **dyads** along. When the cell divides during the first meiotic division, it produces daughter cells, known as **secondary spermatocytes,** which have only 23 double chromosomes. A synthesis of new genetic material does not take place during the interphase preceding the second mitotic division, so the second prophase involves the same dyads that were present in the telophase of the first meiosis. There is centromere duplication during the second metaphase, and when the centromeres move to the poles during the second anaphase, they pull only a **monad,** or single chromosome, with them. The daughter cells produced by the second meiotic division are called **spermatids,** and each daughter cell has 23 single chromosomes. Thus, four haploid spermatids are produced from each diploid primary spermatocyte through the two divisions of meiosis. Each spermatid is then converted into a sperm.

In meiosis there are two cell divisions, but only one duplication of chromosomes and centromeres. Hence, each of the daughter cells will have only one-half the number of chromosomes and centromeres which were present in the parent cell before meiosis. The chromosomes are duplicated in the interphase preceding the first division, but the centromeres duplicate only during the metaphase of the second division. The chromosomes are double in the first meiosis; they separate into duplicate chromosomes at the second meiosis, with one member of each duplicate pair going to each daughter cell.

Human Oogenesis

Egg production, or **oogenesis,** involves meiosis, but cytokinesis in oogenesis differs from cytokinesis in spermatogenesis. The ovaries have an outer germinal epithelium which produces the **primary oocytes.** In the first meiotic division during oogenesis, the spindle figure forms near one edge of the primary oocyte and there is an unequal splitting of cytoplasm. One large cell, a **secondary oocyte,** and one small cell, the **first polar body,** are produced. The secondary oocyte undergoes the second meiotic division, and the cytoplasmic division is again unequal. The second meiotic division produces a **second polar body** and an **egg.** This second polar body can often be found inside the covering which surrounds a human egg. The first meiosis takes place sometime earlier, and the first polar body has usually disintegrated by the time the mature egg is formed. The egg is many times larger than the sperm, but eggs and sperm have the same number of chromosomes, so the heredity of an embryo is equally determined by both parents.

More details on the nature of the sperm and the egg and fertilization are given in Chapter 34.

REVIEW QUESTIONS AND PROBLEMS

1. Cells exposed to a rather high dose of radiation will lose their ability to undergo mitosis. How would such a loss affect these cells and what would be their eventual fate?
2. How do the very long and thin interphase chromosomes become the relatively short rods of the prophase?
3. Why is the centromere such a vital part of a chromosome? What would happen to a chromosome if it lost its centromere?
4. A certain drug has been found which prevents the formation of the microtubules

of protein which make the spindle fibers. What effect would this have on a cell at the time for mitosis?

5. Anaphase chromosomes may be rod-shaped, V-shaped, or J-shaped. What gives them this shape?

6. List the stages of mitosis during which a human cell has 92 chromosomes.

7. During what stages of mitosis would a human cell have twice the number of chromosomes found in an early interphase cell?

8. What is the difference in cytokinesis (division of the cell) between plant and animal cells?

9. Daughter cells are only about half the size of the parent cell, yet we say that daughter cells have all the characteristics of the parent cell. Explain.

10. After a gene has split in interphase, one part of the half-gene has the bases guanine, thymine, adenine, and cytosine, in that order. These can be represented as GTAC. Give the letter designation, in order, of the complementary bases which would be attracted to this group as the gene becomes double-stranded again.

11. Twins of both types are often analyzed in studies on the effect of heredity and environment on human development. Why are twins of greater value in these studies than persons born singly?

12. Cancer research is directed toward the factor which stimulates genes to undergo duplication during interphase. Why would an understanding of this factor be of such importance in cancer control?

13. Compare the first meiosis with mitosis and tell how they are similar and how they differ.

14. Compare the second meiosis with mitosis and tell how they are similar and how they differ.

15. A human egg has about 10,000 times the volume of a human sperm, yet a child inherits equally from both parents. Explain this.

16. Suppose a chemical is discovered which prevents cells from undergoing mitosis. How might this chemical be used to control pests or for other purposes?

FURTHER READING

Cohn, N. S. 1964. *Elements of Cytology*. 2nd ed. New York: Harcourt, Brace and World.

Du Praw, E. J. 1970. *DNA and Chromosomes*. New York: Holt, Rinehart and Winston.

Watson, J. D. 1968. *The Double Helix*. New York: Atheneum Press.

Winchester, A. M. 1972. *Genetics*. Boston: Houghton Mifflin.

———. 1973. *The Nature of Human Sexuality*. Columbus, Ohio: Charles E. Merrill Books.

9

Gene Control of Cell Activity

As research proceeded on the molecular structure and molecular processes of cells, the results of this research raised two important questions. The discovery that proteins are synthesized in the cytoplasm by ribosomes, while the genes, which control protein synthesis, are located in the nucleus, posed the question: How are the directions for protein synthesis transmitted from the genes in the nucleus to the ribosomes in the cytoplasm? Certain observations about gene activity posed another, even more intriguing question. It is known that genes are not active all the time and that all genes in one cell are never activated during the life of that cell. For example, genes in cells on the bottom of the feet never direct the synthesis of proteins which belong in eyes or kidneys, even though these epidermal cells contain genes which could direct these types of proteins to be synthesized. During the development of the human embryo, something induces certain cells in the embryo's head region to begin producing the proteins which are used to form the eyes. After the eyes are formed, the genes which direct the production of these eye proteins are apparently turned off because only two eyes are produced. These pieces of information have led biologists to ask: What causes genes to be activated only at a particular time and in a particular location?

THE METHOD OF GENE CONTROL

The genes within the nucleus contain an encyclopedia of information. Each gene may be

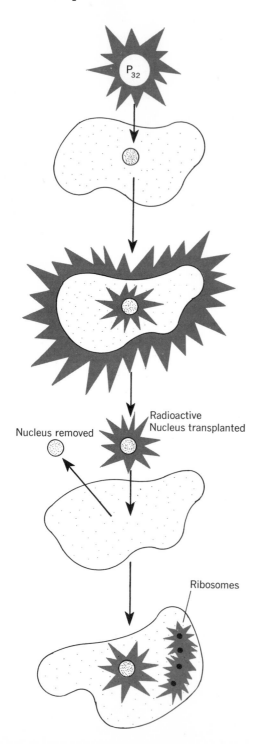

Nucleus removed

Radioactive
Nucleus transplanted

Ribosomes

thought of as an entry in this encyclopedia; there are many thousand entries. As a corollary to this concept, suppose you need some information on the construction of a model airplane, but you do not live near a library. You call the librarian and he makes a photocopy of the entry in the encyclopedia which gives the necessary directions. The photocopy is sent to you by mail and you then make the airplane model from the information. The ribosomes in the cytoplasm need the information on protein synthesis contained in the genes. Copies of this information, which is contained in the genes, are produced and sent out as messages to the ribosomes. Let us now turn our attention to the method by which these messages are produced and utilized.

Distinctions Between DNA and RNA

We know that **DNA** (deoxyribonucleic acid) is shaped into a long double helix of deoxyribose sugar and phosphate with cross-connecting paired bases. **RNA** (ribonucleic acid) differs from DNA in three important respects. First, the sugar in RNA is ribose rather than deoxyribose; ribose has one less oxygen atom. Second, RNA has only a single strand of sugar and phosphate molecules with single attached bases. Third, the pyrimidine base thymine is replaced in RNA by **uracil.**

Three major kinds of RNA are found in a cell. One kind, **ribosomal RNA,** makes up about 60 percent of the ribosomes. Ribosomal RNA is usually abbreviated as **r-RNA. Transfer RNA, t-RNA,** helps move amino acids to the ribosomes. **Messenger RNA, m-RNA,** carries information from the genes to the ribosomes. We shall first

FIG. 9.1 Proof that messages pass from the nucleus to the ribosomes. Radioactive phosphorus-32 makes an entire amoeba radioactive. The nucleus from this organism can then be transferred to an enucleated amoeba. Radioactivity is soon detected in the region of the ribosomes of the amoeba.

learn how m-RNA is produced and passed to the ribosomes.

The Messengers of the Genes

When a gene receives the stimulus to send out information, the gene DNA separates at its weak hydrogen bonds as if it were going to duplicate. Instead of each strand of the helix forming a complimentary strand of DNA, however, one of the two strands forms a strand of RNA, while the second strand remains inactive. The same strand of DNA always forms RNA. Bases are attracted to the active strand: thymine attracts adenine, guanine attracts cytosine, cytosine attracts guanine, but adenine attracts **uracil** rather than thymine. These bases are attached to a backbone of ribose sugar and phosphate, and messenger RNA is formed. Messenger RNA contains a copy of the information which was in the DNA which produced it. Messenger RNA separates from the DNA and passes into the cytoplasm through the thin spots in the nuclear membrane.

Messenger RNA cannot be produced when the chromosomes are tightly coiled during mitosis. During mitosis, therefore, there is little growth or enzyme activity within the cell. The effect of mitosis on m-RNA production was demonstrated through **autoradiographic studies.** Uracil was made radioactive by the incorporation of **tritium** (hydrogen-3) and placed in a culture of hamster cells. Uracil is one of the bases in RNA but not in DNA. A few minutes after the radioactive uracil was applied, the cells were washed and autoradiographs made. The area where uracil had been utilized was characterized by bright spots in the cells. Cells in interphase showed bright, radioactive spots, but those cells which were undergoing mitosis showed no such spots.

Another period of reduced RNA output is found during interphase at the time of gene duplication. During the S phase the genes are producing more DNA and cannot at the same time produce RNA. There is some RNA output during the S phase because not all of the DNA is duplicating at the same time.

Protein Synthesis

When m-RNA leaves the nucleus, it is attracted to the ribosomes, possibly because it has an affinity for r-RNA. The ribosomes have two parts of different size; m-RNA is attracted to the smaller of the two parts. The m-RNA is shaped like a long strand and it seems to associate with a string of ribosomes, known as **polyribosomes.** Sometimes m-RNA is compared to a computer tape which moves along sensitive receptors, conveying its information. This information is a code for the order of amino acids which are to be assembled into a long polypeptide chain. A polypeptide chain may be a protein molecule in itself, or two or four polypeptide chains may combine to make a protein molecule. The human hemoglobin molecule, for instance, is formed of four polypeptide chains which make two different pairs. The alpha chain has 141 amino acids and the beta chain has 146 amino acids. Since there are two alpha and two beta chains, the total number of amino acids in the molecule is 574.

We must still explain how amino acids are

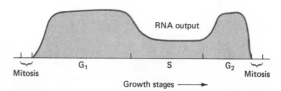

FIG. 9.2 Fluctuations in the amount of RNA output during different stages of the cell growth cycle. When the chromosomes are tightly coiled in mitosis there is no m-RNA output. Output of m-RNA is also restricted during the S phase because the genes are duplicating and cannot produce m-RNA at the same time. There is some m-RNA output during this phase, however, because all the genes do not duplicate at the same time.

moved to the ribosomes. This is the function of **transfer RNA.** Transfer RNA is a relatively short strand of RNA which twists back on itself so that it looks like a twisted wire. There seem to be 87 bases in each t-RNA molecule, but only three of these determine the kind of amino acid which the t-RNA molecule will transfer. These three bases lie at the closed end of the twist, but the amino acids which they attract become attached to the open end where a part of one strand of RNA protrudes. Transfer RNA carries its amino acid to a ribosome. At the ribosome, t-RNA gives up its amino acid and goes back into the cytoplasm to pick up another amino acid molecule which it transfers to the same or different polyribosomes where the amino acid it carries is needed for polypeptide synthesis.

Since each t-RNA molecule can carry only one kind of amino acid, at least twenty different kinds of t-RNA are needed to carry the twenty basic kinds of amino acids. Actually, there are more than twenty kinds of t-RNA because some amino acids are carried by more than one kind of t-RNA.

The method of protein synthesis which has been described here has certain advantages. The genes remain in the nucleus, where they are protected from the destructive forces of metabolism in the cytoplasm. Expendable copies of genetic information are sent into the cytoplasm. These copies are soon destroyed, but more copies can be produced in the nucleus from the master copy of the DNA. This DNA might be compared to an architect's plans for building a house. Blueprints (m-RNA) are prepared from the plans and sent out to the builders (ribosomes). Trucks (t-RNA) bring the materials needed. Each truck brings only one particular material, and from these materials the builders construct the house according to the information in the plans on the blueprints. The blueprints are destroyed from the rough handling they get on the job, but fresh ones can be sent out from the master plans, safe in the architect's office.

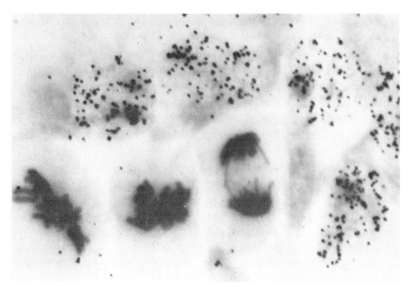

FIG. 9.3 Autoradiograph of a tissue culture of human cells show that there is no RNA output during mitosis. Radioactive uracil was placed in the tissue culture and the dark dots show where it was taken up by cells. The dots are found primarily in the cells which are not undergoing mitosis. (Autoradiography by David Prescott in *Progress in Nucleic Acids:* Academic Press, 1961.)

An enzyme, **RNA-ase,** which is found in the cytoplasm, destroys the m-RNA as soon as enough polypeptide chains have been synthesized. This is necessary because only a certain amount of an enzyme or a structural protein is needed. If the m-RNA were not destroyed, it would continue to direct the production of protein long after the need for protein was satisfied. The cell would be clogged with excess protein and would die.

An enucleated cell can live for a time, but cell growth stops, and as the enzymes are exhausted, the life processes will stop. Mature human red blood cells have no nucleus, although their precursors in the bone marrow do. It can be shown by autoradiography that no protein synthesis takes place in mature red blood cells. Radioactive amino acids can be placed in a blood sample; autoradiographs of this blood are taken. Autoradiographs show that white blood cells, which have nuclei, contain radioactive particles —which means the white blood cells have used the radioactive amino acids to construct proteins —but the red blood cells, which do not have nuclei, do not contain radioactive particles. It can be concluded from these studies that a cell nucleus is necessary for protein synthesis.

Protein synthesis can take place outside a cell. Ribosomes and m-RNA can be removed from fragmented cells by centrifugation. Mixed amino acids can then be added to the ribosomes and m-

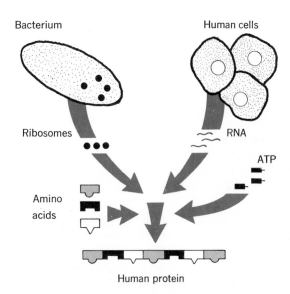

FIG. 9.4 Ribosomes from bacteria will construct human protein if they are added to human m-RNA, amino acids and ATP, which is needed to furnish the energy for the reaction.

RNA along with ATP to supply the energy needed for protein synthesis. Whole proteins, characteristic of the organism which furnished the m-RNA, will soon be produced. The m-RNA can be taken from cells of one species and the ribosome from the cells of another. Ribosomes taken from bacteria and m-RNA taken from human cells have been combined with amino acids and ATP. Human protein was constructed.

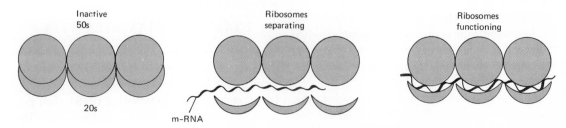

FIG. 9.5 Detail of the union of ribosomes and m-RNA. The ribosomes are made of two parts; m-RNA is captured between the two.

THE GENETIC CODE

Espionage agents are experts in cracking codes used by foreign governments to send secret messages. The genes produce coded messages which direct protein production; this code was difficult to crack, but biologists have succeeded in breaking it.

Triplet Codons

The genetic code is contained in the sequence of the bases which make up DNA molecules. Three consecutive bases in a molecule of DNA form a **codon,** which is a code word directing the production of one amino acid in a polypeptide chain. Since there are four kinds of bases, there are sixty-four possible triplet combinations ($4^3=64$). This is more than enough to code for the twenty different kinds of amino acids.

In addition to codons for the twenty amino acids, there are also triplet codons known as **initiator** and **terminator codons.** The DNA molecule is a long chain which contains many individual genes. The DNA molecule might be compared to a page in a book and the genes to sentences on the page. A way is needed to indicate the beginning and the end of the individual sentences. In writing a sentence in English we use a capital letter to start a sentence and a period or other punctuation mark to end it. In the same way genes in the DNA molecule are set apart from each other. There is an initiator codon at the beginning of a gene and a terminator codon at the end. The presence of initiator and terminator codons ensure that information from an individual is correctly transcribed.

Table 9.1 shows the codons for each of the amino acids and the initiator and terminator codons. Several codons code one amino acid in some cases. In this table the codons are given in m-RNA bases which are complimentary to those in DNA.

TABLE 9.1. The Genetic Code

Codons in m-RNA	Amino Acid Coded
CCU GCC GCA GCG	Alanine
CGU CGA CGG CGC AGG	Arginine
AAU AAC	Asparagine
GAU GAC	Aspartic acid
UGU UGC	Cysteine
CAA CAG	Glutamine
GAA GAC	Glutamic acid
GGU GGC GCA GGG	Glycine
CAU CAC	Histidine
AUU AUC AUA	Isoleucine
CUU CUC CUA CUG	Leucine
AAA AAG	Lysine
AUG	Methionine
UUU UUC	Phenylalanine
CCU CCC CCA CCG	Proline
UCU UCC UCA UCG AGU AGC	Serine
ACU ACC ACA ACG	Threonine
UGG	Tryptophan
UAU UAC	Tyrosine
GUU GUC GUA GUG	Valine
UAA UAG UGA	Terminator codons
AUG	Initiator codon

Comparison to Letters and Words

A better understanding of the genetic code may be obtained by comparing it to the letters which make up the words of a sentence. Since U, A, C, and G, the first letters of the bases found in m-RNA, do not form English words, let us substitute the letters N, E, A, and T for them. Suppose you were to be asked to read the following sequence of letters:

NTAANNATETNTNATATEANTTTN

This sequence makes no sense until you know the code. If you are told that the letters are divided into triplets and that NTA is an initiating triplet and TTN is a terminating triplet, then you would find that the sequence makes sense. The decoded sequence reads:

(NTA) ANN ATE TNT NAT ATE ANT (TTN)

This is a much shorter sentence than the "sentences" in an m-RNA code for a polypeptide chain—the m-RNA code might have hundreds of triplets—but the principle is the same.

Alteration of the Genetic Code

The discovery of the genetic code made it possible to develop a theory which defines **mutations** as small alterations of the bases. Genes can replicate themselves millions of times, making perfect copies each time, but sometimes an error occurs. A substitute base may be inserted, a base may be omitted or added, or larger sections of the DNA sequence may be altered. If a substitu-

tion in a gene occurs, then during the protein synthesis directed by that gene, a different amino acid is inserted into the polypeptide chain. A change in protein structure can have far-reaching effects on the function of that protein. To illustrate this, let us suppose that a typist is copying the sentence described above. He makes perfect copies hundreds of times, but on one copy he strikes N instead of T for the last letter of the last word. The sentence would then read:

(NTA) ANN ATE TNT NAT ATE AN*N* (TNN)

This one letter substitution, which is equivalent to a change in one base in the gene, changes the meaning of the sentence drastically. The addi-

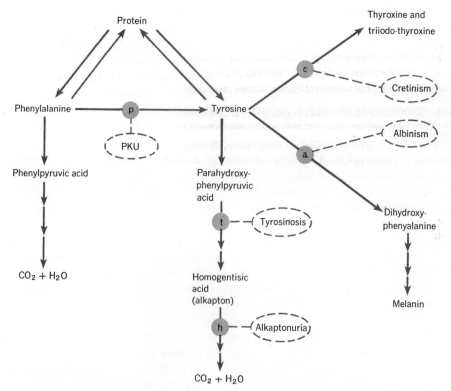

FIG. 9.6 The chain of enzyme-stimulated reactions involved in the breakdown of phenylalanine and tyrosine, two amino acids. Each arrow represents an enzyme. If a mutant form of the gene which produces one of the enzymes is present, it can produce abnormalities in function. Five human abnormalities which are the result of such breaks are shown.

tion or deletion of a letter causes an alteration in the sentence, and a switch in sequence of bases causes an alteration in the amino acid sequence of the protein. Suppose the second and third base of the third codon in our m-RNA sequence were switched. This triplet would now become the terminating codon, TTN, and the rest of the sentence would not be transcribed. This shortened strand of m-RNA would direct the production of only a part of a polypeptide chain. The structure of a protein can be drastically changed by the alteration of a single base.

METABOLIC BLOCKS

The metabolic processes in cells are generally mediated by **enzymes.** Enzymes are proteins whose production is directed by genes through the construction of m-RNA. Enzymes generally have an active site at which enzymatic reactions occur. A genetic mutation which courses an amino acid change in the nonactive region of the enzyme may not have any effect on cell metabolism. These mutations are called "silent" mutations. The fact that proteins of each individual are antigenically different shows that there have been many such mutations in the past. Natural selection neither favors nor disfavors them because they have no effect on physiological functions. Thus, many become established in populations by chance. A mutation which causes an amino acid substitution at the active site of an enzyme, however, can cause severe problems. Mutations which have harmful effects tend to disappear from the gene pool over long periods of time because affected individuals usually do not survive or reproduce as well as normal individuals.

Metabolic Blocks

A series of enzymes control many cell reactions. If, because of a mutation, one enzyme in

such a series fails to do its part, the entire series of reactions can be upset. The result of an enzyme alteration is a **metabolic block,** which can cause great harm. First, the product which the enzyme normally breaks down may accumulate in the cell, and an accumulation of materials can be harmful to the cell. Second, there may be a deficiency of the end product of the reaction. An inability to change dietary nutrients into substances which can be used by the cell can be harmful. Third, an intermediate product, which cannot be transformed because of a defective enzyme, may be acted upon by other enzymes, and another, possibly harmful product may result. Sometimes a single enzyme defect can cause two or three of these harmful effects.

Phenylalanine block. When you eat any food containing protein, it will probably contain the amino acid phenylalanine. As you digest the protein and it is broken down into its component amino acids, the amino acids, including phenylalanine, are absorbed at the intestine. When phenylalanine reaches body cells, it is absorbed and incorporated into the cell proteins being produced by the ribosomes. We usually eat far more phenylalanine than is needed for protein synthesis. Some phenylalanine is broken down by an enzyme into **phenylpyruvic acid.** Most of it, however, is acted upon by another enzyme which converts it into the amino acid **tyrosine.** One mutant gene, however, fails to direct the synthesis of this liver enzyme. This is no problem before birth because the mother's liver can take care of excess phenylalanine. Once a baby is separated from its mother, however, the amount of phenylalanine in the baby's blood begins to rise. The amount of the alternate product, phenylpyruvic acid, also goes up. The baby with this mutant gene is said to have **phenylketonuria,** or **PKU.** Without treatment the baby will become mentally defective and will develop defects of the muscles. These defects are caused by excessive blood levels of phenylalanine. The defects can be prevented if the condition is discovered early and the baby is put on a low-phenylalanine diet. Most states now require that all babies be tested for PKU when they are about three days old. If the test is postponed until the mental and muscular defects become apparent, it will then be too late to restore normal mentality to the child.

Tyrosine blocks. Some **tyrosine** comes from protein taken in from dietary sources and some comes from the conversion of phenylalanine. Several enzymes may act on tyrosine to convert it into various products. One series of reactions involving tyrosine ends in the production of the hormones from the thyroid gland, the primary one of which is **thyroxine.** If a gene which determines the production of an enzyme involved in this series of reactions fails to produce a functional enzyme, the body cannot produce thyroxine. Since this is the hormone which stimulates metabolism, the body is stunted physically, sexually, and mentally without it. A person with this enzyme defect is said to have **cretinism,** and he is called a **cretin.** The damage caused by thyroxin defi-

ciency can be corrected by administering thyroxine extracted from the thyroid glands of other animals.

A second series of reactions involving tyrosine leads to the formation of **melanin,** the brown pigment found in skin, hair, and the iris of the eyes. If an enzyme in this series is defective or missing, no melanin is formed and the person is an **albino.** In this condition the skin is very fair, the hair is almost white, and the eyes are usually a very pale blue.

A third reaction pathway converts tyrosine into carbon dioxide and water through a series of steps. Two enzyme defects can affect this series of reactions. One enzyme defect causes the accumulation of excess tyrosine and is known as **tyrosinosis.** A high concentration of tyrosine is found in the urine, but it seems to cause no difficulties. Another enzyme defect causes the accumulation of an intermediate product known as **alkapton.** Alkapton shows up in the urine and the condition is known as **alkaptonuria.** Alkap-

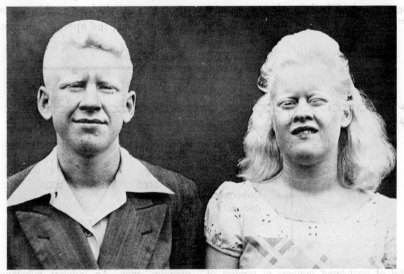

FIG. 9.7 Albinism in a sister and brother. Both children received two recessive genes which cannot direct the production of one of the enzymes in the series which is needed to produce melanin. The parents and three other children have melanin because they have the dominant gene which directs the production of the enzyme.

ton turns dark when exposed to light, so the cartilages of the body may darken as a person with this condition grows older. This darkening will be seen mainly on the earlobes and the tip of the nose.

We have considered only two of the twenty amino acids found in the body, and these amino acids represent only a part of the many products which are acted upon by enzymes. The number of enzyme-mediated reactions in cells is enormous, and an alteration of a single base in a single gene which affects the production of any of the enzymes involved can have far-reaching effects. It leads one to wonder how anyone can be normal when there are so many things which could go wrong.

REGULATION OF GENE ACTION

As you sit reading these lines, probably no more than 5 percent of your genes are actively producing m-RNA. A little later some of these genes may be "turned off" while other genes are "turned on." During fetal life some of your genes which are now permanently shut off were active. What regulates the activity of genes? In 1961 Jacob and Monod reported on the results of investigations on bacteria which documented a method by which gene activity could be regulated in these organisms. Evidence indicates that the regulation of gene activity may be similar in other species. This research led to the **operon theory** of gene regulation.

The Operon Theory

This **operon theory** states that there are **structural genes** which lie adjacent to one another and which produce the m-RNA that codes the production of a related group of proteins. of operator and structural genes is known as an **operator gene** which can stimulate the structural genes to produce m-RNA. The combination of operator and structural genes is known as an **operon.** In addition, there is a **regulator gene** which produces a **repressor** which inhibits the operator gene. As long as this repressor is present in active form, the operon does not function. An

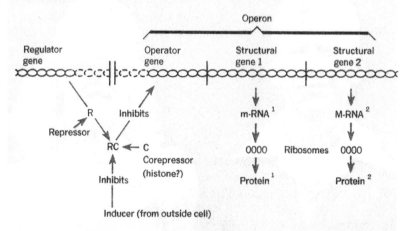

FIG. 9.8 Model of postulated method of control of gene action. A cluster of structural genes are stimulated to open and produce m-RNA by an adjacent operator gene. A regulator gene produces a repressor which works with a co-repression to inhibit the operator gene. An inducer substance, such as a hormone, may enter the cell and block the union of the repressor and co-repressor, freeing the operator gene.

inducer substance may enter the cell, combine with the repressor, and render it inactive. The operator gene is then free to stimulate the structural genes of the operon to produce m-RNA, which passes into the cytoplasm and directs the production of certain types of proteins.

As an example of the possible way in which this system operates, let us consider the digestion of **lactose** (milk sugar). You have genes which direct the production of a series of enzymes to break down this sugar. If you do not drink milk, however, the enzymes are not needed and the repressor keeps the operator gene from stimulating the strutural genes to produce this enzyme. If you drink a glass of milk, the presence of lactose will act as an inducer. The inducer binds the repressor so that it can no longer inhibit the operator gene. The operator is then free to stimulate the structural genes, and the structural genes direct the production of the enzymes which digest lactose. After the lactose has been broken down by the enzyme, none remains to act as an inducer, and their repressor again becomes active and inhibits the operator gene and enzyme production ceases.

Hormones are powerful inducers. When **estrogen,** the female hormone, is applied to a tissue culture taken from the uterus, a sharp increase in m-RNA production can be detected in the cells of the tissue culture. The male hormone, **testosterone,** has a similar effect on cultures of cells taken from the prostate gland. The flowering hormone, **florigen,** causes a similar increase in m-RNA production when it is applied to plant buds, and soon after its application, flowers appear.

When antibiotics are applied to cultures of certain bacteria, the m-RNA output by the bacteria is reduced. It has been suggested that the antibiotics destroy bacteria by binding or destroying inducers which control bacterial m-RNA production and, thus, bacterial growth. Antibiotics must be specific if they are to have any medical value. When you take an antibiotic to overcome an infection, the antibiotic seems to bind or destroy only the inducers which promote the growth of proteins produced by the bacteria and not the inducers needed to maintain protein production of your own cells. Your own cells must continue to produce m-RNA to maintain your life processes.

We shall consider other examples of gene control of cell activity as we continue our study in the following chapters.

REVIEW QUESTIONS AND PROBLEMS

1. Describe the three major structural distinctions between DNA and RNA.
2. List the three major kinds of RNA, tell where each is found, and tell briefly what each does.
3. An enzyme which breaks down the sugar galactose is produced in the liver cells. The code for the production of this enzyme is in DNA in the cell nucleus. Trace the steps involved in the production of this enzyme, beginning with the DNA.
4. Describe an experiment which shows that it is not the kind of ribosomes but the kind of m-RNA which determines the type of protein which will be synthesized.
5. Why is the output of m-RNA greater during some parts of the interphase than during others?
6. Would you expect to find transfer RNA and messenger RNA to be the same in

cells from a horse and cells from a human being? Give reasons for your answer for each type of RNA.

7. What difficulties would arise if there were no enzymes to destroy m-RNA after it had produced one or several polypeptide chains?

8. Suppose the codons for amino acids consisted of only two bases rather than three. What problems would arise?

9. **Galactosemia** is a serious human affliction characterized by an excess of galactose-phosphate in the blood. Galactose-phosphate is an intermediate product in the chain of reactions involved in the breakdown of milk sugar, lactose. Why do you think some babies develop this affliction? How might they be treated to overcome it?

10. One hypothesis holds that certain agents are carcinogenic (cancer causing) because they block repressors in cells. Explain how such a blockage might affect the cells.

11. The leukocytes (white blood cells) in human blood normally do not grow or divide, but when an extract of the kidney bean is added to a sample of human blood these cells begin growth and division. In terms of the operon theory, explain how the kidney bean extract might work.

12. A man has all the genes necessary to produce fully developed feminine breasts, yet normally such breast development does not occur. Explain this in terms of the operon theory.

FURTHER READING

Fraser, S. 1966. *Heredity, Genes, and Chromosomes.* New York: McGraw-Hill.

Hamerton, J. L. 1967. *Human Cytogenetics.* New York: Academic Press.

Watson, J. D. 1970. *Molecular Biology of the Gene.* 2nd ed. New York: W. A. Benjamin.

Winchester, A. M. 1972. *Genetics.* Boston: Houghton Mifflin.

10

The Capture
of Energy

Energy has come to be a very important word in our vocabulary in recent times because we are experiencing the consequences of a world-wide energy shortage. We never knew energy was so important until we found that there was not enough of it. The number of people on the earth has been increasing at an explosive rate and the energy demands made by each person have escalated greatly during this century. It has always been inevitable that the fossil fuels upon which we depended for so much of our energy would eventually be exhausted. This exhaustion has been accelerated by excessive and often wasteful utilization of fossil fuel products. Actually, there is no shortage of energy. There is plenty of energy bound within the atoms on the earth and coming to us in radiant form from space. The problem is to find ways to convert energy from these sources into usable forms. In this chapter we shall learn something about the sources of energy and how energy is captured and transformed into a useful form.

THE SOURCE OF ENERGY

In the final analysis, all energy comes from the conversion of mass into energy. This reaction releases the **nuclear energy** of atoms. Man has only recently learned how to control this type of reaction. Using atomic fission and fusion, we can convert a small amount of mass into great amounts of energy. The destructive force of nuclear weapons has demonstrated the potential power of nuclear energy. Controlled release of nuclear energy is used in the nuclear generating plants which now provide some of our electricity.

The conversion of mass into energy is expressed by Einstein's famous equation, $e = mc^2$. This equation means that energy, in ergs, is equal to the mass of matter, in grams, times the velocity of light, in centimeters squared. If one gram of matter were completely transformed, it could be converted into energy equal to that obtained by burning 23,000 tons of coal. A few grams of atomic fuel could power an automobile for years if we could discover a safe method to produce this kind of energy.

Energy from the Sun

The amount of energy we obtain directly by the conversion of mass here on earth is infinitesimal when compared to the amount of energy we use. A source outside our planet furnishes most of the energy we use. It is the sun. On the surface of the sun hydrogen fusion takes place constantly, converting a small number of the hydrogen atoms of the sun into energy. Like giant hydrogen bombs which hurl their flames over 100,000 miles out into space, these reactions release energy as light, heat, and solar radiation. A tiny bit of this energy reaches us across 93 million miles of space, yet this tiny bit supplies almost all the energy we use and we use only a fraction of it.

Capturing the Sun's Energy

The sun's energy reaches the earth primarily as light and heat. We utilize some of it directly. Some homes have reflectors on their roofs which focus the sun's light onto pipes of water. The water in the pipes is heated by sunlight and circulated around the house to supply heat and hot water. This is a clean, nonpolluting, and unending source of energy. Solar batteries which convert light from the sun directly into electricity have

FIG. 10.1 The hydrogen fusion explosions on the surface of the sun release a great amount of energy in the form of light, heat, and high-energy radiation. The surface of the sun has been shielded in this photo. The flames extend into space for about 132,000 miles. The sun is the primary source of energy which enables life to exist on the earth. (Mount Wilson Observatory Photo in cooperation with California institute of Technology.)

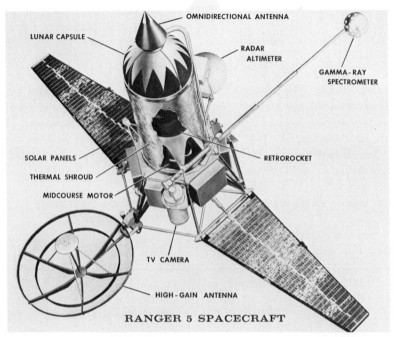

LUNAR CAPSULE

OMNIDIRECTIONAL ANTENNA

RADAR
ALTIMETER

GAMMA-RAY
SPECTROMETER

SOLAR PANELS

THERMAL SHROUD

MIDCOURSE MOTOR

RETROROCKET

TV CAMERA

HIGH-GAIN ANTENNA

RANGER 5 SPACECRAFT

FIG. 10.2 Solar batteries on the winglike structures of an orbiting space craft convert light energy into electrical energy which can be used to broadcast weather information from high above the earth. (NASA photograph.)

been developed. Solar batteries are now being used to furnish power for space laboratories which will orbit the earth for years. The cost of materials that go into the construction of these batteries is too great at present for practical use, but more research may reduce the cost.

Variation in air temperature in adjoining regions accounts for the generation of winds, which can be converted into useful energy. Windmills, turned by wind, generate electricity and pump water. Water power is also created by the release of the sun's energy. The sun evaporates water and winds move the evaporated water around in the atmosphere. As evaporated water contacts cooler air, it condenses into rain, which falls into lakes and rivers. As this water runs downhill it can turn powerful generators which produce electricity.

All these means of energy production are minor, however, when compared to the major method by which the sun's energy is captured: **photosynthesis.** Plants with chlorophyll have the unique ability to convert light energy into a bound form of energy. Even our fossil fuels, which may seem far removed from the sun, are the product of the photosynthetic capture of solar energy.

PHOTOSYNTHESIS

The energy for food synthesis can be obtained by the breakdown of inorganic compounds, and there was a time when **chemosynthesis** was probably the dominant form of energy gathering on earth. Today, however, **photosynthesis** is by far the most common form of energy capture by autotrophic organisms; only a few bacteria use chemosynthesis.

The Discovery of Photosynthesis

The beginning of man's realization that plants had the power to manufacture food from light energy dates back to the early part of the seventeenth century. Before this time it was thought that plants absorbed nourishment from the soil and used this as food in much the same way that animals use their food. In 1646 a Belgian biologist, **Jan Baptista van Helmont,** felt that this might not be a valid assumption and devised an experiment to test it. He dried some soil in an oven to remove all the water. Then he weighed out 200 pounds of this soil and placed it in an earthenware tub. Into this soil he set a small willow plant that weighed 5 pounds. The plant was watered with rain water and the soil was covered so that its weight would not be altered by possible dust blowing in or out. After five years he removed the plant, scraped off the soil, and found that it weighed 164 pounds. The soil was again thoroughly dried and found to weigh only 2 ounces less than 200 pounds. This showed conclusively that the increase in plant substance could not have come from the soil alone. Van Helmont suggested that it must have come from the water. Today we know that water was only a part of the answer. Most of the substance had been derived from the air, that part of the air known as carbon dioxide.

Materials Necessary for Photosynthesis

We often think of **carbon dioxide** as an injurious gas because it is given off from the human body as a waste product, yet this gas is absolutely necessary for the continuation of life on the earth. It is the major ingredient in the

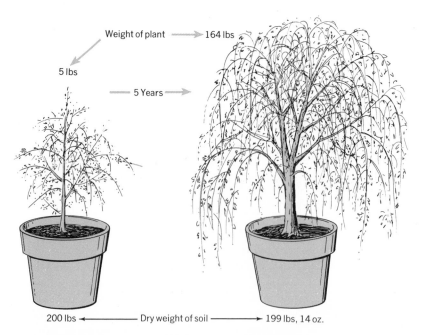

Weight of plant ⟶ 164 lbs

5 lbs

5 Years ⟶

200 lbs ⟵ Dry weight of soil ⟶ 199 lbs, 14 oz.

FIG. 10.3 The experiment of Van Helmont which demonstrated that plants do not obtain their substance from the soil. In five years the willow tree gained 159 pounds, but the soil lost only two ounces in dry weight.

production of food for all forms of life. The normal atmosphere is only about 0.03 percent carbon dioxide, yet this is sufficient to supply the plants in photosynthesis.

This is not to say that plants could not function more efficiently if the concentration of carbon dioxide in the air were higher. Greenhouse operators have found that it is possible to get increased growth of their plants if the carbon dioxide content of the air is artificially increased. During the winter, when greenhouses are normally kept closed, the concentration of carbon dioxide may drop as low as 0.012 percent. Studies by Holley and Goldsberry at Colorado State University showed that carnations and chrysanthemums would show much improved growth if the carbon dioxide level was raised to 0.05 percent. Roses reached their maximum growth rate at 0.1 percent concentration. It was also found that movement of the air was an important factor. The moving air kept a fresh supply of carbon dioxide available to the plant leaves. Without air movement the carbon dioxide content of the air would drop in the vicinity of the leaf during photosynthesis, and fresh carbon dioxide would move in only at the slower rate of diffusion in the air. We can conclude that some wind is favorable for the food manufacture of plants in natural environments.

There are reasons to believe that the carbon dioxide content of the air may have been greater in past ages when there were periods of intense volcanic activity, since such activity generates large quantities of this gas. This could have been a factor in the phenomenal growth of the great fernlike plans whose remains now serve as a base for the great coal deposits of the earth.

In addition to carbon dioxide, **water** is a necessary ingredient in the manufacture of food by a green plant, as Van Helmont surmised. Water contributes the hydrogen which goes into the food. Since hydrogen is a very light element, however, it makes up only about $\frac{1}{15}$ of the weight of the glucose molecule which results

from the photosynthetic process. That is why we say that most of the food comes from the air.

A plant needs more than carbon dioxide and water for photosynthesis. There must be **chlorophyll** and certain other elements which are necessary for the formation of this catalyst. The two ounces which were lost from the soil in Van Helmont's experiment included such things as nitrogen, potassium, phosphorus, manganese, magnesium, iron, and other elements in smaller quantities.

Magnesium and nitrogen have a place in the construction of chlorophyll. Plants grown in soil deficient in either of these elements will show leaves of a pale yellow-green color and will grow poorly. When fertilizer containing these elements is added, there will be an almost overnight change to the bright green color characteristic of normal chlorophyll content.

We know also that potassium, iron, and manganese are needed in the formation of chlorophyll, although they are not a part of the chlorophyll molecule in its final form. Phosphorus, as we soon shall learn, plays a very important part in the capture of energy.

The Nature of Chlorophyll

Since chlorophyll plays such a vital role in photosynthesis, it has been studied extensively by plant physiologists in an effort to determine how it accomplishes its transformation of carbon dioxide and water into food. Two slightly different kinds of chlorophyll have been identified in higher plants. These can be separated rather easily by means of **paper chromatography**. Chlorophyll is first extracted from the leaves of a higher plant; spinach is a favorite of the plant physiologists. Then concentrated chlorophyll extract is placed on a strip of absorbent paper. The chlorophyll moves up the paper and accumulates in two distinct bands. These two bands are found to have slightly different chemical formulas. One, chlorophyll *a*, has the formula

$$C_{55}H_{72}O_5N_4Mg$$

The other, chlorophyll *b*, has the slightly different formula

$$C_{55}H_{70}O_6N_4Mg$$

The proportion of the two varies in different plants, but the higher plants usually have about three parts chlorophyll *a* to one part chlorophyll *b*. Magnesium lies in a central position in these molecules. (Hemoglobin, which is the oxygen-carrying red pigment in the blood of higher animals, has a molecule very similar in structure to that of chlorophyll, except that the central ele-

ment is iron rather than magnesium.) Two other variants of the chlorophyll molecule, *c* and *d*, have been found in some of the algae, and two others in certain bacteria which have been found to carry on photosynthesis.

In all but a few of the simplest algae the chlorophyll is contained in chloroplasts suspended in the cytoplasm of the cells. Electron photomicrographs of individual chloroplasts at very high magnification show that they contain large numbers of layered substances, somewhat resembling stacked coins. These layers are known as **grana.** They seem to have the chlorophyll and enzymes necessary for photosynthesis arranged in the proper sequence for the most efficient handling of this process.

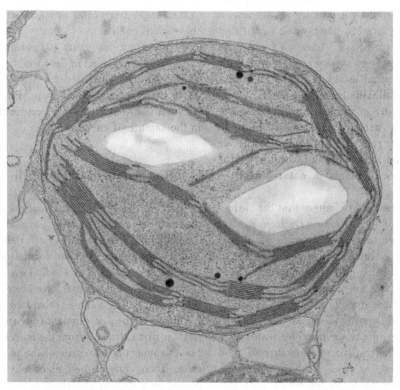

FIG. 10.4 Electron photomicrograph of a chloroplast from an oat leaf showing the layers of grana. Two starch grains are also seen in the chloroplast.

The Method of Photosynthesis

The more we learn about photosynthesis, the more we realize what a complicated reaction it is. There is much yet to be learned, but ingenious techniques devised in recent years have given us some insight into what happens. Much of this information has been gained through the use of isotopes, which permit us to trace the course of an element through the process.

For instance, through the use of an isotope of oxygen we first learned that water was split and that only the hydrogen entered into the formation of glucose, while oxygen was given off as a waste product. When heavy oxygen (oxygen-18) occurred as the oxygen part of the water given to a plant, then heavy oxygen came off from the leaves when the plant was exposed to light, but no O^{18} was found in the glucose which was produced. On the other hand, when heavy oxygen was in the carbon dioxide, it was found in the glucose manufactured by the leaf.

Before we can go into the process of photosynthesis, we need to learn something about the energy "currency" of cells. This exists in the form of **adenosine triphosphate, ATP.** Within all cells there is a substance known as **adenosine diphosphate, ADP.** This is adenosine to which two phosphate groups are bonded. A phosphate group is PO_4 and is usually represented simply as P in equations concerned with energy transformations. A third phosphate group may be bonded to the first two. Such a bonding results in ATP. This is a high-energy bond and considerable energy is required to accomplish the bonding. Likewise, a considerable amount of energy is released when the bond is broken. Hence, through the breaking of these bonds the cell can use the ATP as a source of energy. Energy going into a cell builds ATP; when energy is needed in the cell, the third phosphate bond is broken and energy is released. This process can be represented by the reversible equation:

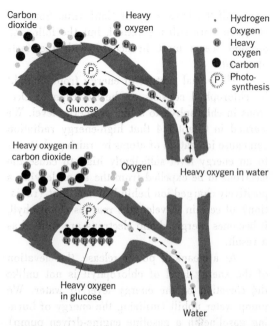

FIG. 10.5 By using an isotope of oxygen (oxygen-18) it can be shown that the source of oxygen in the formation of glucose during photosynthesis is carbon dioxide rather than water.

$$ADP + P + energy \rightleftharpoons ATP$$

The role of chlorophyll in capturing light energy has been clarified considerably in recent years as a result of the work of botanists all over the world and especially by the work of Daniel

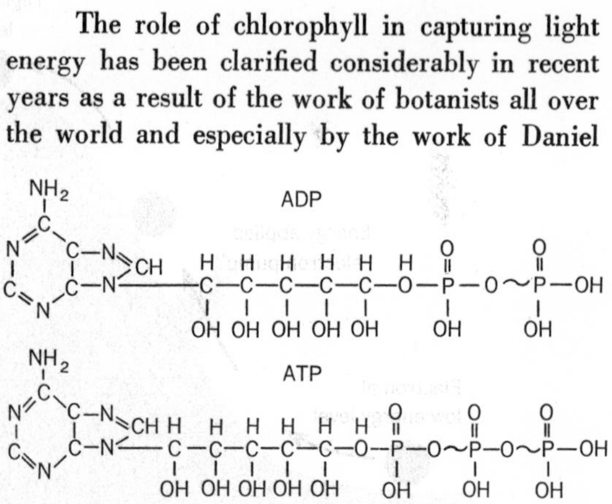

FIG. 10.6 Formation of ATP from ADP. An extra phosphate group is connected to ADP with a high-energy bond. When this bond is broken, energy is released.

Arnon of the University of California. Some of the events are still theoretical, but the following account seems most in accord with the facts observed.

The actual capture of energy from sunlight by chlorophyll consists of the elevation of electrons in chlorophyll to a higher energy level. We learned in Chapter 4 that high-energy radiation can cause ionization of atoms by raising electrons to an energy level sufficiently high to cause the electrons to be expelled from the atom, leaving a positively charged ion behind. When light (radiation) of certain wavelengths reaches chlorophyll, it becomes energized and electrons are emitted as a result.

As a means of power release, the elevation of the energy level of chlorophyll is not unlike the elevation of the energy level of water. We pump water uphill (utilizing the energy of burning gasoline in a gasoline engine-driven pump) to a reservoir for storage. Here it has **potential energy.** When the dam gates are opened, the water is released and flows downhill. It now has **kinetic energy**—it can do work: it can turn

water wheels which furnish power to grind grain, or it can turn the wheels of a generator and produce electricity. Chlorophyll is the "pump" in living plants. It uses the energy of sunlight to "pump" electrons "uphill." Such electrons contain the potential energy which can be transferred to other substances within the cell and eventually be used to do work as the electrons move once more "downhill" to a lower energy level.

For many years it was known that photosynthesis in higher plants would not take place without chlorophyll *b*, and yet it was shown that this form of chlorophyll did not participate in the actual manufacture of food. Now we know why. When a photon of light is absorbed by a molecule of chlorophyll *b*, the chlorophyll becomes excited. This results in the emission of an energized electron from the chlorophyll molecule. Nearby on the grana of the chloroplast there is an iron-containing compound, **plastoquinone,** which is similar to vitamin K. This acts as an electron acceptor and receives the energized electron, which is now at a high energy level. The loss of an electron leaves a molecule of chlorophyll *b* as

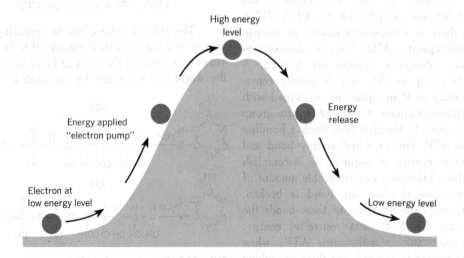

High energy level

Energy applied "electron pump"

Energy release

Electron at low energy level

Low energy level

FIG. 10.7 The "electron pump." When an electron at a low energy level absorbs energy, it is raised to a higher energy level in which it has potential energy.

When the potential energy is released, the electron returns to its former low energy level.

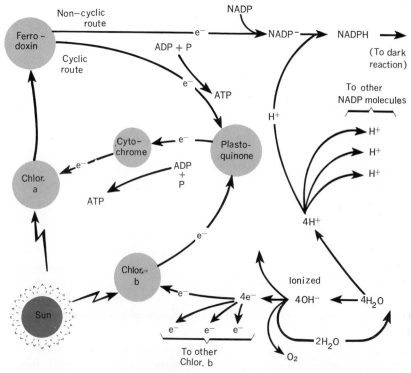

FIG. 10.8 The complex series of events involved in the light reaction of photosynthesis. When **chlorophyll** a and b absorb solar energy, some of their electrons are raised to such a high energy level that they leave the molecule. The pathway of these electrons is shown as they pass from one acceptor to another. Both ATP and NADP are produced. NADP is then used in food manufacture in the dark reaction.

a positively charged ion; it cannot capture more energy of the sunlight until its lost electron is restored. But there is a source of electrons readily available. Water, you remember, is always ionized to some extent, and there are **hydroxide ions** (OH−) near the chlorophyll b molecule. The extra electron is picked off a hydroxide ion, and chlorophyll b is restored to its original state and can again absorb energy of the sunlight. The OH which remains after giving up its electron then combines with other OH molecules and forms water and oxygen as follows:

$$4\ OH^- \rightarrow 4\ e^- + 2\ H_2O + O_2$$

The removal of hydroxide ions from the water will tend to leave an excess of hydrogen ions and the water would be expected to become progressively more acid, but the hydrogen ions are needed in a later part of photosynthesis, so there is no buildup of them in the chloroplasts or surrounding areas.

Now let us return to the plastoquinone which has received the electron raised to a higher energy level. This electron is passed "downhill" to another iron-containing compound, **cytochrome.** In this transfer some of the energy of the electron is used to hook a phosphate group onto ADP, and a molecule of ATP is formed. This ATP is going to be needed in some of the final processes of photosynthesis. From the cytochrome the electron is passed to chlorophyll *a*. During this transfer some of the remaining extra energy of the electron is used to form another molecule of ATP. The joining of a third P onto ADP in this manner is known as **photosynthetic phosphorylation.** The process of ATP formation which takes place in the mitochondria of aerobic organisms is known as **oxidative phosphorylation** and will be discussed in the next chapter.

Chlorophyll *a* is ready to receive the electron because it has lost an electron. The sunlight has also excited chlorophyll *a*, and an energized electron is thrown off, leaving the chlorophyll *a* molecule as a positive ion. The energized electron from chlorophyll *a* is received by a third iron-containing compound known as **ferrodoxin.** From the ferrodoxin, the highly energized electron may follow two courses. One, it may be passed "downhill" to plastoquinone, and a molecule of ATP is generated in the process. Then it is passed on to cytochrome and back to chlorophyll *a*, as already described. Thus, a cycle is completed, ATP being produced at three points of the cycle. This is known as the **cyclic phase** of photosynthesis. The second pathway of the electron, the **noncyclic phase,** carries it to an acceptor known as **NADP** (nicotinamide adenine dinucleotide phosphate). It is produced by the cells from one of the B vitamins, **niacin.** Plants can synthesize this vitamin. The negative charge

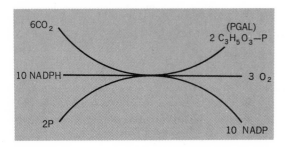

FIG. 10.9 The dark reaction of photosynthesis. NADP from the light reaction is combined with carbon dioxide to form PGAL, which can then be used to form glucose.

on the NADP then causes it to pick up a free hydrogen ion from the water and NADPH is formed.

$$NADP + e^- \rightarrow NADP^- \rightarrow NADP^- + H^+ \rightarrow NADPH$$

All the events described so far make up the **light reaction** of photosynthesis. It includes the transfer of some of the energy of the sunlight to ATP and some to the formation of NADPH.

Although it has taken some time to describe these reactions, they all happen very rapidly—in a split second they are all completed, but no food has been manufactured yet. This takes place in what is called the **dark reaction,** which can proceed without sunlight. The energy has been captured and the balance of photosynthesis is carried on by enzyme-stimulated reactions.

In 1962 Dr. Melvin Calvin, of the University of California, received the Nobel Prize for his long, painstaking work in determining the events which take place in the dark reaction of photosynthesis, and the events described here are largely as a result of his discoveries. **Carbon dioxide** is needed for the formation of food from NADPH. In order to determine the intermediate steps in the process, Calvin used radioactive carbon (C^{14}) in the carbon dioxide he administered to plants, and then he studied the compounds containing the C^{14} at various intervals. For instance, after only five seconds' of exposure to radioactive C^{14} the plants were plunged into boiling alcohol to stop all reactions. C^{14} was then found in an intermediate compound, **phosphoglyceric acid.** When the plants were allowed to function a little longer, the C^{14} was found in **phosphoglyceraldehyde, PGAL.** After about thirty seconds, the C^{14} was found in the sugars, such as **glucose.** A complete accounting of all the intermediate products will not be undertaken here because of the extreme complexity, but we shall discuss some of the end products.

The overall reaction in the production of PGAL ($C_3H_5O_3-P$) is:

10 NADPH + 6 CO_2 + 2 ATP →

2 $C_3H_5O_3$—P + 10 NADP + 2 ADP + 3 O_2
PGAL

The PGAL might be considered to be the end product of photosynthesis, for it is a food and can be used as such by the cell. It is far too reactive, however, to be transported or stored, so it is usually converted into glucose almost immediately after it is formed. In this conversion two molecules of PGAL combine with hydrogen from a molecule of water as follows:

2 $C_3H_5O_3$—P + H_2O → $C_6H_{12}O_6$ + 2 P + $\frac{1}{2}$ O_2
PGAL   **Glucose**

We express the oxygen as one-half O_2 rather than O because the single oxygen atom released in the process immediately combines with another oxygen atom and makes the oxygen molecule, O_2.

If we add up everything that goes into photosynthesis and everything that comes out, omitting those things which go in at one part of the process and come out unchanged at some other part of the process, we obtain the following overall equation:

$$6\ CO_2 + 6\ H_2O + \text{light energy} \xrightarrow{\text{chlorophyll}} C_6H_{12}O_6 + 6\ O_2$$

In the next chapter we shall learn that the overall equation for aerobic metabolism is almost the exact reverse of this equation.

If food is to be stored in the leaf, the glucose is converted into **starch.** If you wish to test leaves to determine if photosynthesis has taken place, you can test for the presence of starch, since some of the glucose will go into this product. For instance, the necessity of light for photosynthesis can be demonstrated by covering a part of a leaf with black paper. After a day or two the leaf can be removed and the chlorophyll extracted by hot alcohol. When iodine solution is added to the leaf, it turns dark blue on the areas which received light but not on the areas which were covered with the black paper. Since iodine reacts with starch to produce the dark blue color,

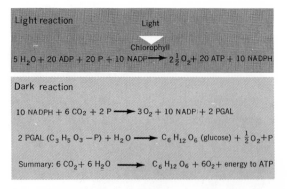

FIG. 10.10 Summary of the light and dark reactions of photosynthesis, together with a summary of the overall equation of both reactions.

we know that food has been manufactured only in the portion of the leaf exposed to light.

Likewise, we can demonstrate the necessity of carbon dioxide for photosynthesis by excluding carbon dioxide from a part of a leaf. This can be done by covering part of the leaf with petroleum jelly. The part which is covered will not turn dark with the iodine test. Variegated leaves will not turn dark in those parts which are not green, thus showing the necessity of chlorophyll for photosynthesis. Starch is easily formed from glucose by combining a large even number of glucose molecules and extracting water:

$$n(C_6H_{12}O_6) \rightarrow (C_6H_{10}O_6) + n(H_2O)$$
Glucose   **Starch**

FACTORS AFFECTING THE RATE OF PHOTOSYNTHESIS

The **dark reaction** acts as a limiting factor on photosynthesis. When there is bright illumination, the NADPH can be formed much faster than it can be utilized in the dark reaction. A plant exposed to a series of bright flashes of light will manufacture food just as fast as a plant continually exposed to light of the same intensity as the flashes. The plant in the continuous light

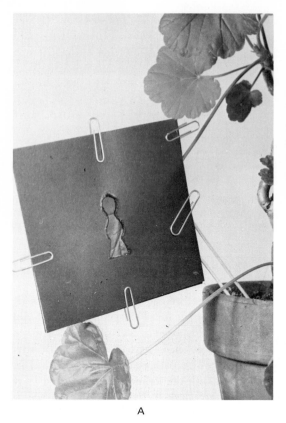

A

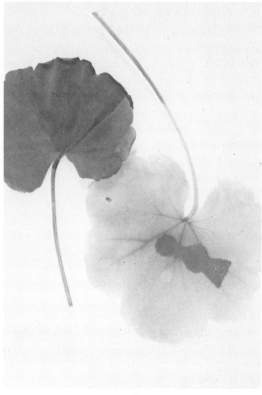

B

FIG. 10.11 Demonstration that food manufacture takes place only in the parts of a leaf exposed to light. A leaf of a geranium plant was covered with black paper with an irregular opening in it (A). After a day in light, the leaf was removed and its chlorophyll was extracted with hot alcohol. When it was placed in iodine, the part of the leaf exposed to light turned dark, showing the presence of starch (B).

may receive ten times as much total illumination, yet the dark reaction is proceeding at its maximum rate in both cases.

Further verification of this limitation is demonstrated by a study of the efficiency of photosynthesis at various degrees of illumination. If we start with a very low light intensity on a plant and gradually increase the amount of light, there will be a corresponding increase in the rate of food manufacture up to a certain point. Then, no matter how much the light intensity is increased, there will be no further increase in the amount of food manufactured. In the subdued light of the dawn, about 30 percent of the light striking a leaf is being utilized in photosynthesis. As the day gets brighter, however, this percentage drops until only about 2 percent of the energy is being utilized in the bright sunlight of midday.

The **humidity** of the air is also a limiting factor in photosynthesis. Air is admitted to leaves through tiny openings, the **stomata.** Very dry air will tend to remove water from the leaves at an excessive rate. Such a loss is prevented to some extent by a closing of the stomata, and less air is admitted to the interior of the leaf. This closure also prevents the leaf from receiving carbon dioxide to some extent, and photosynthesis slows down. As we have seen, plants can manufacture more food when they are supplied extra carbon dioxide, so reduction in the supply of carbon dioxide by a closing down of the stomata can be an important factor in limiting the rate of photosynthesis. Hence, it is on the humid days that plants, in general, are most efficient.

Temperature is also a factor, since photosynthesis, like most other chemical reactions, takes place more rapidly at warm temperatures. Plants are living systems, however, so there is an upper limit; life is destroyed by temperatures above a certain point. For most plants the most favorable temperatures for photosynthesis lie between 85° and 100°F. As temperature drops from this optimum, photosynthesis becomes progressively slower and usually stops altogether at about the freezing level of water. Some desert plants are so adapted that they can thrive in temperatures as high as 130°F.

Much of the land area near the equator is covered with tropical rain forests. The temperature in this region is high the year around and the rainfall is heavy. Thus the humidity stays very high and conditions are ideal for the growth of plants of many types. Some plants will actually grow as much as a foot in a day under such conditions.

PHOTOSYNTHESIS AND OXYGEN PRODUCTION

About 20 percent of the atmosphere of the earth today is oxygen. This important gas is necessary for the existence of all animal life and most plant life. Oxygen is necessary for cellular respiration, and both plants and animals must carry on this vital reaction, with the exception of the few forms which use anaerobic respiration. Of what importance is the oxygen given off as a byproduct of photosynthesis in the maintenance of the oxygen content of the air? It is very impor-

tant. Without photosynthesis there would be a gradual reduction of oxygen in the air. We learned in Chapter 5 that the atmosphere of the early earth probably did not contain oxygen; only after photosynthesis developed was this gas liberated in any appreciable quantities; only then did the process of aerobic respiration become possible. Photosynthesis was and remains practically the only source of free oxygen in the atmosphere. It is estimated that the entire oxygen supply of the air is renewed about every 3000 years through this important process. For this reason alone, animal life would always be dependent upon plant life, even though some means could be devised for synthesizing an adequate food supply.

PHOTOSYNTHESIS AND THE LIGHT SPECTRUM

We say that chlorophyll is green in color; that is, the light reflected from a leaf enters our eyes and affects the bodies in the retina which are sensitive to green. Where does the green color come from? Sunlight striking the leaf contains the entire spectrum of colors—violet, blue, green, yellow, orange, and red. We see only the green because chlorophyll does not absorb this part of the spectrum and it is reflected back. We would not expect the green part of white light to be of any great importance in photosynthesis because it is not absorbed.

An interesting means of testing the utilization of the light of various wavelengths by chlorophyll was devised by a German, **T. W. Engelmann**, in 1882. By means of a prism, light from the sun or any artificial source can be separated into its component colors. We frequently see such a separation when drops of water in the air act as prisms and give a rainbow in the sky. Engelmann used a glass prism to separate light and allowed the separated colors to fall on certain green algae, simple water plants which have a body in the shape of long threads or filaments. Since oxygen evolution is an indication of photosynthesis, there should be more oxygen given off from the regions falling under light of the colors most utilized in photosynthesis. Small oxygen bubbles can be seen to come off when the light is bright, but this is not a very accurate method of testing for photosynthesis. Engelmann used certain aerobic bacteria which were known to swim to regions in the water of greatest oxygen concentration. When the light was on, the bacteria soon accumulated in such numbers that the water became cloudy around the algae filaments in the orange-red area of light and also in the blue-violet area. There were almost no bacteria around the filaments under the green light.

This experiment led to the theory that the orange-red and blue-violet portions of the light spectrum are of primary importance in photosynthesis. In more recent times this fact has been confirmed by experiments in which leaves were exposed to single colors of light and then tested for food production.

EFFICIENCY OF CHLOROPHYLL IN ENERGY CAPTURE

The amount of energy captured by plants will have more meaning if we express it in terms of actual food harvested from a crop. A study of a cornfield in Illinois showed that the corn required about 100 days of growing time and yielded about 100 bushels of corn from one acre of land. When the amount of energy contained in this corn was compared with the amount of radiant energy from the sun which fell on this acre of land during the 100 days, it was found that about 0.4 percent of the radiant energy was represented in the corn harvested. This represented only about one-fourth of the food which the plants manufactured. The other three-fourths went into the leaves, stalks, roots, and the life processes of the plants.

There have been some very interesting studies in recent years on ways of increasing the efficiency of energy capture. This is of more than

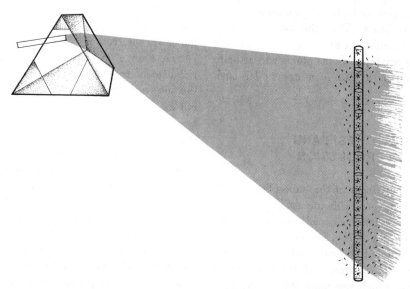

FIG. 10.12 Engleman's experiment which showed that the rate of photosynthesis was greatest in the orange-red and blue-violet portions of the spectrum. Bacteria which need oxygen congregated in much greater numbers around the part of the algal filament which was illuminated by these colors.

theoretical interest. There are large areas of the world where most of the people live in a constant state of semistarvation because of insufficient food supplies. It has been found that one of the most efficient plants is a small one-celled alga, *Chlorella*. This plant grows in water and may be so abundant as to give the water a dense green color. If land is flooded and this plant is allowed to grow in the water, one acre will produce as much as 40 tons of dry weight of food per year. Our best field crops can produce no more than 1 ton per acre. *Chlorella* can be grown anywhere that water and minerals for fertilizing are available, and is rich in proteins, vitamins, and minerals essential for human nutrition. Under favorable growing conditions it can double its volume every twelve hours; it has no growing season; it is not subject to damage by wind, frost, or hail. The entire plant can be eaten—no roots, stems, or leaves need be thrown away. The big problem which would arise in any attempt at widespread use of this plant for food would be in accustoming people to its taste. It is different in taste from most foods and there would surely be a considerable psychological objection to its use for human nourishment. People are notably contrary about changing their eating habits. Research is being conducted along this line and recipes are already available for quite a variety of dishes, even including ice cream, which use liberal quantities of *Chlorella* without creating great taste variations. Perhaps the time will come when this plant will help solve international problems arising from insufficient food supplies.

REVIEW QUESTIONS AND PROBLEMS

1. Elaborate on the statement that the energy used by living matter originates as atomic energy.
2. Differentiate between potential and kinetic energy.
3. How did van Helmont prove that the primary substance of a plant does not come from the soil?
4. In a tightly closed greenhouse, plants will not grow as well as in a well-ventilated greenhouse. Explain all the factors which might be involved in this difference in growth.
5. If magnesium is deficient in the soil, plants will not develop the normal green color. Explain exactly why this is true.
6. How was it proved that the oxygen in glucose comes from carbon dioxide rather than from water?
7. Explain how ATP is formed from ADP and tell why ATP is such a good source of energy for cells.
8. What feature is common to plastoquinone, cytochrome, and ferrodoxin?
9. When chlorophyll *b* becomes energized by the sunlight it gives off an electron. How is this lost electron replaced?
10. Hydrogen ions from water are used in the formation of NADP. This would seem to leave an excess of hydroxide ions and cause the water to become basic, but this does not happen. Explain why.
11. What products are obtained from the cyclic phase of the light reaction of photosynthesis and what products from the noncyclic phase?

12. It has been shown that plants manufacture just as much food when they are exposed to brief flashes of a bright light as when they receive the same light intensity continuously. Explain.
13. Most plants are more efficient as food producers when the air is humid than when it is dry. Explain.
14. If all the green plants on the earth were destroyed and man found a way to produce synthetic food, what would eventually happen to animal life and why?
15. It has been suggested that we might solve some of the world's food problems if we would use the alga *Chlorella* for human food. Explain.

FURTHER READING

Conn, E. E., and Stumpf, P. K. 1972. *Outlines of Biochemistry*. New York: John Wiley and Sons.

Goldsby, R. A. 1967. *Cells and Energy*. New York: Macmillan.

Lehninger, A. L. 1971. *Bioenergetics*. 2nd ed. New York: W. A. Benjamin.

Rabinowich, E., and Govindjee. 1969. *Photosynthesis*. New York: John Wiley and Sons.

11

Energy Conversion in the Cell

Most of the food produced by green plants is stored in one form or another and is available to meet the energy requirements of the plants when food is not being manufactured. Some of the food produced during daylight is retained in the leaves as starch and is used for the energy needs of the plants at night. Some of the food produced during the spring and summer months is carried to the roots and is used for energy and for building new tissues in the following spring, when chlorophyll may be lacking. Some of the food is stored in seeds and is available for the growth of the embryonic plants before they begin manufacturing food.

Not all of the food stored by plants, however, is used by the plants producing it or by their offspring; much of it serves as food for animals and other organisms which do not have chlorophyll. You may eat foods, such as bacon, eggs, butter, and beefsteak, that do not come directly from green plants, but the nourishing qualities of these foods had their origin in plants. Animals eat the stored food of plants and convert a part of it into the various parts of their bodies. Thus, their bodies contain some of the bound energy of the sunlight which was obtained from the plant food. Other animals eat these animals and utilize this energy, and they in turn may serve as food for still other animals. This may extend through quite a number of distinct species; it forms what is known as a **food chain.** Each animal in such a food chain releases a part of the energy which was captured by the green plants at the base of the chain. Fig. 11.1 shows a greatly simplified food chain. The total amount of energy available grows less as the animals become further removed from the source of food. Hence, the animals at the top represent the smallest total amount of protoplasm even though they are the largest animals in individual size. More details of this are to be found in Chapter 38.

Within the living cell there is much activity. There is movement, transformation of energy, chemical synthesis, chemical decomposition, and transfer of materials. The sum total of these living reactions is known as **metabolism.** For convenience we refer to that phase of metabolism that includes the building of new protoplasm as **anabolism. Catabolism,** on the other hand, is the decomposition reactions, which usually involve a release of energy. In this chapter we shall

be concerned primarily with the catabolic phase of metabolism.

ENERGY REQUIREMENTS OF LIVING THINGS

Life cannot continue without an expenditure of energy. To begin with, energy is needed for most forms of **chemical synthesis** in the cell. When atoms, ions, and molecules are combined to form more complex chemical substances, an input of energy is required in most cases. Without such chemical synthesis there could be no formation of new protoplasm, and no growth would occur.

Motion utilizes much energy. The contraction of muscle tissue which makes it possible for

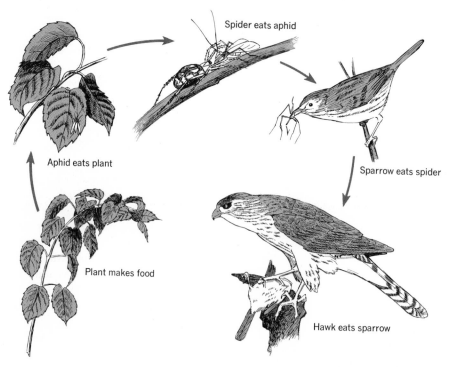

Spider eats aphid

Aphid eats plant

Sparrow eats spider

Plant makes food

Hawk eats sparrow

FIG. 11.1 A food chain. This greatly simplified energy relationship shows how food manufactured by a plant forms the base of a food chain which involves several types of organisms.

you to hold this book and move your eyes in reading uses some of the energy released in cells. Even when you are asleep, some movement of muscles must continue. The beating of your heart and the movement of muscles which enable you to breathe continue throughout the night. There is also considerable movement of the muscles in your digestive organs which continue processing the food you have eaten. There is likewise considerable motion within the cells of plants and in the plants as a whole, although the movement of an entire plant is not generally as rapid as animal movement. **Active transport** of materials across the plasma membrane can also be classified as a utilization of energy for movement, and we have already learned that a great portion of the energy expenditure of cells goes for this purpose.

Heat production is another way in which energy may be utilized. All cells release some heat during their metabolic activities and heat production can be very important in certain animals, those which are **warm-blooded.** Such animals have a body temperature which remains constant at a high level. On cool days much of their food is used to furnish the heat energy needed to maintain this temperature. Warm-blooded animals generally eat much more in proportion to body size than cold-blooded animals do. Cold-blooded animals have a body temperature which fluctuates with the outside temperature, but even these release some heat. In a very cold environment the heat release will be sufficient to keep the body temperature above freezing level.

Electricity is a form of the energy utilized by living cells. The transmission of nerve impulses in animals involves the alteration of electrical potential of nerve fibers, and the alteration is possible only through the utilization of energy from ATP. Changes in electrical potential also characterize muscle activity. Studies of the activity of the heart are made by measuring the electrical changes of the beating heart, a procedure called an **electrocardiogram.** The **electroencephalogram** measures the variations in electrical impulses in brain wave patterns and helps in the diagnosis of brain disorders. Plants also have a pattern of flow of electric currents that may be correlated with physiological processes. Some animals can generate electricity in amazing quantities; the electric eel, which inhabits the Amazon river, can give out such a powerful shock that a person wading in the water nearby can be knocked off his feet. Needless to say, this eel is not disturbed by any animal that might wish to prey upon it. There are also electric rays and electric catfish. If electrodes are attached to these animals and the electrodes are connected to a light bulb, stimulation of the animal will light the light bulb.

FIG. 11.2 A firefly photographed by its own illumination. Only a few species use part of their energy to produce light. Reflectors around the insect furnished enough light to make this photograph.

Light is still another form in which energy is utilized by some life. Fireflies flash their lights on summer evenings as signals between the sexes to help them locate one another in the dark. In deep sea regions where sunlight cannot penetrate, many animals have light-producing organs. Great areas of the ocean are often illuminated at night by concentrations of large numbers of luminescent bacteria. Decaying organic matter often will glow in the dark because luminescent bacteria are active in the decaying process. The so-called foxfire in a dark forest is nothing more than decaying wood or leaves rich in luminescent bacteria.

AEROBIC RESPIRATION

The process of making the energy of food available to the cell is known as **cell respiration,** In most forms of life, oxygen is needed for this process, which is accordingly called aerobic respiration. In some cells, however, energy can be obtained from food without free, atmospheric oxygen, and this process is known as anaerobic respiration.

In **aerobic respiration** glucose serves as the primary energy-yielding food. The entire breakdown of a molecule of glucose can be represented by the following overall equation:

$$C_6H_{12}O_6 + 6\ O_2 \xrightarrow{\text{enzymes}} 6\ CO_2 + 6\ H_2O + \text{energy (38 ATP)}$$

This is almost the exact reverse of the photosynthesis reaction. In photosynthesis the energized electron from **chlorophyll a** plus the ATP energy obtained from light are used to join the hydrogen ion from water with carbon dioxide and to join smaller molecules together to make the larger molecule of glucose. Oxygen is given off as a byproduct of this process. In aerobic respiration, enzymes replace the same amount of oxygen and break the glucose molecule down into its original units of carbon dioxide and water. In the process, the energy which was used to form the

glucose is transferred to form thirty-eight molecules of ATP from ADP and P. ATP can then be used for the energy needs of the cell.

Glucose can be burned outside living cells, a process known as **combustion**; the energy contained in the combusted glucose is released with a considerable amount of heat. A cell would be destroyed if there were a direct breakdown of glucose and an immediate release of its stored energy. Instead, within the cell a gradual, stepwise release of the energy occurs and the released energy is transformed into potential energy in the form of ATP. There are actually about twenty intermediate products between the glucose and its final conversion into carbon dioxide and water. Each step is catalyzed by a specific enzyme. We

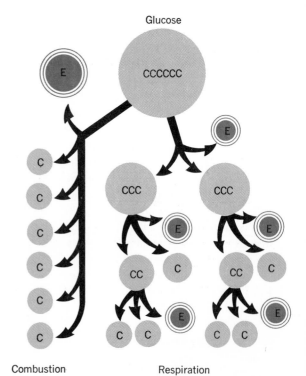

FIG. 11.3 Comparison of energy release in combustion and in cell respiration. In respiration there is a step-by-step controlled release of energy, while in combustion there is a single-step, complete release.

shall not attempt to consider all twenty steps, since these involve very complex reactions, but we can survey some of the more important intermediate products. Two distinct stages are recognized, **glycolysis** and the **Krebs cycle.**

Glycolysis

Glucose is first phosphorylated by a phosphate from ATP and the molecule is rearranged. Then the glucose molecule is split into two 3-carbon compounds with additional phosphorylation by another molecule of ATP. Glycolysis ends with the formation of two molecules of **pyruvic acid.** (PGAL, the compound so important in photosynthesis, is one of the intermediate products.) During these various reactions, four atoms of hydrogen are split off and four molecules of ATP are formed from the energy released. The entire process of glycolysis can be summarized as follows:

$$C_6H_{12}O_6 + 2\ ATP + 2\ ADP + 2\ P \rightarrow$$
Glucose

$$2\ C_3H_4O_3 + 4\ ATP + 4\ H$$
Pyruvic Acid

You can see that two ATP are needed to get the reaction going, but four ATP are formed from the energy released, so there is a net gain of two ATP. Most of the energy from glucose, however, is still bound in the pyruvic acid. If the cell is to obtain the maximum energy from glucose, the pyruvic acid must be further broken down in the Krebs cycle. Later we shall learn what happens to the four atoms of hydrogen given off in glycolysis.

The Krebs Cycle

There are ten stages in this cycle, which serves to remove the remaining hydrogen from pyruvic acid. The cycle begins when pyruvic acid is partially broken down and combined with co-enzyme A to form **acetyl CoA.** Coenzymes are

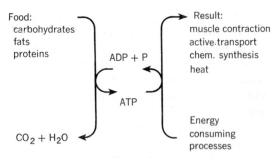

FIG. 11.4 The breakdown of food within the cell releases energy for the addition of a phosphate group to ADP to form ATP. ATP can be broken down to ADP to release energy needed by the cell.

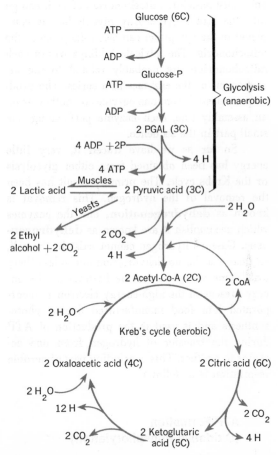

FIG. 11.5 Some of the major steps in glycolysis and the Krebs cycle.

not proteins, as enzymes are, and they do not catalyze reactions but hold some of the products of the metabolic cycle temporarily. Most coenzymes are formed from vitamins; coenzyme A is formed from pantothenic acid, one of the B vitamins. The reactions of the Krebs cycle can be summarized as follows:

$$2 \ C_3H_4O_3 + 6 \ H_2O \rightarrow 6 \ CO_2 + 20 \ H$$
Pyruvic Acid

In eukaryotic cells, the Krebs cycle seems to take place exclusively in the mitochondria of the cells. Within the free cytoplasm the enzymes for glycolysis appear to be present, but the pyruvic acid must pass into mitochondria before it can go into the Krebs cycle. Any glycolysis not completed in the cytoplasm can also take place in the mitochondria. The intricate foldings within each mitochondrion are probably related to the arrangement of the enzymes in a series. The products are passed from one enzyme to another as on an assembly line, each enzyme performing one small part in the sequence.

So far as we have traced it, very little energy has been obtained from either glycolysis or the Krebs cycle. The primary result has been the removal of the hydrogen. This removal is known as **dehydrogenation,** and the enzymes which accomplish it are known as **dehydrogenases.** Gaseous hydrogen cannot exist in the cell, so there can be no dehydrogenation unless there exist some acceptors for the hydrogen. The energy present in the high-energy electron is incorporated into food manufactured during photosynthesis and is used in the production of ATP during the transfer of hydrogen from one acceptor to another. This concluding part of aerobic respiration is as follows.

ATP Formation (Oxidative Phosphorylation)

Oxygen is the final acceptor for the hydrogen, and this is why oxygen is needed for aerobic respiration. A sudden union of these two gases, however, creates an explosive release of energy. Within the cell the hydrogen is passed from one acceptor to another with a release of a part of its electron energy at each step. Thus, there is a gradual and controlled release of energy.

When hydrogen is first split off, either in glycolysis or in the Krebs cycle, it may be picked up by the acceptor NADP, the same acceptor that picks up the energized electrons and hydrogen ions in the light reaction of photosynthesis. As in photosynthesis, the acceptor first takes up an electron. In this case the electron comes from hydrogen, leaving a hydrogen ion remaining:

$$2 \ H \rightarrow 2 \ H^+ + 2 \ e^-$$
$$NADP^+ + 2 \ H^+ + 2 \ e^- \rightarrow NADPH + H^+$$

For simplicity we can think of the product as $NADPH_2$ since the hydrogen ion is carried along with the NADPH.

Next the hydrogens are passed to a flavoprotein, **flavine adenine dinucleotide, FAD,** and some of the energy of the electrons is used to produce a molecule of ATP:

$$FAD + H_2 + ADP + P \rightarrow FADH_2 + ATP$$

Then the hydrogens are split off and the electrons from the two atoms are passed through a series of iron-containing red compounds known as **cytochromes,** which, you will remember, serve as electron acceptors in photosynthesis. As the electrons are degraded to a lower energy level in the cytochrome series, two molecules of ATP are formed:

$$Cyt + H_2 + 2 \ ADP + 2 \ P \rightarrow$$
$$Cyt^{--} + 2 \ H^+ + 2 \ ATP$$

The final acceptor in the series is **oxygen.** Oxygen receives the two electrons from the cytochromes and then adds the two hydrogen ions and forms water.

$$\tfrac{1}{2} \ O_2 + 2 \ e^- + 2 \ H^+ \rightarrow H_2O$$

The ATP production is now completed; the

energy which was bound up in glucose during photosynthesis has been converted to ATP. During both glycolysis and the Krebs cycle, from one molecule of glucose, 24 hydrogen atoms are given off (12 from glucose and 12 from added water). From these, 36 molecules of ATP are formed. We can summarize the ATP formation as follows:

$$24 \text{ H} + 12 \text{ acceptors} \rightarrow 12 \text{ acceptors} - \text{H}_2$$
$$12 \text{ acceptors} - \text{H}_2 + 6 \text{ O}_2 + 36 \text{ ADP} + 36 \text{ P} \rightarrow$$
$$12 \text{ acceptors} + 12 \text{ H}_2\text{0} + 36 \text{ ATP}$$

To this we must add the net gain of two ATP in glycolysis, making a total of 38 ATP formed from one molecule of glucose. The ATP is available for any of the energy requirements of the cell.

Pathways of Other Foods in Aerobic Respiration

Our discussion of the events in aerobic respiration has been based upon glucose as the food being acted upon. This is appropriate because glucose is the primary source of energy for cells, and most other foods can be converted into glucose before being used for energy. There can be a more direct utilization of some other foods, however, and some of the first steps of respiration can be bypassed. The place where the food enters the process depends upon the number of carbon atoms in its molecules. We have seen that the breakdown in glycolysis involves the breaking apart of the carbon linkages as well as the splitting off of some of the hydrogen atoms. Glucose has six carbon atoms, PGAL and pyruvic acid have three, and acetyl-CoA has two.

When fats are broken down in digestion, they form glycerol and fatty acids. The **glycerol,** with its three carbons, may be converted into the three-carbon PGAL, and glycolysis continues from there. The two-carbon **fatty acids** may be converted into the two-carbon acetyl-CoA and enter the Krebs cycle, thus bypassing glycolysis.

Or the fatty acids may be converted into glucose and start at the beginning of glycolysis. The two-carbon **ethyl alcohol** can also go to the two-carbon acetyl-CoA and enter the Krebs cycle from there.

When proteins are digested, they are broken down into **amino acids.** When amino acids are used for energy, the amino group (NH_2) is split off; and if they are 3-carbon amino acids, pyruvic acid, which is also 3-carbon, is produced. If they are 2-carbon amino acids, 2-carbon acetyl-CoA is produced. Some amino acids may have more than three carbons and some of these may enter directly into the Krebs cycle, after chemical changes.

Carbohydrates in foods generally have six or more carbons, and they tend to enter the respiratory cycle as the 6-carbon glucose.

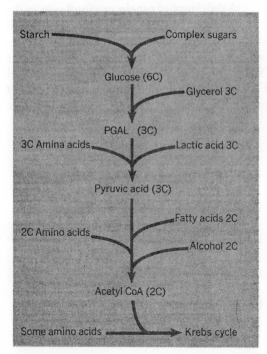

FIG. 11.6 Pathways of various foods showing where they enter the metabolic cycle. Each food enters at a point where there is the same number of carbon atoms.

Respiratory Blocks

There are several ways in which aerobic respiration can be blocked. The most obvious one is the **deprivation of oxygen.** Since oxygen serves as the final acceptor for the hydrogen, there must be a continuous supply of oxygen; otherwise the other acceptors will fill to capacity and will have no outlet for the hydrogen. When a person goes to a high altitude, he may feel dizzy and listless because there is less oxygen in the air and his cells do not receive sufficient oxygen to operate at maximum catabolic efficiency. In the course of time, the body undergoes homeostatic adaptations to the lessened oxygen concentration of the air, and a person who remains at a high altitude for a period of weeks will be able to supply the oxygen needs of his cells satisfactorily.

A deficiency of **niacin** or **riboflavin** in the diet will also slow ATP production. These are B vitamins needed to form some of the intermediate acceptors for hydrogen. Niacin is needed for NADP, while riboflavin is an essential part of FAD. Since **iron** is an important part of the cytochromes, an iron deficiency will reduce

energy output because of reduced numbers of cytochromes. Iron is also needed for the formation of hemoglobin of the red blood cells, and hemoglobin is the primary oxygen-carrying part of the blood. Thus, the cells suffer both from cytochrome deficiency and from oxygen deficiency when iron is not supplied in sufficient quantities. **Cyanide** is a highly poisonous substance which brings death quickly to aerobic forms of life because it forms a chemical union with the cytochromes and prevents them from serving their normal function as electron acceptors.

EFFICIENCY OF AEROBIC RESPIRATION

The energy yield of food is measured in **Calories.** We hear much these days about the intake of Calories, especially in reducing diets. The food we eat in excess of that needed to supply the energy requirements of the body is generally stored as fat. When the diet contains fewer Calories than are needed to supply these requirements, the stored fat is used to supply the extra quantity of energy. We measure Calories by the amount of heat energy that can be realized from a given amount of food. A Calorie is the amount of heat required to raise the temperature of a liter of water by 1 Celsius degree. A slice of bread contains about 100 Calories of potential energy; if this energy were applied to a liter of water at 0°C (freezing), it would raise the temperature of the water to 100°C (boiling).

We can easily measure the caloric content of specific foods by burning them in a laboratory furnace and measuring the total amount of heat given off. When 1 gram of glucose is so tested, it yields about 6.7 Calories. We get about the same from other carbohydrates and from proteins. Fats, however, have a higher caloric value; they yield about 9 Calories per gram.

These methods measure the maximum potential energy available from foods. How much of

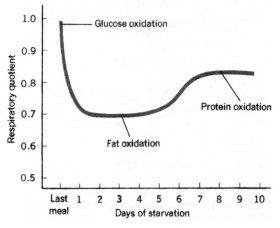

FIG. 11.7 Variation of the respiratory quotient under conditions of starvation. Glucose is utilized first for energy. When glucose is exhausted, fat and finally protein are used as an energy source.

the potential energy is actually captured in the production of ATP? Not all of it, but about 57 percent is available to the cell in ATP formation. One gram of glucose taken into cells will yield sufficient ATP to release about 3.8 Calories. The other 43 percent of the energy represents a heat loss during the process of reducing the size of the molecules and in passing the hydrogen down the series of acceptors. Even this energy is not entirely lost, because heat is of value in keeping the animal body warm in many circumstances. This is a remarkably high level of efficiency of utilization of potential energy. Man's best machines are considered highly efficient if they use the energy of 40 percent of the fuel fed into them.

THE BASAL METABOLIC RATE

When a person first awakens in the morning, it is probable that his body metabolism will be functioning at a low level. Assuming that he has had a good night's rest, does not have a fever, is comfortably warm, has not been out on a spree the night before, and is not wakened by some discordant shock to the nervous system like the jangling of an alarm clock, his body will be using energy at a rate just sufficient to maintain life in the cells. His breathing rate will be low, since comparatively little oxygen is required. We say that his body is functioning at the **basal metabolic rate, BMR.** This rate will vary according to age, weight, height, and sex. It could be measured by putting a person in a chamber and measuring the amount of heat being given off, but this would be a cumbersome and difficult procedure. The BMR can be more conveniently ascertained by measuring the amount of oxygen used in a given period of time. Oxygen is not stored in the body, so all the oxygen that is being used is going to receive hydrogen.

Today a more efficient means of measuring the BMR is available. The rate of cellular metabolism is regulated by the hormone **thyroxin,** which comes from the thyroid glands. A low BMR could indicate a deficiency of this hormone, or a high BMR could indicate an excess. Thyroxin is derived from an amino acid and has iodine in a central position in its molecule. It is bound to a protein and no other protein in the blood contains iodine. Through a quantitative test of the amount of protein-bound iodine in a given sample of blood, it is possible to determine if there is a normal rate of metabolism.

The number of Calories required to maintain normal metabolism of the human body varies with age, sex, body weight, height, physical activity, and the temperature of the surroundings. A man of average size with average physical activity in a temperate climate will use about 3000 Calories per day. A lumberjack cutting timber in the north woods in wintertime may require twice this amount from his food for his metabolic needs. A woman's metabolic needs are usually lower than that of a man; under similar conditions she will require about 10 percent less food per pound of weight.

THE RESPIRATORY QUOTIENT

If we measure not only oxygen consumption but also carbon dioxide release, we can determine what kinds of foods are being used. The ratio of carbon dioxide released to the amount of oxygen taken in is known as the **respiratory quotient.** Each of the main classes of foods has a characteristic respiratory quotient. In the overall equation for aerobic respiration of a molecule of **glucose,** six molecules of oxygen are used and six molecules of carbon dioxide are released. This 1 to 1 ratio gives a respiratory quotient of 1.0. **Fats,** on the other hand, have a lower respiratory quotient. They have a much lower oxygen content than carbohydrates, and before they can enter the respiratory process they must be converted into intermediate products by the addition of oxygen. The complete breakdown of the fat tripalmitin will illustrate:

$$C_{51}H_{98}O_6 + 72.5\ O_2 \rightarrow$$
$$51\ CO_2 + 49\ H_2O + \text{energy to ATP}$$
$$RQ = \frac{51}{72.5} = 0.70$$

Proteins lie between carbohydrates and fats, with a respiratory quotient of about 0.80. When proteins are used for energy, nitrogeneous wastes are given off and removed from the body by the kidneys. Hence, by measuring the quantity of nitrogen in the urine and comparing it with the amount of oxygen consumed, it is possible to determine the amount of protein being used for energy. The nitrogen in protein averages about 16 percent of the total weight, so it is easy to calculate the amount of protein which was used in respiration.

After a person has eaten a meal rich in carbohydrates, his respiratory quotient will rise almost to 1.0, since carbohydrates tend to be used first in respiration. If he eats no more in about twelve hours, the RQ will drop to near 0.70, which indicates that his body is using fat reserves for respiratory needs. Should he continue to abstain from eating, the RQ will gradually rise as the body fats are depleted and some of the protein of the muscles and other body organs are used. In a week or two, depending upon the amount of stored fat, the respiratory quotient will be established at about 0.80. The average RQ reflects a utilization of some of all three classes of foods if the diet is balanced. It will fall between 0.80 and 0.90.

ANAEROBIC RESPIRATION: FERMENTATION

If yeast cells are placed in a carbohydrate solution, such as molasses and water, and air is bubbled through the mixture, supplying plenty of oxygen, there will be aerobic respiration in the cells as we have described it. If a bottle of the yeast-carbohydrate mixture is tightly corked, excluding oxygen, respiration continues and bubbles of carbon dioxide can be seen rising to the top of the mixture. A pressure will soon be generated which is so strong that the stopper will be blown out, or if the bottle is stoppered very tightly, it will break from the internal pressure. This happens because yeast cells contain enzymes which permit an extraction of energy from the carbohydrate without oxygen, a process called **anaerobic respiration.**

Such anaerobic respiration is known as **fermentation.** You will recall that glycolysis does not require oxygen: glucose is broken down to pyruvic acid with the formation of some ATP. Hydrogen is released, however, and there must be some acceptor for this hydrogen. In the absence of oxygen to act as a final acceptor, pyruvic acid itself serves as the acceptor. Ethyl alcohol is formed as a result of this union of the four molecules of hydrogen given off in glycolysis with the pyruvic acid:

$$2\ C_3H_4O_3 + 4\ H \rightarrow 2\ C_2H_6O + 2\ CO_2 + \text{energy}$$
Pyruvic Acid   Ethyl Alcohol

The Krebs cycle does not follow, and as a result, the energy yield from a molecule of glucose is much less than that obtained in aerobic respiration—only about 5 percent as much. The rest of the energy remains in the alcohol. This energy is not necessarily lost to living organisms, however. If apple juice is fermented, it first forms hard cider, which has an alcoholic content. If oxygen is permitted to reach this, however, along with certain bacteria which are present in the air and which were on the skins of the apples, the cider turns to vinegar. These bacteria have enzymes which can convert the alcohol into acetic acid with the extraction of some of the energy in the alcohol:

$$C_2H_6O + O_2 \rightarrow C_2H_4O_2 + H_2O + \text{energy}$$
Ethyl Alcohol   Acetic acid

If the vinegar is left exposed to the air, other forms of yeasts have enzymes which can carry the acetic acid through the Krebs cycle with a release of the remaining energy:

$$C_2H_4O_2 + 2\ O_2 \rightarrow 2\ H_2O + 2\ CO_2 + energy$$

If we add up all three equations, we come out with the same overall equation as we have for aerobic respiration:

$$C_6H_{12}O_6 + 6\ O_2 \rightarrow 6\ H_2O + 2\ CO_2 + energy$$

Some anaerobic respiration takes place in higher forms of life, including man. Anaerobic respiration is not sufficient to sustain life for an extended period of time, but it can be a valuable supplement to aerobic respiration. You are able to be more active than would be possible otherwise because your muscle cells can obtain energy through anaerobic respiration. You could never climb two flights of stairs to a third floor lecture room if you had to depend upon the supply of oxygen reaching your muscles fast enough to permit aerobic respiration. You would have to make several stops along the way to provide the needed oxygen. You could never see a football game because the players could not stand the exertion needed in such a strenuous sport. These activities are possible because the muscles under stress turn to anaerobic respiration. Pyruvic acid is used as an acceptor for hydrogen, as in the yeast cells, but the pyruvic acid is then converted into **lactic acid** instead of ethyl alcohol and carbon dioxide. It would not do for a person to become intoxicated with alcohol each time he exercised vigorously. Instead the enzymes in animals carry the pyruvic acid along another pathway:

$$2\ C_3H_4O_3 + 4\ H \rightarrow 2\ C_3H_6O_3$$
Pyruvic acid   **Lactic acid**

When you exercise vigorously, much of the energy you expend comes from this form of anaerobic respiration. Thus, lactic acid accumulates in the muscles. A little of this acid actually makes your muscles function more efficiently. That is why muscles work better after a period of warmup. In time, however, the buildup of lactic

acid is so great that you must rest. You continue to breathe heavily for a time after you have stopped exercising. You have built up an **oxygen debt** and the oxygen now being carried to your muscles is used to repay this oxygen debt. The lactic acid is first converted back to pyruvic acid by the extraction of hydrogen, since oxygen is now available as a final acceptor for hydrogen. The pyruvic acid can now be carried on through the Krebs cycle if more ATP is needed, or it can be carried back to glucose by using energy from ATP. Some of the lactic acid is carried from the fatigued muscles to the liver, where it is converted into glucose when oxygen becomes available. We shall learn more about energy utilization by muscles in Chapter 28.

Since yeast cells can obtain energy through either aerobic or anaerobic respiration, they are known as **facultative anaerobes**. Some bacteria are **obligatory anaerobes**; they can obtain energy only through anaerobic respiration and in fact cannot even grow when exposed to free oxygen. The waste products of such bacteria can be extremely poisonous. The bacterium *Clostridium botulinum* produces resistant spores that can remain alive under unfavorable conditions. If some of these spores are sealed in a can of food, they can become active and the bacteria grow very rapidly, generating one of the most powerful poisons known to man. **Botulism**

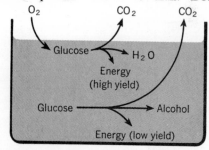

FIG. 11.8 Anaerobic and aerobic respiration of yeast. When oxygen is available (at the top of a glucose solution, for example), aerobic respiration takes place with a maximum release of energy. Anaerobic respiration can take place without the presence of oxygen to produce alcohol and low energy yield.

poisoning is not common today with our modern canning methods, but a few deaths occur every year because of it. (See Chapter 15.)

A closely related species, *Clostridium tetani,* causes the disease of **tetanus** in man and other higher animals. Spores of this bacterium may gain entrance to the body through puncture wounds which leave them beneath the skin, sealed off from the oxygen of the air. In such an environment, the bacteria can grow and generate their own toxin. The toxin affects the muscles, causing them to contract involuntarily. Since the disease is usually fatal, many physicians recommend vaccination as a protection.

CELL RESPIRATION IN PLANTS

The green plants, as well as animals, must carry on cell respiration if they are to make use of the energy within the food which they manufacture. During daylight hours there is a plentiful supply of oxygen as a byproduct of photosynthesis. At night, however, plants are dependent upon the oxygen of the atmosphere, just as animals are. Plants also give off carbon dioxide as a product of their cell respiration, just as animals do. Some hospitals have a ruling that all plants must be removed from the patients' rooms at night. There is a feeling that when a patient has a lowered body resistance, he should not have plants competing with him for the available oxygen in his room or adding to the carbon dioxide content of the air. This would not be the case in the daytime, of course; the conditions would be reversed, provided the room was well lighted. Carefully contrived tests, however, have shown the the composition of the air is altered so slightly in both cases that it would not have any significance for the welfare of the patient.

During daylight hours the rate of photosynthesis is so much greater than the rate of respiration that an average of about ten times as much oxygen is produced as is used. Likewise, about

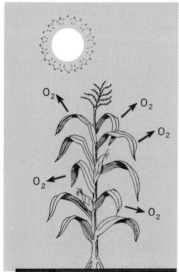

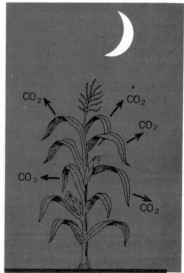

Daylight   Twilight   Night

FIG. 11.9 Green plants give off oxygen during the daytime when photosynthesis is taking place. At twilight the carbon dioxide output from metabolism balances oxygen production. At night, when photosynthesis does not occur, carbon dioxide is given off.

ten times as much carbon dioxide is used as is produced. At night the oxygen production stops and the oxygen consumption continues, but for a twenty-four period the plant still produces far more oxygen than it uses. When the light intensity falls to about 1 percent of that found in bright sunlight, we reach what is called the compensation point. Here the rate of respiration exactly balances the rate of photosynthesis; neither oxygen nor carbon dioxide is given off from the leaves. When plants are kept in the light at this level experimentally, there is neither gain nor loss in weight—food is used at exactly the same rate at which it is produced and none is left over for growth.

One plantlike group, which is known as the fungi, do not have the chlorophyll. Hence, they are much like animals in their requirements of oxygen for cellular respiration. A mushroom can grow as well in the dark as in the light, since it cannot carry on photosynthesis.

INTERDEPENDENCE OF PLANTS AND ANIMALS

As the many facets of the energy relationships of living things become apparent, we can see that there is an interdependence of plants and animals. One aspect of this was demonstrated by an interesting experiment performed about two hundred years ago by **Joseph Priestley,** an English chemist. He placed a mouse in an airtight glass container. After several hours it was breathing with difficulty and finally it died of suffocation. Another mouse placed in the same container became unconscious within a few minutes but recovered when it was removed to fresh air. Then a growing potted plant was placed in a transparent container and allowed to remain for several hours in the sunlight. A live mouse then placed in the container lived several hours without showing any respiratory difficulties. Somehow or other the green plant had restored air which had been

altered by the breathing of an animal. Priestley proposed the hypothesis that the animal had given off some sort of poison into the air and that the green plant had removed this poison. Within a few years, however, the French chemist **Antoine Lavoisier** showed that an animal in such a container actually uses up some of the oxygen in the air and gives off carbon dioxide. A green plant does just the opposite, and this is what restores the air.

These studies gave man his first clue to the dependence of animals and plants upon one another. If a mouse and a plant of the proper size were both sealed in a transparent container and kept in the light, both could continue their respiration, and photosynthesis would continue without difficulty. If we think of the earth as one huge container with a limited volume of air, we can

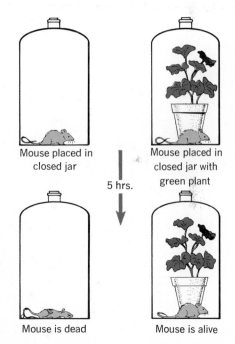

FIG. 11.10 A mouse placed in a closed jar will die of suffocation when it has exhausted the oxygen, but if a green plant is placed in the jar and light is supplied, the oxygen output of the plant supplies oxygen for the mouse and it stays alive.

see that a similar situation exists on a much larger scale. Our massive pollution of the oceans is killing many of the photosynthetic organisms, and some people fear that this process could eventually cause a lowering of the oxygen content of the air.

The interdependence of plants and animals does not stop with atmospheric balance by any means. Plants serve as the sole source of food for animals; this is a clearly recognized dependence of animals upon plants. There is also a part played by the animals and the fungi which is just as important for the growth of the green plants. This is the reconversion and making available of certain vital minerals which are needed by the plants in their growth. Man has long recognized this role of the higher animals and uses the waste products from many kinds of animals as fertilizers for his crops. The bacteria play a particularly important role in these conversions. Through

decay they break down the dead bodies of both plants and animals, and the mineral matter which these bodies contained becomes available for use by growing green plants. Bacteria are also able to take the mineral nitrogen out of the air and convert it into a form which plants can use in building the vital protein food.

THE ENERGY CYCLE

As we sum up cell activities related to energy transformation, we can see that living and nonliving matter on the earth is all tied together in a giant cycle kept in operation by the energy of the sunlight. Plants take carbon dioxide from the air and water and minerals from the soil, and with the aid of the sunlight they convert these substances into foods of different kinds. Some of these foods are broken down by the plants for

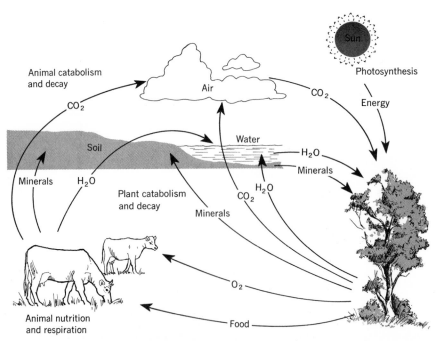

FIG. 11.11 The energy cycle. This diagram shows the relationships between living and nonliving materials on the earth. Solar energy is required for the maintenance of the cycle.

energy, and carbon dioxide and water are released. These again become available for green plants to use in capturing more energy of the sunlight. This is the shortest phase of the energy cycle.

Some of the food produced by the plants is used in building their bodies and may not again become available for energy capture until the plants die. Then the bacteria of decay come in and break down the substance of the plants, liberating the bound energy for their own metabolism. Carbon dioxide and water are released, and the minerals go back into the soil where other plants can absorb them.

Animals which eat some of the food produced by plants utilize a part of the energy in cell respiration with a release of carbon dioxide and water, which again become available for the green plants. Animals also excrete some of the vital minerals which are in the food and these become available again. When animals die, their bodies decay and all the materials which were used in their construction are restored to a state which can be used over again. The bodies which we possess today consist only of a number of rather common chemical substances which have probably been used many, many times in the past as construction material for other forms of life, both plant and animal.

There is only one thing in the energy cycle which does not get back to its source—the energy itself. Energy does not go back to the sun and become available again for photosynthesis. Yet it is not lost in the true sense of the word. Studies on energy show that it cannot be destroyed; it can only be transformed. The kinetic energy of sunlight is transformed into the potential energy stored in the food materials which are constructed by green plants. The potential energy is released in cell respiration and is used in various ways, but this does not destroy the energy. The heat which comes from your body when you have been exercising vigorously is never actually lost. It radiates out and soon becomes widely distributed in the surrounding air. It will continue to spread and in time will be lost from the earth as it escapes into outer space, but it is never actually destroyed. It will go on and on in the future eons of time as an infinitesmal elevation of the temperature of the universe. Energy is constantly being lost from the earth into outer space in this way, a process known as entropy, yet the sunlight keeps it replenished.

There are also nonliving reactions which play a part in the energy cycle. The burning of organic matter by fire, a process known as combustion, consumes oxygen and releases carbon dioxide. Joseph Priestley, the English chemist whom we mentioned earlier in this chapter, demonstrated the fact that combustion was equivalent to animal respiration in its effect on air. He found that a burning candle placed in a closed container would soon go out. Then a mouse placed in this container could not live. The burning candle had altered the air in the same way as if a mouse had been kept in the container until he died of suffocation. Thus, so far as the balance of oxygen and carbon dioxide is concerned, the plants are benefited by combustion just as they are benefited by cell respiration.

REVIEW QUESTIONS AND PROBLEMS

1. When you eat a T-bone steak, you certainly assume that you get some nourishment from it. How can you square this fact with the statement that green plants are the source of all foods?

2. Animals at the top of a food chain may be larger in individual size than animals closer to the plant source of the food, yet they represent a smaller total volume of protoplasm in the aggregate. Explain.

3. List the various ways in which energy is expended within the human body.

4. Give the overall equation for photosynthesis and the overall equation for aerobic respiration. Explain all the ways in which these two equations differ from each other, beginning with the fact that one is the reverse of the other.

5. How does energy release in combustion differ from energy release in aerobic respiration?

6. Explain the events which take place in glycolysis. You need not give chemical formulas.

7. Why is more energy released in the forms of life which can carry on aerobic respiration than in those which have only anaerobic respiration?

8. Glycolysis can take place without oxygen, but the Krebs cycle cannot. Explain why.

9. A person whose diet is deficient in B vitamins is likely to be listless and lacking in energy for normal activities. Give a scientific reason for this.

10. NADP serves as an electron and hydrogen ion acceptor in both photosynthesis and aerobic respiration. How does the source of electrons and ions of hydrogen differ in the two processes?

11. Explain exactly why a person dies if his breathing is stopped for several minutes. Be specific.

12. The energy from the electrons of the hydrogen atoms of a glucose molecule yields 36 molecules of ATP, yet we say that the total yield of ATP from a glucose molecule is 38. Explain.

13. How does the role of cytochromes in aerobic respiration compare with their role in photosynthesis?

14. Explain the general rule governing the entrance of foods other than glucose into the respiratory cycle. Illustrate with fats, proteins, and starches.

15. Why is cyanide such a deadly poison?

16. When estimating the total number of Calories required by a person in a day, we consider such items as sex, amount of activity, and the temperature in which the person lives. Why are these factors necessary?

17. Explain the three ways in which a person's basal metabolic rates may be determined as discussed in this chapter.

18. How can the respiratory quotient be used to determine when a person is in an advanced state of starvation?

19. Wine makers are usually careful to exclude air from the fermenting fruit juice. Explain why this would be done.

20. Explain how the anaerobic metabolism of yeast cells differs from the anaerobic respiration of human cells.

21. Why do you continue to breathe heavily for several minutes after you have climbed several flights of stairs?

22. How does a facultative anaerobe differ from an obligatory anaerobe?

23.   At a certain time between daylight and dark, green plants neither take in nor give off oxygen. The same is true for carbon dioxide. Explain.

24.   Certain chemicals absorb carbon dioxide from the air. Had Priestley put such a chemical in the container along with a mouse and a green plant, what would have happened and why?

25.   Why are bacteria such an important part of the energy cycle?

FURTHER READING

Baker, J. J., and Allen, G. E. 1970. *Matter, Energy and Life.* 2nd ed. Reading, Mass.: Addison-Wesley.

Giese, A. C. 1973. *Cell Physiology.* 4th ed. Philadelphia: W. B. Saunders.

Kennedy, D. 1965. *Living Cell: Readings from Scientific American Magazine.* San Francisco: W. H. Freeman.

Lehninger, A. L. 1971. *Bioenergetics.* 2nd ed. New York: W. A. Benjamin.

McElroy, W. D. 1971. *Cell Physiology and Biochemistry.* 3rd ed. Englewood Cliffs, N. J.: Prentice-Hall.

12

Systematic Classification of Living Things

The variety of living things which share this planet with us is enormous. Over one and one-half million species have been identified and named, and there are countless others which have not yet been named. Without some systematic method of classification, this variety would be a bewildering array. When we examine the species which have been identified, we find that they tend to fall into logical groups. For example, you would probably have no trouble classifying a particular organism as a fish, although you may never have seen that particular species before. All fish have certain characteristics in common which make their recognition comparatively easy. Biological classification takes advantage of these natural groupings and arranges living organisms in a systematic manner.

The classification method is similar to that used by a library. Books are grouped first according to broad subject areas and then are subdivided into progressively more narrow categories. Once you are familiar with the library classification system, you can go to the shelves where books on a particular subject are located and find a particular book. In the same way, by becoming familiar with the classification system used for living things, you can find information about a particular kind of organism by consulting these systematic listings. The area of biology which deals with classification is known as **taxonomy.** In this chapter we shall consider some of the methods used by taxonomists to arrange living things into logical groups.

TAXONOMIC METHODS

Man has a logical mind and tends to classify all things he encounters. When you meet a person for the first time you almost unconsciously begin to classify him. Sex is probably the first classification category: you decide if the person is male or female. Next you might consider age: is the person in babyhood, childhood, youth, middle age, or old age? Then you might notice facial features and skin pigmentation which would indicate racial background. All of this takes place very quickly. It is reasonable to assume that primitive man must have begun grouping living things on the earth in terms of their similarities and differences.

Ancient Systems

The book of Genesis in the Bible dates back over 4,000 years, yet it refers to a well-defined system of classification. Plants were grouped into grasses, herbs, and trees, while animals were classified as fish, creeping things, fowls, beasts, and cattle. Man was placed in a group by himself. In about 350 B.C., the Greek philosopher **Aristotle** devised a system based on three groups of plants—herbs, shrubs, and trees—and three groups of animals—those which lived on the land, water, and in the air. Each of these was divided into subgroups. The narrowest classification groups were the *genos* and *eidos,* which correspond roughly to the **genus** and **species** in modern classification systems. All the early systems recognized two major groups, plants and animals, which were distinctly different from each other.

Beginning of Modern Methods

Aristotle's system became the standard in the Western world and was used for about 2000 years. As scientific investigation intensified, however, a more precise system was needed. In Sweden during the early eighteenth century, Carl von Linné, undertook the enormous task of devising a new system based upon more scientific observations. He attempted to classify all forms of life on the earth with this new system. He is better known by his Latinized name, **Linnaeus,** and he is regarded as the originator of the modern system of classification.

Linnaeus used similarities of structure and function as a basis for his classifications, rather than using living habits and superficial similarities, as had been done previously. The Linnean system was consistent with possible evolutionary relationships. Porpoises and alligators, as an example, both live in the water, while cows and lizards both live on the land. If classification were based on habitat, we would put the first two together and the second two together. A closer examination of these animals, however, shows that, according to structure and function, the cows and porpoises are alike in many respects. Both are warm blooded, both have young born alive, and both nurse their young with milk. Therefore, both cows and porpoises are placed in a classification group known as **mammals.** Lizards and alligators also have many similarities in spite of a considerable size difference. Both are cold blooded, both have a scaly skin, both have hearts with similar structure, and both share many other common characteristics. They are placed in a group known as **reptiles.**

Linnaeus published his system in 1737 in a book called *Systema Natureae.* This book was expanded in succeeding editions and culminated in the twelfth edition, published in 1758, in which 4378 species were classified. This book is still recognized as the standard, and the names in it are given priority over any names that have been given since. A **law of priority** has been established which states that when a new species is discovered, the name which will be recognized is the first one properly published since 1758. This system avoids the confusion which arises when different investigators discover the same species independently and each gives it a different name. The first species name which is published is the one chosen.

Taxonomic Groups

In a favorite guessing game, seen on television and played at parties, the contestants may first ask, "Is it animal, vegetable, or mineral?" Plants and animals are so different that it is logical to separate the two. Linnaeus recognized this grouping and began his system with the **plant** and **animal kingdoms.** He felt that it would be easy to place all living things into one of these two kingdoms. As microscopes were developed, however, many smaller forms of life were found which did not fit easily into either of these kingdoms. New kingdoms have been created, therefore, to accommodate these newly discovered organisms.

The next classification group below kingdom is the **phylum,** plural **phyla.** Unfortunately, many botanists selected the designation **division** as the subdivision of the kingdom of plants. Since it is very confusing to use different words for the same major groups, many biologists now accept phylum as the first subdivision beneath kingdom for all kingdoms, and it will be used in that sense in this book. As we become familiar with the characteristics of the various phyla, we gain a comprehensive overview of the entire world of living organisms. Each phylum is further subdivided into **classes,** the classes into **orders,** the orders into **families,** the families into **genera,** and the genera into **species.** Even the species may be divided into **subspecies,** which are known as races, varieties, or breeds.

Examples of Classification

This classification arrangement is best understood by some examples. Let us suppose you have a cocker spaniel dog which you wish to classify. It is evident that it is an animal, so we place it in the kingdom **Animalia,** which may also be known as **Metazoa.** A survey of the characteristics of the different phyla in this kingdom shows that one includes animals with a backbone. Since a dog has a backbone, we place it in the phylum **Chordata.** Placing it in this phylum narrows the field of possibilities tremendously, but this phylum still includes such diverse animals as frogs, fish, snakes, birds, monkeys, and cows, to name a few, in addition to dogs. The classes of the Chordata include the **Mammalia,** animals which possess mammary glands and nurse their young. This is the class to which dogs belong. Among the mammals, there is an order, the **Carnivora,** which includes those which live primarily on meat and which have long canine teeth for killing their prey. The dog falls into this category. Families within the carnivores include **Filidae** (the cat family), **Ursidae** (the bear family), **Otaridae** (the sea-lion family), and **Canidae** the dog family).

Now we have narrowed the field down to a relatively small group which includes the foxes, jackals, wolves, and dogs. There are two genera in this family: *Vulpus,* the foxes, and *Canis,* the dogs. There are five well-known species of *Canis.* These are *Canis lupus* (the European wolf), *Canis occidentalis* (the timber wolf of North America), *Canis latrans* (the prairie wolf or coyote), *Canis aureus* (the jackal), and *Canis familiaris* (the domestic dog). Within the domestic dogs, there are many breeds: collies, dachshunds, Great Danes, bloodhounds, and cocker spaniels, to name a few.

Table 12.1 shows the classification of dog, man, and the white oak tree according to modern methods. You can see that man and the dog share many groups in common and separate only at the level of the order, but that the white oak is different beginning with the kingdom.

Sometimes the complexities of organisms are such that it is difficult to fit them into this orderly scheme of organization. We find sub-phyla, subclasses, suborders, and other subgroupings which extend the number of classification groupings. These are primarily of value to the specialist in the specific area of study and will not concern us greatly in this book.

KINGDOM ANIMALIA

Phylum chordata

Class mammalia

Order carnivora

Family canidae

Genus canis

Species familiaris

Variety cocker spaniel

FIG. 12.1 The classification of a cocker spaniel. Beginning with the animal kingdom, each subgroup becomes more restricted.

TABLE 12.1. Taxonomic Groups of Three Organisms

	Dog	*Man*	*White Oak*
Kingdom	Animalia	Animalia	Plantae
Phylum	Chordata	Chordata	Trachaeophyta
Class	Mammalia	Mammalia	Angiospermae
Order	Carnivora	Primata	Fagales
Family	Canidae	Hominidae	Fagaceae
Genus	*Canis*	*Homo*	*Quercus*
Species	*familiaris*	*sapiens*	*alba*

Reason for Scientific Names

Why do biologists use long double names for organisms which have much simpler common names? Why not use a simple three-letter word, dog, instead of *Canis familiaris*. Why do scientific names have strange endings and plurals in the Latinized forms? There are several reasons for using scientific names. First of all, they are universal. The same names are used in articles written in French, Spanish, German, or even Japanese. This is important because of the exten-

A

B

FIG. 12.2 Which is corn? This may seem to be a ridiculous question to people in the United States, but in some parts of the world, the plant on the left would be called corn. Common names are not reliable when accuracy of identification is needed.

sive exchange of information by biologists who live all over the world. Mistakes in translations cannot occur when the same name is used in all languages. When the rules for scientific names were formulated, Latin was the language of the scholars. Not only names but all scientific publications were in Latin. This was an attempt to have a universal language of science. Many common English words use Latinized plurals, so it should not be hard for you to learn that the plural of **amoeba** is **amoebae** and the plural of **fungus** is **fungi,** to mention two examples.

Then too, common names are often unreliable. Even within the same language, the same name may be applied to two or more entirely different organisms, and the same organism may have several common names. In some of the southeastern sections of the United States the word "scorpion" refers to a type of lizard, but in the Southwest "scorpion" is used to name a relative of the spiders which has a long curved tail with a stinger on the end. In the Appalachian Mountain regions there are three different plants which may be called "rhododendrons" and many arguments ensue when different people defend their particular interpretation. The plant which we call "corn" is known as "maize" in England, and the English use the word "corn" to refer to wheat and other cereal crops. The scientific name *Zea mays*, however, can refer to only one plant, the one with the big ears which Americans call corn. You can see how important it is to use scientific names in scientific literature or discussions where accurate identification is necessary. There are many small and little-known forms of life which have no common name and we have no choice but to use scientific names.

Common names, of course, have their place. Many are descriptive and intriguing. Baby-blue-eyes, jack-in-the-pulpit, Indian paintbrush, bluebonnet, lady slipper, and black-eyed Susan—all common names of wild flowers—have a romantic appeal which does not characterize scientific names. No one would say, "I saw the first *Turdus migratorius* of spring this morning." Or, "Look at the beautiful field of *Ranuculus acris.*" It is much better to use the common terms "robin" and "buttercup" on such occasions, but at a scientific meeting the scientific names would be appropriate.

Scientific names are often formed by combining shorter words which refer to certain characteristics of the organisms to which they apply. The phylum name **Platyhelminthes** comes from two Greek words, *platys,* meaning flat, and *helmins,* meaning worm. This phylum includes the flatworms. In the plant kingdom, the word Gymnospermae comes from two Greek words, *gymnos,* meaning naked, and *sperma,* meaning seed, and it refers to a group of plants which bear naked or exposed seed.

Rules for Usage of Scientific Names

When we give a scientific name, we should include both the genus and species. The genus name is always capitalized and the species name is always begun with a lowercase letter. The dog is *Canis familiaris* and man is *Homo sapiens.* When printed, these names are in italics. When they are typed or written, they are underlined, which to a printer means italics. The genus name is a noun, while the species name is an adjective which descrbes the noun. *Canis* means dog and *familiaris* means common or familiar. *Homo* means man and *sapiens* means wise, a designation which we may sometimes question when we see how man has wantonly used his intellectual powers to pollute and destroy other forms of life on the earth. The oak tree has the genus name *Quercus,* and the various species names are descriptive. *Quercus alba* is the white oak, *Quercus niger* is the black oak, and *Quercus suber* is the cork oak.

The same species name may be used under a number of different genera, but the genus name is unique. The catbird is *Dumatella carolinesis* and the white-breasted nuthatch is *Sitta carolinensis.* Both birds were discovered in the Carolinas and so they were given the same species names, but *Dumatella* and *Sitta* are not used to refer to any other organisms.

In scientific literature the name of the person who first described a species may be included in the scientific name. You may read *Rana catesbeiana Shaw*. This is the genus and species name of the bullfrog and the name was assigned by a man named Shaw.

When one refers to a particular species over and over again, it is common practice to use only the first letter of the genus after it has been given in full in the first usage. Thus, you may see *E. coli* used in reference to the bacterium *Escherichia coli*.

THE DISTINCTION OF A SPECIES

The species is the fundamental unit of taxonomy, but it is not always easy to draw a firm line of distinction around a species. The traditional definiton is "a freely interbreeding group, a natural population, which is reproductively isolated from other groups." This definition is easy to apply to higher animals and many higher plants which have reproductive methods easy to observe, but we must turn to other criteria for many forms of life which are not so easily observed or controlled. First, let us see how this definition is applied to some of the larger animals and plants.

Higher Animal Species

The cheetah, leopard, and lion are members of the cat family which are found in the same regions of Africa, but they are different species and are reproductively isolated. A male cheetah is not attracted to a female lion in heat. The sexual odor of the lion may be strange to the cheetah and may not excite a male cheetah to the point where he would attempt mating with a female lion. Odor is an important sexual attrac-tant in many mammals. Even if the male cheetah should attempt to mate with the female lion, he would probably be repulsed by her because his mating habits would be strange to her. And even if mating did take place, there would be another barrier to fertilization. Females of a particular species produce eggs which have an affinity only for sperm of their own or closely related species. It is doubtful, therefore, that the sperm of the cheetah would enter the egg of the lioness even if artificial insemination were performed. The leopard and the lion are more closely related than the lion and the cheetah, and the sperm of a leopard might fertilize the eggs of a lion, but the embryos would probably be abnormal and soon die. When eggs of a bullfrog are artifically fertilized with sperm from a leopard frog, the embryos develop for a while and then become abnormal and die.

When the species are very closely related, a hybrid can result from a cross-species mating, but the hybrid will usually be sterile. Different species of fruit flies may inhabit the same garbage can and some may mate and produce hybrid off-spring. The genes and chromosomes of the two species are different, however, and hybrids can-not produce functional gametes. In fact, some of the different species of fruit flies look so much alike that they cannot be distinguished from one another by appearance. It is only when they are crossed and the offspring prove to be sterile that the distinction between the two species can be made. A study of the chromosomes of such hy-brids shows that the chromosomes do not match in meiotic pairing and, therefore, no gamete gets a complete set of chromosomes, a necessity for survival.

Geographical separation may be the only method of reproductive isolation for some species. Lions and tigers are very closely related, and in captivity they sometimes do mate. The offspring is known as a liger when the male parent is a lion and the female a tiger. Offspring of the reverse mating are known as tigrons. These hybrids have been known to produce offspring when mated to either parental species, so the chromosomes must match sufficiently well to produce viable gametes. In their natural habitat there is generally no such hybridization, so the two species remain distinct. Lions live primarily in Africa and tigers, in Asia. Up until about 1880 there were Asiatic lions which lived in regions overlapping those of the tigers. The lions, however, preferred the open ranges and lived in groups known as prides, while the tigers were more solitary and preferred the jungle regions. Hence, there was little opporunity for contact between the two species, although there were rare reported sightings of ligers and tigrons. Today there are only a few hundred Asiatic lions left and they are restricted to a small area.

The Alaskan brown bear and the polar bear are classified as different species even though they sometimes produce fertile hybrids when kept together in a zoo. Brown bears live in forested regions, while polar bears prefer the open regions of the tundra and the ice floes.

Different breeding cycles can also provide reproductive isolation. Two species of frogs may inhabit the same pond yet remain as distinct species even though hybrids can be produced in the laboratory by controlling the hormonal cycles of the two species. One species may come to the peak of their cycle in May and complete reproduction by the time the second species is ready for reproduction in June. This means that males of the first species will have lost the hormone stimulation which made them responsive to females in May and will not be attracted to the sexually active females of the second species in June.

Higher Plant Species

Hybridization experiments are often used to distinguish between species in higher plants. Whenever two plants can be crossed and yield fertile offspring, they are generally considered to be of the same species. Geographical isolation, however, can prevent reproduction and the two will be considered different species, even though they can produce fertile hybrids when brought together. The season of reproduction can also provide a barrier between two species in the same locality. One species may release its pollen at one time of the year, and its female receptacles can be fertilized only at this season. Another species which is similar may release pollen at a later season, so there is no chance of hybridization.

Importance of Reproductive Isolation

Without reproductive isolation there could be no such thing as a species. Suppose cats could breed with rabbits and produce fertile offspring. Soon there would be no cats or rabbits but rather animals with various mixtures of the characteristics of each species. There are cases in which humans have sexual relations with such animals as cows, sheep, and dogs. If these matings resulted in fertile hybrids, the human species would gradually lost its identity. In time there would be no dogs, sheep, cattle, or people. In fact, all the mammals would be a conglomeration of individuals of almost every conceivable description.

We can see what would happen if there were no reproductive isolation from the example of mongrel dogs which appear when dog breeding is not controlled. The same blending of characteristics would occur between different species if reproductive isolation did not exist. These reproductive barriers have value because they permit each species to undergo genetic changes which make that species more adapted to its environment. New mutations and chromosome changes occur and become established where the breeding population is limited, and these changes make the species more likely to survive. Selection may favor one species over others if it adapts more successfully or more rapidly to a particular habitat.

Subdivsions of the species can occur when

there is some degree of reproductive isolation. The many breeds of dogs which have been derived from a common ancestor as a result of many generations of selection show how artificial control of mating can maintain breeds. Man has developed many varieties of corn from a wild ancestral plant, **teosinite.** These varieties can be maintained as long as they are not planted too close together. Popcorn, sweet corn, and flint corn will freely hybridize, however, if they are

A

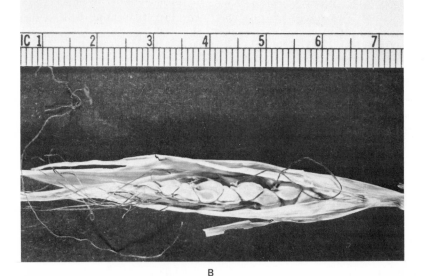

B

FIG. 12.3 Teosinte, wild corn. The small ears, only 4 cm long were the kind eaten by early man before the development of agriculture. Through cultivation and selection, we have established the modern varieties of corn which have a much better yield than the ancestral variety.

planted close enough for the pollen of one variety to reach the tassels of another. In nature subspecies develop when geographical isolation separates populations. The red-shouldered hawk, *Buteo lineatus,* exists in at least four subspecies in the United States. One lives east of the Rocky Mountains and north of the Gulf states. It is given a subspecies name of *lineatus,* so its complete name may be written, *Buteo lineatus lineatus.* The western subspecies is slightly smaller and more reddish in color and is called *B. l. elegans.* In the Gulf states the subspecies is smaller and paler in color from the eastern variety. It is called *B. l. alleni.* Still another subspecies is found in the Keys of Florida and is called *B. l. extimus.* At some points where the ranges of these subspecies overlap, there will be some inbreeding and the distinctions between them blend together. These groups are not different enough to be called different species, although if they remain separated for a long time, the differences might become much more distinct and they would then be classified as different species. The physical differences might be accompanied by genetic differences which would make hybridization impossible.

Use of Other Criteria

When organisms have no detectable forms of sexual reproduction, some other criteria must be used to determine species' classifications. Certain parasitic worms mate only within the intestine of their hosts and it is very difficult to try to cross two worms together outside their hosts' bodies. As a result, most of these worms are classified according to their morphological features and their host relationship. One species of round worm which infects the hog but will not infect man and another which infects man but will not infect the hog look almost identical, but the two species are reproductively isolated because they can never mate, since they are never found in the same intestine.

Some of the different species of bacteria look alike under the microscope, but they can be distinguished by their habits of growth on different types of media and by their pathogenicity when they enter certain animal bodies. Most of the microorganisms are classified by their variations in reaction to different nutrients and by other physiological distinctions which are indicative of genetic variation.

Organisms which have become extinct—and this includes the majority of the species which have inhabited the earth—cannot be hybridized to determine species distinctions. We have primarily fossil remains, from which we can determine many morphological distinctions which permit us to assign species names to the extinct organisms.

Artificial Hybridization

A hybrid is any offspring of parents which are genetically different. A cross between a cocker spaniel dog which is **particolored** (has white and colored spots) and one which is **solid colored** yields hybrid pups even though the parents are of the same breed and the same species. Only one gene is responsible for the variation of pigment distribution in the dogs' coats. A cross between a black Angus and a Holstein, two breeds of cattle, also produces a hybrid. These breeds differ with respect to many genes. Hybrids can also be obtained by crossing different species which are closely related. Horses and donkeys belong to different species, yet they cross readily and produce a hybrid, the mule. The mule is sterile, however. Domestic cattle can be crossed with the bison (American buffalo) to produce a cattalo, another sterile hybrid.

We may hear of more extreme and fictional hybrids. One example is a "cabbit"—a hybrid which is the result of a cross between a cat and a rabbit. Cats and rabbits belong not only to different species but also to different genera, different families, and even different orders. Such great differences in classification groups are indicative of extreme differences in genes and chromosomes and rule out any chance of successful hybridization, even with artificial insemination. There are even stories of gorillas capturing native women in Africa and carrying them off to the jungles for mating. Some unverified reports even tell of sightings of human-gorilla hybrids, but we can attribute such reports to sensational journalism rather than to actual events. The genetic differences between gorillas and humans would be too great for hybridization to occur. Dog-human

FIG. 12.4 The mule is a hybrid offspring of parents from two different species, the donkey and the horse. The mule exhibits hybrid vigor and has greater strength than either parent, but it is sterile because its chromosomes did not match properly during meiosis.

hybrids have even been reported in some sensation-seeking papers, but the genetic differences between these two species is even greater than between humans and gorillas. Jokes are often made about even more extreme hybrids. A typical one is the cross of an abalone and a crocodile by a man who hoped to get an abadile but ended up with a crockabalone.

Sometimes several varieties are included in complex hybridization to produce new varieties. On the great King ranch in south Texas, it was desirable to have a strain of cattle which would flourish on the vegetation of the area and which would be resistant to the heat and tropical diseases which are found there. At the same time, these cattle had to be good beef cattle. Using Brahman and shorthorn breeds, the ranchers conducted crosses and backcrosses while selecting those cattle which met the criteria best and came up with a very desirable breed for that region. It is known as the Santa-Gertrudis breed and it is now being raised in many parts of the world.

Hybridization is widely used by plant breeders to obtain new and improved varieties. A cross between a grapefruit and a tangerine, for instance, created the **tangelo,** which pleasantly combines the sweetness of the tangerine with the tartness of the grapefruit. Most commercial flowers and vegetables used today are hybrids. Hybrids are generally more sturdy and more productive than their parents. Without hybrid corn we could not supply our needs today. Sometimes hybrids are bred from hybrids, so that the final hybrids have actually come from four varieties, which is shown in Fig. 12.6

Hybrid Vigor

The hybrid between two different varieties or species is frequently more vigorous and stronger than either of the parents. A possible reason for this is that hybrids tend to express the most desirable traits of both parents. Most of the

FIG. 12.5 Santa Gertrudis cattle on the King ranch in Texas. These cattle were bred by hybridization of the Brahman and shorthorn breeds. The desired characteristics of each breed were established by interbreeding and selection.

traits which contribute to strength and vigor are the result of the action of dominant genes, and the less desirable traits result from the expression of recessive genes. In a hybrid the dominant, good characteristics of both parents would tend to be expressed, while the recessive, bad characteristics are covered. Close inbreeding has just the opposite effect. The mating of closely related animals tends to bring out many harmful traits. When an animal breeder crosses the equivalent of brothers and sisters, he will find many defective offspring because the harmful recessive genes in their common ancestry stand a greater chance of being expressed. Marriages between human cousins have a much higher proportion of defective children than those from nonrelated marriages. Cousins are more likely to carry the same harmful recessive genes which can come together in their children.

Heterozygote superiority is another possible explanation for the vigor which characterizes many hybrids. Sometimes an organism is more vigorous when it has genes which are different, even though some of them may be for

harmful traits. The human gene for sickle-cell anemia is harmful when a person has two of these

FIG. 12.6 The tangelo at the bottom is a hybrid produced by crossing a grapefruit with a tangerine. It combines the juiciness and tartness of the grapefruit with the sweetness and thin skin of the tangerine.

A   B

FIG. 12.7 Crossing different varieties of corn produces hybrid plants which grow taller and produce larger ears. The hybrid's characteristics are in between those of two parental varieties.

genes, but if he has only one abnormal gene and one normal gene he has a superior chance of surviving if he lives in a region in which malaria is prevalent. Hence, those organisms which have parents of different genetic backgrounds may enjoy a superiority over those with parents of the same genetic stock.

Hybrid vigor does not last through succeeding generations. A farmer must buy newly created hybrid seed each year to maintain a maximum yield from his corn crop. If he tries to beat the game by saving seed from the hybrids, he will find a gradual decline in yield each year for about seven years because the genes will become more alike each year. Fig. 12.6 shows how the height of stalks declines with continued plantings for seven years. At this point the height

of the stalks stabilizes but this corn produces a much lower yield than either parental variety.

When plants are propagated asexually, the hybrid vigor can be maintained by using budding, grafting, rooting, and bulbs to keep the hybrid gene combination from generation to generation. Even hybrids between different species can be maintained this way.

It is possible to create new plant species by hybridization. By treating the seedlings with a chemical which causes a doubling of the chromosome number, a hybrid between two different species which otherwise would be sterile will produce fertile seed, and this seed can be used to start a new species. **Karpechencho,** a Russian botanist, was one of the first to do this. He crossed a radish with a cabbage. The hybrid was fully fertile with others like itself but sterile when crossed back to either the radish or cabbage. A more practical hybridization was performed by this technique recently in the United States. A new species of grain, *Triticale*, has been produced by hybridizing wheat and rye. Such grain has a high yield and other desirable features which may help solve some of the world's food problems.

This method works well for plants because the chemical can be applied to the growing tip of the plant from which all new growth comes. It cannot be used for animals because animals grow all over their bodies and the chemical is too poisonous to be applied to all tissues.

Genetic Basis of Speciation

All populations are constantly changing genetically. The genes undergo modification through mutation and selection. When a small population is separated from the main population and lives under different conditions for a long time, these changes can add up to a considerable divergence. At first the changes will be minor and will result only in different varieties of the same species, but with continued modifications, the populations will in time become so different that they may be classified as different species.

Use of Serology

Serology is a method of determining the similarities of proteins in closely related species. Since genes code the production of proteins, protein differences are indicative of gene differences. Serology is possible because vertebrate animals produce antibodies in their blood plasma which will react with foreign proteins which may be introduced into the body. The part of the protein which stimulates antibody production is known as an antigen. Different proteins have different antigens, so antibodies react primarily with the antigens which stimulate their production.

Serology is one of the best indicators of the closeness of taxonomic relationships. As an example, some of the serum (defibrinated plasma) from a dog can be injected into a rabbit and the rabbit will respond by producing antibodies to react with the proteins in the dog serum. After about ten days or longer, to allow time for full antibody production, some of the rabbit serum can be removed and mixed with dog serum. A cloudiness will appear in the mixture as the proteins in the dog serum are precipitated. The antibodies have caused the protein molecules to aggregate into large groups which precipitate out of the mixture and become visible as small specks. The precipitation reaction causes the cloudiness. If rabbit serum which has developed antibodies to dog proteins is mixed with serum from a wolf, some cloudiness will form, but not as much as when dog serum is mixed with the rabbit serum. These reactions show that the wolf has plasma proteins which are similar, but not identical, to those of the dog. If we assume a common ancestor from which these two species have diverged, we then know that mutations have caused slight differences in the amino acid sequence of the proteins of the dog and the wolf. If serum containing antibodies to dog proteins is mixed with serum from a fox, less cloudiness will appear, so we can conclude that the fox and the dog have been separate species for a longer time than have the wolf and the dog and have developed more gene differences. Mixing serum from other families or order, such as cats, cows, or monkeys, with serum containing antibodies to dog proteins produces no precipitation at all and points out the lack of close relationship between these groups.

The method can be used in other species. The entire bodies of one kind of fruit fly, for instance, can be ground up and injected into a rabbit. This rabbit serum can then be mixed with the ground bodies of other kinds of fruit flies.

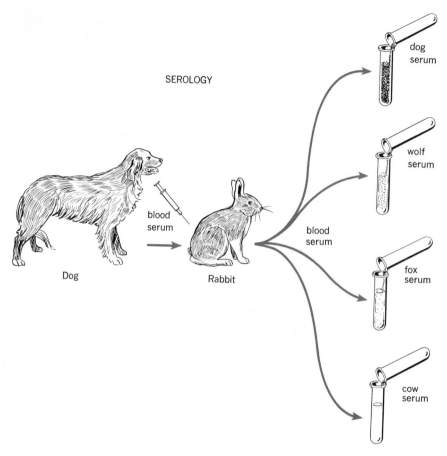

SEROLOGY

blood serum

Dog

Rabbit

blood serum

dog serum

wolf serum

fox serum

cow serum

FIG. 12.8 The classification of a dog. Note how members of successive groups become more and more alike and how each group includes fewer animals as you proceed in the classification.

The degree of reaction between the serum and the fruit fly bodies indicates the closeness of relationship between the different kinds of fruit flies.

DIFFICULTIES IN TAXONOMY

When Linnaeus proposed his system of classification, he accepted the belief of the time that all species were fixed and unchanging, existing then in the same form in which they had been created. He further assumed that they all fell into natural groups and that man had only to discover these groups. Since that time, however, we have learned that living things do change, and over a long period of time the changes can be very extensive. As a result, species do not always fall neatly into man-made taxonomic units. Taxonomists have had many headaches as they try to arrange them in a logical order. Just as they think they have established criteria for one group, some organisms are found which do not fit exactly, and subgroups or new groups must be established.

When the class Mammalia was first established, certain characteristics were used to identify members of his class. These included mammary glands to nourish the young, hair, and live birth. Then a strange creature was found in Australia which had hair and produced milk to nourish its young like a mammal, but which laid eggs and incubated them like a bird. The duckbill platypus, as it is called, was placed in the Mammalia class, in a separate subclass from the other mammals which give live birth to their young.

More difficulties arose when the microscope was developed and its use revealed a great host of organisms which had never been seen before. Some of these microscopic organisms did not fit readily into either the plant or animal kingdom. For example, the one-celled water organism *Euglena* has green chlorophyll like a plant, but it moves about freely and absorbs and digests food like an animal. Some botany books listed it as a plant—one of the green algae—while some zoology books listed it as an animal—one of the protozoa. The slime molds created another dilemma. At one stage of its life, these molds look like a mass of jelly. They creep among decaying wood and leaves, digesting and absorbing food very much like the amoeba, which was considered an animal. When time for reproduction arrives, however, the slime molds dry out, and from their remains, sporangia arise and release spores into the air which start the life cycle over again in a new location. This form of reproduction is a common feature of the fungi, which were placed in the plant kingdom at that time.

In an attempt to establish a grouping which would provide a place for these organisms, some systems of taxonomy which enlarged the number of kingdoms were devised. This has resulted in confusion, because while some biologists still prefer the two kingdom system, others prefer to enlarge the classification system to include three, four, and five kingdoms. Thus, taxonomy becomes more complex as we learn more about living things. These questions are primarily of concern

FIG. 12.9 The duckbill platypus, shown swimming under water, has been a taxonomic puzzle. Although it possesses fur and nourishes its young with milk like mammals, it lays eggs and incubates them like birds.

FIG. 12.10 A slime mold crawling on a tree trunk appears to make animallike responses, but when it is ready for reproduction it produces sporangia which are similar to those produced by some fungi. Classification of the slime mold was difficult until new kingdoms were established to accommodate such organisms.

to the specialists in the various fields of biology, however, and should be of little concern to the beginning student. Nevertheless, in order to show the methods used and to provide a reference, the following chapter will list some of the major taxonomic units and the kinds of organisms included in each.

REVIEW QUESTIONS AND PROBLEMS

1. Why was the system of classification proposed by Linnaeus a better indication of true relationships than the systems previously used?
2. The following are classification groups: family, genus, kingdom, order, phylum, species, and class. Rearrange these in correct sequence beginning with the largest group.
3. Give the advantages of scientific names over common names in scientific reports.
4. Genus names are sometimes used alone, but species names must always be used in conjunction with a genus. Explain why.
5. Why is reproductive isolation, which exists between most species, so important in the maintenance of taxonomic groups?
6. Suppose you were given two cages containing little ratlike animals which were similar in many respects, but had characteristic differences. How would you determine if the two were the same or different species, assuming you had plenty of time?
7. Why does a farmer continue to buy expensive corn seed each year when he could save seed from his crop of the year before? Why does he get a better yield by buying new seed?
8. Why is the mule sterile?
9. Why are more defective offspring born to parents who are first cousins than to those who are not so closely related?
10. New plant species can be created by the hybridization of existing species, but this method cannot be extended to animals. Explain why.
11. Why do we say that serological differences are indicative of genetic differences?
12. Why do modern taxonomists have problems in classification which were not encountered by Linnaeus?

FURTHER READING

Benson, L. 1957. *Plant Classification*. Boston: D. C. Heath.

Cain, J. A. 1971. *Animal Species and their Evolution*. 3rd ed. New York: Hutchinson University Press.

Jaques, H. E. 1946. *Living Things: How to Know Them*. Dubuque, Iowa: W. C. Brown.

Lawrence, G. H. M. 1951. *Taxonomy of Vascular Plants*. New York: Macmillan.

Mayr, E. 1969. *Principles of Systematic Zoology.* New York: McGraw-Hill.

Simpson, G. G. 1961. *Principles of Animal Taxonomy.* New York: Columbia University Press.

Stanier, R. Y., et al. 1970. *The Microbial World.* 3rd ed. Englewood Cliffs, N. J.: Prentice-Hall.

13

The Kingdoms of Living Things

The living things on the earth are not always organized into neat groups which can be easily arranged into taxonomic units. Just when it seems as if the criteria for a unit have been well defined, some new organism is found which has some characteristics of one unit and some of another. What can we do? We can place it into one or the other with the notation that it is a transitional form which lacks some of the features of others in the group. Or we can create an entirely new taxonomic group for these new forms. Biologists may be classified as lumpers and splitters. The lumpers tend to keep the number of taxonomic groups down by including more in each group, even though there may be some diversity and exceptions in each. The splitters tend to add more taxonomic groups for those which do not fit exactly into existing groups. This results in increasingly complex classification systems. Since there is no complete agreement on the proper extent of lumping or splitting, there is some variation in all the systems of classification which have been proposed. Since those systems which are accepted by most are in a state of flux and will certainly be changed in the future, it is not possible in a textbook to present a system which is universally acceptable. Still, the classification of organisms is such an important part of biology that the method which seems to be most widely accepted at this time will be presented. This system is based on five kingdoms plus a separate category for those entities known as viruses. It is not suggested that you try to memorize all these groups, but you should become familiar with the kingdoms and some of the larger subgroups and the kinds of organisms included in each. Then you can refer back to this chapter when questions on classification arise. An aid to pronunciation of the group names is also included.

THE VIRUSES

The viruses are not accepted as living organisms by some biologists, even though they have some of the characteristics of living matter. They are not placed in a kingdom, therefore, but they must be listed somewhere. Viruses can cause serious infections in many cellular forms of life. They can grow and multiply only in living cells, so most of them are parasitic, but a few viruses have established a pattern of growth which does not harm the host. Viruses consist primarily of a nucleic acid core of DNA or RNA which is surrounded by a protein coat, although some have additional compounds. Three types of viruses are recognized according to the hosts which they infect.

1. **Bacteriophages** which infect bacteria.
2. **Higher plant viruses** which infect plants primarily.
3. **Higher animal viruses** which infect primarily vertebrate animals, although some infect the arthropods.

KINGDOM MONERA

The kingdom **Monera** (mow-nee'-ruh) includes the most primitive organisms which have cellular organization. They are **prokaryotic,** so they do not have a definite nucleus, but they do have a nuclear area which contains a single long, circular strand of DNA which some call a chromosome but which others call a **genophore** because proteins as well as DNA are included in the chromosomes of eukaryotic cells. In addition, the Monera lack some of the cellular organelles, such as mitochondria and lysosomes, which are found in eukaryotic cells. They also lack centrioles and do not form a spindle figure. Instead, the circular genophore replicates and one duplicate goes to each daughter cell. Some monerans have chlorophyll, but the chlorophyll is not contained within chloroplasts.

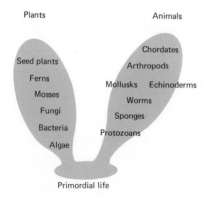
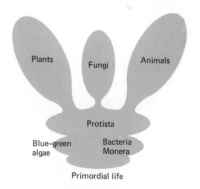

FIG. 13.1 Two methods of classification. One puts all organisms into the plant and animal kingdoms, while a second, recently proposed method uses five kingdoms.

Phylum Schizomycetes

(skit′-zo-my-see′-teez)

Class Myxobacteria (mix′-oh-back-ter′-i-uh), the slime bacteria. These have a layer of slime outside the cell and they often form a slippery coating over damp, shaded surfaces.

Class Rickettsiae (rick-et′-si-uh). These organisms are very small spheres or rods which seem to lie between viruses and bacteria. They are parasites on arthropods such as fleas, lice, and ticks but some can be spread to humans.

Phylum Schizophyta

(skitz-of′-i-tuh), the bacteria

Class Eubacteria (you-back-ter′-i-uh), the true bacteria. These are one-celled organisms which reproduce by fission. Many Eubacteria are parasitic and can cause serious diseases in man and other animals and in some plants. Many are **saprophytic** and live off the remains of dead organisms. A few can manufacture their own food.

Class Actinomycota (ak-tin′-oh-my-cot′-uh). These organisms are filamentous forms which resemble certain fungi. They reproduce by fragmentation and typically live in the soil. They cause some diseases. This class includes *Streptomyces,* which is the source of some of our antibiotics, such as streptomycin and aureomycin.

Class Spirochaetae (spy′-row-keet′-ee). These are spiral-shaped forms which can cause some of man's most serious infections, such as syphilis. Some are free-living forms which are found in stagnant water.

Phylum Cyanophyta

(sy′-an-of′-i-tuh), the blue-green algae

Although these algae are called blue-green, some may be yellow or red because they contain other pigments. They are one-celled, but some adhere to each other after division to form filaments. Others produce a slime layer which holds the cells together in clusters after division.

KINGDOM PROTISTA

The kingdom **Protista** (pro-tiss′-tuh) includes both unicellular and multi-cellular species. The Protista are eukaryotic, have a definite nucleus bounded by a membrane, and have chromosomes which include protein as well as DNA. They have mitochondria and some have chlorophyll contained in chloroplasts.

Phylum Euglenophyta

(ewe′-glee-nof′-i-tuh)

These organisms are small cells which have one or two hairlike flagella. They may absorb soluble food or they may engulf larger food particles. Most have chlorophyll and can manufacture their own food, so they can live in either light or dark. *Euglena* is a typical genus with chlorophyll. An *Astasia* is similar in appearance but lacks chlorophyll.

Phylum Pyrrophyta

(py-rof′-i-tuh), flame colored algae

These algae have a brown pigment in addition to chlorophyll. Some are called **dinoflagellates,** a group which includes *Gymnodinium,* the genus which sometimes multiplies in such large numbers that it causes large rust-colored areas in the sea. *Gymnodinium* are the cause of the red tide which appears off the Gulf coast of Florida at times.

Phylum Crysophyta

(cry-sof′-i-tuh), golden-yellow algae

These have yellow pigment in addition to chlorophyll. This phylum includes the algae which form a yellowish scum on the surface of stagnant water. They are widespread in fresh-water and marine habitats. This phylum includes the diatoms which are present in enormous numbers in nearly all waters of the world. Diatoms are an important source of oxygen in the atmosphere. Members of the phylum have silicon, glasslike shells.

Phylum Chlorophyta

(klor-of′-i-tuh), green algae

The green algae are abundant in fresh and salt water and they also grow on shaded, moist soil. This phylum includes single-celled forms, filamentous forms, and large colonies which have specialization of different cells.

Phylum Phaeophyta

(fay-of′-i-tuh), brown algae

These organisms are brown in color because they contain a red pigment which blends with the green of chylorophyll. Most of the Phaeophyta are large marine algae. This phylum includes most of the seaweeds.

Phylum Rhodophyta

(row-dof′-i-tuh), red algae

Considerable quantities of red pigment give many of these algae a bright red color. Most Rhodophyta are marine. Many species have beautiful branched bodies.

Phylum Protozoa

(pro-tow-zoh′-uh)

These organisms are one-celled forms without chlorophyll. They have animallike methods of nutrition and they show a high degree of specialization. They have many organelles for feeding, excretion, water balance, and protection. This phylum includes over 30,000 species, some of which are the most serious infectious organisms that affect man. Many are free living in both fresh and salt water.

Class Mastigophora (mas′-tig-of′-oh-ruh) or **Flagellata** (fla-jel-ah′-tuh). Organisms of this class have at least one flagellum for locomation. They may live in foul water and in moist soil. Members of this class include some serious human parasites, such as tripanosomes.

Class Sarcodina (sark′-oh-dee′-nuh). Sometimes these organisms are called blobs of glob. These protozoans can change their shape and move by means of pseudopodia, or projections from the cell. *Amoeba* is an example.

Class Sporozoa (spoh′-row-zoh′-uh). These are parasitic protozoans which produce sporelike reproductive bodies at some time in their life cycle. This class includes the malarial parasite.

Class Ciliata (sil′-i-ah′-tuh). Members of this class bear many tiny cilia which are used in locomotion. In some, such as *Paramecium*, the cilia are distributed over the entire cell, but in others, such as *Vorticella* and *Didinium*, they are localized in rings.

KINGDOM FUNGI

The kingdom **Fungi** (fun′-jy), singular fungus (fun′-gus), includes some highly developed organisms. The Fungi are sometimes included in the plant kingdom, even though they lack chlorophyll, or they are placed with the Protista in some systems of classification. They range in size from one-celled organisms to very large forms.

Phylum Myxomycota

(mix'-oh-my-kot'-uh), slime molds

Slime molds are found in moist soil, wood, feces, and decaying vegetation. The typical form is an acellular mass of slime, a **plasmodium,** which can creep about like a giant amoeba. Slime molds engulf food particles. They tend to move away from light until the time of reproduction when they come out into light and dry out. Sporangia are produced from the dried mass.

Phylum Eumycota

(ewe'-my-kot'-uh), true fungi

Class Phycomycetes (fy'-koh-my-see'-teez), algae-like fungi. These generally have long filaments. Some, like *Saprolegnia*, live in fresh water and may parasitize fish. Many of the molds are found in this group. A typical genus is *Rhizopus,* the black bread mold.

Class Ascomycetes (as'-koh-my-see'-teez), sac fungi. These fungi have spore sacs containing eight spores typically. Yeasts are one-celled examples. *Penicillium,* which belongs to this class, is multicellular blue-green citrus fruit mold.

Class Basidiomycetes (buh-sid'-i-oh-my-see'-teez), club fungi. These fungi bear spores on clubs called **basidia.** This class includes mushrooms, puffballs, and many wood-decaying forms. Rusts which live on grains are members of this class.

KINGDOM PLANTAE

This kingdom is sometimes listed as **Phyta** or **Metaphyta.** It includes the plants with 250,000 known species alive today and countless others which have become extinct and are found only as fossils. Most are multicellular and have **eukaryotic** cells. Most have chlorophyll in plastids. Some call the subgroups divisions, but we will use the term phyla to be consistent.

Phylum Bryophyta

(bry-of'-i-tuh), liverworts and mosses
Class Hepaticae (hep-at'-i-see), the liverworts. These have flat, lobed bodies.
Class Musci (mus'-see), the mosses. These have root, stem, and leaf organization, but they do not have the conducting tissues characteristic of true roots, stems, and leaves.

Phylum Tracheophyta

(tray'-kee-of'-i-tuh), vascular plants

These plants have small tubes in a continuous system which conduct water and dissolved substances. This phylum includes the large land plants.
Subphylum Psilopsida (sigh-lop'-sid-uh). These have bare stems which come up from up from underground rhizomes. There are only a few living species but many fossils.
Subphylum Lycopsida (ly-kop'-sid-uh). These have upright stems with small spirally arranged leaves. They were large treelike plants in the carboniferous period and furnished much of the deposit for coal. Only four genera remain
Subphylum Sphenopsida (fee-nop'-sid-uh). The horsetail subphylum includes one living genus and many coal-age fossils.
Subphylum Pteropsida (ter-op'-sid-uh). The ferns and seed plants are the most abundant of the tracheophytes. They have leaves which are well developed.
Class Filicineae (fee-li-sin'-i-ee). These are true ferns with broad fronds above ground and underground stem. They reproduce by spores.

Class Gymnospermae (jim'-no-sperm'-ee). This class includes the many coniferous trees which bear naked seeds on some sort of cone.

Class Angiospermae (an'-ji-oh-sperm'-ee). Commonly called the flowering plants, these plants have enclosed seeds.

KINGDOM ANIMALIA

Sometimes called the **Metazoa,** this kingdom includes those forms which we commonly call animals.

Phylum Porifera

(poh-rif'-er-uh), the sponges

These organisms have many tiny pores on their bodies. There is no differentiation into distinct organs.

Phylum Coelenterata

(see-len'-ter-ay'-tuh) or **Cnidaria** (nee-day'-i-uh)

These are usually cylindrical animals which have stinging cells on tentacles. Examples of this phylum are jellyfish, sea anemones, and corals.

Phylum Platyhelminthes

(plat'-i-hel-min'-theez), flatworms

This phylum includes many freshwater forms, such as *Planaria,* and many parasites, such as tapeworms.

Phylum Aschelminthes

(ay'-shel-min'-theez), roundworms and rotifers

This phylum includes many roundworm parasites which infect man, as well as freshwater, saltwater, and soil organisms. It is a very abundant phylum. It may also be called **Nemathelminthes.**

Phylum Annelida

(an-nel'-i-duh), segmented worms

The bodies of organisms in this phylum are formed like many small rings, joined together. The earthworm, sandworm, and leech are members of this phylum.

Phylum Acanthocephala

(ay-kan'-tho-self'-uh-luh), spiny-headed worms

It includes many vertebrate parasites.

Phylum Mollusca

(mol-usk'-kuh), the mollusks, often called shellfish

This group includes oysters, clams, snails, squid, and octopus.

Phylum Ctenophora

(ten-of'-oh-ruh), comb jellies or sea walnuts

Phylum Bryozoa

(bry'-oh-zo'-uh), moss animals

Phylum Nemertea

(nee-mer'-tee-uh), ribbon worms

Phylum Onychophora

(on'-i-ko-fore'uh), segmented animal with claws on its legs

There appears to be a link with annelids and higher forms. A member of this phylum is *Peripatus.*

Phylum Arthropoda

(ar-throp'-oh-duh), animals having jointed appendages

This phylum has more species than all the others. It is a very important phylum economically. It includes spiders, scorpions, insects, centipedes, millipedes, crabs, shrimp, and many others.

Phylum Chordata

(kor-dot'-uh), the chordates

Organisms in this phylum have a notochord. **Subphylum Hemichorda** (him'-o-kord'-uh), acorn worms with notochord only at anterior end.

Subphylum Urochorda (your'-oh-kord'-uh), sea squirts. They have a full notochord as larvae, but they become soft bags as adults.

Subphylum Cephalochorda (self'-uh-low-kord'-duh), *Amphioxus*. This organism has a full notochord extending into the head.

Subphylum Vertebrata (vur'tee-bra'-tuh), the vertebrates. This subphylum includes the great majority of the chordates. Members of Vertebrata have a vertebral column of joints of cartilage or bone and have a well-developed brain. This group includes the most intelligent animals.

14

The Viruses

Viruses are the orphans of the taxonomic system. They are left out of the established kingdoms because they seem to lie somewhere between the world of the living and the nonliving. They consist primarily of a central core of nucleic acid which may be either DNA or RNA but not both as a rule. It has been discovered recently, however, that *Vaccina*, cowpox virus, does have both of these nucleic acids, as do cellular forms of life. In addition, a few viruses also have a protein-lipid membrane on the outside of the coat and a few include complex polysaccharides in association with the coat. Viruses are all parasitic in nature and can grow and multiply only when they are in cells of the proper type. In some instances, they can be reduced to dry crystals and still retain their infective power for many years. With all these properties, they can hardly be called nonliving, so we either must recognize them as living, or as lying in a twilight zone somewhere between living and nonliving. We can refer to viruses as biological entities without making a decision either way.

DISCOVERY OF VIRUSES

Filterable Infectious Agents

During the latter part of the nineteenth century, **Robert Koch** and **Louis Pasteur** found that many diseases were caused by microorganisms which could be spread from one infected organism to another. They were able to isolate bacteria from people and other animals suffering from a disease and culture these bacteria in nutrient media in the laboratory. Koch and Pasteur found that each type of disease they studied was caused by a particular kind of bacterium and they identified the bacteria by genus and species. In the case of smallpox, however, microbiologists could not find any bacteria which were associated with the disease. Pasteur also failed to find any bacteria in the saliva of rabid dogs, yet when this saliva was injected into laboratory animals, they developed rabies. Pasteur assumed that the bacteria must be present but that they were too small to be seen under his microscope. He also found that it was impossible to develop any sort of culture medium upon which rabies or smallpox organisms could be grown and he could maintain these organisms only by growing them in living

animals. He grew the rabies agent in rabbits, for instance, and in this way was able to develop a vaccine for rabies.

In 1882 **Dmitry Iwanosky,** a Russian, was trying to find the cause of the **tobacco mosaic disease.** In this disease, the leaves of tobacco plants would develop transparent spots. He suspected that these spots might be caused by a bacterial disease similar to the bacterial diseases found in animals by Koch and Pasteur. He found that if he squeezed juice from leaves with the mosaic spots and placed the juice on healthy leaves, the mosaic disease would appear. When he examined this juice under the microscope, however, there were no signs of bacteria. Iwanosky thought that the organisms involved must be too small to be seen, so he tried to remove them by passing the leaf juice through a fine porcelain filter. This method had previously proved effec-

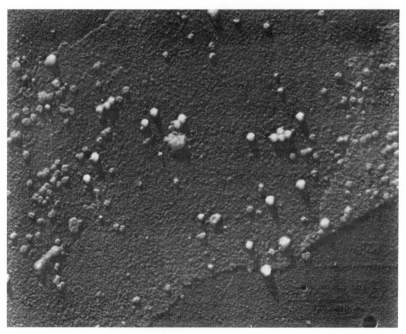

FIG. 14.1 Virus particles attached to a cell membrane. Viruses inject their nucleic acid core into the cell, but the protein coat will remain outside the membrane.

(Electron photomicrograph by Keith Porter, Rockefeller University.)

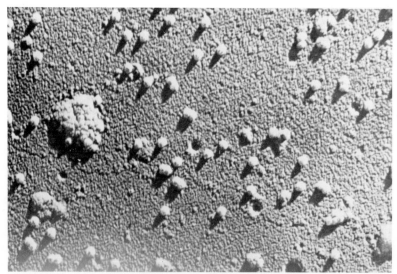

FIG. 14.2 Virus particles which cause polio, seen with the electron microscope. It was only proven that viruses were particles after the development of the electron microscope, since viruses are so small that they cannot be seen with an optical microscope.

tive in removing bacteria from liquids. To his surprise, however, the juice remained infective after being passed through the filter.

Discovery of Virus Particles

It was discovered that other diseases of higher plants and animals were caused by similar agents and these became known as the **filterable viruses.** Speculation about the nature of these ubiquitous entities arose. Some suggested that they must be tiny bacterialike organisms which could pass through the filters, but all attempts to culture them on nutrient media, which were effective for bacteria, failed. Others suggested that they might be nonliving chemicals which had the power of replication when placed in certain living organisms. Support for this concept came when some of the juice from infected tobacco plants was evaporated and crystals were observed. These crystals could be kept in a jar on the shelf for years and they would still cause infection. One jar of tobacco mosaic virus which was produced in 1935 is still infectious today. The puzzle surrounding filterable viruses was solved during the 1930's when the electron microscope was developed. With this microscope virus particles could be seen. Virus particles had distinctive shapes by which they could be distinguished. Each distinctive virus was called a **viron.**

Naming Viruses

Since viruses do not have all the properties of cellular life, they have not been given genus and species names, but they are generally named for the diseases which they cause. There is the tobacco mosaic virus, the smallpox virus, and the polio virus. Some microbiologists have created confusion, however, by giving viruses names similar to the genus and species names given to other organisms. The fever blister virus is known as *Herpes simplex* and its close relative, which causes chickenpox in children and shingles in adults, is *Herpes zoster*. Another close relative is *Herpes keratitis*, which infects the eyes and is a major cause of blindness. Still another, which causes infection of the genital region is known as *Herpes genitalis*.

THE NATURE OF VIRUSES

The Bacteriophages

In recent years we have learned much about viruses by using chemical and genetic techniques,

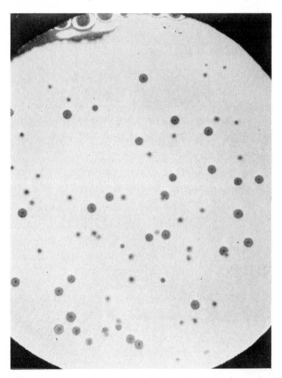

FIG. 14.3 Lysis of a culture of bacteria, taken from the human colon, by the T₂ bacteriophage. Many bacteria were placed on this plate of nutrient agar and they grew into a confluent lawn. When phage particles were placed on the bacterial growth, however, they caused lysis of many bacteria. Each phage particle is surrounded by a circular plaque of lysed bacteria.

as well as by observation of viruses through the electron microscope. Viruses are all composed of an outer protein coat surrounding a nucleic acid core. Much of our knowledge of viruses has come through the study of viruses which infect bacteria. The first virus of this type was discovered in 1915 when **Frederick Twort,** an English microbiologist, found that some of the bacteria he was culturing were dying in certain spots on the culture medium. The culture would be growing well, and then the bacteria in a circular area would be **lysed,** or broken open, leaving a liquid behind. When this liquid was transferred to a healthy culture, it would cause a similar lysis. The liquid remained infective even after it was filtered through a porcelain filter, so it appeared that the death of the bacteria was caused by a virus. Two years later, **Felix d'Herelle,** a Canadian, did extensive work on this virus and gave it the name **bacteriophage,** a word which means "bacteria eaters." The word is often shortened to **phage** by those working with them.

Detailed study of the phage viron has shown that it has a hexagonal head with an outer covering of protein and an inner core of double-stranded DNA along with a little protein which probably helps hold the DNA together. The DNA is present as a single long molecule containing enough information to produce about one hundred different kinds of protein molecules. Extending from the viron head is a tail, made of protein, which terminates in six spikes. All together, this viron looks very much like a space vehicle designed to land on the moon. It is incapable of motion under its own power, but it can be carried about by the movement of liquids generated by the processes of diffusion, osmosis, and thermal gradients.

Most studies on phages have been done on *Escherichia coli,* which is a very abundant bacterium in the human intestine. Different strains of the phage have been identified and designated as T-1, T-2, and so on. Those with even numbers are highly virulent and are known as the T-even phages. The T-2 phage has been investigated extensively. The bacterium has protein sites on its surface which are recognized by the tail fibers of the phage. Other bacterial species do not have the appropriate receptor sites and will not be infected by this virus. Upon contact, the tail fibers contract and bring the end plate of the tail down against the surface of the bacterium. The nucleic acid contents of the phage are then injected into the bacterium. The tail of the virus is somewhat like a tranquilizing dart which is shot into wild animals. When a dart hits an animal, the contents of the syringe are injected.

The bacterium has a single, long, circular molecule of DNA which may be called a chromosome. Because it does not have the protein component which characterize chromosomes in more highly developed organisms, some prefer to call this DNA a **genophore**. The DNA of the bacteriophage also forms into a circle after it enters the cell and may be called a chromosome or a genophore. The DNA of the phage genophore begins to produce m-RNA, which codes the production of phage protein at the site of the bacterial ribosomes. One of the first phage proteins to be synthesized is an enzyme which attacks and destroys the DNA of the bacterium. The DNA of the phage has a slightly modified form of cytosine, which apparently protects it from degradation by this enzyme. From this point on, therefore, all metabolic processes in the bacterium are directed by m-RNA produced by the phage DNA and only phage proteins are synthesized. The phage DNA undergoes repeated replication and each molecule forms a mass known as a **prophage**. The prophages have no protein coats, however, and are not infective. If the cell is broken open at this stage, these prophages cannot enter other cells. About one hundred or more new phage particles develop within a single bacterial cell. At this time the phage DNA directs the production of a lysogenic enzyme, one which dissolves the outer

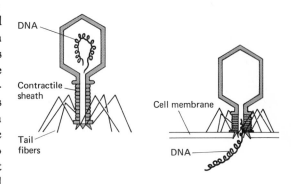

FIG. 14.4 Diagrammatic representation of the T₂ phage. The T₂ phage looks somewhat like a space vehicle ready to land on the moon (left). It is also shown after making contact with the outer cell membrane (right). As the neck collapses, its DNA is injected into the cell.

membrane of the cell and allows the contents to spill out. This cell breakage is known as **lysis.** Each phage viron is now free to infect other bacterial cells which it may contact.

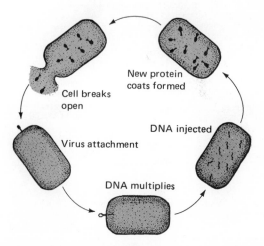

FIG. 14.5 The cycle of infection of a bacterium by the T₂ phage.

The Tobacco Mosaic Virus

Viruses can infect most of the flowering plants, the angiosperms, but very few of the cone-bearing or lower plant forms. The tobacco mosaic virus is typical of the higher plant viruses and can be used to illustrate this group. The tobacco mosaic virus is much simpler in structure than a bacteriophage. A viron of the tobacco mosaic virus is a single rod with an outer protein coat and an inner core of nucleic acid. At the time of infection, the entire viron may work its way into the cell, but the nucleic acid core alone is sufficient to cause infection. The nucleic acid is RNA rather than DNA. The two components of the viron, the coat and the core, can be separated by osmotic shock, by placing the virus in alternate hypotonic and hypertonic solutions. Then the two viral components can be recombined. It is even possible to combine the coats of one strain with the core of another strain. Doing this has shown that the viron coat determines the type of cell which the viron will infect. The protein coat must match the receptor sites on the cell wall if infection is to occur. The viral core determines genetic traits, such as the degree of virulence and rapidity of multiplication.

The higher plant viruses can be spread from cell to cell without the need for entry through the outer cell wall. Many higher plants have cytoplasmic connections between cells, and the prophage particles as well as the fully formed virons can pass from one cell to another through the natural flow of cytoplasm across these connections.

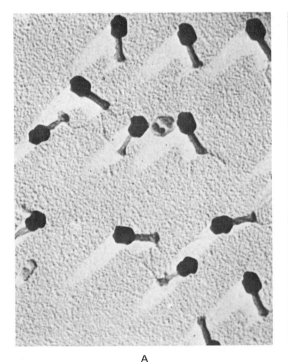
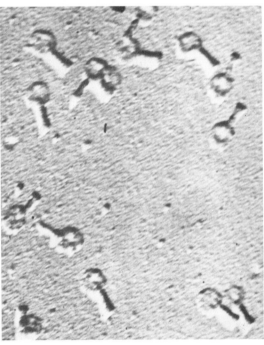

A

B

FIG. 14.6 Photographs of intact bacteriophage particles (left) and particles which have received osmotic shock (right). The contraction from the shock has caused the bacteriophages to eject their nucleic acid cores, leaving only the protein coats behind. (Courtesy of R. N. Herriot, Johns Hopkins University.)

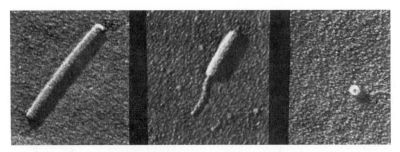

FIG. 14.7 Tobacco mosaic virus. An intact particle is shown at left. In the center photograph the nucleic acid core is coming out of the protein coat. An end view showing the hollow core after the extrusion of RNA is shown at right. (Courtesy of Wendell Stanley, University of California.)

RNA Replication

One important problem faced by those doing research with the tobacco mosaic virus was to discover the method of RNA replication. No cases of RNA replication have been found in cellular forms of life. Some investigations indicate that the virus RNA can direct the production of an enzyme, **reverse transcriptase,** and that this enzyme causes a reversal of the usual transcription of RNA from DNA. Instead, DNA is produced from RNA. This DNA can then produce m-RNA, which directs the production of viral protein within the cell. In time it can also produce the RNA prophages which serve as the core which fits inside the newly forming protein coats. When the cell is lysed, the viral DNA is lost, but the RNA can make more DNA when it enters another cell.

Viruses of Higher Animals

Many diseases of higher animals are caused by virus infections. Some human examples are measles, mumps, smallpox, influenza, rabies, polio, encephalitis, chickenpox, fever blisters, shingles, hepatitis, mononucleosis, and the common cold. You can see that these diseases include some of the most serious ones which affect us. Both DNA and RNA viruses are involved in higher animal infections. The method of replication and cell lysis is about the same as for the phages and plant viruses, although there are a few which can leave an infected cell without destroying it. These include the DNA type of herpes virus and the RNA type of influenza virus. These viruses tend to bud off from the outer part of the cell, carrying with them a surrounding envelope of the plasma membrane. When they infect a new cell, this outer membrane fuses with the plasma membrane of the host and the virus particles are inside the cell.

The higher animal viruses are not only specific about the species which they will infect, but they are specific about the type of cells which they will infect. The hepatitis virus attacks primarily cells in the liver, the encephalitis virus enters only certain types of nerve cells, and the common cold virus infects the cells lining the nasal cavity, the sinuses, and the throat.

RESISTANCE TO VIRUS INFECTION

Antibody Production

Virus infections induce antibody production in vertebrate animals. These antibodies react with the protein coat of the virons and prevents the virons from penetrating cells. Antibody production not only brings an infection under control but also provides immunity to future infections. The period of immunity varies with different diseases. Usually we have mumps, measles, and chickenpox only once and are thereafter immune. Unfortunately, however, there can sometimes be a mutation of a viral gene which results in a slight alteration in the protein coat. Thus, a person may have antibodies against the previous strain but may be susceptible to infection by the mutant strain. We have seen a succession of influenza epidemics sweep around the world because of these mutations. These viruses are named for the region in which they originate. We have had the Hong Kong flu, the Spanish flu, and the London flu. Three different strains of the polio virus are known, so when we receive polio vaccine we must be inoculated against all three strains because immunity to one will not protect against the other two.

The comparatively short period of immunity to the virus of the common cold may be due to the fact that the nasal membranes do not have a ready supply of antibodies to resist infection. Another suggestion which has been made is that there are many different viruses which can infect the mucous membranes of the nose, sinuses, and throat. You may have a cold and develop antibodies against the invading virus, but another, dissimilar virus may cause you to develop another cold within a few weeks after the first. This has made it impossible to develop a vaccine which will protect against the common cold. Our best protection at present seems to be the maintenance of a good general body condition, which makes us more resistant to all infections. Cold viruses are so widespread that it is not possible to avoid contact with them. They are most likely to cause an infection when your resistance is low as a result of excessive fatigue or poor nutrition. It has recently been contended that a high intake of vitamin C will help prevent colds, but there are differences of opinion about this. It is definitely known though that an insufficient intake of this vitamin in foods makes one more susceptible to colds.

Interferon

Sometimes when a person has a severe outbreak of fever blisters on the lips, he will be given a smallpox vaccination. How could a vac-

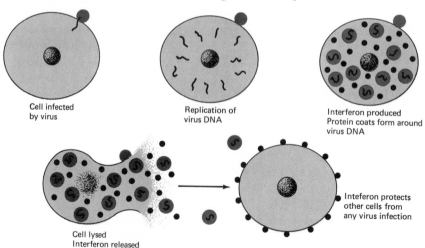

Cell infected
by virus

Replication of
virus DNA

Interferon produced
Protein coats form around
virus DNA

Cell lysed
Interferon released

Inteferon protects
other cells from
any virus infection

FIG. 14.8 Interferon is produced in a body cell when it is infected with a virulent virus and it is released when the cell is lysed. The interferon then becomes attached to the cell membrane of other cells and protects them from infection by the virus.

cination against an entirely different virus help overcome this infection? We have learned that antibodies are specific and immunity to one strain of an organism does not protect you against other strains. Still, we also know that whenever an animal is infected by one virus, it cannot be infected by another virus of the same or different kind for several weeks as a rule. There must be some other protective mechanism involved besides the antibody mechanism. Infected cells produce a substance known as **interferon,** which inhibits further infection. Fluid taken from macerated infected cells will confer resistance to virus infection when the fluid is applied to these cells. This seems to be a natural method by which the body overcomes virus infections. The first cells which are infected may be destroyed, but as they are lysed, they release interferon, which is carried over the body by the circulation and makes other body cells more resistant. Interferon appears in the body within twelve hours after a viral infection begins. Inoculation with the weakened virus of smallpox, therefore, will cause interferon production and this may prevent continued infection of cells of the lips by the herpes virus which causes fever blisters.

When it was first discovered, interferon was hailed as a possible wonder treatment which might cure all viral diseases. Unfortunately, however, it was found that interferon is species specific. Interferon produced by a rabbit will not help humans overcome infections. Interferon effective against human diseases has to come from humans, and there are not many who would allow themselves to be infected with virus disease and then give up large quantities of blood from which interferon may be extracted. Even if large quantities of blood were available, the extraction of interferon is somewhat difficult. Given our success in overcoming other obstacles which have stood in the way of medical advances, however, it is possible that this obstacle may fall and that treatment by interferon may become commonplace.

Temperate Viruses

Some viruses establish a **commensal** (harmless) **relationship** with their host cells and do not destroy them. **Temperate viruses** among the bacteriophages have been studied extensively. A temperate phage viron injects its nucleic acid into a bacterium, but instead of destroying the DNA of the host cell, the phage DNA becomes incorporated into the circular genophore of the bacterium. As a result, the phage DNA replicates only when the bacterial DNA replicates, which is once during each cell division. Bacteria which contain these integrated genophores are said to be **lysogenic** and they can undergo many divisions while carrying the phage DNA. In time, however, this phage DNA may separate from the host DNA and begin independent replication. New phage particles are generated in the bacterium cell and this cell is lysed. When these phage particles infect other bacteria, their DNA becomes integrated with the host DNA and the cells are lysogenic.

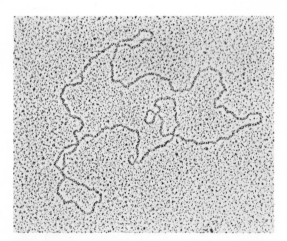

FIG. 14.9 The circular genophore of the bacterium *Escherichia coli.* The phage viron also has a circular genophore which may become incorporated into the bacterial genophore, making the bacterium lysogenic. A lysogenic bacterium is not destroyed by the invading DNA. (Courtesy of D. A. Wolstenhohue and I. B. David, Carnegie Institution of Washington.)

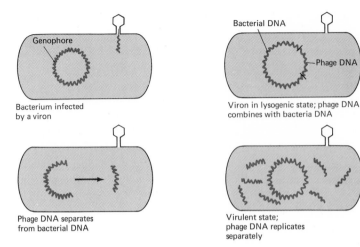

Genophore

Bacterium infected
by a viron

Phage DNA separates
from bacterial DNA

Bacterial DNA

Phage DNA

Viron in lysogenic state; phage DNA
combines with bacteria DNA

Virulent state;
phage DNA replicates
separately

FIG. 14.10 The genophore of a temperate viron becomes incorporated into the genophore of a bacterium. There is only one replication of the virus geno-phore to each replication of the bacterial genophore. When the viron DNA breaks loose, however, it replicates faster and the cell is lysed.

Some viruses are temperate in one strain of a species and virulent in others. Practically all potatoes raised in America carry the "healthy potato virus" in a lysogenic state. This virus seems to do no harm. It is spread easily because potatoes are generally propagated from "eyes" cut from potatoes raised the previous season. When American potatoes are brought in contact with European potatoes, however, the European varieties develop a serious viral disease. The European potatoes lack some factor which enables them to carry the virus in its lysogenic state.

People carry many viruses in their lysogenic state and occasionally some of them become virulent. The herpes virus is carried in a lysogenic state in the cells of the lips by most people. The herpes·virus is acquired early in life because of the habit of kissing. Occasionally some factor can trigger it into virulence and the virulent virus causes fever blisters to develop on the lips. The precipitating factor which causes the virus to leave the lysogenic state can be a fever, too much sunlight, a common cold, low-ered general resistance, or sometimes there may be no apparent reason at all.

Cancer Virus

Certain viruses have been isolated from cancerous tissue, and it has been suggested that a virus may be involved in the complex reactions which initiate malignant growth. In higher plants there are viruses which initiate excessive growths which result in galls. The *Rous sarcoma* virus induces malignant tumors when it is injected into domestic chickens. Since human cancer does not seem to be infective, there is considerable doubt that human cancer is caused by viruses. The possibility has not been ruled out, however. A virus might be carried in a lysogenic state by most people, and a carcinogenic agent might cause it to become virulent and to alter the growth pattern of the infected cells. Viruses with such ability are known as **oncogenic viruses.** Much remains to be learned about these viruses and their possible role in human cancer because we also have found that genetic factors in the host cells play a role in determining susceptibility to cancer development.

GENETIC ALTERATION BY VIRAL INFECTION

Bacteriophage DNA Transfer

Sometimes when viral DNA separates from its bacterial host after a lysogenic association, it will carry with it some of the DNA of the bacterium. This DNA will be included in the protein coat formed around the prophage particles. When these virons infect other cells, they can transfer some of the genes from the first bacterial cell to the new host. This is known as **transduction,** the transfer of genes from one cell to another through a virus. One strain of a bacterium may lack the ability to produce an enzyme needed to synthesize the amino acid threonine. A second bacterial strain may have the gene for the production of this enzyme. When bacteriophages from this second strain infect cells of the first strain, the bacteriophages may carry along with them the gene for the production of the enzyme. The first strain then acquires the ability to produce the enzyme and synthesize threonine. Transduction has occurred.

Correction of Human Defects

There is the hope that someday we may utilize this method of transduction to correct certain human genetic defects. Many serious human defects are caused by the lack of a particular gene which directs the production of some vital cellular protein. If a normal gene could be transduced into the cells of a person who lacks the gene, the affliction could be corrected. This sort of transfer has been accomplished in cells in tissue cultures. A certain virus, the **Shope**

popilloma virus of rabbits, also infects human cells in tissue cultures without harming them. Some people have **argininaemia,** a defect caused by the lack of a gene which produces the enzyme **arginase,** which can break down the amino acid arginine. Arginase is a liver enzyme which removes and chemically alters arginine when its concentration in the blood gets too high. Excess levels of arginine in the blood can cause mental retardation, epilepsy, and early death. Cells from a person with argininaemia can be grown in tissue culture and infected with a papiloma virus which has been taken from rabbits which have the gene to produce arginase. Some of these cells will develop the power to synthesize arginase, evidently because of the transduction of the rabbit gene to the human cells. Perhaps the injection of this virus into the liver of persons with argininaemia would cause a similar transduction to take place and enable that afflicted person to produce the vital enzyme. We must be certain first, however, that the virus cannot become pathogenic and destroy liver tissue and that other harmful rabbit genes will not be introduced along with the desirable gene.

ORIGIN OF VIRUSES

How did viruses begin? One theory says that viruses may have originated in the thin soup of organic compounds in the seas of the early earth. Under these conditions viruses could exist and multiply outside host organisms. Later, after the organic soup began to disappear and cellular organisms had developed, viruses adapted themselves to a parasitic existence in cellular hosts. Another theory holds that viruses represent an extreme form of parasitic specialization. This theory says that viruses were once cellular organisms which became parasitic on another cellular forms. All parasites tend to lose those parts and functional abilities which are provided to them by the host, so viruses then lost their cytoplasmic components and retained only the bare essentials, the nucleic acid core and the protein coats needed to penetrate host cells. Still another theory holds that viruses arose as detached pieces of chromosomes within cells. We know that chromosomes do break sometimes and pieces may end up in the cytoplasm. Usually cytoplasmic enzymes break down stray pieces of genetic material, but pieces which managed to resist a breakdown might begin replicating independently of the nuclear genes. There are independently replicating, DNA-containing particles in the cytoplasm of cells. The mitochondria and the plastids, which are normal organelles and play a vital role in cellular metabolism, are such particles. Some chromosome pieces, however, might have become harmful and caused lysis of the cell. These chromosome pieces might then have been engulfed by other cells of the same kind and the process of replication and lysis of the host cell repeated. In time, the chromosome pieces might have acquired the ability to penetrate the cells without being engulfed. This theory would explain the specificity of viruses. Viruses would find the right protein configuration which would allow them to attach to and penetrate the cell membrane only on these cells from which they originated. The viruses could utilize the cytoplasmic components only in cells like the ones which were their original hosts.

REVIEW QUESTIONS AND PROBLEMS

1. The bacterial diseases were brought under control before the viral diseases in general were controlled. Explain why this is true.
2. Why was the term filterable viruses used by the early workers who studied the viruses?
3. What special provision does the T-2 bacteriophage have for introducing its DNA into a bacterium?
4. Why do some biologists prefer to call the circular strand of DNA in a bacterium or a virus a **genophore** rather than a **chromosome**?
5. RNA is generally produced from DNA, but some viruses have only RNA in their cores. How do such viruses replicate?
6. Why does a particular viron infect only one kind of cell, as a rule?
7. Some virus diseases or vaccinations seem to confer a lifetime immunity, but others, such as the common cold, seem to leave only a very short period of immunity. Give possible explanations.
8. How does interferon differ from antibodies as a means of protection from viral infections?
9. How does infection by a temperate virus differ from that of a virulent virus as illustrated by bacteria?
10. Since viruses must kill the host cell in order to spread to other cells, how can transduction benefit a cell by bringing in desirable genes?
11. What problems would be involved in the use of transduction for the correction of human defects?
12. Which of the theories of the origin of viruses do you think has the greatest amount of evidence behind it? Explain your reasoning.

FURTHER READING

Fenner, F. J. 1968. *Biology of Animal Viruses.* New York: Academic Press.

Fraenkel-Conrat, H. 1968. *Molecular Basis of Virology.* New York: Van Nostrand Reinhold.

Hayes, W. 1968. *The Genetics of Bacteria and Viruses.* New York: John Wiley and Sons.

Luria, S. E. 1967. *General Virology.* 2nd ed. New York: John Wiley and Sons.

Stanley, W. M., and Valens, E. G. 1961. *Viruses and the Nature of Life.* New York: E. P. Dutton.

Stent, G. 1963. *Molecular Biology of Bacterial Viruses.* San Francisco: W. H. Freeman.

15

Bacteria and Related Organisms

Organisms which cannot be seen with the naked eye are known as **microorganisms,** or **microbes.** The study of these organisms is known as **microbiology.** In this chapter we shall discuss those microorganisms known as **bacteria** and some related organisms which are found in the kingdom Monera. (For classification, see Chapter 13.)

THE NATURE OF BACTERIA

Bacteria are among the smallest of the one-celled organisms, but they are a very important group. Some bacteria cause disease, suffering, and death in human beings; others are highly beneficial to man. In fact, in spite of the damage they do, the cycle of life on earth could not continue without bacteria. Bacteria can be found in almost every conceivable environment. Each swallow of water you take is swarming with hundreds of bacteria; each breath you take includes many of them; they are on almost all the objects you touch, and in the food you eat, they flourish in your mouth and intestine. Fortunately, the great majority of bacteria are harmless.

Discovery of Bacteria

A typical rod-shaped bacterium will be about 3 μ long and some ball-shaped forms are only about 1 μ in diameter. In comparison, a grain of salt is about 500 μ across. Over 10,000 bacteria could be placed on the head of a pin without overlapping. Hence, it is not surprising that the existence of bacteria was not suspected until microscopes were invented.

The first records of the observation of bacteria date back to 1683 when **Anton van Leeuwenhoek** ground glass lenses and made primitive microscopes. When he looked at a bit of matter taken from between his teeth and gums, he was astonished by the great number of organisms he saw. A drop of stagnant pond water was found to contain many varieties of living organisms. He drew some of these organisms, and from these drawings we can recognize some bacteria which are known today. At first the small organisms were thought of only as curiosities. After about two hundred years, however, it became evident that bacteria played a very important part in human life and death.

In 1864 **Louis Pasteur** suspected microbes of causing undesirable changes in stored wine. This was an important problem in his native France, where wine making was a major indus-

FIG. 15.1 Anton van Leeuwenhoek explains to his wife the little microscope he has built. In this artist's conception, he is holding the microscope in his left hand. (© Bausch and Lomb Optical Company.)

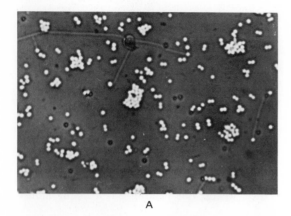

A

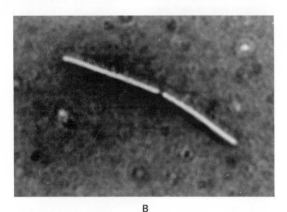

B

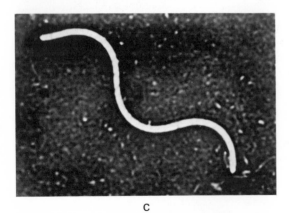

C

FIG. 15.2 The three major morphological types of bacteria: (A) coccus (staphylococcus); (B) bacillus (streptobaccillus); (C) spiral (spirillum).

try, so he set out to find ways to prevent the souring of wine. His studies led to the discovery of the principle of **pasteurization** which we use extensively today. By heating grape juice to a temperature which was high but not boiling, he could kill the microbes which soured wine, and then by adding microbes from good wine, he could get predictable results. When the wine was heated again, it would keep without souring.

At about the same time **Robert Koch,** a German, proved that there was a relationship between bacteria and disease. He found that he could isolate and grow bacteria from sheep with anthrax. Some of these cultured bacteria could then be inoculated into healthy sheep and they would contract anthrax. He worked out a vaccination technique to protect sheep against anthrax. He gave the sheep inoculations from aged cultures in which the bacteria had lost their virulence. Pasteur developed a vaccination for rabies, even though he never found the organisms which caused this disease because the organisms were viruses.

At the same time in England, **Lord Lister** was laying the foundation for modern antiseptic surgery. He suspected that microorganisms were the cause of the infections which almost always followed surgical operations and led to many deaths. By washing his hands and instruments in weak carbolic acid in order to destroy the organisms that might gain entrance to the body during surgery, he was able to reduce infections. His patients had clean wounds which healed quickly, while those patients whose surgeons did not observe these precautions continued to develop gangrene, blood poisoning, and tetanus and to have a high mortality rate.

Morphology of Bacteria

Bacteria fall into three broad classifications according to shape.

1. Coccus (spherical bacteria). The **cocci** are subdivided according to their arrangement with respect to one another.

a. Single
b. Diplococcus (a pair of cocci)
c. Streptococcus (a chain of cocci)
d. Tetracoccus (a flat sheet of cocci)
e. Sarcina (a cube of cocci)
f. Staphylococcus (an irregular bunch of cocci)

    2. Bacillus (rod-shaped bacteria). The **bacilli** fall into two subdivisions.

a. Single
b. Streptobacillus (a chain of bacilli)

    3. Spirillum (spiral-shaped bacteria).

Some of the bacilli and most of the spirilla have **flagella** which are used for locomotion. The flagella are so slender that they may be overlooked with the optical microscope, but certain stains will build up on them and thus thicken them so they can be seen. Against a dark field they may also be visible. Those bacteria without flagella are generally dependent upon other forces for transportation. Under the microscope they can be seen vibrating back and forth very rapidly. This motion is **Brownian movement,** which is brought about by the bombardment of molecules in a liquid. This phenomenon was explained in Chapter 3.

Reproduction in Bacteria

Bacteria are said to reproduce by **fission,** which is a simple splitting in two. No mitotic figure such as that seen in eukaryotic cells appears before this splitting, but the circular strand of DNA undergoes replication and one of each of the two circles formed moves to one end of the cell before it splits.

    The rate of reproduction of bacteria is amazing. Under favorable conditions, some bacteria may undergo fission as often as once every twenty to thirty minutes. Because of their small

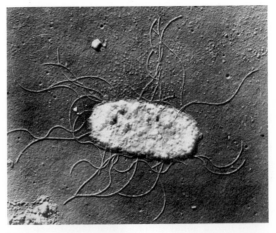

FIG. 15.3 The electron microscope brings out the flagella surrounding the single *Proteus vulgaris* bacillus.

size, bacteria can absorb food very rapidly and double their size in this short period of time. At the thirty-minute rate, a single bacterium could multiply and produce 281,500 billion bacteria in two days, and within a week it would produce a mass of bacteria equal to the weight

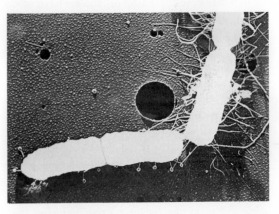

FIG. 15.4 Conjugation in bacteria. These two have formed a protoplasmic connection through which part of the donor's genophore will pass into the recipient. (Electron photomicrograph, courtesy of T. F. Anderson, The Institute for Cancer Research, Philadelphia, and E. L. Wollman and F. Jacob, Institut Pasteur, Paris.)

of the earth itself. This cannot happen, however, since the rate slows down as the bacteria exhaust the food and build up waste products. The tremendous potential rate of growth, however, makes it easy to understand how a few bacteria which enter your body today can make you deathly sick the day after tomorrow. Each bacterium is very small, but all together they can make quite a volume.

Some of the rod-shaped bacteria, the **bacilli,** produce **spores** which enable them to live for long periods in unfavorable conditions. The protoplasm in a bacillus cell becomes concentrated into a small sphere which then develops a hard protective coating. Bacillus spores can withstand immersion in boiling water for up to two hours or the application of strong solutions of poisonous chemicals. Fortunately, very few of the disease bacteria produces spores, but bacterial spores do make food preservation difficult.

For many years after their discovery, it was believed that bacteria were organisms without any form of **sexual reproduction.** Then some well-planned experiments by **Joshua Lederberg** and **E. L. Tatum** at Yale University showed that even these tiny organisms have some way of exchanging genes. These men combined two strains of *Escherichia coli*, the common human colon bacillus, in one tube and found that some of the descendants had characteristics of both strains. One strain had the power to synthesize all the amino acids except two, which we will designate A and B. The other strain could synthesize all amino acids except C and D. The two strains were grown together in tubes which contained all the amino acids. After a time, some of the bacteria from these tubes were placed in a culture medium lacking all four of these amino acids. Some of the bacteria grew, showing that they could synthesize A, B, C, and D. Somehow genes had been passed from one strain to the other. Later the electron microscope made it possible to see this process. Two bacteria come together and form a protoplasmic conjugation tube between them. One of the two conjugated forms serves as the donor. The donor's genophore, which has become disjoined to form a long strand, begins to pass into the recipient. After the genophore is part way in, as a rule, the two bacteria break apart. The recipient has now acquired some of the genes of its partner in conjugation. These genes may become incorporated into the recipient's own genophore and give it a new combination of traits. The duplicate genes from the recipient's genophore are thrown off during succeeding cell divisions. The donor usually duplicates its genophore before donating it to the recipient; therefore, the donor ends up with a complete set of genes. The donor also throws off any excess genetic material.

In this manner the basic purpose of sex, the recombination of genes, is accomplished. Sexual reproduction provides the variety of traits which is necessary for natural selection. In this case, we have sex without reproduction. We started with two organisms and we ended with two, but reproduction can be accomplished very efficiently through fission at a later time.

Transformation

Bacteria are unique in that they have two methods of achieving gene recombination. They can also take in genes from other organisms. This process is known as **transformation.** It was first demonstrated by **Fred Griffith** in 1928. He was studying the bacterium which can cause pneumonia in man and other animals. This bacterium is a diplococcus, which is usually encased in a polysaccharide capsule and is commonly called the pneumococcus. When sputum from a person who has this form of pneumonia is injected into mice, they sicken and die within approximately twenty-four hours. The polysaccharide capsule seems to protect these bacteria from the natural body defense of the host. Another strain of this organism, however, lacks the capsule and is not virulent. Mice injected with the unencapsulated strain are not injured.

Griffith hypothesized that the nonvirulent strain might pick up the gene for capsule production from the virulent strain. To test this hypothesis, he injected some mice with heat-killed encapsulated forms and living nonencapsulated forms of pneumococcus bacteria. The mice injected died. From the dead mice he recovered living encapsulated bacteria. Some of the normally harmless nonencapsulated forms had picked up some of the genes for capsule formation from the DNA of the dead encapsulated forms. This process is called **transformation.**

It was then found that transformation could take place in culture dishes as well as in living organisms. When DNA was extracted from the virulent form of pneumoccocus and placed on culture media upon which the nonvirulent form was growing, some of the nonvirulent colonies which developed had capsules. The nonvirulent strain had taken up the gene for capsule formation from the surrounding medium. Phagocytosis was the probable method of DNA intake.

Later it was found that bacteria could even take up genes of other species. This discovery caused concern about the use of antibiotics in farm-animal feeds. It had been found that adding a small amount of antibiotic to the animal feed kept down the incidence of infection and that the animals grew faster. It was also found that spreading a little antibiotic on the surface of animal products to be consumed by man would retard spoilage. Some strains of bacteria, however, have genes which enable them to resist antibiotics. When people ate the antibiotic-treated food, they might ingest some of these resistant bacterial strains. Even though these strains would be heat-killed during food processing, there might be a residue of antibiotic-resistant DNA which could be taken up by other bacteria in the body. This DNA could transform some bacteria which cause human disease into antibiotic resistant forms which could not be killed by antibiotic treatment.

LABORATORY CULTURE OF BACTERIA

Many bacteria can be grown in the laboratory in culture media which contain all required nutrients. Beef broth was one of the first culture media used. Later sugars, starch, blood, and other nutrient media were used to accommodate the needs of specific bacteria. To these nutrients a solidifying agent is often added so that the bacteria will remain in one place and form a colony. Gelatin has often been used, but usually an extract of a red alga, **agar,** is used. When a single bacterium contacts this surface, it sticks. Then, as it grows and divides over and over again, a large mass of many thousand cells develops into a visible colony.

Such cultures are used in determining the number and kind of bacteria in a given substance. Suppose we open a sterile dish of nutrient agar for several minutes. Then we cover it and keep it warm for about twenty-four to forty-eight hours. By that time each bacterium which settled down onto the agar from the air will have grown to form a colony. The colonies on the dish will be different shapes, sizes, colors, and textures. We can then remove a small bit of one of the colonies, stir it in water on a microscope slide, dry it, stain it, and determine its morphology. If we find that a particular colony is interesting, we can transfer a bit of the colony to nutrient agar in a test tube and thus have a pure culture to study. By such techniques we can find the number and kind of bacteria in milk, water, and food or the number and kind of bacteria on a person's fingers, lips, or other body parts. Some of the colonies may turn out to be yeasts and molds, because these organisms also produce spores which are air borne. Public health officials regularly make cultures from dishes, silver-

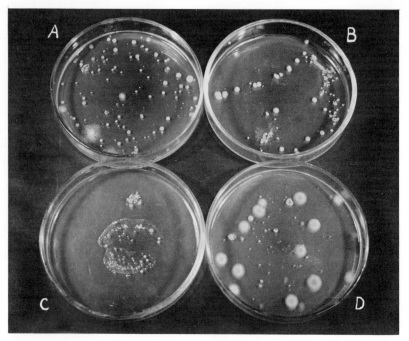

FIG. 15.5 Colonies of bacteria growing on nutrient agar. The plates were exposed as follows: (A) opened in the air for five minutes; (B) a fly was allowed to crawl across the agar; (C) a kiss on the agar (note where the nose touched); (D) a person's hair was shaken over the agar (note the large, fuzzy mold colonies).

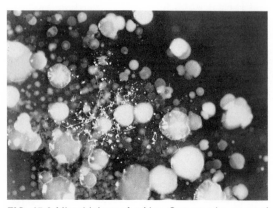

FIG. 15.6 Microbiology of a kiss. One may be amazed at the number of bacteria which are transferred by lip contact. Most of these bacteria are harmless species. Each colony is composed of hundreds of thousands of bacteria, all of which have grown from a single organism.

ware, and glasses used in restaurants to see that the restaurants maintain the proper standards of cleanliness. Milk and water are also tested to be sure that the bacterial count is not too high and that no contaminating disease organisms have gained entrance.

Maintaining **sterility** is a great problem in preparing cultures for studies of bacteria. All media, glassware, and inoculating needles must be absolutely free of all life before cultures are made. Only then can one feel sure that only those organisms will grow on the media which were purposely placed there. Since some spores can withstand boiling for as long as two hours, complete sterility is difficult to obtain. Most laboratories turn to the pressure sterilizer, the **autoclave**, which will give sterility in about twenty minutes at fifteen pounds pressure. At this pressure the steam in the autoclave reaches a temperature of 120°C, as compared to 100°C for boiling water at atmospheric pressure. An open **flame** may be used to sterilize small metal objects, such as inoculating needles, which can withstand this great heat. Only a few seconds are required in this method. A **hot-air oven** can be used to sterilize glassware and other materials which can stand the dry heat. About one hour at 170°C is used in this method.

IMPORTANCE OF BACTERIA

We have already said that life as it exists on the earth today would not be possible without bacteria. What makes bacteria so important? Let us summarize some of the functions of bacteria.

Nitrogen Fixation

Nitrogen is necessary for the building of protoplasm in all forms of life. It is an essential element in all amino acids and the amino acids form the proteins which are an essential part of protoplasm. Nitrogen is an abundant element. There are about 1500 pounds of it in the air above a single square food of earth, but many of the plants and animals, including people, suffer from nitrogen deficiency. The reason for this is that nitrogen in atmospheric form cannot be used in the synthesis of animo acids. Certain species of bacteria which live in the soil can combine this atmospheric nitrogen with oxygen and produce nitrates ($-NO_3$). The nitrates combine with sodium to form sodium nitrate and with potassium to form potassium nitrate. Bacteria which can convert atmospheric nitrogen to nitrates are known as nitrogen-fixing bacteria. The higher plants use nitrates for their growth. We sometimes add nitrates to the soil in fertilizer, but most of the nitrates used by plants come from the bacteria in the soil.

Some of the nitrogen bacteria live free in the soil, but one species has formed a mutually beneficial association with one family of flowering plants. This family includes the **legumes,** a group including peas, beans, clover, and alfalfa. These plants are all effective nitrogen producers because nitrogen-fixing bacteria actually live in their roots. If a healthy soybean plant is pulled up, wartlike nodules will be seen on the roots. This might appear to be a harmful infection, but the more nodules there are present, the better it is

FIG. 15.7 Nodules on the roots of a soybean plant. Each nodule contains millions of nitrogen-fixing bacteria. Such nodules are characteristic of the plants known as legumes.

for the plant. Each nodule contains millions of nitrogen-fixing bacteria. Legumes are always rich in protein because of their association with nitrogen-fixing bacteria. When we cannot afford meat, we may turn to beans or peas as a protein substitute. Alfalfa is fed to cattle because of its high nitrogen content. Sometimes a farmer will plant a field of legumes and then plow them under as a green manure. As legumes decay, they release their rich storehouse of nitrogen for other plants to use. Sometimes farmers will mix their legume seed with cultures of nitrogen-fixing bacteria to ensure a maximum infection of the root system.

Decay

Decay is essential to the maintenance of life on earth. It would be difficult to find space to live, if you had to pick your way around the dead bodies of dinosaurs, mastodons, and the many millions of other animals and plants which lived and died in the past. Bacteria serve as the scavengers which clean up this mass of dead things so that there is room for the living. Bacteria are useful in another way as well. Contained within the bodies of the living plants and animals is a sizable amount of the mineral wealth of the eatrth. It is important that these minerals be returned to the earth after one form of life has finished with them so they can be used by new forms. When leaves fall from the trees in a forest, the bacteria of decay soon begin to work on them and the minerals which they contain are released to filter down into the soil. The minerals are picked up by plant roots and contribute to· plant growth the following spring. Dead animals also undergo decay, and minerals from their bodies are returned to the soil where they can be used by plants. The American Indians taught the Pilgrims that their corn would grow better if they put a fish in each hill of corn. The decaying fish increased the fertility of the soil.

Food Preservation

Although decay is essential to the continuation of life on the earth, it has its disadvantages. We must go to great lengths to prevent decay of our foods. Great amounts of the food we produce are lost because of spoilage which is the result of the action of bacteria and related organisms. The supermarket manager knows that as much as one-fourth or more of all his fresh produce will have to be discarded because of spoilage, in spite of all his efforts to prevent it. The following methods of preservation are commonly used.

Cold. Most bacteria grow best at warm temperatures, and as the temperature drops they grow more slowly. Refrigeration keeps foods at a temperature of about 35° to 40°F. This slows spoilage considerably. Freezing is widely used. At about 0°F bacterial activity practically stops, and foods can be kept in a fresh condition for a much longer period of time.

Heat. High temperatures of sufficient duration will destroy all forms of life. This makes it possible to preserve foods by first applying heat and then preventing other organisms from entering the food. Our widely used canning methods are examples of this method of food preservation. It is imperative that sufficient heat be applied, because some bacteria can grow even in a sealed can.

Osmotic pressure. Bacteria can grow only within certain limitations of concentration of dissolved substances around them. When the concentration becomes too high, their growth stops and there is no decay of food. Salt and sugar are commonly used. Meat that is thoroughly **salted** can be kept without refrigeration. Jelly does not ferment because the sugar concentration is too high. The **drying** of foods works on the same principle. When most of the water is removed from grapes, plums, peaches, and apricots through drying, the natural fruit sugar is so highly concentrated that bacteria cannot grow.

Honey does not ferment because much of the water from the nectar is evaporated before it is stored in the hive.

Chemical preservation. We add chemicals to foods in order to prevent spoilage. Acetic acid in the form of vinegar has been used for centuries in the process which we call **pickling.** Sodium nitrite and sodium nitrate are often added to canned meats as an extra precaution against spoilage. Benzoic acid and benzoate of soda are commonly added to apple and cherry cider and to many other bottled beverages to inhibit growth of microorganisms. Sulfur dioxide is used to aid in the preservation of many dried fruits. (There are strict laws against the use of formaldehyde for this purpose, and public health authorities often check for the presence of this chemical in milk, because it is poisonous to human cells as well as to the bacteria.) **Smoking** is another method of food preservation which depends upon .chemical action. As the wood burns, certain chemicals are released in gaseous form which aid in preservation. Beef, pork, and fish are the chief foods which are smoked.

Radiation. Certain high-energy radiations have been found to be effective in food preservation. **Ultraviolet rays** prevent spoilage of hams and beef in the "tenderizing" process. Meat becomes more tender when held for a time at a temperature higher than ordinary refrigeration, but it would spoil without the use of ultraviolet rays to inhibit bacterial growth. Orange juice is now being shipped from Florida in tank cars. This is possible because of the use of ultraviolet rays on the juice when it is squeezed. The ultraviolet rays kill most of the organisms of decay and the juice retains its fresh qualities and can be delivered in cartons to consumers many hundreds of miles away. Recent experiments have been conducted on the use of high-energy radiation from the neutrons of atomic fission. Foods sealed in plastic containers were radiated and kept for months at room temperature without spoilage. The time may not be far distant when we can buy foods which have been prepared in this way for home consumption.

The consumption of food which has not been properly preserved can lead to unpleasant and sometimes dangerous reactions. In addition to simple decay which makes food unpalatable, there are a number of bacteria which may grow in food that produce toxins powerful enough, to cause **food poisoning.** Several species of *Staphylococcus* are the most common offenders. They develop best in foods rich in protein, such as meat, milk, eggs, and many other foods which have not been properly preserved. After eating food thus contaminated, a person will become violently ill within several hours with nausea, diarrhea, and general body weakness. Most of us have experienced at least a mild case of this type of food poisoning.

There is another type of food poisoning which is rare, but so dangerous when it strikes, that everyone should know about it. According to public health records, about 65 percent of the people who develop **botulism** die from its effects. The poison is produced by a bacterium, *Clostridium botulinum,* and is so powerful that one gram could kill 25,000 people if evenly distributed among them. A person may ingest a fatal amount by simply tasting the contents of a can that appears suspicious. This is an **anaerobic** organism and develops only within containers tightly sealed from the air. It is a gas former and causes a positive pressure to build up within a can so that the ends bulge out and gas escapes to the outside when the can is punctured. Food should be discarded whenever it comes from a swollen can, when it shows gas bubbles, or has an odor of rancid cheese. The nonacid, home-canned foods, such as string beans, have caused most of the cases of botulism in the past, although carelessness in commercial canning has taken its toll. Recently several thousand cans of mushrooms had to be recalled by a cannery after several deaths had resulted from consumption of these mushrooms. A worker had failed to leave the cans in the pressure cooker long enough. Bacterial toxins are destroyed by thorough cooking: five minutes at boiling temperature is generally enough. When contamination is suspected, it is better to discard the food than try to save it through cooking. Botulism does not develop until about twelve to forty-eight hours after the intake of the toxin, and the infection worsens gradually for about a week. Death, if it occurs, comes at about the tenth day after ingestion of the food.

Salmonella infection is usually classified as food poisoning but this is actually a disease brought on by the ingestion of live bacteria and not by the toxins produced by bacteria in food. Salmonella is an intestinal infection. Workers who have the infection can infect food they handle with the bacteria. Meat is the primary source of infection. It can be contaminated by the intestinal contents of the infected animals when slaughtering is not done properly. Salmonella is sometimes called sausage poisoning because it is frequently obtained by eating sausages, such as bologna and liverwurst. These sausages are often eaten without further cooking. Thorough cooking destroys the bacteria and makes the meat safe to eat. Symptoms such as vomiting and diarrhea appear in about eight to twelve hours after eating infected meat. A few deaths have been reported, but most people recover after about a week of unpleasant sickness.

DISEASE

The word "disease" is a very broad term. It is used to refer to almost every deviation of the human body from a normal state. Bacteria and other microorganisms, of course, are a major cause of disease, but their importance is dwindling as new methods of prevention and treatment are developed. Today more and more the physicians are called upon to treat diseases which arise from other causes. Many of these diseases have a genetic background. We must not relax our vigilance, however, or the infectious diseases may make a quick comeback. The epidemic of venereal infections which has appeared in recent years is an example of the way in which infectious diseases can spread rapidly.

Diseases can be classified into two basic types; those caused by microbes and those which are not caused by microbes. First let us consider those which are not infectious.

Noninfectious Diseases

Most of the diseases which arise without any infectious agent can be divided into three categories: organic, allergic, and deficiency.

Organic diseases. These arise as the result of some malfunction of one of the body parts. **Coronary thrombosis,** heart attack, is caused by a clot in the coronary artery leading to the heart muscle. This disease is a major cause of disability and death in the United States. **Hemophilia,** the bleeding disease, results from an inability of the body to produce one of the factors which are necessary for normal blood clotting. **Epilepsy** or seizures can be caused by abnormal pressure on regions of the brain. **Cancer** is caused by uncontrolled cell division in a region of the body.

Allergic diseases. Many people are plagued with allergic reactions to various substances in their surroundings. Respiratory allergies are caused by the inhalation of pollen, dust, animal dander, or many other substances which may be in the air. When the offending substance reaches the lungs, allergic asthma can result. Other people react to certain protein foods, such as beef, eggs, and milk, by developing irritating hives on the skin or by vomiting after the food is eaten. Allergic diseases are the result of the antibodies developed by the body against substances. Some people may develop antibodies against the poison of bees or wasps. Quite a number of people die each year from an allergic response to stings by these insects. Many people become sensitized to penicillin and can die from an injection of this antibiotic if they are not promptly treated. Treatment for allergic diseases may involve avoidance of the offending substance, and if this is not possible, participation in a desensitization program. Desensitization involves injection of very small quantities of the substance into the body. The amount is gradually increased. Through this treatment, the body "learns" to recognize the antigen in the substance as a part of its body and reduces its antibody output against it.

Deficiency diseases. These diseases are the result of the failure of the body to receive a sufficient amount of substances necessary for normal body functioning. **Food deficiency diseases** are well known. If a person does not receive a sufficient amount of a certain vitamin he may develop **rickets, scurvy, pellagra** or **beriberi.** Anemia can be the result of an insufficient amount of iron in the diet, and **goiter** can be caused by insufficient iodine. Deficiency diseases are becoming less common in the developed countries because many of these vital substances are added to foods. Goiter, for instance, has practically disappeared in the United States because of the addition of iodine to table salt. **Hormone deficiency diseases** are common. **Diabetes** is caused by insufficient insulin output from the pancreas. We have also overcome many of these hormone deficiency diseases by the regular administration of the proper hormone. Many people with diabetes live normal lives today, although they would have suffered greatly and probably died in the past, when insulin was not available. **Enzyme deficiency diseases** are another very important group. When certain vital cellular enzymes cannot be produced as a result of defects in the genes, serious complication can result. Some of these were discussed in Chapter 9.

Infectious Diseases

Infectious diseases arise as a result of the invasion of the body by microorganisms. Some **infectious** diseases destroy body tissues. An example is **paresis**, a type of insanity, which results when syphilis organisms destroy brain tissue. The majority of infectious organisms, however, produce toxins which cause the illness. These toxins include some which are the most powerful of all poisons known. We recognize two kinds of infectious diseases: contagious and noncontagious.

Contagious diseases. These are easily spread from one person to another by casual contact. As children we "catch" diseases like mumps, measles, and chickenpox through association with other children who have the infection. The organisms of most contagious diseases enter the body through the nose or mouth where they first infect the mucous membranes and then spread to other body parts. They may be coughed or sneezed into the air, but mostly they are spread by the hands. An infected person rubs his nose and gets the disease organisms on his hands, then he handles objects and the disease organisms are transferred. A second person handles the same object and gets the disease organisms on his hands, then puts his hands to his nose or mouth and becomes infected. Of course, kissing is another habit which makes it easy for the infectious agents to pass from one person to another.

Noncontagious Diseases. Many infectious diseases are not contagious. You can associate with a person who has the disease without "catching" it. You may have close associations with a person who has malaria and be quite safe but you must be careful of the mosquitoes that spread the organisms which cause this disease. Drinking water can be a source of typhoid fever and dysentery. Blood poisoning and tetanus can result from the entry of bacteria through breaks in the skin.

Kinds of Infectious Organisms

Bacteria are the major infectious organisms, but there are other agents as well. A few of the **yeasts** are **pathogenic** (disease producing). Some yeasts can cause skin infections and others infect the lungs. Some **molds** can cause irritating infections of the skin, the ears, and the lungs. **Protozoans** cause some very serious diseases, such as malaria and sleeping sickness. The **viruses** have been mentioned in Chapter 14 as the cause of some of the infectious diseases which are the most difficult to control. The **rickettsias** cause tick fever and Rocky Mountain spotted fever. Some of the larger organisms, such as the **worms,** can also cause serious malfunctions of the human body which may be known as disease.

Factors Influencing Infection

You do not succumb to a disease each time you get the organisms in your body. A person varies from time to time in his susceptibility to infection, and one person may be different from another. Let us survey some of the factors which are involved.

Virulence of the organisms. Some infectious agents are much more likely to cause infection than others. They may be better adjusted to the body than other strains of the same organism, and they may be able to grow more quickly. The vaccine for polio may be composed of a strain of the virus with a low virulence that does not cause a generalized infection. Some strains of the polio virus may be so highly virulent that nearly all who are exposed to it come down with the disease. The virulent strains are also likely to cause more severe reactions. We think of the flu as an inconvenient but not dangerous infection, but the 1918 epidemic of influenza killed more people than were killed during World War I. This epidemic was caused by a particularly virulent strain. There is always the danger that existing strains of low virulence may undergo mutation and be changed into highly virulent strains.

Number of invading organisms. You may be able to destroy a small number of infectious organisms which may get into your body, but if there is a massive invasion, your chances of resisting them is greatly reduced. All of us come in contact with the bacteria which cause tuberculosis and usually resist them, but if you live with a person who has the disease, you will come in contact with many more of the organisms and you are more likely to get the disease.

Body resistance. Your chance of resisting infection is much greater when your body is in good physical condition. It is practically impossible to avoid exposure to the viruses of the common cold at certain seasons when many people around you have this infection. Your chance of avoiding infection, therefore, depends more upon your body resistance than upon the intake of the virus. Proper nourishment is important in maintaining resistance, as are avoidance of extreme fatigue and avoidance of exposure to extremes of weather. Some infectious organisms are opportunists. They cannot gain a foothold in a healthy body, but when the body is invaded by other organisms and weakened, these opportunists may also get started. Pneumonia may develop after a bad case of the flu, or you may develop a strep throat after a bad cold.

Immunity. People have various degrees of immunity to specific infectious agents. This important point is discussed later in the chapter.

The Course of a Disease

When infectious agents enter the human body, the development of the disease follows a pattern. The organisms multiply slowly at first, and then rapidly until enough toxin is produced to make the person feel ill. The time between invasion by the infectious agent and the first symptoms of the disease is known as the **incubation period.** This may vary from twenty-four hours for the common cold to several months for rabies. After the incubation period, the disease may worsen as the organisms continue to multiply until a crisis is reached. The crisis is a critical time because at this time the disease may follow one of two courses. The disease organisms may continue to overcome the body defenses until a sufficient amount of toxin is produced to cause death, or the body may begin to destroy the invaders faster than they can multiply and a gradual recovery occurs.

The body has two major defenses to combat disease agents once they have entered the body. First, the phagocytic leukocytes can engulf the invaders and digest them. Second, the body can produce antibodies which react with the antigens of the invaders. This may cause the invading organisms to clump together and be inactivated, or the disease organisms may be lysed or burst open, and destroyed.

The Carrier State

Sometimes the resistance of the body may be balanced against the virulence of the invading organisms, so the organisms stay alive but do not cause symptoms of the disease. A person with such a balance will be a carrier. He can spread the infection to others in whom it may cause serious symptoms. A classic case of this kind was Typhoid Mary, a cook who carried typhoid bacteria in her intestine without showing any symptoms of typhoid fever. She worked in several homes when the people came down with the disease without suspecting that their cook was the source of the disease organisms.

IMMUNITY

When a person can withstand repeated exposures to a disease without developing it, we say that he is immune to that disease. There are degrees of immunity, however. Some people have complete immunity; others are immune most of the time, but when their resistance is low they may develop the disease. Others have no immunity and develop the disease whenever they are exposed. One type of immunity comes to you through the genes you inherited from your parents. The other is acquired.

Inherited Immunity

Species immunity. We are immune to many diseases simply because we are human beings. We do not develop hoof-and-mouth disease, which is so deadly to cattle, or hog cholera or dog distemper. Conversely, a cow, hog, or dog is immune to many human diseases. It seems as if the pathogenic microorganisms are so adapted to the physiological balance of body fluids of one species that they cannot develop in the body of a different species. Closely related species may develop the same diseases because their physiology is enough alike to permit existence of the same organisms. Gorillas are hard to raise in captivity because they are susceptible to the same respiratory infections as man. Monkeys and other primates are the only animals other than man which can be used in experimental work on the polio virus.

Racial immunity. Different subspecies, races, or breeds within the same species also show variation in their degree of immunity. They have physiological differences, but these are less pronounced than those between different species and the immunity is not likely to be complete. Most of you have had chickenpox, which is a minor children's disease for some racial groups, but for some races it is highly fatal. Many Indian tribes in the Americas were wiped out by diseases such as chickenpox which were introduced by the Europeans who came to these continents. The Tahitians, Hawaiians, and other South Sea Islanders suffered many deaths from tuberculosis, smallpox, measles, and other diseases when the missionaries arrived. The entire population of one island perished from tuberculosis after being visited by English sailors. Likewise, a Caucasian cannot live in parts of Africa without taking many precautions against the diseases which the native people are highly immune to. Natural selection establishes such racial immunities.

Individual immunity. Even among members of the same race there is a difference in the degree of immunity. People vary in the protein content of their protoplasm and in other physiological functions, and these variations can confer differing degrees of immunity to certain disease organisms on different individuals. Certain diseases tend to run in families because of an inherited low degree of immunity to that disease.

Acquired Immunity

One of the many homeostatic responses of the body is the development of tolerance for the disease organisms to which we are exposed. There are two ways we can acquire immunity to a specific disease.

Active acquired immunity. Once a person has the measles, he is usually immune for life. The presence of a foreign antigen in the body stimulates the production of antibodies which react with and neutralize the antigen. Since the pathogenic microorganisms and the viruses have proteins which are antigenic, antibody production is a natural reaction to disease organisms which invade the body. These antibodies continue to be produced even after the antigen is gone and thus are present and can destroy the invaders which may enter the body in the future. This accounts for the immunity which is acquired after a person has recovered from a disease.

A person can acquire immunity without having a disease. In 1776 **Edward Jenner,** an English physician, noticed that milkmaids had beautiful complexions, while most other people

FIG. 15.8 The first vaccination for smallpox in the western world. In this old woodcut, Edward Jenner is shown inoculating a boy with pus taken from a cowpox infection on the hands of a milkmaid. (Courtesy of Fisher Scientific Company.)

had the scars of smallpox on their faces. He investigated and found that the milkmaids usually had scars on their hands which were caused by cowpox infections which the milkmaids acquired from the udders of the cows. He associated the two and began putting pus from cows with cowpox onto scratches on the arms of people. A pustule, or cowpox, formed and the persons were then immune to smallpox. Since the cowpox was known as **vaccina,** this treatment became known as **vaccination.** Through widespread vaccination, smallpox has been practically wiped out in the developed countries of the world. The protein composition of the coat of the cowpox virus is so close to that of the human smallpox virus that antibodies produced against the first also protect against the second.

We now have vaccines for many other diseases. Polio is one of the more recent diseases which has been practically eliminated by the oral vaccination. Another recent one is the measles vaccine. This is especially valuable for those races which are highly susceptible to this disease. Most vaccines consist of organisms which have been killed or weakened in such a way that their protein structure is not disrupted. Some vaccines consist of organisms which have a natural low virulence.

The period of immunity after having a disease or being vaccinated for it varies considerably. The antibody production slows down after a time without contact with the antigen, but antibody production can be quickly reactivated by a booster shot. A small amount of the antigen causes the body to go into full production of antibodies again. Tetanus and typhoid booster shots are recommended about every three years.

Passive acquired immunity. The antibodies produced in one animal can be transferred to another animal and confer an immunity. We often take advantage of this passive acquisition. Horses may be inoculated with disease organisms or with toxins of various diseases. After allowing about ten days or longer for the generation of antibodies, some of the blood serum from the immunized animal can then be injected into people to help them overcome infections which they already have or to protect them from infection. Many people have been saved from death by tetanus by an injection of immunized horse serum which contains antibodies against the tetanus bacteria. Blood serum from people who have had a disease is sometimes used to save others suffering from the same disease. There is a certain type of *streptococcus* blood poisoning which is usually fatal and for which we cannot establish antibodies in other animals. When a person survives this infection, his blood is very valuable for saving the lives of others who take it. We have learned that the **gamma globulin** fraction of the blood contains the antibodies. Hence, we can borrow a pint of blood from the immune individual, remove the gamma globulin fraction and return the rest of the blood so that he will not lose the red blood cells and other blood proteins by his donation.

Passively acquired immunity does not last as long as that acquired actively. The body is constantly replacing the blood elements and within about six weeks after introduction of antibodies, the concentration of acquired antibodies will drop so low that the immunity is lost. Vaccination causes the body cells to start production of antibodies and to continue producing them for years. Some people become sensitized to horse serum and cannot take injections of such serum. More details about antibody production are given in Chapter 30.

SOME COMMON HUMAN INFECTIOUS DISEASES

Diseases Caused by Bacteria

Diphtheria. This disease is caused by a bacillus, *Corynebacterium diphtheria,* and spread by discharges from the throat of an infected person. The incubation period (the time between contact with the germs and development of symptoms of the disease) is from two to five days. This serious disease of children tends to occur in epidemic form. Vaccination has greatly reduced the incidence and severity of the infection in the United States. It is most common in the winter months.

Gonorrhea. This disease is caused by a diplococcus, *Neisseria gonorrhoeae*, spread through discharges from infected body parts, primarily through sexual contact. The bacteria grow best in the tubes of the reproductive system, but may infect the eyes. In previous times many cases of blindness in infants resulted from infection of the eyes during birth by the genital tract of the mother. Today this blindness has largely been eliminated because modern hospitals follow the practice of dropping antiseptic solutions in the eyes of all newborn babies. Permanent damage and sterility often result from advanced infections of the genital tubes of either sex. Antibiotic treatment has been very effective, but increasing numbers of resistant strains of the bacteria are making treatment more difficult, and the disease is increasing in incidence.

Pneumonia. This disease may be caused by either bacteria or viruses. The most common bacterium causing it is *Diplococcus pneumoniae*. The incubation period is from one to three days. Pneumonia is a serious disease of the lungs that usually follows a lowering of the body resistance due to exposure or infection with some other disease, such as influenza. It occurs commonly in older people who may be confined to bed for long periods because of broken bones or other causes. It was formerly a major cause of death, but antibiotics have greatly reduced its toll of human life.

Syphilis. This disease, caused by *Treponema pallidum*, is spread by active lesions of the body and has an incubation period of about three weeks. It is a widespread, very serious, and, in advanced cases, highly fatal disease of man. Most cases arise through venereal contact, but some are spread by other forms of contact, such as the lips. Final effects may follow initial infection by many years. These final symptoms include disorders of the brain, spinal cord, heart, blood vessels, bones, muscles, liver, kidneys, and other body organs. Heavy doses of penicillin and arsenic compounds seem to cure most cases in about a week.

Tetanus (lockjaw). This disease is caused by *Clostridium tetani*, which usually gains entrance into body tissues through deep wounds. It has an incubation period of from four days to three weeks. The bacterium grows in the intestine of many animals without doing harm. Horses are excellent breeders of it; the number of cases of tetanus has greatly diminished with the decline of the horse as a means of transportation. Tetanus is highly fatal; death results from most cases. Tetanus can be prevented by antitoxin given after serious wounds or by previous *vaccination.*

Tuberculosis (pulmonary). Tuberculosis is caused by a small rod-shaped bacterium, *Mycobacterium tuberculosis*, spread by sputum from infected persons. The incubation period is variable. These bacteria are so widespread that most persons develop a tubercular infection of their lungs at some time during their life, usually during youth. Most people overcome minor infections without knowing they had the disease, but if body resistance is low, the infection may spread and damage the lung tissue extensively. Tuberculosis is a disease that requires treatment in its early stages for best results. Young people, especially young women between the ages of seventeen and twenty-five, are most susceptible to infection.

Whooping cough. This disease, caused by *Hemophilus pertussis*, is spread by discharges from the throats of infected persons and has an incubation period of from ten to sixteen days. It is primarily a disease of childhood and is usually not serious in older children, but often is fatal to babies.

Diseases Caused by Molds

Ringworm and athlete's foot. These are skin infections that may be caused by any of a number of genera of the filamentous fungi. They may be transmitted by spores of these organisms from infected persons. They develop best in parts of the body where the skin is likely to be moist from perspiration. Children often get ringworm from pet cats or dogs.

Diseases Caused by Protozoa

Amoebic dysentery. Caused by *Entamoeba histolytica*, this disease is spread through ingestion of cysts from infected persons. The incubation period is about one week. It causes destruction of the lining of the large intestine, diarrhea, and general body weakness.

Malaria. Caused by protozoan *Plasmodium*, it is spread by the female *Anopheles* mosquito. The incubation period is about three weeks.

Diseases Caused by Viruses

Chickenpox. Caused by a virus, chickenpox is spread by contact with sores on the skin of an infected person. The incubation period is from two to five weeks. It is a common, comparatively mild disease of childhood in most races. In adults this same virus can cause shingles, which can be very painful and sometimes does permanent damage.

Influenza. The virus is spread by discharges from nose and throat. The incubation period is from twenty-four to seventy-two hours. It tends to occur in epidemics with varying virulence.

Measles. The virus is spread by discharges from nose and throat. The incubation period is from twelve to fourteen days. One attack gives permanent immunity.

Mumps. The virus is spread by discharges from the throat of an infected person. The incubation period is from twelve to twenty-six days. It usually attacks the salivary glands, causing the typical swelling characteristic of the disease; but it may settle in other glands of the body, including the testes. This may cause a serious infection and result in sterility.

Poliomyelitis. This disease is caused by a virus that is spread by secretions from nose, throat, and possibly body excrement. Insects are suspected of playing a part in its spread. The incubation period is from seven to fourteen days. Many probably have the disease in a mild form, often without knowing that it is polio, and they may transmit the infection to others. Paralysis occurs in a small proportion of the cases. The disease is most prevalent during hot, dry weather. Vaccination has almost eliminated this disease in the United States today.

Rabies. This is spread through the saliva of an infected animal, most commonly by the bite of a dog. The incubation period ranges from about one to six months. This long incubation period makes it possible for a rabies vaccination after a bite to develop immunity and prevent the disease from occurring. Death results from almost every case of the disease that develops in animals, including man.

Smallpox. Smallpox is spread by exudate from pustules on the skin of infected persons. The incubation period is from one to two weeks. It is a very serious, disfiguring, and highly fatal disease but is now largely controlled in civilized countries through vaccination.

Diseases Caused by Rickettsias

Epidemic typhus. This disease is transmitted by the body louse. It is a serious infection which tends to occur in epidemics whenever sanitation measures break down, such as during wars.

Rocky Mountain spotted fever. This disease is spread to man by the bite of ticks which may obtain it from wild animals which are carriers. It is a serious recurring fever which causes spots on the skin.

Psittacosis, parrot fever. This disease is spread to man by infected birds. It causes a pneumonia type of infection of the lungs.

Lymphogranuloma venereum. This is a venereal disease which is especially prevalent in the tropics. It causes extensive destruction of the tissues in the region of the reproductive organs unless properly treated.

REVIEW QUESTIONS AND PROBLEMS

1. Can the very small size of bacteria have anything to do with their very rapid rate of growth? Explain your answer. (Refer to Chapter 7 for helpful information.)

2. When the theories of bacterial cause of disease originated, some people said, "How can things so small that one cannot see them cause illness in anything so large as a human being?" What answer can you give to this question?

3. Flagella are so small that they usually cannot be seen on living bacteria. When you see a mixture of bacteria in water under the microscope, how can you tell which have flagella and which do not?

4. List the three major morphological types of bacteria and tell which may have flagella and which may bear spores.

5. Of what advantage is spore production to bacteria?

6. What experiments indicate that there is sexual union in bacteria?

7. What are the advantages of the cultivation of bacteria on artificial media?

8. What is the importance of the nitrogen-fixing bacteria?

9. Why are the plants in the legume family usually richer in proteins than plants in other families?

10. Why is decay an important beneficial activity of bacteria?

11. Fish is a food widely used by man. Name as many ways of preserving fish for food consumption as you can, and tell why each method prevents spoilage.

12. Honey will not spoil, but if we add water to honey the mixture will ferment. Explain.

13. Fresh Florida orange juice is now delivered to many homes in all parts of the country along with milk. What development in food preservation has made this possible?

14. Food poisoning appears to be more common after outdoor picnics than after meals in the home. What do you think could be the cause of this?

15. Under what conditions does the botulinus toxin develop, and how can you recognize food contaminated by this toxin?

16. At noon a third-grade student eats a chicken salad sandwich which his mother made early that morning from canned chicken. When he arrives home from school in the afternoon he begins vomiting and feels very ill. His distraught

mother asks you for advice. What type of food poisoning would you say the child had and why?

17. Tell how contagious diseases differ from infectious diseases.

18. In the United States during recent years the number of deaths from infectious diseases has shown a steady decline, but the number of deaths from noninfectious diseases has increased. What do you think has brought this about? (Consider such factors as the public health program, increasing medical and hospital care, increasing average age of the population, and so forth.)

19. A group of five men on a fishing trip drink from a polluted stream. Two of them develop typhoid fever, the other three do not. List the factors which might have been involved in this difference between the men.

20. What is meant by the incubation period of a disease?

21. How does the body combat the organisms which cause infectious disease?

22. What is meant by a "carrier" of a disease? Explain how this condition comes about.

23. Many Americans who go into the undeveloped regions of Africa soon develop malaria, dysentery, or other tropical diseases. Without prompt medical treatment they may die of these infections, yet many African natives live their entire lives in these regions without medical treatment. Explain.

24. List three ways in which a person can acquire an immunity to a disease.

25. Describe the conditions under which you would recommend serum inoculation and those under which you would recommend vaccination.

FURTHER READING

Bisset, K. 1970. *The Cytology and Life History of Bacteria.* Baltimore: Williams and Wilkins.

Bold, H. C. 1970. *The Plant Kingdom.* 3rd ed. Englewood Cliffs, N. J.: Prentice-Hall.

Burdon, K. L., and Williams, R. P. 1968. *Microbiology.* 6th ed. New York: Macmillan.

Carpenter, P. L. 1972. *Microbiology.* 3rd ed. Philadelphia: W. B. Saunders.

Delevoryos, T. 1966. *Plant Diversification.* New York: Holt, Rinehart and Winston.

Fraser, D. 1967. *Viruses and Molecular Biology.* New York: Macmillan.

Volk, W. A., and Wheeler, M. F. 1973. *Basic Microbiology.* Philadelphia: J. B. Lippincott.

16

The Algae

Mention the word algae to many people and they think of the green pond scum which is so common in stagnant pools of water, but the algae include a much wider variety of organisms. Some are one-celled and microscopic in size and some have millions of cells and may be as much as one hundred feet long. According to the two-kingdom method of classification, the algae are listed as plants because they have chlorophyll and most of them have cell walls composed of cellulose such as those found around the cells of higher plants. In a newer system of classification, such as the one given in Chapter 13, the phyla of algae are split between two kingdoms, Monera and Protista. The algae are of great economic importance, as will be apparent in the survey which will be presented in the balance of this chapter.

THE BLUE-GREEN ALGAE

The blue-green algae are in the phylum **Cyanophyta** under the kingdom **Monera.** They contain a blue-green pigment in addition to chlorophyll. Some have other pigments as well which cause them to show other colors such as purple, yellow, black, and red. The Red Sea derives its name from one of these species of blue-green algae which occasionally become so abundant that they give the water a reddish tinge. Some species of these algae have adapted themselves to life in water of hot springs which have temperatures up to 167°F. There are many such springs in Yellowstone National Park, where boiling water comes up out of the ground in beautiful pools and terraces. Around the edges of such springs the water is often a reddish color as a result of the abundant growth of blue-green algae in water which is hot enough to scald the hand of unwary tourists who reach in to feel the algae. This is a good example of the ability of living things to become adapted to extremes of environment through natural selection.

The blue-green algae are among the oldest organisms, from an evolutionary point of view. They appear to have undergone little change from their earliest form as one of the first kinds of cellular life which developed on the earth over two billion years ago. Some of the oldest known fossilized deposits of living organisms include blue-green algae.

Most of the blue-green algae have body forms consisting of cells joined together end to end to form long, threadlike **filaments.** The cells are also primitive in their organization. The chlorophyll and other pigments are not contained in definite plastids, as they are in other algae and higher plants, but are diffused through the protoplasm. These algae have no definite nucleus, as found in most cells, but in a central region in each cell there is a circular strand of DNA, a **genophore,** like the type found in the bacteria, which are also classified as Monera. Reproduction takes place by the duplication of the genopore, which is followed by fission of the cell.

Oscillatoria is a common genus of this phylum. It is found floating abundantly in freshwater ponds and lakes and there are also saltwater forms. Under the microscope it can be seen as slender filaments which move back and forth in an oscillating motion.

FLAGELLATED ALGAE

The phylum **Euglenophyta** includes single-celled forms possessing **flagella,** which are organelles used for locomotion. The flagellated algae, along with the other algae (with the exception of the blue-greens), are placed in the kingdom **Protista.** They have a definite nucleus and those species which have chlorophyll have it

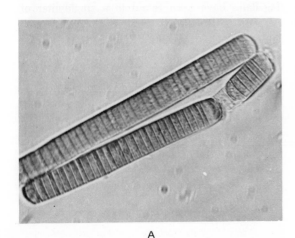

A

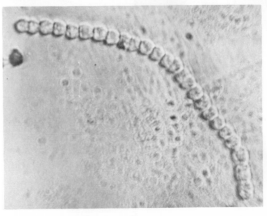

B

FIG. 16.1 Blue-green algae: *Oscillatoria* (A) and *Nostoc* (B). These are prokaryotic organisms which do not have a nucleus, chloroplasts, or other cellular inclusions found in more advanced algae.

contained in chloroplasts. The flagellated algae do not have rigid cell wall of cellulose, but they are rather surrounded by a flexible membrane which enables them to change shape considerably. Many are both autotrophic and heterotrophic. They may have chlorophyll and carry on photosynthesis, but they also may produce digestive enzymes which can digest food in their surroundings and absorb it. Some of them engulf food particles.

Euglena is the best-known genus in the phylum. This organism has a single elongated cell with a long flagellum attached at one end. It pulls itself along in the water by beating the flagellum down to the sides, somewhat like a swimmer using the breast stroke. It has a small red eyespot near the point of attachment of the flagellum, and this spot is sensitive to light. Nearby is a contractile vacuole which picks up excess water and empties it to the outside. Reproduction is conducted by longitudinal fission after a mitotic separation of the chromosomes.

A very similar genus, *Astasia*, has all the characteristics of *Euglena* except that it lacks chlorophyll and is entirely heterotrophic.

FLAME-COLORED ALGAE

The algae in the phylum **Pyrrophyta** contain a flame-colored pigment in addition to chlorophyll. Most are known as **dinoflagellates.** Each is a one-celled organism which bears two flagella. The red tide which occasionally appears in the Gulf of Mexico off the coast of western Florida is caused by excessive reproduction of the genus *Gymnodinium*, which is in this phylum. As many as sixty million of these dinoflagellates have been found in a single liter of water taken from such an area. These high concentrations clog the gills of fish and also release toxins into the water. Millions of fish may be killed and washed up on the beaches. The red tides seem to be positively correlated with years of heavy rainfall and it has been suggested that the heavy runoff of water over the heavily fertilized land of the costal area may carry so much fertilizer into the Gulf that the growth of this form of algae is excessively stimulated.

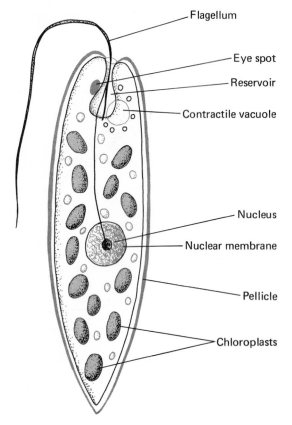

Flagellum

Eye spot

Reservoir

Contractile vacuole

Nucleus

Nuclear membrane

Pellicle

Chloroplasts

FIG. 16.2 *Euglena* is a strange organism which has chlorophyll contained in chloroplasts and produces digestive enzymes. *Euglena* is both autotrophic and heterotrophic.

GOLDEN-YELLOW ALGAE

The phylum **Crysophyta** includes the algae which have a yellowish-brown pigment in addition to chlorophyll. The most important members

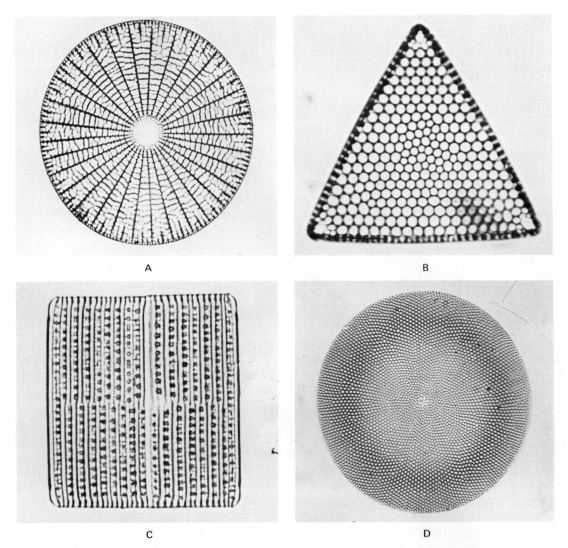

A

B

C

D

FIG. 16.3 Diatoms are one-celled algae with "glass" cell walls which are finely etched in beautiful patterns. They are so abundant, as a group, that they are very important in maintaining the world's supply of oxygen.

of the phylum are the one-celled **diatoms,** which are distinguished by their "glass" walls, formed primarily of silica (SiO_2), the most common ingredient of commercial glass. The diatom walls are covered by characteristic markings which, in beauty of pattern, rival the finest cut glass that man can produce, even though the diatoms are among the smallest of organisms. Diatoms are frequently used to test the resolving power and aberrations of microscopes because their markings are much smaller and more exact than man could ever produce.

In spite of their small size, the diatoms are among the most abundant plants on the earth by volume. They may be found floating in or lying on the bottom of almost every body of water in the world, both fresh and salt. They furnish much of the food for water animals, and the rich vitamin content of fish-liver oils is derived mainly from the vitamin-rich oil produced by the diatoms. The walls are highly resistant and do not decay when the protoplasm dies. The "shells" settle to the bottom of the water and may form great deposits. In California, deposits more than 1400 feet deep have been found. These are the remains of diatoms which flourished in an ancient sea which covered this area. These deposits are used as a base for cleansers, toothpaste, and automobile polish. Diatomaceous earth is also used for insulating and soundproofing material. The fats and oils in ancient diatoms contribute to the great oil deposits which have fueled our industrial complex for many years.

The diatoms are an important source of atmospheric oxygen which animals depend upon for their existence. The numbers of diatoms in the oceans are being reduced by continuing pollution on a grand scale, and some people are concerned that this may eventually lead to a reduction in the oxygen.

The cell wall of a diatom consist of two halves fitted together like the two halves of a pill box. During asexual reproduction, the protoplasm undergoes mitosis, the two halves of the cell wall separate, each carrying one-half of the protoplasm along with a full set of genes, and a new half shell is formed by each. Sexual conjugation has also been observed.

THE GREEN ALGAE

Phylum **Chlorophyta,** the green algae, includes more species than the other phyla of algae and also includes a greater variety of body forms. Some green algae grow in shaded, damp places on the land, but most live in the water, both in fresh and salt. Green algae have cell walls of cellulose and most of the cells contain large vacuoles, so the cells are similar to those of the plant kingdom. This is such a varied phylum that we shall survey several representative genera to give you an idea of their variety.

Protococcus is a one-celled form which may be found growing on trees, fences, and buildings where it is damp and shaded. In the Northern Hemisphere *Protococcus* is generally found on the north side because this is the side which never receives the drying rays of the sun during the middle part of the day. The individual plants cannot be seen with the naked eye, but they are present in such great numbers that they give a green color to the surface on which they are growing. When some of this green substance is scraped off and viewed under a microscope, it is seen to be composed of many one-celled plants; these often cling together after mitosis and occur in groups of two, three, four, or more. Each plant has a cell wall and a single large *chloroplast* which fills most of the cell. The cytoplasm fills the rest of the interior of the cell except for the small nucleus which lies within the cytoplasm.

Several genera of the green algae are motile. They possess **flagella** resembling those found in *Euglena* except that there are usually two flagella on each cell. *Chlamydomonas* is a one-celled form that lives in fresh water. The body consists of an elliptical cell with a single large cup-shaped chloroplast, a reddish eyespot, and two contractile vacuoles. The two flagella extend out from one end of the cell. The eyespot is sensitive to light, and the plant can respond to variation in light intensity by swimming toward regions with the best conditions for photosynthesis. Reproduction is more specialized than that found in *Protococcus*. In asexual reproduction the protoplasm within a cell divides twice in most species, thus giving rise to four new cells within the old cell wall. Eventually the old cell wall breaks open, releasing the new daughter cells, each of which now has developed its own cell wall and flagella.

Chlamydomonas also has a simple form of sexual union with a resultant mixing of genes from two individuals. In this process two individuals of opposite mating types unite at the region of the flagella. The mating types are designated as the plus and minus strains. The cell walls are then shed and the protoplasm of the two cells mingles into one body, known as a **zygote.** The nuclei unite to form a single nucleus within the zygote. The zygote then undergoes a series of two divisions, known as **meiosis** (see Chapter 8).

The cells of *Chlamydomonas* are normally **haploid (n)** in their chromosome number. This means that each cell has one of each kind of chromosome and one of each kind of gene. When two cells unite, however, each cell contributes a complete set of chromosomes, so there are two of each kind of chromosomes; the cells are now **diploid (2n).** In the two divisions of meiosis the chromosome number is reduced back to the haploid (n) state and remains so during the repeated asexual divisions by mitosis. Of the four cells produced by meiosis of a zygote, two will be plus and two will be minus. Any two cells that unite in sexual reproduction are known as **gametes,** and when the gametes are of the same size, as in this case, we call the process **isogamy.**

The zygote which is formed secretes a thick wall around itself and undergoes a period of dormancy before producing the four new haploid cells. In this condition it can withstand such unfavorable conditions as the drying up of a pond. Such thick-walled zygotes are known as **zygospores.** Sexual union is stimulated when conditions in a pond become unfavorable: for instance, a decrease in available nitrogen is one factor which brings about such union. Thus, the sexual union and formation of zygospores can have a survival value to the organism.

Volvox is another motile green alga with cells somewhat similar to *Chlamydomonas,* but the cells of *Volvox* cohere after cell division and form a rather large spherical aggregation of cells. The individual cells are thus joined together to form a colony. Each cell is connected to the others in the colony by small strands of cytoplasm. New colonies are produced when some of the cells in the hollow sphere divide and form new spheres, which grow within the parent colony. Sometimes as many as twenty or thirty small colonies may be seen inside a single parent colony. Eventually the parent colony breaks open and releases the daughter colonies.

Sexual reproduction is also found in *Volvox,* but in this case the gametes are of different sizes (**heterogamy**). The smaller gametes are known as **sperm;** they are motile and swim to the larger gametes, **eggs,** which remain stationary. When a sperm penetrates the egg, **fertilization** has been accomplished and a zygote is formed. The zygote forms a heavy wall around itself and thus becomes a zygospore.

Spirogyra is one of the most common genera of the green algae. By the naked eye it can be seen as a light-green, slimy scum on the surface or beneath the surface of stagnant ponds of water. Under the microscope this slimy material is quite beautiful. The plant body consists of a filament of long cells attached end to end, with a delicate ribbonlike chloroplast coiled around the periphery of each cell. In the center of the cell is a large vacuole within which the nucleus is suspended by delicate strands of cytoplasm that connect with the cytoplasm at the outer part of the cylindrical cell. The reason for the slimy look and feel of *Spirogyra* can be detected under the microscope; around the cell wall there is a gelatinous sheath which envelops the entire filament. Numerous spherical bodies on the chloroplasts, called **pyrenoids,** are centers of starch formation or starch storage. There is no specific asexual means of reproduction. The filaments become longer through mitosis of the cells but increase in number only when they are broken accidentally, as by a nibbling fish. A type of sexual reproduction known as **conjugation** occurs at definite seasons of the year—in the spring and in the fall. At the beginning of this process two filaments come to lie opposite each other, and small protuberances grow out from opposing cells and unite. These form a conjugation tube connecting the two cells. In the meantime, the protoplasm within each cell rounds up and forms a single large gamete. One of the gametes then migrates through the conjugation tube into the opposite cell and unites with the gamete in that cell. This forms a diploid zygote which then develops a thick coat around itself and becomes a free **zygospore** in the water as the old cell wall distintegrates. In this condition it can withstand much more unfavorable conditions, such as the drying of a pond in the summer or freezing in the winter, than can the active cells. When conditions become favorable, the zygote wall breaks open, meiosis occurs, and a new haploid filament grows from it.

Oedogonium is another filamentous green alga which may float or become attached to debris in ponds, lakes, and quiet streams. It is somewhat similar to *Spirogyra*, but it does not have the slime around the filaments and it feels like wet threads when it is rubbed between the fingers. The chloroplasts form a lacelike pattern within the cells. It reproduces asexually by **zoospores,** which are small bodies that can swim about actively by means of the hairlike cilia attached to one end. Each zoospore is formed by a single cell in the filament. The cell wall breaks open, and the zoospore is released into the water where it swims around freely for a time, but it then begins dividing and a new filament is formed. Sexual reproduction also occurs. Certain cells, **oogonia,** swell and form eggs, one egg in each cell. Sperm are produced in other cells, **antheridia.** Each antheridium produces two ciliated sperms which break out and swim freely in the water. Sperms are attracted to the eggs, possibly by some chemical substance produced by the eggs which diffuses out into the water. When a sperm reaches an oogonium, it penetrates the cell and fertilizes the

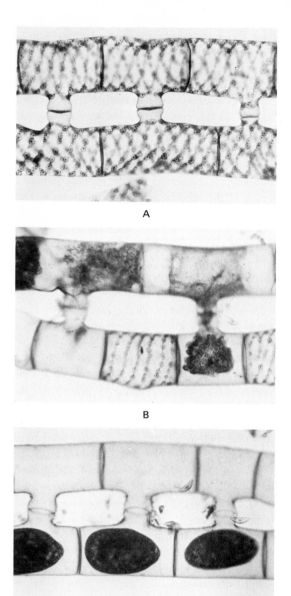

A

B

C

FIG. 16.4 *Spirogyra* is a green alga with a spiral chloroplast. It is shown here in conjugation. Two filaments of opposite mating types have united and formed a bridge between the cells (top). Protoplasm from the cells at the top move into cells at the bottom (center). The gametes have rounded up into zygotes which can withstand unfavorable conditions (bottom).

egg within. The zygote which is formed by the union of sperm and egg becomes a resistant **zygospore.**

Sometimes you may see a soft, green, feltlike growth on the damp soil at the edge of a pond or stream. Some of the growth may extend down into the shallow water. This growth is very likely a green alga, *Vaucheria.* When viewed under the microscope, this feltlike mass can be seen to be composed of many branched filaments. The filaments do not have crosswalls; each filament is like a hollow tube with many nuclei floating in the cytoplasm within the tube. This is known as a **coenocytic** body form. Asexual reproduction is by zoospores, and sexual reproduction is by means of sperms and eggs. The gametes are borne on specialized sex organs, **gonads.** A gonad is any organ, male or female, which produces gametes, and this terminology applies to animals as well as plants. In *Vaucheria* the male gonads are known as *antheridia* and the female gonads are **oogonia;** both names are used commonly in the plant kingdom.

THE BROWN ALGAE

The brown algae, phylum **Phaeophyta,** have a brown pigment in addition to chlorophyll. The largest of the algae are in this phylum; one of the kelps may grow to a length of over one hundred feet. All these algae are multicellular and most are marine. They can be seen attached to rocks along the seacoast and are called seaweeds.

The rockweed *Fucus* is a typical brown alga. It has a holdfast at one end which is usually attached to a rock. The rest of the body is a flat branched thallus formed by a series of Y-shaped branches, an arrangement called **dichotomous branching.** There are gas bladders along the thallus which keep the plant upright in the water. The tips of some of the branches are swollen and contain the sex organs. The antheridia produce sperms and the oogonia produce eggs. Both eggs

and sperms are released in the water and fertilization takes place there. The zygote which is formed grows into another plant.

Sargassum is another common genus of brown algae. It is of more than passing interest because it forms great masses of seaweed floating in the Atlantic Ocean. The area is known as the **Sargasso Sea,** a huge oval lying between the Bahamas and the Azores. The ships of Columbus, on his fateful voyage to the New World, are reported to have passed through a part of this sea, and many of the crewmen feared that they might be trapped in the seaweeds. Old sailors' yarns tell of small ships caught in the Sargasso Sea and slowly drawn into the center of the area by a spiral current. The ocean currents around the Sargasso Sea keep the algae moving in a giant oval. When hurricanes churn through the waters of the Atlantic in this region, many of these plants are scattered. The beaches of Florida's east coast may be littered with *Sargassum* after a hurricane.

The **kelps** are brown algae which are among the giants of the plant kingdom. They consist of one or several expanded portions, similar to leaves, connected by a stalk to a holdfast which is anchored to some solid support. Air bladders on the stalks keep the expanded portions

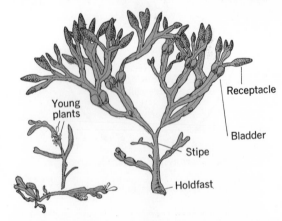

FIG. 16.5 *Fucus,* the rockweed, a marine brown alga. This drawing shows how young plants can grow out from branches of older plants.

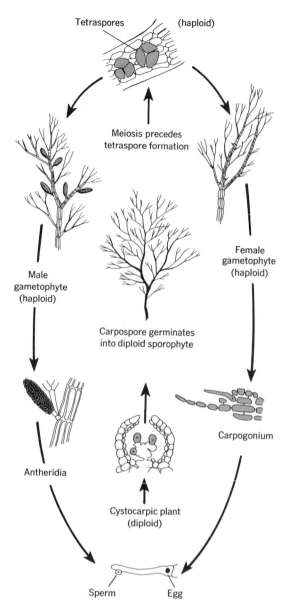

Tetraspores (haploid)

Meiosis precedes tetraspore formation

Male gametophyte (haploid)

Female gametophyte (haploid)

Carpospore germinates into diploid sporophyte

Antheridia

Carpogonium

Cystocarpic plant (diploid)

Sperm Egg

Union of gametes results in diploid zygote

FIG. 16.6 Reproductive cycle of *Polysiphonia,* a red alga. Note the three generations, one sexual and two asexual.

up near the surface. The kelps have considerable economic importance; a valuable commercial product, **algin,** is extracted from them. This is used by many ice cream manufacturers to impart a smooth consistency to their product and to prevent the formation of large crystals during freezing. Algin is also used in many ointments, cosmetics, shaving creams, drugs, and dental plasters. Since kelp grows in the mineral-rich sea water, it is rich in minerals in which land plants may be deficient. It is especially rich in iodine and potassium chloride. The leaflike portions of the kelps are sometimes collected, dried, and pulverized, and in this form they may be used to supplement a human diet liable to certain mineral deficiencies. Potassium chloride may also be extracted from kelp for commercial purposes. The Japanese people have long used kelp as a major vegetable in their diet.

THE RED ALGAE

This final phylum of the algae, **Rhodophyta,** includes forms which are mostly red because they possess a red pigment in addition to their chlorophyll. Nearly all are marine in habitat. Many genera in this phylum have beautiful body forms, with many delicate feathery branches. Some people collect them and mount them on cardboard as a hobby.

The red algae have considerable economic importance. They are eaten by people who inhabit the rocky coasts of Europe and North America, as well as by many Asiatics. Extracts from red algae are used as stabilizers in ice cream, puddings, candy, preserves, and pie fillings. A most important product is the substance known as **agar,** a gelatinous material which can be extracted from certain red algae. It is sold in a dried form; this can be dissolved in hot water and will cause the water to "jell" when it cools. Agar is **a** complex carbohydrate, and there are no enzymes in the human body for its digestion. Hence, it can be eaten to add bulk to the intestinal contents and thereby aid sluggish elimination. To the biologist, however, agar has a much more valuable use: it is added to culture media upon which bacteria are grown. This causes the media to solidify, somewhat like gelatin, but agar is much more satisfactory than gelatin for this purpose.

Reproduction in the red algae is more complex and more advanced than in any of the other algae. *Polysiphonia,* a beautiful, feathery marine form will illustrate it. The sexes are sepa-

rate; both male and female plants exist. On the male plant large clusters of **antheridia** develop, but each antheridium produces only a single sperm. Strange to say, this is a nonmotile sperm. When mature, the antheridia break open and release the sperms which float passively in the water. A female plant produces female organs, known as **carpogonia,** and each carpogonium produces a single large egg at its base. A long neck extends up from each carpogonium, and when a sperm chances to contact the neck, it adheres. The sperm contains enzymes which dissolve

A   B   C

D   E   F

FIG. 16.7 The red algae form many beautiful patterns and are often collected and mounted as shown here.

away the cell walls of the cells on the neck, and the nucleus of the sperm is then free to enter the neck and fertilize the egg within the carpogonium. Thus, a diploid zygote is formed; but meiosis does not follow, as in the other algae we have studied. Instead, the zygote divides by mitosis and produces a small filamentous plant, the **cystocarp,** which has diploid cells. Through continued mitosis the cystocarp produces spores, the **carpospores,** which are diploid. This is unusual because most spores produced by plants are haploid.

Each carpospore can grow into a diploid plant, a **tetrasporic plant.** On this plant meiosis takes place, and four haploid **tetraspores** are formed. Two of these will grow into male haploid plants and two into female haploid plants. Since these plants will produce gametes, we call them **male** and **female gametophytes.**

This complex cycle is known as **alternation of generations,** and it is a common phenomenon in the higher plants. In *Polysiphonia* there are three generations to the cycle. First, there is the large haploid generation, the gametophyte. Second, the diploid cystocarp is a generation which grows from the zygote; it is a **sporophyte generation** and produces the diploid carpospores. Third, there is **tetrasporic plant,** also a sporophyte generation, which is diploid also and which produces tetraspores by meiosis. The cycle is complete as the haploid tetraspores grow into the large gametophyte generation. The sexual gametophyte generation alternates with the two asexual sporophyte generations. *Polysiphonia,* along with some of the other red algae, is unusual in that it has two sporophyte generations in the

cycle. In nearly all plants there is only one sporophyte alternating with one gametophyte.

EVOLUTION OF BODY FORM

The algae show considerable variation in complexity of body form, and it is interesting to speculate on the possible evolutionary sequence of the development of the various body forms. Without doubt, the blue-green algae are the most primitive and represent some of the earliest kinds of cellular life on the earth. The one-celled flagellated forms like *Chlamydomonas* show an advance in the development of definite nuclei, chloroplasts, and locomotor attachments. From such plants there appear to have been two main lines of development. The spherical colonies probably arose because of adherence of the cells after division. *Volvox* represents a rather high stage of development of this series; there are other colonies of cells which are intermediate between the one-celled forms and the large colonies. Another line of development could have led to filamentous forms. If flagellates adhered together after division and if division was always in one plane, a filament would be formed. The flagella have been lost in these filamentous forms. The flagellated zoospores which are formed by many of the filamentous algae, such as *Oedogonium,* are very much like the one-celled flagellated algae. Branched filaments represent another step in increasing complexity. From this it is a short step to the sheetlike layers of cells, such as those found in *Ulva.*

REVIEW QUESTIONS AND PROBLEMS

1. Sexual reproduction has not been observed in the blue-green algae. Could this have anything to do with the fact that these algae have remained in their primitive prokaryotic state? Explain.

2. Under the two kingdom system of classification, *Euglena* was very hard to assign to either plant or animal kingdom. What characteristics made this so difficult?

3. *Gymnodinium* can always be found widely scattered in the water of the Gulf of Mexico, but at certain times they become so abundant that they form a red tide which kills millions of fish. What is a possible explanation of this great increase?

4. The algae all have chlorophyll which is green, but they are of many other colors. Explain how this can be.

5. The dumping of various pollutants into the ocean off our East Coast has resulted in a great decline in the number of diatoms in these waters. Ecologists have become concerned about this decline. Why should we worry about the number of these little algae which seem to be of no use to man?

6. Why is *Protococcus* found most abundantly on the north side of trees, fences, and buildings in the United States?

7. Use *Chlamydomonas* to illustrate the principle of alternation of generations with respect to chromosome number.

8. How could a colony of cells, such as *Volvox*, have arisen from one-celled organisms, such as *Chlamydomonas?*

9. Why is the term isogamy used to describe sexual reproduction in *Spirogyra,* and heterogamy for sexual reproduction in *Oedogonium?*

10. The Japanese people have never been troubled with the iodine-deficiency type of goiter. Explain.

11. Some of the brown algae have a holdfast on rocks deep under the water, but the upper part of the plant stays near the surface. How is this managed?

12. In what respects does *Polysiphonia* differ from *Chlamydomonas* in its alternation of generation cycles?

13. Trace the possible steps in the evolutionary development of body form in the algae.

FURTHER READING

Alexopoulos, C. J., and Bold, H. C. 1967. *Algae and Fungi.* New York: Macmillan.

Brock, T. D. 1969. *Biology of Microorganisms.* Englewood Cliffs, N. J.: Prentice-Hall.

Prescott, G. W. 1968. *Algae: A Review.* Boston: Houghton Mifflin.

Wilson, C., et al. 1971. *Botany.* New York: Holt, Rinehart and Winston.

17

The Protozoans

We never cease to marvel at the great variety of life which becomes visible when a drop of murky pond water is viewed under the microscope. Organisms of many shapes and sizes can be seen, many of which move back and forth rapidly, while others move at more moderate speeds, and still others seem to be immotile. Some organisms are reproducing; others are being devoured by the carnivorous forms. Life in all its complicated relationships goes on here as it does among the larger organisms in a forest.

Many of these small organisms are protozoans. The phylum **Protozoa** is a very important group under the kingdom **Protista.** Protozoa was considered a phylum under the animal kingdom when a two-kingdoms classification system was used because the protozoans do show many animallike characteristics. We often think of them as very simple forms of life because they are so small. Yet, since certain vital life processes must be carried out by all forms of life, there must be some organelles which perform all these functions in the one cell. The protozoan cell must function as a nerve cell when it receives stimuli; it must function as a muscle cell when it moves; it must produce all the enzymes necessary for food digestion; it must function in reproduction. Hence, a protozoan must be considered a rather complex organism.

In spite of their small size, the protozoans have considerable economic importance. Protozoans are food for larger water animals which, in turn, are eaten by still larger forms. At the

232

top of this food chain are many fish and other water animals which are food for man. Protozoans are also valuable for research studies. For example, *Tetrahymena geleii* is a protozoan with nutritional requirements very much like those of human beings. This protozoan is used in research on the absorption and use of various foods, the effects of poisons, the agents stimulating cell division, and so on. The results of such research have given us clues about the activities of human cells and have been of value in cancer research.

On the debit side of the economic ledger, we find that some protozoans have accommodated themselves to a parasitic existence, utilizing man and other animals as hosts. Many serious human diseases are traceable to protozoan parasites. **Malaria, amoebic, dysentery,** and **African sleeping sickness** are all protozoan infections, and the control of diseases caused by protozoan infection is one of the great public health problems today. The World Health Organization and other agencies have made great progress in many developing countries, but much remains to be done.

PROTOZOA WITH FALSE FEET— THE SARCODINA

The class **Sarcodina** includes those protozoans which form **pseudopodia** (false feet). These are temporary projections which can be pushed out from any part of the body and withdrawn just as readily. They may be either slender and fingerlike or broad and spreading.

Amoeba is the genus most frequently studied as an example of the class. It may be found at the bottom of freshwater ponds and streams among the debris formed by decaying vegetation. There are also some marine forms which are found in the seas. The large freshwater form, *Amoeba proteus,* is a typical species. When viewed under the microscope, this protozoan appears as a small mass of protoplasm surrounded

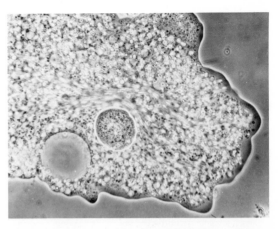

FIG. 17.1 A portion of an *Amoeba proteus* showing the nucleus and the clear contractile vacuole. Note the clear ectoplasm around the outside. The flowing movement of the protoplasm is evident in the blurred region, just above the nucleus.

only by a thin plasma membrane. When in an active state, it constantly changes shape as it moves about. A fingerlike pseudopodium may be pushed out in the direction of movement. Then the rest of the body flows into the pseudopodium. This movement is made possible by changes in the sol-gel relationship of the protoplasm in different parts of the cell. The inner portion of the protoplasm is more fluid than the outer portion, and a flow occurs when a weakness develops in the outer portion. To illustrate, let us think of an amoeba at rest; all pseudopodia are withdrawn and the animal is in a spherical shape. Then a weakness develops in the thicker outer protoplasm, the thinner protoplasm in the center flows through this weak spot, the plasma membrane is forced out, and a pseudopodium is formed. As the protoplasm spreads out at the tip of the pseudopodium, it assumes more of the gel consistency and forms the outer portion of the protoplasm. The amoeba has control over these changes in consistency and thus can control the formation and direction of movement of the pseudopodia.

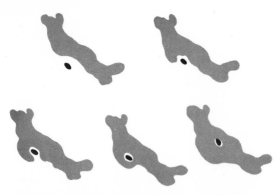

FIG. 17.2 An amoeba ingests a food particle by surrounding it with pseudopodia and engulfing it.

This means of locomotion is known as **amoeboid movement** and is found in other types of cells as well. The white blood cells of the human body, for instance, use pseudopodia for movement and also for engulfing disease organisms which may have gained entrance to the body.

The amoebae use the process of **engulfing** to obtain their food. Various plants and animals, as well as bits of decaying organic matter, serve as food. The amoeba approaches food, and may actively pursue it if it happens to be a motile organism. Pseudopodia are thrown out on either side of the food and gradually close in on it. We might say that the amoeba just wraps itself around the food. This is the means of **ingestion.** There is no mouth—the food can be taken in at any point of the body. Once inside, the food is contained within a **food vacuole,** which might be thought of as a temporary stomach. Here digestion takes place. There is a thin membrane around the food and the enzymes diffuse through this membrane into the vacuole and break the food down into soluble substances which can diffuse through the membrane and be utilized by the cell. When digestion has been completed, there will still be an indigestible residue remaining in the vacuole. This must be expelled from the amoeba, and it is done by the process known as

egestion, which is somewhat like ingestion in reverse. The vacuole moves to the outer edge of the cell and bursts through the plasma membrane, thus carrying the waste out.

Amoeba is an aerobic organism and therefore must have a constant supply of free oxygen. This is a small problem, since the water in which amoebae live contains dissolved oxygen from the atmosphere. This diffuses readily into the cell through the plasma membrane. The carbon dioxide generated in **cellular respiration** easily diffuses out through this membrane into the surrounding water. The importance of dissolved oxygen in the water to amoebae can readily be demonstrated. When water is boiled, the dissolved gases are driven off. If amoebae are placed in water which has been boiled and cooled rather quickly, they will die because of oxygen deficiency. In time, however, more oxygen from the air will dissolve in the water and then the water will support living amoebae.

Excretion is another vital life process which is accomplished by simple diffusion. The excess minerals which remain after the catabolic phase of metabolism must be expelled. They are in solution and pass out through the plasma membrane.

The amoeba has another biological problem which must be solved—the elimination of excess water. While water is a necessary part of any living cell, it cannot be allowed to accumulate in excessive quantities. The amoeba obtains water in three ways; some is generated in metabolism, some is ingested along with the food, and some enters through the plasma membrane by osmosis. These three sources provide much more water than is needed; without some method of eliminating the excess, the amoeba would swell from internal water pressure and might even burst. This event is forestalled by the functioning of a **contractile vacuole.** It can be seen as a clear, spherical body which gradually grows larger as it accumulates more cell water. Then suddenly it seems to burst and disappear. Careful observation shows that when it has grown large it moves to the edge of the cell and contacts the outer plasma membrane which then bursts. This allows the water to flow out while the inner membrane of the vacuole becomes the outer membrane of the amoeba at this region. A small new vacuole then forms near the center of the cell and the process is repeated. When amoebae are placed in water where the concentration of minerals is equal to or greater than the concentration of minerals in the protoplasm, no vacuoles are formed. This indicates that osmosis is the main source of the excess water. Saltwater forms of amoebae do not form contractile vacuoles unless they are removed to fresh water.

Under favorable conditions an amoeba will continue to grow until it reaches a certain size; then it undergoes mitosis and forms two cells of equal size. Thus amoebae never grow old and die of old age; they may live indefinitely, even though they split into many pieces. But death is very frequently caused by destruction by other forms of life and extreme changes in environment. We may say, then, that amoebae die frequently by accident, but never from old age. An amoeba is potentially immortal: those that you see under a microscope have lived for millions of years through countless fissions.

Amoebae in the active state which we have been describing are said to be in the **vegetative condition,** but, as such, they are very delicate things and easily destroyed. Some species of amoebae are known to pass through unfavorable periods, such as the drying up of a pond, by developing a protective coating or **cyst** around themselves and reducing metabolism to the barest minimum necessary to maintain life. In this encysted form they are able to withstand unfavorable periods and return to the vegetative state when conditions become normal again.

Amoebae display **irritability** even though they have no distinguishable areas of sensitivity. They will move away from a strong light source; this can be watched by viewing them under the microscope while shining a light from one side of the slide. It is also demonstrated by the fact that they are most abundant in the darkest portions of laboratory culture jars. They will react to various chemicals, moving toward some that attract them and away from others that repel them. They can distinguish between food material and other matter in the water; they engulf only that which they can use as food. They will actively pursue small water animals in an effort to capture them for food. They respond to heat and cold, always moving to the area of the water which is nearest their optimum temperature. They respond to vibrations in the water by pulling in extended pseudopodia. In their various reactions to their environment, these single cells display sensory perceptions akin to those of the highest forms of animal life. Senses similar to touch, taste, smell, sight, temperature sensitivity, and even hearing are represented in these perceptions.

There is another genus in the Sarcodina which is amoebalike in its body form. This is the genus *Entamoeba* (Greek *entos,* within) so named because it lives within the body of other forms of life. The first species of the genus that we will consider is *Entamoeba coli,* which lives in the body of man and other vertebrates. This organism may be found in the colon (large intestine) of about half of all people, but fortunately it is harmless, since it feeds on the indigestible food or feces which pass through the colon. It is considerably smaller in size than *Amoeba proteus* and somewhat more sluggish in its reactions.

A closely related species, *Entamoeba histolytica,* attacks the lining of the intestinal wall, and thus is a serious parasite of man. In the intestine of man, these organisms attach themselves to the cells and secrete a tissue-dissolving enzyme that destroys the cells. This permits the entamoebae to go deeper into the wall of the intestine and destroy more cells, until they may produce a serious ulceration of the colon. In extreme cases the wall of the colon is perforated and millions of bacteria, which are always present in a human colon, are liberated. They pass out into the body cavity and cause an infection known as peritonitis, which may be fatal. Fortunately, there is a tough muscular layer in the intestine that normally checks the penetrations of these organisms and thus prevents such serious results.

The disease caused by this infection is known as **amoebic dysentery.** The irritation of the colon usually causes the victim to suffer from severe diarrhea accompanied by general weakness of the body. Examinations of the watery stools reveal thousands of the entamoebae in the active vegetative state. Many of the cells will also be seen to have formed cysts by means of which the disease may spread to other persons. Cysts are produced from the vegetative cells by the elimination of water, which concentrates the protoplasm, and by the formation of a protective coating around the outside. The vegetative cells die rather quickly after they leave the protective warmth and moisture of the body, but the cysts can live a much longer time.

The cysts may be spread from one person to another through contaminated food or drinking water. The food may be contaminated by flies that have come in contact with fecal material from infected persons or by food handlers that harbor the organisms in their bodies. Many persons may carry the organisms in their intestines without showing any symptoms of the disease. In such a person a balance has been struck between the virulence of the organism and the resistance of the body, so that neither is able to overcome the other; yet the parasites are able to live and produce their cysts, which may infect others. Drinking water may be contaminated from sewage seeping into surface wells or streams which are used as drinking water without proper purification. There was even one famous case of confused plumbers making the wrong connections in a hotel and allowing some sewage to get into running ice water in the hotel rooms. Many

fatal cases of amoebic dysentery resulted from this mix-up.

There are different strains of *Entamoeba histolytica* which vary in their effects on the body. Some are so mild that many persons are infected without knowing it. In tropical countries the infection is more serious.

Any organism that causes a disease is known as **pathogenic.** *Entamoeba histolytica* is accordingly a pathogenic protozoan. A third species of *Entamoeba, E. gingivalis,* is found in the mouths of as many as 70 percent of the population. These organisms are especially abundant around the base of the teeth where they join the gums. If you scrape some of the material from this region of your mouth and stir up the material with a 1 percent salt solution and keep the slide warm while studying it under the microscope, you are very likely to find that you are harboring this organism in your own mouth. They are more sluggish than the free-living amoebae, but you can see them throw out broad pseudopodia. Human customs permit this organism to spread from mouth to mouth in the vegetative condition and it is not known to produce cysts. It is not generally pathogenic, but there is some evidence that it may play some part in the damage done by the disease of pyorrhea. For the average person with a healthy mouth, however, there is no need to worry about the presence of these organisms and, human nature being what it is, there is little likelihood that we shall ever be able to control their spread.

PROTOZOA WITH FLAGELLA— THE MASTIGOPHORA

The members of the class **Mastigophora** have a more definite shape than the Sarcodina and are characterized by the presence of a whip-like attachment known as a **flagellum.** The flagellum is used in swimming. For this reason they are known as flagellates and the class name is sometimes given as **Flagellata.**

Some of the Mastigophora invade the blood stream of higher animals and cause serious diseases. Among the many serious diseases found in Africa is **African sleeping sickness.** This disease is generally caused by *Trypanosoma gambiense,* one of the Mastigophora that lives in the blood stream of man. It is spread through the bite of the **tsetse fly,** a blood-sucking insect which is abundant in many sections of Africa. Death usually terminates untreated cases of African sleeping sickness, but drugs have been discovered that will kill the organisms and effect a cure when properly used. (There is a virus-caused form of sleeping sickness in the United States which afflicts domestic animals, especially horses, and sometimes human beings. This should not be confused with the protozoan-caused African sleeping sickness.)

Another species of this genus, *Trypanosoma brucei,* infects domestic animals, and in sections of Africa where tsetse flies abound it is impossible to keep such animals. Wild animals, such as zebras and antelopes, serve as carriers of this parasite without themselves being seriously affected by it, but death usually follows infection for horses, cattle, and pigs.

Dumdum fever, or **kala-azar,** is a serious disease in India, northern Africa, and parts of South America. It is caused by a torpedo-shaped flagellate, *Leishmania donovani,* and it causes enlargement of the liver and spleen as well as numerous sores on the skin. This often fatal disease is spread from person to person by sandflies. Great epidemics of kala-azar occurred in parts of India from 1890 to 1900 and decimated whole villages. Only in recent years has it been brought under control.

There are several other less dangerous flagellates that often inhabit some part of the human body. One called *Trichomonas vaginalis* has its home in the human vagina, where it is usually benign, but it may cause inflammation

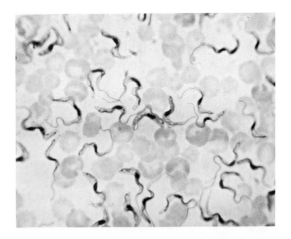

FIG. 17.3 *Trypanosoma gambiense* in human blood. These pathogenic protozoans cause African sleeping sickness.

under certain conditions; another species of *Trichomonas* lives in the tartar of our teeth; and a third, in the human intestine. Another protozoan in this class, rather large and bearing several flagella, is *Giardia lamblia,* which has been found in 15 percent of all human beings. It lives in the large intestine and sometimes causes a type of diarrhea which is known as giardiasis. It is difficult to eradicate and infections may last for years.

SPORE-BEARING PROTOZOA— THE SPOROZOA

The members of this class are all parasitic and produce spores as a part of their reproductive life cycle. Spores are small, asexually produced reproductive bodies that are usually quite resistant to unfavorable surroundings. Many plants produce spores, but only this class of animals has this mode of reproduction. Only one genus in this class infects the human body; but this one, *Plasmodium,* is a very serious pathogenic protozoan because it causes the disease of

malaria. The word "malaria" (Italian) means "bad air," a name that goes back to a time when the disease was attributed to the damp night air. We now know that it is mosquitoes, which are likely to be out in the night air, that spread malaria. This is one of the best understood diseases in the world today. We know how it is spread, how to treat it successfully, and how to prevent it. But it still ranks as the most serious infectious disease on our globe.

At one time there were large sections of our own country where malaria was prevalent, but today its spread has been reduced to the vanishing point in the continental United States. In spite of new cases brought in from foreign countries, the vigorous program of the public health departments plus the prompt use of new antimalarial drugs keeps the disease suppressed.

There are several types of malaria caused by different species of malarial parasites, but the one most prevalent and once common in our southern states is tertian malaria, which is caused by *Plasmodium vivax.* The word "tertian" means "third," and this name was chosen because the chills and fever tend to reappear every forty-eight hours. By an old Roman method of reckoning, it was customary to count the day that something happened as the first day, the day following as the second, and the day following

that as the third. Thus, even though the chills were only two days apart, they were said to recur on the third day.

Infection in a person may result through the injection of spores that are present in the saliva of an infected mosquito. The injected **sporozoites** enter cells lining the capillaries of the liver, spleen, and other internal organs. There they multiply for about a week or ten days and then break out into the blood stream, where they attack the red blood cells. Each sporozoite may enter a red blood cell. There it rounds itself up into a ball with the nucleus at the edge and a vacuole in the center that makes it look like a signet ring. This is called the **signet-ring stage.** It gradually takes on an amoebalike form that fills the center of the red blood cell. Then it breaks up into fifteen to twenty spores, the **merozoites.** The weakened cell then bursts, approximately forty-eight hours from the time it was first invaded, and liberates the merozoites into the blood plasma. Each of these spores may enter another red cell and the cycle will be repeated until literally billions of red cells are being destroyed every forty-eight hours.

The first symptoms of the disease appear within two or three weeks after infection. When the many red cells burst at about the same time, liberating the spores and the accumulated wastes of the parasites, the body reacts to this poison with a chill. This is followed by a rather high fever as the body speeds up metabolism in response to the chill. Then, as the fever subsides, the victim will feel better, but will be weakened and anemic from loss of red blood cells. Without treatment, the same cycle will be repeated forty-eight hours later, and so on, until the victim dies or his natural body resistance brings the disease under control. There are many malarial carriers that harbor the parasites and serve as a reservoir of infection but do not show serious symptoms themselves.

This cycle spreads the parasites within the body, but if the parasite is to survive, it must have some means of spreading to other

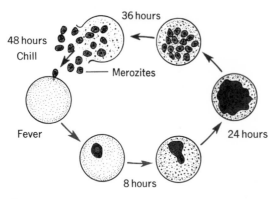

FIG. 17.4 The cycle of tertian malaria in human blood cells.

persons, because every host must die sooner or later, taking all his parasites to the grave with him. We have already mentioned the mosquito as the agent of spread. This mosquito is not just any kind of mosquito, but the female *Anopheles* mosquito. It must be a female because male mosquitoes are vegetarian in their diets and suck only the juices of plants. The females also suck plant juices, but they will suck blood when they can get it. It must be an *Anopheles* mosquito because this is the one in which *Plasmodium* can continue its life cycle; the physiology of other mosquitoes is different and the parasite cannot live in their bodies.

Certain of the merozoites grow in human blood plasma and become male and female **gametocytes.** If the proper kind of mosquito bites and sucks up at least one of each kind of gametocyte, the cycle continues in the mosquito's stomach. The female gametocyte becomes an egg without much change, but the male gametocyte

gives off from four to eight slender sperms which break off and swim to the egg. One unites with the egg and produces the **zygote.** This zygote becomes amoeboid in nature and squeezes itself between the cells of the stomach wall and forms a cyst on the outside of this wall. The zygote divides within this cyst until large numbers of spores, the sporozoites, are produced. The cyst then breaks and the sporozoites migrate through the body cavity, some of them reaching the salivary glands of the mosquito. This cycle requires ten to twelve days, so a mosquito is not infective until at least this long after it has bitten a person with malaria. Once it becomes infective, however, it is likely to remain so for its entire life, which may be several months. Each zygote produces about 10,000 sporozoites.

The cycle is completed when the infected mosquito bites a human being. As she bites she injects a little saliva into the wound, which prevents the blood from clotting. This saliva will

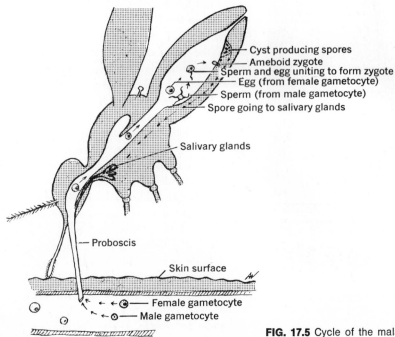

FIG. 17.5 Cycle of the malarial parasite in the female *Anopheles* mosquito.

carry the sporozoites into the body of the person she is biting, and the cycle begins again.

Malaria control is one of the major health problems of the world. Many resource-rich regions of the world remain undeveloped because of the prevalence of malaria. Some of these are being opened to development by programs to fight malaria. A campaign to find and treat persons carrying the malarial parasite will do much to eradicate the disease.

PROTOZOA WITH CILIA— THE CILIATA

Members of the **Ciliata** class are abundant in the water of ponds and streams which is rich

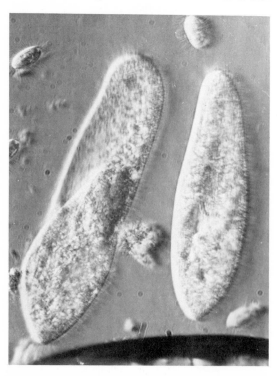

FIG. 17.6 *Paramecium caudatum,* a protozoan ciliate. The oral groove, which shows clearly in the organism on the left, leads to the mouth. The large macronucleus can be seen as a bulge to the right of the oral groove.

in decaying organic matter. They can be cultured rather easily in the laboratory by placing dead grass or other leaves in pond water and allowing the mixture to stand for a wek or two. As the water becomes turbid with suspended particles of decaying matter, the ciliates will usually be rather abundant. All members of this class bear **cilia.** Cilia are very small hairlike structures which can beat back and forth in locomotion. By controlling the beating of the cilia, these organisms can move in any direction.

Paramecium is a good representative genus of this class, often selected for laboratory study to illustrate a complex protozoan. A number of different species are commonly available. One, *Paramecium caudatum,* is found abundantly in many ponds rich in organic matter. It is a slipper-shaped animal, its body covered by a tough yet flexible **pellicle.** This pellicle enables the paramecium to bend its body and then return to its original shape when it relaxes the bending force. The plasma membrane lies under the pellicle. Cilia are embedded in the pellicle and are arranged in the form of a spiral down the longitudinal axis of the animal. This arrangement causes the paramecium to revolve and take a spiral course through the water when swimming. A single animal will have about 2500 cilia on its body. There are also tiny bulbs of a gelatinous liquid, the **trichocysts,** embedded in the pellicle. When irritated, the paramecium may squirt this liquid out into the water, where it hardens and forms sticky threads which serve in a protective capacity; enemies are likely to get tangled in these threads and thereby be rendered harmless.

Many of the life processes of *Paramecium* are similar to those found in *Amoeba,* but there are noteworthy differences. For one thing, there is a definite opening for ingestion and another for egestion. The entire process of **digestion** can be demonstrated if we place some yeast cells stained with congo red on a microscope slide with some paramecia. Under the microscope we can see the masses of red yeast cells being caught up by cilia which line the oral groove and being forced through the gullet into the body. There the yeast cells are rounded up into food vacuoles. As time passes the color of the yeast within the vacuoles will change from a deep red to blue. This indicates that enzymes have passed into the vacuoles and are digesting the yeast. The color shift means that the contents have changed from alkaline to acid. A similar change takes place in the human stomach. The vacuoles become smaller as the food is digested, and the soluble part diffuses out into the protoplasm. When only an indigestible residue remains, it will be expelled through the anal pore which lies slightly below the gullet. This pore is quite small and cannot be seen except when the animal is in the act of egestion.

There are two **contractile vacuoles** in *Paramecium caudatum*, and these remain in a fixed position. Radiating canals bring water to the vacuoles from other parts of the cell. When the vacuoles become full, they contract rather quickly and force the water to the outside. A small pore connects with each vacuole having this function. It requires only about fifteen seconds for a contractile vacuole to fill and empty its contents.

Respiration and **excretion** are carried on by diffusion through the plasma membrane as in *Amoeba*. The pellicle is highly porous and offers no hindrance to this diffusion.

In stained specimens of *Paramecium caudatum* two nuclei can be seen—a larger **macronucleus** and a smaller **micronucleus.** The micronucleus has a full set of genes, but they do not produce RNA. The macronucleus, on the other hand, has many sets of genes and they produce the RNA. This multiple condition of the genes results from repeated gene duplication without nuclear division.

Asexual reproduction occurs by transverse fission in which both nuclei divide along with the cell. The micronucleus divides by mitosis, but there is a simple splitting of the macronucleus. Since there are many sets of genes in the macronucleus, however, it is certain that each of the two nuclei will get a number of diploid sets of genes. Fission takes place about every ten to twelve hours when conditions are favorable.

It appears that the micronucleus is unnecessary; it can be removed and the paramecium continues to live, grow, and reproduce by fission. The organism cannot have sexual reproduction, however, because the micronucleus is the only part which can undergo meiosis. Conjugation is the means of sexual union. As worked out in *P. aurelia,* two animals of opposite mating types unite at their oral grooves. The macronucleus of each disintegrates and the micronucleus undergoes the two divisions of meiosis, forming four haploid nuclei in each cell. Three of the four disintegrate, leaving a single haploid nucleus remaining in each cell. This one divides into a larger immotile micronucleus and a somewhat smaller motile micronucleus. The small one of each moves across a connecting protoplasmic bridge and unites with the larger micronucleus of the other animal. Then the two animals separate. So far there has been no actual reproduction, but there will be two divisions immediately following conjugation, accompanied by some nuclear phenomena which are shown in Fig. 17.7. Conjugation is not necessary for the continuation of the species in *Paramecium.* One research worker isolated a single individual and followed 15,000 generations of descendants over a period of twenty-five years without conjugation ever occurring. It does, however, accomplish an important objective. In conjugation there is a trading of genes between two individuals. This brings about new combinations of hereditary traits and results in a variety in the descendants that would not be possible without such a genic exchange. Variety is important, because through variety there can be a natural selection of the most fit individuals which will continue the propagation of the species.

Although there are no morphological distinctions between the two paramecia which undergo conjugation, it has been shown that there is some physiological difference which is akin to sex. Not any two paramecia can conjugate. There will never be conjugation among the descendants

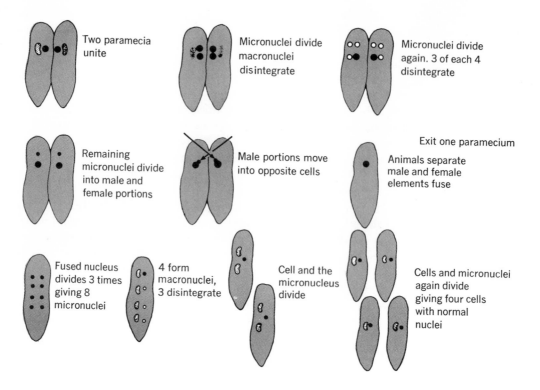

Two paramecia unite

Micronuclei divide macronuclei disintegrate

Micronuclei divide again. 3 of each 4 disintegrate

Remaining micronuclei divide into male and female portions

Male portions move into opposite cells

Exit one paramecium

Animals separate male and female elements fuse

Fused nucleus divides 3 times giving 8 micronuclei

4 form macronuclei, 3 disintegrate

Cell and the micronucleus divide

Cells and micronuclei again divide giving four cells with normal nuclei

FIG. 17.7 Conjugation in *Paramecium caudatum*. Two protozoans of different mating types unite at their oral grooves and a part of the micronucleus of each passes to the partner.

of a single individual; the conjugating animals must be of different **mating types.** We cannot call one of them male and the other female, however, for in some species of paramecia several different mating types have been found and any one can conjugate with several of the others. Still we can think of these as a beginning of the physiological differences which distinguish sexes in higher forms.

There are quite a number of other genera of ciliates, including *Tetrahymena*, which we mentioned in the early part of this chapter. One species in this class is parasitic on human beings and also lives in pigs and chimpanzees. This is *Balantidium coli*, which may be found in the intestine of about 80 percent of the domestic swine in some localities. It is a large ciliated organism which forms numerous round, thick-walled cysts. These occasionally find their way into the human digestive system. Here they lose their thick walls and start to multiply. They penetrate the inner lining of the intestine and often form ulcers which cause an attack of dysentery. Infected persons, if not promptly treated, pass numerous cysts which may spread the disease.

STALKED PROTOZOA— THE SUCTORIA

A number of protozoans which are ciliated in the early stage of life lose the cilia and become attached to some object by means of a long stalk when they grow older. They capture food, which often consists of other protozoans, by means of tentacles. Some of these tentacles have small suction cups on the end, a characteristic which gives the class its name. Young animals are formed from a portion of the parent cell. They develop cilia, swim around for a time, and then form a stalk and turn into the adult form. An example is *Podophyra*.

REVIEW QUESTIONS AND PROBLEMS

1. Describe some of the ways in which human life is affected by protozoans.
2. Describe the changes in the protoplasm which are involved in the formation of a pseudopodium.
3. We say that amoebae die frequently by accident but never from old age. Explain what is meant by such a statement.
4. Describe an experiment which would demonstrate that amoebae can react to unfavorable stimuli. Use one different from those given in this chapter.
5. One experiment showed that the rapidity of the filling and emptying of the contractile vacuole was negatively correlated with the concentration of dissolved materials in which amoebae were kept. Explain why.
6. Of what advantage is the cyst formation by members of the Sarcodina?
7. People who have visited regions of central Africa may return to the United States with the trypanosomes of sleeping sickness in their blood, yet we do not worry about the disease being spread here. Explain.
8. Many ecologists have recommended the abandonment of the use of DDT as an insecticide because it is not easily degradable and has a harmful effect on the eggs of birds. Some health authorities feel that its use should not be abandoned in regions where malaria is prevalent. Explain.
9. A group of drug addicts in New York City were sharing the same needle for heroin injections. One had just returned from Vietnam. Several in the group developed malaria. Explain what happened, giving scientific names.
10. Suppose *Plasmodium* had no sexual reproduction. How would this affect its spread?
11. What are trichocysts and how are they used?
12. What results would you expect if you removed the macronucleus from a paramecium?
13. How does food ingestion in *Paramecium* differ from that of *Amoeba*?
14. Conjugation does not occur among paramecia which have all descended from a single individual. Explain why.

FURTHER READING

Jahn, L. 1949. *How to Know the Protozoa*. Dubuque, Iowa: W. C. Brown.

Kudo, R. R. 1971. *Protozoology*. 5th ed. Springfield, Ill.: Charles C. Thomas.

18

The Fungi

The fungi have traditionally been classified as plants without chlorophyll because many of them have definite plantlike characteristics. More recently, the tendency has been to place them in a kingdom of their own. They appear to have been an offshoot of some primitive plantlike protistan ancestors, but they have followed a line of development different from that of the plant kingdom. Some may have had chlorophyll at one time, but as they developed other ways of getting food, they lost it. Fungi are sometimes called **absorptive heterotrophs** because they absorb food from their surroundings. They produce enzymes which pass out of their cells and digest food which is thus rendered soluble and which they can then absorb. Some can also engulf food particles.

The fungi have an advantage over green plants; they can grow in the dark. Most of them are **saprophytic;** that is, they obtain nourishment from dead organic matter. Many are **parasitic,** or obtain nourishment from living organisms and harm the host organism in the process. **Facultative parasites** can use both methods of nutrition, while **obligatory parasites** can derive nourishment from living things only.

THE SLIME MOLDS

As the name suggests, slime molds exist as a mass of slime at some time during their life cycle. In some species of slime molds cells unite to form a large **plasmodium** which creeps about like a giant amoeba, engulfing food particles as it moves. Slime molds avoid the light and dry air, so in the daytime they are found underneath decaying leaves, in the crevices of decaying wood, or underneath the surface of soil which is rich in organic matter. They are also abundant in the feces of larger animals. Cow dung in a damp place will probably harbor extensive slime mold growth. At night when the air contains high humidity, they come out on the surface. You can see them best with a flashlight at night. Some biologists feel that the slime molds, because of their animallike motion, arose from animallike protistans and are not related to the other fungi.

In a typical species, such as *Dictyosteium discoideum*, the plasmodium is a large multi-nucleated body. When food is in short supply or when the conditions begin to become dry, the plasmodium changes its usual habitat and moves out into the air and light. As it dries into a crust-like surface, **sporangia** grow up and each sporangium produces thousands of **spores**. This method of reproduction is very much like that of the other fungi, and this is the primary reason for including the slime molds in the fungi. The spores are carried by air currents and widely scattered. When they reach an area with the proper conditions of moisture, they will germinate and produce amoebalike organisms. Many of these spores come together from a wide area and unite to form a large plasmodium. The cells fuse, making the plasmodium one unit with many nuclei. You might wonder what force draws these widely scattered cells together. Laboratory experiments show that when a few cells happen to contact one another, they unite and begin releasing **cyclic AMP** (adenine monophosphate), which is similar to the energy-giving ATP but which lacks two of the phosphate groups attached to ATP. Cyclic AMP acts as an attractant to other cells of the species, and they move along a concentration gradient toward the greater

concentration of this substance. As they come together, they fuse and even more cyclic AMP is released and more cells are attracted to the plasmodium from greater distances. As many as 200,000 cells have been observed to fuse into one very large plasmodium.

A few of the species of slime molds are parasitic. One of these is the **clubroot fungus** which invades the roots of certain plants. The

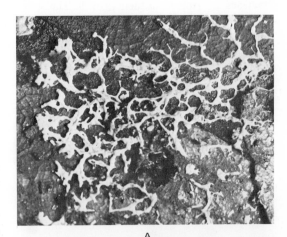

A

B

FIG. 18.1 Slime molds. The active vegetative stage is shown in (A) and the reproductive, spore-bearing stage in (B). The two different phases of the life cycle have made molds difficult to classify.

clubroot disease of cabbage is caused by this slime mold. The amoebalike cells produced by the spores in the soil penetrate the roots and then unite to produce the slimy plasmodium. This eventually dries out and produces spores which remain in the soil ready to infect the cabbage plants the next season. The presence of the parasitic plasmodium causes the roots to grow abnormally large and eventually to rot. As a result, the heads are stunted in their growth. The spores will infect only the cabbage, and the amoebalike cells from the spores will die if no cabbage plant is available.

THE ALGAELIKE FUNGI

The algaelike fungi are so called because some of them resemble algae in body form and methods of reproduction. They are in the class **Phycomycetes,** which is the first class of the

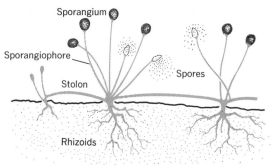

ASEXUAL REPRODUCTION

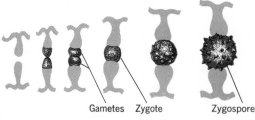

Gametes   Zygote   Zygospore

SEXUAL REPRODUCTION

FIG. 18.2 *Rhizopus,* the bread mold, reproduces asexually by spores and sexually by conjugation.

true fungi, phylum **Eumycophyta.** Typical members of this class have a vegetative body composed of filaments that run through and on the substance upon which they are growing. This vegetative body is known as the **mycelium.** The mycelium bears **sporangia** which produce **spores.** There is also sexual reproduction in most of these fungi. This takes the form of conjugation with **isogamy.** The common bread mold, *Rhizopus,* is typical of this class.

The **bread mold** gets its name from the fact that it grows very commonly on bread. Today in the United States we see less of it than we did in past years because most bread from bakeries today contains a mold inhibitor. Without such inhibitors bread that is kept in a warm, moist place for two or three days is likely to be covered with this mold, which is found on many foods besides bread. When the mold first begins to grow, it can be seen as a mass of delicate, cottonlike filaments on the surface of the food. Then little stalks can be seen extending upward from the filaments, and each stalk terminates in a little white ball. Finally, the white balls become larger and turn black as the black spores mature inside. A detailed examination of the mature mold growth with a magnifying glass or wide-field binocular microscope shows that the **mycelium** consists of small rootlike **rhizoids** that extend down into the bread and absorb nourishment for the entire plant, horizontal **stolons** that extend out over the surface of the bread, and the **sporangiophores** that extend up and bear the balllike **sporangia.** Filaments of all three types may be referred to as **hyphae.** These filaments are **coenocytic;** that is, they have no crosswalls and therefore form long tubes containing many nuclei. This condition we first describe in the alga *Vaucheria.*

You might sometimes wonder where the mold comes from that appears to arise by spontaneous generation wherever food is kept under conditions favorable for mold growth. It is not hard to understand when we learn of the enormous number of spores produced. A growth of *Rhizopus* on a small scrap of bread thrown in the garbage can be covered by thousands of sporangia, and each sporangium may bear as many as 70,000 **spores.** When the thin sporangium wall ruptures, these tiny spores are caught by the air currents and can thus be widely distributed. There are probably some spores of the bread mold floating in the air around you now, and no doubt you have inhaled some of them while you have been reading about it. It is small wonder, therefore, why we have so much trouble with the molds. They develop whenever the spores find suitable food, moisture, and warmth.

Under certain conditions **conjugation** can be seen in the bread mold. Two stolons grow side by side and each puts out a protuberance which forms a **gamete** on the end. The gametes from the opposite filaments fuse and form a **zygote.** After a period of dormancy the zygote may germinate and produce a new mycelium. It has been found that there are two strains of mold, designated as plus and minus. Whenever growths from spores of these two strains are brought together, conjugation takes place. This is a primitive sort of sexual differentiation. We cannot call one strain male and the other female because this is isogamy and the gametes are alike, but there is some physiological difference between the two strains.

This class of fungi includes many molds which invade stored fruits and vegetables, and they are the most important cause of rotting of such foods. It also includes a group of plants known as the **water molds.** These grow on dead animals and plants in the water, but they may also be parasitic and attack living fish and other water animals. In the tropics some of these infect the skin of human beings making it dangerous to wade or swim in fresh water. Such skin infections are difficult to cure.

Other parasitic forms in this class include the **downy mildews** and **white rusts** which infect many plants.

THE SAC FUNGI

Members of the class **Ascomycetes** are characterized by the formation of saclike structures which usually contain four or eight spores. This is the second class of the true fungi, phylum Eumycophyta. It is a very large class and one of considerable economic importance. The fungi in this class cause many important plant diseases. Examples are the **chestnut blight,** which has almost wiped out the chestnut trees in North America, and the **Dutch elm disease,** which has been so destructive to our shade trees. It also includes the **powdery mildews** which are destructive to many species of higher plants.

The **cup fungi** are ascomycetes which are mostly saprophytic and grow on decaying organic matter. We see them as cup-shaped bodies, but these appear only after there is a considerable growth of hyphae in the organic matter upon which they feed. The cups range in size from a fraction of an inch to several inches, and in some species they are brightly colored. The **asci,** spore-bearing sacs, appear within the cups. There is a sexual union of male and female gametes, and a zygote is formed. This divides three times and produces the eight spores found within each ascus. When the spores are mature, the ascus suddenly contracts and the spores are forced out through a hole in the end. When many asci contract at about the same time, as a result of changes in temperature or humidity, the spores may appear as a white cloud above the fungus.

The **morel fungi** are sometimes called sponge mushrooms because they look like mushrooms with spongelike caps. The caps are folded and convoluted, and the asci are borne within the folds. Morels grow in soil rich in organic matter, usually in densely forested regions. Growth starts from spores which send out the mycelium underground, and the reproductive body then grows up and the spores are produced. Morels are edible and are highly prized by many for their flavor.

Truffles are also edible fungi, highly prized

FIG. 18.3 The blue-green mold, *Penicillium,* has infected the orange on the left. This has caused the fruit to soften and it has collapsed. The spore masses appear at the top. An uninfected orange is shown at right.

by gourmets. They grow entirely underground and produce fruiting bodies as large as three inches in diameter. In France, where truffle hunting is a profitable sideline for many farmers, hogs are trained to detect the presence of these underground fungi by odor. Having detected them, the

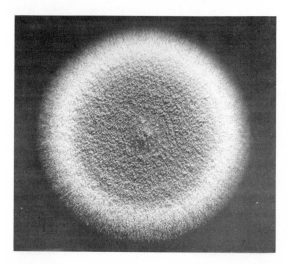

FIG. 18.4 *Penicillium* growing on a culture medium. The cottony filaments spread out from the center and the blue-green powdery sporangia develop in the center. We obtain the antibiotic penicillin from this mold.

hogs then root them up. They are a delicacy in French cuisine.

The **blue-green molds** include a group of ascomycetes that produce powdery spores which are usually blue-green, although some of them may be red, yellow, or black. They grow on nearly all kinds of organic material when moisture and warmth are sufficient. A pair of shoes left in a closet during a damp, warm summer may come out in the fall completely covered with a greenish powder as a result of the growth of one of these molds. Much of the spoilage of fruit and vegetables in shipment and storage is caused by blue-green molds. The citrus mold *Penicillium* causes millions of dollars worth of damage to citrus fruits each year. One of these citrus molds, *Penicillium notatum,* has provided us with one of mankind's most valuable drugs, **penicillin.** The flavor and color of Roquefort or other blue cheese is due to another of these fungi.

Ergot is a parasitic ascomycete which grows on wheat, rye, and wild grasses. The spores are produced in dark bodies, the ergots, which protrude from the head of the grain. This fungus is of particular significance to man because it is poisonous when ingested in sufficient quantities. There have been many cases of ergot poisoning resulting from eating bread made with ergot-infected wheat or rye. It causes a contraction of the smooth muscles and the walls of the blood vessels. Gangrene may develop in regions of the body deprived of blood due to the contractions of the blood vessels, and abortions may occur in pregnant women because of the contractions of the smooth muscles of the uterus. Death sometimes results in cases of heavy intake of the fungus. Modern milling methods remove any ergot which may be present on grains, but cases of ergot poisoning still occur where the milling is done locally. Domestic animals may also suffer from eating ergot-infected grasses. Extracts of ergot have medical uses, for example treatment of migraine headache and stimulation of the smooth muscles of the body.

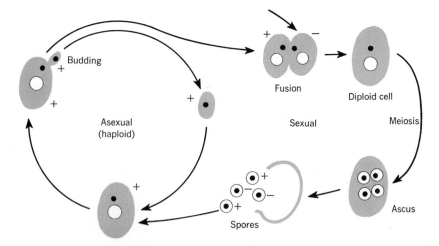

FIG. 18.5 Sexual and asexual reproduction in baker's yeast, *Saccharomyces cerivisiae.*

The **yeasts** are ascomycetes which form no fruiting bodies. They differ from the majority of the fungi in that they are one-celled organisms. These cells reproduce by **budding** rather than by fission. In fission there is an equal division of the cytoplasm, but in budding the spindle figure of mitosis is formed near the edge of the cell, and one set of the duplicated chromosomes is incorporated into a small bud. In some species this bud soon breaks away and grows into a full-size yeast cell, but in other species the cells may adhere after budding and form a long chain resembling the myceia of the other fungi.

The common baker's and brewer's yeasts are the best-known examples of these plants. A cell of such yeasts is an oval containing a large vacuole and a small nucleus within the cytoplasm. The buds break off shortly after they are formed, so single cells are generally seen when these yeasts are viewed under the microscope. The yeasts are classified as ascomycetes because at times the entire yeast cell may become an ascus and spores are produced within. These spores can withstand unfavorable conditions and become active when conditions turn more favorable. They may be carried in air currents like the spores of bread mold. There is also a fusion of cells in a sexual union as a part of the life cycle of the yeasts.

As we learned in Chapter 11, the yeasts have the power to obtain energy from carbohydrates anaerobically in a process known as fermentation. Carbon dioxide and alcohol are liberated in this reaction. We use yeast in bread to make it rise. The carbon dioxide forms tiny bubbles which are trapped by the dough; heat causes these bubbles to expand even more, and the result is light and tasty yeast-raised bread. We need not fear becoming intoxicated by the alcohol thus formed; the heat of baking drives most of this off before bread is ready for eating. Alcohol production, however, is the prime purpose of fermentation in other uses. In almost every region of the globe today and far back in the historical past, we find fermented carbohydrates used by man for beverage. The most abundant source of carbohydrate is generally used. Fruit juices are universally popular, and corn, rye, rice, and wheat are used in proportion to their relative abundance. In some sections of the world even milk is fermented. In certain parts of the Near East, it is an insult to a host for a visitor to refuse to partake of fermented camel's milk.

Strong concentrations of alcohol are inimical to the growth and normal activities of living tissue; a person will die if the concentration reaches as much as 0.5 percent in the body fluids and even the yeast cells that produce alcohol will ordinarily die when the concentration reaches about 5 percent. Some special yeast strains have been selected, however, that can withstand up to 16 percent alcohol. Higher concentrations are obtained by distilling; alcohol, being lighter than water, distills off first. Thus, whiskey containing 50 percent alcohol can be produced by distilling a fermented mash that contains no more than 8 or 10 percent alcohol.

A

B

C

D

FIG. 18.6 Some of the larger club fungi. A puff ball is shown in (A) and an earth star in (B). A coral fungus is shown in (C) and a shelf fungus is shown in (D).

Opinions may vary on the desirability of human use of alcohol in beverages, but even if this use were discontinued, alcohol production would still be a very important commercial process. Alcohol is indispensable in many chemical industries; it is invaluable in scientific laboratories; it is an important preservative and an excellent antiseptic and has many other valuable qualities.

THE CLUB FUNGI

The club fungi, class **Basidiomycetes,** are in the third and final class of the true fungi, phylum **Eumycophyta.** Most of the saprophytic forms in this class have large and conspicuous fruiting bodies. The spores are borne on small clubs, **basidia.**

The **gill fungi** include many species commonly known as **mushrooms** or **toadstools.** They grow in rich soil where there is an abundance of organic matter and plenty of moisture. They begin their development as a branching network of underground hyphae, which later send up the umbrella-shaped fruiting bodies. At first the fruiting body appears as a small ball or button on a stalk. Later the lower portion of the ball pulls away from the stalk and spreads out and forms the umbrella. The gills radiate out from the stalk and the spores are borne within the gills. When mature, the spores break off the clubs on which they are formed and sift down through the gills. Beautiful spore prints can be obtained from mushrooms. Cut off the cap close to the stalk and place it on a piece of paper. Cover it with a bowl to prevent air currents from disturbing the spores. After several hours lift the cap. The spores which have fallen on the paper will have left an outline of the gills. The spores of different mushrooms differ in color, so use two caps and place one on black paper and one on white paper, since some spores are light in color and some are

dark. Colors such as white, pink, yellow, brown, purple, and black will be found.

Many mushrooms are edible and furnish an important source of food for man, but unfortunately some are quite poisonous. No person should attempt to collect and eat wild mushrooms unless he is thoroughly familiar with the species which are edible. Only about seventy species out of several thousand are poisonous, but there is no easy way to recognize the poisonous forms. One of the most poisonous mushrooms is also one of the most beautiful. It is the destroying angel, *Amanita verna,* a snow-white species with sparkling drops of liquid on its top surface. This spe-

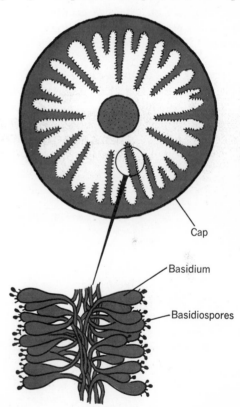

Cap

Basidium

Basidiospores

FIG. 18.7 Cross-section through a mushroom cap showing the gills bearing the clublike basidia with spores at their tips. Detail of the basidia is shown below.

cies can be fatal if only one cap is eaten, although the symptoms of poisoning will not appear until six to sixteen hours after it has been eaten. It is so poisonous that one may become ill by absorption of the poison through the skin. In the author's laboratory two students once became ill the day after they had handled some of these mushrooms in dissections.

The **pore fungi** can be distinguished by the small holes (pores) on the underside of the fruiting bodies. These are common as the **shelf** or **bracket fungi** which grow from the trunk and limbs of living and dead trees. Like the other fungi, this fruiting body appears only after there has been an extensive growth of hyphae. The hyphae penetrate the bark and wood of living trees and are serious parasites. By the time the

shelves appear, a tree is already heavily infected. The spores get on the exposed wood of trees injured by cutting, falling branches, fire, or other causes. The hyphae from the germinating spores produce enzymes which digest the cellulose, and thus they can continue to spread. These fungi are largely responsible for the decay of living trees. Many species continue to grow on a tree after it is dead, and some fallen trees may be literally covered by the fruiting bodies. A new ring of growth is formed each year by most species, and the age of a shelf fungus can be determined by counting these rings on the upper surface.

The **puffballs** bear spores within a ball. Some of them, like the **earthstar,** develop a small pore at the top through which the spores emerge in a cloud whenever the ball is pressed. In others

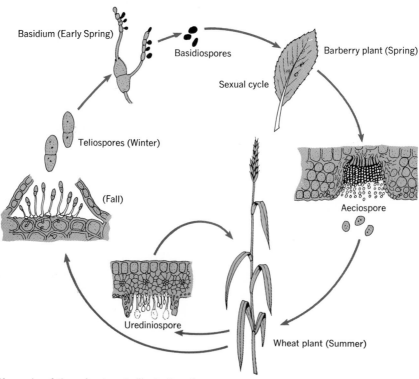

FIG. 18.8 Life cycle of the wheat rust, illustrating the alternation of generations on two different hosts.

the walls of the ball break down and the spores escape. Some of them resemble golf balls, and when they are seen growing on a golf course are likly to cause some confusion in golfers. When young, the puffballs are edible.

The **stinkhorn fungus** should be mentioned because its disagreeable odor is sure to attract attention. Its fruiting body is a hollow stalk with a swollen tip. When first formed, the tip is covered with a membrane, but when this membrane breaks away it exposes a sticky material with a strong odor resembling that of a putrefying animal body. This odor attracts certain flies and other insects. Some of the spores adhere to the bodies of these visiting insects and are carried away and spread to other localities.

The **coral fungus** is a fleshy form which looks very much like some of the branching corals. The fungi show beautiful pastel shades of yellow, pink, purple, and gray, as well as white. The spores are borne on clubs on the outside of the branches.

A large group of the club fungi are called **rusts** because of the rust-colored mass of spores that they produce at one stage of their life cycle. They are all parasitic on higher plants, and some of them infect two different species in an alternation of generations. **Wheat rust** is one of the most important economically. Rust infection may reduce a wheat crop by as much as 50 percent when it is widespread.

The wheat plant is infected when it is young and the fungus sends its hyphae through all parts of the plant, stunting its growth and interfering with the production of normal heads of wheat. When infection is well established, reddish spore masses appear on the leaves and stems of the wheat plant. These are the **uredinia,** and they bear many tiny **urediniospores.** As the uredinia break open, the urediniospores are carried by the wind and may start the infection in other wheat plants. As the wheat plants reach maturity, the fungus produces black spore masses, the **telia,** which bear **teliospores.** These are the winter

spores and pass through the winter months on the stubble from the harvested wheat. In early spring they germinate and each spore produces a small **basidium** which bears four **basidiospores.** Strange to say, the basidiospores cannot infect other wheat plants. If they are to germinate, they must be carried by the wind to the common barberry bush. Within the barberry leaves this fungus carries on a sexual phase of its life cycle, centered in small yellowish spots, **spermagonia,** on the upper surface of the leaf. Finally, large dark spore masses, the **aecidia,** appear on the lower surface of the leaves. These bear the **aecidiospores** which break out and may be carried by the wind to the young wheat plants, where the cycle begins again.

This complex cycle is worth the paragraph of description that we have given it because it exemplifies the parasite with two different hosts. In one host the parasite has a form of sexual reproduction. We shall find other parasites, both plant and animal, which have similar alternation of hosts.

Wheat rust infection may be controlled by widespread destruction of barberry bushes in localities surrounding wheat fields. This must be a cooperative enterprise, because the spores from the barberry may be carried as far as ten miles. In the southern areas of wheat production this cannot entirely eliminate the rust, because some of the urediniospores can live through the winter and infect the spring growth of wheat. The planting of rust-resistant varieties of wheat will reduce the loss from this infection.

The **smuts** are similar to the rusts. **Corn smut** is probably the best known of this group of fungi, which infect many members of the grass family. The spores appear as a large black mass, covered by a membrane, which develops on the ears of corn. The number of grains borne on the ears is greatly reduced by such growths. There is no alternate host. The spores fall to the ground, germinate, and form **basidiospores,** and these infect corn seedlings the next spring. Corn seeds are often treated chemically before planting to destroy spores that may be on the seeds, and crop rotation may be practiced to avoid infection from spores in the ground.

A MUTUALLY BENEFICIAL RELATIONSHIP—THE LICHENS

The **lichens** are a large group of plants that grow in nearly all parts of the earth. They are able to withstand extremes of temperature, drought, and other unfavorable conditions. In arctic regions they even survive beneath the snow and furnish an important item of diet for the reindeer, which paw the snow to get at the lichens beneath. We probably know them better as the crustlike, gray-green growth found on the north side of nearly all trees. They may also be found growing on barren soil and even on bare rock.

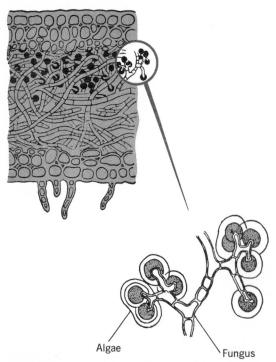

FIG. 18.9 An enlarged portion of a lichen showing the relationship between the algal and fungus portions.

The success of the lichens in living under conditions of great hardship is due to their makeup. They are formed from not one but two different plants living together in harmonious association with mutual benefit. One of the plants is a fungus which bears its spores in sacs (in most lichens) and is a member of the Ascomycetes. The other plant is an alga, usually a green alga such as *Protococcus*. The plant body is composed of the tightly interwoven filaments of the fungus within which are the cells of the alga. The fungus has rhizoids which serve to attach the plant and help in absorbing water and minerals. It can also absorb water like a sponge during a rain and hold it for future use. Minerals can also be absorbed from the dust that blows onto the plant. These features enable the lichens to live on bare rock which cannot support other plants. The algae benefit from the association by being protected by the fungus and furnished with the water and minerals necessary for food manufacture. The fungi also benefit by using some of the food manufactured by the algae.

This is a good example of the biological relationship known as **mutualism,** in which two organisms live together in a mutually helpful relationship. The fungus in this case is not a parasite, since the alga is not harmed but rather is benefited by the association. We will find many examples of mutualism in both the plant and animal kingdoms as we continue our study.

Lichens are pioneers in the plant kingdom. They can grow on bare rock where other plants cannot exist. As they send their rhizoids down into the tiny crevices of the rock, they actually help break the rock up into small particles. When lichens die, their decaying bodies mix with the small particles of broken rock and give it organic content. This is the beginning of soil upon which other plants can grow and continue the process of soil formation.

Lichens are commonly grouped according to their general appearance into three types: **crustose,** crustlike; **foliose,** leaflike; and **fruitcose,** shrublike.

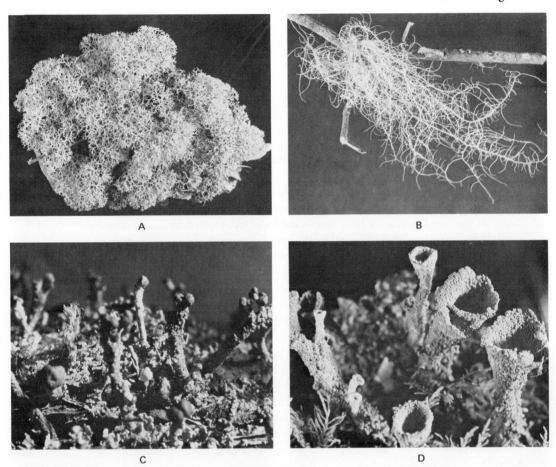

FIG. 18.10 Lichens are found in a considerable variety of shapes. Shown here: fruiticose (A), old man's beard lichens (B), the match head (C), and trumpet lichens (D).

FIG. 18.11 Lichens are pioneers in the formation of soil. This photograph shows crustose lichens growing on bare rock. Lichens begin the breakdown of rock and the production of soil upon which other organisms can grow.

THE IMPERFECT FUNGI

There are some true fungi for which we have never been able to demonstrate sexual reproduction. These are lumped together and called the imperfect fungi (*fungi imperfecti*). Many of the plant **blights** and **wilts** are caused by fungi in this group. A number of them are parasites of man. **Ringworm** is an infection of the skin which may be found on young calves, dogs, cats, and human beings. It is sometimes even found on the comb of chickens. Many people think this infection is caused by a worm because it frequently forms a ring-shaped area of skin irritation. When it occurs on the feet, we call it **ath-lete's foot.** It is spread by small pieces of the fungus growth which break off and are transferred to new areas where growth can take place.

REVIEW QUESTIONS AND PROBLEMS

1. What advantages and disadvantages do the fungi have in comparison with chlorophyll-bearing plants?
2. Distinguish between parasitic and saprophytic organisms.
3. What is a facultative parasite? Give an example.
4. Some of the fungi appear to have descended from chlorophyll-bearing ancestors. Explain how this loss of chlorophyll might have taken place.
5. Describe the life history of a slime mold, pointing out the phases which are animallike and those which are plantlike in nature.
6. Suppose you take three pieces of home-made bread and place several drops of water on each. You put one in the refrigerator, you wrap one in wax paper and leave it at room temperature, and you leave the third piece at room temperature without wrapping. Which piece is most likely to develop mold growth first and why?
7. In many regions of the country, leather goods may mildew during the summer months. Analyze each of the following places in terms of suitability for storage of leather goods during warm, humid weather: a refrigerator; an attic where the sun causes a very high temperature but a lower relative humidity; a basement storage room where the temperature is cooler than in the rest of the house but the humidity is very high; an air-conditioned room where both temperature and humidity are lower than in the rest of the house.
8. Describe conjugation in the bread mold.
9. What is the relation between the words **mycelium, stolons, rhizoids, sporangiophores, sporangia, spores,** and **hyphae?**
10. How do the yeasts differ in body form from the other sac fungi?
11. What are the commercial uses of yeasts?
12. How does budding compare with fission as a means of reproduction?
13. Compare the method of spore formation in the sac fungi and the club fungi.
14. How can you distinguish the gill fungi from the pore fungi?
15. How are trees injured by the shelf fungi?
16. Trace the life cycle of the wheat rust.
17. Wheat rust is more easily eradicated in the northern wheat-growing states than in the southern states. Why is this?
18. What is meant by alternation of hosts as applied to parasites?
19. What is mutualism, and how is it illustrated by the lichens?
20. Explain how the lichens may act as pioneers in soil formation.
21. What is the cause of ringworm infection?

FURTHER READING

Alexopoulos, C. J. 1962. *Introductory Mycology.* New York: John Wiley and Sons.
———, and Bold, H. C. 1967. *Algae and Fungi.* New York: Macmillan.
Christensen, C. M. 1965. *Molds and Man.* New York: McGraw-Hill.
Funder, S. 1968. *Practical Mycology.* New York: Hafner Publishing.
Hale, M. 1961. *Lichen Handbook.* Washington, D. C.: Smithsonian Press.
Webster, J. 1970. *Introduction to Fungi.* London: Cambridge University Press.

19

Liverworts and Mosses: The Bryophytes

The phylum **Bryophyta** includes the simpler members of the plant kingdom. These organisms lack an elaborate system of vascular tubules for conduction of materials from one part of the plant to another. The bryophyta are limited in size because they are dependent upon movement of materials from cell to cell by osmosis, diffusion, and active transport. Bryophytes can live on the land, but only in moist places, because they do not have the extensive root systems necessary for water absorption. Their reproduction depends upon water since they produce motile sperm which must swim to the female sex organs. There are two major classes in this phylum. They are known commonly as liverworts and mosses.

THE LIVERWORTS

If you had lived in ancient Rome and fallen sick, the doctor might have diagnosed your ailment as "liver trouble" and would probably have given you some small liver-shaped plants to eat, in acccordance with an old belief that plants with the shape of some human body organ were good for treatment of disorders of that organ. We know today that there is no such relationship, but we still give these plants the name "liverworts" because of their former use as liver medicine and their rough resemblance to the human liver. They are in the class **Hepaticae.**

Botanists have identified about 9000 species

of liverworts. The typical liverwort body consists of a small green **thallus** that grows flat on the ground with delicate rootlike **rhizoids** extending down into the soil. The rhizoids are short; the water must move from cell to cell by osmosis since there are no little tubes (as there are in higher plants) which carry the water. The liverworts, therefore, must live in a very moist environment—in fact, the soil on which they grow must be so damp that we might call it mud. The direct rays of the sunlight are drying, so the liverworts live mostly in shaded habitats. The most common place to find them is along the shaded banks of creeks, rivers, and similar locations that combine high humidity, shade, and abundant moisture. This need for moist environment confirms the view that the liverworts have not completed the transition to land.

For reproduction, the dependence of the Hepaticae upon water is even more pronounced. The reproductive organs are characteristic of water plants, not land plants; sperm are produced which must swim to the egg through water. There are a few liverworts that have returned to the water habitat and live floating on the water surface, with rhizoids extending down into the water.

A typical liverwort, such as *Marchantia*, has a body that is a small thallus about an inch in length. Its upper surface is divided into polygonal areas, each of which contains an air pore that looks like a light dot in the center of each area. This air pore admits air to the porous upper surface of the thallus. At times little cups can be seen on the large veins of liverworts. These are **gemmae cups**, which produce small multicellular bodies having the power to grow into new plants. These new plants may extend the vegetative growth of the plants where they were produced, or they may be carried to new locations by water to start new growths.

Special organs, known as **receptacles**, function in sexeual reproduction. The male receptacles and female receptacles are produced on different plants, so we say that *Marchantia* is **diecious,** in contrast to **monecious** plants that bear both sex organs on one plant. The male receptacle consists of disk borne on a stalk which rises well above the thallus. The male sex organs, **antheridia,** are formed near the top of the disk. The female receptacle is also stalked and looks somewhat like a miniature palm tree. The female sex organs, **archegonia,** are borne on the underside of the fingerlike processes at the top of the stalk. Only one egg is produced in each arche-

FIG. 19.1 A liverwort, *Conocephalus conicus.* This is a thallus from a plant which grows flat on the ground.

gonium, but many sperm are produced in each antheridium. When the plants are wet with rain or dew, the sperms come out on the disk. A drop of water falling on the disk at this time may scatter the sperms over a distance of several feet, enabling them to reach female plants where they can swim up the female receptacles and fertilize the eggs.

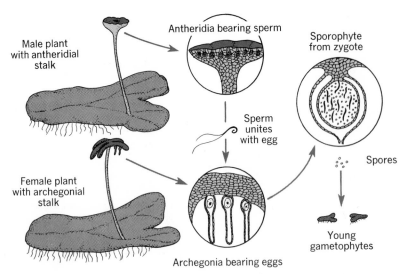

Male plant with antheridial stalk

Antheridia bearing sperm

Sporophyte from zygote

Sperm unites with egg

Spores

Female plant with archegonial stalk

Young gametophytes

Archegonia bearing eggs

FIG. 19.2 Life cycle of a typical liverwort. The male and female gametophytes produce sperm and eggs which unite to produce zygotes. A microscopic sporo- phyte grows from the zygote and produces spores. New male or female gametophytes grow from the spores.

Since the thallus produces sex organs which in turn produce gametes, this part of the life cycle of the liverwort is called the **gametophyte.** Then after fertilization the zygote produces a microscopic body, called the **sporophyte** because it produces a number of spores. The sporophyte remains attached to the female receptacle where the egg was fertilized until the spores are shed. Each spore then grows into another thallus, the gametophyte. Thus, we have the gametophyte generation alternating with the tiny sporophyte generation. Such alternation of generations is studied more thoroughly in Chapter 20.

THE MOSSES

The word "moss" is to the plant kingdom what the word "bug" is to the animal kingdom. Whenever an American sees a small animal with wiggly legs and doesn't know what it is, he is likely to call it a bug. The same person seeing a mass of green plant material which he cannot name specifically is likely to call it moss. The name may include algae (water moss), lichens (reindeer moss), and seed plants (Spanish moss), to cite a few cases of misplaced applica- tion of the word.

Biologically speaking, the word "moss" ap- plies to only one of the classes of the Byrophyta, the **Musci.** Typical moss plants are vegetatively advanced over a liverwort because they have a **root, stem, and leaf organization.** In their sexual reproductive cycle they are quite similar to the liverworts, although the sporophyte genera- tion is much larger and many more spores are produced. Mosses are more widely distributed than the liverworts and somewhat more successful in existence on the land.

The main part of a typical moss plant, such as *Mnium*, consists of rhizoids that anchor the plant and an upright stem that bears delicate leaves coming out in a spiral arrangement This is the **gametophyte** of the moss because it bears the gametes. There are male and female moss plants; the **antheridia** are borne on the upper tip of the stem of male plants, and the **archegonia** are borne on the upper tip of the stem of female plants Sperms are produced in the antheridia. Scattered by falling drops of water, they reach the female plants and fertilize the eggs.

After fertilization the zygote grows into a comparatively long stalk that extends up from the top of the female plant and bears a capsule filled with many spores at its upper tip. This **sporophyte** of the moss plant is much larger than the liverwort sporophyte, but it is still attached to the gametophyte. The spores are light and may be carried quite a distance by the wind. If they reach favorable conditions, each spore will germinate into a small filament, the **protonema,** which grows along the surface and pro-

duces a number of buds which grow into new moss plants.

Mosses are important in soil production and conservation because they often form a dense mat with the rhizoids so firmly entwined around the soil particles that there can be little erosion, even on a steep hillside. Then, as their bodies die and decay, forming humus, the soil becomes more favorable for the growth of higher plants.

The peat moss, *Sphagnum*, grows in wet, boggy areas, often forming a thick mat over the ground. This plant continues to grow from year to year from the terminal portion while the older parts die and decay. Sometimes this growth may extend from the land out over water in swampy areas. The moss plants may be so intertwined

A

B

FIG. 19.3 Moss plants, *Mnium*. The plants often grow close together to form a green carpet on damp soil (A). A single plant which shows the root, stem, and leaf organization is shown in (B).

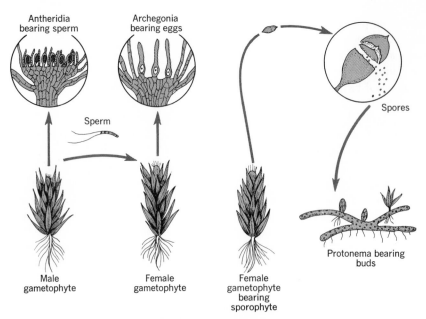

Antheridia
bearing sperm

Archegonia
bearing eggs

Sperm

Spores

Sperm

Male
gametophyte

Female
gametophyte

Female
gametophyte
bearing
sporophyte

Protonema bearing
buds

FIG. 19.4 Life cycle of a typical moss plant. The sporophyte grows from a zygote at the tip of the female plant. Spores from the sporophyte grow into protonemata which bear buds which grow into new moss plants.

as to give a surface appearance of solid soil, and many people have gotten wet by venturing out on this growth only to fall through into the water below. Such areas are called "quaking bogs." Sometimes other plants, even trees, may grow on the moss. When such a growth breaks loose from its moorings it may form a floating island. As the moss plants die, they settle to the bottom and form peat. This helps to convert the swamp into a region of solid soil.

Sphagnum moss grows naturally in a tangled porous mass and, thanks to its porosity, has many uses. Nurserymen use it as a moist packing around the roots of plants that have to be taken up for replanting. It is sometimes used as a surgical dressing; it costs less than absorbent cotton and will absorb up to twenty times as much liquid as an equal weight of cotton. In regions of northern Europe, where wood and coal are scarce and expensive, the decaying moss plants, peat, are taken from the water, dried, and burned as fuel.

REVIEW QUESTIONS AND PROBLEMS

1. Why are the Bryophytes generally smaller than the vascular plants?
2. Why do we say that the liverworts have not completed the transition to land?
3. Describe the habitat of liverworts and mosses.
4. Describe reproduction in the liverworts.
5. What is a gametophyte? A sporophyte?

6. Compare the vegetative structure of liverworts with that of mosses.
7. Describe reproduction in the mosses.
8. What are some economic uses of peat moss.
9. What led to the superstitious use of liverworts for treating liver trouble?
10. The water plants and animals which once lived on the land and returned to the water are generally more complex than those that stayed in the water all along. Why?
11. Why are the rhizoids of liverworts and mosses less efficient than the roots of higher plants for transporting water?

FURTHER READING

Bold, H. C. 1973. *Morphology of Plants.* 3rd ed. New York: Harper and Row.
Conrad, H. S. 1966. *How to Know the Mosses and Liverworts.* Dubuque, Iowa: W. C. Brown.
Doyle, W. T. 1970. *Biology of the Higher Cryptograms.* New York: Macmillan.
Harris, R. M. 1969. *Plant Diversity.* Dubuque, Iowa: W. C. Brown.

20

The Vascular Plants

The vascular plants, phylum **Tracheophyta,** show an outstanding advance over more primitive land plants. They have special vascular or conducing tissue in their roots, stems, and leaves. It consists of elongated tubes which transport water and dissolved substances from one part of the plant to another. It is a much more efficient means of transportation than can be provided by osmosis and diffusion alone. The advantage of this vascular transportation has allowed plants in this phylum to become true land plants and to grow to a very large size. The Tracheophyta is the largest phylum in the plant kingdom, with more than 200,000 identified species. It includes the ferns and the seed plants as well as several phyla which were once quite abundant on the face of the earth but are now mostly extinct. The latter phyla are of interest primarily in studies of plant evolution.

THE FERNS

The ferns, class **Filicineae** of the subphylum **Pteropsida,** may be found growing mainly in moist, shaded places in most parts of the United States. Many of them are cultivated and grown as pot plants. Others are used as greenery by florists, although the much-used and very fine-leaved "asparagus fern" is not a true fern, but a seed plant. The leaves of most ferns are typically flat and finely divided; they are called **fronds.** The stems of most ferns growing in temperate regions are found underground, horizontal to the surface, and the leaves come out of the buds by uncoiling somewhat like the merry-makers of a New Year's Eve party. These underground stems are called **rhizomes,** and the coiled leaves, as they first push themselves up above the surface of the earth, are sometimes called "fiddle heads." The fronds are killed by the winter frost, but the rhizome continues to live underground and new fronds appear the following spring. In the tropics the stems of some species of ferns grow above the ground in a vertical position, producing new fronds continually from the upper tip. This produces the tree ferns which may reach a height of sixty feet. Tree ferns were a common part of the vegetation of North America some 250,000,000 years ago, when the climate there was tropical.

The life history of a typical fern reveals an outstanding difference from that of the liverworts and mosses; in ferns the sporophyte is the dominant generation. In the liverworts and mosses we found the sporophyte insignificant and growing on the gameotophyte. In the ferns the rhizome, the roots, and the fronds make up the sporophyte; and the gametophyte is a small, usually unnoticed part of the life cycle. The **spores** of the ferns are produced in **sporangia** which are usually borne on the underside of the fronds in clusters called **sori.** In some species of ferns the clusters of sporangia are protected by an **indusium** which is shaped like an inverted umbrella. Each sporangium is surrounded by an **annulus** which breaks open and releases the spores when they are mature. The spores are small and light and may be carried by the wind for some distance.

Upon reaching favorable conditions, a spore will germinate and produce a small heart-shaped plant that grows flat on the ground and has short rhizoids for anchorage and absorption. It resembles the liverwort in general appearance but is smaller—usually no more than half an inch long. This inconspicuous plant is the **gametophyte**

of the fern. Since both male and female sex organs are borne on this one plant, we say that the fern is **monecious.** The **archegonia** are produced up near the notch of the gametophyte and the **antheridia** are produced near the opposite end. An antheridium will burst when mature, releasing the many sperm that have been produced within. Water is necessary for fertilization—the sperm must swim to an archegonium to enter it and unite with the **egg** within. A number of archegonia are produced on each gametophyte, but usually only one egg will be fertilized. It is thought that the plant produces a chemical which repels the sperm after this one zygote is

FIG. 20.1 The fiddle-heads of the uncoiling fronds of the cinnamon fern may be seen in forests in the spring. The plants come up from underground rhizomes which have survived the winter.

A   B

FIG. 20.2 Fern fronds and a portion of a frond showing the sori underneath. The dark dots on the sori are sporangia.

formed thus preventing more eggs from being fertilized.

The zygote is the beginning of a new **sporophyte.** It begins division while still in the archegonium and for a time derives nourishment from the gametophyte, but soon a rhizome, roots, and a leaf are produced and the gametophyte withers

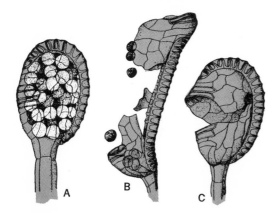

FIG. 20.3 Fern sporangia. (A) unopened, filled with spores. (B) annulus has straightened, throwing out the spores. (C) empty sporangium remaining.

away as the dominant sporophyte generation develops.

The principle of alternation of generations was first discussed in connection with our study of the algae. At this point we shall learn more about this interesting principle as it is related to the number of chromosomes in the cells. In sexual reproduction we have the problem of keeping the chromosome number constant from generation to generation. Whenever two gametes unite, a zygote is formed with twice the chromosome number of each gamete. Sometime before the gametes for the next generation are formed there must be a reduction of the chromosome number to one half or there would be a doubling of the chromosome number each generation. In animals this reduction typically comes just before the gametes are formed. As a result, the cells of the animal body all contain the double or **diploid** number of chromosomes with the exception of the gametes. These contain the single or **haploid** number of chromosomes. This is true of plants also, but the reduction comes much earlier in some of the more primitive plants. As a result there may be a considerable amount of tissue in the plant life cycle which is haploid. This part of the plant we call the **gametophyte,** whereas the part which is formed by the union of gametes is diploid, produces spores, and is called the **sporophyte.** As we survey the plant kingdom we note that there tends to be a gradual change from a dominant gametophyte generation to a dominant sporophyte generation.

Going back to the algae, we find that in *Spirogyra* the entire filament is haploid and would therefore be classified as gametophyte. A gamete is produced from a filament cell and is, of course, haploid. When two gametes unite, a zygote is formed which is diploid, and this is all there is to the diploid, or sporophyte, generation. When the zygote germinates, the nucleus divides twice, but the chromosomes divide only once, so we end up with four haploid nuclei. Three of these degenerate, but the fourth remains as the nucleus of the cell and divides by regular mitosis each time the cell divides. This produces the cells of the filament, which are all haploid. In the red alga *Polysiphonia* we learned that the haploid plant is the largest of the three generations, but there are two small sporophyte generations.

In the liverworts we found that gameto-

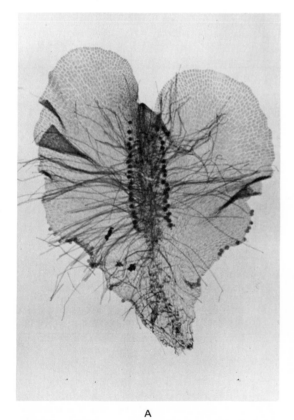

A

B

FIG. 20.4 Fern gametophyte shown in (A) bear antheridia which appear as dark spots near the edge of the plant. The dark areas near the center are archegonia. In (B) a young sporophyte grows from the gametophyte.

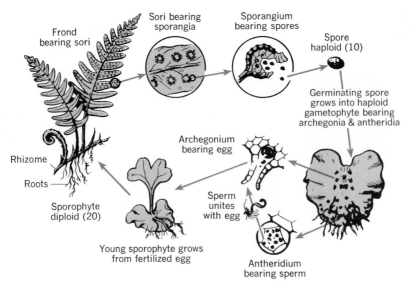

FIG. 20.5 Life cycle of a typical fern.

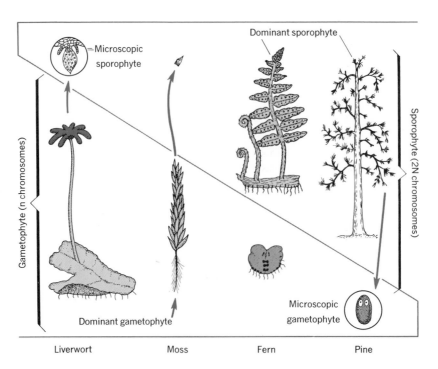

FIG. 20.6 Alternation of generations in plants. There is a gradual shift from a dominant gametophyte gener- ation to a dominant sporophyte generation as the plants become larger and more complex.

phyte is still the dominant generation and the sporophyte is only a microscopic group of cells living parasitically on the gametophyte. Most of the sporophyte is used in spore production. The spores are produced by a reduction division of the cells, so they have only the haploid chromosome number and represent the beginning of the gametophyte generation.

In mosses the gametophyte is still dominant and the sporophyte is still attached to the gametophyte, but the sporophyte is much larger than the sporophyte in the liverworts and a smaller proportion of it produces spores. The sporophyte has chlorophyll and can manufacture food, but it is dependent upon the gametophyte for water and minerals.

In the ferns, as we have just seen, the gametophyte is the smaller of the two generations, and only a small part of the sporophyte becomes spores. The final step in this evolutionary progression comes in the seed plants. The sporophyte is the dominant generation and the gametophyte exists as a tiny, microscopic body that is parasitic on the sporophyte. Thus, we find a complete shift of emphasis from the gametophyte to the sporophyte as the primary plant.

Note that the spores always germinate to form **gametophytes,** and the **zygotes** always germinate to form **sporophytes,** with the exception of *Polysiphonia*, where the spores from one sporophyte germinate to produce a second sporophyte. Spores are reproductive cells that are produced asexually: there is no union of cells to produce a spore. Zygotes are reproductive cells that are produced sexually: two gametes of opposite types must unite in order to form a zygote. Therefore, there is alternation of the method of producing reproductive bodies—one generation, the gametophyte, reproducing **sexually** and the other generation, the sporophyte, reproducing **asexually.**

This may be clarified by a survey of the chromosomes of a fern. Let us start with a zygote that has just been produced and assume that it

has 20 chromosomes. The first division of the zygote is by mitosis, which produces a division of the chromosomes before cell division, so each daughter cell also has 20 chromosomes. The second division is also by mitosis, as well as the third, the fourth, and so on. Therefore, all the cells of the sporophyte will have 20 chromosomes. However, when spores are produced within a sporangium, the cells have meiotic divisions and each daughter cell receives only 10 chromosomes. Thus the spores have only half as many chromosomes as the large sporophyte cells that produced them.

When the spores break from their confinement, each spore that falls in a favorable place for germination will divide by mitosis and give two daughter cells with 10 chromosomes each, and so on until the entire gametophyte is formed —each cell containing 10 chromosomes. The gametes are produced in the antheridia and archegonia and they also have 10 chromosomes. When a sperm carrying 10 chromosomes unites with an egg also carrying 10 chromosomes, the resulting zygote will have the diploid number of 20 chromosomes and the cycle is completed.

Since the chromosome number varies with different species of plants we sometimes say the gametophyte has the n number of chromosomes and the sporophyte has the $2n$ number of chromosomes and allow n to represent the haploid number of the plants under consideration.

Fig. 20.6 shows how the proportion of tissue shifts from a predominance of n-chromosome cells to a predominance of $2n$-chromosome cells in the plant kingdom.

MINOR SUBPHYLA OF THE VASCULAR PLANTS

Psilopsida

This subphylum includes some of the most primitive vascular land plants, pioneers that flourished on the earth over 300 million years ago. Most of them have long since become extinct and exist only as fossils, but two genera have survived. One of these may be found in Florida; it is commonly known as the "whisk fern" and has the scientific name *Psilotum nudum.* It has an underground rhizome, like a fern, and upright stems come off from this and extend above ground for several inches. These stems form a series of Y-shaped branches—**dichotomous branching,** as it is called. There are no true leaves on these stems, only tiny scales. Some of the branches bear sporangia which contain the spores. Wind spreads the spores and each grows into a small gametophyte which bears gametes of both sexes. After fertilization a new sporophyte develops. This living plant probably reprresents but little change from the ancient forms in this subphylum.

Lycopsida

This subphylum includes the genus *Lycopodium,* which contains the commonly called

ground "pines," running "cedar," and the club mosses. It also includes the genus *Selaginella,* often called the little club moss. The lycopodia have dark green, needlelike leaves that make some of them look quite like miniature pine or cedar trees. They are much used for making wreaths and other decorations at Christmas time. They produce spores—all about the same size—which germinate into gametophytes bearing both antheridia and archegonia on the same plant, a characteristic of nearly all ferns.

Selaginella is quite interesting from a botanical viewpoint because it produces two distinct kinds of spores, a characteristic of higher plants. Vegetatively, this plant resembles the mosses, but it has well-developed conducting tubes and the reproductive organs are distinctly of a higher type of plant. The sporangia are produced on the sides of the stems on specialized leaves, the sporophylls. Some of the sporangia are conspicuously four-lobed. When these are broken open, they are found to contain only four relatively large spores; these are called **megaspores.** The other sporangia are smooth in outline and when broken open are found to contain

A   B   C

FIG. 20.7 *Lycopodia.* These plants, which are related to the ferns, are often used for ornamentation at Christmas. They are, from left to right: the ground pine, *Lycopodium obscurum;* the running cedar, *Lyco-* *podium complanatum;* and the shining club moss, *Lycopodium lucidulum.* Note the conelike arrangement of the sporangia.

many very small spores, the **microspores.** This production of two distinct types of spores by a plant is called **heterospory,** in contrast to **homospory,** the production of only one morphological and physiological type of spore.

The gametophytes of *Selaginella* are greatly reduced in size, a characteristic considered as a sign of advance, since this is also the case in the more highly developed seed plants. The female gametophyte develops entirely within the outer covering of the megaspore. Archegonia are produced which bulge from one end of the spore case, permitting fertilization. A small, male gametophyte is produced within the microspore wall. A single antheridialike structure develops which produces sperm; and a single vegetative cell is produced, the **prothallial cell,** which represents the entire vegetative part of the fern gametophyte. When moisture is available, the sperm may be liberated and swim to the egg within the archegonia of the female gametophyte. The zygote then begins growth into a young sporophyte while still in the female gametophyte.

Sphenopsida

This subphylum includes an interesting group of plants, the **horsetails,** which bear little outward similarity to the ferns, but the life history of the two is much alike. The horsetails, *Equisetum,* have underground stems, rhizomes, which send up aboveground stems of two types —vegetative which bear the slender, needlelike leaves, and reproductive. The sporangia are borne on specialized leaves which are grouped together to form cones at the tips of the reproductive stems. The spores are all of approximately the same size, but the spores give rise to two different kinds of gametophytes—one bearing antheridia and one bearing archegonia; hence, it shows heterospory.

The horsetails are found in a wide variety of habitats—many grow in ponds and swamps; others, in damp, shaded places, and, strange to say, many of them seem to thrive on the embankments along the sides of railroad tracks. At one time during the earth's history this was a very important order of plants. Huge relatives of our present-day horsetails once formed a conspicuous feature of the earth's vegetation. They are quite abundant as fossils which were deposited in the Upper Carboniferous period some 250 million years ago.

THE SEED PLANTS

The seed plants represent the climax in the development of the plant kingdom and are by far the most successful of the land plants. They produce seeds as their primary means of reproduction and distribution. Seeds are extremely important structures for mankind. Modern civilization would be impossible without them because of two seed characteristics. First, seeds always contain a considerable amount of concentrated food material. This supplies nourishment for the developing plant embryo but often is diverted for the use of man—either directly, for personal consumption, or indirectly, for consumption by animals that ultimately will furnish food for man. The seeds of wheat, corn, oats, rye, barley, rice, beans, peas, and nuts furnish the great bulk of the food consumed by man and his food producing animals. Second, seeds are capable of living at a reduced rate of metabolism for a comparatively long, dormant period and then germinating under favorable conditions. This makes it possible for man to gather seeds from the harvest and put aside a portion of them for the next year's crop.

There is great variation in the size, structure, and other characteristics of seeds of different species of plants, but certain features are common to nearly all of them. A typical seed consists of a young plant, the **embryo;** a protective covering, the **seed coat;** and a supply of stored food, the **endosperm.** Most seeds have developed some method of **distribution;** otherwise, all the

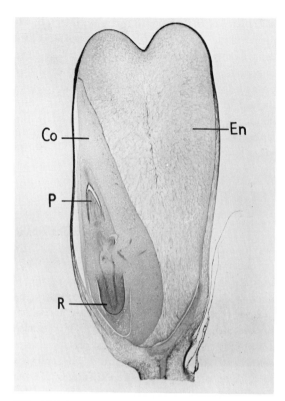

FIG. 20.8 A seed consists of a young plant embryo together with stored food and a protective covering. This section through a grain of corn shows the embryo, which consists of a plumule, which will develop into the stem and leaves, and the radicle, which will produce the roots. The cotyledon (covering) and the endosperm (food storage) can also be seen.

the waste from the animal's body. Many nuts are actually planted by various rodents that bury them underground as a food cache. Bean seeds are developed in a pod; when the pod becomes dry it suddenly breaks open with the force of a coiled spring and throws the seeds for a distance of several feet. These are but a few of the many means of seed distribution.

CLASS GYMNOSPERMAE— PLANTS WITH NAKED SEED

The seed plants are divided into two classes. One class bears its seeds in a naked, or exposed, state; the other class bears its seeds enclosed in some sort of covering, a fruit. We call the first group **gymnosperms,** a word which means "naked seed," and the second group **angiosperms,** a word which means "enclosed seed."

FIG. 20.9 Each dandelion seed is attached to a fluffy parachute which can carry it for great distances, much to the annoyance of those who wish to keep dandelions out of their lawn.

seeds would fall beneath the parent, germinate, and compete with one another for survival. The pine seed has a wing that sends it spinning through the air for some distance as it falls. Each dandelion seed has a dainty parachute and may be borne for great distances by the wind. The cocklebur seed is surrounded with sharp barbs that stick in the skin or fur of passing animals, eventually to drop off in a new locality. The strawberry produces a delicious fruit which is eaten by animals. The seeds of the fruit resist digestion and pass through the animal's body unharmed, to germinate where they are deposited in

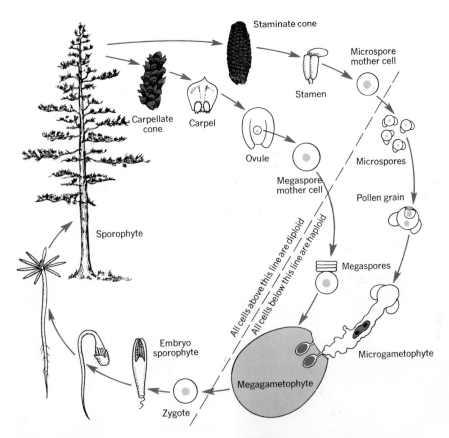

FIG. 20.10 Life cycle of a pine tree.

The gymnosperms are chiefly woody plants and include the largest and the oldest living trees on the earth.

The Life Cycle of a Pine Tree

The most abundant of the gymnosperms is the pine, found in many sections of the United States and known to almost everyone for its tall, graceful beauty and fragrant odor. Economically valuable, pine is our most abundant source of lumber and is a source of turpentine and resin. We will study the life cycle of the pine as a typical seed plant.

By far the major part of the pine tree is the **sporophyte**—the extensive root system, the towering trunk (which may reach a height of two hundred feet or more), the needlelike leaves, and most of the cone are sporophyte tissue and therefore contain the diploid or *2n* chromosome number. The **gametophyte** is reduced to a tiny microscopic body found in the cone and entirely dependent upon the sporophyte for nourishment. This completes the trend toward sporophyte dominance that we noted earlier.

Pine cones appear on the tree in the early spring. There are two types of cones produced on each tree—male, **staminate** cones, and female, **carpellate** cones. As in *Selaginella*, two kinds of spores are produced—one destined to form a male gametophyte and the other destined to form a female gametophyte. The staminate cone produces spores that will form the male gametophyte. Since these spores are the smaller, they are called **microspores** and are also known as **pollen grains**. The male cone is composed of many scales, **microsporophylls,** and each bears two pollen sacs filled with pollen. As the staminate cones mature, the pollen sacs rupture, releasing a cloud of winged pollen grains. The air around a thick pine forest may show a distinct yellow cast at this time. This is the first plant studied that does not require water for movement of the male gamete to the female.

In the meantime, the carpellate cones have been developing. They are larger than the staminate cones, and their surface shows a series of diamond-shaped **megasporophylls.** Two are **ovules** produced on the inner, upper surface of each megasporophyll, and a **megaspore** is produced within each ovule. The megaspore germinates, producing the small **female gametophyte.** Some of the pollen grains sift down into the cone and enter the ovule through a special opening, the **micropyle.** These later germinate into small pollen tubes which grow down to the female gametophyte. This tube, together with the nuclei within, forms the **male gametophyte**— a tiny remnant of the once dominant generation

A

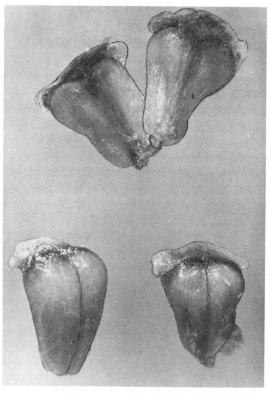

B

FIG. 20.11 The staminate (male) cone of the pine tree. (A) shows an enlarged portion of the cone, while (B) shows several of the microsporophylls which have been removed.

FIG. 20.12 The carpellate (female) cone of the pine tree. The small spring cone is shown at left. This is how the cone appears during the first year. A year later the cone looks like the center picture. It now contains developing seed. The two-year cone, at right, has opened and the winged seeds have escaped.

as first studied in the liverworts. When one of these nuclei fuses with a female nucleus, the diploid condition is restored and the next sporophyte generation begins. Cell division follows and a small pine embryo is produced, the outer part of the ovule becomes a seed coat, and the rest of the ovule becomes the endosperm. We now have a pine **seed.** This develops a winglike extension that aids in distribution. When the seed is ripe, the scale of the cone opens up, and since the seed is not enclosed, it drops out. The wing causes it to take a spiraling course that carries it for some distance. The entire cycle requires about two years; pollination occurs in the first spring, but the pollen tube growth and union of male and female nuclei do not occur until the following spring, about a year from the beginning of the carpellate cone. Then the seeds are released in the fall after fertilization and they germinate the following spring. However, the staminate cones grow, mature, release their pollen, and die all within the space of a few weeks in the spring.

ORDERS OF GYMNOSPERMS

Order Cycadales

The cycads are widely distributed in tropical regions, but their numbers are quite small compared to their great abundance in the prehistoric past. They were so abundant during the Triassic period, about 175 million years ago, that this period is sometimes called the age of the cycads. These plants have fernlike leaves and bear a close resemblance to the tree ferns. But their method of reproduction is much like that of the pine. Micropores and megaspores are borne on staminate and carpellate cones respectively, but these cones are produced on separate male and female plants rather than both on the same plant, as in the pine. More concisely we can say they are **diecious,** and the pine is **monecious.** Pollen is transported by air, but motile, flagellated sperms are produced within the pollen tubes which grow down carrying the male

A

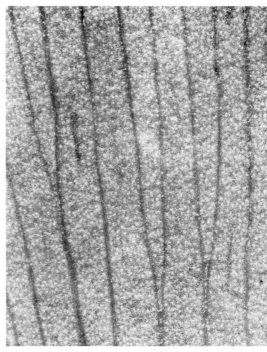

B

FIG. 20.13 *Cycas,* a cycad. The male plant bears a single large staminate cone (A). The female plant bears a cluster of megasporophylls in a loosely arranged cone (B).

A

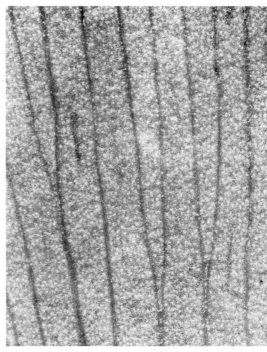

B

FIG. 20.14 *Ginkgo,* the maidenhair tree (A). This is the only living species of an ancient order of gymnosperms which was once very abundant. The leaves are somewhat like the fronds of ferns and they have veins with dichotomous branching. An enlargement of the leaf is also shown (B).

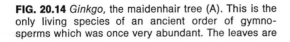

gametes to the female gamete. There seems to be no useful purpose for this motility, for the sperm have nowhere to swim except within the narrow confines of the microscopic pollen tube. The nonmotile gametes within the pollen tube of the pine fertilize just as efficiently, so we may assume that this sperm motility is an evolutionary carryover linking the gymnosperms with some primitive fernlike ancestors. This assumption receives further support when we examine the staminate cones; in them the microsporangia sometimes occur in sori on the lower surface of the microsporophylls very much like those of the ferns.

Only two genera of cycads are commonly seen in the United States—the genus *Zamia*, which is native to Florida, and the genus *Cycas*, which is native to Australia but grown in greenhouses in the United States and used for decorative purposes by florists. Both genera have attractive, dark green leaves and resemble small palm trees.

Order Ginkgoales

There is only one living representative of this ancient order of gymnosperms. This is the **ginkgo tree,** sometimes called the maidenhair tree because of the resemblance of the leaves to those of the maidenhair fern. The leaves have an unusual dichotomous arrangement of the veins that is not found in any other seed plant but is found in some of the ferns. This plant has the distinction of being the oldest genus of the living trees. It can be traced back to the Permian period of the late Paleozoic era, about 200 million years ago. Fossilized leaves have been found in deposits from this period which are almost identical with leaves on the living tree of today. It is strange that this plant has been able to persist almost unchanged throughout this great period of time, whereas plant and animal life all around it was undergoing very extensive changes. The living representatives are native to western

China, but have been transplanted to the United States as ornamental trees. One or more specimens of this living fossil may be found on most college campuses.

Like the cyads, the ginkgo is strictly diecious and has swimming sperm; but the sperm motility is again unnecessary, because the pollen tube carries the male gametes directly to the location of the female gametes.

Order Gnetales

This order includes three strange living representatives. The best known genus is *Ephedra*, for it produces the valuable drug ephedrine, which has an effect on the human body somewhat like adrenalin. It is of great value in shrinking the nasal passages in colds and hay fever, thus permitting free breathing. It must not be used excessively, however, because of its effect on the nervous system. *Ephedra* is the only genus that grows in the United States. A second genus, *Welwitschia*, found only in South Africa, is a very curious plant that has a large conical stem which spreads out into a saucer-shaped crown bearing two long, leathery leaves that sprawl over the ground and continue to grow during the entire lifetime of the plant. The genus *Gnetum* consists of several species that grow as woody vines or small trees in a number of tropical regions. They have broad leaves that grow opposite one another on the stem and closely resemble some of the more advanced trees.

Order Coniferales

The great majority of the living gymnosperms are in this order, which gets its name from the cones that produce the spores and seed. It includes our common evergreen trees with needlelike leaves, such as the pine, spruce, hemlock, cedar, redwoods, sequoias, and firs. There are also a few that shed their leaves, such as the cypress.

CLASS ANGIOSPERMAE—
PLANTS WITH ENCLOSED SEED

The height of evolutionary development of the plant kingdom is reached in the angiosperms. They are the dominant plants of the earth's flora today and may be found in almost every type of ecological habitat. Practically all of man's food crops are angiosperms. The roots, stems, leaves, fruits, and seeds that we consume are almost all from plants in this class. They also have great esthetic value because they bear their reproductive organs in flowers, many of which add color and beauty to our homes and gardens as well as to the natural landscape. Not all flowers have beautiful petals and a fragrant odor, however. These are characteristics of the insect-pollinated flowers that attract insects to visit them and feast on the nectar found at the flower base. On such a visit an insect becomes coated with sticky pollen which it may transfer to other flowers, thus accomplishing cross-fertilization. However, the

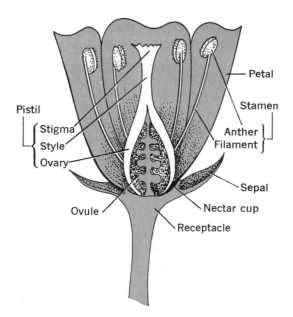

FIG. 20.15 A typical angiosperm flower cut away to show the ovules contained in the ovary.

pollen of many angiosperms is carried by the wind, and the flowers from such angiosperms have greatly reduced petals or may lack petals altogether. The pollen is very light and powdery and produced in great quantities, increasing the probability that some of it will reach the female part of a flower. In many cases people become sensitive to pollen, so abundant in the air we breathe, and develop hay fever or asthma when they inhale it.

The Life Cycle of a
Typical Angiosperm

The structure of a typical flower is shown in Fig. 20.15. The male reproductive organs are the **stamens.** These are the microsporophylls of the angiosperms, since they produce the microspores, **pollen grains.** A stamen is composed of a long, stalked **filament** and an **anther** in which the pollen develops. When mature, the anther ruptures, exposing the pollen to the wind or insects. The female reproductive organs, or megasporophylls, are the **pistils.** A pistil is composed of the **stigma,** which is usually a sticky, upper tip of the pistil which receives the pollen; a **style,** a slender tube; and the **ovary,** which is a swollen organ at the base of the pistil enclosing the **ovules.** The class name, Angiospermae, is derived from this enclosed condition of the ovules. The entire flower is contained in a **receptacle,** and there are several **sepals** that extend up from the receptacle and support the showy **petals** of the flower. The sepals are usually green and the petals colored.

Most angiosperms are **monecious** and bear both female and male organs in the same flower, but there are many ingenious arrangements which prevent self-pollination and allow cross-pollination. The pistils of many of the wind-pollinated plants mature earlier. Thus, by the time the stamens mature and begin to release pollen, the ovules have already been fertilized by pollen from another plant. However, in case such pollen has not arrived, there can still be self-pollination, which is better than no pollination at all. The insect-pollinated plants usually have the pistil extending out beyond the stamen. An insect entering the flower will brush by the pistil first and transfer pollen from another flower before it contacts the pollen in the flower that it is visiting. Some flowers have pistonlike plungers, trapdoors, and other special devices that cause the visiting insects to be showered with pollen.

Certain plants, such as buckwheat, rye, cabbage, and corn, have stigmas and anthers that mature at the same time, but for some reason the pollen will not germinate (or germinates too slowly to accomplish fertilization) when it reaches the stigma of the same plant. On the other hand, there are some, such as the garden pea, in which the stamens and pistils are both enclosed in the petals so that only self-pollination can occur in normal circumstances.

In some plants pollination is such a close mutual relationship with certain insects that neither plant nor insect can survive without the other. For instance, the wild yuccas of the southwestern United States depend upon a certain genus of small white moth, *Pronuba*. When the flower opens, the female moth enters and collects a ball of the very sticky pollen. She then visits another flower and lays her eggs within the ovary, where the larvae will develop. Before leaving the second flower, she fastens the ball of pollen directly on the stigma. No other natural means of pollination of this flower is possible and no other place for the development of the larvae

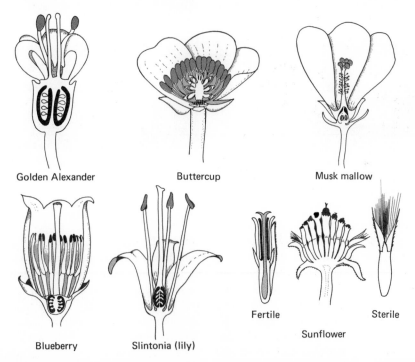

Golden Alexander   Buttercup   Musk mallow

Blueberry   Slintonia (lily)

Fertile   Sterile

Sunflower

FIG. 20.16 Variation in the arrangement of flower parts in different kinds of flowers. Top row: golden Alexander, buttercup, and musk mallow. Bottom row: blueberry, slintonia (a lily), and sunflower (a compound flower with a single, fertile, disk flower at the left and a sterile, ray flower at the right).

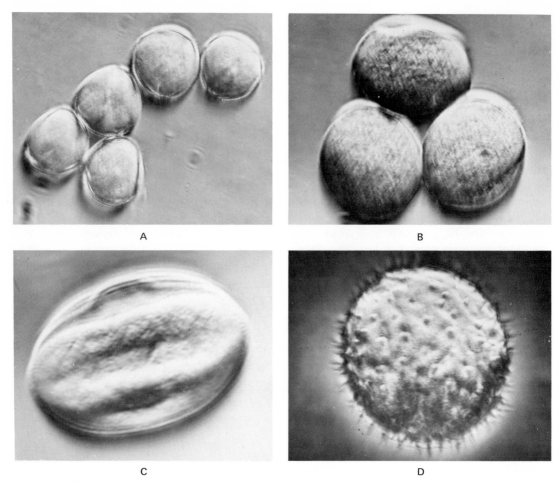

FIG. 20.17 Pollen is found in a variety of shapes and sizes. These pollen grains are: *Plantago major* (A); *Pedicularis groenlandica* (B); *Zygodenus elegans* (C); *Althea rosea* (D).

is known; each organism is completely dependent upon the other for its continued existence.

Thus, we see that there are various methods by which pollen is transferred in the different species of plants. However, once it is transferred, the story is much the same in all the angiosperms. The stigma contains a moist, viscid material which stimulates the germination of the pollen grains. The absorbed water increases the turgor within the pollen causing a slender pollen

tube to be pushed out. This tube pushes directly into the stigma and forces its way through the style, pushing cells aside or crushing them in its advance toward the interior of the ovary where the ovules are located. Ultimately, through the influence of some guiding force which is not clearly understood, the tube arrives at the ovule. Each ovule has a micropyle, as we found in the pine ovule, through which the pollen tube may enter. Once inside, the tip of the pollen tube dissolves, allowing the discharge of a nonswimming male nucleus containing the haploid chromosome number. This unites with the haploid female nucleus within the ovule, thus restoring the diploid number and starting the **sporophyte** generation again. The **gametophyte** of both sexes is limited to tiny nuclei which are formed by reduction division within the pollen grains and the ovules.

Each ovule within the ovary is fertilized by a male nucleus from a different pollen grain. Thus, there must be at least as many pollen grains

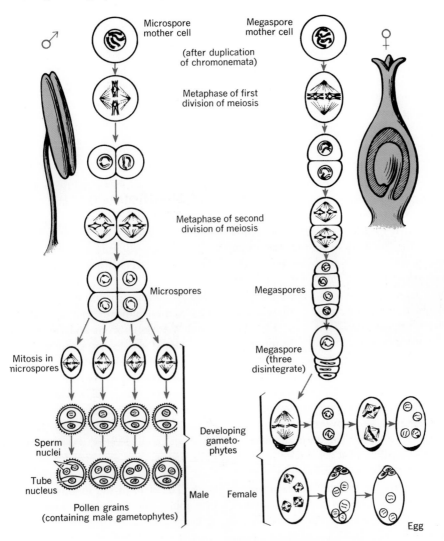

Microspore mother cell

(after duplication of chromonemata)

Metaphase of first division of meiosis

Metaphase of second division of meiosis

Microspores

Mitosis in nicrospores

Sperm nuclei

Tube nucleus

Pollen grains (containing male gametophytes)

Male

Megaspore mother cell

Megaspores

Megaspore (three disintegrate)

Developing gameto- phytes

Female

Egg

FIG. 20.18 Method of formation of pollen grains and eggs in angiosperms. (Redrawn from Winchester, *Genetics*, Houghton Mifflin.)

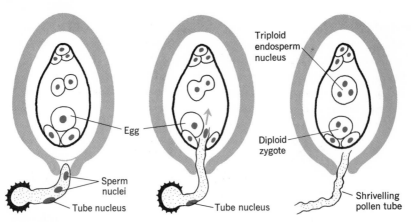

FIG. 20.19 Fertilization in a typical angiosperm. A pollen tube from a pollen grain grows into the ovule and a sperm nucleus unites with the egg to form the diploid zygote. Another sperm nucleus unites with a diploid cell to form the triploid endosperm nucleus.

as there are ovules in order for each ovule to be fertilized. Any ovules that are not found by pollen tubes or any pollen tubes that are too late to find unfertilized ovules soon wither and die. This explains the empty places that are often found on ears of corn or in bean pods—the ovules were not fertilized and hence no seed developed.

Each ovule which has been fertilized normally develops into a **seed.** We now have the foundation for a **fruit.** In common terminology the word **fruit** is often restricted to certain succulent, edible plant outgrowths—the apple, orange, peach, pear. Botanically speaking, however, a fruit is the mature ovary together with the seeds it encloses and any remnants of the flower associated with it. The term applies to structures that we call berries, vegetables, melons, nuts. The fruit of the bean is produced by a great enlargement of the ovary and the seed contained therein. The petals wither and drop off, but the receptacle and the sepals can still be seen attached to "snap beans." The fruit of the apple is produced by a great enlargement of the receptacle and the ovaries within form the core. The stamens, styles, and sepals can be seen at the end of the apple opposite the stem.

SUBGROUPS OF THE ANGIOSPERMS

The angiosperms are divided into two subclasses—**Dicotyledoneae** and **Monocotyledoneae.** The term "cotyledon" refers to food storage sections within the seed, and the subclass names indicate that the first group has two of these cotyledons and the second group has only one. If we compare the seedling of a bean plant (**dicotyledon**) with that of a corn plant (**monocotyledon**), we can easily see this distinction. The mature plants may be rather easily distinguished also. The monocotyledons have leaves that are usually rather long and slender and have parallel veins, whereas the dicotyledons have leaves that are broader in proportion to their length and the veins are arranged in a network. The stem of the monocotyledon is divided into joints and it may be rather easily broken at the nodes separating the joints, but the dicotyledon stem does not have such joints.

The important orders of the living angiosperms are listed below, together with the common names of some of the well-known plants found in each order for reference purposes.

Subclass Dicotyledoneae

Order Ranales, the buttercup order. Buttercups, larkspurs, laurels, magnolia, barberry, water lilies.

Order Malvales, the mallow order. Mallows, hollyhock, cotton, elm, nettles, hibiscus.

Order Geraniales, the geranium order. Geranium.

Order Papaverales, the poppy order.

A	B	C	D
E	F	G	H

FIG. 20.20 Distinctions between monocotyledon and dicotyledon seedlings. Corn seedlings (monocotyledon) push up from a single cotyledon which remains underground. Bean seedlings (dicotyledon) push up two cotyledons which open and allow the two seed leaves to expand.

Poppies, mustard, cabbage, bleeding heart, blood-root.

Order Caryophyllales, the carnation order. Carnations, pinks, buckwheat, rhubarb, willow, poplar.

Order Primulales, the primrose order. Primroses, plantain.

Order Ericales, the heath order. Heaths, heather, rhododendron, mountain laurel, wintergreen.

Order Gentianales, the gentian order. Gentians, olive, lilac.

Order Polemoniales, the phlox order. Phlox, morning-glories, sweet potato, forget-me-not, heliotrope.

Order Scrophulariaceae, the snapdragon order. Snapdragon, foxglove.

Order Lamiales, the mint order. Mints.

Order Rosales, the rose order. Rose, cherry, strawberry, pear, peach, legumes.

Order Myrtales, the myrtle order. Myrtle, clove, eucalyptus, evening primrose.

Order Cactales, the cactus order. Cacti.

Order Celastrales, the bittersweet order. Bittersweet, grape.

Order Sapindales, the maple order. Maple, walnut, hickory, oak, beech, chestnut, birch, hazel.

Order Umbellales, the parsley order Parsley, parsnip, carrot, dogwood.

Order Rubiales, the madder order. Madder, coffee.

Order Asterales, the aster order. Aster, dandelion, sunflower, burdock, ragweed, chrysanthemum.

Subclass Monocotyledoneae

Order Alismales, the water plantain order. Water plantain, arrowhead, eelgrass, cattail, pondweed.

Order Liliales, the lily order. Lily, tulip, hyacinth, onion, asparagus, trillium.

Order Arales, the duckweed order. Duckweed, jack-in-the-pulpit, skunk cabbage, calla lily.

Order Palmales, the palm order. Palms.

Order Graminales, the grass order. Grass, oats, wheat, barley, rice, corn, bamboo.

Order Iridales, the iris order. Iris, gladiola, crocus, amaryllis, narcissus.

Order Orchidales, the orchid order. Orchid, ladyslipper.

REVIEW QUESTIONS AND PROBLEMS

1. What feature enables the ferns to grow to a much larger size than the mosses?
2. Distinguish between a rhizome and a rhizoid.
3. Describe the life cycle of a typical fern plant.
4. What is meant by the term "monecious"?
5. Trace the change from a dominant gametophyte to a dominant sporophyte generation.
6. If a moss plant had 16 chromosomes as its diploid number how many chromosomes would there be in a cell of a leaf? In a cell of the sporophyte? In a spore? In a sperm? In an egg? In a zygote?
7. If a fern has 20 chromosomes as its diploid number, how many chromosomes would there be in a cell of a frond? In a cell of the rhizome? In a cell of the rhizoid of the gametophyte? In a sperm? In a spore? In a zygote?

8. In alternation of generations which generation reproduces sexually and which asexually?
9. How does *Psilotum* differ from the ferns?
10. Which is considered the more advanced method of reproduction, heterospory or homospory? Why?
11. Evidence indicates that the rainfall was considerably greater during the Carboniferous period when the tree ferns, horsetails, and *Lycopsida* flourished. Could a reduction in rainfall have had anything to do with their decline as dominant plants on the earth? Explain. (Consider the reproductive cycle of these plants in formulating your answer.)
12. Why are seeds of such importance in modern civilization?
13. Describe the parts that make up a typical seed.
14. Why is seed distribution of great importance? Give some of the methods of distribution.
15. Describe the reproductive cycle of the pine.
16. What interesting reproductive characteristic of the cycads is of particular evolutionary significance? Explain.
17. What vegetative feature of ginkgo is of particular evolutionary significance? Explain.
18. What is the economic importance of *Ephedra*?
19. What characteristics of insect-pollinated flowers make them rather easily distinguished from the wind-pollinated flowers?
20. Describe some of the methods which ensure cross-pollination.
21. What part of an angiosperm constitutes the gametophyte? The sporophyte?
22. Distinguish between a dicotyledon and a monocotyledon.

FURTHER READING

Bold, H. C. 1973. *Morphology of Plants.* 3rd ed. New York: Harper and Row.

Salisbury, F. B., and Parke, R. V. 1970. *Vascular Plants.* Belmont, Calif.: Wadsworth.

Scagel et al. 1965. *Evolutionary Survey of the Plant Kingdom.* New York: Harper and Row.

Sporne, K. R. 1966. *Morphology of Pteridophytes.* London: Hutchinson.

Torrey, J. G. 1967. *Development in Flowering Plants.* New York: Macmillan.

21

Structure and Function of the Seed Plants

Seed plants dominate the environment on most of the earth's surface and they have great economic importance. In this chapter we shall learn something about their structure and function. Seed plants are generally divided into three parts: roots, stems, and leaves. To these we can add the reproductive bodies which were discussed in Chapter 20.

ROOTS

Because of their larger size and greater complexity of structure we often refer to the seed plants as the higher plants. Since they occupy such a large area of the earth's surface and are of such great economic importance, we shall study their structure and function in this chapter. For convenience of study the body of a seed plant is divided into roots, stems, and leaves. To this list might also be added the flowers and fruits, but since we have already considered these in Chapter 20, we shall restrict our study in this chapter to the roots, stems, and leaves. Let us begin with the root system.

In general the root system of a higher plant is the part of the plant that grows underground, but this statement must be qualified by numerous exceptions. We see aerial roots in the orchid, which obtain water from the moist air; in the common ivy, roots attach the plant to its aboveground support; and in corn, there are prop roots aboveground. We see underground stems in

Irish potatoes, iris plants, onions, and dahlias. There are even underground fruits: the peanut plant flowers aboveground and the stems then push underground to produce the seed in an automatic planting process. However, we may think of these cases as exceptions and consider the root system as the underground part of the plant for higher plants as a whole.

Anchorage is a primary function of the root system. A higher plant must maintain its upright position in the soil, and it is dependent upon its root system for this. **Absorption** is another important root function: the water and minerals necessary for the life processes are absorbed through the roots. **Food storage** is the final important root function. Food is manufactured in the leaves, but it may be transported to the roots and stored. Turnips, carrots, beets, and sweet potatoes are good illustrations of such root storage. Man often takes advantage of this and uses the roots for his own food.

Extent of the Root System

A root system is likely to be much more extensive than is generally supposed. The roots of a corn plant growing in good deep soil may have a system that extends about seven feet deep and three feet laterally from the base of the plant. This explains why plants must be well spaced in the field or garden in order to reach maximum development. It is sometimes said that the spread of a root system underground is equal to the spread of the plant aboveground. This is a good average, but there is considerable variation. The extensive root system is necessary because of the quantities of water required by the plant. A mature apple tree will lose about 250 pounds of water from its leaves every day through evaporation. This amounts to about 18 tons per season and takes no account of the water used in the process of photosynthesis. When we consider that this water must enter the plant by osmosis

through tiny root hairs, we can understand why the root system must be extensive.

The roots of land plants need both moisture and air and tend to establish themselves at a depth where the combination of the two is most suitable. If soil conditions are greatly altered, the plant may die because its roots are established at an unsuitable level. If a lawn is watered every day and the practice is suddenly discontinued, the grass is likely to die because the roots have been developed near the surface. If we pile several feet of dirt around such a tree as a maple, it will probably die because we have changed the depth of the roots in relation to the top of the soil, and insufficient air will penetrate through the thicker layer of earth.

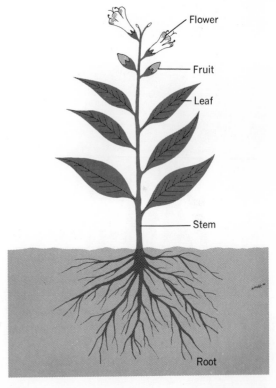

FIG. 21.1 The major systems of a higher plant. The flower and fruit are grouped together under the reproductive system.

If an area is flooded for a time, the trees on it may die as a result of lack of air around the roots. Yet certain plants can withstand flooding. Willows grow along the banks of streams right at the water line, and cypress trees actually grow in the water. Their roots send up growths that extend above the water line. These are called "cypress knees."

At the other extreme we find plants like the mesquite tree that can live in very dry regions where only a few inches of rain fall each year. This is possible because of the extensive root system; it may penetrate as far as seventy feet underground and pick up the moisture that is more abundant in these lower regions. Above-

FIG. 21.2 Prop roots of corn. These roots can be seen coming out from the first two nodes above the ground. They help to anchor the plant in the soil.

ground the tree will not be more than ten feet high. Cotton is able to withstand the hot, dry weather of the southwestern summers because it sends a taproot far below the surface to regions where the water is available.

Structure of Roots

If we pull a young seedling from the ground we cannot tell much about its root system, because the soil particles cling to the roots and we cannot remove them without destroying much of the system. However, we can germinate seed in moist air and get a good impression of the structure. By studying such a seedling in conjunction with a longitudinal section through its root, we can identify the parts of a young root. At the lower tip will be a **root cap,** made of tough cells that serve to protect the delicate growing tip of the root, much as a thimble protects the finger of a seamstress. Above this is an **embryonic region.** Here the cells are rapidly dividing, and in the longitudinal section, many mitotic figures may be seen. (Root tips are good material for mitosis studies because of the great number of dividing cells found in this region.) Above this is the **region of maturation,** where the cells are enlarging and undergoing differentiation. Here the delicate **root hairs** may be seen growing out horizontally. The lower root hairs will be short, but longer hairs are found as we go up the root until the hair shows a uniform length. At the point where they reach their maximum length the mature region of the root begins. Here the cells have become differentiated.

Details of the mature region can best be seen through a cross section from this region. The outer layer of cells, in contact with the soil, is the **epidermis.** Each root hair is an outgrowth of a single epidermal cell, and the major portion of the absorption of the plant takes place through the root hairs. Absorption can take place through any of the epidermal cells, but the development of root hairs as outgrowths of these cells increases their absorptive capacity about twenty times. Underneath the epidermis is a large cylinder, the **cortex.** The cortex consists primarily of large, thin, walled **parenchyma** cells, but there is an inner ring of cells of another type, the **endodermis.** The central portion of the root is occupied by the **stele.** The outer portion of the stele is called the **pericycle,** and branch roots arise from cells in this region. The inner portion of the stele contains cells of two important types —the large **xylem** cells, which form tubes through which water rises to supply the needs of the parts of the plant aboveground, and the **phloem** cells, which transport food in solution. The xylem cells are often in the shape of a four-pointed star in the center of the stele, and the phloem cells lie in between the points of the star. There will also be some parenchyma cells in the stele, filling in the extra space not occupied by the cells just described.

In many plants, root growth is limited to the embryonic region at the tip of the roots. The roots become longer by the addition of new cells at this region, but after these cells enlarge and mature, there is no further growth. Such plants tend to have a large number of slender roots, known as a **fibrous root system.** Other plants, especially the larger woody plants, have a secondary thickening that enables the roots to increase in diameter after they have been formed. A young root of this kind is similar to a fibrous root, but as it grows older it becomes evident that there is a layer of cells outside the xylem and inside the phloem in which new cells are produced. The root increases its diameter as the plant grows. This layer of cells that retains the power of division is the **root cambium.** As the root grows larger, the epidermal cells lose their absorptive capacity and are replaced by a **corky layer** that protects and waterproofs the outside of the root. There is a special **cork cambium** that produces new cork cells around the expanding root. Such a root is no longer of any value

for absorption and there must be a continual growth from the ends of the roots if there is to be a plentiful supply of epidermal cells with root hairs.

Absorption and Transportation

Absorption takes place at the roots through a combination of osmosis, diffusion, and active transport. In normal soils the moisture content exists in the form of a film adhering to the tiny soil particles. Water has a tendency to adhere to solid particles; when we wash dishes, some water will adhere after the final rinsing and must be dried off with a cloth or through evaporation. If we moisten two flat pieces of glass and press them together, we find that considerable force is necessary to pull them apart because of

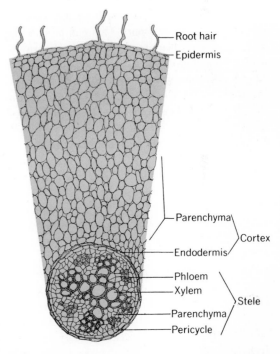

FIG. 21.3 A portion of a cross section of a buttercup root.

the adhesion of the water. In the same way, after a rain some water will drain down through the soil, but some will adhere to the small particles of rock and organic matter that make up the soil. Mineral matter of the soil becomes dissolved in the water. The delicate root hairs twine themselves around the soil particles so that they are in direct contact with this film of water and dissolved mineral matter. Absorption takes place here. The total concentration of dissolved substances is greater in the protoplasm of the root hairs than in the water outside, so water moves into the root hairs by osmosis. On the other hand, the concentration of certain minerals is greater outside than inside the cell and there will be absorption of these minerals by diffusion. In addition, as we learned in Chapter 7, there is some movement of minerals against the tendency of simple diffusion. Some minerals vital to the plant will be present in a lesser concentration outside the cell than they are inside the cell, yet they will be absorbed by recourse to active transport. This can be done by a living plasma membrane. It requires energy to achieve this movement against the tendency of simple diffusion. Once inside an epidermal cell, the water and minerals will alter the concentration there and there will be a continued movement to other cells toward the center of the root. Eventually the water and minerals reach the xylem cells. These cells are in the form of small continuous tubes extending up through the stem into the leaves.

The water and minerals are now in the conducting tubes, but they still must be lifted up to the leaves. This may require some force, for some trees reach a heigh of more than three hundred feet, and the combined diameter of the xylem tubes in their trunk may exceed that of a city water main. Water and minerals are lifted by a combination of three forces. (1) As water continues to pour into the xylem cells by **osmosis** from the surrounding cells, there will develop a pressure that forces the water up somewhat like the rise of the water in a thistle tube, as described in Chapter 7. But osmosis alone does not suffice, for the force of gravity would balance the osmotic pressure before the water could reach the necessary height. (2) **Capillary attraction** is the second force in the lifting process. We have already called attention to the tendency of water to adhere to solid objects. If we touch a very small glass tube about the size of a human hair to water, the water will immediately rise up the tube to a height of several feet. The attraction of the glass walls of the tube causes this lifting power. Since the xylem of plants consists of very fine tubes, capillary attraction is an important factor in preventing water from being drawn down the tubes by gravity. (3) Finally, **transpiration** plays a part in lifting the water up a tree. Water is constantly being evaporated from the leaves as the air circulates through them. As the water is thus removed at the top, there is a suction that helps to pull up more water to replace that which is lost. This works somewhat like a pump on a well—as water moves out at the top, other water molecules move into the leaves and replace those which have been lost. Water has a high degree of cohesion; just as water molecules tend to adhere to solid objects, they also tend to adhere to one another and resist separation. When some water molecules leave the plant at the top of the columns of water in the leaves, other molecules move in, and this movement exerts a pull which draws up the entire column of water. This explanation is known as the **cohesion-tension theory.** The energy for this pulling force comes from the heat, usually of the sun, which brings about the evaporation of water in the leaves. Thus, with osmotic pressure from below, cohesion-tension pull from above, and the capillary attraction of the walls of the tubes along the way, water, along with the dissolved minerals it contains, can be supplied to the leaves of the highest trees.

When the concentration of dissolved minerals in the film of water around the root hairs becomes greater than that within the protoplasm, there is reverse osmosis and the water is drawn out of the root hair and the entire process just described is reversed. Water is drawn down, the leaves wither, and soon the entire plant dies unless normal conditions are restored. Salt placed around the base of a plant will soon kill it for this reason, and heavy applications of concentrated commercial fertilizers may retard the growth of the plants they are supposed to benefit.

It is said that cut flowers will stay fresh longer if some of the lower part of their stems is cut off under the water in which they are placed. Whenever a flower stem is cut, the column of water in the small xylem tubes quickly moves upward a short distance. This leaves air spaces in the lower part of these tubes. When placed in water, these columns of air will interfere with the movement of the water up to the petals of the flower. When the lower part of the stems is cut off under the water, these air columns are removed, no air can enter the xylem tubes, and there is a continual flow of water up to the flower as moisture evaporates from the petals.

Roots are very important as an anchorage of the soil particles. When the plant cover of a region is removed, there can be extensive erosion from both water and wind.

THE STEM SYSTEM

We often restrict the use of the word "stem" to the small twigs of a tree, but in botanical terminology the stem is all of the higher plant that is above ground except the leaves and reproductive structures, such as flowers and fruits. This is generally true, but there are exceptions. There are underground stems in the Irish potato, the iris, the bulbs, Johnson grass, and other similar plants. The aerial roots of ivy and orchids have already been discussed. However, the stem has certain characteristic structures that make it easily distinguishable even though it may not always grow above ground.

The stem produces many valuable commercial products. Lumber leads the list; the wood of trees is an invaluable product in the construction industry of the entire world. Cork is obtained from the bark of the cork oak. Quinine is extracted from the bark of the cinchona tree. Rubber is made from the sap which flows from the rubber tree. Chewing gum owes its distinctive properties to chicle from the chicle tree. Turpentine is distilled from the resin which flows from the cut trunk of the pine tree. We use the underground stem of the Irish potato and the aboveground stem of the sugar cane for food. These are but a few of the many ways that man has learned to use plant stems to his advantage.

Function of Stems

The stem **supports and displays the leaves** in a manner that gives exposure to the light. The stem forms a connecting link between the two regions of raw material intake—roots and leaves—and serves the important function of **transportation** of materials back and forth between the two. Water and minerals must be conducted from the roots to the leaves. It is also

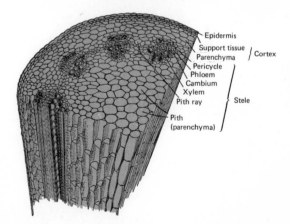

Epidermis
Support tissue
Parenchyma } Cortex
Pericycle
Phloem
Cambium
Xylem
Pith ray } Stele
Pith (parenchyma)

FIG. 21.4 A portion of a young sunflower stem showing both cross and longitudinal sections.

just as important that some of the food produced in the leaves be conducted down to the roots to nourish the living cells existing underground. Stems also function in **food storage:** potatoes are underground stems, and the sugar cane stalk stores large amounts of concentrated food. Finally, there is some **photosynthesis** in young stems. If you examine the twigs of a shrub or tree out near the ends of the branches, you will usually find that they show the typical green color of chlorophyll. This indicates that they manufacture food. Such stems will bear many small openings, **lenticels,** that admit the air necessary for photosynthesis. Without lenticels, there could be no food manufacture, because the stems are covered with a layer of cork through which air cannot penetrate. Photosynthesis in cactus plants is carried on entirely in the stems, the leaves having been reduced to the small spines that are found on the expanded stems. This is an advantage in the dry habitat characteristic of this group of plants. Every phase of this plant's activities conserves water. If the plant contained large numbers of delicate leaves, the rate of trans-

piration would cause excessive loss of moisture. It is probable that the cactus plants at one time had leaves, and the conversion into spines may have come about as an adaptation to desert life. This conception is confirmed by what happens when certain species of cactus plants are raised in a moist environment; many of the spines grow into small leaves that contain chlorophyll.

The Herbaceous Stem

Dicotyledon plants may be divided into two groups according to stem structure. One group, **herbaceous plants,** are not able to withstand frost. Hence, in temperate climates they are killed to the ground each winter. **Woody plants,** on the other hand, have stems which may withstand winter frost and continue to live above-ground from year to year. A cross section through a typical herbaceous stem, such as a sunflower or a clover plant, reveals certain similarities to the structure of a young root. There is an outer **epidermis** with the **cortex** immediately to the inside. However, the vascular tissue is not in the center but is arranged as a series of **vascular bundles** in the form of a circle just inside the cortex, Each vascular bundle consists of the **phloem** on the outside and the **xylem** on the inside, separated by the **cambium** in between. It will be remembered that the cambium has the power to divide and produce new cells, so the stem can continue to increase in diameter as long as it remains alive. The inner portion of the stem is filled with large, thin-walled **pith** cells.

Herbaceous plants may be **annuals,** which complete their life cycles in a single growing season; **biennials,** which store food in the roots and come up a second year from the roots; or **perennials,** which have roots and sometimes also underground stems. They remain alive for a number of years and new growth comes up from them each year.

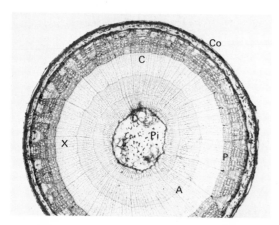

FIG. 21.5 Cross section through a three-year-old stem of *Tilia*, basswood, showing the parts of a young woody stem. Pi—pith, X—xylem, C—cambium, P—phloem, Co—corky layer.

The Woody Stem

Growth of a woody stem during the first season is like that of a herbaceous stem, but important differences are evident after several years. Around the outside there is a durable **corky layer** which protects the stem. Just beneath this is the **cork cambium.** As the stem increases in diameter, the outer corky layer cracks and sloughs off, but the cork cambium provides an abundant supply of new cork cells to take care of this loss and growth of the stem as well. The **cortex, phloem,** and **cambium** are like those parts of the herbaceous stem. The **pith** is relatively smaller, and the **xylem** occupies the rather extensive region between the pith and the cambium. The xylem cells, grouped to match the phloem, form vascular bundles, and cells which extend out into the space between the xylem groups form the **vascular rays.**

Annual rings. As the stem increases its diameter, new xylem cells are formed by the cambium, but these vary somewhat in their size and the thickness of their walls. The xylem that is laid down in the spring, when conditions for growth are favorable, usually has much larger and thinner-walled cells than that laid down in the summer, when conditions generally are less favorable. At the point where the small, thick-walled, summer cells meet the large, thin-walled cells of the following spring, a definite line of demarcation is formed. The term **annual ring** is given to the combined spring and summer wood laid down in a year. It is possible to tell the age of a stem after it is cut by counting these annual rings. In many areas of the tropics annual rings are formed even though there is no frost to stop the growth during the winter months. These regions have alternate periods of plant growth due to the wet and dry seasons that result in annual ring formation in the stems.

The mature woody stem. The woody stem continues to increase its diameter each year

of its life. The pith in the center of the stem degenerates. The cortex also loses its identity owing to the pressure of the expanding phloem. All the area outside the cambium is called the **bark** and includes the phloem, the cork cambium, and the corky layer. All the area inside the cambium is called the **wood** and consists of xylem cells together with the vascular ray cells. However, there is usually only a small cylinder

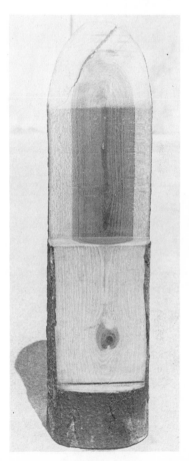

FIG. 21.6 Portion of a chestnut trunk showing both radial and tangential cuts. The radial section exposes the dark heartwood in contrast to the light sapwood. A knot, which shows in the tangential cut, shows where a side branch came off the tree.

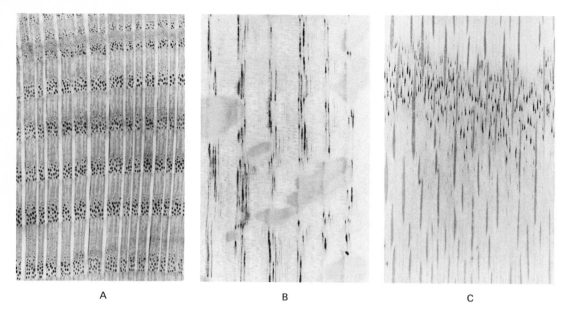

A   B   C

FIG. 21.7 The grain which shows in lumber is influenced by the way in which it is cut. These sections were all made from the red oak. At left is a cross cut made at right angles to the main axis of the trunk. In the center is a radial cut made in a vertical plane passing near the center of the trunk. This may be called quarter-sawed oak. At right is a tangential cut also in a vertical plane, nearer the edge of the trunk. The vessels of the annual rings and the medullary rays look different in each cut.

of xylem cells next to the cambium that remains functional as a water conductor. This cylinder is lighter in color than the rest of the wood and is called the **sapwood.** As new xylem cells are laid down by the cambium on the outside of the cylinder, the inner cells become filled with **resin** and assume a darker color. They then become a part of the **heartwood** which occupies the central region of a mature woody stem. Heartwood cells do not serve as conductors, but they do add strength to the stem, and the resin acts as a preservative that tends to inhibit decay and insect damage.

We sometimes think of the bark of a tree as having no vital function in the life of the plant; it is only a comparatively thin layer of apparently dead tissue around the wood. Yet, if a ring is cut through the bark all the way around the tree, the tree will die because the

bark contains the phloem. The necessary transfer of food materials cannot occur if the phloem is cut in a complete cylinder. Hogs sometimes kill trees this way. As they rub against the bark to scratch themselves, they eventually wear away the bark down to the wood.

The wood of the tree is the part sought by the lumber industry. Trees have different types of wood due to their habits of growth. In general, the slow-growing trees produce small, thick-walled cells that make a very hard wood. Lumber from such trees is greatly valued because of its superior wearing qualities. Oak and mahogany are typical. The teak tree, of India, produces wood so hard and solid that it will not float on water. The rapidly growing trees produce large annual rings containing large, thin-walled cells that result in a soft wood. Even within the same tree the wood varies in hardness depending on the speed of growth. Summer wood is always harder than spring wood, and when termites attack woods they invariably begin eating away the soft spring wood, leaving long slivers of the harder summer wood. When a floor becomes badly worn, the spring wood is worn down first, leaving the summer wood protruding so that it easily breaks loose as splinters.

The grain of lumber results from the annual rings and vascular rays that run through the planks. Grain patterns of different types may be produced by cutting the log so as to bring out the most desirable type. Fig. 21.7 shows some of the different wood patterns and how they are produced.

The twig. The stem, like the root, increases its length only by growth at the tip end. If we drive a nail through a four-foot sapling two feet from the ground, it will be halfway between the ground and the tallest point of the sapling. Thirty years later the sapling may have grown into a tree, perhaps thirty feet tall, but the nail will still be two feet from the ground and not halfway between the ground and the upper tip of the tree as might be supposed. The diameter of the tree will be much greater and will have completely surrounded the nail, but this growth has been in girth and not in length.

A typical twig during the dormant season shows a **terminal bud** at the tip end. This bud is composed of delicate embryonic tissue, the **meristem,** in the center surrounded by the **bud scales,** which are usually impregnated with a waxy secretion that seals the bud and helps protect the meristem from desiccation and frost. Embryonic leaves, branches, and possibly flowers may also be seen coming out from the meristem ready to enlarge and make functional structures when growth is resumed. The meristem contains the only tissue which produces cells that increase the length of the stem; hence, growth in the height and spread of a tree is always through addition of cells at the tips of the branches. This

contrasts with typical animal growth, where height is increased by the production of new cells in various regions of the body.

Lateral buds will be found on either side of the twig. Most of these produce side branches and leaves, but some also produce flowers and fruit. The buds are formed just above the point where the leaf joins the stem, and after the leaves have been shed, there will be a **leaf scar** just below each of the lateral buds. The vascular bundles that ran out into the leaf may often be distinguished in this scar. The outer portion of the bark is perforated with many small openings,

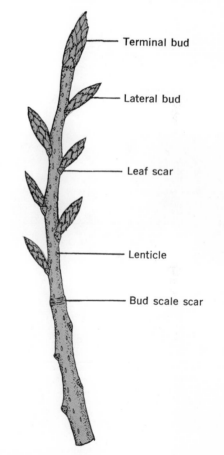

FIG. 21.8 A typical twig.

the **lenticels,** that admit air to the cells containing chlorophyll.

Most twigs also show **bud scale scars** at the region where the terminal bud spent the dormant season during previous years. These appear as a series of bands that encircle the twig. These scars are valuable as indicators of the twig growth. By measuring the distance between the scars, it is possible to see just how far the twig has grown in each of the past several years. Thus, a rapidly growing plant can easily be distinguished from a slow growing species. Also, variations in growth of the same plant from year to year can be determined and correlated with weather, soil fertility, and other conditions.

The Monocotyledon Stem

The preceding description of stem structure applies to the dicotyledons and the woody gymnosperms. Monocotyledons show certain differences in stem structure that warrant consideration. It will be remembered from the survey of the plant kingdom that the monocotyledons include such plants as the grasses and the cereal grains. The stems of this group of plants are often divided into sections or joints, each of which has a disk of embryonic tissue at its tip. Thus, the joints are able to increase in length as the plant becomes older. However, there is no cambium in the stem, and the diameter of the stem increases only by enlargement of the existing cells after they have been formed by the meristematic region at the nodes, the region where the joints come together.

The structure of a monocotyledon stem in cross section is somewhat like the herbaceous dicotyledon stem; there is the outer **epidermis** followed by the **cortex** and **pith** on the inside. However, the area of distinction between the cortex and pith is not sharply defined. The **vascular bundles** have no cambium, as already mentioned; neither do they have the arrangement of the herbaceous stem. There will be an outer ring of them, but they are also scattered throughout the central portions of the stem.

FIG. 21.9 Longitudinal section through a bud of a lilac. Note the meristem, which contains the embryonic cells at the top, and the primordial leaves and flowers. The bud scales surround the bud.

Stem Growth

The woody plants tend to assume a characteristic form or shape as they grow. We have the broad, spreading maple; the tall, stately pine; the weeping willow; and the prostrate juniper. These variations are brought about as inherited traits which influence growth. In the tall trees with small side branches, the growth from the terminal bud at the top of the tree is much more rapid than the growth from the terminal buds at the ends of the side branches. This produces the **excurrent** type of branching which is characteristic of the evergreens that we cut for Christmas trees. Other trees, such as the maple, have growth from the buds on the side branches which is just about as rapid as the growth from the bud at the upper tip of the tree. This results in the **deliquescent** or spreading type of tree which usually has several main branches. Such trees are very nice for shade.

Some of the tropical monocotyledons, such as the palms, grow year after year and form trees. The trunks of such plants usually have about the same diameter from the ground up to the leafy portion because there is no cambium, the diameter of the stem remaining fixed once cell enlargement has stopped. Since dicotyledons have a cambium which adds a new ring of wood each year, the trunks of trees in this group tend to taper from the ground up to the twigs.

Many environmental factors can cause an alteration of the typical forms of tree growth. Two trees growing side by side will tend to be lopsided, because each will grow away from the other. Pine trees in a forest have long trunks with small branches and leaves concentrated in a crown at the top, whereas those growing out in the open retain their lower branches and leaves are found almost from the ground up. Uneven illumination will cause trees to grow unevenly, with the greater growth toward the side receiving the greatest amount of light. Man often prunes and trims woody plants in order to make them grow in a form which suits his convenience or fancy. Many of these characteristics of stem growth are brought about by the action of plant hormones secreted by the cells at the growing tips of the plants. We will learn more about them in the next chapter.

Since growth of the dicotyledon plants is dependent on the meristem contained in the buds on the twigs, one might wonder how such plants continue to live and grow when all of the bud-bearing branches are removed from a tree, as sometimes happens in severe pruning. Continued growth is possible only because these plants produce **adventitious buds** at the site of the cut. These buds must arise from the cambium, since this layer contains the cells that have the embryonic power to divide and produce other cells of all types. When a large tree is cut down, a number of buds may arise from the cambium, which lies just between the bark and the wood, and a series of shoots will appear in the form of a circle. Sometimes one of these becomes dominant and forms the main branch of the newly formed tree, but quite often a number of the branches will continue growth to produce a tree with several trunks and only one root system.

Herbaceous dicotyledon plants also regenerate new meristems when they are cut, and therefore their growth is not stopped when the tips of the stems are cut. However, growth is slowed and a continued cutting of these tips will result in eventual death.

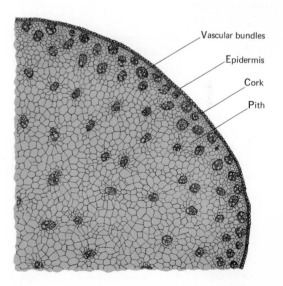

FIG. 21.10 Cross section of a young corn stem.

Higher animals typically grow through a period of development until they reach adult size and then stop growing. New cells are produced only to replace those that are destroyed. Finally, a period of old age arrives when the growth capacity of the body cannot keep pace with the cell death, and death of the organism eventually results. The higher plants that are adapted to live more than a single season have no such cycle—they must continue to grow as long as they live. Only the newly formed root tips can absorb the necessary water and minerals; only the newly formed twigs bear the leaves necessary for photosynthesis; and only the newly formed xylem functions in water conduction. Therefore, a tree must increase its root spread, its height, and the diameter of its trunk and branches every year of its life. Since the functional part of the tree is always newly formed tissue, there is no such thing as old age and natural death from age. Trees die through the influence of external forces. Diseases, insects, fire, wind, and man are the usual agents that terminate the life of a tree. Those trees that withstand these forces best live to be the oldest.

Stem Propagation

The cambium is a remarkable part of the stem; it retains the power to produce any type of cell found in the entire plant. Because of this property of the cambium, plants can be propagated asexually and retain exactly the same characteristics possessed by the parent plant. Seeds are usually produced by the union of pollen from the male organs of one plant with ovules within the female organs of a different plant. This produces a mixed parentage, and the exact combination of traits expressed in the offspring may be different from that expressed by either parent. Cross pollination is of great value in producing new types of plants from which those commercially desirable may be selected;

however, once we have a type that is of value, we want to keep it that way. Seeds from a desirable type of plant can grow into plants that may be entirely different—a seedling pecan tree grown from a big paper-shell pecan may bear small thick-shelled pecans. Desirable types are maintained by vegetative propagation. A twig cut from a paper-shell pecan tree may be put in the ground; roots will be produced from the cambium underground, adventitious buds will appear on the stem aboveground, and in time a tree will be produced that will bear exactly the same type of pecans that the single parent bore. Many plants are propagated entirely by their stems to maintain the variety.

Budding and Grafting

It is possible to attach the growing point of one tree onto another tree by budding and grafting. Then, since all growth in the trees comes from the tip of the twigs, all the new growth will be of the type that was transferred. Suppose you have an apple tree that produces only a few small sour apples. It is possible to place buds from a number of varieties of apple trees on several of the main branches. A slit is made in the bark of the branch and the bud is slipped under the slit so that the cambium layers of the two are in contact. Later, the branch is cut off just beyond the point where the bud has been inserted so that all the strength of growth in the branch will go into the new bud. Within a year or two each bud will have grown into the dominant branch of that region of the tree and this tree may produce Delicious, Winesap, Jonathan, and Roman Beauty apples from a single trunk and root system. There will be no mixing of the qualities of the new types of fruit with the old, for all of the new cells are from the meristem of the bud and all that they receive from the trunk and root system is water and minerals. Budding, therefore, gives us a method of converting a tree bearing undesirable fruit into one bearing fruit of high quality.

It is the general policy of nurserymen to plant or root the stem of a native fruit tree that bears undesirable fruit and, then, when it is several years old, to place a bud from a desirable type of fruit on it. Experience has shown that such a plant is much healthier than one grown entirely from a cutting of a desirable fruit tree. In the process of selecting the best qualities of fruit, it is likely that a plant that will not be as strong as a native plant will be produced. By budding we can obtain the root system of the hardy native tree and the crown bearing the desirable fruit.

Grafting is another way of accomplishing the same thing. In this process the entire twig from one plant is grafted onto the twig of another plant. The twigs should be of about the same size so that the cambiums of the two are in contact, but it is possible to place a small twig on a large branch of a tree provided it is placed near the edge in the cambium. Grafting is not as popular as budding because it uses up more of the desirable stock. There are a number of buds on the twig that is used for grafting and each of these buds could be used as a growing point through the budding method.

THE LEAF SYSTEM

The leaf system is a very conspicuous and important part of a higher plant. Leaf tissue must be thin and spread for an optimum exposure to the light. In spite of the conspicuous nature of the leaves, we seldom realize the great extent of the exposed surface area. If we collected the leaves from one mature maple tree and spread them on the ground side by side, they would cover a surface approximating the size of a football field. Thus, even though the amount of sun's energy captured in a single leaf is small, the com-

bined effects on this entire leaf surface are quite significant.

Function of Leaves

Photosynthesis is the primary function of leaves. In the cactus plant leaves have abandoned this function, but in the great majority of plants all other functions are secondary to photosynthesis.

Protection is an important secondary function of leaves in many plants. Leaves are generally tender, succulent, and tempting to many forms of animal life as a source of food. Many leaves have sharp spines on their surfaces to discourage the larger animals from eating them. Some of these, such as the stinging nettle, even have an irritating acid on the end of the spines, which give a sting like an insect sting. Some, such as the holly, have leaves that roll at the edges, thus forming sharp spines. Others, such as the geranium, produce a volatile oil on small hairs on the leaf surface that acts as an insect repellent. In some desert plants, such as the agave, the leaves are enlarged and store water. Desert wanderers have sometimes saved themselves from death by cutting into these plants and drinking the liquid within. The Mexicans produce a powerful beverage, tequila, from the fermented liquid of a species of agave.

There is one strange group of plants, the **carnivorous plants,** whose leaves trap small animals as prey. The leaves of the **sundew** have sparkling drops of a viscid secretion that attract and catch insects which are digested by the plant. The **pitcher plant** has pitcher-shaped leaves with a fluid at the bottom containing the digestive enzymes. Insects enter the plant from the top but are impeded from leaving by an array of bristles. Eventually they end up in the fluid at the bottom of the leaf. The **Venus's-flytrap** has leaves which fold around insects that happen to alight on them. Some tropical carnivorous plants

A   B

FIG. 21.11 Some leaves have a protective function. The Spanish bayonet (A) has leaves which are very sharp on the end. The cactus (B) has its leaves re- duced to small, sharp spines which carry an irritating poison.

are large enough to capture small animals and birds, but none are large enough to endanger the life of a larger animal.

Structure of Leaves

A simple leaf consists of a thin, broad **blade** attached to the stem of a **petiole.** Some sort of a lateral bud appears just above the point where the petiole joins the stem. Sometimes there will be two small lateral projections at this same point, the **stipules.** Small **veins** through- out the leaf carry materials to and from the cells. These veins contain xylem and phloem cells which are continuous with those found in the stem and root.

Photosynthesis is possible because of the cell organization which can be studied in a cross section of the leaf. At the top of the leaf the **upper epidermis** covered by a **cuticle** water-

FIG. 21.12 Poison ivy. This innocuous-looking plant contains a substance which is highly irritating to the skin of persons who have become sensitive to it. Poison ivy can be either a climbing vine or a ground vine.

bean-shaped cells, the **guard cells,** which expand and contract and regulate the rate of gaseous exchange. On a hot, dry day the stomata tend to close down, preventing the excessive transpiration which would extract large amounts of valuable moisture from the plant. This somewhat impairs the efficiency of photosynthesis, yet conserves the water supply of the plant. For this reason plant growth is most rapid in moist, tropical regions where the humidity is so high that the stomata can open and allow a maximum circulation of air through the leaves without losing much water through transpiration. The guard cells contain a few chloroplasts, but there are none in the epidermal cells.

proofs the leaf and prevents evaporation of water from the surface. Beneath this is a layer of columnar cells, the **palisade layer,** where most photosynthesis takes place. These cells contain many chloroplasts which move about in the cell. Below the palisade layer is the **spongy mesophyll.** This layer contains cells of irregular shape fitted together loosely, allowing for many air spaces between. This arrangement allows the free circulation of air through the leaf that is so necessary for the exchange of gases. The **veins,** consisting of xylem, phloem, and sheath cells, run through the mesophyll bringing in the water and dissolved minerals and carrying away the manufactured food. There are some chloroplasts in the mesophyll cells, and therefore some photosynthesis occurs here. Most of the light is absorbed by the palisade cells above, which carry on most of the photosynthesis in the leaf. The **lower epidermis** is found on the lower surface of the leaf and is also covered by a cuticle. Air passes into and out of the leaf through small openings, stomata, found on the lower surface of the leaf. Each **stoma** is surrounded by a pair of

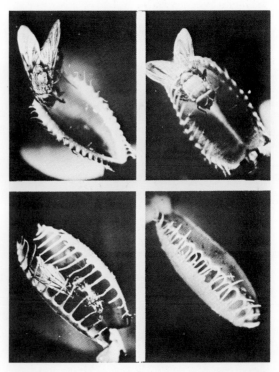

FIG. 21.13 The Venus's flytrap. This plant has leaves modified into traps which catch insects. When a fly touches the trigger hairs in the center of the leaf, the leaf closes around it and produces enzymes which will digest the fly.

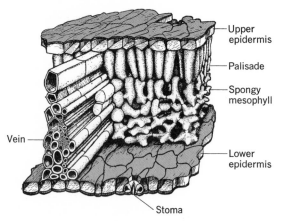

Upper epidermis

Palisade

Spongy mesophyll

Vein

Lower epidermis

Stoma

FIG. 21.14 Cross section through the leaf of *Syringa* (lilac), shown in perspective.

Leaf Venation

The veins of leaves arise at the petiole, or if the leaf does not have a petiole, they arise at the junction of the leaf and stem. From this origin they run out to all parts of the leaf, supplying water and minerals. Veins also add rigidity to the leaf because of the firmer construction of their cells. Two major types of venation may be recognized—parallel and netted. In **parallel venation** the veins are all of about the same size and run parallel to one another. It is characteristic of the monocotyledons. In **netted venation,** typical of most dicotyledons, the veins branch and rebranch, forming a fine network of tiny veins. **Pinnately netted leaves** have one major vein running down the center of the leaf with numerous side branches coming off from it. Such leaves somewhat resemble a feather and take their name from this resemblance. **Palmately netted leaves** have several major veins with side branches coming out from these. Such leaves are usually broader than the pinnately netted leaves and their margins are often lobed to correspond with the major veins.

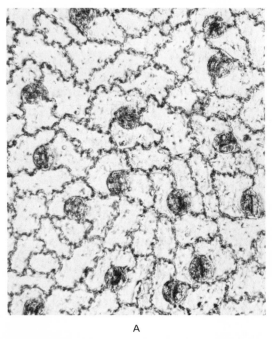

A

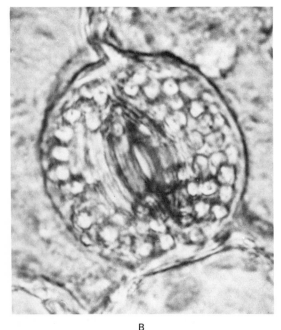

B

FIG. 21.15 Stomata on the underside of a *Smilax* leaf, a running vine (A). The enlarged view shows the chloroplasts in the bean-shaped guard cells on either side of the stoma (B).

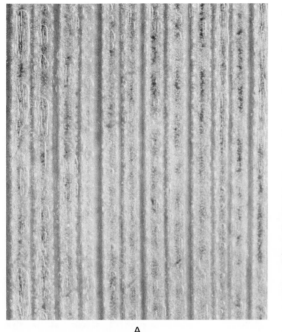

A

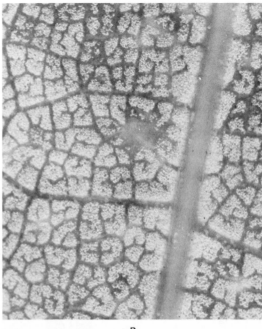

B

FIG. 21.16 The two major types of leaf venation. Parallel veins are typical of the monocotyledons (A). Netted veins are typical of the dicotyledons (B).

Leaf Complexity

The margin of a leaf shows various degrees of indentation, as illustrated in Fig. 21.16. The leaf may also be divided into smaller individual leaflets. This type is called a **compound leaf,** in contrast to the **simple leaf,** which has no division of the blade into leaflets. Compound leaves may be **pinnately compound, palmately compound,** or **double compound,** as illustrated in Fig. 21.17.

It may sometimes be confusing to decide just where the stem ends and the leaf begins. Some simple leaves may be composed of a single blade only a fraction of an inch long; some compound leaves may be composed of hundreds of leaflets and be several feet long. Just how are we going to know where the leaf starts? In woody plants our most reliable clue lies in the lateral buds; a lateral bud of some description always lies in the axil of the leaf where it joins the stem. The bud may be small and covered by the petiole of the leaf, but it is there and usually can be seen easily. If there are any leaves that have fallen from the tree on the ground nearby, the extent of the leaf is also easily determined. When a leaf falls, the petiole breaks off at the point where it joins the stem, and even though the leaf is several feet long, it will be shed from the point where it joins the stem.

In herbaceous plants the lateral buds are usually small and inconspicuous. In many cases this makes it difficult to judge the extent of the leaf except by pulling the leaf from the stem. The petiole of an entire leaf will usually break clean from the stem, but a leaflet will not make a clean break.

Leaf Arrangement on the Stem

Leaves arise from the stem according to a regular pattern characteristic of the species. **Opposite** leaves come out in pairs, each pair arising from the same level on the stem. In the **alternate** pattern, each leaf arises from the stem singly and the next leaf arises from the opposite side of the stem higher up, and so on as shown

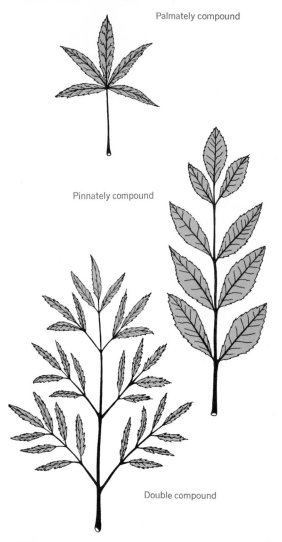

Palmately compound

Pinnately compound

Double compound

FIG. 21.17 Types of compound leaves.

in Fig. 21.18. **Spiral** leaves come out from the stem in a spiral order. In most cases there are four leaves in one spiral turn around the stem. In the **whorled** pattern three or more leaves arise at the same level on the stem. Then perhaps a little higher up on the stem there will be another whorl, and so on up the stem.

Leaf Fall

Leaves, unlike roots and stems, have no embryonic region to produce secondary growth that would enable them to enlarge after they are matured. The tiny embryonic leaf, which can be seen folded up within the bud, contains all the cells that the mature leaf will contain. At the proper time the bud opens, the embryonic leaf unfolds, and the leaf grows through enlargement of the cells. When the cells have reached the limit of their enlargement, growth stops; the leaf is mature and, because of this characteristic, is doomed to a short life. In a great many plants this life is no longer than a few brief months from the time of unfolding in the spring until frost in the fall.

Deciduous Trees and Shrubs

In herbaceous plants the stems as well as the leaves are killed by frost. Among the woody plants, there are some whose leaves are killed by frost but whose stems are not. These are called deciduous trees or shrubs, according to their habit of growth. Since a group of dead leaves hanging on a tree would be disadvantageous, there is a special layer of cells at the point where the leaf joins the stem that cuts the leaf cleanly from the tree. This **abscission layer** consists of thin-walled cells; when frost comes, the cell walls disintegrate and the leaf is left hanging by the tiny vascular bundles. Then, when a breeze breaks the final connection, the leaves fall.

Sometime between the first cold snap of fall and the final shedding, the leaves assume an autumnal coloration that makes the outdoor landscape a riot of color in regions abundant in deciduous trees and shrubs—yellow, orange, red, and brown leaves, often standing out against the deep green of evergreen foliage. Many of these colors are present in the leaves all of the time, but in summer there is so much chlorophyll that they cannot be seen. Chlorophyll is a very unstable substance which disintegrates rapidly when the cells die. As it fades, the more stable colors show and the leaves seem to develop new colors overnight. Some changes are associated with chemical reactions occurring within the leaf.

Evergreen Trees and Shrubs

Woody plants of another type retain their leaves throughout the winter months. These are the evergreens. Their leaves show characteristic distinctions that make it possible to recognize evergreens even at a time of year when all the plants have their leaves. Evergreen leaves are thicker and have a heavier coating of cuticle on the surface which causes most of them to be shiny in comparison with deciduous leaves. Both of these characteristics help protect the leaf from the cold, but with some sacrifice of their capacity for food production. The leaves are too thick for light to penetrate, and only the topmost cells are able to carry out much photosynthesis. Because of their weight, only a smaller number of leaves can be borne on a limb, and this reduces overall food production. This disadvantage is offset by the fact that evergreens do not need to accumulate the great reserve of food characteristic of the deciduous plants, for they do not have to replace such a large mass of leaf tissue each year.

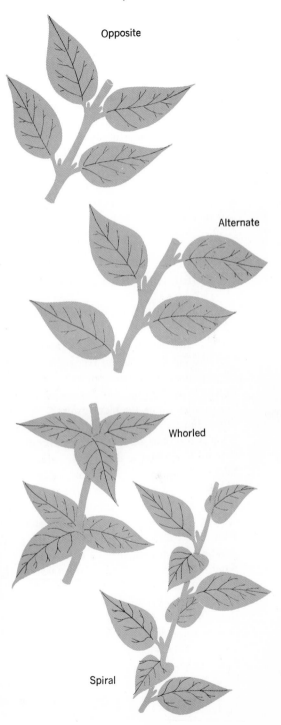

FIG. 21.18 Variations in the type of arrangement of leaves on the stem is illustrated by these diagrams.

Thick, well-protected leaves in themselves would not be sufficient to withstand freezing temperatures. A physiological characteristic of evergreen leaves makes this possible. When fall approaches the leaves may change some of their stored starch to sugar. Starch exists in the protoplasm in the form of a colloid and, as such, does not alter the freezing point of the water in the leaf. However, sugar dissolves to form a true solution, and any substance dissolved in water will lower its freezing point. We put salt in water to make brine that can withstand a temperature far below the usual freezing point of water alone.

In the same manner, this extra sugar in solution lowers the freezing point of the water in the leaf and allows it to live through the winter.

Evergreen leaves fall into two general classifications—the broad-leaved, with a tough, leatherlike leaf, and the needle-leaved, with the leaf rolled into a compact needle. Evergreen plants do shed their leaves—a walk through a needle-carpeted forest will confirm this—but they are shed gradually and new leaves are produced each year. Hence, evergreen plants are never completely bare of leaves.

REVIEW QUESTIONS AND PROBLEMS

1. Why do the leaves of most plants wilt when the plant is transplanted, even though most of the large roots are preserved intact?
2. List the different regions which can be seen in a longitudinal section of a young root and tell how you can recognize each of the regions.
3. What parts of a mature root are primarily concerned with transportation?
4. In what way does the root system of a plant give a clue about the weather conditions under which the plant grows?
5. What characteristic of growth produces the fibrous type of root system?
6. What is the function of a root cambium?
7. Explain how water and minerals enter the root hairs.
8. Why does a plant die when salt is placed around the roots?
9. Describe the three forces that lift water from the roots to the leaves of plants.
10. If we cut a growing plant off near the ground, we can often see drops of water coming out on the surface of the cut stem. What forces bring this water up?
11. What forces bring the water up into the petals of a cut flower in a vase of water?
12. What are the principal functions of stems?
13. What is the primary distinction between herbaceous and woody plants?
14. What tissues are found in a vascular bundle of a herbaceous dicotyledon stem?
15. Explain how an annual ring is formed.
16. How does heartwood differ from sapwood with respect to color and function?
17. Why is wood from a slow-growing tree usually harder than that from a tree that grows rapidly?
18. What tissues are found in the bark of a tree?
19. What is a meristem and where may it be found?
20. How can you determine the rapidity of growth of a plant by an examination of the twigs?

21. How does the growth of a monocotyledon stem differ from the growth of a dicotyledon stem?
22. What difference in manner of stem growth causes some trees to assume the deliquescent shape while others are excurrent?
23. Where do adventitious buds appear and why are they of such importance?
24. A citrus grove owner in Florida plants seeds of a wild, sour orange. When these have produced saplings several feet tall he buds a sweet variety of orange onto them. Why did he not plant seeds of the sweet variety in the first place?
25. Mr. Brown wants to produce some new varieties of iris in his garden, while Mr. Smith has a variety with very beautiful flowers and he wants to produce more of these for commercial sale. How would you advise each of these men to proceed?
26. The cambium layers of the two stocks must be in contact before there can be any successful budding or grafting. Why?
27. How is the exchange of gases regulated in a leaf?
28. Why is photosynthesis at a comparatively low level when a plant is in direct sunlight and the air is very warm and dry?
29. Describe the major types of leaf venation.
30. What are the differences between evergreen and deciduous leaves?
31. Describe the ways in which leaves may be arranged on the stem.

FURTHER READING

Bold, H. C. 1973. *Morphology of Plants.* 3rd ed. New York: Harper and Row.
Eames, A. J. 1961. *Morphology of Angiosperms.* New York: McGraw-Hill.
Esau, K. 1965. *Plant Anatomy.* 2nd ed. New York: John Wiley and Sons.
Galston, A. W. 1964. *The Life of a Green Plant.* Englewood Cliffs, N. J.: Prentice-Hall.
Greulach, N. A., and Adams, J. E. 1967. *Plants.* New York: John Wiley and Sons.
Kozlowski, T. 1964. *Water Metabolism in Plants.* New York: Harper and Row.
Meeuse, A. D. J. 1966. *Fundamentals of Phytomorphology.* New York: Ronald Press.
Meyer, B., ed. 1973. *Introduction to Plant Physiology.* New York: D. Van Nostrand.
Ray, P. M. 1963. *The Living Plant.* New York: Holt, Rinehart and Winston.

22

Regulation of Growth in Plants

The response of a plant to its environment often appears to be regulated by some sort of intelligent comprehension of its needs. No matter what the position of a seed when planted, the roots will grow down and the stems will grow up. When a plant is illuminated from one side only, the leaves will turn toward the light and the stems will grow toward the light. When the top of a tree is cut off, a smaller branch will turn up, enlarge, and become the dominant growing top of the tree. All these reactions accomplish something for the plant, but the reactions are based on chemical and mechanical processes and not on intelligent or instinctive responses such as those made by animals.

Plant hormones control many of these reactions. In this chapter we shall learn more about these important substances and their influence on plant reactions.

HORMONAL REGULATION OF GROWTH

Hormones in animals are well known, and we have also become aware of their importance in plant physiology. A hormone is a substance produced in one part of the body which is transported and influences cell activity in other parts of the body. The most widely known plant hormones have been given the name of **auxins** and are produced primarily in the growing tips of plant stems and in the young, growing leaves.

From this point they pass down, influencing cell growth and other activities. The main auxin produced is **indole-3-acetic acid.**

An auxin functions, in part, by controlling cell elongation. Under the influence of certain concentrations of this auxin, the thin cell walls in the actively growing parts of the plant become quite plastic. This permits a considerable degree of elongation of the cells and growth is stimulated. This effect can be demonstrated very force-

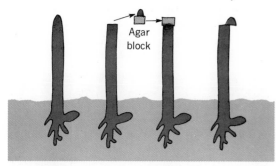

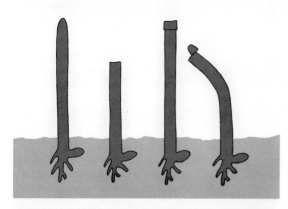

FIG. 22.1 Experiments to demonstrate the production of a growth-stimulating hormone in the tip of the coleoptile of the oat seeding. The bottom row shows the reactions which take place several hours after treatment, as shown in the top row. The undisturbed control seedling on the left shows normal growth. The one with the tip removed does not grow. When the tip is placed on an agar block for a time, however, and this block placed on the cut coleoptile, growth continues. When the cut tip is replaced off center, there is growth on the side under the tip, but no growth on the other side. The plants were kept in the dark to eliminate any influence of light.

fully by a series of experiments with seedlings of oat plants. The first leaf of such a seedling is known as the **coleoptile.** When the tips of the coleoptiles are cut from seedlings, growth stops. Control seedlings with intact tips will go on and grow normally. If a cut tip is replaced on the plant from which it was taken, growth resumes. But the tip itself is not needed to restore growth. The cut tip can be set on a block of agar jelly for about two hours. Then the agar jelly can be put on the cut portion of the seedling. Growth resumes. If the coleoptile tip is cut off and replaced so that it is off-center, then the surface of the coleoptile which is under the tip will grow, but the surface not under it will not grow. This results in a curved growth as shown in Fig. 22.1.

All these reactions can be explained on the basis of the auxin which is produced in the tip of the coleoptile. When this growing tip is removed, the supply of auxin coming down the plant is cut off, the cells are not stimulated to elongate, and growth stops. When the tip is replaced, or an agar block which has absorbed some of the auxin is put on the cut tip, growth resumes. When the tip of the coleoptile is replaced off center, the auxin goes down one side of the plant only and the cells on that side elongate, but those on the other side remain unaffected. This causes the plant to bend to the side. Now let us see how this response can explain some of the reactions of plants to their environment.

Response to Gravity—Geotropism

Suppose we have some young bean plants growing in a pot and we turn the pot on its side. Within an hour the stems will begin to turn upward by a curvature at the region of cell elongation. Within a day the stems will have executed a complete 90-degree turn and will be growing upward. This is what is known as a **negative geotropic response** (growing away from the pull of gravity) which is one of the tropisms exhibited by plants during their growth. We can explain this tropism as an auxin reaction. When a stem is turned on its side, the auxin produced by the growing tip of the plant tends to flow downward in response to the pull of gravity. This causes a greater concentration of auxin on the lower surface of the stem with a consequent greater stimulation of the cells in the region of elongation. These cells grow faster than their counterparts on the upper surface of the stem and the upward turn is executed. When the growing tip becomes erect, the flow of hormones be-

comes equalized all around the stem, and the cell elongation becomes equal. This is illustrated in Fig. 22.2. Note that the part of the stem affected is limited to the region of cell elongation. Once the cell walls have become fixed, there is no further chance of elongation regardless of auxin concentration.

Roots also respond to gravity in their growth pattern, but they are **positively geotropic** (grow toward the pull of gravity). An auxin is again responsible, but how does it cause a reaction in roots which is just the opposite of the reaction seen in stems? All plant cells are

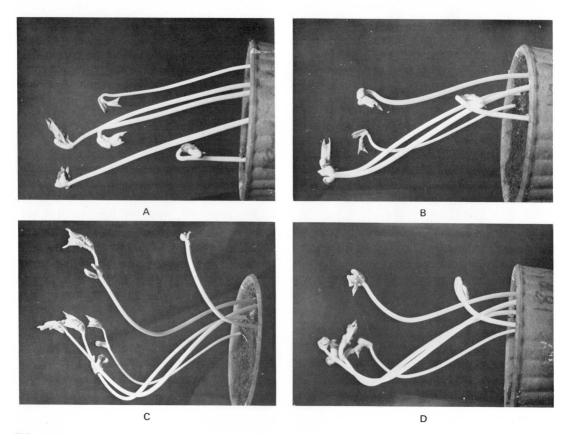

FIG. 22.2 Negative geotropism is demonstrated in these bean seedlings turned on the side. At the start (A) the stems are straight, but after one hour (B), four hours (C), and twenty-four hours (D), a progres- sive upturn can be seen. Note that the effect is exhibited only in the region of the stem where the cells are elongating, the region of differentiation.

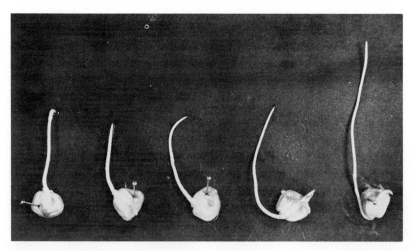

FIG. 22.3 Positive geotropism is demonstrated by a root growing from corn seed. Each of these seeds is in a different position, but the roots have all grown downward.

inhibited in growth by high concentrations of auxin and are stimulated by low concentrations. The cells of the stems and those of the roots differ in the level of auxin which will cause inhibition. Stems require higher concentration of auxin than roots for inhibition to take place. The cells in the embryonic region of roots are inhibited from growing by concentrations of auxin as low as 1 part in 50 million, while stems are not inhibited until the concentration of auxin reaches about 1 part in 500. On the other hand, roots are maximally stimulated to grow by concentrations of 1 part of auxin to 10 billion parts of water, while stems show maximum growth from the application of 1 part of auxin to 100,000 parts of water.

Such variations in sensitivity cause the opposite reactions which are observed. When a plant is turned on its side the auxin accumulates on the lower surface of the root in sufficient concentration to cause inhibition of growth. This causes the cells on the upper surface to grow faster and the root turns downward. The same concentration in the stem would give the opposite reaction—the cells on the lower surface would be stimulated.

Were it not for this opposite geotropic response of roots and stems, agriculture as we know it today would be impossible. A farmer drops his seeds in the ground, confident that the stems will grow up and the roots will grow down, no matter what position the seeds fall in. Think how complex it would be if each seed had to be planted by hand in just the right position so that some of the roots would not come out from the tops of the seeds and grow above ground while the stems and leaves grew down.

After studying the chemical nature of indoleacetic acid, biochemists have made synthetic products with similar chemical properties. Some of these have proved to be quite valuable economically. One, commonly known as **2,4D,** is used extensively as a **weed killer** for lawns, pastures, and fields and unwanted water plants in rivers, lakes, and streams. This synthetic auxin has been found to have a powerful growth-stimulating action, but different plants respond to different concentrations. The dicotyledons are generally more sensitive to this growth-stimulating chemical than the monocotyledons. Most of our common weeds are dicotyledons, while our lawns, pastures, and most of our field crops are monocotyledons. When the 2,4D weed killer is applied in proper concentrations to an area where both types of plants are growing, the dicotyledons are stimulated and they outgrow their reserve food supply and die. The monocotyledons, on the other hand, are not stimulated as much and they remain alive. It is much easier and cheaper to spray a corn field with a low concentration of this chemical than it is to pull up or dig up the weeds by hand. The same is true of a lawn. In Florida and other southern states 2,4D has recently been used to kill water hyacinths. These are beautiful plants, but they grow so abundantly in warm waters that they clog rivers, irrigation ditches, and ship channels. A light spray of 2,4D kills them.

Response to Light—Phototropism

The response of growing plants to light is one of the most striking adaptations that we find in the plant kingdom. A plant growing in a box with a small hole in it will grow to the hole and emerge into the light. Even if there are several partitions in the box with holes at different places, the plant will still find its way to the outside. This apparently purposeful activity can be explained by an auxin reaction. When the light which strikes the growing tip is stronger on one side, there seems to be a migration of auxin from the brighter side of the tip to the darker side of the tip. There is also evidence that light may destroy auxin to a certain extent. As a result of these effects, we find that when a plant tip is illuminated from one side only, the amount of auxin which passes down the stem will be greater on the shaded side. This was first demonstrated by experiments performed by Charles Darwin, who is better known for his theories on evolution. Darwin reported that when he placed little black caps over the tips of oat seedlings light from one side did not cause the plant to bend, even though the part of the plant which does the

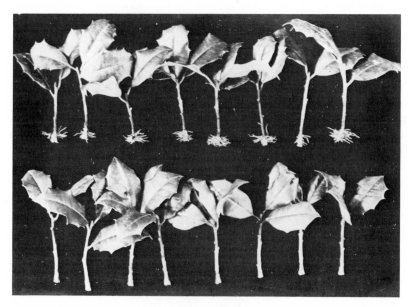

FIG. 22.4 The American holly twigs at the top were dipped in indolebuturic acid twenty-one days before this photograph was taken. They were then placed in moist soil. All the treated twigs have put out roots. The twigs at the bottom had no hormone treatment and no roots have appeared. (Paul C. Marth, United States Department of Agriculture.)

bending was fully exposed. This proved that light must have an effect on some substance in the tip, which then moved down the plant. In the reverse experiment black tubes were placed on all of the coleoptile except the tip; bending took place. When a tip was cut off, the plant showed no response to uneven illumination. These experiments pinpointed the effect of light on the tip of the coleoptile of the oat plants, which is the place where the auxin is produced, as we have learned.

The influence of light on auxin accounts for a number of commonly observed plant reactions. A potted plant growing in a window will bend and grow toward the window if it is not turned regularly to equalize the differential effects of auxin concentrations. A plant growing in a darkened area will grow toward a small opening which admits light. Any time that the growing tip tends to point in a different direction, there will be an unequal distribution of auxin traveling down the stem and the plant will turn so that it grows toward the light source. This enables many plants to survive by growing out into the light when seeds happen to sprout in a darkened area.

The fact that light somehow inhibits the production or function of auxin is demonstrated by growing plants in total darkness. Seedlings which have never been exposed to light show a great elongation of the stems, a condition known as **etiolation.**

Auxins and Root Production

The rooting of cuttings is a common method of propagating woody plants. A twig cut from a shrub or tree can be put in the ground and in many cases it will develop roots and form a new plant with characteristics like the plant from which it was taken. There are many woody plants, however, which are very difficult to root in this manner. Yet rooting can be accomplished by the use of an auxin with ease in many of them. Indo-leacetic acid, which is produced naturally in the growing tips of plants, and which can be produced synthetically or obtained from other organic sources such as human urine, stimulates root formation. Roots will even appear on the side of a plant stem if the surface is scraped and this auxin is applied. Sometimes this method is used for rooting difficult plants. The stem is scratched and the auxin applied. Then the area is wrapped in damp sphagnum moss or similar water-absorbing material and covered with a watertight wrapping. In the course of time a vigorous growth of roots will develop within this covering. The stem can then be cut from the plant and put in the ground with a root system already established. In many cases, it is easier and just as effective to cut the twigs from the plants and soak them in a solution of this hormone or dip them in a powder containing the hormone before planting.

Auxins and Spring Growth

Many songs have been composed about the beauties of the spring growth of plants, the sudden resumption of plant growth which transforms the drab winter landscape into a fresh green set off by the colorful blossoms of the spring flowers. As with so many other plant activities, we find that hormones are involved in the spring renewal of growth. The terminal buds of the woody plants are stimulated to activity by the warming temperatures. As they burst open, auxin is produced in the tip region, and diffuses down and stimulates the stem to resume growth. Buds are more sensitive to auxin than stems. The concentration of auxin which stimulates stem growth is strong enough to inhibit the growth of the buds lower down on the stem. The auxin which diffuses down from the bud at the top of a tree slows the growth of the buds below it. Thus, the top branch tends to grow faster than the side branches. If the top bud is cut off, however, the next lower bud will be released from this inhibition and will begin more rapid growth. Some of the buds at the lower part of the tree will receive so much auxin from the combined production of the buds above that they will not begin to grow at all. If the buds above are removed, these lower buds will become active and put out branches. If all buds are removed, the growth of the cambium cells will form new buds—adventitious buds—which will burst out through the bark and start new branches. Whether a tree is tall and slender or broad and spreading depends, to some extent at least, on the amount of auxin produced by the buds and the sensitivity of the buds to auxin concentration coming down the stem.

We take advantage of this principle in the process of **budding** woody plants. We can take a bud from the side of a stem of an orange tree —a bud which would have made only a leaf or a minor side branch—and place it on the stem of another tree. If we leave this stem intact, the transplanted bud may never open. But if we cut off the stem just above the bud, we remove the source of the inhibiting auxin and the bud will produce a shoot which will become the dominant growing branch.

Auxins and Fruit Production

One of the great uncertainties which make fruitgrowing a gamble in many parts of the world is the chance of late frost which may wipe out a fruit crop. The year during which these lines were written has seen the peach crop of Georgia and South Carolina almost destroyed by a late cold spell. Experiments have been tried in which auxins and related chemicals were used to inhibit bud opening to delay growth until all danger of frost was past. This technique has been successful with many fruit trees, but more investigation is required on its harmful side effects.

In the normal course of fruit formation, the ovary containing fertilized ovules enlarges and forms the fruit. If the ovules have not been fertilized, the ovary will form an **abscission layer** and fall from the plant without forming fruit? What is there about fertilization which causes the ovary to remain on the plant and form fruit. Hormones again seem to be the answer. Pollen grains contain an auxin which is carried down into the ovary in the pollen tube. The presence of this auxin seems to liberate bound auxin which is already within the ovary, and this auxin prevents the formation of the abscission layer. In many cases, however, as the fruit develops, the amount of auxin is lowered, the abscission layer forms, and there is a premature dropping of the fruit. Apple growers often lose more than half of their crops in this way. Premature dropping can be prevented to a certain extent by the application of a synthetic auxin (**naphthalene-acetic acid**) which is cheaper than the natural product. When this is sprayed on the apples in a very weak concentration, the fruit remains on the tree until it is fully ripe and ready for picking. This auxin can also be used to increase the yield in annual plants such as string beans and tomatoes.

This synthetic auxin can be used for another very interesting and possibly valuable application. The pleasure of eating many fruits is often diminished by the presence of seeds. How nice it would be if we could produce seedless watermelons! Such a thing is possible and has been done. Normally, a watermelon flower will fall from the plant if it is not fertilized, but if we spray it with the auxin which prevents the abscission layer from forming, it will remain on the plant. Thus, we can spray with this auxin to prevent flower-drop and cover the flower to prevent fertilizations. A seedless watermelon will then develop—no seeds can form from infertile ovules. This requires a considerable amount of hand work for each watermelon, but we might be willing to pay a higher price for a melon if we could eat it without being bothered with seeds. Tomatoes, cucumbers, and squash are other fruits which have been produced without seeds by this method. Female holly trees may be sprayed so they will develop red berries even though there are no male plants around. Thus, a person can have all his holly trees be female and have all of them bear red berries. As it is now, he must have at least one male tree around or he will get no berries from any of the trees. When the holly is cut and brought indoors the leaves can be prevented from falling by this same auxin. Living leaves produce the auxin which prevents abscission, but when leaves die there is no auxin and the abscission layer forms. Auxin-sprayed leaves will stay on a cut branch until they are brown and shriveled.

PHOTOPERIODISM

Anyone afflicted with hay fever caused by ragweed pollen can usually mark the date on the calendar when the sneezes, sniffles, and watering of the eyes will begin. In the area around Chicago and Indianapolis this is almost exactly August 15 of each year. In more southern regions of the United States it is later—about September 1 in central Texas, for instance. The date in the spring when the plants begin growing will vary because of variations in the onset of suitable weather conditions, but no matter when they begin growing they mature and produce pollen at the same time each year. This, and other phenomena of plant growth related to season, is explained by the effect of the length of day and night, known biologically as the **photoperiod.**

Photoperiodism was discovered in 1906 when a variety of tobacco, known as Maryland Mammoth, was produced. It grew to a great height, ten to fifteen feet, during the summer months, but frost would always come before it could flower and produce seeds. This created a great problem in propagation. To solve the problem, workers at the Bureau of Plant Industry near Washington, D.C., transferred some of the plants

to a greenhouse so they could continue growth after the first frosts. Within several weeks the plants bloomed and seed was produced. But why? Was it the intensity of the light out of doors that prevented the flowering? Experiments with various light intensities showed that this was not a factor. Perhaps it was the variations in temperature outside which were not found in the green-

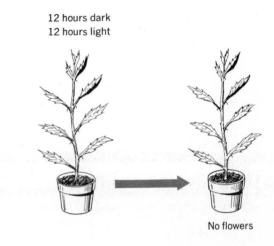

12 hours dark
12 hours light

No flowers

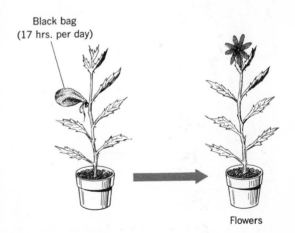

Black bag
(17 hrs. per day)

Flowers

FIG. 22.5 Experiment which demonstrates the effect of a long dark period on the production of flowers in the tobacco plant.

house? Experiments with many various temperature conditions had no effect. Only then did the experimenters turn to the one factor that appeared to remain—variations in the periods of day and night to which the plants were exposed. In these investigations they discovered the principle of photoperiodism and its relationship to flowering.

During the early summer some seedlings were covered each day at 4 P.M. and uncovered the next day at 9 A.M. This gave them only seven hours of daylight and seventeen hours of darkness, while the normal amount for this season

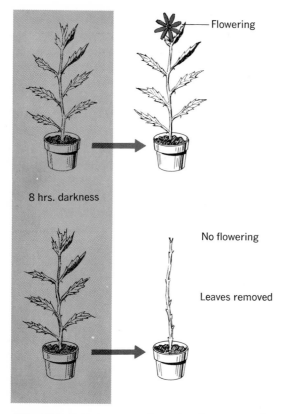

8 hrs. darkness

Flowering

No flowering

Leaves removed

FIG. 22.6 Experiment to show that leaves are the source of florigen. The leaves are removed and, even though the proper dark period is provided, no flowers are produced. This experiment was done with a cocklebur plant.

and latitude is about fifteen hours and nine hours respectively. Soon these young plants bloomed and produced seeds. Since then it has been found that the photoperiod influences flowering in many different plants.

At first it was assumed that it must be the length of day which brought about the flowering, but as experiments continued, it became evident that it was actually the period of darkness which was the critical factor in triggering the response. When some of the tobacco plants were exposed to seven hours of light alternating with seven hours of darkness, there was no flowering. When they were given seventeen hours of each, however, there was flowering. This showed that it was the long period of darkness rather than the short period of light that stimulated flower production. Further experiments showed that if the seventeen hours of darkness were interrupted by even one very short yet brilliant flash of light, there would be no flowering. A full seventeen hours of uninterrupted darkness were required.

Not all plants have the same requirements. For the cocklebur eight hours of darkness suffice to induce flowering. This plant will flower after receiving just one period of darkness of this length, even though the preceding and succeeding periods of darkness are much shorter.

How does the period of darkness bring about flowering? We have much to learn yet, but the evidence indicates that hormones are involved. If the cocklebur is exposed to a period of darkness greater than the required eight hours and the leaves are then removed, there will be no flowering even though the flower buds are left intact. From this we might conclude that the dark period stimulates the leaves to produce a hormone which migrates to the flower buds and stimulates them to open. Further support for this concept is provided by experiments on the tobacco plant. One leaf on a growing plant was covered by a black bag a part of each day so that it had the long-night photoperiod. The rest of the plant had the typical midsummer short-night exposure. In a short time flowers appeared from buds nearest the leaf which had been covered. Similar experiments on the cocklebur confirmed this. When one branch was covered, the blooms appeared on this branch and the effect gradually spread to the rest of the plant. To further confirm the hormone concept, a stem from a plant which had received the long-night treatment was grafted onto a plant living in the short-night environment and was thus induced to begin blooming.

It appears obvious that there must be a hormone which spreads from one part of the plant to another and induces flowering reactions. The hormone was given the name **florigen,** and for years its existence remained a hypothesis. Then in 1961 it was isolated from the cocklebur by biologists at the California State College. Commercial production of this hormone could have many economic uses. Plants could be induced to flower at a time when flowering was economically desirable without the laborious process of artificial regulation of the length of darkness in their nights.

As the studies were extended, it was found that the time of flowering of many plants is regulated by the photoperiod, but each variety may have its own critical period of darkness. In the Northern Hemisphere the length of day gradually becomes shorter after June 21 and there is a corresponding increase in the length of night. Many plants respond by blooming when the night period increases to a specific length. These are called the long-night plants. The aster, chrysanthemum, cosmos, dahlia, goldenrod, and poinsettia are examples. Poinsettias require an extremely long night which normally is not reached until about December so they must be raised indoors in most parts of the United States. (It should be mentioned that the beautiful red structures on the poinsettia are not flower petals but leaves; they turn red when the small yellow flower opens at the tip of the plant.)

Florists make a practice of regulating the photoperiod of many of their plants. The beautiful chrysanthemums we see at the early fall football games and the Easter lilies that always come to full bloom at Easter would not be possible without this practical application of the results of biological research.

Some plants are stimulated to flower by a short period of darkness every night. These short-night plants bloom in the spring when the period of darkness each night drops below the critical level. Such plants as clover, delphinium, gladiolus, hollyhock, iris, lily, and radish are in this group. To make them bloom prematurely, it is necessary to expose them to artificial light for a time after the sun has gone down in order to shorten the period of darkness to which they are exposed at night.

A few of our very earliest spring flowers are long-night plants. They grow and begin blooming just as soon as they are established, because the nights are long enough in early spring to stimulate the production of the florigen. Later in the spring, as the nights become shorter, the blooming stops. Sometimes, however, they will bloom again in the late fall if they are not killed by frost before the nights

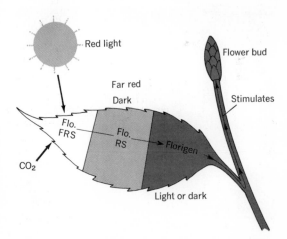

FIG. 22.7 Relationship of red light and far red light to the release of florigen in a long-night plants. See text for details.

reach a sufficient length. Violets are in this group. They are among the very early spring flowers, but sometimes we see them again in the fall before winter's arrival brings an end to the plants as well as the flowers.

A number of plants show no response to the photoperiod. They bloom whenever they reach a certain state of vegetative development. We can be thankful for this, because we can enjoy tomatoes, string beans, and fresh corn at all seasons of the year. Wherever the climate permits, these plants will grow and produce their flowers whenever they are at the proper stage of growth. Among the plants with aesthetic appeal because of their flowers, we find the carnation, nasturtium, rose, and sunflower in this group along with that perpetual lawn pest, the dandelion.

Leaves contain a protein, **phytochrome,** which exists in two shades of blue. One absorbs red light at a wavelength of around 660 mμ. The other shade absorbs far red rays at about 735 mμ. Red light is part of the sunlight spectrum, but far red rays are invisible to human eyes and can be present at night. When exposed to red light the red sensitive phytochrome (RS) is con-

verted into the far red sensitive (FRS) form. At night, when leaves are exposed to far red rays, the phytochrome is reconverted into the (RS) form. If we assume that florigen exists in a bound form in the leaves and is converted into an active form by the enzymatic action of phytochrome (RS), we would have an explanation of flower induction in the long-night plants. During the day the leaves receive red light and cannot build up phytochrome (RS). At night, under far red rays, the phytochrome (RS) is built up, but a specific period of darkness is required before the concentration becomes sufficient to release the florigen. Even a brief interruption by light puts the phytochrome back to the (FRS) form. In short-night plants florigen may be inhibited by phytochrome (RS) and would bloom only when the nights are too short to allow phytochrome (RS) to accumulate.

VERNALIZATION

Something more than photoperiod is involved in the blooming of flowers. An unusually

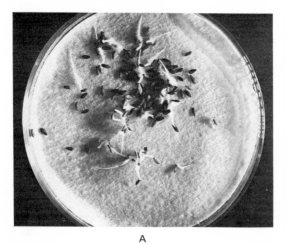

A

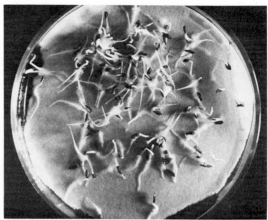

B

FIG. 22.8 Coumarin inhibits growth of radish seeds. Those seeds which were treated with coumarin in a 10⁻³ concentration are almost completely inhibited (A). Control seeds have almost all germinated and grown well (B).

FIG. 22.9 Seed of the shadscale, a desert plant, will not germinate unless they stay wet for a long time. A fungus must grow and penetrate the seed coat, which then absorbs water and swells to several times its former size. The young plant emerges after this.

protracted cold winter will result in the delay of the appearance of spring blossoms. Cherry trees which blossom out so beautifully in Washington D.C., will grow, but will not bloom, when transplanted to southern Florida. Seeds of many plants must be exposed to rather cold temperatures during their dormant period or they will not germinate. We speak of the conditioning of plants and seeds by exposure to cold as **vernalization.** The embryo is the part of the seed which is sensitive to this exposure. After the temperature has reached a certain critical low point, the embryo is stimulated to produce a hormone which permits germination.

In biennial plants—plants which do not flower in their first year of growth—vernalization occurs in a more advanced state of growth. Beets, for instance, store food underground during the first year of growth, but produce no flowers. The winter cold acting upon the underground growth causes vernalization, and flowers appear during the second year's growth. In trees and shrubs, vernalization occurs in the terminal buds. This is the reason why the flowering cherry trees will not put out blossoms unless a minimum temperature is reached during the winter.

THE GIBBERELLINS AND PLANT GROWTH

Plant geneticists in the United States have long known of an inherited characteristic in corn which causes the plants to be dwarfed in size. Plants receiving genes for this characteristic are no more than one-fourth the size of normal plants grown under the same conditions, and there is a corresponding reduction in yield. In Japan plant biologists have long known of a disease in rice which causes the plants to grow ridiculously tall. They grow so tall that the stems break and they usually never flower and produce seed. The Japanese biologists called this the "foolish seedling" disease and found that it appeared when the seedlings bcame infected with a fungus, *Gibbarella fujikuroi.* These two apparently unrelated facts led to the recent discovery of an important plant hormone which affects growth.

The Japanese biologists succeeded in extracting the active substance from the fungus. This substance, **gibberellic acid,** caused excessive growth when applied to normal rice plants. Here is where the geneticists stepped into the picture. If this chemical could stimulate the growth of normal plants so that they became abnormally tall, could it stimulate the growth of dwarf plants so that they would become normal? To answer this question, gibberelic acid was applied to the seedlings of corn plants which had inherited the characteristic of dwarfism. They did indeed grow to a normal size and gave a normal yield of corn. Continued investigation showed that the normal corn plant produces certain quantities of gibberellic acid; dwarf plants inherited genes which do not direct production of normal quantities of this substance. The geneticists added what nature failed to provide and thereby overcame an inherited defect in growth.

This and closely related chemicals, all known as gibberellins, have been found in many higher plants, but few of the fungi produce it and *Gibberella* is nearly unique in this regard. When applied experimentally to higher plants, the gibberellins have varied effects. The most common reaction is the rapid lengthening of the stems. In some experiments the stems of citrus trees have been stimulated to grow at a rate six times greater than normal. Other effects are being discovered. For instance, when gibberellins are applied to the young fruit of seedless grapes, the grapes become larger and stay on the vine longer. No doubt, future research will bring out many practical applications of this discovery.

THE KININS

The **kinins** are recent discoveries among the plant hormones. They promise to greatly

FIG. 22.10 Gibberellic acid stimulates extreme growth on cabbage plants. The two at left are untreated, while the three at right received 100 parts per million of the acid. (Courtesy of S. H. Wittwer, Michigan State University.)

extend man's control over plant growth and development. Normally, the only cells of plant tissue which have the power of division are those located in the cambium and in the meristematic regions at the tips of the stems and roots. Other cells—such as xylem, phloem, and cortex—have all of the genes necessary to form all types of plant tissue, but after they reach a certain degree of specialization they lose their power to form new cells. With the discovery of kinins, however, we find that this is not an irreversible loss. Under proper stimulation specialized cells can undergo division and produce other cells and even produce an entire new plant.

This can be illustrated by some experiments on carrots. Let us take a carrot and, using sterile techniques, cut small plugs from it in the regions which do not include any cambium tissue. The plugs are placed on sterile agar medium containing several mineral salts and vitamins. To this we add some coconut milk (the liquid inside a coconut) and a very small concentration of indoleacetic acid. Soon the cells of the carrot plug will begin division and a considerable growth of carrot tissue will form. We can break the carrot plugs into individual cells and each cell functions in the same manner as a zygote and produces an entire carrot plant.

Just what did the coconut milk do to restore this embryonic power to cells which normally would never have divided again? The coconut milk contains an ingredient, known as **myo-inositol,** which stimulates cell division, but it requires the presence of an auxin to have its maximum effect. Indoleacetic acid or the synthetic auxin, 2,4D, serves very well in this capacity. Auxins stimulate cell growth when they are present in proper concentrations, and kinin stimulates cell division. The interplay of the two results in normal plant development.

Other kinins have been extracted from immature corn kernels, from the immature fruit of apples and plums, and from tumorous plant tissue. The presence of kinins in plant tumors

might lead us to important discoveries on the reasons for the rapid cell divisions and the uncontrolled overgrowth in tumors and cancers even among animals.

The kinins contain **adenine,** an essential ingredient which seems to give these substances their biological activity. Adenine is a nucleic acid base found in genes and it is necessary for gene duplication. This discovery might also lead to a better understanding of the process of gene duplication.

Other experiments with kinins show that tobacco stems, when placed on a medium containing kinin and auxin, will show a great increase in the number of roots and buds that are formed. Varying the proportions of the hormones affects the proportions of roots and buds: more auxin produces more roots; more kinin produces

FIG. 22.11 Grape size can be increased by spraying with gibberellic acid. Those at right received 50 parts per million of gibberellic acid during their early growth stage. (Courtesy of Robert J. Weaver, University of California at Davis.)

more buds. Cut carnations placed in a kinin solution of ten parts per million will be fresh and beautiful after a month, while they normally would wither and lose the petals within a week. Fresh vegetables sprayed with similar dilutions will remain fresh and green much longer than unsprayed controls.

ETHYLENE—A GASEOUS HORMONE

Gas lights were used extensively for street lighting back around the turn of the century. It was noticed that the leaves of branches of trees near the lamps would droop and be shed. Later it was found that tomato plants which were stored in a warehouse alongside ripening bananas would have leaf drooping, or **epinasty.** Then orange growers in Florida found that oranges, picked while still green and sour would ripen and become sweet when they were warmed for a few days with heat from oil heaters.

FIG. 22.12 The leaves of the coleus plant in (A) show epinasty caused by the ethylene gas released by ripening bananas enclosed with them. Those in (B) show the normal leaf position.

What do all these situations have in common? In all of them **ethylene,** a gas which affects the leaves and fruit of plants, is produced. Since a hormone is a chemical messenger produced in one place and acting at other places, ethylene might be called a gaseous hormone. Natural gas and the gas given off by an oil flame are rich in ethylene; ethylene is produced naturally in fruit which is ripening. Ripening bananas and pears are especially rich sources of it. Today many fruits are picked and shipped green, then exposed to ethylene gas for a short time and ripened for market.

Ethylene also stimulates leaf epinasty and stimulates the development of the abscission layer which causes leaves to fall. As long as there is a normal flow of auxin down the leaf, no ethylene is produced, but if a leaf is injured or begins to wither in the fall, the amount of auxin dwindles and ethylene is released. This promotes the abscission of the leaves.

Abnormally high quantities of auxin also cause ethylene production and leaf abscission. Various synthetic auxins act as defoliants for this reason. Defoliants like these were used extensively in Vietnam to defoliate the jungles so troop movements could be seen from the air and also to destroy rice crops in enemy hands. Unfortunately, the solvent used was **dioxan** and this chemical is a **teratogenic agent**—one which causes embryonic abnormalities. Many women in these zones bore defective babies as a result of the use of dioxan.

COMMERCIAL USES OF PLANT HORMONES

A number of other plant hormones have been discovered and the time may be near when they will be more widely employed than they are today. One hormone group, known as the **morphactins,** will inhibit growth of a plant but will not disturb the life process of the plant which has already grown. Think how nice it would be to mow a lawn once in the spring, spray it with morphactin, and then enjoy a neat green lawn all summer without any mowing or worries about the growth of crabgrass or dandelions. Seedlings exposed to this hormone lose all response to light and gravity and extend out in all directions. The hormone can also be used to prevent potatoes from sprouting while in storage. If you plan to buy potatoes to use for planting, be sure you get some which have not been sprayed with morphactin. Other hormones have been found which stimulate pineapple plants to flower and produce fruit several weeks earlier than they would otherwise.

With the currently available herbicides and hormones, it is possible to spray a field with a combination of pesticides which will eliminate all but one species. This reduces the hours spent in weeding crops by hand. One spray, which can be applied from an airplane, will do the job. These compounds make it possible to control the time of maturation of fruits and vegetables, to control the size of the product, and to cause abscission of the leaves so that mechanical harvesters can be used to gather the crops.

REVIEW QUESTIONS AND PROBLEMS

1. Suppose someone told you that plants seek out the light so their leaves will be better exposed for photosynthesis. Tell how you could demonstrate by experiments that this was not a scientific explanation of this phenomenon.
2. Tell how a hormone differs from a cellular enzyme. (Consider the origin of the substance, the function, and the place of function in answering this question.)
3. When a plant is turned on its side, it bends in only one portion of the stem as it turns away from the pull of gravity. Why is the area of response limited to this one region of the stem?
4. Describe an experiment that would demonstrate that the growth-stimulating auxin is produced in the tip of the stem.
5. Why do roots grow down and stems grow up when they are both influenced by the same auxin?
6. When a certain synthetic auxin, 2,4D, is applied to a lawn in proper concentration it will kill most plants that we classify as weeds, but the grass is not killed. Explain how there can be such a selective destruction.
7. Describe an experiment which demonstrates that the effect of light on a stem is limited to the tip of the stem.
8. Some fruit trees, such as peach trees, produce more fruit and the fruit is easier to pick if there is a spreading growth of the branches. To produce this desirable shape, the trees are pruned by cutting the tips of the stems of the highest branches. How does this encourage the spreading growth of the trees?
9. How can auxins be used as an aid to the rooting of cuttings?
10. A bud on the stem of an orange tree will produce a small side branch if left on the stem. If this bud is removed and placed on the stem of a seedling and the terminal portion of the seedling is cut off, however, this same bud may grow to form the entire crown of a tree. Explain.

11. How can auxins be of value in increasing the yield of apples?

12. How can auxins be used to produce seedless fruit? Explain the principle involved.

13. In Florida chrysanthemums normally do not bloom until near the first of December, yet in greenhouses the florists have the blooms ready for the October football games. How can they do this?

14. Explain how it can be demonstrated that it is the period of darkness rather than the period of light which causes the response of those plants which respond to the photoperiod.

15. Most spring flowers are short-night plants, but the violets which bloom in the early spring are long-night plants. Explain this.

16. Suppose you are shown two groups of corn plants of about equal size. One group has inherited the genes for dwarfism, but these plants have been treated with gibberellic acid and have grown to a normal size. How would you determine which group was normal and which had been so treated? (If you have difficulty with this refer to Chapter 36.)

17. Suppose you devise a cheap and practical method of propagating carrots by breaking them into cells and applying kinin and auxin to produce small carrot plants which then can be set out in the soil. What advantage could you claim for your technique as compared to the traditional method of planting seeds as a means of raising carrots? (If you have difficulty on this, refer to the section on budding and grafting in Chapter 21.)

18. Some plants are desired as ornamental shrubs if they have many small branches that give them a bushy foliage. How might plant hormones be used to increase or decrease the bushiness of a plant? (Consider the fact that the branches arise as buds from the stem.)

19. An oil refinery is built near a pecan grove in Louisiana. The leaves of the trees soon begin to fall even though it is midsummer. Explain.

20. When a hand of bananas is kept in a plastic bag, the fruit ripens more quickly than when the hand is kept in open air. Explain.

FURTHER READING

Galston, A. W., and Davies, P. J. 1970. *Control Mechanisms in Plant Development.* Englewood Cliffs, N. J.: Prentice-Hall.

Leopold, A. C. 1964. *Plant Growth and Development.* New York: McGraw-Hill.

Philips, I. D. J. 1971. *Biochemistry of Plant Growth Hormones.* New York: McGraw-Hill.

Salisbury, F. B., and Ross, C. 1969. *Plant Physiology.* Belmont, Calif.: Wadsworth.

Steward, F. C. 1968. *Growth and Organization in Plants.* Reading, Mass.: Addison-Wesley.

23

Some Simpler Invertebrate Phyla

We often speak of the **invertebrate** animals which do not have a vertebral column, and the **vertebrates** which have a vertebral column. We shall start our survey of the different kinds of animals with a consideration of some of the simpler invertebrate phyla.

THE SPONGES

Sponges belong to the phylum **Porifera,** a name given to them because their bodies are perforated with many tiny pores. Because they are permanently attached to a solid object and do not show active movements, they were thought of as plants for many years. Then it was discovered that they have active flagellated cells which beat and keep a current of water circulating through their pores. This is definitely an animallike characteristic. We now consider them a part of the animal kingdom, although some would place them in a separate category all their own. They represent an evolutionary blind alley with adaptations to a type of life which more highly developed animals have abandoned. Most sponges are marine, but there are some fresh-water forms.

A Simple Sponge—Scypha

The genus name *Scypha* is given to a small vase-shaped sponge which grows to about an inch in height and which tends to be found in clusters attached to rocks in shallow water near the seashore. *Scypha* shows **radial symmetry.** Radial symmetry is the kind of symmetry shown by an apple. If you divide an apple into two equal halves by making a cut which passes through the stem end and the rounded end, the two halves will be alike. If you cut the apple in half by making a cut which passes through the apple in any other direction, however, the two halves will be different. The body of *Scypha* is covered externally with many sharp spines, the **spicules,** which protect and support the animal. One end of the body is attached. At the other end is an

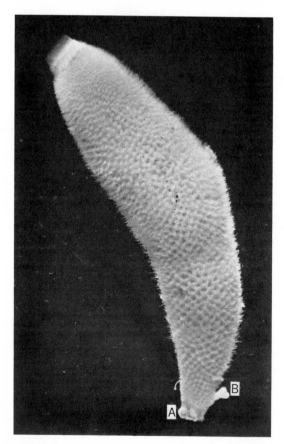

FIG. 23.1 *Scypha,* a simple sponge. A—attachment, B—bud.

opening, the **osculum.** The canals which perforate the body are lined with flagellated **colar cells** which keep a constant flow of water circulating through the body into a central **cloacal cavity,** and out through the osculum. This flowing water is a necessity, for it brings in dissolved oxygen which is absorbed by the cells and carries out the carbon dioxide and excretory wastes which diffuse out of the cells. This water also contains bits of organic matter which are engulfed by the cells lining the canals.

Reproduction is accomplished by both asexual and sexual methods. Buds grow out near the base of the sponge and gradually enlarge until they form a full-sized animal. The new animal remains attached, however, and this is the reason why *Scypha* is usually seen in clusters. At certain times, some cells of the body undergo meiosis and form sperm, while others form eggs. The sperm are released in the water, where they swim freely. They are attracted to eggs, which are usually in other animals since the sperm and eggs of one animal usually do not mature at the same time. After union of the gametes, a zygote is formed. The zygote grows into a ciliated larvae which is free to swim around and find a new location where it settles down and becomes permanently attached. Since both gametes are produced by each animal, the sponges are said to be hermaphroditic.

Regeneration in Sponges

Sponges have a remarkable power of regeneration. The cells do not show much differentiation and almost any cell can produce an entire new animal. If *Scypha* is cut into many parts, each part will regenerate the lost portions to produce an entire animal. The power of regeneration has been strikingly demonstrated by an experiment in which pieces of the red sponge, *Microciona,* were forced through heavy silk bolting cloth. This broke up the sponge into separate cells or groups containing a small number of cells. When these cells were placed in a dish, they began to clump together and form a number of small sponges. Soon the colar cells began to beat their flagella and each clump became a functioning individual.

The Bath Sponge

To the average person the word "sponge" means **bath sponge,** although the use of sponges for bathing has been almost wholly supplanted by rubber and cellulose sponges. The Mediterranean region was long an important source of sponges for the world market. In the United States, some Greek immigrants settled around Tarpon Springs, Florida, to continue their sponge fishing. With continuing pollution of the water in this region, however, the sponges have become scarce, and little remains of what was once a flourishing industry.

SACLIKE ANIMALS— THE COELENTERATES

The phylum name for this group of animals, **Coelenterata,** means "hollow intestine." It was chosen because representatives of this phylum have a rather large hollow area in the middle of the body which serves as an intestine. They have **radial symmetry.** They include more than 9000 species, but they are not very well known to the average person because most species are marine, and the few freshwater forms are small. The jellyfish seen on the beach at the seashore often represents the extent of our familiarity with this phylum.

A Typical Coelenterate—Hydra

The genus *Hydra* is not typical in its habitat, because it is a freshwater form, but it does

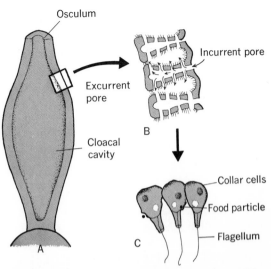

FIG. 23.2 The internal structure of a simple sponge.

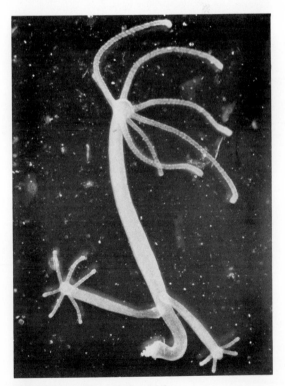

FIG. 23.3 A hydra with two buds.

illustrate the life processes of this phylum well. It can be found in most sections of the United States attached to submerged rocks and leaves in ponds, lakes, and streams where the water does not become too murky and does not get too warm. The slender, cylinder-shaped body attaches itself at the base end to some solid object and bears a mouth, surrounded by from five to seven tentacles, at the other end. The entire body, including the tentacles, will probably not be more than half an inch long. The hydra can move along slowly by sliding on its base or more quickly by reaching out and attaching the mouth and then pulling the base up near the mouth, like a measuring worm. Some have also been seen turning end over end like an acrobat doing hand-springs. If you touch one of the tentacles of an extended hydra, you will probably see it jerk it-

self up into a ball less than one-tenth as long as the extended form. When these body activities are compared with the passive condition of *Scypha*, it is evident that this metazoan has some connection between the cells which causes the entire body to react as a unit. This connection is the **nervous tissue.** It cannot be seen in the living hydra, but by treating a preserved specimen with a stain known to be absorbed by nerve tissue, it is possible to bring out nerve cells all over the hydra's body. There is no brain, but the nerve cells are more concentrated at the oral end (mouth end) where most of the activity occurs. While this nerve tissue is not sufficiently complex and differentiated to be called a nervous system, it does represent a beginning of specialization of tissue for reception of stimuli and transmission of impulses.

Hydra uses its tentacles to capture its food, much as an octopus does. When some small water animal, such as a water flea, swims within reach, *Hydra* may suddenly extend itself and wrap its tentacles around the animal. At the same time tiny poisonous arrows are shot from the stinging capsules, **nematocysts,** which cover the tentacles. The poison from these arrows tends to subdue the prey, which is then carried to the mouth by the tentacles and **ingested.**

The stinging reaction is automatic; the tentacles will wrap around any object, such as a small piece of wood, that comes within reach, and the nematocysts will eject the poisonous threads. The mouth will not open, however, and the object will not be ingested unless it is food. How can the hydra distinguish between food and other objects? There must be a sense of taste. If pieces of wood are treated with an extract from crustaceans, the wood will be ingested as readily as the crustaceans. If the juice extracted from crustaceans is placed in the water where a hydra is living, the mouth will open and, in the absence of a solid object, will turn down and open wider and wider on the dish to which it is attached, as if it were trying to swallow the dish itself. So

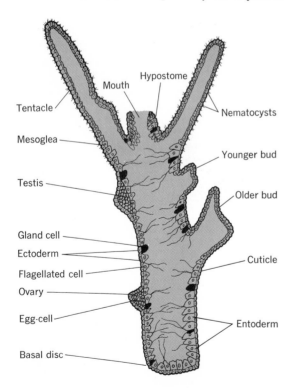

Mouth
Hypostome
Tentacle
Nematocysts
Mesoglea
Younger bud
Testis
Older bud
Gland cell
Ectoderm
Cuticle
Flagellated cell
Ovary
Egg-cell
Entoderm
Basal disc

FIG. 23.4 Longitudinal section of a hydra.

strong is this stimulation that the animal may turn itself inside out. This is another good example of how we must avoid teleological explanations of observed facts. It is easy to say that the hydra opens its mouth in order to swallow the food, but this is the end result and not the cause of the mouth opening. The mouth opens in response to definite chemical stimulations of sensory cells which, under normal conditions, do result in the intake of food.

Hydra is a transitional form with respect to the process of **digestion.** The ingested food passes into the **gastrovascular cavity,** where it is exposed to **extracellular enzymes** flowing out from the cells lining this cavity. These enzymes break the food down into smaller particles which will be engulfed by the cells lining the cavity in much the same way as an amoeba would engulf food. **Intracellular enzymes** within the cells complete the process of digestion.

Another feature of the digestive system of *Hydra* is rather primitive. There is only one opening to the gastrovascular cavity and this must serve both as a port of entry for the food and as a port of exit for the waste which remains after digestion is completed. It is called a mouth, but it also serves the same function as the anus in higher forms of animal life.

Practically every cell of the body of a hydra is exposed to water, either on the outside or within the gastrovascular cavity. The oxygen needed in **respiration** can easily diffuse into the cells and the carbon dioxide can diffuse out. **Excretion** involves only the diffusion of metabolic wastes out of the cells and into the surrounding water. There are only two layers of cells, the outer **epidermis** and the inner **gastrodermis,** in the main body of *Hydra*. Between these is a thin layer of a jellylike substance known as the **mesoglea.** Since there is no true mesoderm—middle layer—this is a **diploblastic** animal.

Asexual reproduction in *Hydra* is accomplished by **budding,** as in *Scypha*. A small

protusion appears on the side of the body; this enlarges and soon develops tentacles. Within two or three days it breaks off from the parent as a small but fully developed animal.

Sexual reproduction is also found in *Hydra*. Some species are **hermaphroditic,** producing both male and female gonads on one animal; others appear to produce only one type of sex organ on each animal. The male gonads, **testes,** appear on the side of the hydra and release sperm through nipples at their tips. **Ovaries** appear as spherical bodies near the base of the animal. Only one egg is produced in each ovary. Fertilization occurs when a sperm penetrates the egg. The zygote then begins division and forms a small embryo in the form of a ball with an inner **endoderm,** which will form the gastrodermis, and an outer **ectoderm,** which will form the epidermis. At this stage the embryo ceases development, forms a hard outer covering around itself, and drops from the parent. In this dormant state it can survive unfavorable conditions which would cause the death of the active adult.

Gonads appear, in some species at least, when there is a lowering of the water temperature. This is an adaptation which allows the re-

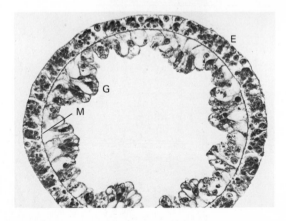

FIG. 23.5 Cross section of a hydra. This photograph shows the two body layers, epidermis (E) and gastrodermis (G), separated by the thin jellylike mesoglea (M).

sistant embryos to be formed when winter is approaching and the adults may be killed. Gonads will also appear at other times of the year when the hydras grow in crowded conditions. The crowding itself is not a factor, however. This was demonstrated by an experiment in which it was shown that a single hydra would produce gonads if it was placed in water taken from a crowded culture. Somehow the water surrounding a crowded culture stimulates the development of sex organs. Could it be the lowered oxygen content of the water that gives it this power? Experiments in which a hydra was put in water from which most of the oxygen had been removed showed that this did not stimulate gonad production. Could it be the increase in the carbon dioxide content of the water which would accompany crowded living conditions? Individual hydras were placed in water into which carbon dioxide was bubbled and it was found that they produced gonads. It was concluded that the active agent in this water is carbon dioxide. The end result of this reaction is the stimulation of

gonad production whenever growing conditions become unfavorable and there is a good chance of death of the adults.

Regeneration in *Hydra* is extensive and proverbial; *Hydra* gets its name from its regenerative powers. According to a Greek myth, there was a great sea serpent named Hydra that Hercules went out to slay. The monster had nine heads, and if one of these was cut off, two new ones would grow back in its place. A "twoheaded" coelenterate hydra can be produced by splitting the oral end back for a short distance. Each half of a "head" will regenerate the other half and there will be two complete oral ends attached to a single aboral end. The animal can also be cut into several pieces and each piece can regenerate the missing portions and complete hydras are formed. Individual tentacles can regenerate lost portions. Forked tentacles are sometimes created when a tentacle is crushed on one side; this stimulates a new portion of a tentacle to grow from this region. A final illustration of how adaptable this animal is to physical damage is that it can

FIG. 23.6 A small jellyfish, *Gonionemus*. This is the medusa stage; the polyp stage is quite small and looks somewhat like *Hydra*.

restore itself to a normal state even when turned inside out. The cells of the gastrodermis and the cells of the epidermis migrate past one another and re-establish themselves in their proper position.

A Small Jellyfish—Gonionemus

The **jellyfish** take their name from the fact that the mesoglea is highly developed, and if the outer covering is broken, jellylike material will ooze out. Some jellyfish are quite large and bear tentacles with powerful nematocysts. They are attractively beautiful in the water, and many a newcomer to the seashore has picked one up to examine it more closely, only to drop it quickly as the powerful nematocysts came in contact with the skin. We choose a small one, *Gonionemus*, as an example of the jellyfish because it illustrates the interesting principle of **alternation of generations.** We will recall this principle in the plant kingdom—the ferns, for instance, have a sexual and an asexual generation alternating with each other. A similar process occurs here.

Gonionemus is umbrella-shaped, with tentacles bearing nematocysts hanging down from the outer rim, and a mouth at the lower end of the handle of the umbrella. Its body shape might be compared roughly to a hydra that has been turned upside down and given a hard slap on its base, spreading it out to form the umbrella shape. This laterally expanded portion of the body is the bell. The jellyfish swims by rhythmic pulsations of the bell. Water is slowly sucked into the under surface, like air swelling a parachute, and then is suddenly expelled as the bell contracts. This propels the animal in the opposite direction through the principle of jet propulsion, which the jellyfish were using for thousands of years before man thought of it for aircraft.

Dr. Jekyll and Mr. Hyde have nothing on *Gonionemus*; this little animal also leads a double life. The form we have been describing is known as the **medusa** stage and is the one most commonly seen and known. Its lesser known self is the **polyp** stage and looks more like *Hydra* in shape and size than the medusa stage. The medusa may produce eggs or sperm, which are united in the water. It seems logical that the zygote thus produced would grow into another medusa, since it is a general rule in the animal kingdom that the zygote grows into an animal resembling the ones that produced it. But this zygote grows into something that looks like neither parent; it looks like an entirely different species. This is the polyp stage, which produces buds asexually. The bud forms a medusa, which breaks off and continues the unusual cycle.

Some of the small jellyfish live in fresh water, and inland lakes may be filled with them at certain seasons of the year but at other seasons may seem to have none. However, a careful search will usually reveal them in the less conspicuous polyp stage.

A Colonial Coelenterate—Physalia

Another well-known coelenterate is the large marine form commonly called the **Portuguese man-of-war,** with the scientific name of *Physalia*. Its common name is derived from the fact that it has an air-filled membrane, acting as a

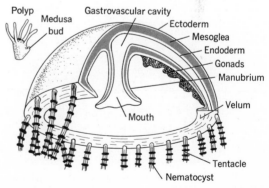

FIG. 23.7 A cut-away view of *Gonionemus* to show the body parts. At upper left is the small polyp stage with a bud which will grow into the more conspicuous medusa stage as shown below.

float, that stays on top of the water and resembles the sails of the old men-of-war. No fighting ship in history, for its size, was more heavily armed than is *Physalia*, with its thousands of powerful nematocysts which are able to inflict an extremely painful and sometimes dangerous sting on persons that happen to brush against its tentacles in the water. This is a colonial form in which the individuals adhere to one another after budding and different individuals assume different tasks in the group. This specialization of individuals is somewhat similar to the specialization of cells in multicellular animals. Just under the float there are quite a number of feeding polyps that have become specialized in digesting and absorbing food for the entire colony. Some polyps are specialized as producers of the gametes of sexual reproduction. Others trail out in the water bearing

FIG. 23.8 The Portuguese man-of-war, *Physalia*. This colonial coelenterate has stung a fish into insensibility and has pulled it up close to the feeding polyps, where it is being digested and absorbed.

the nematocysts. These capture the food and defend the colony against enemies. There will often be thousands of these colonies left on the beach as the tide goes out.

An average-sized Portuguese man-of-war will have a float about six inches long, with tentacles trailing down for several feet. The float glistens with iridescent shades of pink, blue and purple—a beautiful sight which belies the deadly nature of the tentacles hidden beneath. A fish that happens to touch the tentacles is enclosed and stung into insensibility; then the tentacles contract and draw the fish up under the float where the feeding polyps can go to work on it. These reach down and spread their lips over the body of the fish, secrete their digestive juices on it, and suck up the digesting bits of food as they break off.

One of the most interesting relationships between two animals in the entire animal kingdom is that between the Portuguese man-of-war and a small fish, *Nomeus*. To venture within reach of the deadly tentacles of the Portuguese man-of-war means sudden death to most small animals, yet *Nomeus* not only ventures within reach, but actually swims in and out among the tentacles in perfect safety and is protected from enemies that do not enjoy this privilege. The association is not a one-sided affair; the little fish will sometimes swim out and, if a larger fish gives chase, *Nomeus* will dart through the tentacles. As the pursuer follows, he finds himself ensnarled and soon he is serving as a square meal for *Physalia* and *Nomeus*, the latter swimming around and picking up the pieces that come off as digestion causes the fish's body to disintegrate. This is a case of **mutualism,** the association of two living things in a mutually helpful relationship.

The Sea Anemone

The sea anemone is named after a flower, the anemone, but seen in its natural habitat, it seems more closely to resemble a chrysanthemum, a marigold, or a dahlia in its different forms and colors. It is abundant in shallow, rocky regions in tropical waters. When seen in large numbers, sea anemones present a display of color harmony and body forms that rivals the most gorgeous flower gardens. This animal is of the polyp type in body form but much larger than the other polyps that we have studied. Sea anemones are entirely marine in habitat and abundant in tropical waters but are also found in temperate regions. One genus, *Metridium,* is abundant along our north Atlantic coast, where the animals attach themselves to piers and other solid objects near the shore. *Metridium* is a dull brownish animal, in contrast to the beautifully colored forms of the tropical waters.

Coral

Another small polyp in this phylum, an animal not much larger than *Hydra,* produces a hard, protective skeleton around itself. This is the little coral polyp, which is not much more than half an inch long. Through the combined action of many of these polyps, however, islands and reefs of tremendous size have been constructed. The Great Barrier Reef which runs along the eastern shores of Australia is more than 1200 miles long; it was all constructed from the deposits made by these little animals. Their bodies are soft, and they secrete little protective cups of limestone around themselves. As they bud and produce more individuals and these produce more limestone, the colony continues to grow. The growth of the reef is slow, only about half an inch a year, but during the eons of time in which they have been growing, the huge masses have been produced. The South Pacific waters are dotted with many thousands of coral islands which have been built up from the bottom of the ocean, and the coral reefs that lie just under the surface of the water are a great hazard to ships

traveling uncharted areas in this region. The Florida Keys are examples of such islands in North America. These are surrounded by coral reefs upon which many ships have been wrecked, especially during the centuries when small sailing vessels went around the southern tip of Florida to reach the cities on the Gulf of Mexico.

Though there is a great abundance of coral of many varieties, one, called precious coral, is rare and commands as much as six hundred dollars an ounce. This is found primarily in Mediterranean waters and off the coast of Japan. It is usually pink or red and makes beautiful bracelets, necklaces, and similar ornaments.

THE FLATWORMS

Several phyla include animals which are known as worms. The group which we shall now consider, phylum **Platyhelminthes,** is distinguished from the other worms by a flattened

FIG. 23.9 A group of sea anemones and their colorful, spreading tentacles are a beautiful sight, resembling an underwater flower garden.

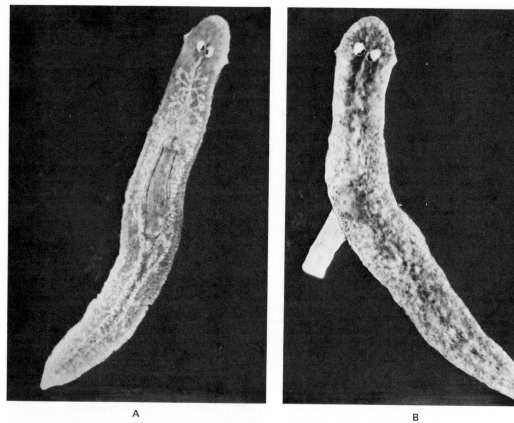

A

B

FIG. 23.10 The planarian, *Dugesia,* a free-living flat-worm. The proboscis is retracted in (A), and fully extended in (B). Some of the branches of the intestine show near the anterior end of the animal in (A).

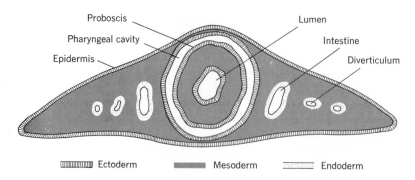

Proboscis   Lumen

Pharyngeal cavity   Intestine

Epidermis   Diverticulum

Ectoderm   Mesoderm   Endoderm

FIG. 23.11 Cross section through a planarian worm in the region of the proboscis, showing the body parts formed by the three germ layers.

body form. They have two distinct sides and therefore show **bilateral symmetry.** Most flatworms live as parasites in the bodies of higher animals, but parasites tend to become rather highly specialized. Some structures may be lost and others overdeveloped in the adaptation to a parasitic existence. As a result, a free-living (nonparasitic) form is usually chosen as a representative of the general characteristics of this phylum. The little freshwater planarian flatworms of the genus *Dugesia* are good examples.

Planarians—Free-Living Flatworms

Planarians are usually found under rocks and leaves at the bottom of freshwater ponds and streams. They are small animals, usually less than an inch in body length when fully extended. The planarian body has a head at one end which bears a pair of **eyes.** The pigment spots of these eyes are both on the inner side, and this gives planaria a cross-eyed appearance. These eyes are merely light-sensitive spots; they are not capable of picking up definite images. A planarian can probably see about as well as you can with your eyes closed. You can still detect changes in light intensity with your eyes closed and thus become aware of movements which cause variations in this intensity.

When you watch a planarian moving, it will appear to be gliding along over the surface without any noticeable body contractions that characterize the crawling of a snake. This happens because the body is covered by microscopic **cilia** which beat and propel the animal. From time to time the planarian will change its body shape—sometimes stretching out very long and slender and other times drawing itself up into a compact mass. It may also raise the head and turn it from side to side. These movements are possible because of the presence of a **muscular system.** There are muscle fibers running in different directions which lie just under the outer

epidermis. They are formed from a **mesoderm,** a middle layer of embryonic cells. Thus, a planarian has three main body layers: mesoderm, ectoderm, and endoderm. This makes it a **triploblastic** animal, in contrast to the **diploblastic** condition, two body layers, found in the coelenterates. All higher animals are triploblastic. Since the mesoderm fills the space between ectoderm and endoderm, there is no body cavity or **coelom.**

If you put a little piece of liver or other fresh meat in shallow water at the edge of a pond in which there are planarians, you may find it covered with these animals within a few hours. If you remove them to a dish of clear water, you can observe **ingestion.** The mouth of a planarian is located on the ventral surface of the body about one-third of the body length from the posterior end. From this mouth the planarian extends its **proboscis,** a tubelike structure, and attaches it to the meat. Digestive juice is poured onto the meat through the proboscis, and the bits of partially digested food are sucked up as they break off. These food particles then enter the **intestine,** which has three major branches, one anterior and two posterior, with many smaller branches called **diverticula.** These branches carry the food to all parts of the body. The cells engulf the incompletely digested food particles and digestion is completed inside the cells, thus combining **extracellular** and **intracellular digestion** as *Hydra* does. Small animals and smaller bits of food will be taken directly into the intestine through the proboscis without preliminary digestion outside the body, but the planarian will usually take time to secrete mucus on the food to ensure its smooth passage into the intestine.

When the time comes for **egestion,** the planarian is faced with the same problem that *Hydra* has; there is only one opening to the digestive tract. So the mouth must serve as an anus also, and the indigestible food residue is expelled through the mouth.

Respiration is still not represented by any system. There is direct absorption of oxygen from the water by individual cells and carbon dioxide is given off in the same way.

For **excretion,** however, a system has been developed. Scattered over the body are many **flame cells** that pick up the waste products of metabolism as they are thrown off by the individual cells. A flame cell is shaped like a funnel with cilia around the mouth; these beat and suck the excretory products down into the funnel. When seen under the microscope, these beating cilia resemble the flickering of a flame—hence the name, flame cells. After being sucked into these cells, the waste passes into one of two **excretory tubes** located on either side of the body. These, in turn, empty it into the surrounding water through **excretory pores**—very small holes opening on the dorsal surface, the first pair lying directly posterior to the eyes and several other pairs farther down the body.

The extensive branching of the digestive and excretory system is necessary in the planarian if food is to reach all parts of the body and if the excretory wastes are to be brought back from all regions, since there is no circulatory system to transport materials from one part of the body to another. More advanced animals with blood coursing all over their bodies have the digestive and excretory systems localized and depend on the blood to distribute the food to the more distant parts of the bodies and to bring back the wastes of metabolism to the excretory organs for removal and elimination.

In *Hydra* we find a nervous system consisting of a group of nerve cells without any brain, but with a concentration of cells at the oral end. In the planarian this concentration of nerve tissue forms a "brain" (cerebral ganglia) in the anterior end which exercises a degree of control over the entire body. The "brain" has two distinct ganglia with a **nerve cord** from each running the length of the body on either side. Large commissures run across to connect these two nerve cords at regular intervals and give a structure resembling a ladder, which is frequently called a **ladderlike nervous system.** Many smaller nerves branch off from these and run to all parts of the body to pick up sensations resulting from various stimuli and to transfer impulses to the muscles for the proper reaction to these stimuli.

The planarian has a well-developed **reproductive system** with many of the organs similar in structure and function to those found in the higher animals. Both male and female organs are found in each animal, so planarians are said to be **hermaphroditic,** or **monecious,** animals. Since planarians have bilateral symmetry, a complete set of sex organs of both sexes is found on both sides of the body. In spite of the fact that each planarian has within itself all the possibilities for fertilization, this does not happen; instead, they have a sexual union, **copulation.** The advantages of the blending of genes from two different organisms in producing variety in the offspring have been emphasized previously. Self-fertilization results in offspring very much like the single parent organism, but in cross-fertilization genes from two different parents are blended to produce something new. Such a variety of organisms is an essential ingredient for natural selection.

In planarians during copulation, there seems to be a mutual exchange of sperm, both animals serving as male and female at the same time. Sperm are produced in the numerous **testes** and make their way through the slender **vasa efferentia** (sing. **vas efferens**) into one of the two **vasa deferentia** (sing. **vas deferens**) and thence down into the **seminal vesicles,** which serve as a storage chamber for the sperm until the time comes for their discharge. In copulation the **penis** of each animal is thrust through the **genital pore** of its partner and the seminal vesicles contract. This forces sperm into the seminal receptacle of the other. The animals then separate, and a short time later eggs are produced by the ovaries. The eggs pass down the oviducts, where they meet the sperms which come out of the seminal receptacles and journey up the oviducts. The **zygotes** which result from the union of the sperms and eggs continue down the oviducts. The **yolk glands** secrete many **yolk cells** which also pass out into the oviducts and will serve to nourish the young embryos. When these reach the **genital atrium,** a group of from four to twenty zygotes and thousands of yolk cells will be surrounded by a cocoon which then passes out of the genital pore and is usually attached to the underside of a rock, leaf, or similar underwater structure. The embryonic planarians develop within the cocoon, deriving nourishment from the yolk cells for several weeks; then they break this confining wall and crawl out to start life on their own.

Regeneration in Planarians

Sexual reproduction seems to be the dominant method in planarians, but they also have an asexual method, **transverse fission,** which has a very interesting physiological explanation. As we mentioned earlier in the description of the nervous system, the "brain," which is at the anterior end of the body, acts as a center of control and coordination of all parts of the body. As the animal grows longer, however, under favorable living conditions, it seems that the posterior end of the body gets so far away from the brain that the brain gradually loses its control over it and it begins to act somewhat independently. A new center of coordination develops at the posterior end and the planarian literally pulls itself in two as a tug of war develops between the two ends. The lost portions of each half are then regenerated. This method of reproduction can be prevented if the planarians are kept in containers greased with petroleum jelly. On a slick surface there can be no tug of war and the two ends are not pulled apart.

Although the planarian shows considerable

specialization of tissues, it still has regenerative powers and is an excellent animal to experiment with in this connection. If you slit the head down the middle with a sharp razor blade, each half will regenerate the missing half; the result is a two-headed animal. These two can be split to give four, and an eight-headed planarian has been produced by a splitting of the four. Grafting is also possible, just as it is in plants; a part of one animal can be cut off and placed in an incision in another animal and it will grow there.

Some Parasitic Flatworms

The **flukes** resemble planarians in body shape. Different species of them may be found in many parts of the bodies of higher animals. There are blood flukes, lung flukes, liver flukes, intestinal flukes, and even flukes that cling to the external surface of some water animals. Many of the flukes have two hosts and alternate between the two. The snail serves as one of the hosts for a large number of flukes.

The **Chinese liver fluke,** *Clonorchis sinensis,* is found in large numbers of people in China, Japan, and Korea because of the living and eating habits of the people of these regions. The adult flukes live in the human liver and sometimes cause a partial breakdown of that organ called the "liver rot." A type of jaundice may be caused by flukes which clog the passageways for the bile. The animals copulate in the liver and their eggs are released to make their way down the bile duct, into the intestine, and eventually out of the body along with indigestible food or feces. The eggs must reach water in a short time in order to continue the cycle. This they frequently do because many of the houses are built over the water or on the water in these crowded countries, and the rivers and lakes offer a convenient means of disposal of feces. If snails of the right species eat these eggs, the eggs will hatch and the larvae go through a series of asexual reproductive phases within the snail's body. Eventually a large number of larvae possessing tails are produced. These break out of the snail and swim around in the water until they come near a fish; then they burrow through the skin of the fish and encyst in its muscles. People ingest the parasite by eating raw or poorly cooked fish. It is a common practice in many Oriental countries to slice the raw fish and put it on top of a bowl of steaming rice to warm it a little, or perhaps to dip it in a hot sauce. The encysted larval flukes are not destroyed by such treatment. When they reach the stomach, the cyst wall is digested, and fluke larvae pass into the small intestine and the bile duct, which they swim up to the liver, where they grow into adults.

We do not have the human liver fluke in the United States, but we do have a liver fluke that is very common in sheep and is sometimes found in cattle. This is called the **sheep liver fluke.** Instead of going to a fish, the larvae that comes from the snail crawls up a blade of grass or other succulent plant growing up out of the water and forms its cyst where it may be devoured later by its mammalian host.

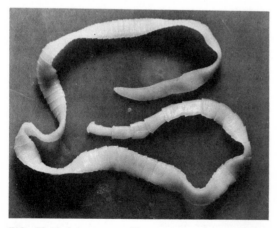

FIG. 23.12 A tapeworm. The scolex is at the center. The many proglottids are narrow at first but gradually become longer and more swollen as they near the posterior end. The proglottids are filled with ripe eggs and soon break off.

The liver fluke offers another good example of **alternation of generations:** the sexual generation in the liver of one host and the asexual generation in the snail.

The **tapeworms** are abundant parasites in practically all mammals and many of the other vertebrates. Their bodies are more specialized than those of the flukes. The body of an adult tapeworm consists of a head, or **scolex,** which is attached to the intestinal wall of its host with the aid of **suckers** and sometimes **hooks** and a series of many sections, or **proglottids,** which arise from the neck of the scolex and trail out into the intestine, sometimes for several feet. It somewhat resembles the tape measure, from which it derives its name. Tapeworms have no **digestive** system; they live in digesting food and absorb food through their body walls. Their **muscular** system is poorly developed; they usually lie passively in the intestine soaking up the food needed for their growth and metabolism. The **reproductive system** of each tapeworm contains a complete set of male and female organs.

The **pork tapeworm,** *Taenia solium,* is common in sections of Europe and other parts of the world and at one time was quite prevalent in the United States, but an efficient system of meat inspection has made it comparatively rare here today.

The adult lives in the intestine of man and normally reaches a length of six to ten feet, with a width of about two-fifths inch at the posterior end. This may seem like a great length for man to accommodate, but we have about twenty feet of small intestine, so there is plenty of room for several of them without crowding. They do some harm by absorbing food intended for the body, but the main damage seems to be due to our absorption of their waste products, which may cause anemia and nervous disorders. An obstruction of the intestine may develop when several tapeworms are present at the same time.

New proglottids arise from the neck throughout the life of the tapeworm, but they do not increase in length after they reach maturity. Mature proglottids break off at the posterior end as fast as they are produced at the anterior end. A newly formed proglottid is small, but it grows in size as it is pushed away from the neck and develops sex organs. The male organs mature first and release sperm into the intestine. They swim about and fertilize mature eggs in other worms, if any are present. As more eggs are produced, the uterus swells until it fills the proglottid and it is bulging with young embryos. By this time the proglottid has reached the end of the tapeworm and is said to be ripe. It then breaks loose, perhaps with several other segments, and is carried from the body in the feces.

A hog must eat these ripe proglottids if the cycle is to continue. Once inside the intestine of the hog, the shell of each egg is digested off and the little embryos, which now have six hooks on their body, escape. They immediately burrow into the intestinal wall and get into the blood stream through the little capillaries which are so abundant in this region. They are then carried over the body by the blood and leave the blood vessels when they reach some of the muscles. They bore their way into the muscle fibers, round themselves up into a ball, and encyst, This encysted stage is called a **bladderworm** and for many years was thought to be an entirely separate animal. It consists of an inverted scolex inside a rounded bladder. The bladderworms live in this dormant condition throughout the life of the hog until they are eaten by some person. These cysts are rather easy to see with the naked eye, and trained meat inspectors can easily detect them, so it is unusual to find them in pork sold at the markets. Thorough cooking will kill them.

The **beef tapeworm**, *Taenia saginata*, has a similar life history, except that cattle serve as the intermediate host. These worms are extremely rare in the United States, but in some countries, such as Tibet, nearly all the people have them, because of the habit of broiling large chunks of beef over an open fire, which thoroughly sears the outside but does not carry the heat to the center.

The **fish tapeworm** has the distinction of being the longest parasite to infect man. Worms have been measured with a length of sixty feet and a width of about an inch, and may have as many as four thousand proglottids in a single worm. The cycle is similar to the one just studied, with the fish in the place of the hog or cow.

Most warm-blooded animals have certain species of tapeworms that may live in their intestines; it is possible for man to harbor the bladderworm of some of these species that live in some other animal as an adult. One of these is the **dwarf tapeworm of dogs**, *Echinococcus granulosus*. It is no more than one-third inch long as an adult in the intestine of the dog, but there may be hundreds present in one dog. It is not difficult for a person to get the eggs when the association between dog and human is intimate, since dogs often get the embryos on their noses and tongues and seem to have a great desire to show their affection by licking a person on the mouth. Once inside the human body, the embryo bores through the intestine and finds its way to a suitable spot to encyst. It does not form a small cyst, like *Taenia*, but buds and rebuds and may eventually produce a cyst as large as a coconut with scoleces inside it which number in the thousands. Such cysts must be removed from the body surgically. Of course, it is not likely that a dog could obtain these scoleces from a human body, but the cysts will also develop in cattle and sheep, and dogs can become infeced by eating the offal from slaughtered animals.

ROUNDWORMS

Members of the **Aschelminthes** phylum have rounded bodies. They are among the most abundant of all animals, and there is hardly a spot on the earth that does not contain them. A little debris taken from the bottom of a pond is likely to contain hundreds of tiny roundworms which can be seen under the microscope threshing around vigorously. A spadeful of garden soil may contain millions of them, but the aid of a microscope is needed to see them because of their small size. If you like homemade pickles and similar foods prepared with apple-cider vinegar, it might be best if you do not examine the scum that forms on the top of the liquid, for you are very likely to find the same little worms. These are called **vinegar eels** and do man no harm. One species found in Germany lives a life that some people would envy; they inhabit the mats upon which mugs of beer are placed and live entirely on the beer that slops over the brims of the mugs.

Besides these very abundant free-living forms, the roundworms include by far the greatest number of the metazoan parasites. Some are plant parasites, which crawl around among the cells, sucking the cell sap, causing the plants to wither and otherwise interfering with their normal growth. Vertebrate animals usually carry from one to half a dozen species in their bodies at all times, with the exception of civilized man, who, thanks to sanitation and medical vigilance, manages to avoid the universal infection. There are comparatively few persons who have not harbored at least one of these parasites at one time or another, perhaps unknowingly. The free-living roundworms are so small that we will turn to one of the large parasitic forms as a typical animal of the phylum.

The Large Roundworm of Man—Ascaris

Ascaris is a large roundworm that lives in the intestine of man. It is probably the oldest known parasite of man because it is so large that it could hardly be overlooked when passed from the body, even by prehistoric people. It may be about a foot long and normally it stays in the small intestine, but it sometimes crawls into the liver, pancreas, appendix, or occasionally all the way up the esophagus and out the nose. These worms eat the digesting food floating in the intestine and sometimes bite the intestinal wall and suck blood. They may become so numerous that, if not removed surgically, they block the intestine and cause death.

Ascaris shows a body organization that may be called a **tube-within-a-tube** type of body. There is an outer tube, the body wall, around a cavity which contains another tube, the digestive tube, which runs the length of the body. The digestive tube opens anteriorly with a **mouth** and posteriorly with an **anus.** The digestive tube of *Ascaris* is not differentiated into organs, such as a stomach, along its length, but there is a slight enlargement, the **pharynx,** just posterior to the mouth. This pumps the digesting food through the rest of the tube, the **intestine,** where absorption takes place. There are no digestive glands; enzymes from the human digestive glands accomplish the function of food digestion. Some roundworms, those which must do their own digesting, do have these glands. It is believed, therefore, that this is a case of degeneration in *Ascaris*, brought about by its specialized parasitic existence. As the food is absorbed through the intestine wall it is spread over the body by a fluid which circulates in the body cavity around the intestine. Thus, there is no need for an extensively branched intestine, such as is found in the planarian. This circulating fluid might be thought of as a forerunner of a **circulatory system.** The body cavity is not a true coelom, since it is lined internally with endoderm and externally with ectoderm. Hence, this is a pseudocoelomate animal.

Respiration is a little difficult to account for since there is no oxygen in any quantity in the intestine. Some of the other roundworms that inhabit the intestine suck into their bodies blood containing oxygen that was absorbed at the lungs of the host. *Ascaris* does not do this with any regularity, so it seems to derive its energy in the same manner as the anaerobic bacteria that are able to live without any contact with free oxygen.

Excretion is carried out by a pair of excretory tubes that run the length of the body on either side and unite and empty through a pore at the anterior end a short distance behind the mouth. There are no noteworthy changes in the **muscular system** in comparison with that of the planarian.

The **nervous system** consists of a nerve ring which circles the pharynx and two nerve cords running the length of the body as in the planarian, but these two are dorsal and ventral rather than lateral. Connecting nerves circle the intestine at intervals, and smaller branches are given out to all parts of the body.

The **reproductive system** is represented by organs similar to those found in the planarian; the one significant difference is the separation of the sexes. There are definite male and female *Ascaris* and they can be recognized externally rather easily. The female body is about one-fourth longer than the male body and is somewhat thicker, and the male has a distinct crook at the posterior end with a pair of small hairlike projections called **penial setae** near the end of the crook. Because the sexes are distinct, there must be at least two animals in a person's body before the infection can be spread to others. The male and female copulate in the intestine, and the female soon becomes an egg-producing machine with a daily egg production that would make the world's champion hen seem poor in comparison. Something like 200,000 eggs in twenty-four hours is a good estimate for a healthy *Ascaris* female. These pass from the body along with the feces, and the main method of diagnosis for *Ascaris* infection consists of an examination of the feces for the eggs. If a person has the worms, even a small sample of the excrement from his body will show many of these eggs. The eggs may get on people's hands and eventually into the mouth. When swallowed, the eggs

reach the intestine, where the embryos hatch out as tiny larval forms. A newly hatched *Ascaris* larva is apparently not adapted for life in the strong digestive fluids of the small intestine, and so for the first ten days of its life it lives in other parts of the body. First it bores through the intestine and gets in the blood stream, which carries it all over the body, and it finally ends up in the lungs. There it bores through the lung tissue into one of the many little air sacs found in the lungs. This breaking through the tissue causes a little bleeding and a blood clot is formed around the little larva. The presence of this foreign material in the lungs causes the infected person to cough and the blood clot containing the larva is coughed up. If a person coughs hard enough, he may get the larva all the way up into the mouth, where it may be spit out. On the other hand, a polite little "hack" is likely to bring it up only to the throat, where it is immediately swallowed back down into the stomach and intestine and rapidly grows to its adult size and begins the vital process of reproduction.

If a person has ingested a large number of eggs and there is extensive injury to the lung tissue, a type of pneumonia may result which may be quite serious. The adult *Ascaris* which live in the intestine do not seem to do the body great harm, but may cause anemia and nervous disorders because their poisonous waste products are absorbed. A number of drugs are effective against them, for example, oil of chenopodium and hexylresorcinol.

The Hookworm

There is a much smaller roundworm which causes body symptoms which are much more serious than those caused by *Ascaris*. This is the hookworm, which gets its name from the hooked body of the male. The scientific name, *Necator americanus*, literally means American killer, and a study of its depredations will justify the name.

The hookworm lives in the small intestine but does not absorb the digesting food like *Ascaris*; instead, it bites the intestinal wall and sucks blood. Since it moves around quite a bit and there is quite a bit of bleeding at a spot after the worm has moved to another place, it does extensive damage to the intestine. The excretory wastes of a number of these worms cause a general physical lassitude that for years affected many people living in rural areas in the southern states, where the proper conditions of moisture, temperature, and lack of sanitation made the spread of this worm easy. Today it has been largely brought under control through a vigorous program of education and sanitation.

The life cycle of the hookworm is identical with that of *Ascaris*, except for its mode of entry into the body. When the hookworm eggs leave a human body, they hatch into small larvae on the ground and are able to infect man by boring through his skin, the most common site being through the bare feet. The larvae stick to the skin when they are stepped on and bore through. Once inside the body, they are carried by the circulatory system to the lungs, where they bore through, are coughed up and swallowed, and end up in the intestine.

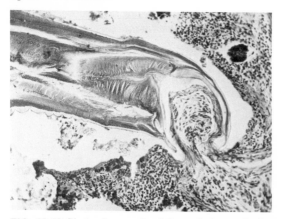

FIG. 23.13 Photomicrograph of the anterior part of a hookworm attached to the intestine. Note the part of the intestine extending down into he hookworm's mouth.

The Trichina Worm

The trichina worm, *Trichinella spiralis*, is another of the serious parasites of man. This is another parasite obtained from pork and unfortunately our efficient system of meat inspection is not a guarantee against infection. The larvae are most numerous in the diaphragm of the slaughterhouse animals and can be detected if the infection is heavy, but a light infection may be overlooked. Thorough cooking of all pork to be eaten is the only sure way to avoid possible infection.

When a human being eats pork containing live *Trichinella* larvae, the digestive juices soon digest the pork and release the tiny worms in the intestine. They become quite active and mature into adults in a short time. These adults are quite small; females are only about 3 mm long and males are about half as long. Within two or three days after being ingested, they copulate and the males die a short time later. The females retain the eggs in their bodies until they hatch. Then they deposit the tiny living larvae within the lining of the human intestine. The larvae reach the blood stream and are carried throughout the body by this medium. They invade the voluntary muscles and finally settle down and encyst. It is this working around through the muscles that causes the serious symptoms of this infection. A single female may deposit as many as 1500 larvae in the intestine. The movements of the worms through the muscles cause extreme pain and cramps which may result in death if the infection is heavy enough. The females will die within two or three months, so the disease will play out if you can survive these pains, but you will always carry the little encysted larvae in your muscles as a souvenir of your experience.

Since hogs do not have access to human flesh, we must still explain how the larvae happen to be in the hog muscle. The larvae seem to be acquired primarily through an accidental cannibalism. Hogs are fed on garbage which may contain some scraps of uncooked pork that have been discarded during the preparation of the meat for cooking. If these scraps contain the encysted larvae, the hogs can readily become infected. Because of repeated cases of human infection, some regions require thorough cooking of all garbage before it is fed to hogs.

The Pinworm

One of the most widespread of the roundworms is the pinworm, *Enterobius vermicularis.* It is found in all parts of the world and is especially abundant among children in the United States. The adult worms, about half an inch long, live in the large intestine. The females migrate down and lay their eggs around the anal opening. This causes an intense itching and desire to scratch, so children easily get the eggs on their hands and thus can reinfect themselves. The eggs make a direct trip through the digestive system, hatch, and grow to maturity when they reach the large intestine. They live only a few weeks. If reinfection can be prevented, they will disappear from the body within this period of time. It is very difficult to prevent reinfection, however, because the eggs are very small, light in weight, and easily spread; medical treatment to expel them is advisable.

Elephantiasis

One of the most repulsive diseases of tropical regions is known as **elephantiasis.** In this disease certain parts of the body increase in size tremendously. An ankle may swell until it is as big as a person's waist; a single finger may enlarge until it is larger than the wrist; the male genital organs may swell until a man cannot walk. These symptoms are caused by an accumulation of small roundworms in the lymph spaces and the consequent interference with the free flow

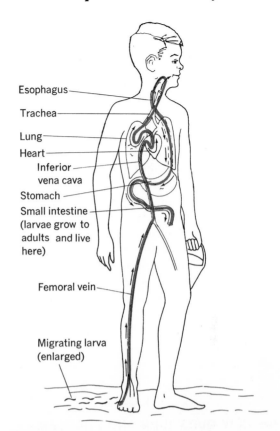

Esophagus
Trachea
Lung
Heart
Inferior vena cava
Stomach
Small intestine (larvae grow to adults and live here)
Femoral vein
Migrating larva (enlarged)

FIG. 23.14 The cycle of hookworm infection.

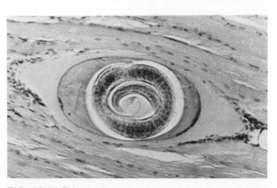

FIG. 23.15 *Trichinella* larvae encysted in pork muscle. If pork containing *Trichinella* larvae is eaten without thorough cooking the cyst dissolves and releases the larvae which bore through the muscles of the consumer causing a painful condition known as trichinosis.

of lymph through these regions. This causes the lymph to accumulate and brings a temporary swelling which is followed by a growth of connective tissue that makes the swelling permanent. *Wuchereria bancrofti* is the name of the worm causing this condition. The females may be about three inches long, but only about as thick as a coarse sewing thread, while the males are about half this length. They are found in great tangles in the lymph spaces when infected regions of the body are cut open. After mating, the females release live larvae into the lymph. As is the case with *Trichinella*, the eggs hatch before they are laid. These tiny larvae are only about a hundredth of an inch long and are carried to the blood. The bite of a mosquito may spread them to another person. In most parts of the world, the common *Culex* mosquito is the intermediate host. After being sucked into the mosquito's body, the larvae undergo further development and in a few days are ready to infect a second person.

THE ROTIFERS

The rotifers are biological misfits when it comes to classification. They are wormlike in appearance and have some of the characteristics of the flatworms, such as flame cells in their excretory system, but they also have a mouth, digestive tract, and anus similar to those in roundworms. We shall consider them as a class of the **Aschelminthes,** but they are often placed in a separate phylum. They are microscopic forms which are frequently found in freshwater cultures along with protozoans. Their name means "wheel-bearers," and they are sometimes called wheel animals because many of them have cilia on disk-shaped projections from the anterior end of the body. These cilia beat so that the disk seems to be revolving, and a rotifer's movement resembles a wheel turning.

REVIEW QUESTIONS AND PROBLEMS

1. Is the osculum of *Scypha* comparable to the mouth of more advanced forms of animal life? Explain.
2. What is a hermaphroditic animal? What are the advantages and disadvantages of hermaphroditism?
3. Regeneration generally becomes less extensive in the more complex animals. Explain why.
4. What advantage does *Hydra* have over *Scypha* in obtaining food? Explain.
5. What is an animal colony and how does a colony differ from an individual animal? Give an example from the Porifera and the Coelenterata.
6. Suppose you have a culture of *Hydra* and you want them to produce gonads. What would you do? How is this related to survival in nature?
7. What would you do to make *Hydra* open its mouth and ingest a small pebble?
8. Describe a case of **mutualism** among the Coelenterata.
9. A type of starfish has been found to be increasing its numbers greatly in the South Pacific. It eats coral. What long-term effect might this increase in population have if it continues?
10. Of what advantage is extracellular digestion to the planarian worm?

11. Why is it necessary for the digestive and excretory systems to be so extensively branched in the planarian worms?

12. Even though planarians are hermaphroditic, they have copulation. Why has this reaction developed when both types of gametes are present in the body to begin with?

13. What advances in body organization are found in planarians as compared to *Hydra*?

14. The snail is important in the life history of many flukes. Why?

15. The tapeworm has neither mouth nor digestive tract, yet it manages to continue life very well. How is this possible?

16. Parasitic worms, as a rule, produce many more eggs than other worms. Explain why.

17. *Ascaris* does not have the extensive branching of the intestine and excretory system which is found in planarians. What adaptation has it made which makes such branching unnecessary?

18. Even though the hookworm is smaller than *Ascaris*, it is generally more harmful. Explain.

19. How does the hookworm differ from *Ascaris* in its mode of entry into the body?

20. Pinworm infection is an occupational hazard of teachers in the primary grades of public schools. Explain why.

21. What is the cause of the painful symptoms of *Trichinella* infection?

FURTHER READING

Buchsbaum, R., and Milne, L. 1960. *Lower Animals*. New York: Doubleday.

Chandler, A. C. 1961. *Introduction to Parasitology*. New York: John Wiley and Sons.

Lenhoff, H. M., and Loomis, W. F. 1961. *Biology of Hydra and Some Other Coelenterates*. Coral Gables: U. of Miami Press.

Sasser, J. N., and Jenkins, W. R., eds. 1960. *Nematology*. Chapel Hill: U. of North Carolina Press.

24

Invertebrates of Increasing Complexity

In this chapter we shall discuss some representative organisms belonging to some of the invertebrate phyla which have well-developed organ systems and which lie somewhere between the simpler and the more complex invertebrates.

SEGMENTED WORMS

The phylum **Annelida** includes worms which have bodies divided into segments. The phylum name means "little ring," and some annelids look like a series of little rings put together side by side. Since body segmentation is characteristic of all the higher forms of animals, this feature is considered an important mark of advance in body structure. The earthworm is the most abundant representative of the phylum, and we shall use it to illustrate the group.

The Earthworm

Earthworms may be found everywhere that there is moist soil. They are valuable in agriculture because they loosen the soil as they work their way through it. Some gardeners buy earthworms to place in their soil. The organ systems of earthworms are well developed in comparison to the organ systems in the invetebrate phyla we have already surveyed. Let us use the species *Lumbricus terrestris* as an example.

Respiration is carried on by direct absorption of oxygen through the moist skin; carbon dioxide is released from the skin. The skin is richly supplied with small blood vessels which make this gas exchange possible. The earthworm moves up and down in the soil so that it is always at a level where the soil is damp and contains many air spaces. Earthworms frequently come out on the surface at night when the humidity is very high. They crawl back at dawn, leaving little castings behind as evidence of their nocturnal prowlings.

The **digestive system** is divided into distinct organs. It is basically a tube which runs the length of the body but has been enlarged and specialized into organs. Just posterior to the mouth is the **pharynx,** a muscular organ which expands, creating suction to bring in food. From the pharynx, the food passes into and along the **esophagus** by means of rhythmic contractions known as **peristaltic movements.** From here it passes into the **crop,** which is a storage chamber, and then into the **gizzard,** which is a grinding chamber. The gizzard pulverizes the food by muscular contractions. This process is aided by the grains of sand found in this organ. Birds also have a crop and a gizzard and they swallow bits of gravel from time to time to improve the effectiveness of the gizzard. Then the food passes into the **intestine,** where it is exposed to enzymes secreted by the intestinal wall. Digested food is absorbed into the blood vessels, which liberally supply the intestine. The absorbed food is then transported all over the body. The indigestible part of the food is egested through the anus. The dorsal surface of the intestine is folded downward to form a **typhlosole,** which provides a greater exposed inner surface for absorption. The earthworm also eats dirt as it moves along the ground; it is able to digest and absorb the nutrients which are in the dirt.

The **circulatory system** consists of two large blood vessels with smaller vessels which connect the two main ones. At the anterior end

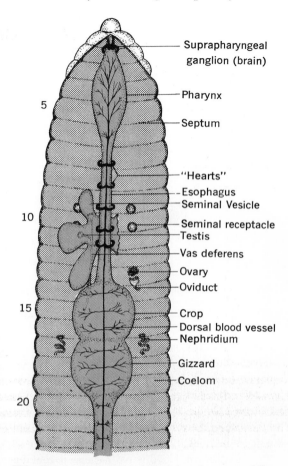

FIG. 24.1 A dorsal view of the internal organs of the earthworm in the anterior region of the body. Nephridia are shown in only one segment. The seminal vesicles have been removed on the right side to show the organs beneath.

of the body there are five pairs of so-called **hearts,** which are little more than valves which draw blood up from the ventral vessel and pump it out into the dorsal vessel. The blood moves forward to the head region and then backward along the dorsal body wall. You can see this blood movement in a live earthworm. Its movement is aided by peristaltic contractions of the dorsal vessel. Small blood vessels branch off in each segment and become **capillaries** going to individual

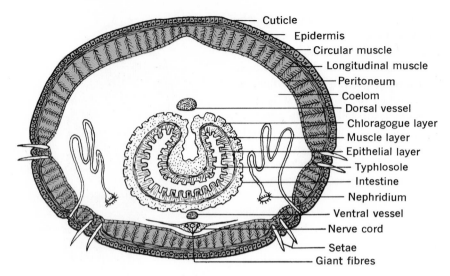

Cuticle
Epidermis
Circular muscle
Longitudinal muscle
Peritoneum
Coelom
Dorsal vessel
Chloragogue layer
Muscle layer
Epithelial layer
Typhlosole
Intestine
Nephridium
Ventral vessel
Nerve cord
Setae
Giant fibres

FIG. 24.2 Cross section of an earthworm in the region of the intestine.

body parts. The capillaries fuse and empty the blood into the ventral vessel which carries the blood forward to the hearts. Blood from the head region passes posteriorly, of course, to the hearts. This is a **closed blood system.** The blood never leaves the blood vessels in the circuit of the body.

Earthworm blood is red because it contains **hemoglobin,** which is the oxygen-absorbing part of the blood. Earthworm hemoglobin has a somewhat different chemical make-up from the hemoglobin of man, but it has the same function. Unlike human hemoglobins, it is not carried in red blood cells. Circulation is aided by movements of the coelomic fluid in the **coelom.** The coelom is the cavity which lies between the intestine and the outer muscle layer. In the earthworm the cavity is lined with mesoderm, so it may be called a true coelom. The fluid has no hemoglobin, but it can carry dissolved products very well. Each segment is partitioned off from the other segments by a septum, but the septa are perforated so that fluid can flow freely from one segment to the others.

The **excretory system** consists of a pair of **nephridia** in each segment except the first three and the last. Each nephridium consists of a little funnel surrounded by cilia which collect excretory waste from the coelomic fluid. The nephridia empty into a slender coiled tube that runs backward and empties to the outside through a pore on the ventral surface. Water may be absorbed from the tube so that the waste is concentrated before it is expelled.

Earthworms have a **muscular system** consisting of circular and longitudinal fibers. The contraction of the circular fibers squeezes the worm and makes it longer, while contraction of the longitudinal fibers makes it shorter and fatter.

The **nervous system** of the earthworm is organized very much like that of the higher animals, including man. There is a **brain** at the anterior end of the body and a **nerve cord** running the length of the body, giving out nerve branches at different levels that run to all parts of the body. Some of these are **sensory nerves** which bring impulses from sense organs to the nerve cord and then continue up the cord to the brain. Others are **motor nerves** which lead from the brain and nerve cord out to the muscles and stimulate them to react. The nerve cord is on the ventral body wall and the reactions are more or less of the reflex type, the brain having comparatively little to do with correlating the body reactions. A cross section of the nerve cord reveals three **giant fibers** in the dorsal region of the cord. These fibers contain nerves running the length of the body and connecting the different segments, so that they work in a coordinated manner, and strong stimulation brings response from all segments even though the brain has been removed. This is possible because each segment contains an enlargement of the nerve cord called a **ganglion** which acts somewhat like a little brain for that segment and makes the proper connections for response to stimuli. An earthworm can be cut into several parts and each part will go right on crawling around for a time and reacting to stimuli much as if nothing had happened, for the nerves in the giant fibers connect the different ganglia of each part together and give it coordination.

Reproduction in the earthworm is very interesting. They are **hermaphroditic,** but they also **copulate** with a mutual exchange of sperm. The **testes** are contained within the large **seminal vesicle** that extends from segments 9 to 13, and the two **vasa deferentia** lead from the testes to a pair of external openings in segment 15. The female organs consist of a pair of **ovaries** in segment 13, a pair of **oviducts** which open in segment 14, and two pairs of **seminal receptacles** in segments 9 and 10.

During copulation two worms come together facing in opposite directions with their ventral surfaces in contact. The anterior ends of their bodies become fused together by a layer of mucous secretions so that the openings of their sex ducts are close together. Sperm are discharged from the openings of the vasa deferentia and pass down two pairs of grooves in the slime until they come to the openings of the seminal

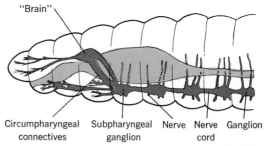

FIG. 24.3 The position of the brain, nerve cord, and connecting nerves in the anterior part of the earthworm.

receptacles, which they enter. It is a mutual exchange, each worm receiving sperm from the other. When insemination is completed, the worms separate.

Later a swollen, bandlike portion of the body, called the **clitellum,** which occupies segments 32 to 37, secretes a series of cocoons. Each cocoon slides over the anterior end of the earthworm, and as it passes the openings of the oviducts, eggs are laid in it. Then as it passes the pores of the seminal receptacles, sperm which have been received from the other worm are discharged and fertilize the eggs. The cocoon passes over the head and the ends close up tightly. It is left just beneath the surface of the ground by the worm. After completing their embryological development, the young worms crawl out of the cocoon.

FIG. 24.4 Copulating earthworms. The picture was made outdoors on a rainy night.

There is no asexual means of reproduction, but the power of **regeneration** is possessed to a great degree. If an earthworm is cut in two between segments 15 and 18, the head piece will regenerate a new tail and the tail piece, a new head. If the cut is made posterior to segment 18, the tail piece cannot regenerate a new head, but instead usually develops another tail to form a worm with a tail on either end. This earthworm can very easily take care of egestion but is at a loss for a place of ingestion and must slowly starve to death.

A Marine Annelid—The Clam Worm

The clam worm, *Neanthes*, is representative of the many marine annelids. This worm lives in the sand of shallow water near the shoreline of many of the sandy beaches of the world. It burrows down into the sand, leaving only its head and tentacles protruding. Whenever some small water animal ventures too close to this innocent-looking protrusion, the clam worm may throw out its proboscis, grasp the animal, and pull it down into the burrow, where it will be devoured. In external appearance the clam worm differs rather strikingly from the earthworm because of the presence of a row of paddlelike **para-podia** on each side of the body. There is a pair of these appendages on each segment. They are used in swimming and also to keep a current of water flowing through the burrow when the animal is at rest. This is necessary for proper respiration.

The body systems of the clam worm are very much like those of the earthworm except for reproduction. The clam worm has separate sexes, and the reproductive cells develop from the wall of the coelom in the posterior region of the body. During the reproductive season this portion of the body becomes greatly swollen. As the sexes mingle in an annelid courtship, these swollen regions burst open, liberating the sperms and eggs, and fertilization takes place in the open water. The worms then usually die, but in some species the uninjured anterior end may regenerate a new posterior and life continues. In one species of marine annelid, the palolo worm of the South Seas, the posterior end of the animal breaks off and goes swimming to the surface, leaving the anterior end, which begins regeneration of its lost posterior. This reproductive process always takes place on the first day of the last quarter of the October–November moon. At this time the surface of the water near the shore may be filled with these swimming worm segments. Natives of the regions have

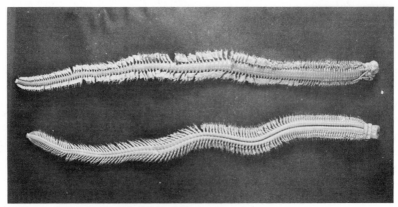

FIG. 24.5 *Neanthes,* the sand worm, seen from the dorsal surface above and the ventral surface below.

learned of this and gather these in large numbers and have a feast.

Parasitic Annelids—The Leeches

Leeches are a kind of segmented worm which has adapted to a parasitic existence. They do not live in the bodies of other animals like many of the flatworms and roundworms, but suck blood through the outer skin of their hosts. Thus, they are **ectoparasites** rather than **endoparasites,** like the others. If you put a living leech in an aquarium, it will attach itself by a posterior sucker, and if you put your finger in the water, it will stretch and wave the anterior portion excitedly in an effort to reach the finger. If it succeeds, it will attach itself to your finger by the anterior sucker and release its hold at the posterior. There are three sharp teeth on the mouth within the anterior sucker and the skin is punctured by these teeth, which are so sharp that there is usually no pain at all. Then the leech will gorge itself on blood, sucking in about three times its body weight if undisturbed. It will then drop off in a stupor and will not need to feed again for several months.

The practice of medicine was at one time spoken of as "leechery" because doctors always carried a supply of large leeches in their little black bags. There is one leech that was well adapted to their purposes and it is still called the medical leech, *Hirudo medicinalis*. It is about four inches long and capable of ingesting a considerable quantity of blood. When a person was sick, it was believed that there was an upset balance between the body fluids and that removal of some of the blood would restore the balance and health would return. The leeches were a very convenient method of accomplishing this, which accounted for their popularity. More recently, the colorful barber shops of the early part of the century kept a supply of leeches on hand, and when a man got a black eye or other conspicuous

black and blue spots on his body, a lean hungry leech was applied to the skin and blood from the discolored area was sucked out. This practice has not entirely disappeared today, and it is possible to buy these large medical leeches at some drug stores for this purpose.

A particularly vicious land leech lives in some densely vegetated tropical regions where the rainfall and humidity of the air are great enough to allow it to live out of the water. A person walking through jungle regions where they abound must be careful to pick these leeches off when they become attached to the skin or suffer the loss of an appreciable quantity of blood. There are even some cases of death from loss of blood by persons injured and unable to keep the leeches off. A few such leeches are found in the Everglades of south Florida.

MOLLUSKS

Animals in this phylum are highly specialized for a type of existence which is different from that of higher animals. They form a side branch which goes no further than this group on the evolutionary tree.

The phylum **Mollusca** is very extensive, including nearly 80,000 species, a greater number than is found in any other phylum except one. The name is derived from the Latin word *mollis*, which means "soft," and is appropriately chosen in view of the soft body parts of these animals. Most of them have a shell on the outside of the body that protects their delicate organs. Because of the presence of this shell, many of the marine forms are called shellfish. The phylum includes such widely divergent forms as the snail, octopus, squid, oyster, and clam. These may not seem to have many characteristics in common that would cause them to be grouped together in the same phylum, but a survey shows them to be quite similar in fundamental structure. They all have a **mantle** surrounding a part of the body called the **visceral hump,** which contains important internal organs. Extending down ventrally from the visceral hump, in all the species that are able to move about, is a **foot,** which is a locomotor organ. The mantle secretes a liquid which hardens to form a **shell** in most species. Some Mollusca, like the squid, have no external shell, but the mantle is somewhat tough and offers a degree of protection. The soft bodies of many of them make delicious eating, and the mollusks (common name) are an important source of food.

The Freshwater Clam

There are marine as well as freshwater clams. The freshwater species are found widely distributed in the rivers and lakes of the United States wherever there is sufficient calcium carbonate in the water to enable them to produce their shells. They are abundant in the Missis-

sippi Valley, where much of the water filters through limestone deposits and is rich in calcium. In this region clams are collected and their shells are used to make "pearl" buttons, such as those found on men's shirts, and pearl-handled knives, revolvers, and ornaments.

The outer surface of the shell is made up of numerous concentric rings; these are **rings of growth,** since the rings are deposited by the mantle around the outer rim of the shell, thus increasing its size. A number of rings will be deposited each year, but in regions where there are great fluctuations of water temperature, the rings representing the dormant periods of the winter months are usually more pronounced; thus, it is possible to estimate the age of a clam by these rings. Near the hinge and slightly anterior is a little hump in the shell called the **umbo.** The rings radiate out from this region, which is the oldest portion of the shell.

The outer surface of the shell is covered by a tough and somewhat eroded horny layer, but

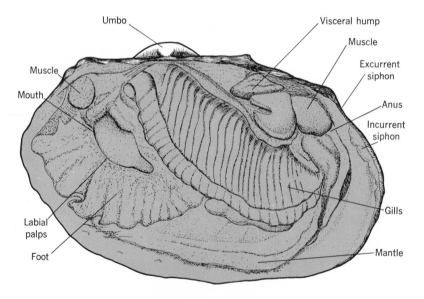

FIG. 24.6 A freshwater clam with the shell and mantle removed from the left side to show the body organs within the shell.

the inside is a beautiful pearly layer. This inner layer has an opalescent quality, reflecting the light waves in their various colors, owing to the presence of thousands of tiny prismatic bodies embedded in the mother-of-pearl.

The two halves of the shell are closed by a pair of strong muscles that make it quite difficult to pry the shell open. When the clam opens the shell, it relaxes these muscles and the elastic hinge pulls the valves apart. With the shell open, the clam can extend its foot and pull itself along rather awkwardly in the mud at the bottom of the river or lake where it lives.

Some unusual developments in the organ systems of the clam allow it to carry on its life processes in its sluggish state. At the posterior end of the animal there are two openings which allow water to enter and leave the **mantle cavity,** which is the cavity lying between the mantle and the visceral hump and foot. The ventral opening is the **incurrent siphon,** through which water enters the mantle cavity; the dorsal opening is the **excurrent siphon,** through which the water leaves. A pair of **gills** on either side of the mantle cavity bear cilia that keep up the circulation of water through the cavity. Respiration is carried on by the gills as the water flows through them.

The water coming into the cavity will contain bits of organic matter which can be used as food. A pair of **palps** on either side of the **mouth** at the anterior end of the body sort out particles that are edible and carry them into the mouth. **Digestion** takes place in a small **stomach** and absorption through the **intestine.** **Egestion** is through the **anus,** which opens just dorsal to the excurrent siphon. **Excretion** is taken care of by a pair of **nephridia** which lie in the dorsal portion of the visceral hump and empty into the water that is on its way out of the dorsal siphon.

The **circulatory system** of the clam includes a single heart and connecting arteries and veins which transport the blood over the body.

Some of the arteries, instead of forming small capillaries which connect with the veins, empty into **blood sinuses,** which are cavities that connect directly with the veins. The blood has no hemoglobin but does have white corpuscles.

The **nervous system** consists of three paired **ganglia** with connecting nerves, smaller nerve branches, and rather poorly developed sense organs. A pair of ganglia above the mouth, the cerebropleural ganglia, are the equivalent of the brain, with paired nerves running posteriorly to the visceral ganglia and another pair of nerves running ventrally to the pedal ganglia in the foot.

Reproduction involves no contact between the two sexes. The male releases sperm through the excurrent siphon; some of them will be taken into the incurrent siphon of the female. In the meantime the female has released her eggs, which are held in the gills where they are fertilized by the sperm passing through. The resulting zygotes are held here until small larvae, called **glochidia,** are produced. The glochidia are expelled from the mother and must find a fish in order to continue their existence. If a fish comes near, these larvae immediately clamp onto its skin or gills and live there as parasites until they have developed into small clams. This seems to be a distribution adaptation as well as a means of protection and nourishment. A fish may carry these little parasites a great distance before they drop off. Since the clams are so sluggish in their movements, this is probably the only way that the worms could be well distributed, especially upstream in a river.

Certainly there is no one who has not heard of the renowned clam bakes and clam chowder of the New England states, which testify to the good eating qualities of the marine clam. Many varieties of edible clams are found in the ocean, ranging from tiny ones no bigger than the end of your finger, which make delicate hors d'ouevres, to huge ones from which you can get a good-sized steak.

Oysters furnish us with an important source of food and are cultivated in many places in order to supply the great demand for them. The oyster is not able to move about like the clam; the left side of the shell becomes attached to a solid object when the oyster is in the larval stage and it remains there for the rest of its life. The oyster resembles the clam in structure except for the fact that it has no foot. The foot is a locomotor organ and is not needed by the sessile oyster.

Whenever a foreign body, such as a grain of sand or a small parasite, gets between the mantle and the shell of one of the bivalves, a layer of mother-of-pearl will be secreted around it. Additional layers of mother-of-pearl will be formed around this as time goes on, and at the end of several years a **pearl** will be formed. The most valuable pearls are found in the pearl oyster, a large oyster of the West and South Pacific.

In Toba, Japan, a man named Mikimoto discovered a method of artificially introducing foreign bodies between the mantle and the shell of the pearl oyster, thus stimulating pearl formation. Several years are required for the formation of a pearl of commercial value.

The **shipworm**, *Teredo*, is a highly specialized relative of the clam which is very destructive to wood in salt water. It has a slender body with a small anterior shell. The shell is used

FIG. 24.7 A snail. Note the stalked eyes and the slimy appearance of the body.

to burrow in the wood of ships or wharves. The many burrows which shipworms leave can greatly weaken these structures.

Snails and Slugs

Most snails have somewhat disrupted their bilateral symmetry by twisting their visceral hump in a spiral and secreting a shell around it. They creep along on the foot at the traditional "snail's pace," which averages about 2 inches a minute at full speed. Their progress is slow because they slide along in a layer of mucus which is secreted by a gland at the anterior end of the foot. The pathways of snails can be recognized by the silvery trails of dried mucus which they leave behind. Land snails usually do their traveling at night when the air is cool and moist. They withdraw their foot inside the shell and seal the opening with mucus during the day so their bodies do not dry out.

The snail has its eyes on stalklike tentacles which can be extended and retracted. The eyes do not see definite images but can only detect relative light intensity.

A large species of snail is raised extensively in western Europe for food. They are considered a delicacy in France and can be obtained in some parts of the United States.

Some people think that snails can crawl out of their shells and return to them because they sometimes see animals that look like snails without shells. These are the **slugs,** which do not have noticeable shells as adults. However, a study of the embryonic development of the slug reveals that a shell is formed in the embryo, just as it is in the snail, but fails to continue its development to a functional size.

The Cephalopods

The class name **Cephalopoda** means "head foot." The foot is wrapped around the head, causing some confusion of the parts of each.

The **squid** is a good example of the class. The squid has a somewhat torpedo-shaped body, with the visceral hump drawn up into a long pointed shape and covered by the mantle but not surrounded by a shell. There is a vestigial remnant of the shell, the pen, embedded in the mantle at the anterior edge, and this supports the softer parts of the body. The foot bears ten arms which are lined with suckers that can give the animal a good grip on anything around which it wraps them. The squid has two eyes, with iris, lens, and retina like those found in vertebrate animals. This does not mean that the squids are very close relatives of the vertebrates; there are many other points of difference between the two. But it is thought that the eye of the squid developed independently along the same lines as the vertebrate eye. It is a case of analogous development.

The squid sucks water up into its mantle cavity during respiration and may use this same water for locomotion. Leading from the cavity is a little tube called the siphon which can expel this water forcefully and propel the animal in the opposite direction, on the principle of jet propulsion. By pointing the siphon in different directions, the squid can control the direction of movement, aided somewhat by a pair of fins that project out at either side of the mantle. With this method of locomotion, it can move in a series of rapid darts when in a hurry, but it can also use the arms to crawl along in a more leisurely fashion.

The squid not only has a pen, it also has ink. The ink sac opens into the mantle cavity; when in danger, the squid will squirt a cloud of black ink out into the water. This throws down an underwater "smoke screen," giving the squid an opportunity to escape. This device is not always effective, however, for the squid is very commonly found in the stomachs of fish and furnishes them with a source of food. The squid is also an important source of food for man in the Orient.

An average-sized adult squid will be no more than a foot long, but there is a species of giant squid (*Architeuthis princeps*) which has the distinction of being the largest of the invertebrates. These squids have ben found with a body length of about twenty feet and a total length, including their long arms, of fifty-two to fifty-five feet.

One species of squid, called the **sepia**, produces a sepia-colored ink and is an important source of this pigment for use in artists' paints. It is also called the cuttlefish and has a large pen, the cuttlebone, which is removed and placed in the cages of captive birds like canaries to supply calcium.

One of the most feared creatures of the South Pacific is the giant **octopus,** or devil fish, that has been known, on rare occasions, to ensnare divers with its long snakelike tentacles. It does not compare with the giant squid in size, but some have been found with tentacles that could be spread in a radius of 15 feet. This is easily large enough to overcome a person. The cold, wicked-looking eyes, the soft, slimy body, and the long, wiggling tentacles combine to make this a fearsome animal; yet the great majority of octopuses are far too small to do man any bodily harm, and they furnish him an important source

FIG. 24.8 An octopus. Note the suction cups on the tentacles. These suction cups enable the animal to attach itself firmly to solid objects.

of food. An average-sized octopus will have a body no bigger than a grapefruit, and in the Mediterranean countries, South America, the South Pacific, and the Orient, octopus is a much better known and more frequently eaten dish than lobster.

The octopus is very much like the squid in its body structure, but it does not have a pen for support, and when taken out of the water, its body collapses like a mass of jelly. It has eight arms instead of ten.

The **chambered nautilus** is another member of the class. Unlike the squid and octopus, its mantle secretes a shell, which is laid down in a series of spiral chambers to form a beautiful coil. A new chamber is formed as the animal outgrows the older one, so each new chamber is larger than the preceding one.

FIG. 24.9 A section through the shell of the chambered nautilus showing the chambers. The animal moves forward each season, making a new and larger chamber.

Primitive Mollusks—The Amphineura

This class of mollusks seems to show a lesser degree of specialization than any other in the phylum and is therefore of interest in establishing possible relationships between this phylum and others. These mollusks have a ladderlike nervous system, such as the one found in the planarian, and a ciliated larva which is similar to that produced by *Neanthes*. Embryonic development of the flatworms, annelids, and Amphineura shows that the cells divide in the same way and form similar structures up to a certain point before beginning development in different directions. These facts indicate a rather close relationship between these three phyla.

A typical genus is *Chiton*, which crawls around among the rocks on the seashore eating the algae growing there. Its body looks like a broad flattened worm but is covered on the dorsal surface with a series of eight plates with articulating surfaces that give more freedom of movement than is found in other mollusks with rigid shells.

The Toothshells

This is a minor class of marine mollusks that have a tubular shell open at both ends. Its members are commonly called **toothshells** because of the similarity of appearance of the shells to the teeth of some large animals.

A Living Fossil—Neopilina

One class includes a number of primitive, extinct mollusks which lived in Paleozoic times and are known today only by their fossil remains. In 1957, however, a Danish expedition collecting off the west coast of Costa Rica found a living member of this class from material brought up from a depth of about 12,000 feet. It was given the name of *Neopilina galatheae*. It

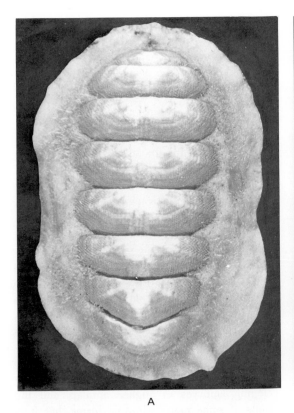

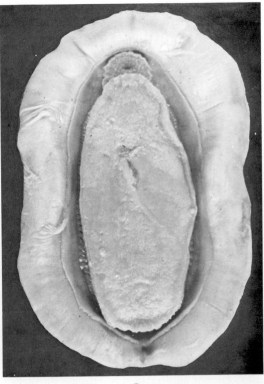

A B

FIG. 24.10 Chiton, dorsal and ventral views. This is the most primitive of the modern mollusks. Note the row of articulating plates which take the place of the less flexible shell of other mollusks. The broad creeping foot resembles that of the snail and the slug.

is segmented internally with several pairs of organs such as nephridia, which are annelidlike characteristics, but there is a mantle, a single shell, and other characteristics which place it in the Mollusca. This was considered a very important biological find because it is another "missing link" or "living fossil" which helps us to trace the possible past course of biological evolution.

SPINY-SKINNED ANIMALS: THE ECHINODERMS

The phylum name *Echinodermata* means "spiny skins," and if you have ever examined a starfish or perhaps stepped on a sea urchin at the seashore, you know that the name is appropriate for some members of this phylum at least. Most of them have a calcified outer covering of the body which bears spines. They are all marine and very abundant in tropical waters.

The echinoderms are rather advanced in some systems, but in others they seem to be more like the simplest Metazoa, especially the coelenterates. For instance, they have radial symmetry like *Hydra*, but they also have a coelom. In still other systems they are neither primitive nor advanced, but specialized in ways that are peculiar to this phylum. They have a system of circulating sea water in their bodies which no other phylum has. Some clue to this strange mixture of characteristics may be found through critical studies of the embryonic development of these forms.

Importance of Embryos in Establishing Phylogenetic Relationships

Although the adult echinoderms are radial in symmetry, the embryos first develop into larvae with bilateral symmetry. This condition is not found in any of the other animals with radial symmetry as adults. This is of particular significance because many studies have shown that the embryonic characteristics are usually more reliable than adult characteristics in determining animal relationships. Adults often become so altered by their adaptations to varying environments that they may show little resemblance to adults of closely related species which may have become specialized along other lines. The embryos, however, are not subject to this extreme fluctuation in structure and tend to develop in the same way that they did before the adults became so specialized. Hence, when we find two animals with a very similar course of embryonic development, we may suspect that they are rather closely related, even though they may show considerable morphological difference as adults. On this basis of reasoning, it is logical to conclude that the echinoderms were probably bilateral in symmetry at one time but that they have become radial in the adult condition. The embryos continue to develop in the bilateral direction as they did before the change in the adult.

Detailed studies of starfish embryonic development show that the three germ layers (endoderm, ectoderm, and mesoderm) and the coelom form in almost exactly the same way as they do in the primitive chordate animals. This similarity is the basis for the theory of echinoderm ancestry of the chordates, which is plausible in spite of the great diversity of the adults of these two phyla. Considering all these studies of echinoderm embryos, we conclude that this phylum is rather advanced in its phylogenetic position and the living representatives have become highly specialized.

A word of caution is in place about the reliability of embryos as a means of establishing animal relationships. In 1828 the great German embryologist Von Baer brought forth his observations on the similarities among the embryos of different animals. He emphasized the point that the embryo of one animal may resemble the embryo of a closely related animal as we have pointed out. Less scientific workers, such as Ernst Haeckel, seized upon this idea and formulated a **theory of recapitulation** which received wide acceptance for a time. This theory holds that an embryo tends to retrace all the changes which have occurred in the ancestry of the animal. As this theory has been subjected to careful scientific analysis, it has become apparent that this is not true. There are a few similarities in the developing embryos of higher animals and the adults of some more primitive forms, but it is now believed that these are incidental to the development of the embryo and not a recapitulation. The embryo tends only to reproduce those embryonic characteristics which it had in the past as an embryo and not the characteristics of the past adult forms. Thus, comparative embryological studies are of great importance in establishing animal relationships but not in showing the changes undergone by the adults in their evolutionary development.

The Starfish

The most commonly known and perhaps the most widely distributed of the echinoderms is the starfish, *Asterias*. It is not shaped like a star nor is it a fish, but it was given this name when it was thought that stars had five points and that every animal which lived in the water was a fish. We now know that stars are spherical in shape and that fish are only one type of the many animals that inhabit our ponds, rivers, lakes, and seas. We retain the name starfish, however.

The starfish is well protected by a **skeletal system.** The outer surface of the body is covered by skin through which hard knots of lime protrude and form **spines.** This gives excellent protection and support for the soft body parts within. At the same time the arms are quite flexible because of soft skin between the spines which allows the arms to bend easily. Small pinchers, the **pedicellaria,** come out from around the base of the spines. These pinchers keep off foreign matter that might adhere to the body of the starfish and interfere with respiration and other life processes.

In the living starfish there are delicate little fingerlike projections coming out from the skin between the spines. These projections are extensions of the body cavity, or **coelom,** which contains the coelomic fluid. They are called **skin gills** and absorb oxygen from the water and return the carbon dioxide to it. The coelomic fluid circulates over the body, but there are no well-organized vessels as there are in the earthworm. The skin gills also aid in **excretion.** The coelomic fluid contains amoeboid cells which engulf the excretory wastes as they diffuse out into this fluid from the body. Eventually these cells become filled with the waste and squeeze their way out of the body through the skin gills, carrying themselves and the waste out into the water.

The starfish has a **muscular system** which enables it to move the arms, to pinch with the pedicellaria, and to make certain other simple movements. However, the major movements are accomplished by an entirely new system that is found in no other phylum of animals. This is the **water vascular system,** which uses sea water under a sort of internal hydraulic pressure. On the aboral surface of the central disk of the star-

fish is a hard round structure called the **madreporite,** which is the beginning of this system. The madreporite admits the proper amount of sea water into the system. A little calcified canal, called the **stone canal,** leads from the madreporite to the **ring canal,** from which a **radial canal** radiates out into each of the five arms. Alongside the radial canals are little **ampullae** which connect to **tube feet** which can be extended through the oral surface of the body. These can be seen externally lying in rows in the **ambulacral grooves** of each arm. To extend a tube foot, the starfish squeezes the connecting ampulla. This forces the sea water out into the tube foot and makes it much longer. On the bottom of the tube foot, in most starfish, is a little suction cup which may be attached to some solid object. Once this attachment is made, the muscular wall of the tube foot contracts, forcing the excess water back into the ampulla. This contraction of the tube foot will pull the starfish toward the point of attachment. Through cooperative action of many tube feet, the animal draws itself along in a rather slow and awkward fashion. Because the tube feet cannot get a grip on sand, starfish are seldom found along shores with a sandy bottom but are very abundant where the shoreline is rocky.

FIG. 24.11 A starfish, showing the use of the tube feet in locomotion.

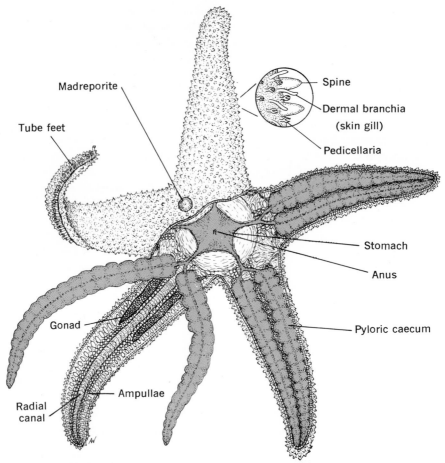

FIG. 24.12 A cutaway view of the starfish showing the internal organs in one arm and in the central disk.

The starfish has one of the most interesting methods of **digestion** found in the animal kingdom. Its food consists almost completely of mollusks which have hard protective shells around their bodies. If you have ever tried to pry open the shell of a clam or oyster you may wonder how such a small and clumsy animal as a starfish could possibly feed on these mollusks. The starfish, however, can open these shells with comparative ease through a principle of applied physiology. As an illustration, we will say that it finds an oyster to eat. It will settle iself down over the portion of the shell that opens, spread its arms down on either side of the shell, attach the tube feet by suction, and begin to pull. The oyster has very powerful muscles which close the shell and it would appear to be secure against the slight pull which the starfish is able to exert. However, the starfish continues pulling, and when one arm gets tired, it will rest it while pulling with another. It is a well-known principle of muscle physiology that continuous contraction cannot be maintained for a very long period of time, and so it is only a few minutes before the muscles of the oyster must relax and the shell opens. Then the starfish everts its stomach through its mouth, wraps it around the soft body parts of the oyster, and pours out powerful digestive juices that digest the oyster in its shell. The digested oyster is absorbed and the liquefied food passes into the coelomic fluid by which it is distributed.

A single starfish may destroy ten or twelve oysters a day in this way. Starfish are great enemies of the oyster fishermen, who take steps to combat their presence in oyster beds. One common method is to drag the ends of a frayed rope or similar structure over the bottom. The starfish will grasp the rope ends with their pedicellaria and are thus lifted to the surface. At one time they were chopped in two and thrown back in the water, but because of regeneration, this actually doubled the population. Now they are collected and dried so that death is certain.

There are many tiny mollusks which may also serve as food for the starfish; these are taken into the stomach through the mouth, and after digestion the shells are spit out through the mouth.

The **anus** of the starfish is used very little, since for the most part only digested or easily digested particles are taken into the digestive system. When small mollusks are ingested, their shells are more easily discharged through the mouth. The ancestors of the starfish, in all probability, had a functional anus; now, although still present, it is so greatly reduced in size that it is regarded as a vestige. The secretion of a plentiful supply of digestive fluids reduces the time that a starfish must sit on an oyster with its stomach protruding. The supply comes from five large pairs of **pyloric caeca** that fill the coelom of each of the five arms.

The nervous system includes a **nerve ring** around the mouth and a nerve running out into each arm, with small branches connecting with the different parts of the body. **Sense organs** are poorly developed, but at the end of each arm there is a pigmented eyespot which is sensitive to light, and both the skin gills and pedicellariae are sensitive to tactile stimuli.

Reproduction is quite simple; sexes are separate and a pair of **ovaries** or **testes** lies in each arm between the pyloric caeca and the ampullae. Their size varies considerably with the degree of maturity of the animal and the season of the year. These gonads open to the outside by small pores at the points on the central disk between the arms. Sperm and eggs are shed into the water during the reproductive season and the sperm find the eggs and fertilize them. A bilaterally symmetrical larva develops. Bands of cilia around it enable it to swim freely in the water, and this aids in its distribution.

Regeneration of the starfish is extensive. A single arm, as long as it contains a portion of the central disk attached to it, can regenerate all the missing parts. Another interesting process, called **autotomy,** is illustrated by the starfish. If a portion of an arm is crushed, the entire arm may be automatically severed from the body near the central disk and a new arm regenerated.

Other Echinoderms

A number of other common echinoderms will be surveyed briefly. The **brittle stars** have a central disk and five arms like the starfish. They get their name from the fact that their arms come off so easily; if you merely grasp one of the arms with your fingers, it is very likely that the animal will cut it off at the central disk. It can easily regenerate a new one.

The **sea urchins** are globe-shaped echinoderms with very long spines projecting out from their bodies. Some species have poison on their spines, and you should be careful when you wade in water where these may be found. When the brittle skeleton is cracked, the yellow gonads may be seen almost filling the interior. The gonads are sometimes used as food.

FIG. 24.13 A living sea urchin. The long, sharp spines provide good protection to this echinoderm.

The **sand dollars** are small disk-shaped echinoderms that have the characteristic star-shaped outline on both oral and aboral surfaces. From a distance an observer might mistake one lying on the sand for a silver dollar.

The **sea cucumbers** are found abundantly along the bottom of the sea in shallow tropical waters. They somewhat resemble the cucumber in appearance and differ from the other members of this phylum in having a flexible outer skin that feels more like a piece of soft leather than the hard spiny skin of the starfish. A number of tentacles around the mouth are waved in the water until sufficient food particles adhere to them; then they are turned down into the mouth, the mouth is closed, the tentacles are withdrawn, and the food is sucked off, in very much the way a child sticks his fingers in a can of molasses and then sucks the molasses off. In Oriental countries sea cucumbers are in great demand for soup making.

The **sea lilies** are other animals that are somewhat plantlike in superficial appearance. Most of them have a stalk which they attach to a rock or similar underwater structure. It expands into five arms with branches coming out around the mouth. The sea lillies are abundant animals, but they are seldom seen because many of them inhabit the deep seas where human beings cannot venture. They are also abundant as fossils, especially in limestone deposits. These are often found far inland and give us definite evidence of the evolution of the earth's surface. Their presence indicates that these regions were at one time at the bottom of the sea.

FIG. 24.14 Sand dollars which have washed up on the beach.

OTHER INVERTEBRATE PHYLA

There are a number of comparatively small groups of invertebrates which do not fall readily into any of the major phyla. In the past there was a tendency to place them in one of the existing phyla anyway, even though it was understood that their characteristics were not in line with those of the other animals in the phylum. Modern methods of classification place these misfits into separate phyla, and this adds cons'derably to the number of phyla in the animal kingdom. These phyla are listed here together with a brief description of some of their more impor-

FIG. 24.15 The sea cucumber differs from most echinoderms in that it has a soft skin. A close-up view of the skin shows the projecting tube feet.

tant features. Some of them do have considerable biological significance; the very characteristics which make them misfits may be important clues to the relationships among the other phyla. They may be links which show the direction of changes in the past.

Mesozoa

This phylum includes a small group of wormlike animals with the simplest body structure of any of the multicellular forms. They resemble the colonial protozoans; they are only about 8 mm long, and the body consists of only about twenty-five ciliated outer cells around one or several long internal reproductive cells. The outer cells take in food and there is intracellular digestion. They could be very primitive metazoans, or possibly they are extremely degenerate forms related to the flatworms. A typical genus is *Dicyema,* which lives as a parasite in the nephridia of squids and octupuses.

Ctenophora

This phylum includes the **comb jellies, or the sea walnuts.** The first common name is given because they bear eight rows of small ciliated bodies which closely resemble combs and their bodies are jellylike in consistency. The shape of some of them has given them the name of sea walnuts. They are marine forms which float and swim near the shore, and great numbers are sometimes cast up on the beach by the breakers.

They have many features in common with the coelenterates, such as a gastrovascular cavity and a mesoglea, but there are long muscle fiber cells in the mesoglea. This indicates a higher degree of development. Again, they are not placed in the Coelenterata because they have a higher organization of the digestive system and no nematocysts.

Nemertina

This is a phylum of worms which are almost all marine and not commonly seen. They are long and flattened and are commonly called **ribbon worms.** They have a long proboscis which may be extended from its sheath at the anterior end of the body. They have a circulatory system with blood and blood vessels and two openings to the digestive tract. In many respects they are similar to the free-living flatworms; they have flame cells, a flat, unsegmented body, no coelom or respiratory system. However, they are different in that they possess a complete digestive system with mouth and anus, a circulatory system with blood and vessels, and a simpler reproductive system. They commonly live buried in the mud or other material at the bottom of the water and feed on other animals. They have great powers of regeneration and one, *Lineus,* can be cut into as many as a hundred pieces and each piece will regenerate a tiny entire worm. These may then be cut and yield still smaller worms and so on until the regenerated worms are thousands of times smaller than the original.

Entoprocta

These small animals have a vaselike body with a stalk by means of which they attach themselves to seaweed or other objects in shallow coastal waters. Many live in a commensal relationship with marine annelids.

Gastrotricha

Another group of microscopic forms living in habitats common to many freshwater protozoans and rotifers. They resemble some ciliate protozoans, but they are multicellular and have organ systems such as a blood vascular system. A typical genus, *Chaetonotus*, has a slender, flexible body with a forked posterior end. The flat, ventral surface has two longitudinal bands of cilia.

Kinorhyncha

These are very small marine worms not over 1 mm in length. They live in the mud or sand at the bottom of various bodies of salt water. They resemble the nematodes but differ in their excretory and reproductive organs.

Acanthocephala

These are commonly known as the **spiny-headed worms** because they bear a proboscis covered with spines. They are parasitic animals living in the intestine of vertebrates as adults and as larvae in the bodies of various anthropods. They have a long, flattened body which may have a superficial resemblance to a tapeworm when they are removed from an animal's intestine. They have many similarities to the Aschelminthes, but they differ in the absence of a digestive tract and the presence of the proboscis and circular muscle fibers, as well as in other features.

The abundance of this parasite may be illustrated by an account of a study of squirrels near De Land, Florida. About 50 percent of all the squirrels killed showed infection with the acanthocephalan *Monoliformis moniliformis*. Even though the worms reach a size up to 14 inches, there was an average of 13.8 worms in each infected squirrel. Investigation showed that the larval stage of the worm is passed in the larva of a beetle that lives in acorns, which are a common food of the squirrels.

Bryozoa

Animals in this phylum are sometimes called **moss animals** because they resemble mosslike plants and are often mistaken for seaweed in their marine form. They develop a hard covering, somewhat like coral polyps, and extend themselves from this when they feed, but can withdraw into the covering in times of danger. They have a polyplike body which gives them a superficial similarity to some of the coelenterates, but closer study shows the presence of a brain and an anus, which make them more advanced.

Bugula is a common bryozoan that is found attached to objects in shallow sea water. It is a colonial form, with small individuals closely united in longitudinal rows.

Brachiopoda

These are commonly called the **lampshells.** They bear an external resemblance to the bivalved mollusks. There are two valves to the shell, held together with a hinge, but there are many morphological differences. The two halves of the shell represent dorsal and ventral surfaces rather than lateral, as in the mollusks; and within the shell there is a pair of spirally coiled arms which have no counterpart in any other phylum. They are all marine and are able to move about like clams but have a stalk coming out near the hinge of the shell which they can use for a temporary attachment.

Phoronidea

This is a small group of about fifteen species of sessile, marine, wormlike animals that secrete a leathery tube in which they dwell. They have tentacles at the anterior end which they extend from the tube when feeding. The tentacles lie on the bottom of the water and capture small organisms or organic debris. They can withdraw into the tube in times of danger or when they are exposed to the air by the receding tide.

Chaetognatha

This phylum includes only a few genera of marine animals which feed upon small marine forms of life. They dart about through the water like arrows as they capture their food, and hence are given the common name of **arrow worms.**

Sipunculoidea

This is another phylum of marine worms, sometimes called **peanut worms** or **spoon worms** because of their shape. Some make a mucus-lined burrow in the sand and extend their tentacles. The beating cilia on the tentacles draw small organisms into the mucus, where they are trapped and ingested.

Priapuloidea

Only three species of this phylum are known. They are cylindrical, wormlike, marine animals living in the mud or sand of cold, shallow waters.

Echiuroidea

These are rather short, fat marine worms with a very large coelom and a very long proboscis. *Bonellia* is a genus which has become well known to biologists because of the extreme difference between the sexes and the method of sex determination. The female body is about the

FIG. 24.16 *Peripatus*, an animal which has characteristics both of the annelids and the arthropods.

size of a walnut, but with a forked proboscis which can be extended for several inches. The male is a minute ciliated form which lives in the reproductive tract of the female. The larvae are potentially of either sex. If they develop alone they become females, but if they come near a mature female they settle upon the proboscis and develop into males.

Onychophora

This phylum is of special interest because it shows a combination of annelid and arthropod characteristics. Hence, it is thought to show the relationship between these two groups. The best-known genus, *Peripatus*, is somewhat like a caterpillar in superficial appearance. It has a thin cuticle over the body, paired nephridia in each segment, a dorsal blood vessel, and ciliated reproductive organs. These are annelidlike characteristics not found in the arthropods. On the other hand, it is similar to the arthropods because it has claws on the appendages, tracheae for respiration, and blood sinuses.

REVIEW QUESTIONS AND PROBLEMS

1. Keepers of the greens on golf courses usually put down poison to kill the earthworms which may be in the soil beneath the putting surface, but gardeners actually buy earthworms to place in the soil. Why the difference?

2. The **typhlosole** increases the surface area of the earthworm intestine. Of what value is this increased surface area?

3. Suppose there were no perforations of the septa of the earthworm. What problems would this cause?

4. The earthworm can only extend its length or shorten its length, yet it can move either forward or backward by these movements. Explain.

5. Both earthworm blood and human blood are red, but under the microscope a distinct difference in the distribution of the hemoglobin can be observed. Explain.

6. If the posterior part of the earthworm is cut off, it can continue to move about and respond to stimuli for days, yet if a person loses a leg it will cease activity and response almost immediately. Explain.

7. Why are seminal receptacles present in earthworms and not in people?
8. How is *Neanthes* conspicuously different from the earthworm externally? Why do you suppose this difference exists?
9. **Ectoparasites** often have a temporary association with their hosts, but **endoparasites** usually remain with one host all their lives. Explain.
10. How are leeches adapted to their type of parasitic existence?
11. The octopus and the clam are quite different in appearance, yet both are in the same phylum. Show what characteristics they have in common which would cause them to be placed together.
12. Clams move very slowly, yet they are widely distributed in rivers. What feature of their life history makes this possible?
13. A general principle of evolution holds that an organ which is not used tends to degenerate. Illustrate this principle in the oyster and the starfish.
14. Suppose you want to induce pearl formation in oysters. Where would you place the material which will serve as a nucleus for a pearl, and why would it be placed there?
15. Snails actually breathe by gills, yet they can be land animals. Explain how this can be possible since gills can function only when they are very moist.
16. A slug has a shell as an embryo but not as an adult. Why is this fact of biological interest?
17. How does the squid use its ink?
18. The eye of the squid and the eye of man are similar in structure, yet the other body organs are very different. How might this correlation in eye structure have come about?
19. Why is *Chiton* of importance in the study of mollusks?
20. Why are the echinoderms considered to be more closely related to the chordates than are other invertebrate phyla?
21. The larva of the starfish has bilateral symmetry. Why is this of significance in classification studies?
22. How can a starfish open an oyster, a feat which often defies human efforts?
23. What unusual method does the sea cucumber have of defending itself?
24. What is unusual about the method of reproduction and sex determination of *Bonellia?*
25. Why is *Peripatus* of importance in establishing relationships of other phyla?

FURTHER READING

Barnes, R. D. 1968. *Invertebrate Zoology.* Philadelphia: W. B. Saunders.
Fingerman, M. 1969. *Animal Diversity.* New York: Holt, Rinehart and Winston.
Gardiner, M. S. 1972. *The Biology of Invertebrates.* New York: McGraw-Hill.
Mayr, E. 1969. *Principles of Systematic Zoology.* New York: McGraw-Hill.

25

Animals with Jointed Legs: The Arthropods

We commonly think of the vertebrate animals as the most successful and dominant of all the animal groups, but if the Arthropoda had the power of reasoning and could communicate their thoughts, they would probably claim this distinction—and they would have good arguments in support of their claim. They could point out that more than three-fourths of all the animals on the earth are arthropods. If success is measured in total number of representatives, there would be no question of their superiority. They could show that over half of the species of animals that have been named and classified are arthropods and that the phylum is very extensive, including crayfish, crabs, shrimp, spiders, scorpions, centipedes, and insects. If success is measured in the diversity of types within the phylum we could not dispute their claim. Arthropods are found everywhere that vertebrate animals are found and in many places where vertebrates cannot live. They may be found at the equator and in the arctic regions; they may be found flying high in the air and in the depths of the ocean; they inhabit the humid jungles and the dry sandy deserts. No extreme of environment seems to discourage them. The Great Salt Lake contains about 27 percent salt and is so dense that a man's body floats on it like a cork, yet it contains many little shrimp which spend their entire existence in this concentrated solution. When the point of adaptation to environment is brought up, we must concede the advantage to the arthropods. Finally, there is no other group of animals on the earth so well armed for offense and defense. It is true that man, as one of the vertebrates, has been able to forge terrible weapons of destruction, but the arthropods have forged from their own bodies weapons that are highly efficient and always ready for use. Of course, there would be some points in favor of the vertebrates, such as larger size and greater development of the brain and its functions. In the end, however, we would have to concede that the arthropods are at least the equal of the vertebrates in success as a group, although they have followed a different pattern of development.

Because of their great numbers and wide distribution, it is evident that this group of animals is of considerable economic importance to man. They are probably more important than any other phylum. On the harmful side, they are serious competitors for our food supply; we must constantly fight them in order to secure food for ourselves. They spread serious diseases which cause untold human suffering and death. Many of them have unpleasant bites and stings which can be serious. On the beneficial side, they provide us with quantities of food, medicine, clothing, and chemicals; they kill many of our enemies, including some of their own group; they cross-pollinate many of our important plants.

CHARACTERISTICS OF THE ARTHROPODS

The phylum name **Arthropoda** comes from Greek words meaning **"jointed feet,"** and this is one of the most distinguishing features of the group. The legs consist of a series of articulating joints which may be highly modified in form for functions such as swimming, respiration, and seizing food. Members of this phylum have **segmented bodies,** such as those found in the annelids, but the segments are not all similar to one another as they are in the annelids. They have an efficient **exoskeleton** of **chitin.** The

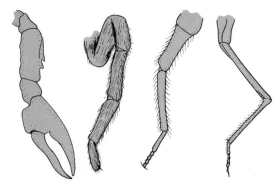

FIG. 25.1 Arthropod legs are characterized by their numerous joints. These are the legs of a crayfish, a spider, a cockroach, and a praying mantis.

term "skeleton" is commonly associated with an assortment of bones including a skull, ribs, vertebrae, and arm and leg bones, an **endoskeleton.** The outer covering of a grasshopper is just as much a skeleton, but it is found on the outside of the body. Both kinds of skeletons serve their purpose very well, but each has certain advantages and disadvantages.

The exoskeleton has a great advantage in protection. The soft body parts are found inside the skeleton and are much better protected than when some of them are exposed to injury on

FIG. 25.2 An exoskeleton of chitin is characteristic of the arthropods. The cast-off exoskeleton of a cicada is shown here. Note the slit down the back through which the insect crawled out.

the outside of the body. However, the endoskeleton has an important advantage in growth: the skeleton and the body parts can grow together; the skeleton places no limitation on the size of the animal. In contrast, the exoskeleton encloses the remainder of the body, and once it is formed, there can be no further increase in the animal's size. Some of the arthropods partially avoid this disadvantage by shedding their skeletons completely, in a process called **molting,** and then growing larger ones. This method has its drawbacks, however, and even with molting the body size is definitely limited. Since, in general, a large animal has an advantage over a small one, the animals with an endoskeleton enjoy a point of advantage. We may well be thankful that we have this advantage in size; without it we would not have a chance in competition with arthropods. At their present small size, they are still quite formidable.

CRUSTACEA

This class name means **"hard shell"** and refers to the hard, crustlike exoskeleton found on members of this group. The **Crustacea** are also characterized by a large number of paired appendages which typically are branched into

FIG. 25.3 The crayfish is a typical aquatic arthropod.

two terminal portions. They use **gills** for respiration, although a few of them live on the land and absorb their oxygen from damp air circulating over the gills. They have a body composed of many segments, as do all arthropods, but these are combined to form two or sometimes three main body parts. There are 20,000 species of Crustacea ranging in size from tiny microscopic water fleas to huge crabs that have a leg spread of 10 or 12 feet.

The Crayfish

We will study the crayfish, *Cambarus*, as a typical representative of this class as well as of the phylum as a whole. "Crayfish" is the name of this animal found in books, but the names "crawfish" and "crawdad" are common in speaking of them. They live in fresh water and seem to prefer ponds and sluggish streams where the bottom is somewhat muddy. Many a child has spent happy hours fishing for crawdads with a slice of bacon tied on a piece of string. The animal will grab the bacon with its pinchers and, with just the right degree of pull on the string, will hold on until it is out of the water. The abdomen, or "tail," contains edible white meat, and crayfish are often caught in large numbers for human consumption.

The appendages of the crayfish are diversified and meet a number of different needs, although in the embryo they all arise according to the same plan. We can think of the crayfish in its very early embryonic condition as resembling the clam worm, with each segment bearing a pair of appendages. The first four segments fuse to form the head and the next eight to form the thorax. The head and thorax are so closely connected that they form a single part of the body, the cephalothorax. The final six segments form the abdomen. Thus, there are eighteen segments to the whole body, but since there is an additional pair of appendages, the antennules, which are

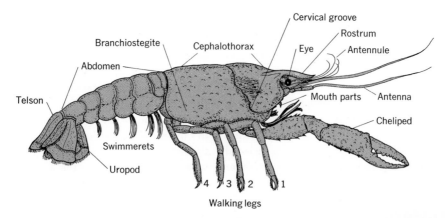

FIG. 25.4 External features of a crayfish, seen from a lateral view.

apparently not associated with a segment, there is a total of nineteen pairs of appendages. These include a pair of antennae; six pairs of mouth parts; a pair of **chelipeds** (pinchers), the first of the five pairs of walking legs; five pairs of **swimmerets** on the abdomen; and a final **uropod** at the tip of the abdomen.

The appendages of the crayfish illustrate a very important biological principle, **homology.** Body structures which arise in similar ways in the embryo are said to be **homologous.** Thus, we can say that the appendages of the crayfish are homologous to one another even though they become greatly modified and assume different functions in the adult. The concept applies not only to a series of similar structures on the body of one animal but to structures on different animals that arise in the same way embryologically. For instance, the arm of a man, the wing of a bird, the foreleg of a horse, and the pectoral fin of a fish all arise as a little outpocketing of the ectoderm and mesoderm at the same spot on each embryo. In the early embryo of each of these animals the structure would appear as a little projection without any indication of its ultimate form and function. Therefore, we can say that these four structures are homologous to one another. These all happen to have different func-

tions, but that need not necessarily be so. The arm of a man and the arm of a monkey have similar functions and are homologous also.

Another biological principle, **analogy,** bears a close relation to homology and the distinction between them should be made clear. Body structures are said to be **analogous** when they have the same function but a different embryonic background. For example, the wing of a bird and the wing of a housefly serve the same general purpose—flying—yet the two arise in entirely different ways in the embryo. The bird wing, as just mentioned, bears a definite relation —homology—to the front leg of other animals, yet the wing of the fly arises as a little balloonlike puffing out of the ectoderm on the back of the insect and bears no relation to its legs. Although these two structures have no embryonic relationship, they are used for the same purpose and so are analogous to each other.

The crayfish is a scavenger, eating bits of dead animals that it may find but also catching and eating live water animals that get within reach of its powerful pinchers. After seizing the food, the crayfish uses its pinchers to cut up and carry the bits of food to the mouth. The food is crushed into finer pieces by the mandibles and passed through the short **esophagus** into the

stomach. The anterior, or **cardiac portion,** of the stomach is a storage chamber and is followed by a **gizzard** lined with teeth made of chitin that grind and pulverize the food. The residue is then passed into the posterior, or **pyloric portion,** of the stomach, where it is digested by enzymes from a pair of large digestive glands and then passed on into the **intestine.**

The respiratory organs consist of a series of **gills** that lie in the branchial chambers on either side of the thoracic region of the body. The chambers are formed by an overlapping portion of the exoskeleton, the **branchiostegite.** Water is kept flowing through the chambers by the action of the **bailer,** which is continually bailing the water out at the anterior end as it flows in at the posterior end. This keeps the gills bathed in fresh water as oxygen is absorbed and carbon dioxide is given up. The process is furthered by movements of the gills themselves; most of them are attached to appendages, and when the appendages are moved on the outside, the gills are moved in the chambers.

Circulation in the crayfish is of the **open type system,** where blood flows through the sinuses in the body, in contrast to the closed type system, where the blood does not leave the blood vessels in its circuit around the body. The earthworm and vertebrates have the closed type system, whereas the freshwater clam has a combination of both. The **heart** lies just dorsal to the anterior portion of the intestine. It connects to **arteries,** which lead from it in all directions; there are no veins. Blood is taken into the heart from the surrounding **pericardial sinus** through three pairs of openings in the heart, the **ostia.** When the heart expands, blood flows in through the ostia; when the heart contracts, valves close the ostia and the blood cannot pass out through them but flows out into the arteries instead. The main arteries form smaller branches which empty the blood into the body sinuses. The most important of these is the **sternal sinus,** from which the blood flows out into the gills and then flows back into the pericardial sinus to repeat the circuit of the body. The blood contains corpuscles but is clear because there is no hemoglobin. **Hemocyanin,** an oxygen-absorbing substance in the plasma, is clear in the body but turns blue when removed and allowed to stand awhile.

The body of the crayfish is liberally supplied with muscles which move the body and its appendages. The crayfish can crawl along rather slowly in any direction, using its walking legs, but it can also dart backward for several feet so rapidly that its movements can hardly be followed. It is able to do this because the greater portion of the abdomen is filled with a strong **flexor muscle.** When the fanlike portion at the rear of the body is suddenly pulled forward

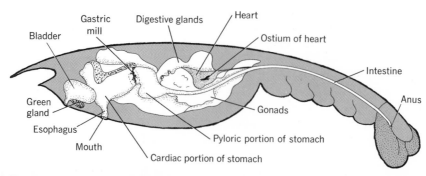

FIG. 25.5 The internal organs of a crayfish.

against the water by the flexing of this muscle, the animal is jerked backward at great speed. A smaller **extensor muscle** lies on top of the flexor and extends the abdomen.

The **nervous system** is very much like that found in the annelids. The **brain** is in the dorsal part of the head region. Two **circumesophagal connectives** circle the esophagus and join ventrally to form the **nerve cord.** There are enlargements of the nerve cord, **ganglia,** located in the different segments, and the **sensory** and **motor** nerves are given off from the ganglia.

The sense organs include a pair of stalked **compound eyes** on the first body segment. The eyes are called compound because each is composed of many individual facets; there are about 2500 single facets in each. A single eye is just a light-sensitive spot, but the combined mosaic image produced by the entire group gives something approximating the image produced by a lens of a higher animal's single eye.

Sense organs of **equilibrium** are found in the base of the antennules. One of these consists of a small sac that contains grains of sand which shift around as the crayfish changes its position. These shifting grains stimulate delicate nerve endings in the sac and the crayfish is able to adjust its position so as to maintain its equilibrium. One research worker confused the crayfish by substituting iron filings for the sand grains in these organs. When a magnet that was placed above the animal caused the iron filings to be drawn to the top of the sacs, the crayfish reacted as if it were upside down and turned over on its back and remained there as long as the magnet was working.

The appendages and other parts of the body bear small sensory bristles of two kinds. One group is sensitive to **touch** and another is sensitive to chemicals and might be called sense organs of **taste.**

The male reproductive organs consist of a pair of **testes** lying under the heart and a **vas deferens** leading from each to an opening in the base of the fifth walking leg. The sperm are removed from this opening and transferred to the female by the first two pairs of swimmerets, which are modified in the male. The female system consists of the paired **ovaries** and the **oviduct** leading from each to an opening in the base of the third walking leg. There is also a **seminal receptacle** on the ventral surface of the body between the fifth walking legs.

During the breeding season, which is usually in the fall, the male approaches the female and a struggle follows, with the female apparently making a great effort to resist his advances. However, her resistance finally weakens and the male throws her over on her back. As he stands over her, sperm flow from the openings of the vasa deferentia and are guided into the seminal receptacle of the female by the first two pair of swimmerets. The sperm are kept here through the winter months and fertilize the eggs in the spring. When the proper time arrives for egg-laying, the female first secretes an adhesive substance from the basal portion of the uropod. This is spread over the swimmerets. Then the crayfish turns on her back and lays the eggs, which are passed backward over the seminal receptacle and receive the sperm which were deposited there. They are then guided back and stuck to the swimmerets by the adhesive that covers them. During the embryonic development of the young crayfish within the eggs, the female waves the swimmerets back and forth in the water, keeping the embryos well supplied with oxygen. Upon hatching several months later, the young crayfish cling to the swimmerets of the mother until after their second molt.

Regeneration in the crayfish is limited to the appendages and eyes. If a part of an appendage is injured, the entire appendage may be cast off at a certain joint and replaced by regeneration. Its effect in the crayfish seems to be to prevent excessive bleeding from crushed and mutilated portions of the appendage, because little bleeding occurs from a break at a joint.

Other Crustaceans

Among the other crustaceans, the **lobster** is probably the most important economically because it is a source of delicious food. It is just an enlarged saltwater edition of the crayfish, and the body organs and habits are very similar in the two.

The **crabs** are also crustaceans and have appendages of the cephalothorax very much like the crayfish, but the abdomen is greatly reduced and folded up under the cephalothorax. There are many varieties of crabs. The edible or blue crab is an important source of food in the United States. Unlike the lobster, the main muscles consumed as food are located in the cephalothorax and are used to control the powerful pinchers. The fiddler crab has one pincher larger than the other in the males and the two are held in a position somewhat like a bow and "fiddle." The hermit crab backs into a deserted mollusk shell and carries it around on its back. The body has become soft, allowing it to curve around inside the shell. The little sand crabs come out on the sand of the beach and can run very fast over loose sand, traveling sideways for better traction.

FIG. 25.6 Pill bugs are crustaceans which roll themselves into balls resembling pills when they are disturbed.

They are so common on some beaches that a sun bather stretching out on the sand must be careful or he may receive a pinch from a sand crab beneath him.

Shrimp cocktails as a dinner appetizer have become standard fare in all parts of the United States. The catching and shipping of shrimp constitute a major industry, especially in the coastal regions of Florida and Louisiana.

The **fairy shrimp** are beautiful little crustaceans often found in great numbers in small freshwater ponds in the spring. They swim on their backs by undulating movements of the appendages, a charming sight. They get their name from the fact that they seem to have power to make themselves disappear in the summer when the ponds dry up and reappear in the following spring. The adults actually die when the ponds dry in the spring, but the previously laid eggs live through the dry summer and are ready to hatch when the rains and warmth of spring return.

Barnacles are crustaceans that bear some resemblance to mollusks. They develop a shell and become permanently attached to some solid support as adults. The young are free-swimming but soon pick a rock, wharf, ship bottom, or similar underwater structure for attachment. They are a great nuisance to shipping; ships often have to be hauled into drydock to have the barnacles scraped from them. An accumulation of barnacles greatly reduces the speed of a ship.

Water fleas are only about one-tenth inch long and do not bite, as the land fleas do. It might be thought that they do not have much possible economic importance, but they breed prolifically and furnish an important source of food for many water animals, including fish which are caught for human consumption. In size and general body shape they resemble fleas, but their body structure is entirely different.

Cyclops was a huge giant with a single eye in the center of his forehead, according to Greek mythology, but according to the zoology textbook the cyclops is just another small crustacean about the size of the water flea. It certainly does not resemble the mythical Cyclops in size, but it does have a single eye in the center of the head region.

The **pill bug,** or sow bug, is another crustacean. It differs from the others studied in that it spends its life on the land. However, like all crustaceans it breathes by gills. It is able to exist on the land by coming out only when the air is moist, which is usually at night. During the day it gets under rocks, logs, or similar structures where the humidity is high and the air cool. Its name is derived from the habit of rolling into a tight little ball when disturbed, which makes it look like a little pill. Sometimes pill bugs are destructive to garden plants, and poison must be put out to kill them.

ARACHNIDS

The arachnids include a motley group of animals, apparently with little in common— spiders, scorpions, mites, ticks, and king crabs. However, they have a number of like characteristics that cause them to be grouped together in one class. They all have two main body parts, the **cephalothorax** and **abdomen,** and four pairs of **walking legs,** which are on the cephalothorax. They have no compound eyes and no chewing jaws.

Spiders are thoroughly disliked by many persons. This aversion is largely unjustified, because most of them are beneficial. True, some of them do occasionally bite, and there are a few with a serious bite, but they greatly benefit us by destroying many harmful insects, and they bite only in self-defense.

The cephalothorax of the spider bears eight simple **eyes** on the anterior dorsal surface. These are not well-developed eyes, and it is doubtful if the spiders can see well for more than 5 inches. A pair of **fangs** at the anterior end of the ventral surface of the cephalothorax is connected internally with poison sacs. The fangs are hollow, with an opening near the end like a hypodermic needle. When the spider bites, the poison sacs contract and inject a small amount of poison into the wound. Just back of the fangs is a pair of appendages, the **pedipalps,** which look like a miniature pair of legs. These are mostly sensory in function but also serve as copulatory organs in the male and help to squeeze the juice out of an insect in feeding. The abdomen is unsegmented and bears three pairs of **spinnerets** on the ventral surface. These are sometimes mistaken for stingers, but no spiders can sting and these organs are used only for spinning the web. Spiders can breathe air, for there is a pair of **book lungs** in the anterior portion of the abdomen. Each lung consists of a single sac with an external opening, through which air is taken in, and fifteen to twenty folds within, which resemble the leaves of a book. Gaseous exchange takes place as the air circulates over these leaves.

The **web** plays a vital part in the lives of the spiders, and its many uses make an interesting study. Inside the abdomen are several silk glands that secrete a viscid fluid from which the web is formed. This liquid is forced out of the spinnerets under rather high pressure and hardens almost instantly on contact with the air to form the web. The principle is the same as that of manufacturing rayon and nylon thread— forcing a viscid liquid through tiny holes to harden in the air. One of the best known uses of the web is as a trap for insects, which provide most of the food for spiders. Funnel-shaped and orb-shaped webs are the most common types of traps used by the spiders. No sight is more beautiful than the geometrically arranged web of the orb weaver sparkling with dewdrops in the early morning sun. When an insect touches the web, the vibrations attract the spider and it rushes out and begins squirting sticky web at its prey. Soon the insect will be helplessly trussed up, awaiting its doom. The spider may hang it up for future use or may eat it at the time. Its mouth parts do not permit the spider to chew an insect up and swallow it. Instead the spider will bite a hole in the body and inject digestive enzymes which digest the soft internal body parts. Then the liquefied digesting food may be sucked up into the body of the spider. The hard exoskeleton of the insect is not digested and is cast aside after feeding. It is often possible to find a large number of these empty skeletons underneath the web of a spider.

Not all spiders use a web to trap their prey. Some, like the wolf spiders, stalk and pounce on their prey like a cat. These spiders can produce web, however, and use it to form a

FIG. 25.7 Black widow spider. The red hour glass on the ventral surface of the abdomen readily identifies the black widow, our most dangerous spider.

soft and dry lining to their homes, which are usually holes in the ground.

Spiders also use their web as a convenient means of transportation. A spider on the ceiling can lower itself to the floor by attaching the web to the ceiling and spinning out the web as it comes down. Should it want to go back up, it can wind the web in and go up very rapidly. It can move from one tree top to another by spinning a little tuft of sticky web and throwing it out in the breeze while spinning a single strand attached to it like a boy flying a kite. When the tuft hits a solid object it will stick and the spider has a cable on which to cross. This accounts for the many single strands of web which lie across woodland paths.

There are two dangerous spiders in the United States. The **brown recluse** is a newcomer, but more common is the **black widow,** one with large poison sacs which extend far down into the body and can inject a relatively large amount of poison into its victim. Quantity, coupled with the fact that the poison is of high potency, accounts for its serious effects on the human body. Bites are rare, considering the wide distribution and abundance of this spider, because the black widow is rather shy and retiring. However, a female protecting an egg case will bite readily when disturbed; most bites are received in this way. The poison is a **neurotoxic** type—a nerve poison—and causes severe systemic reactions for several days after the bite. Some of the symptoms are pain in the muscles, difficulty in breathing, nausea, mental confusion, and general retardation of the body functions. The bite is fatal only in rare instances.

Reproduction among the black widows may be studied as typical for the group. Spiders are ordinarily solitary animals; that is, they do not live in social groups. There is a good reason for this lack of sociability; spiders are cannibalistic and often prey on one another when they are confined to close quarters. For reproduction, however, there must be some association, and

over again, each time getting a little nearer to the female. In preparation for this occasion the male has already removed a ball of sperms from his reproductive organs and is carrying it in his pedipalps. After repeated approaches in which the female does not attempt to capture him, the male finally approaches close enough to tuck the ball of sperm into her seminal receptacle. Then he turns and rushes away, but this time the female pursues him and may overtake and eat him. The black widow gets her name from this behavior and her black color. Later she will lay the eggs and spin a web around them to form an egg case.

The little spiders hatch within the case and soon break out and climb on their mother's back. Female spiders are frequently seen with their backs covered with tiny spiders clinging to the hairs. Distribution among spiders is an important thing. Here, again, the web comes in handy. The little spiders climb the nearest tree or other elevation and each spins a little tuft of web which it throws out into the breeze like a parachute. In this way it can be carried a considerable distance.

Tarantulas are often thought of as very poisonous, but experiences indicate that their

FIG. 25.8 Male and female black widow spiders. The small male is usually eaten by the female after copulation.

the cannibalistic tendencies are repressed temporarily at mating time. A female ready for reproduction will sit in the center of her web and wait for a male. Some studies show that she casts out single strands of web which are carried a considerable distance by the wind. When a male touches one of these strands he picks up some sort of signal which indicates that the female is ready for insemination. He can follow the strand of web and it will lead him to her. The male is much smaller than the female and he approaches her with great caution. He will come in slowly and then turn and run away, only to return and repeat the process over and

FIG. 25.9 A tarantula, the largest of the spiders, does not spin a web, but stalks its prey and pounces on it. Tarantulas are greatly feared by many people, but their bite is not serious.

bites are painful, but not dangerous like those of the black widows. Tarantulas are large and hairy and are jumping spiders that pounce on the prey which they find in their nocturnal wanderings.

The **scorpion** bears a formidable sting on the tip of its long curved "tail" which makes it highly respected by anyone who has been stung by one. Stings may occur frequently when scorpions are around the house, for they come out at night in search of food and stay in some secluded place in the daytime. The author has found that their hiding place may be the inside of a man's trousers. They may make their presence suddenly and painfully known when the trousers are put on. They have an interesting courtship; before mating they grab each other's claws and do a dance, the "dance à deux." Holding claws is probably not so much a sign of affection as it is of mutual distrust, since scorpions are cannibalistic. The eggs are retained in the body of the female until they hatch and the young are brought forth live. These clamber on the mother's back and ride around awhile before departing from the family group.

The **daddy longlegs** look like spiders but have segments to their abdomens and must therefore be placed in a different group. Their very

FIG. 25.10 The scorpion has a powerful sting on the tip of its long curved abdomen. The sting is very painful.

long legs seem to be a disadvantage since they move rather awkwardly and readily pull off if grasped. Daddy longlegs are also called **harvestmen** because they are so abundant at harvest time.

Mites have a name that is a byword for minuteness, but they cause trouble out of all proportion to their small size. **Itch mites** burrow into the skin and the females lay eggs as they go. When these eggs hatch and the young mites begin burrowing in all directions, their host has the "itch." The mites are too small to be seen with the naked eye, but if scrapings are taken from the infected skin and placed under the microscope, the small arachnids will be seen. The term "seven year itch," which is often given to this infection, indicates that it is not easy to cure. The mange of cats, dogs, and other domestic animals is caused by similar mites that burrow around in the skin.

The tiny **chiggers,** or red bugs, are mites that may be so abundant in the southern and north central parts of the United States that a person will look as if he has the measles after walking through grass in the spring or early summer. These mites attach themselves to the skin and inject a digestive fluid which digests a little burrow through which they feed. This fluid is quite irritating to the skin of some persons. It causes large, raised, red blotches, and an almost unbearable itching. Chandler quotes a description of their effects that can hardly be improved upon. The chigger is a "small thing, but mighty; a torturer—a murderer of sleep; the tormenter of entomologists, botanists, and others who encroach on its domains; not that it bites or stings —it does neither; worse than either, it just tickles." Considerable discomfort may be avoided by dusting sulfur on the legs before going out where chiggers are abundant. Sulfur may also be dusted on the lawn to rid the grass of these pests.

The **ticks** are blood-sucking arachnids that not only cause inconvenience to their hosts but also spread serious diseases. When a young tick hatches, it will climb the nearest bush and begin a vigil that ends either in starvation or a full meal from some passing animal. When there is the slightest rustle of the leaves of the surrounding bushes that might indicate the approach of a possible host, the little tick waves its legs frantically in the air. If a suitable host gets close enough the tick grasps it, crawls around on its skin for a while, pierces the skin, and begins sucking. It buries its head in the wound and holds so tightly that the body may be pulled off without releasing the hold. After gorging itself for several days the young tick will drop off on the ground, rest for a week or so, molt, climb another bush and again take up its vigil. This may be repeated several times during the life of the tick, and disease organisms ingested along with the blood of one host may be transferred to a later host. Texas cattle fever is a serious disease spread by ticks, and expensive dipping of cattle to eradicate ticks is necessary in many cattle-raising regions of the country. Rocky Mountain spotted fever is a serious human disease spread through bites from ticks which have fed on infected rodents. The disease of tularemia may also be spread by ticks.

The **king crab,** or horseshoe crab, is an arachnid that lives in salt water and is not a true crab. It is an exception to the rule in that it has five pairs of legs rather than the four pairs commonly found in the arachnids; but it breathes by book gills and has other arachnid characteristics that place it in this class.

CHILOPODA

The name **centipede** means "hundred legs," but the actual number may range from 30 to more than 400 in different species of centipedes. They have a long, loosely jointed body with a pair of legs on each segment except the first and the last. The first segment bears a pair of vicious looking fangs that are connected with poison sacs and can inflict a very painful bite. Specimens from the southwestern United States are often as long as 8 inches, and tropical forms grow to more than a foot in length. However, most of them are smaller in the United States and are beneficial because of the large number of insects which they eat. One species, the house centipede, is common in human dwellings. It is only about an inch long and has extremely long

A

B

FIG. 25.11 A centipede (A); and a millipede (B).

legs, even longer than the body itself near the posterior end. Like other centipedes, it comes out at night in search of food. Its presence is often unknown until one is discovered in the bathtub some morning where it ventured and could not climb the slick sides to get out. Since it eats cockroaches, bedbugs, crickets, and other household pests, it is quite beneficial and should not be destroyed.

DIPLOPODA

The **millipedes,** "thousand-legged" worms, have so many legs that it would seem difficult for them to move them all in a coordinated manner, but they manage slowly. Like the centipede, the body is divided into segments, but there are two pairs of legs per segment rather than one. Since one pair of appendages per segment is the usual maximum among arthropods, this condition would be hard to explain were it not for the embryonic development. Developing millipedes have only one pair of legs per segment, but later the segments fuse in pairs so that the adults show only one segment for each two embryonic segments. However, the two pairs of legs are retained. Millipedes are harmless; they have no fangs or poison sacs.

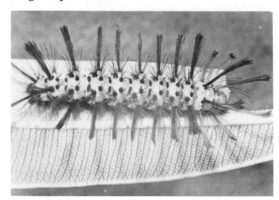

FIG. 25.12 Insects are extremely important competitors for the world's food supply. This caterpillar is methodically eating a leaf.

INSECTA

Insects have three distinct body parts—**head, thorax,** and **abdomen.** Embryonic, or larval, insects have regular body segments that are similar to one another and resemble the annelid plan of body organization. Insect larvae are so similar to annelids that they are often called worms; a caterpillar or a maggot looks more like a worm than like an adult insect. However, during typical insect development the anterior six segments fuse to form the head, the next three form the thorax, and the remainder of the segments form the abdomen. The dividing lines between the segments cannot be seen in the head, are visible as indentations in the chitin surrounding the thorax, and remain distinct with independent movement in the abdomen.

Insects have **three pairs of legs.** This is the most easily recognized characteristic of the insects because no other group of animals has six legs. There is one pair attached to the ventral surface of each segment of the thorax.

Insects breathe through the **spiracles.** These are tiny breathing pores found on the side of the abdomen, one pair on each segment, and usually two additional pairs on the posterior thoracic segments. They connect internally with tiny tubes, **tracheae,** which carry the air directly to the parts of the body where needed. They distribute oxygen much more efficiently than can be done with lung or gill respiration and help give insects the energy which enables them to do so much in spite of their small size. However, this method of respiration makes them highly susceptible to foreign substances in the air, for they are distributed over the body. We take advantage of this by spreading insecticides in the form of sprays or dusts. A beekeeper can make a hive of bees quite docile by blowing a few puffs of smoke in at the top of the hive. This causes the bees to settle to the bottom, allowing him to remove the honey from the top. Chickens can often be seen "delousing" themselves by sitting in a nest of fine dust and working it through their feathers. The lice are killed by a clogging of their spiracles by the fine dust particles.

Insects have **compound eyes.** All but a very few primitive insects have this type of vision, which was described in connection with the vision of the crayfish. Some species of insects have larger compound eyes with a correspondingly greater number of individual facets than others, and accuracy of vision seems to depend on the number of facets. The eyes tend to curve around the head and there are some facets pointing in almost every direction, so it is quite difficult to approach an insect without being seen by some of these facets. About the only way a person can approach a fly without being seen is from below and to the rear, but it is a little difficult to get in this position.

Insects usually have **two pairs of wings,** one pair attached to each of the two posterior segments of the thorax. There are a few insects, however, that have no wings at all and others with only one pair, so this is not a universal characteristic of the group.

Economic Importance of Insects

Insects, although small individually, are certainly large in aggregate. They are abundant almost everywhere on earth; there are probably more of them in your backyard than there are people in your city. They seem to have a persistence and aggressiveness that are unequaled in any other animal group. Think of a mosquito buzzing around your head. No amount of slapping and swatting discourages it; it will persist for hours if necessary until either a full meal or death rewards its efforts. Insects have abilities that challenge the best that man has to offer. They were making paper from wood pulp ages before man was chiseling crude characters on stone tablets. They domesticated animals and cultivated plants long before man had the idea. They have highly developed social organizations that seem to function more smoothly than the best that man has devised. They have a strength all out of proportion to their small size; an ant commonly carries loads several times its own size and weight, and a flea easily jumps a hundred times its own body length. We could continue this list of accomplishments to a great length. Insects are certainly not a group to be belittled in spite of their small size.

The insects include some of man's greatest enemies. They destroy our food plants. It is hard to find a peach or an apple without a "worm" in it unless the tree was carefully sprayed during the time the fruit was developing. Ears of fresh corn bought at the market may have "worms" that have eaten some of the tender kernels at the end. Grasshoppers destroy and damage many different kinds of crops. Even after a crop is harvested, it is by no means exempt from insect damage; large quantities of stored grain, fruit, and other foods are destroyed each year. Insects are a nuisance with their bites and stings, and some of the most serious infectious diseases of man are spread by insects. Our clothing is destroyed by insects; furs, woolens, mohair, and feathers furnish food for the larvae of the clothes moth. Some insects do not wait for the material to be made into clothing before destroying it; the cotton boll weevil may take as much as one-fourth of an entire potential cotton crop each year. Books and valuable papers are injured and destroyed by the silverfish; termites destroy all types of wooden construction and may destroy books and other stored paper products. We spend millions of dollars each year in an attempt to control the destructive insects.

After such a survey of only a few of the harmful effects of insects, we may sometimes wish that there were no such things on the earth. Yet some of the insects are of such great economic value that this would be a very different world to live in if they were eradicated. Most of our fruits, many vegetables, and many of our grains would

FIG. 25.13 Bumblebee on red clover. Many insects are highly beneficial to man. As they visit flowers to obtain nectar, they transfer pollen from one flower to another. Without cross-pollination, many of our important food plants would disappear.

disappear with the insects. There would be no cotton or silk; most of the land birds and most of the freshwater fishes would die of starvation. Many commercial products, such as dyes and hair tonics, would disappear from the market, for certain insects form important ingredients in such products. Many people of the world would be without a direct source of food. Fried ants and grasshoppers are widely eaten in India; fried caterpillars are sold like hot dogs on the streets of some cities in Mexico; certain African natives eat the large tropical termites with great relish.

Metamorphosis

We think it natural that newly born or newly hatched offspring should closely resemble their parents, except for size and that they should gradually grow into adult size. This is what happens among most of the larger animals. It is by no means what always happens among insects,

however; many young insects are so different from their parents that they would be placed in entirely different phyla if their entire life histories were not known. Only through long observation do we learn that a green, wormlike caterpillar can undergo transformation into a beautiful butterfly. Such a thoroughgoing change from the young to the adult type is called metamorphosis.

Insect metamorphosis may be divided into three major types. (1) In **complete metamorphosis,** the individual starts as a fertilized **egg** which develops into a **larva,** usually wormlike. Common names of insect larvae indicate their similarity to worms: bagworms, grubworms, silkworms, measuring worms, and fuzzy worms are all insect larvae. Insect larvae have voracious appetites and eat nearly all the time. The mature females instinctively lay their eggs near a plentiful supply of the proper kind of food, so the newly hatched larvae have no difficulty finding it. Their gluttonous feeding habits cause the larvae to grow very rapidly; a fly maggot will double its size in a few hours and increase its size several hundred times in a few days. A single tomato worm may eat half of a large tomato in a day and increase its size accordingly.

Finally, when the larva seems to have eaten its fill, it goes into a quiescent stage called the **pupa.** It may spin threads of silk around itself to form a cocoon in which it spends its pupal state, or its outer body covering may simply harden to form a pupa case, or both may take place. The metamorphosis into the adult condition comes about within this covering, and a fully developed **adult** emerges as the fourth and final stage in the development of an insect with complete metamorphosis. The adult is as large as it will ever be when it emerges; big flies do not grow from little flies; they are big or little flies when they leave the pupa case and retain their size throughout their lives.

FIG. 25.14 Complete metamorphosis. The monarch butterfly undergoes complete metamorphosis. The egg hatches into a caterpillar which becomes a pupa and then an adult emerges.

(2) **Gradual metamorphosis** may be illustrated by the grasshopper. The egg hatches into something that resembles the adult, and this must undergo a series of gradual changes before it becomes the adult type. The immature stages of such metamorphosis are called **nymphs.** A newly hatched grasshopper nymph can be recognized as a grasshopper—it feeds in the same way as the adult and lives in the same type of environment —but its body proportions are somewhat different. It has no wings, its body is proportionately shorter, and its hind legs are not as well suited to jumping as those of the adult. It increases in size and becomes like the adult through a series of **molts.** When a nymph is ready to molt, a fluid will form between the body and the skeleton. As the two separate, the skeleton splits down the back, letting the insect emerge. It has a soft body covering at this time and takes in air, which stretches it while the skeleton is hardening. When the new skeleton hardens, which may be within an hour, the insect may be twice its original size, and it seems impossible that it could have come from the cast-off exoskeleton. The grasshopper goes through five molts, but other insects with gradual metamorphosis may have as many as twelve. The **adult** emerges at the final molting.

(3) **Incomplete metamorphosis** is similar to gradual metamorphosis in that the adult is produced after a series of molts of immature stages. The primary difference is that the immature insects with incomplete metamorphosis live in different conditions and feed differently from their parents. They usually have special body parts, not possessed by their parents, adapting them to this different type of life. In the immature stages, they live in the water and are called **naiads.** The dragonfly is a good example of this type of metamorphosis. The adult flies in the air, catching other insects for food. The eggs are laid in the water and hatch into naiads that roughly resemble the adult, but have a very long tongue which they use to catch small water animals for food. Thus, the method of feeding is quite differ-

ent from that of the adult. Respiration is also different; the adult breathes air by means of spiracles, but the naiad of the dragonfly sucks water into the rectum and expels it through the anus. From this water it absorbs oxygen. When ready for its final molt the naiad crawls up out of the water and emerges with fully formed wings, ready to take up its life in the air.

(4) Finally, there are a few primitive insects that have **no metamorphosis.** In these the egg hatches directly into the *adult* form without intervening stages of development. The silverfish and some of the lice are examples of this method.

Wings

The wings of insects show great variation, and wing structure is a frequent basis for subdividing the class into orders. A few insects, such as the fleas and fishmoths, have no wings at all. Others, such as flies, have only one pair, but the second pair is represented by a rudimentary projection at the place where the second pair should be. Some, such as the aphids, have generations without any wings at all, but eventually produce a winged sexual generation that flies away and spreads the species. The majority of insects, however, have two pairs of wings as adults.

Wings which are used for flying are thin and light, with delicate veins running through them. These are **membranous wings.** In some insects, such as the beetles, the front pair of wings has been greatly thickened and hardened to form **horny wings** used as wing covers to protect the

FIG. 25.15 Incomplete metamorphosis. The naiad of the dragonfly undergoes a series of molts as it increases in size and becomes more like the adult. The dragonfly naiad lives in the water and feeds in a different manner than the adult. The lower photographs show the mouth parts of the naiad and those of the adult. Both mouth parts are adapted to their feeding habits.

A

B

C

D

delicate second pair of membranous wings which are used in flying. During flight the horny front pair of wings is held out to the side of the body like the wings of an airplane. In other insects, such as the grasshopper, the two front wings are tough, yet flexible, and they are called **leather-like wings.** They also act as wing covers and are held rigid in flying. Some insects, like the butter-flies, have membranous wings covered with scales which rub off like a fine powder when they are handled. Some, such as bugs, have a leatherlike front part of a wing and a membranous hind part. With so many variations it is often possible to tell the order of an insect by glancing at its wings.

Food and Feeding Habits

Almost every conceivable kind of organic matter furnishes food to one or another species of insect. A small pile of feces dropped from the body of a cow, after the cow has extracted all possible nourishment from it, provides food for hundreds of maggots to live and grow into flies. An inch of rainwater standing in a tin can may contain organic matter that furnishes all the nourishment needed for dozens of mosquito wigglers to live and grow. Even solid, dried wood, dry feathers, paper, and starched clothing may contain sufficient food for some insects to live on for their entire lives. There seems to be almost nothing that has an organic origin that the insects have overlooked.

Some insects are **restricted** in their feeding habits and can eat only one or a small group of foods. Termites eat only wood or cellulose products; silkworms stick to mulberry or similar leaves; the cotton boll weevil restricts itself to the cotton square or boll; fleas must have the blood of a mammal; certain wasp larvae live only on spiders. Other insects are **generalized** and can have a varied diet. The housefly is certainly not particular about what it eats: it may have an appetizer from the body of a dead dog, an entree from the fried chicken on your dining table, and dessert from a pile of manure. Grasshoppers certainly do not restrict themselves to grass; they may damage almost any kind of crop, as farmers have learned from experience.

The **mouth parts** of insects are highly specialized to accommodate the kind of food eaten and the method of obtaining it. In general, there are two types—**biting** and **sucking** mouth parts. Insects with biting mouth parts have hard chitinous jaws that work from side to side to bite off and crush the food as it is taken into the mouth. In the other group the mouth parts are fused, forming a proboscis through which liquid food may be drawn. This is often sharp on the end and may be thrust through the skin of animals to suck blood, or through the epidermis of leaves to suck the plant sap. In attempting to control insect pests of plants, it is important to know the kind of mouth parts that the insects possess. A poisonous spray put on the leaves will kill those with biting mouth parts, because they will eat the poison along with the leaves. But those with sucking mouth parts will not be affected, since they stick their proboscis right through the poison and suck the unpoisoned sap beneath. A contact spray that will be absorbed by the bodies of these insects then becomes necessary for their control.

Protective Attributes

All animals have enemies, and if they survive as a group they must develop means of protection against these enemies. Insects have many methods of protecting themselves. The body may have **protective coatings** that discourage predators. The hard exoskeleton of many insects in the adult stage is quite effective. Such a protection is often lacking in the fast-growing larvae, but they may live in a protected environment. Some, such as the caterpillars, that must live in an exposed environment have developed prickly hairs that will pierce the mouth of any animal that attempts to molest them. A few even have poison on the hairs and can give quite a sting to a person that handles them. Spittle bugs suck plant juice, expel part of it from the posterior end of the body, and beat it into a froth with which they cover the entire body for protection.

Locomotion is a means of protection to some insects. By rapid flight and maneuverability they elude their more sluggish enemies. Some jump or run to escape.

Other insects protect themselves by **combat;** they fight with a ferocity and aggressiveness that is unexcelled in the animal kingdom. Some bite with their jaws and others have a powerful sting on the tip of the abdomen. There is an ant that can spray its poison from the sting so that intruders are greeted with a cloud of poisonous vapor when they come too close.

Some insects have **odors** and **tastes** that protect them from predators. These may leave an odor that remains for some time on any object they touch, and often a delicious looking berry will have a most disagreeable taste from its prior contact with a stink bug. The monarch butterfly probably looks as if it would taste good to a bird, but a bird will attempt to eat only one; after that, other monarchs do not tempt him. The blister beetles have blood that will irritate and raise a blister on the skin of any animal that crushes them.

Protective mimicry is another very effective means by which insects escape detection by their enemies. They develop body shapes and colors which resemble their surroundings so closely that detection by sight is very difficult. A tomato worm may be on a tomato plant and provide evidence of its presence by the damage done by its feeding, but a careful examination of the plant may fail to reveal a worm even though it is 2 or 3 inches long. Some butterflies look so much like leaves that they are indistinguishable from them when at rest. Walking sticks look very much like twigs and assume poses identical with the appearance of the twigs that they are on. Other forms of life also show protective mimicry, but it seems to reach its height of development in the insects.

Insect Voices

When we hear the loud singing of the cicada, the chirping of the cricket, and the buzzing of the bees, we might get the impression that insects are a noisy group of animals, but the great majority of them are mute. None of them have any organs like vocal cords which vibrate when air is expelled from their bodies. The few that do make sounds must do so by other means. A common method is scraping together the wing covers or legs. Male crickets scrape their wing covers to produce their singing and strum them to produce the chirping. It is said that the timing of the chirps of a tree cricket is in direct relation to temperature and that the temperature can be ascertained by counting the number of chirps per minute. The katydids also use their wing covers and produce a sound resembling their name. During the mating season, male grasshoppers fly in the air above the females clicking their wing covers together like castanets to produce a very characteristic sound.

The cicadas, also called locusts or dog-day harvest flies, climb trees in the summer and produce a lonesome, continuous singing that can be heard a great distance. These insects have a cavity formed by a part of the thoracic exoskeleton that extends back over the abdomen. Within this cavity is a membrane that can be stretched and relaxed to make it vibrate at a variable pitch, and the cavity acts as a resonance chamber to amplify and reflect sound.

The humming of insects is an incidental sound accompanying the vibration of the wings in flying, but should be mentioned under insect voices because it is sometimes used as a means of communication. The pitch of the hum depends on the rate of vibration of the wings. A housefly hums the note of F in the middle octave of a piano. When this piano key is struck, the piano string vibrates 345 times per second. That means that the housefly's wings must beat up and down 345 times per second. This has actually been checked by allowing the fly's wings to beat against a rapidly revolving smoked drum and counting the strokes made in one second. The humming of bees varies with the temper of the hive, and a beekeeper can tell by the tone of the hum whether a hive is going about its business or is possibly ready to swarm. Recent research shows that mosquitoes use the hum of their wings as a means of communication. Greatly amplified recordings have been made which show that they have danger signals, mating calls, and numerous other sounds with a definite meaning.

Reproduction

Reproduction among the insects is highly specialized. Many of the females attract the males by their odor. Moths are active at night, and the males might have a hard time finding their mates were it not for a special gland which the females protrude from their bodies when they are ready to mate. This gland has a characteristic scent and can attract a male, which then flies in the direction from which the odor is coming until he finds the female. The antennae seem to be the olfactory organs and are much more extensively branched in the male than in the female moth. Some day-flying insects, such as the flies, use sight to find their mates, and we can usually distinguish these by their large, well-developed eyes. There is one night-flying insect that also uses sight, the firefly. Fireflies have their own lanterns with them which give off flashes of light that aid them in locating each other in the dark. Many of the sounds made by insects as described under insect voices are made by the males. In some cases they seem

to attract the attention of the females. In most cases only the males have the sound apparatus and it serves solely for sexual attraction.

The mating act usually consists of direct copulation, the male inseminating the female, usually only once during her lifetime. Seminal receptacles are frequently present that store the sperm in the female's body, fertilizing the eggs as they are laid, sometimes as long as three or four years after insemination, as in the queen bee.

Social Organization

Some insects are solitary, but some are organized into social groups. The solitary insects, such as the grasshoppers, usually get together only for reproduction and live individualistic lives the rest of the time. Of course, they may be seen in large groups sometimes, but the association is casual; they happen to be together because they are after the same food supply. On the other hand, the social insects have a highly developed organization with a distribution of tasks among members of the group and a sharing of the benefits of their labors. In some of these, such as the bumblebees, the groups are small and the organization lasts only one season; in others, such as the honeybee, the group numbers high in the thousands and may be maintained almost indefinitely.

The insects have power through social unity. A single termite is a soft, weak insect that cannot stand exposure to light or dry air, yet a social colony of these insects can undermine the wooden structure of a house within a few years. A single hornet seems to be of little consequence, but woe betide the person who trespasses on the privacy of a social group (nest) of these insects. The honeybees have reached a high degree of social organization including thousands in each social group. Some details of this organization are presented in Chapter 37.

REVIEW QUESTIONS AND PROBLEMS

1. In what one respect is the development of the vertebrate animals superior to that of the arthropods?
2. Suppose you find some strange little wiggly animals in a trip to a desert island. How would you decide if these animals were arthropods or not?
3. Assuming that you find that the above animals are arthropods, how would you decide if they are insects or not?
4. Suppose you find two rather different animals which both have a certain body part which is similar. How would you determine if this body part is homologous or analogous?
5. The crayfish has an open type of circulatory system. What does this mean and how does it compare with the closed type of circulatory system?
6. Why is the flexor muscle in the abdomen of the crayfish so much larger than the extensor muscle?
7. Compare the function of the swimmerets in the male and female crayfish.
8. Fairy shrimp sometimes appear in a pond which was formed by rainwater but has been dry for months. How did they get there?
9. Why are barnacles of such great economic importance?
10. Describe the uses of spider web by spiders.
11. Tell how the male spider inseminates the female spider.
12. Why do the itch mites cause the itch?
13. Ticks do not suck enough blood to do a person any harm, but their bites can be quite dangerous. Explain why.
14. No more than one pair of legs per segment is the rule in arthropods, but the millipedes have two per segment. Explain.
15. What are the advantages and disadvantages of the method of respiration carried on by the insects?
16. Insects do much damage which affects people, but we could hardly live without them. Explain why.
17. Suppose you find some insect eggs which have been laid on a twig. You bring them inside for study of the development. How would you decide which type of metamorphosis the developing immature insects were undergoing?
18. What are the advantages and disadvantages of horny front wings in insects?
19. You find some insects which are harming some of your garden plants. By examining them how would you decide on the type of treatment to be used to eradicate them?
20. Explain how the protective mimicry of the tomato worm might have developed through natural selection.
21. Some moths can be quite harmful to plants. How might you take advantage of their reproductive habits to destroy them?
22. Some insects, such as the honeybees, have a highly organized social group where each contributes and each shares in the fruits of their labor. Many at-

tempts have been made to organize groups of human beings into such socialized groups, but most of these have met with failure. Why does it work so well in insects and not with people?

FURTHER READING

Borror, D. J., and De Long, D. M. 1971. *Introduction to the Study of Insects.* 3rd ed. New York: Holt, Rinehart and Winston.

Cloudsley-Thompson, J. L. 1958. *Spiders, Scorpions, Centipedes, and Mites.* New York: Pergamon Press.

Farb, P. 1962. *Insects.* New York: Time-Life Books.

Fox, R. M., and Fox, J. W. 1964. *Introduction to Comparative Entomology.* New York: Reinhold

Wilson, E. O. 1971. *The Insect Societies.* Cambridge: Harvard University Press, Belknap Press.

Winchester, A. M. 1970. *Concepts of Zoology.* New York: Van Nostrand Reinhold.

26

Cold-Blooded Chordates

The phylum **Chordata** includes such a variety of animals—fish, snakes, birds, people—that it may seem difficult to list generalized characteristics that apply to all. Yet all possess certain morphological characteristics which cause them to be grouped together.

CHARACTERISTICS OF THE CHORDATES

A dorsal tubular nerve cord. In the chordates the nerve cord runs the length of the body against the dorsal body wall, in contrast to the ventral nerve cord found in the higher invertebrates. Embryologically this cord is formed from a strip of ectoderm lying on the dorsal surface of the embryo. This strip first sinks inward, forming a groove, and then the upper walls of the groove come together and fuse, forming the tube. The anterior part of the tube enlarges to form the brain, whereas the remainder forms the spinal cord, which retains its tubelike characteristics throughout the life of the animal.

A notochord. Just under the spinal cord is a flexible tubelike body called the notochord. The phylum name is taken from this characteristic. In the higher chordates the notochord is present in the embryo only.

A postanal tail. A part of the body, containing the notochord or verterbral column, projects beyond the anus to form a tail at some time during the animal's life. The last part of the statement is necessary because there are a few chordates that have dispensed with their tails in the adult stage. These include frogs, guinea pigs, apes, and man. All of these, including man, have tails in their early development, but they fail to develop in proportion to the rest of the body and are rudimentary in the adult animals. Man has a tail bone, consisting of several small vertebrae fused together, as a remnant of his embryonic tail. Occasionally a baby is born with a definite tail when this rudimentary structure develops too far; generally this is quietly and inconspicuously removed at birth.

Pharyngeal gill clefts. The external pharyngeal region of all embryonic chordates bears a series of gill clefts. In those forms that develop gills for respiration these clefts extend inward and perforate the pharynx to produce gill slits with both internal and external openings. In those that use lungs for respiration, either the clefts never completely penetrate the pharynx and form no gill slits or, if the slits are formed, they become closed during embryonic development.

Blood flows in a posterior direction in the dorsal vessel. As the blood leaves the heart, it flows down the dorsal body wall just under the notochord or vertebral column in a vessel called the dorsal aorta and supplies all the body parts posterior to the heart. It is then returned to the heart by ventral vessels. This is just the reverse of the condition usually found in the higher invertebrates. The blood is carried back to the heart in vessels in the chordates and they, therefore, have a closed type of circulatory system, rather than the open type of system as found in the Arthropoda.

An endoskeleton. The skeleton of the chordates, when present, is internal to some of the muscles and other body structures and therefore is called an endoskeleton. The advantages and disadvantages of this type of skeleton have been discussed in Chapter 25 in comparison with the exoskeleton of some of the higher invertebrates. Some vertebrates have developed a type of exo-

skeleton in addition to their endoskeleton. The turtle is well protected by an enlargement and fusion of scales which form its "shell."

SUBPHYLA OF THE CHORDATES

The phylum Chordata is divided into four subphyla. The first three of these are small and little known and are ordinarily not considered in detail in a general course in biology, but they are listed here in order that our survey may be complete.

Subphylum I. Hemichorda. This group includes only one animal, a primitive, wormlike, burrowing animal. Its body is divided into a head, proboscis, and trunk, and the notochord is found is found in only the first two divisions.

Subphylum II. Urochorda. This group includes the sea squirts or tunicates, which are not much more than flexible little bags attached at one end and sucking water in and squirting it out at the other end. They do not seem to have much in common with the other members of the phylum, but their larvae are free-swimming. forms and have a notochord in the proper relation to the dorsal nerve cord and the other chordate characteristics.

Subphylum III. Cephalochorda. The most important genus of this group is *Amphioxus*, a little marine animal that is found near the shoreline of sandy beaches in certain tropical regions of the world. It is an excellent specimen for advanced study in working out the beginnings of vertebrate characteristics and is an animal with which all students of comparative vertebrate anatomy soon become thoroughly familiar.

Subphylum IV. Vertebrata. This group includes the remainder of the chordates. They take their name from the vertebral column, which consists of bony articulating vertebrate surrounding the spinal chord, found in all but the most primitive vertebrates. This is such an important subphylum and of such great interest because of its relationship to man that we shall consider its classes in some detail.

ROUND-MOUTH EELS— CLASS CYCLOSTOMATA

It may seem peculiar that an animal would have a mouth which it cannot close, yet this is exactly what is found in the cyclostomes. The mouths are round and permanently open, for there are no jaws to close them. The marine lamprey eel is a well-known representative of this class. It is a long, slender, eel-like animal, but it should not be mistaken for the true eel, which is a bony fish. Lampreys are common along the Atlantic Coast and may also be found in fresh-water rivers and streams where they go to reproduce. Many have gotten into some of our larger lakes, especially the Great Lakes. The lamprey has an unusual method of feeding which is very efficient in spite of its lack of jaws. Teeth are present in circles around the sucking mouth and help the animal to hold on to its prey. Teeth are also found on the tip of the firm tongue. When a hungry lamprey attaches itself to the body of a large fish by suction, the tongue is moved back and forth like a saw until a hole is cut in the body of the fish. Then the blood and small particles of flesh are sucked into the digestive system of the parasitic lamprey. The hole in the fish is so large and deep that the fish usually dies as a result of this attack.

A lamprey has seven pairs of gill openings. This is a greater number than is found in members of other groups of vertebrates. In many ways the lamprey is either primitive or degenerate—it has no paired appendages, there is only one nostril on the top of the head, and there is no stomach, pancreas, or spleen. If we cut into the body of the lamprey, we find that there is no bony skeleton, but there is a large notochord which runs the length of the body and supports the softer body tissues. There is also a basket

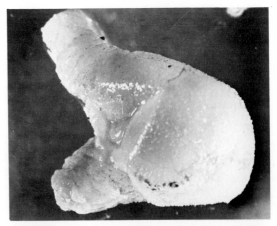

FIG. 26.1 A sea squirt, tunicate. This primitive chordate is little more than a bag with siphons for sucking and squirting out water. Its larva has many chordate characteristics.

composed of cartilage which supports the gills, and there are small pieces of cartilage in the head. When ready to reproduce, lampreys swim up into small, clear streams and build nests by moving rocks with their sucking mouths. The females then lay their eggs in the nest, and the males release sperm over them. The lampreys of many species die soon after reproduction.

An egg hatches into a small fishlike creature that looks very different from an adult. In fact, it is so different that it was once classified as a different animal under the name *Ammocoetes*. This larval stage resembles *Amphioxus* in many ways and was for a time grouped with it in the Cephalochordata. Then it was discovered that it underwent a metamorphosis after several years and became a lamprey. This resemblance indicates that the lamprey and *Amphioxus* are probably closely related through a common ancestor.

FIG. 26.2 A lamprey eel; note the absence of paired appendages.

A

B

FIG. 26.3 Fishes with cartilage skeletons. The tiger shark (A) has a small shark sucker attached to its ventral surface. The sting ray (B) rests in the sand with several lemon sharks.

CARTILAGINOUS FISHES— CLASS CHONDRICHTHYES

The members of this class have a much better developed skeleton than that of the cyclostomes, but the skeleton is composed of cartilage—there is no true bone. **Cartilage** is a material which is flexible to a certain degree, but is rigid enough to maintain its shape and support softer body parts. The ear flap of man which projects on either side of the head is composed of cartilage covered with skin. We can bend our ears in various directions, but they go back to their original shape when we release the pressure.

The best-known member of this class is the **shark.** Its body shows a number of distinct advances over the lamprey. It has a **mouth** with movable jaws lined with several rows of sharp teeth. It has a pair of **nostrils** at the front of the head. These are used for the sense of smell only and play no part in respiration. Sharks have a very keen sense of smell and become highly excited when they detect the scent of blood in the water. Five gill pouches on each side of the head open to the outside by means of five **gill slits.** Water is taken in through a pair of **spiracles** on top of the head, passes over the gills which line the gill pouches, and goes out through the gill slits.

A shark bears two appendages, one in the shoulder region, the **pectoral fins,** and one in the hip region, the **pelvic fins.** Internally these are attached to the **pectoral girdle** and the **pelvic girdle,** respectively. There is a segmented vertebral column composed of cartilage and vestiges of the notochord. The **brain** is completely enclosed in a cartilaginous skull.

We ordinarily think of **ears** as organs of hearing, but the shark has ears that do not function in this sense. **Balance** is the primary sense localized in the ears; the secondary function of hearing did not appear until the land animals developed. The shark has a pair of internal ears embedded in the cartilage of the skull. Each ear consists of three **semicircular canals** and a central chamber. If these ears are removed, the shark cannot tell whether it is right side up or upside down. It is just as likely to swim on its back as it is in the more normal position with the back up. The sense of hearing is localized in a row of sense organs contained in a **lateral line** which runs along both sides of the body. These sense organs pick up vibrations in the water.

With respect to reproduction the sharks seem to be quite advanced, for the **young are born alive,** a condition that we commonly associate only with the highest class of vertebrates, the mammals. However, closer study reveals that sharks have eggs like the other lower vertebrates, but these are retained in the body until they hatch and the young pass from the mother's body alive and active. We say that animals that lay eggs are **oviparous** and animals that have the young born alive, such as the mammals, are **viviparous,** so this condition, in which they have eggs and have the young born alive also,

might be called **ovoviviparous.** In any animal in which the young are born alive there must be copulation and internal fertilization. The median parts of the pelvic fins of the male sharks are modified into **claspers** which aid in this process.

Sharks have a bad reputation because they do attack man on certain occasions. The attacks are comparatively rare in the coastal waters of the United States, however. In a typical recent year there were only eleven attacks reported and only three of these were fatal. This is far below the number of deaths from drownings and other water accidents. Still, it is well to know that the danger exists and not to take chances. Skin divers who spear fish under water should make it a habit to bring their catch to the boat as soon as possible, for the scent of blood may attract and excite nearby sharks. In tropical waters sharks are more numerous and attacks more common.

The cartilaginous fishes also include another group of animals called the **rays.** Their general characteristics are about the same as those of the sharks, but the body shape has been greatly modified. They are flattened out dorsoventrally into a broad, fanlike body with a long slender tail. One of these is called the **sting ray** because it has one or several sharp spines that project up from its tail. It has a habit of resting on the sand at the bottom where a bather may step on it and receive a rather ugly wound from the sharp, barbed spine. The **electric ray** possesses organs that generate electricity, and when disturbed, it can give a powerful shock to any intruder on its privacy. The **sawfish** has an extension from the front end of the head armed with sharp spines on either side that make it resemble a saw.

BONY FISHES— CLASS OSTEICHTHYES

An external examination of a bony fish seems to indicate that there is only one pair of gill slits because there is an **operculum,** or gill cover, that covers the real gill slits which lie underneath. Upon raising the operculum, it will be seen that there are actually four pairs of gills with five pairs of gill slits. There is a **lateral line,** as in the sharks. The bodies of most fish are covered by **scales** which overlap like shingles on a roof. A close examination of scales reveals that they bear concentric rings of growth similar to those found on the shell of a clam. The **pelvic fins** of the fish have **moved anteriorly** to occupy a position near the center of the body or, in some forms, just under the pectoral fins. Internally this is accompanied by a complete **disappearance of the pelvic girdle.**

The skeleton is **composed of bone,** a fact which hardly needs mentioning since everyone who has eaten fish knows it from experience. Another noteworthy internal structure is the **swim bladder,** an air-filled bladder in the dorsal region of the body cavity which serves to make the fish more buoyant. By regulating the amount of air in this sac, the fish is able to maintain itself at a desired level in the water. In some fish this bladder has a connection with the pharynx, and air can be gulped down into it. In these fish the air bladder is an accessory organ of respiration, for oxygen can be absorbed through its walls when the oxygen content of the

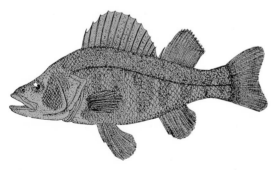

FIG. 26.4 The perch, a typical bony fish Note the operculum at the posterior part of the head, the lateral line, and the pelvic fins which are just under the pectoral fins.

water is reduced. The alligator gar is such a fish, and observation shows that it can live in foul water long after other fish have died from lack of sufficient oxygen. The swim bladder, thus, has a significance as a possible forerunner of the lungs of the land vertebrates.

Life in the water is rugged and dangerous, and there are numerous methods of protection. Universal is **protective coloration.** Fish are colored so that they are hard to see in their natural environment. The dorsal surface is colored darker than the ventral surface. An enemy swimming over them would view them against the dark bottom of the body of water and, while swimming under them, would view the light under surface of the body outlined against the light upper surface and sky and would overlook them in either case. This characteristic has probably been developed through natural selection. The flounder or sole shows that this is not accidental. This fish spends much of its time lying on its side against the ocean floor, and it even swims on its side. The side which is usually underneath, the left side, has become light colored, whereas the side which is usually exposed to light is darker. Curiously enough, the left eye, which was formerly on the left side in a normal position, migrates around the skull during embryonic development and comes out on the right side along with the right eye.

The colors and patterns of certain fishes sometimes change and blend with their surroundings. The flounder is another good example of this; if it is placed on sand the specks on its body will be quite small, but if it is placed on gravel the specks will be larger and blend with the larger rock particles. A flounder was placed on a black-and-white tile and developed a rough pattern of checks corresponding in size to the tile. This happens because small color bodies are present in the skin. These expand or contract and can make any area of the skin dark or light.

Many small tropical fish show brilliant coloration and, when they are seen in a fish bowl

FIG. 26.5 The flounder has the ability to change its skin pattern to correspond to its surroundings. On a plain surface its skin becomes a uniform blending shade (A). On a dotted surface its skin color turns into a dotted pattern (B).

or aquarium, might appear quite conspicuous. However, when seen in their natural surroundings of brilliantly colored coral, seaweed, and other tropical forms of life, they blend nicely and the brilliant coloration is a protection.

In one group of marine fishes the paired fins are enlarged and may be used in gliding through the air over the surface of the water. These are the **flying fish,** but since there is no flapping of the fins, they are actually gliders rather than fliers. When an enemy goes after one, it swims rapidly to get a start and then flips itself out of the water and sails through the air for distances as great as an eighth of a mile.

Another interesting means of protection is found in the **electric eel.** This large eel, which grows to a length of several feet, is found in South American rivers and lakes. When irritated, it is capable of giving out a shock that will knock down a person standing in the water nearby. Wires can be run from one of these eels to an electric light bulb and when the eel is disturbed the bulb will light up.

The **porcupine fish,** found in tropical

waters, does not need any of the means of protection described. Its body is covered by sharp spines, which normally lie flat on its body. But when in danger, the fish will inflate itself with either water or air and expand into a tight ball. The spines will stand out to discourage any animal that might wish to eat it. If a large animal, such as a shark, should be so hungry as to swallow it anyway, it is just too bad for the shark. The porcupine fish secretes a red fibrous material around itself, which protects it from the digestive juices of the shark, and then pro-

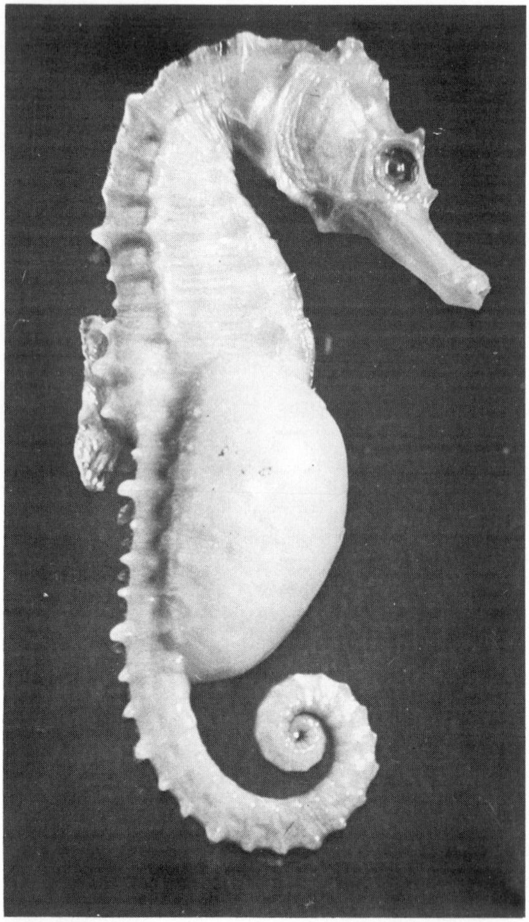

FIG. 26.6 This male sea horse has a brood pouch filled with hatching young about ready for "delivery."

ceeds to eat its way through the shark's stomach and into the water outside, none the worse for its experience.

Fish Reproduction

The reproductive habits of fishes show great variation in the number of eggs and the degree of care or protection given to the eggs and young. It is a general rule of nature that these two factors are in inverse proportion to one another. At one extreme, we find fishes like the **sturgeon** that deposit the eggs indiscriminately in the water, and release sperm in the same way. The fertilized eggs sink to the bottom and receive no protection. As a result, they are eaten by many animals; few of them hatch, and many of the young that do hatch are eaten before they can mature. The sturgeon compensates for this great destruction by laying an enormous number of eggs. A female will lay approximately three million eggs each year for about ten years, or a total of about thirty million eggs. From this huge number only two, on the average, will survive and live to reproduce, one male and one female. Even man is numbered among the animals that eat the sturgeon eggs, for caviar is made from them. Much of our caviar comes from the paddlefish, or spoonbilled catfish, a relative of the sturgeon which is common in the southern Mississippi River.

At the other extreme are quite a number of ovoviviparous fishes, such as our common **freshwater minnow**, *Gambusia*. Like the sharks, this little fish retains the eggs in its body until they hatch. Inasmuch as the young are born alive, they can swim away from their enemies from the first. The anal fin of the male is modified into a long slender tube which serves as a copulatory organ in transferring sperm to the female. A much smaller number of eggs is formed, and there may be no more than a dozen young minnows produced at one reproductive cycle.

The **grouper** is somewhere between. These fish scoop a nest out of the mud along the side of the river in which they live and the female lays the eggs in the nest and goes on her way. The male releses sperm over the eggs and then, waving his fins near them, causes a fresh current of water to flow over them which supplies plenty of oxygen. When the young hatch, he stays with them for a time and he may even open his mouth so that the young can swim inside for safety when there is danger. This illustrates an interesting principle of fish reproduction. We are accustomed to thinking of care of the eggs and young as a maternal instinct, but in some fishes it is the male that gives care to the young, if any is given.

The **sea horse** is one of these fish. The name is not descriptive of size, for the animal is seldom more than five inches long, but the head does somewhat resemble the head of a horse. The male sea horse has a brood pouch located on the ventral surface of the tail. During the reproductive season, he swims behind a female and, as she lays the eggs, he catches them in his pouch. As he swims around in an upright position with his pouch distended with eggs, he looks very much like a pregnant animal about to give birth to offspring. It is sometimes said that this is the only species of animal in which the male becomes pregnant. When the eggs begin to hatch, the male will hold on to a twig with his tail and appear to go through all the pangs of labor as he brings forth the living baby sea horses from his pouch.

The **Pacific salmon** has one of the most complicated reproductive cycles. The adults are found in salt water, but they never reproduce there. When they are ready to reproduce, they go into one of the freshwater rivers that empty into the Pacific. They are one of the few fishes able to make the transition from salt to fresh water without fatal reactions. They continue up the river until they reach shallow, spring-fed streams far inland in which to spawn. The remarkable part about the whole process is that they return to the identical stream where they were hatched. This can be determined by marking the young salmon and checking their return. As they proceed up the river they come to many forks and turns, waterfalls, and rapids, but they return to their place of origin.

During their journey upstream they eat no food and their mouths become modified for digging their nest. Upon arriving at the breeding places, they dig a nest over a spot where a spring is bubbling up through the gravel. A spot like this will not freeze during the severe winter which occurs in the northwestern United States, western Canada, and Alaska. The eggs are then laid, the sperm released over them, they are covered with gravel, and the parents flounder around awhile in the shallow water and die.

CLASS AMPHIBIA

The class name **Amphibia** comes from Greek words meaning "both forms of life," an appropriate name because the representatives of this group are both land and water animals. We borrow the name to speak of other things—amphibian planes, tanks, jeeps—which operate on both land and water. The class includes the frogs, toads, salamanders, and a few primitive, tropical, wormlike forms. All the vertebrates studied up to this time have been water animals, and the amphibians start life as if they were going to follow this pattern of existence. They hatch out of the egg as fishlike animals with three fully developed and functional gills. After a time, however, certain changes occur which transform them into animals adapted to live on the land. The gills gradually disappear and lungs appear. Small limb buds appear which grow into legs. These animals are now able to crawl out on the land and take their place as terrestrial vertebrates. Some of them have made a success of their life on land; others have returned to the water habitat, although they retain some of the land characteristics and must come to the surface to breathe. Some of this latter group do not lose the larval gills and use both gill and lung respiration.

In making the transition to land, the amphibians have failed to develop a new method of reproduction, and they continue to use a method characteristic of water animals. Some of them are able to live on the land far from water, but when the time comes for reproduction most of them return to the water to reproduce. There are two reasons for this. First, the male has no copulatory organ to transfer sperm into the body of the female. This is a necessity for reproduction on land. The amphibians must be in the water during the reproductive period so that the sperm can swim to the eggs released by the females. Second, the amphibians do not have a land egg. An egg which lives and hatches on land must have a protective outer covering and special membranes which allow respiration in the air. Since the amphibian

FIG. 26.7 Amphibians with tails. Top: newt, *Triturus viridescens.* Center: slimy salamander, *Plethodon glutinosus.* Bottom: hellbender, *Cryptobranchus.*

egg has neither, it must develop in the water. We might mention an interesting exception to this rule; in South America there is a large Surinam toad whose young develop on the back of the mother. The female lays the eggs and, with the aid of the male, spreads them on her back. The male releases sperm over the eggs and fertilization is accomplished. The skin of the female then grows over the eggs to completely enclose them. The embryo develops into a miniature form of the adult without going through a tadpole stage. These young toads then break the skin and emerge from the back of the female. Although this is not at all like the land method of reproduction used by higher forms of vertebrate life, it does indicate an interesting method of avoiding dependence on water for reproduction.

In one group of amphibians, the adults retain the tail. Animals in this group are commonly called **salamanders** or **newts.** Some of these live almost entirely in the water and may keep the gills throughout life, but also develop lungs and use both gill and lung respiration. Others live mostly on land, but they must crawl under rocks and logs during the daytime or their skin will dry out.

There is one, the **tiger salamander,** that has a larva that often does not change into the adult type. It will mature sexually and reproduce in the larval body form with three pairs of gills. However, if kept in a comparatively dry environment, the gills may disappear and the adult form develop.

The **mud eel,** another interesting salamander, has a long, slender, slimy body and front legs but no hind legs. Its movements are somewhat like those of a snake, and the legs are of no particular use. In fact, the hind legs would actually be a hindrance. The slipperiness of the eel is a means of escape from its enemies, and hind legs would just interfere with the smoothness of its body.

Amphibians that lose their tails as adults are either **frogs** or **toads.** The toad lives more

as a land animal than the frog and has developed a dry warty skin. It is sometimes thought that a person may get warts by handling a toad but there is not the slightest foundation for such belief. The eggs of toads may be distinguished from those of frogs when seen in the water: toad eggs are laid in a long string and look like beads on a string, but frog eggs are laid in a gelatinous mass.

CLASS REPTILIA

The transition from water to land habitat, started by the amphibians, is completed by the reptiles. It was accomplished by the development of a copulatory organ in the male and the development of an egg that can hatch on the land. There are some reptiles, such as turtles and crocodiles, that spend a great part of their time in the water, like some amphibians. Most reptiles are typical land animals, however, and many inhabit the most arid regions of the world.

The water egg of a fish or amphibian consists of an outer membrane containing the developing embryo attached to a yolk sac. The land egg of a reptile or bird must have a thick, protective shell on the outside. This **shell** may be tough and leatherlike, like that of a reptile, or brittle, like that of a bird. Beneath the shell another membrane is developed, the **allantois,** which is necessary for respiration. This membrane is supplied with many blood vessels from the embryo, and the circulating blood absorbs oxygen through the porous shell and releases carbon dioxide. The embryo is surrounded by another membrane, the **amnion,** which contains the **amnionic fluid,** in which the **embryo** floats. Thus, the embryos of land animals develop in a liquid medium after all, even though the egg remains on the dry land. Reptiles' eggs are not incubated. They are cold-blooded animals that usually bury their eggs in the ground and leave them to hatch.

A

B

FIG. 26.8 A frog and a toad. The photographs of the leopard frog (A) and the common toad (B) show the external distinctions between these two tailless amphibians.

Prehistoric Reptiles

At one time, reptiles were the dominant animals of the earth. There were the huge land dinosaurs—much larger than any land animal of today—as well as many smaller land reptiles that dominated life on the land. There were flying reptiles in the air and fishlike reptiles that were abundant in the waters of the earth. Through studies of prehistoric and modern reptiles, we have been able to identify 125 families in this class. Of this great number, however, only eighteen are found alive on the earth today. These show particular adaptations that enabled them to survive while their many relatives became extinct.

The decline of the reptiles was accompanied by a rise of the mammals. There are a number of theories to account for this change in the animal world, but one of the most plausible is based on climatic changes which are known to have occurred in the past. The climate of the United States, for instance, has varied from tropical to frigid during past geological eras. Reptiles are cold-blooded animals and may be quite active in warm weather. When the temperature drops, their muscles become sluggish, and they become entirely inactive when it gets down around 45 degrees or below. A rattlesnake placed in a refrigerator becomes quite harmless until he gets warmed up again. The opposite is true of mammals; they may feel sluggish in very warm weather, but as the weather becomes brisk they become more and more active. Thus, the mammals had a great advantage as the climate became cooler.

Turtles

These reptiles have traded activity for protection. They are slow and awkward but so well protected by their outer "shell" that they have been able to survive while their more active relatives became extinct. Although amphibians typically have a smooth skin, the reptiles have scales on their bodies. In this group of reptiles some of the scales have become greatly enlarged, thickened, and fused together, forming the "shell." In many turtles this is hinged. The head and legs may be drawn inside and the "shell" tightly closed. Land turtles are usually referred to as **tortoises,** whereas the name turtle commonly refers to those that live primarily in the water. Turtles hold the world's record for holding their breath under water, for they have been known to stay submerged for as long as three hours.

On the Galapagos Islands there are huge tortoises weighing several hundred pounds and estimated to be up to four hundred years old. They are probably the oldest living animals on the earth.

Crocodiles and Alligators

These are the giants of the living reptiles; they frequently grow to a length of twenty feet and have actually been measured thirty feet long. Their huge bodies, armed with powerful jaws lined with sharp teeth, make the crocodiles antagonists to be respected by any living creature. They are always found in or near the bodies of water and may be seen sleeping on the bank of a tropical river.

There are only slight differences between the crocodiles and the alligators, but the crocodile seems to be the more vicious and aggressive of the two. Fortunately, the alligator is the type most abundant in the coastal regions of the southern United States, although there is one species of crocodile in Florida. Crocodiles infest rivers and lakes in tropical regions all over the world and are especially abundant in the rivers of Africa.

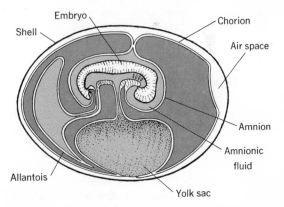

FIG. 26.9 The relationship of the parts within an egg of a reptile or a bird. The large yolk sac supplies the food for the embryo and the allantois takes care of respiration.

A

B

FIG. 26.10 Alligator, above, and crocodile, below. These large reptiles have changed very little in millions of years.

Lizards

The lizards are an extremely varied group of reptiles adapted to a wide range of environments. They range from tiny wormlike burrowing forms to animals more than twelve feet long with long sharp claws which can tear their prey apart. They are among the few animals able to exist on the hot, dry, sandy deserts; they are also found in moist forests and even in water. Some move so swiftly that they look like a streak to the eye; others move so slowly that they seem to be in slow-motion movies. Some have well-developed limbs; some have none at all and may be mistaken for snakes.

Many lizards, at one time or another, have been accused of being poisonous, but there is only one species in the United States that deserves this accusation. This is the **Gila monster,** a clumsy, thick-bodied lizard with a heavy tail that serves as a food reservoir to tide it over periods of food scarcity. It is found primarily in

FIG. 26.11 The Gila monster is the only poisonous lizard found in the United States. Its habitat is restricted primarily to the desert region of Arizona and California.

Arizona and is especially abundant along the Gila River. It moves awkwardly but, when molested, is capable of quick movements of the head accompanied with a snapping of the jaws which may easily fasten on to a hand held too near. Gila monsters have grooved teeth in the lower jaw with poison glands at their base. Once a Gila monster has fastened its grip, it holds on for a time, chewing a little and working the poison down into the wound. It also tries to turn on its back so that the poison can flow down into the wound. The bite makes a man quite ill, but the poison does not seem to be injected in sufficient quantities to cause death.

The **horned lizard** of Texas and other southwestern states has a broad flattened body resembling that of a toad and is often called a horned toad. When disturbed it may shoot a fine stream of blood out of its eyes for a distance of several feet. This does not seem to accomplish any particular purpose, but it might startle an attacker and give the lizard a chance to escape.

The American **chameleon** is abundant in the southeastern United States and is remarkable for the way its skin changes color according to its surroundings. It has the same type of color bodies under the skin as the flounder, and the changes are mainly brown, gray and green, with intermediate shades.

The **glass snake** is worthy of mention because it has given rise to the superstition that snakes sometimes become unjointed in times of danger, only to reassemble themselves and crawl away whole when the danger is past. This is not a snake at all, but a legless lizard. Like many lizards it is able to throw off a part of its tail when greatly excited. This is a protective measure; if it is being pursued by an enemy the tail is most likely to be caught first. The tail may be broken off and the main part of the lizard escapes while its enemy is satisfied with the tail. If the tail is thrown off during the chase, it will attract the enemy because it will wiggle actively for a time after being thrown off. The glass snake often does this, and the rest of the superstition was pieced together from this one fact. A new tail will be regenerated.

FIG. 26.12 The Texas horned lizard. In this close view it looks like a model of an ancient dinosaur.

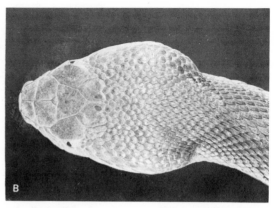

FIG. 26.13 Pit viper head characteristics. This copper-head shows the eliptical (slitlike) pupils of the eye; the pit is anterior and ventral to the eye (A). The arrow-shaped head is characteristic of these poisonous snakes (B).

Snakes

There is probably no single group of animals on the earth more feared than the snakes, yet an objective study of their habits shows that generally they do not deserve such a bad reputation. As a whole, snakes are highly beneficial to

FIG. 26.14 Rattlesnake coiled, ready to strike. The rattles of this Eastern timber rattlesnake were moving so fast that the eye could not see them, but the movement was frozen with a high-speed flash.

mankind, because they destroy great numbers of rodents, which are among our worst enemies.

Unfortunately, some snakes have fangs connected to poison sacs and are capable of inflicting a poisonous bite which in some cases may be venomous enough to cause the death of a human being. In spite of whatever benefit they might offer otherwise, such snakes are undesirable. However, they make up only a small portion of the total number of snakes in the world, and it is certainly unwise to condemn an entire group of animals just because there are some undesirable species in their number.

The poisonous snakes in the United States number only four, and these have easily recognized characteristics. Three of them are **pit vipers.** These take their name from the pit which lies just anterior and ventral to the eye; the pupil of the eye is slitlike, like the pupil of a cat's eye, rather than round as in the other families of snakes. Some readers may think that it does no good to know what kind of eye the pit vipers have because they are not going to get close enough to see the pupils anyway. However, it is safe to get near enough to see the pupil clearly; snakes can strike no farther than half their body length. The third characteristic of the pit vipers is the arrow-shaped head, caused by an enlargement of the head just where it joins the neck. Other snakes show this characteristic to some extent, but it is quite distinctive in the pit vipers.

The **rattlesnake** is by far the most abundant and important of the poisonous snakes in the United States. It is a nervous, temperamental snake that is quick to strike, and it carries a very potent poison. This snake is best known by the rattle, a series of rings of dried skin on the tip of the tail. When a person gets excited and nervous, his hands often tremble; snakes have no hands, but their tails tremble when they are excited. When the tail trembles at high speed, the rattle makes a characteristic whirring sound that is easily recognized after it has once been heard. When the skin is shed, a ring of the posterior part is left behind to form a ring of the rattle.

The fangs are long and curved and fold back against the roof of the mouth in a sheath of skin when not in use. When danger threatens, the snake will probably try to get away first, but if pressed closely will pull itself into loose loops facing its foe, ready to strike. As the snake strikes, the lower jaw is dropped down, the fangs become erect, and the head hits with enough force to drive the fangs into the flesh. At the moment the fangs enter, the poison sacs contract and inject the poison and the snake jerks its head back ready to strike again. All of this takes place with lightninglike rapidity. The poison is a blood poison and begins destruction of the elements of the blood. If the area of the bite is cut and allowed to bleed freely, a good portion of the poison will be washed out; this is the most important first-aid measure to remember. Practically all properly treated victims recover; a few of the improperly treated victims will die. Rattlesnakes account for about 90 percent of the deaths from snake bite in the United States.

There are about fifteen species of rattlesnakes in the United States. Some, like the little pigmy rattlesnake or ground rattlesnake, have only a single button on the tip of the tail, which does not rattle. The timber rattlesnake and the eastern diamondback may exceed six feet. The Texas diamondback is also a large snake.

The **cottonmouth water moccasin** is a dangerous and feared snake of the swamps and bayous of the southern United States. It has the pit viper characteristics and can also be recognized by its dull, irregular, olive brown color and its thick, heavy body. Moccasins are found in and around water at all times and, like the rattlesnake, they prefer to escape danger rather than face it. When approached they may be seen to slip silently into the water and swim away, but when cornered they are quite vicious, and death sometimes results from the bite.

The **copperhead** is the third and final

FIG. 26.15 The beautiful coral snake has a highly potent poison. The letters indicate the sequence of the colored bands on the body.

member of the pit viper family in the United States. It is a smaller snake than the other two and has a much better disposition. In fact, it is so gentle that it seldom bites unless it is stepped on. On their slender, graceful bodies copperheads have pretty copper-colored bands of irregular shape that make them beautiful snakes. They are not more than three feet long and they do not seem to inject enough poison to cause death in a normal, healthy person, but it is enough to make him quite sick, so first-aid treatment should be given following their bite.

The beautiful **coral snake** is the fourth and final venomous type found in the United States. It is not a pit viper. It has a round pupil and no pit. The head is not puffed out where it joins the body. However, it can easily be recognized by the glossy red and black bands, bordered by smaller yellow bands that run around the body. Corals are very gentle snakes, like the copperheads, and the author knows of two cases where they were mistaken for the harmless king snake and were picked up and handled for some time before the handlers realized they were dealing with a poisonous snake. Nearly all bites are received when the snake is stepped on. The coral is a member of the cobra family; the snakes of this family have short erect fangs, rather than long retractable fangs like the pit vipers. Accordingly, the coral snake does not strike, but actually grabs the skin with its mouth and forces the fangs through the skin with a biting motion. Ordinary trousers seem sufficiently thick to protect the legs from its bite. The coral is small, seldom more than eighteen inches long, slender, and has a nerve poison. A person who is bitten may go blind temporarily and have difficulty in breathing and carrying out the other body activities controlled by the nerves. Death may result in some cases.

Death from snake bite in the United States is very low, only twenty to thirty per year—almost negligible in comparison with the many thousands who die from automobile accidents, but in tropical countries deaths are much more frequent. There are many deaths in India from the bite of the **hooded cobra.** This interesting snake, when alert, spreads the ribs at the anterior part of the body to form a hood. It often goes into houses in search of mice, and many people are bitten in their own homes. Unfortunately, the majority of the people hold these cobras in superstitious reverence and do not kill them for fear that they may be the reincarnation of their late grandmothers or other relatives. This cobra may be quite large, but it has the short erect fangs characteristic of the family and has to chew the fangs into the skin. Ordinary clothing is good protection from them.

The **king cobra** of Africa and parts of Asia stands supreme among the poisonous snakes of the world. It grows to a length of twelve feet and has venom which may cause death in a few minutes. When disturbed, it does not show the nervous, threshing movements of many other snakes but gracefully raises its head, spreads its hood, and, weaving from side to side ready to strike, may approach the intruder.

Africa is also noted for the **spitting cobra** that is able to spit its venom six or eight feet. It instinctively aims for the eyes, and temporary blindness will result if even a small quantity of the poison reaches them.

Another member of the cobra family is the **asp,** well known because Cleopatra pressed an asp to her bosom to commit suicide.

In Central and South America the most dangerous poisonous snakes are the **bushmaster** and the **fer de lance,** both pit vipers with very bad reputations that are well deserved.

Among foreign snakes those of the **boa family** are well known because of the great size of some of them. Fortunately, they are not venomous. Boas are constricting snakes. Such a snake will wrap itself around its prey and squeeze it to death before swallowing it. A snake swallowing its prey is a very interesting sight to watch. It can swallow animals with a body diameter at least three times that of its own body. A **python** can swallow a large pig or antelope. After squeezing it to death, it grasps one end of the animal with its jaws. The snake's teeth point backward, and as the jaws are worked backward and forward, the animal is gradually forced into the throat. Once through the throat, powerful peristaltic movements of the esophagus help pull the animal down. After the meal is finished, there will be a huge lump in the python's body, which will gradually diminish as the animal is digested. Huge pythons have been known to eat naked native children, but show aversion to clothed persons.

The most abundant of all snakes in the United States are the striped snakes that have one or several longitudinal stripes running the length of the back. They are commonly called the **ribbon snakes** and the **garter snakes** and may be found in the backyards of city dwellers as well as in the fields and forests. They eat cold-blooded animals, such as earthworms, frogs, and toads, and are harmless to man.

The **water snakes** live in and around water and eat fish, frogs, and toads. They are almost invariably called water moccasins by those who think that any snake around the water is a water moccasin. They are somewhat ill-tempered and able to give a painful bite, but carry no poison.

The **king snakes** are a highly beneficial group because of the large numbers of rodents that they destroy. They also eat other snakes and seem especially fond of poisonous ones such as rattlesnakes. Sometimes the king is bitten during the battle, but it is immune to the poison and suffers no harm. King snakes are very gentle toward man and make nice pets that are entirely safe to handle.

Probably the most interesting of all snakes is the deceptive **spreading adder.** If a person suddenly comes upon one it will rear up and spread its hood, which makes it look very much like a cobra. It forces air rapidly out of its mouth to produce a most vicious hissing noise. It feints as if just ready to strike at any moment. Because of these habits it has a fearsome reputation; even its breath is supposed to be poisonous. But it is all a bluff; the spreading adder cannot bite a person even if he sticks his finger in its mouth. It has no teeth around the jaws. The spreading adder may suddenly roll over on its back and go through the most realistic death agonies and then lie still. It will hold this position as long as a person stays near and can even be picked up and handled without giving any sign of life, but if it is placed on the ground with the ventral side down it will immediately flop over on its back. The spreading adder becomes very gentle after being handled a little and makes a nice pet.

REVIEW QUESTIONS AND PROBLEMS

1. Explain why the nerve cord of a chordate is tubular in shape as a result of its embryonic development.
2. The presence of a postanal tail is cited as a chordate characteristic, but some chordates, like man, do not show a tail. Explain.
3. What is a notochord and how is it represented in the lamprey eel and in man?
4. Compare the chordates with the higher invertebrates with respect to the direction of flow of blood in the major blood vessels.
5. Why are the sea squirts placed in the chordates when they appear more like some of the simpler invertebrates?
6. What evidence indicates a close relationship between the lamprey eel and *Amphioxus,* even though they appear quite different?

7. Describe the function of the lateral line in fishes.
8. Describe the function of the ears of fishes.
9. How does the ovoviviparous reproduction of the shark compare with the viviparous reproduction of the mammals?
10. Describe two functions of the swim bladder.
11. Would you expect the sea horse to lay as many eggs as the sturgeon? Give reasons for your answer.
12. Describe the reproductive cycle in the salmon.
13. What happens to the number of gill slits as the water chordates become more complex?
14. Compare the skeletons of the three classes of chordates discussed in this chapter.
15. Describe the changes that take place in a larval amphibian which result in adaptation to life on the land.
16. Give the reasons why amphibians must have water if they are to reproduce, while the reptiles are independent of the water.
17. Explain how the tiger salamander, through individual adaptation, shows how inherited adaptations of the amphibians may have taken place through natural selection.
18. How may the eggs of frogs and toads be distinguished from one another?
19. Describe the egg of a reptile and contrast this with the egg of fish and amphibians. Explain why the extra parts are needed in the reptile egg.
20. What role might climate have played in the decline of the reptiles and the rise of mammals in past evolutionary history?
21. Explain the nature of the "shell" of turtles.
22. Why is the glass snake often mistaken for a snake when it is actually a lizard?
23. What are the head characteristics of the pit vipers?
24. Name the poisonous snakes of the United States, with a brief description of each.
25. Compare the fangs of the cobra family with those of the pit viper family.

FURTHER READING

Bishop, S. C. 1967. *Handbook of Salamanders*. Ithaca, N. Y.: Constock.

Conant, R. 1958. *Field Guide to Reptiles and Amphibians*. Boston: Houghton Mifflin.

Goin, C. J., and Goin, O. B. 1971. *Introduction to Herpetology*. 2nd ed. San Francisco: W. H. Freeman.

Hardisty, M. W., and Potter, I. 1972. *Biology of Lampreys*. New York: Academic Press.

Lagler, K. F., et al. 1962. *Ichthyology*. New York: John Wiley and Sons.

Oliver, J. A. 1955. *Natural History of North American Amphibians and Reptiles*. New York: Van Nostrand.

Winchester, A. M. 1970. *Concepts of Zoology*. New York: Van Nostrand Reinhold.

27

Warm-Blooded Chordates: Birds and Mammals

Only two classes of organisms in the entire animal kingdom maintain a constant high body temperature by internally regulating their metabolism. These are known as the warm-blooded animals; both classes belong to the chordate phylum. These animals, as a rule, require more food than the cold-blooded animals because of their increased need for energy to produce heat. The body temperature is generally higher than the surroundings. We shall consider the birds first and then the mammals.

THE BIRDS

The birds, class **Aves,** are a very important group of vertebrates so far as man is concerned. They have great aesthetic appeal because of their beautiful colorations and their pleasing songs. They also furnish an important source of food for man. Some have been domesticated and so selected that they lay many more eggs than would be normal in a natural environment where eggs are used solely for reproduction. Game birds challenge many sportsmen who seek recreation in hunting. Birds are also highly beneficial as a group because of their feeding habits. They destroy many harmful insects and rodents, and it would be difficult for man to grow many crops without this assistance.

Some birds are scavengers and are valuable in the elimination of dead animals. The vultures or buzzards glide high in the air. Their keen

A

B

FIG. 27.1 Guana birds on an island off the coast of Peru. Guana birds are Peruvian cormorants which are found in large numbers on the offshore islands. They leave huge deposits of fecal waste which is gathered and sold around the world as fertilizer. (Courtesy of the American Museum of Natural History.)

Peru, where millions of birds, such as cormorants, boobies, and pelicans, congregate; in fact, the islands are literally covered with birds when they come in to roost for the night. They deposit an enormous amount of fecal material which dries and hardens quickly in the sunny, dry climate to form a rich natural fertilizer known as guano. The birds deposit as much as 750 tons per acre per year. One might wonder how the birds can obtain sufficient food to supply such a horde, but the deep, cold waters surrounding the islands abound in fish life, which makes food getting easy.

Of course, birds have some activities which are not beneficial and may be economically harmful. Some eat grain or the tender shoots of growing plants. Some hawks catch chickens, although the great majority of hawks are beneficial because they destroy harmful rodents. Some birds, like the English sparrows and starlings, often live in large numbers near human habitations and may be a general nuisance because of their fecal deposits. Many a statue in a public place is marred by the wastes from birds which find the statues a convenient perching place.

As a whole, however, birds are beneficial and need protection in many cases. Excessive clearing of the land removes the trees and protected places where they might nest. Reforestation and protection of existing forested regions help protect birds. The extensive use of insecticides in some regions has created a problem. Birds have been poisoned by eating insects that had received the poison.

Bird Characteristics

Flight is one of the most distinctive characteristics of birds. There are some flightless birds, such as the ostrich and the penguins, but the great majority fly and have many body adaptations which make them efficient in this activity. Careful observation of birds in flight shows how

eyesight enables them to see dead animals on the ground, and as they circle down to feed, others in the vicinity note the action and join in the descent. Soon there will be dozens of these birds, and it takes them only a short time to pick the flesh from a dead animal.

In some sections of the world, birds furnish a valuable source of fertilizer. An outstanding example is the islands which lie off the coast of the wings function. On the upbeat the wings are partially folded and the feathers are so turned that they resemble the open slats of a Venetian blind, thus offering little air resistance. The feathers are closed again on the downbeat and the air resistance maintains the proper altitude. Forward movement is obtained by a propellerlike action of the terminal portion of the wings.

Some birds, such as the vultures, appear to fly without flapping their wings. Such birds are experts in the art of gliding and soaring and take advantage of all the updrafts of air currents to maintain their position in the air.

A bird's speed of flight varies considerably; most travel at from 20 to 40 miles per hour when in full flight, but some are known to travel at higher speeds. The Old World swift has been clocked at nearly 200 miles per hour, duck hawks reach speeds up to 180 miles per hour when in a dive, and some ducks can go as fast as 90 miles per hour.

The body shape of birds is generally streamlined in such a way as to offer the least resistance to the air. We have learned much about the design of airplanes through a study of the shape of the bird's body.

Feathers are a distinguishing feature of birds. All birds have feathers and no animals except birds have feathers. The birds appear to have a reptilian origin and the scales typical of the reptiles are present on the legs, but on other parts of the body the outgrowths from the skin are in the form of feathers. These are of three types. **Contour** feathers have a central shaft with parallel barbs to the sides. They give shape and protection to the body and those on the wings offer resistance to the air in flying. Those on the tail serve as a rudder which helps control the direction of flight and also can be spread and serve as a brake to slow the speed of flight. **Down** feathers consist of a central shaft with barbs coming out irregularly. They function primarily to conserve body warmth and are very efficient in this regard. We take advantage of this and use

down for making quilts which are very light but which have great heat-holding qualities. The best quality of down comes from the eider duck. This bird plucks down from its body and uses it to line the nest, where man may then collect it. Some feathers may be mixed—down near the body and contour on the outside. The third type of feathers are **filoplumes,** which are little hairlike projections with tufts on the end. They seem to have no useful function.

Feathers are borne on certain regions of the skin known as **feather tracts.** The feathers are shed in a process known as **molting,** which takes place in the late summer as a rule, and are replaced by a new set of feathers. This is a gradual process, however, and the bird is never "naked" (without feathers).

The body temperature of birds is higher than that of man; it ranges from about 105° to 111°F. A few species, for example, the poorwills, undergo a hibernation during cold weather during which their body temperature drops, but most maintain a constant high temperature, even those that remain in cold regions during the winter. The hummingbird is an interesting example. It is very small and therefore has a very high proportion of body surface in relation to its body volume. This is conducive to a great heat loss through radiation from the skin. It can be calculated by the laws of physics that it is impossible for the hummingbird to live through the night without feeding, yet we know that it does. This is possible because of a nightly hibernation. The body temperature begins to drop when the bird comes to rest for the night and throughout the night there is not the high rate of food consumption in metabolism that would be necessary to maintain the constant high temperatures found during the day.

The **songs** and **calls** of birds are distinctive; they are produced in a way not found in any other group of animals. There is a special organ, known as the **syrinx,** which is found at the junction of the trachea and the bronchial tubes in the chest. There is no homologous organ in other animals. This is used to produce call notes and songs which are for communication with one another. Songs are of a longer duration and generally produced only by the males during the breeding season. The calls and songs are so distinctive that species of birds can be identified by them. Some birds, such as the mockingbird, can mimic the songs and calls of other species, and some, such as the mynas, parakeets, crows, and parrots, can mimic human speech rather accurately.

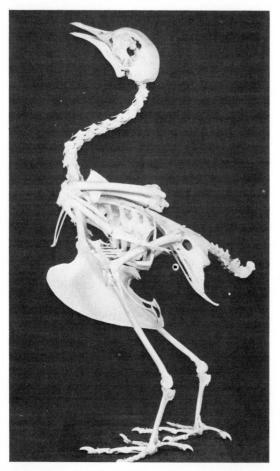

FIG. 27.2 The skeleton of a pigeon. Note the large, keel-shaped sternum which is the point of attachment for the large muscles needed for flight.

Birds have a **keel-shaped sternum,** or breast bone, that accommodates the large amount of muscle necessary for flight. The space on each side of the keel is filled with the attachments of large muscles that are used on the downbeat of the wings, for which most of the power is needed. The muscles to lift the wings are on the back. The back muscles are small. A person that gets the back of a chicken does not get much meat from it. Some of the flightless birds, such as the ostrich, do not have this keel.

Birds have **air sacs** that run from their lungs into the muscles of the chest. This is another adaptation to flight. The chest muscles have their origin on the sternum and during flight the sternum must be kept rigid; therefore, the chest cannot be expanded in breathing. However, the alternate contraction and relaxation of the chest muscles in flying exert a bellowslike action on the air sacs and air is inhaled and exhaled with each beat of the wings. This works very efficiently and gives birds the tireless energy necessary for flying.

How can a bird sleep and still maintain its position on a branch which is swaying in the wind? The answer lies in a special **perching adaptation** which is found in all the birds that live in trees. The tendons of the leg which are attached to the claws are pulled when the legs are bent and this closes the claws. Thus, the weight of the bird's body on the bent legs keeps the claws tightly gripped on the limb without any need for muscle contraction.

The brain of birds is proportionately larger than that of the reptiles. This is due mainly to an increased size of the **cerebellum,** which is the part of the brain associated with the many activities of muscle coordination. This is another adaptation for flying. Watch a bird flying toward a cliff at high speed. It may seem as if he is surely going to dash himself to pieces against it, but at the last moment he throws his wings in reverse, spreads his tail, comes to a dead stop, and settles on a slight projection. This requires a high degree of coordination which is made possible by the large cerebellum with its many folds. The olfactory lobes of the brain are poorly developed, and birds are known to have a very poor sense of smell. A hungry buzzard can be placed near a dead animal that has a very strong odor of decomposition, yet the buzzard will not detect the animal if he cannot see it.

On the other hand, the sense of sight is extremely keen in birds, as evidenced by the highly developed optic lobes of the brain. A buzzard at a thousand feet in the air can detect a dead squirrel on the ground. The eyes are very large; some hawks have eyes as large as human eyes. Thus, a great proportion of the head is occupied by the eyes. The retina of a bird is about twice as thick as that of a man, and in hawks, noted for the keenness of their eyesight, there may be as many as 1,000,000 rods and cones per square millimeter. This gives them a visual acuity about eight times that of man. Accommodation, or focusing, is rapid and very efficient. It is accomplished by changes in the shape of the cornea as well as the lens. Birds can also adjust rapidly to changes in light intensity. A flashlight can be passed back and forth across the eyes of an owl in the dark and the pupils will dilate and constrict as fast as the light moves. In man, several minutes are required to achieve adjustment to sudden changes in light intensity. Birds have a functional third eyelid, the **nictitating membrane,** which can move across and cover the eye when the eyelids are open. This can clear the vision in rain or fog and has been compared to a windshield wiper. Man has a small vestige of this membrane in the medial corner of each eye, but it is not functional.

The sense of taste in birds is poorly developed, probably correlated with the fact that they swallow their food whole. The sense of hearing, on the other hand, is highly developed, a fact which is correlated with their calls and songs. There is an external tympanic membrane which picks up vibrations, and a single ear bone, columella, carries them to the inner ear.

Modern birds do not have teeth and therefore cannot chew their food. The food is swallowed and passes into a soft-walled crop where it is stored. It then passes into a thick-walled gizzard where it is broken down into smaller pieces. Birds swallow gravel and similar hard objects which lodge in the gizzard, and as this organ contracts, the food is ground. Then it passes into the long, slender intestine where digestion and absorption take place.

Courtship and Care of the Young

Courtship among birds is highly developed. In many species the males will strut and perform to attract the females for days or even weeks befor actual copulation.

Since birds are warm-blooded, the eggs must be incubated if they are to hatch. A small embryo cannot generate sufficient heat to main-

FIG. 27.3 Young barn owl. Like many other birds, the barn owl is unattractive when young. Among mammals the young are usually more attractive than the adults.

tain a constant high body temperature. Most birds build a nest in which this incubation is done. Nest building is one of the most remarkable instincts in the entire animal kingdom. Each species tends to use particular materials and to construct the nest in a particular way. A young female bird can be hatched in a commercial incubator and raised in isolation from all other birds, and she will still build a nest typical of her species. She will use the same materials, choose the same type of locality, and construct it in the same shape as her ancestors have done for centuries past. Some birds use saliva for sticking the pieces of the nest together. One, the salangane swift, which lives in southeast Asia and surrounding islands, lines its nest entirely with saliva. These nests are used for making bird's-nest soup.

The number of eggs laid by birds at one nesting period varies from one to twenty or more. In most birds only the female sits on the eggs, but in some species, such as doves and the ostriches, the male will share the brooding. There are even some species in which the male does it all—the kiwi and emu are examples. The time of incubation varies; in general, the larger the egg, the longer the brooding period. The ostrich requires forty-two days, the Canada goose twenty-nine days, the domestic chicken twenty-one days, the crow eighteen days, and the robin thirteen days.

Most birds hatch as ugly, naked, blind bundles of protoplasm, awkward and ungainly, but with very large mouths which open at the slightest sound which might indicate that food is coming. The mother bird and sometimes the father are kept busy providing food for the ravenous appetites of their offspring. When many young are in the nest, how can the parent decide which mouth to feed so that some will not get more than their share while others starve? It appears that there must be a time lapse after the swallowing of one bit of food before a second can be swallowed. Hence, the parent bird places the food in the first open mouth, but then watches to

see if it goes down immediately. If not, she removes it and places it in another mouth and again observes to see if it is swallowed. Thus, the recently fed young will not get food needed by others. A great amount of food is required for the growing young and many trips must be made to supply the needs. One wren was found to make 1217 such feeding trips between sunrise and nightfall of one day.

The young birds which are hatched naked, blind, and helpless are known as **altricial young.** This is typical of most of the perching birds whose nests are built in trees and other protected places. Many birds, however, build nests on the ground and would be an easy prey for predators if they spent their early days in such a helpless state. Such birds usually have **precocial young** which hatch out covered with down, with eyes open, and able to begin walking and taking their own food shortly after hatching. Chickens, grouse, pheasants, ducks, and loons are examples.

Protection of the young reaches a high degree of development in the birds. Most birds will fight ferociously for their offspring. A mother hen with baby chicks can make a dog turn tail and run. A clever means of deception is used by the quail and other ground-nesting birds to draw enemies away from their nests. When a possible enemy approaches, the mother quail will dart from the nest flying and fluttering on the ground like a bird with a broken wing. The predator will most probably give chase, since this appears to be an easily obtained meal. After leading him for some distance, however, the female will fly into the air, leaving the predator a safe distance from the nest. This broken-wing trick is familiar to all who walk in regions where such birds are found.

Migration

The power of flight gives birds a means of rapid transportation which is not possessed by most other vertebrate animals. Many utilize this ability to make migration flights which may take them over thousands of miles each year and enable them to live the year round in surroundings where food is abundant, the days are long, and the climate is favorable. Most birds stop along the way and do not cover more than about 25 miles a day, but some may make long-distance flights of 2000 miles or more at one hop. A number of birds winter in South America and spend the summer in northern Canada, traveling up to 14,000 miles each year. The long-distance record, however, is held by the Arctic tern, which nests in the summer inside the Arctic Circle, then flies down the eastern coast of Canada, across the Atlantic, down the west coast of Europe, down the west coast of Africa, and finally ends up in the Antarctic region where it spends the northern winter months. This makes a yearly round trip of about 25,000 miles. These terns spend both winter and summer near the poles at the time when the sun never sets.

It has generally been assumed that bird migration is an adaptation which permits the avoidance of extremes of weather, and this is certainly one of the end results of most migration habits. Still, birds will leave southern areas in the spring when there is an abundance of food and the weather is favorable and perhaps fly into the teeth of a late spring blizzard on their way north. It seems that the length of day is more important than the temperature factor in triggering the migration instinct. In summer the days are longer in the north than in the south, but in the winter the conditions are just reversed. Most birds feed only when it is light and go to roost when it becomes dark. Thus, they have longer periods for feeding in the regions of long day and will feed for the entire twenty-four hours in regions where the sun never sets. Poultrymen sometimes take advantage of this and keep lights on in the poultry houses at night. The chickens will feed almost continuously and grow much faster and lay more eggs than would be the case if they spent half their time roosting.

Experiments indicate that, in some birds at least, the photoperiod (amount of daylight) initiates hormonal changes which bring about the urge to migrate. A migratory bird in the north in midsummer can be put in a cage which is kept covered so as to shut out the light for a part of each day. This gives him an artificially short day similar in length to those occurring in the fall. When released from the cage he will start his southward migration. This explains why the birds of a certain species arrive at a certain place at about the same time each year even though weather conditions may vary considerably from one year to another. The arrival of the swallows at the mission of San Juan Capistrano in California occurs on almost exactly the same day each year.

There has been considerable speculation about how birds find their way during migration. In some cases it appears that an older bird leads the way and that he follows landmarks along the way. Such birds will stop their flight when the weather becomes too overcast to permit them to follow these landmarks. There are some birds, however, which migrate thousands of miles over open water and find a small island. There are no landmarks on open water, and some have suggested that birds have a sense organ which tells them their position by the magnetic field of the earth and the force brought about by the earth's rotation. Experiments seem to indicate that there are some migratory birds that actually navigate by means of the position of the sun and stars and can make allowance for the changes in relative position of these through an internal time mechanism.

Fossil Birds

As already mentioned, birds appear to have a reptilian origin. This theory is supported by fossil remains of prehistoric birds. In slabs of limestone near Solenhofen in southern Germany the fossilized remains of two birds were found which were quite different from modern birds. They were given the genus name of *Archaeornis* (*Archaeopteryx*). There was a clear imprint of feathers, which indicated that they were true birds and were warm-blooded, but they also had teeth similar to reptile teeth. There was also a long, bony tail with thirteen separate vertebrae. The creatures were about the size of a modern pigeon. The head, neck, and front part of the body did not have feathers. There was not the enlarged breastbone found in modern birds and there were digits with well-developed claws on the wing bones. *Archaeornis* probably had little power of flight, but could run and glide with some degree of efficiency. Other fossils of toothed

FIG. 27.4 *Archaeornis.* This restoration from skeletal remains shows the probably appearance of this ancestor of modern birds. *Archaeornis* had teeth and claws like reptiles, but it also had feathers and wings which it used to glide from tree to tree. (Field Museum of Natural History, Chicago.)

birds have been found in Kansas, Montana, and Europe, but *Archaeornis* appears to be the most primitive.

Typical Birds

There are about twenty-five orders of birds, each containing numerous families, so it will be impossible to give a complete survey of the field. But by taking a few well-known birds and learning something about their habits, we may better understand the class.

The eagles. Perhaps it is appropriate that we start our study with the group that includes the American national bird, the bald eagle. This eagle is not really bald but has a thatch of white feathers on its head that makes it appear bald at a distance. It is a majestic bird, with a wing spread of about six feet, that soars high in the air but can drop like a bullet when it spots its prey. It catches fish and some mammals for food. It is sometimes a robber. It will soar in the air watching the osprey hawks. When the osprey dives in the water and brings out a nice wriggling fish, the eagle will drop out of the sky and attack the smaller osprey, causing it to drop the fish. The eagle then plunges down, catches the fish, and flies away. Eagles build huge nests of sticks in rocky or inaccessible places in which they live and raise their young year after year.

The owls. "Wise as an owl" is an old saying; it would seem that any creature that looks so dignified and sits so long, apparently in deep meditation, must be wise. However, when night comes, owls no longer sit but become quite active as they fly about in search of nocturnal rodents. Their eyes are probably the keenest in the animal kingdom. During the daytime the pupils become reduced to narrow slits, but, contrary to superstition, they can see during the daytime. At night, in the reduced light, the pupils enlarge tremendously, enabling them to see clearly in the darkest forests. The pupils are able to change their size almost instantly to accommodate to changing light intensity, and so an owl can chase a mouse from the bright moonlight into a dark forest without difficulty. The eyes of owls are placed side by side on the front of the head so that they look straight ahead and the owls have to turn the head to see in different directions. When an owl looks at something behind it, the head gives the appearance of being attached backward.

Most owls are beneficial, since rodents seem to be their favorite food, but sometimes the great horned owl finds that it is easier to catch chickens and other domesticated birds. The monkey-faced or barn owl is highly beneficial. The little screech owl gives out an eerie cry that may be frightening on a dark night, but it eats many insects and mice. In the western United States there is a superstition that the little burrowing owl lives in the same burrow with prairie dogs and rattlesnakes. It often lives in a prairie dog hole, but devours the original occupant first and in turn is probably devoured by any rattlesnake that may enter the hole.

The vultures. The vultures, or buzzards, are close relatives of the hawks, but they lack the strong beak and the strong legs of the hawks. Therefore, they do not attempt to catch live prey, but instead devour anything that is dead. No rural landscape is complete without several buzzards gliding high in the air on a sharp lookout for the carcass of a dead animal. Upon sighting anything that looks promising, they circle around and slowly approach the earth. Other buzzards in the distance see this characteristic reaction and come to join the feast. The body of a dead cow will be picked clean— only the bones remain when the buzzards have finished. They seem completely immune to the poisons of bacteria that may be present in such flesh, although other animals may be killed by it. The young are fed on a regurgitation of partly digested flesh.

The loons. Visitors to the lakes in the northern part of the United States never forget the cry of the loon sounding over the water in the middle of the night like the hysterical laughter of a lunatic. However, the loon is anything but a lunatic. It is very hard to get near one; it will dive under the water when you approach in a boat, suddenly pop up behind you, dive again, and appear again in some unexpected direction as if mocking your pitiful efforts to outsmart it. Loons are the champion diving birds, sometimes diving and swimming under water to a depth of fifty feet. A frog does not have a chance against a loon, for the loon can outswim it under water.

The hummingbirds. These are the midgets of the birds; the eggs are exceedingly small, and a young hummingbird will fill only a portion of a teaspoon. Their wings are small and vibrate so rapidly when flying that the characteristic humming sound is produced and only a blur is

FIG. 27.5 Ruby-throated hummingbird. The smallest of birds, the hummingbird, has great maneuverability in the air. (Courtesy of Marlin Perkins, Director, St. Louis Zoo.)

seen, as if a wingless bird were suspended in space without any visible means of support. Since even their tiny bodies would be too heavy for most flowers, their suspended flight is a convenience for their nectar-sucking habits. They suspend themselves in the air while thrusting their long bills down in the flower for a drop of sweet nectar. Small as they are, they are among the most fearless of birds, evidently confident in their speed and maneuverability to escape at a sign of danger.

The ostrich. This is the largest of the living birds. The ostrich may be more than eight feet tall and weigh up to three hundred pounds. It uses only two toes, which are covered with broad pads and are well adapted for travel in the desert. Ostriches easily take twenty-five feet in a stride and travel more than twenty-five miles an hour. The tale about burying the head in the sand to escape detection seems to have no foundation in fact. The female lays a large egg which weighs about three pounds and requires forty-five days of incubation to hatch.

The woodpeckers. It would be hard to believe that any bird has a bill sharp enough and a head and neck strong enough to dig holes in solid wood if we could not actually see it. The feet have sharp claws that cling to slight outgrowths of bark and the tail is held against the tree for firm support, while the long sharp bill drives itself into the wood with a rapid series of blows, like an air hammer digging into concrete. Most woodpeckers are beneficial to man because their pecking is done in search of insects that are boring through the wood. The tongue is long and bears a barb on the tip, so that it can be thrust down a slender hole to stick insects and bring them out. One of the woodpeckers, the yellow-bellied sapsucker, has become harmful, for it digs holes in trees that produce sap, which it sucks out as food.

The cowbird. Cowbirds are one of the villains of the bird class. The female lays her eggs in nests of other birds, usually only one to a nest. A song sparrow may leave her nest of eggs temporarily and find the larger egg of the

FIG. 27.6 The ostrich, the largest of the modern birds, cannot fly, but it can run rapidly to escape enemies. The male, on the right, has a plumage distinctly different from the female, on the left.

cowbird among her own when she returns. It is incubated and the young cowbird duly hatches along with the young song sparrows. When food is brought to the nest the larger cowbird manages to get most of it and may even push the young song sparrows out of the nest. It is quite pathetic to see a small female song sparrow bringing food and stuffing it down the huge mouth of a young cowbird that may be twice her own size.

Quail. The quail, partridge, or bobwhite—so called by different people in different sections of the country—is among the most tasty of the game birds. Its comparatively small wings are vibrated very rapidly during flight, and they produce a characteristic whirring noise that sounds like a miniature airplane taking off.

The male excavates a place in the ground and lines it with leaves and grass and takes turns sitting on the eggs. The young quail usually stay with their parents after they have matured, so quail are usually seen in coveys. When flushed, they scatter in all directions and reassemble after the danger is past, finding each other through a special call. When sleeping, the covey is gathered in a circle facing outward. If they are disturbed during the night, they can take sudden flight without colliding with one another in the darkness. In the spring the coveys separate and the males take up a stand and give their bobwhite call. Bachelor males will keep up the call throughout the summer.

The storks. One of the best-known birds does not live in the United States, although a close relative, the heron, is found along our shores. The storks are abundant in the coastal regions of Europe and have a habit of building their nests on the tops of houses. In many countries this is considered an omen of good luck and they are never driven away. Just how they became an accessory in maternity is not clearly known, but a stork serenely perched on one leg on top of a chimney after a baby was born would offer an easy answer to embarrassing questions from the older children. The storks are wading birds whose long legs make it possible for them to wade in shallow water, picking out fish and other water animals with their long bills.

THE MAMMALS

The mammals include the largest and most intelligent animals on the earth. They are a very successful group, having replaced the reptiles as the dominant large animals of the earth. Many body organs, such as heart and brain, reach their height of development in the mammals, culminating a long series of gradual changes.

Characteristics of the Mammals

Hair. Only the mammals have hair, and most of them have bodies almost covered with hair. Even man, whom we think of as a fairly hairless creature, has hair everywhere on the body except the palms of the hands and the soles of the feet.

Mammary glands. The mammals take their name from the mammary glands, which in the females enlarge and produce milk when they bear young. We commonly express this by saying that the mammals nurse their young. Embryonic mammals have a mammary ridge on either side of the ventral surface of the body. In most mammals nipples develop along the entire length of this ridge; dogs and cats are typical examples. In some mammals, such as horses and cows, nipples develop only at the posterior end of this ridge; in others, such as elephants and human beings, only a single pair of nipples may develop at the anterior end. Occasionally, however, supernumerary nipples appear in addition to those which are usually found in the species. Cows may often be seen with an extra pair of nipples anterior to the two normal pairs. These are usually rudimentary and give no milk. Human beings sometimes have from one to three pairs of nipples posterior to the normal pair, formed in a row similar to those found in the dog.

Diaphragm. This is an internal muscular partition, separating the chest from the abdomen, which can be raised and lowered to exhale and inhale air. Mammals also use chest expansion for respiration, as do the reptiles and birds.

Warmbloodedness. Like the birds, the mammals maintain a constant high body temperature. The rate of metabolism is rather high, and this generates the necessary heat which holds the body temperature at an even, high level when outside temperature fluctuates.

Complex teeth. In lower vertebrates the teeth are usually a row of conelike projections, with a few exceptions such as the fangs of poisonous snakes. In mammals the teeth are specialized to accomplish specific functions. There are **incisors** for cutting, **canine** for tearing, **premolars** and **molars** for grinding. These are developed in varying degrees in different mammals according to different feeding habits.

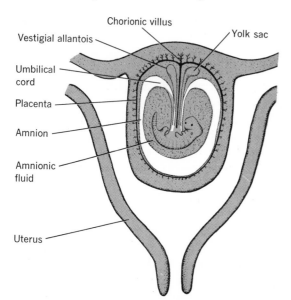

FIG. 27.7 The developing embryo of a mammal in its mother's uterus. The placenta enables it to derive nourishment from the mother.

Placenta. All but the two most primitive orders of mammals have a placenta, that is, an organ that grows out from a developing embryo and becomes firmly attached to the uterus of the mother and absorbs food and oxygen from the blood of the mother. When the mother nourishes the embryo in this way, it is unnecessary to have a large amount of yolk in the eggs, so mammal eggs are much smaller than those of other vertebrates. A rabbit egg is no larger than a dot made with the point of a fine-pointed pencil. This is **true viviparous reproduction.**

Highly developed cerebrum. While the birds have a highly developed cerebellum, for control of equilibrium and muscle coordination necessary for flight, the mammals have developed the cerebrum to the greatest size and complexity of all vertebrates. Since the cerebrum is concerned with the process of learning and reasoning, the mammals can probably be considered the smartest animals on the earth. The cerebrum has complex folds in the higher mammals which increase its efficiency. Man has by far the largest cerebrum, and it bears more convolutions in proportion to his size than the cerebrums of other mammals.

No cloaca. The amphibians, reptiles, and birds have a cloaca, that is, a cavity at the terminal end of the intestine into which the wastes of digestion as well as the urogenital products are emptied. Thus, the anus serves as the external opening for all of these. In the embryonic mammals a cloaca is formed, but tissue grows between the anus and the openings of the urogenital system, so that there are separate openings in the adult. The first order of mammals is an exception, however, and does possess a cloaca as an adult.

Order Monotremata

This order includes the most primitive of the mammals. They have hair, mammary glands, and other characteristics that cause them to be

FIG. 27.8 The duckbill platypus has fur and nurses its young like other mammals but lays eggs and incubates them like birds. It is an endangered species whose habitat is restricted to parts of Australia.

classified as true mammals, but they have a cloaca and lay eggs like the reptiles and birds.

The little **duckbill** of Australia is one of the few examples of this interesting order of mammals. It has a bill like a duck, fur like a seal, webbed feet like a frog, and barks like a dog; it seems to be made up of odd parts of other animals. The female lays eggs which have shells like birds' eggs. She makes an underground nest lined with dried grass and the young hatch out after a short period of incubation. There are no nipples to the mammary glands, but the ventral surface of the body is covered by a broad milk field. When lunch time arrives the mother lies on her back and the young crawl on her abdomen and drink the milk that flows out from the pores of the skin.

Order Marsupialia

The marsupials are advanced over the monotremes in some respects, but are primitive in other respects when compared with the other mammals. There is no cloaca in the adult—openings of the anus and the urogenital tubes are separate. The mammary glands have well-developed nipples, and there are special muscles to pump milk down the throats of the nearly helpless young. The eggs are larger than in most mammals and, in most species of marsupials, contain all the food which will be used by the embryos until they can nurse.

The **opossum** is one of the best known of the marsupials. It is very abundant in our southern states. The opossum eggs are retained in the uterus until they hatch, which is only thirteen days after fertilization, so the young are very tiny and in an early embryological stage of development at birth. They will be no more than half an inch long, but they have sharp claws which they use to climb up the mother's body and enter the pouch on the ventral surface of her body. There they find the nipples, and each little oppossum grabs a nipple which he sucks down his throat, and his mouth actually grows onto the nipple. When the young have completed their embryonic development, they leave the pouch but clamber back in to nurse or to escape danger.

The **kangaroo** and the cuddly little **koala bear** of Australia are other well-known marsupials.

FIG. 27.9 The opossum is the only marsupial (pouched mammal) native to North America.

FIG. 27.10 The kangaroo is one of many marsupials which live in Australia. The young continue to use the pouch of the mother after they are able to leave it.

Order Insectivora

These little animals, which derive their name from their insect-eating habits, have a placenta and true viviparous reproduction. Their teeth have an outgrowth near the base which protects the gums from the hard parts of insect bodies.

The little **mole** shows how animals may be adapted to their living conditions. Moles spend most of their waking hours burrowing along under the ground in search of insects that may be there. They have powerful front legs with sharp claws, and they move along underground with the same movements that a person uses to swim the breast stroke under water. Their eyes have become degenerate and the eyelids remain fused together throughout life. The external ears also are covered with skin.

The **shrew** is an animal which has been described as the fiercest animal in existence, ounce for ounce. Shakespeare showed that he knew his biology when he selected the title "The Taming of the Shrew" for his clever comedy. The shrew is the smallest of the mammals and is about as small as any warm-blooded animal can be. The smaller the mammal, the greater is the exposed body surface in proportion to its size and the more it has to eat to keep warm. When a tiny shrew is placed in a cage with a mouse which is many times larger, the mouse will be so terrified that it will allow the shrew to eat it with hardly any resistance. A shrew must eat about one and one-half times its body weight per day in order to live. Think of a person eating 225 pounds of food per day and you can appreciate the problems facing the shrews in finding and digesting sufficient food to keep alive. A shrew will die of starvation in a few hours if it is deprived of food.

Order Chiroptera

The order **Chiroptera** includes the **bats,** which are the only flying mammals. (We speak of flying squirrels, but they only glide.) The finger bones of the last three fingers of each hand are greatly elongated and skin is stretched between the fingers to form the wings. The first two fingers terminate in claws which the bats use to suspend themselves upside down while they are sleeping during the daytime. Bats come out at dusk and fly around during the night in search of insects. It is an amazing sight to see the bats come circling up out of Carlsbad Caverns by the thousands every evening. Bats orient themselves and avoid collisions with trees, buildings, and other objects by giving out shrill cries that are higher in pitch than the human ear can hear. As the echoes of their cries come back to them, they know the nearness of solid objects and can avoid them.

Not all bats eat insects. The vampire bats of the tropics will bite a person or other large animal, such as a cow, and lap up the blood as it flows out. They are particularly dangerous as transmitters of such diseases as rabies.

Order Carnivora

The order **Carnivora** includes the carnivorous mammals that include meat as an important item of their diet. Most of them are predatory and have strong, agile bodies, with sharp claws and teeth—all adapted to catching and killing prey. All their teeth are pointed, in an adaptation for cutting and tearing flesh; they have no teeth for chewing, so the food is cut up into small chunks and swallowed.

The cats. The members of this family are easily recognized by their similarity to the familiar house cat. They include the lions, tigers, leopards, pumas, wildcats, house cats, and similar animals. They are adapted to stalking their prey and then pouncing upon them with one sudden burst of speed and, as a rule, are not capable of long-sustained high speed.

The dogs. The doglike carnivores often depend on running their prey down in a long chase in contrast to the cats. The **wolves, foxes,** and **coyotes** are included in addition to our common dog.

The hyenas. These strange-looking animals ordinarily do not kill their own prey but wait for some of the bolder carnivores, such as the lions, to do the killing. Then when the lion has eaten his fill, they come in and devour what is left of the carcass. They have the strongest jaws of any mammal and can easily crush large bones with their teeth.

The racoons. These little carnivores are found near the water and often catch crayfish, fish, and clams for food. They seem very particular about their meals, for they thoroughly wash their food before eating it if water is available. This habit seems not so much to be cleanliness as it is an aid to swallowing. The salivary glands are not well developed, and the food is more easily swallowed when wet. Raccoons have

FIG. 27.11 The cheetah is a large carnivore which runs very fast when chasing its prey.

delicate little handlike front paws which they use to handle the food while washing and eating it. The rare but well-known **pandas** from the mountains of Tibet are in the raccoon family.

The civet cats. The civet cats are found in oriental countries, and the mongoose of India is one of the group. The **mongoose** is known for its ability to fight and kill cobras. It is not immune to cobra poison, but is so quick that it can tease a cobra into striking and dart in and grab it behind the neck before the cobra can get back into position to strike again.

The musk-bearing carnivores. The **skunk** is the first animal thought of when animals with an odor are mentioned. It bears special musk glands which can eject an odor, there being none more odoriferous. Confident of this protection, the skunk goes serenely on its way through the forests and is usually given a wide berth by all the other animals. Should some inquisitive animal be rash enough to come too close, the skunk will first raise its tail and wave it as a warning signal, then will give a little jump with its hind legs and squirt a few drops of its musk at the intruder. That will be the last time the intruder will molest any animal even faintly resembling a skunk.

There are many others in this family that do not have such a pronounced odor and do not throw their musk. They are a bloodthirsty group, including **martens, minks, badgers, sables, weasels,** and **otters.** In spite of the rather nauseating odor of some of the live animals, they produce some of the most valuable furs on the market.

The bears. These are the largest of the carnivores. They are more omnivorous in their diet than most of the order and have teeth adapted to do some grinding. The black bears of North America are the ones most commonly seen. No one can make a trip to Yellowstone Park without seeing the **black bears,** which visit the garbage cans in the camps and wait along the highway for handouts from passing cars.

They seem to be very tolerant toward the many tourists and their inevitable cameras, but occasionally people get bitten when they take undue liberties with the bears and forget that these are wild animals. The brown bears and cinnamon bears are only different color variations of the black bear. **Grizzly bears** are also found in the park, but they are quite dangerous. Fortunately, they generally stay away from people and are seldom seen along the highways or in the camps. They get their name from a silvery ring around each hair that gives them a grizzled appearance.

The aquatic carnivores. Like the whales, this group is modified for life in water, although they may spend considerable time out of water. Fish provide the main item of their diet. They are masters at diving and swimming under water. The **sea lions** are the most commonly seen of the group. No zoo of any size is complete without a sea-lion pool, where these interesting creatures seem to get great enjoyment from showing off for the crowds. The trained sea lions of the circus show that they are experts at balance and can be taught to do many intricate tricks if the keepers' supply of fish holds out. The **walruses** are larger than the sea lions and live primarily in arctic regions. They are distinguished by their mustaches and the large tusklike canine teeth that protrude from their upper jaws. The **seals** are also found in cold climates, and are well known as the source of sealskin coats.

Order Ungulata

Members of this order are characterized by the highly developed grinding teeth in their jaws, an adaptation to their herbivorous diet, and the presence of hoofs. Most of them stand on the tips of their toes, with their heels never touching the ground. This makes it look as if they had three joints in the legs rather than the two found in other vertebrates. The toenails, which formed the claws of the carnivores, are enlarged to produce hooves.

The odd-toed ungulates. The **horses** are the most abundant of this group. They use only one toe on each foot, which is greatly enlarged. This is the third, or middle, toe; rudimentary second and fourth toes can be seen higher up the leg, although they do not touch the ground. There were ancient prehistoric horses that used four toes and more recently some that used three toes. A fairly complete history of the development of horses has been worked out from fossils. The donkeys and zebras are very similar in their structure to our modern horse.

The **tapirs** are also placed in the odd-toed group, but they have four toes on the front feet and three on the hind feet.

The **rhinoceros** is a well-known ungulate that has three toes on both front and hind feet. They have one or two "horns" coming out of the median dorsal region of the head. These are not true horns but are formed by a massing and fusion of hairs in this region.

The even-toed ungulates. Most of the ungulates use two of their toes, the second and third, which are incompletely fused together and form the hoof. They are often spoken of as the animals with a **cloven hoof** because of the slit between the two toes. There is one group of even-toed ungulates which have a special four-chambered stomach. Animals in this group are known as **ruminants.** They can swallow their food rapidly when they may be grazing out in an exposed place and store this food in a chamber known as the **rumen.** Then they can retire to a protected place where they can chew the food in leisure. The second chamber, the **reticulum,** forms a mass of this unchewed food, which is regurgitated a mouthful at a time and chewed thoroughly before being swallowed a second time. This time it goes into the third chamber, the **psalterium,** which is filled with folds that strain out the larger particles and return them to be chewed some more. The well-chewed food passes

on into the last chamber, the **abomasum,** which is the true stomach where the digestive glands are located. Such animals are said to chew a cud. The ruminants include **cattle, deer, elks, antelopes, goats, sheep, giraffes, camels,** and **llamas.**

Those even-toed ungulates which are not ruminants include the **hog, peccary,** and **hippopotamus.** The latter uses four toes on each foot. Even our common pig has four rather well-developed toes, but the second and third are usually the only ones that touch the ground.

FIG. 27.12 The porpoises are very intelligent aquatic mammals. This one is shown leaping for fish at the Marineland exhibit in Florida.

Order Cetacea

This order includes the **whales, porpoises, and dolphins.** Whales are sometimes thought of as fish because they live in the water and have a body that is fishlike in shape. However, they are warm-blooded, have a placenta, nurse their young with mammary glands after birth, breathe by lungs, and have the other typical mammalian characteristics. The tail is horizontal, rather than vertical as in fish. This makes them swim with an up-and-down rolling motion in contrast to the side-to-side movements of the fish.

The whales have the distinction of being the largest animals that have ever existed. One measured 103 feet long and weighed 294,000 pounds. Whales have a heavy deposit of fat, or blubber, under the skin that gives them a good insulation from the cold waters in which they are often found. Unfortunately for the whales, this blubber furnishes commercially valuable whale oil, and so they are hunted by whaling vessels. An extremely valuable substance, **ambergris,** is found in the intestine, and from it the most expensive perfumes are made. Some species have long strainers in their mouths, called whalebone, which in the past were used extensively as stays in women's corsets.

The whales have their single nostril on the top of the head. They may be seen "blowing" at a great distance. This spout, which can be seen coming up from the nostril, is formed of condensing moisture and is not a spout of water. If the outside air is cool, the warm air from the lungs will condense and form a visible water vapor, just as a person's breath condenses on a cold day.

The porpoises are a miniature edition of the whales, their structure being about the same. They are an interesting sight following in the wake of boats with their rolling motions, and sometimes they jump entirely out of the water.

FIG. 27.13 The elephants are the largest land animals of modern times. The Indian elephant has smaller ears than the African elephant.

Order Proboscidea

The only living members of this order are the **elephants,** the largest of the living land animals. There are two varieties: the African elephant, which may be 11½ feet tall at the shoulders and has huge ears that completely cover its shoulders, and the Indian elephant, which may be 10 feet high and has smaller ears. In spite of their huge size, they are rather easily tamed and become valuable beasts of burden in tropical countries. In some parts of India, travel by elephant is the only way a person can get through a dense jungle. The tusks, which may be mort than 11 feet long in the African elephant, are a valuable source of ivory.

The woolly **mammoths,** whose bodies have been found in a nearly perfect state of preservation in Siberian ice fields were the prehistoric relatives of our modern elephants. Apparently these were the only members of the elephant family that have been capable of living in cold, arctic regions. When some of them were found, the flesh was so well refrigerated that it was actually fed to dogs, and at a banquet in London mammoth steaks were served which had been in a deep freeze of Siberian ice for several thousand years. The tusks were sold for ivory. Still more ancient were the **mastodons,** which lived in lush, tropical rain forests.

Order Sirenia

Sometimes exhibits are set up at carnivals and circuses to allow the public to gaze on a real live mermaid for a small fee. With visions of a beautiful female creature, humanlike from the waist up and fishlike from the waist down, the ever gullible people buy their tickets. They are disappointed when they walk in to see an ugly **sea cow** floating lazily in a tank of water. However, the sea cows are the real mermaids, although only a drunken sailor on a dark night could mis-

take them for the mythical sirens that give this order its name. They are found along the Atlantic coasts of Africa and South America and are numerous along lagoons of the Florida coast. Their bodies are adapted for life in the water, with the front legs modified into flippers and the hind legs absent. The tail is broadened and flattened out to form a finlike structure for swimming.

Order Rodentia

This is the largest order of mammals; it contains more genera and species than any other, and it probably contains more individuals than all other orders of mammals combined. Mammals in this order have many enemies among snakes, birds, and other mammals, but they are extremely

FIG. 27.14 Rodent teeth. The long, sharp, curved teeth are especially adapted for gnawing. This is the mouth of a deer mouse.

prolific in reproduction. They mature rapidly and the females have large numbers of offspring at frequent intervals during their lifetime. Mice are ready to reproduce three months after they are born, have a gestation period of only twenty days, and bear an average of about eight offspring every eleven weeks. Rodents are omnivorous in their diet and are man's chief competitor among the mammals for food. In fact, their diet is so nearly like that of man that they are used in dietary experiments to determine the effects of the elimination of important elements in the human diet. Rodents have no canine teeth, but the incisors are long and curved. This is the most easily recognized characteristic of the group. The teeth grow continually and the rodents must wear them down on the ends by gnawing on hard objects so they will not become too long and enter the skull.

The squirrellike rodents. This large group includes the **squirrels, ground squirrels, flying squirrels, chipmunks, gophers, ground hogs, prairie dogs, beavers,** and others. The **beavers** are the largest rodents in the United States. They gnaw down trees and place them across streams to form dams, filling in cracks between the logs with mud. They build their homes over the lakes which are formed above the dams. These homes are above the water line, but they have only underwater entrances, and they serve as excellent retreats from enemies. Beavers are beneficial because their dams aid in controlling erosion, but the popularity of their fur threatened their extinction before protective measures were put into effect.

The ratlike rodents. The many species of **rats** and **mice** are found all over the world. They do tremendous damage to grain crops during growth, after harvest, and during storage. They also help spread some of the world's most serious diseases, such as bubonic plague and typhus.

Other rodents. Porcupines have long, stiff, sharp-pointed hairs which can be made to stand erect by the use of special muscles. Although they cannot throw their quills, as is sometimes believed, they have a special mechanism which releases the quills when they become stuck in another animal, and many dogs come away from an attack on a porcupine with quills embedded in their noses and mouths. The **guinea pig** is another well-known rodent. The **chinchilla** is a beautiful little rodent which has very fine and soft fur. Chinchillas have been imported into the United States from the mountains of Chile and Bolivia, where they have been all but exterminated by trappers.

Order Lagomorpha

This order includes the **rabbits.** They were formerly placed in the order *Rodentia,* but recent studies indicate that they have distinctive characteristics which warrant their inclusion in a separate order. They have gnawing teeth, but there are two rows of these teeth rather than the single row found in the rodents. Detailed serological studies comparing the blood of rabbits with rodent blood also indicate that the relationship of the two groups is not close enough to warrant their inclusion in one order.

Order Edentata

This order name means "without teeth." Some edentates have small jaw teeth, but these have no enamel and no roots.

The armadillos. The skin of the armadillo bears bony plates that make this one of the few vertebrates that have an exoskeleton. When the armadillo rolls itself up into a tight little ball, it is about as well protected as the tortoise. Food consists primarily of insects, and the stout legs and sharp claws of the armadillo are perfect equipment for digging insects out of the ground.

The sloths. These strange creatures, somewhat like long-haired monkeys in appearance, are found in the jungles of Central and South America. They spend most of their lives hanging upside down from the limbs of trees and possess long, curved claws which can be hooked over the limbs for support. Their food consists primarily of leaves which provide nourishment and sufficient water so that they do not have to drink. When one is placed on the ground, it makes a ludicrous sight as it awkwardly pulls itself along in this unaccustomed position.

The ant bears. These animals live primarily on ants and termites. They have powerful legs and claws with which they dig open the nests and a long sticky tongue which they put down in the opened nest. When the insects are stuck to it, the tongue is withdrawn into the mouth and the insects are swallowed.

Order Primata

The most intelligent mammals are included in this order. The cerebral hemispheres of members of this order are very large, and their surface is greatly convoluted. These are features which are correlated with great mental capacity. Instead of claws or hooves, primates have nails on the tips of their fingers and toes. Most of them are arboreal animals with hands and feet adapted for grasping, a characteristic which enables them to swing from limb to limb. This grasping ability also makes them adept at handling things, and the hands, rather than the mouth, are used for picking up many objects.

The eyes of primates are on the front of the face, and images seen by the two eyes are superimposed. This produces **binocular vision**; that is, both eyes see the same thing, but from slightly different angles. This type of vision gives depth to images and aids greatly in judging distances. Accurate eyesight is closely correlated with arboreal habits, since this kind of vision is required for leaps from tree branch to tree branch. We see tiny details with our eyes far better than other mammals because of the development of a spot of clearest vision, called the **fovea,** in the center of the retina. In other mammals smell is often the dominant sense; a dog recognizes his friends by smell rather than by sight and some dogs have even learned to follow the scent of an animal while running at a high speed. The sense of smell, on the other hand, is poorly developed in the primates.

The primates have a single pair of mammary glands located in the pectoral region of the body. This is correlated with their upright sitting position. The young are born with a strong grasping instinct and, in most primates, cling to the mother's body while she leaps around in the trees. Even the human baby is born with this instinct and for a week or two after birth can grasp a limb strongly enough to support its own weight; after this time this strength is lost and an older baby is unable to support its body weight.

The **lemurs** are the most primitive of the primates. They are timid little animals found in tropical jungles and look something like a cross between a cat and a monkey. Some of them bear a claw instead of a nail on the second finger.

FIG. 27.15 The armadillo is a mammal which has developed bony plates on its skin for protection. Note the large, powerful claws which are used for digging.

The monkeys. Many tropical regions of the world have swarms of **monkeys** filling the trees. The monkeys of the New World often use the tail as a fifth leg and can swing from a tree by the tail just as they can by their legs or arms. One of the New World monkeys is familiar as the cute, somewhat timid, little **capuchin monkey** that accompanies the organ grinder. These monkeys can be taught to do many tricks. The Old World monkeys are built more stockily than the slender New World monkeys and have a large callused pad on which they sit. The **rhesus monkey** is the best known of this group. This monkey came into prominence when it was discovered that about 85 percent of all people have a factor in their blood that was first found in the rhesus. We call this the rhesus, or Rh, factor; persons having it are Rh positive; those not having it are Rh negative.

The apes. These are the most manlike of all other animals. The arms are considerably longer than the legs; thus, apes walk in a semi-upright position touching their knuckles to the ground for balance as they go. Their feet are more adapted for grasping than walking and they walk primarily on the outside edge of the foot. This gives them a characteristic shuffling, bow-legged gait. They do not have a tail as adults, but, like man, they have an embryonic tail and rudimentary tail vertebrae in the adult.

The **gibbons,** found in the Malay region, are the smallest and most primitive of the apes. They do not reach a height of more than three feet. In one respect they are more manlike than the other apes. They walk erect without touching their arms to the ground, but holding them out to their sides as balancers.

The **orangutans** of Borneo and Sumatra are probably more humanlike in facial features than any of the group. There is no development of the heavy brow ridges, which give the other apes a somewhat ferocious appearance. In an upright position, however, the similarity ceases; the fingers almost touch the ground. Care must be taken at the zoo, for they can reach through the bars of a cage for an unbelievably long distance.

The **chimpanzees,** found widely distributed in Africa, are tamed easily and, when young, seem to enjoy human company. The adults seem to realize that people are just trying to "make monkeys" of them and resent it. This makes the adults quite dangerous. Experiments show that chimpanzees can reason. They will stack boxes to climb on in order to secure food suspended out of reach. They can be taught to do many simple tricks that make them a feature attraction at the zoo or circus. They can dress themselves, eat with a knife and fork, ride a bicycle, waltz, and do acrobatics that put human efforts to shame.

The **gorillas** are the largest of the primates.

FIG. 27.16 The chimpanzee is the most humanlike of the anthropoid apes.

They grow to a height of six feet and weigh up to six hundred pounds. They can be found today only in the highland regions of Africa; those living in lower altitudes have been exterminated. In spite of their huge size and great strength, they are among the most difficult of animals to keep in captivity, for they are susceptible to the many respiratory ailments which affect people.

Man is the most highly advanced of the primates. He has attained a dominant place in the animal kingdom because of the high degree of development of his cerebrum. Physically man is a weakling; a chimpanzee, by actual measurement, has strength in his biceps several times that possessed by man, yet his biceps are smaller in size. It is only by superior intelligence that man is able to survive in a world where biological competition is so keen.

REVIEW QUESTIONS AND PROBLEMS

1. Conservation authorities advise farmers to leave shrubs and small trees along the edges of their farms. These can serve as nesting places for birds. Of what economic value is this to the farmers?
2. Buzzards are often thought of as obnoxious birds because they eat dead and decaying animals, but they play an important role in the balance of nature. Explain.
3. A bird eats much more food in proportion to its size than a snake. Explain the reason for this.
4. It is much more difficult to train a bird to do tricks, yet birds have a much better sense of coordination than the mammals. Explain.
5. In the past men have built wings and strapped them to their arms. They then attempted to fly by flapping them like a bird. Why are such attempts always doomed to failure?
6. A French short story tells of a man who incubated and hatched some chicken eggs while he was in bed with a broken leg. Is there any biological reason why this could not be done?
7. How can a comparison of the brain of a bird with that of a mammal show which of the senses is best developed in the two groups?
8. Bird migration tends to take them to regions where the days are longer. Of what advantage is a long day to the birds, outside of its possible association with warmer weather?
9. Which kinds of birds tend to have precocial young and why is this necessary?
10. Why was *Archaeornis* considered to be a bird in spite of its reptilelike characteristics?
11. Some birds migrate for thousands of miles over open water. What are some of the ways that they find their direction?
12. Why is the duckbill considered to be a mammal even though it lays eggs?
13. What is a cloaca and how is it related to the characteristics of the mammals?
14. How are the marsupials advanced over the monotremes and more primitive than the other mammals?
15. Why is the shrew considered to have the smallest size possible for mammals?

16. If you were shown the teeth from a rodent, a carnivore, and an ungulate, how could you distinguish from which group each tooth came?
17. How is the complex stomach of ruminants an adaptation for survival?
18. The largest aquatic mammals are considerably larger than the largest land mammals. How might this be explained? (Consider the skeleton and the problem of support.)
19. What one characteristic is most highly developed in man, a characteristic which has enabled him to dominate the other animals of the earth?
20. In spite of the fact that they are eaten by many other animals, the rodents are the most numerous of all the mammal orders. What characteristic do they possess which makes this possible?

FURTHER READING

Blair, W. F., et al. 1968. *Vertebrates of the United States.* 2nd ed. New York: McGraw-Hill.

Napier, J. R., and Napier, P. H. 1967. *Handbook of Living Primates.* New York: Academic Press.

Norris, K. S. 1965. *Whales, Dolphins, and Porpoises.* Berkeley: U. of California Press.

Robbins et al. 1966. *Birds of North America.* New York: Golden Press.

Schultz, A. H. 1969. *Life of Primates.* New York: Universe Books.

Winchester, A. M. 1970. *Concepts of Zoology.* New York: Van Nostrand Reinhold.

28

Support, Protection, and Movement in Vertebrate Animals

In our survey of representatives of the animal kingdom, we have considered some of the organ systems which sustain life. Since the vertebrates are of particular interest to us as members of this group, we shall consider, in somewhat greater detail, these systems in the vertebrate body. The human systems will be the primary source of illustrations, but there will be reference to other vertebrates too. This chapter surveys those systems involved with support, protection, and movement, since these are all closely correlated.

THE SKELETON

Did you ever wonder what shape you would be without a skeleton? Without your skeleton, you would become a soft blob without any particular shape because your other body parts are too soft to support themselves. You can see how important your skeleton is to you. In the arthropods and other invertebrate phyla, we found that an **exoskeleton** is the usual skeleton type, but an **endoskeleton** is typical of the vertebrates. The endoskeleton lacks some of the protective features of the exoskeleton, but it has advantages—growth and maneuverability—over the exoskeleton.

The Axial Skeleton

This part of the skeleton includes the skull

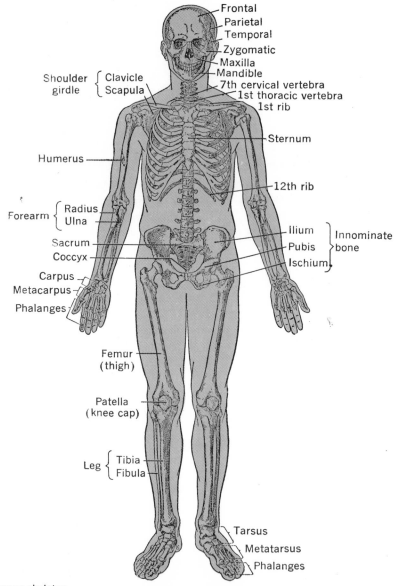

Frontal
Parietal
Temporal
Zygomatic
Maxilla
Mandible
7th cervical vertebra
1st thoracic vertebra
1st rib

Shoulder { Clavicle
girdle { Scapula

Sternum

Humerus

12th rib

Forearm { Radius
{ Ulna

Ilium
Sacrum
Pubis
Coccyx
Ischium

Innominate
bone

Carpus
Metacarpus
Phalanges

Femur
(thigh)

Patella
(knee cap)

Leg { Tibia
{ Fibula

Tarsus
Metatarsus
Phalanges

FIG. 28.1 The human skeleton.

and vertebral column, along with attachments like ribs. The skull includes the **cranium, jawbones, and facial bones.** The cranium encloses and protects the brain, but it is perforated with a number of openings which permit the exit of nerves from the brain. Most of these openings are small and accommodate the small nerves that run to the face and sense organs of the head. There is one rather large opening, the **foramen magnum,** at the base of the cranium, through which the spinal

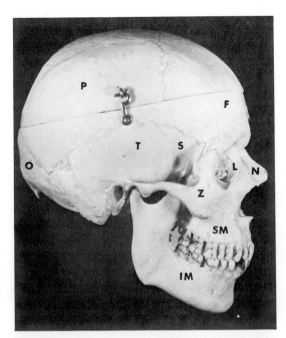

FIG. 28.2 A human skull. The major bones which can be seen are: P—parietal; O—occipital; T—temporal; S—sphenoid; F—frontal; L—lacrymal; N—nasal; Z—zygomatic; SM—superior maxillary; IM—inferior maxillary.

cord passes. Closely associated with the cranium are the **auditory capsules,** which contain the ear bones; the **orbits,** sockets for the eyes; the **nasal cavity;** and the **sinuses.** The skull is made of a number of separate bones, but they are all fused together except the large bone which forms the lower jaw. This bone is separate and is attached to the rest of the skull by movable joints which permit the opening and closing of the mouth. The **teeth** are a part of the skull. They are mostly of the same kind in most vertebrates, with some exceptions such as the fangs of poisonous snakes. Modern birds do not have teeth, which means the mammals are the only class possessing specialized teeth. There are three types

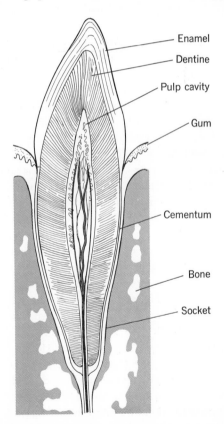

FIG. 28.4 Section through a tooth and its surrounding socket of bone.

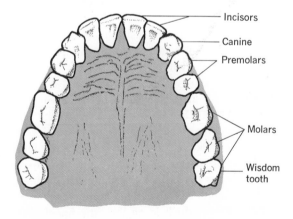

FIG. 28.3 Human upper jaw showing the types of teeth and their arrangement.

of teeth in man—the **incisors,** at the front of the mouth (used in biting) ; the **canine,** or eyeteeth; the **premolars** and **molars** (used for grinding the food). If we start at the center of the front teeth and consider only those found in half of the lower jaw, there will be two incisors, one canine, two premolars, and three molars in the normal adult mouth. This makes eight for half of one jaw or thirty-two for the entire mouth. The third molar, the "wisdom tooth," is often not functional and may cause considerable trouble when it pushes through the skin long after the other teeth are fully formed. During early childhood the molars are lacking entirely, but after the "milk teeth" are shed, the molars come in along with the other teeth that had predecessors in the first set of teeth.

The **spinal column** is made of a series of **vertebrae** with the spinal cord running through them. Disks of cartilage are found between the vertebrae and these permit a limited movement of the vertebral column, thus giving flexibility to the back. In human beings these disks also act as shock absorbers which prevent the jarring of the brain with each step. The upright position of the human body puts a considerable strain on the vertebral column, and "back trouble" is one of the most common of human complaints. Some of the disks between the vertebrae in the lower part of the back may slip out of position, causing adjacent vertebrae to turn in an abnormal position. Breaks in the vertebral column can be very serious because of the danger of injury to the spinal cord within.

In man the vertebrae are divided into seven **cervical** (neck) vertebrae; twelve **thoracic** (chest) vertebrae; five **lumbar** (lower back) vertebrae; five **sacral** vertebrae fused together to form the **sacrum;** and about four **caudal** (tail) vertebrae, fused together to form the **coccyx,** a vestige of the tail. The first cervical vertebra that articulates with the skull is the **atlas,** and the second one is the **axis.**

The structure of a single vertebra can best be studied by removing it from a skeleton and looking at it anteriorly. There will be a ventral, oval portion composed of solid bone. This is the **centrum.** A **neural arch** of bone extends dorsally from the centrum and encloses the **neural canal** which contains the spinal cord. A **neural spine** extends dorsally from the neural arch, and **transverse processes** extend to the sides. There are also small **articulatory processes** where one vertebra joins the vertebra above it in the spinal column.

Each of the thoracic vertebrae bears a pair of **ribs** which extend out and encircle the chest. The ventral tips of the ribs are made of cartilage rather than bone. These are the **costal cartilages** and they provide the flexibility needed to expand the chest. In rare instances these cartilages become ossified, producing a rigid chest which cannot be expanded. A person with this kind of chest may still live since breathing may be accomplished by the diaphragm alone. The first ten ribs are attached to the **sternum,** breastbone, by the costal cartilages; these are the **true ribs.** The last two ribs have no such attachment; these are the **floating ribs.**

The Appendicular Skeleton

The appendicular skeleton consists of the arm and leg bones together with the girdles to which they articulate.

The pectoral girdle and arms. In man the **pectoral girdle** consists of the **scapula,** shoulder blade; the **clavicle,** collarbone; and the **sternum,** breastbone. The scapula contains a hollowed-out socket, the **glenoid cavity,** into which the **humerus,** upper arm bone, fits. There are two bones in the forearm: the **radius,** which is on the thumb side and is the shorter of the two, and the **ulna,** which is on the little finger side and projects out to form the **olecranon** at the point of the elbow. There are eight small wrist bones, the **carpals;** five long hand bones, the metacarpals; and the **phalanges,** which form the finger bones. There are two phalanges in the thumb and three in each of the other fingers.

The frog pectoral girdle has the same essential parts that are present in man, but some of the parts are much better developed in the frog. For instance, the frog has a rather large **coracoid** bone that extends from the scapula to the sternum; whereas in man this bone has a homologue in the small coracoid process projecting out only slightly from the scapula. The humerus is similar to that of man, but the radius and ulna are fused to form the **radioulna.** There are six metacarpals and five carpals, but only four digits. The thumb is rudimentary and possesses no phalanges.

The pelvic girdle and leg bones. The pelvic girdle of the human skeleton forms something of a basket that contains the internal organs of the abdominal region. A part of the dorsal portion of this basket is formed by the sacrum, which is a part of the vertebral column. The rest of the basket is formed by a fusion of three bones on either side of the body. The large crest of the pelvis which projects anteriorly is the **ilium.** This joins the **ischium** posteriorly. The ischium is the part of the girdle that we sit on. Then coming off from the ischium ventrally is the **pubis,** which joins with the pubis from the other side of the body to form the **pubic symphysis** in front of the girdle. Ordinarily this joint is immovable, but in the later stages of pregnancy the tissue joining the two bones becomes soft, and during childbirth there may be some movement at the pubic symphysis allowing enlargement of the pelvic cavity. This circular cavity in the center of the pelvis is surrounded by the bones just described. It is considerably larger in the female skeleton than in the male.

At the point where the three pelvic bones come together there is a socket, the **acetabulum,** into which the **femur,** thighbone, articulates. At the knee there is the **patella,** kneecap, and a junction with the bones of the shank. The shank corresponds with the forearm and bears two bones: the **tibia,** shinbone, which corresponds to the radius of the arm, and the **fibula,** a thinner bone, which corresponds to the ulna. The tibia lies just beneath the skin on the front part of the shank and receives many painful blows because of its exposed condition. It is called the "boneyard of the body" by surgeons; they often go to the tibia, because of its easy availability, to secure bone for grafts to injured bones in other parts of the body. Many a person carries parts of his shinbone in the vertebrae of his back where they were placed to repair crushed parts of his spinal column. Next, there are seven **tarsals,** anklebones, the largest of which is the heel bone, the **calcaneum.** The instep of the foot contains five **metatarsals,** corresponding to the metacarpals of the hand. Finally, there are two **phalanges** in the big toe and three in each of the other toes. The metatarsals often become dislocated from their normal attachment to the phalanges and drop down in the region of the ball of the foot. This puts an undue pressure on the skin at this region and usually results in the formation of calluses in addition to the pain that accompanies this condition. Metatarsal supports may be placed under the fallen bones to relieve both conditions.

The frog pelvic girdle appears quite different from that of man, but the same bones are present and occupy the same relative positions. The ilium is much longer and thinner, and the pelvic cavity is smaller in proportion to size. The femur is the same as in the human skeleton, but the tibia and fibula are fused to form the **tibiofibula.** Two of the tarsals are greatly elongated, giving the frog a long ankle which provides the additional leverage necessary for leaping. The big toe, like the thumb, is reduced in size, but the other digits are long.

Joints Between Bones

The human skeleton is made of 206 bones, and the region where two or more of these bones meet is known as a joint. Many of these are **immovable joints,** such as are found at the junction of bones in the skull and in the pelvic girdle. The points of union of the bones of such joints are known as **sutures,** and they can easily be distinguished on the skull of either the frog or man as irregular lines where the bones fit together somewhat like the pieces of a jigsaw puzzle. **Movable joints** are found where the adjacent ends of the bones can move freely. The ends of the bones at such joints are usually padded with cartilage, and the entire joint is enclosed in a joint capsule, or **bursa.**

The bursa is filled with a lubricant, the **synovial fluid,** which makes joint movements smooth and free of excessive friction. Occasionally a person will receive an injury to a joint and

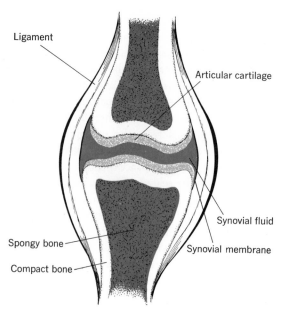

FIG. 28.6 Section through a hinge joint, a typical movable joint.

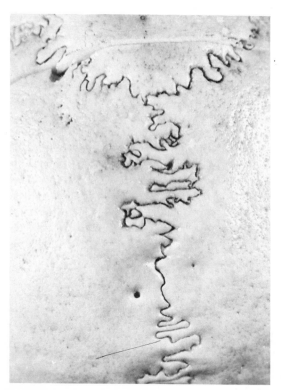

FIG. 28.5 A suture of bones of the human skull, typical of immovable joints.

some of the synovial fluid will leak out and accumulate at the region. This causes a swelling. If the joint is at the knee, the condition is known as water on the knee. Irritation of the cartilages found on the ends of the bones within the bursa is known as **bursitis.** Such irritation is frequently found at the shoulder joint.

The bones must be held together at the joints and the membrane of the bursa cannot do this alone. There are tough **ligaments** extending from one bone to another which hold the bones firmly yet allow for the necessary movement. Ligaments can be pulled loose from the attachment on the bone when there is undue strain; a pulled ligament is an incapacitating injury suffered by many athletes and one which heals slowly.

There are several different types of movable joints. The **ball and socket joint** is found at the shoulder and the hip. The femur, or thighbone, bears a ball-shaped extension which fits into the acetabulum of the pelvis in a manner which permits a free rotary movement at the hip joint. The elbow and the knee are examples of **hinge joints,** where the movement is limited to one plane. **Gliding joints** are found at the wrist, the ankle and between the vertebrate. Here the bones slide on one another and permit a wide degree of movement. Finally, a **pivot joint** is found in the articulation of the skull with the spinal cord. The atlas fits into the axis in such a manner as to permit a rotation of the head.

Histology of Bone and Cartilage

The vertebrate skeleton is composed of two types of tissues—bone and cartilage. A longitudinal section through a shaft bone can illustrate bone structure. The outer surface of the bone is covered by a membrane, the **periosteum,** on all its surfaces except at the joints. The periosteum is rich in blood vessels and nerves which pass from it into the bone. Underneath the periosteum is a layer of **compact bone,** but the central part of the bone is entirely hollow or contains only a spongy network of bone tissue. This structure of a bone provides the maximum of strength with a minimum of weight. The compact bone is in the form of a cylinder, and this represents the most efficient arrangement to give strength. The hollow spaces in the bone are filled with **marrow.** The central part of the shaft contains the fatty, yellow marrow which we see every time we eat a piece of round steak. The spongy area at the ends of the bones may contain red marrow, which bears a vital relationship to the circulatory system; new blood cells are formed in the red marrow.

A microscopic examination of a cross section of the outer, compact bone reveals a series of concentrically arranged layers, the **lamellae.** In the center of each series there is a cavity, the **Haversian canal,** through which the blood vessels and nerves run. Between adjacent lamellae

are small cavities, the **lacunae,** which contain the bone cells, **osteoblasts.** Small canals, the **canaliculi,** extend out in all directions from each of the lacunae into the adjacent lamellae. Processes from the osteoblasts extend out into these canaliculi. The outer part of the bone, just under the periosteum, contains many osteoblasts which lay down new bone, thus permitting increase in bone diameter during the growth period of life. Bone is incapable of further growth, once it is laid down, because of the hardness of the calcified matrix between the cells. If there is an injury to the bone, such as a break, the osteoblasts in the lacunae become active again and lay down new bone tissue, healing the injury. In fact, bone tissue heals more readily than the majority of body tissues.

Cartilage resembles bone in its structure. There are scattered cells in **lacunae** surrounded by a **matrix,** but the matrix is not calcified and cartilage is able to carry on interstitial growth, growth by production of new material throughout the cartilage. A **perichondrium** around the

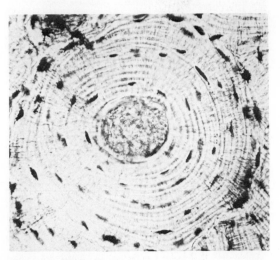

FIG. 28.7 Section of compact bone from the femur. Note the Haversian canal in the center and the small lacunae arranged concentrically around the canal. The processes extending outward from the lacunae are the canaliculi.

outside of cartilage corresponds to the periosteum of the bone. The matrix between the cells varies in nature in the different types of cartilage. **Hyaline cartilage,** such as is found in the costal cartilage, has a clear, homogeneous matrix. **Elastic cartilage,** such as that found in the external ear, has a yellowish colored matrix in which numerous elastic fibers are imbedded. **Fibrous cartilage,** such as is found in the disks between the vertebrae, has such an abundance of fibers that the clear matrix is not even visible. This type of cartilage appears fibrous even when examined with the naked eye.

Formation of the Skeleton

As the skeleton is formed in the embryo, most of the bones are laid down first as cartilage. By the time a human embryo is seven weeks old, rudiments of most of the bones of the body are already present as cartilaginous precursors. This

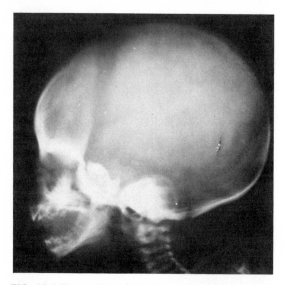

FIG. 28.8 X ray of the skull of a newborn baby. Note the cartilaginous separations between the bones and the fontanel at the top of the head. (Courtesy of H. K. Robinson, University of Colorado Medical School.)

is an advantage so far as growth is concerned. Cartilage is capable of growth by internal expansion, while a bony skeleton could grow only by addition of materials on the outer surface. Such a growth would not permit the many articulations and muscle attachments which are necessary in embryonic development. At the beginning of about the eighth week, ossification centers develop in the cartilages and bone is laid down. Special cells, **osteoblasts,** move out from these centers, depositing calcium carbonate and calcium phosphate which replace the softer tissue with true bone.

A few bones of the adult skeleton are laid down as **membranes** in the embryo. These include most of the bones of the cranium and the clavicles (collarbones). These bones develop as thin membranes in the dermis of the skin. Bone is deposited on either side of these membranes, and they then sink down beneath the skin and become a part of the skeleton. Such **membrane bones** are in the nature of modified scales. At birth, ossification of both cartilages and membranes is rather advanced, but the bones are by no means completely formed. There is a conspicuous soft spot, the **fontanel,** at the top of a newborn baby's head that persists for several months. There is a heavy blood supply to this region and the pulse of a baby may be taken by merely watching the pulsations of the fontanel. Sometimes needles are inserted through this membrane to take blood samples or to give blood transfusions because the veins of the arm are so small in a newborn baby. Although this incomplete ossification of the skull subjects the brain to possible injury at this point, there could be no normal births without it. The opening in the pelvic girdle of the female skeleton is too small to pass the head of a baby ready for birth. The skull must be bent in the proper shape during birth. A newborn baby may have a strangely shaped head as a result of this bending, but the head returns to its normal shape after a few days. Babies delivered through a Caesarian section do not have this misshapen head during their first few days of life.

Since the bones are rigid, they grow by means of regions that remain active after the rest of the bone is completely ossified. In the shaft bones, such as are found in the arms and legs, there is a growth disk near each end of the bone that allows it to elongate during the period of growth of the child. After adolescence these disks become ossified and bone growth ceases. There are certain cartilaginous parts of the skeleton that do not become ossified. The front part of the ribs remain cartilage, as do the external ears, the tip of the nose, and the ends of bones where they come togther to form a joint. Occasionally ossification gets out of control and not only ossifies cartilage that should remain cartilage, but even ossifies surrounding muscles and other body regions. This results in death when a vital body organ is affected. Types of arthritis are caused by the ossification of the cartilages at the joints. This interferes with movement of the joints, causing pain and swollen joints.

Functions of the Skeleton

Perhaps the most obvious function of the skeleton is **support.** Most of the other cells of the body are soft and incapable of supporting the body. Because of their generally soft condition, most animal cells are liable to mechanical injury from external forces. Such injuries occur frequently to the skin, but these are not serious because the skin will grow back and repair the damage. There are many vital organs, however, which must be protected, because injury to these organs will result in death. The skeletal system forms a **protective armor** for these organs. The brain is fully encased in bone, and the rib cage surrounds the heart and lungs. The skeleton also serves as a place for the muscles to **attach.** The voluntary muscles have no value unless they are

attached to a firm structure. If the tendon of Achilles, just above the heel, is severed, a person will lose use of all the calf muscles, because the tendon of Achilles attaches the calf muscles to the bones of the foot. Without this attachment the contraction of the calf muscles results in no movement of the foot.

The skeleton of vertebrates contains tissues which **produce blood cells.** Red bone marrow is found in certain bones and is especially abundant in the sternum, or breastbone. Red blood

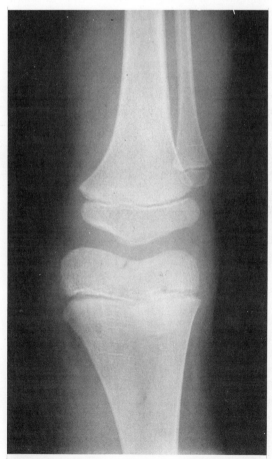

FIG. 28.9 The growth disks of the long bones of a child's leg shows clearly in this X ray of the knee joint. (Courtesy of H. K. Robinson, University of Colorado Medical School.)

cells, most white blood cells, and the blood platelets are all formed in this red bone marrow.

The bones also serve as an important **reserve of certain minerals.** The mineral content of the bones is constantly being renewed. It is estimated that over a fifty-day period, about one-fourth of the minerals in the bones is removed and replaced. Calcium and phosphorus are especially abundant in the bones, and the level of these two minerals in the blood must be maintained at a constant level of concentration. When the diet is low in these minerals, they can be drawn from the bones, while any excess in the diet can be stored in the bones for possible future use. During pregnancy, when the demand for minerals to form the bones of a growing embryo is great, a woman's own bones and teeth may be excessively depleted unless her diet is rich in calcium and phosphorus. An old wife's tale says that a woman will lose a tooth for each child she bears. With our modern knowledge of nutrition, this certainly need not be true. During starvation the skeleton serves as a source of minerals, maintaining mineral concentration in the blood at the proper level, and life is sustained for a much longer period than would be possible without this source.

The exchange of calcium and phosphate ions between the blood plasma and the bones is rapid. When the bones are not under stress, the outgo of these ions from the bones exceeds the intake and there is a gradual depletion. As a result, the bones may become soft and spongy and easily broken. A long period of confinement to bed can bring about such a condition. An even more rapid depletion can take place when an animal is kept in a weightless state. This has been demonstrated with dogs. This fact is of great concern because astronauts experience this weightlessness. A prolonged stay under such conditions might have serious consequences on the skeleton. Some system may have to be devised to create an artificial gravity for men sent on prolonged space missions.

THE INTEGUMENTARY SYSTEM

The integumentary system is closely related to the skeletal system and often studied along with it. Like the skeleton, the integument serves protective functions and helps give the body its shape. This system is composed of the material which we call skin, together with its various outgrowths.

Structure of the Skin

The skin is composed of two main parts—the rather thin, outer epidermis, which is tough and insensitive, and the inner dermis, which contains many blood vessels and nerve endings.

Epidermis. The epidermis is composed of epithelial cells which are columnar in shape when they first form, but as they are pushed to the outside, they form a flat sheet of cells (**squamous epithelium**). These outer cells of the epidermis constantly flake off as new cells grow in and replace them. We do not notice this process on most parts of our body, but the hair may retain these flakes, producing a condition we call dandruff. The epidermis varies in thickness on different parts of the body. In man it is

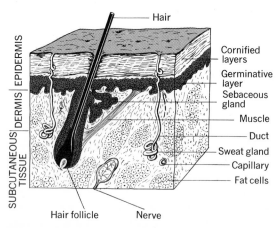

FIG. 28.10 Enlarged diagrammatic section of human skin.

thickest on the palms of the hands and the soles of the feet, and even here the thickness varies according to use. A child accustomed to going barefooted develops tough leatherlike soles on his feet which protect about as well as leather soles on shoes. A man who does hard manual work with his hands has a thickened epidermis on his palms which withstands injury as well as a pair of leather gloves. This is an excellent illustration of the adaptability of living things to the conditions of their environment. The development of corns and calluses on the feet and hands protects the softer parts of the skin underneath. These growths cause pain, however, when they are pressed tightly against the sensitive tissue beneath. On parts of the hands and feet the epidermis becomes pushed up in characteristic ridges which are different for every person. This makes identification by fingerprints possible.

Dermis. The dermis is much thicker than the epidermis and is made up mainly of connective tissue. In the outer part of the dermis this tissue is composed of thickly matted fibers. This is the part which is tanned to make leather from the skin of cattle, pigs, and horses. Beneath this there is a layer composed of more loosely woven fibers of connective tissue, blood vessels, nerves, and sense organs. At the innermost part of the dermis there is **adipose tissue** (fat cells). This fat deposit is a food storage depot and an insulation layer which prevents loss of body heat during cold weather. Whales have a very heavy layer of this fat and are thus insulated from the cold waters in which they may live. As a rule, women have heavier deposits of fat than do men. As a result, they can tolerate exposure to cold better than men as a group. Notice the reaction of the two sexes around a swimming pool of cold water. The women may complain about cold water when they first enter it, but they are usually still comfortable long after their manly escorts are blue with cold. Unfortunately, this deposit of fat under the skin tends to diminish after a person has attained physical maturity. This accounts for the wrinkling of the skin and the greater sensitivity to cold which comes with advancing years.

Outgrowths of the Skin

A number of structures grow out from the skin and therefore are considered a part of the integumentary system.

Hair. Hair is the most abundant skin growth in most mammals. The entire skin of man, except the palms of the hands and the soles of the feet, contains many hair follicles. Hair originates as a growth from follicles in the inner part of the dermis. By the time a hair extends from the epidermis, however, it is dead tissue and it only becomes longer by the addition of new material at its base. As long as the follicles are in good health there is an indefinite succession of new hairs replacing old hairs which are shed. In many men the follicles on top of the head lose their vigor and fail to replace the hairs which are shed. This results in baldness, a condition influenced by both heredity and sex. Each hair has a small muscle in the dermis which can elevate it. Such elevation ("the hair stands on end") comes about automatically when an animal is cold. This hair erection ruffles up the fur and traps many small dead air spaces that provide excellent insulation. Man has this reflex—he gets goose pimples when he gets cold—but this does him little good because of the sparseness of hair on his body. Fright also initiates this reflex. A cat looks twice its normal size when disturbed by a dog, but man looks no more ferocious with goose pimples than without them.

Scales and feathers. Fish and reptiles have outgrowths of the skin known as scales which furnish excellent protection. Birds have feathers, which appear to be homologous with the scales of reptiles because of their similar embryonic origin. The amphibians have no such outgrowths.

Skin glands. The skin contains glands of several kinds. There are small sebaceous glands which secrete an oily substance. They are found in close association with the hair follicles and help keep the hair in good condition. The sweat glands also originate in the dermis. They have slender tubes which carry their products up through the epidermis and empty through pores on the outside. They excrete water and some minerals and are regulated by the autonomic nerves. During very hot weather the loss of minerals through the sweat glands may be so great as to result in heat prostration. The danger of this happening can be reduced by taking extra salt during such times. The mammary glands which are characteristic of the mammals are skin glands specialized for nursing.

Nails and similar outgrowths. The nails appear as outgrowths of the epidermis. They are composed of closely packed dead cells which are somewhat transparent. This allows the blood in the underlying capillaries to show through and give the nails their natural pink color. In some mammals there is an outgrowth of a similar nature from the top of the head which forms structures known as horns. (The so-called horns of the rhinoceros however, are composed of fused hairs and are not homologous to the horns of cattle, deer, and other ungulates.) Teeth should also be included in this category, even though they are commonly studied as a part of the skeleton. There is an infolding of the skin into the mouth, and the teeth are formed as outgrowths of this skin.

Functions of the Skin

Mechanical protection. The skin is a relatively tough, yet pliable, covering of the body. It protects the more delicate tissues beneath it from external friction or blows. Man's skin is less efficient in this regard than that of most mammals because it is thinner, softer, and not covered with hair. The skin of a rhinoceros is so tough that it can deflect a rifle bullet.

Protection from temperature variations. The skin of man serves an important function in regulating the temperature of the body. If the interior of the body tends to become too warm, the small blood vessels in the skin become dilated and heat radiates out from the skin. To cool the blood in the skin when the body is overheated, the sweat glands excrete greater amounts of perspiration. The evaporation of the water from the surface of the skin cools it and this helps cool the blood flowing through. In cool weather, on the other hand, the vessels in the skin contract, and even though the skin may get cold, the body heat is conserved in the deeper tissues. Likewise the sweat glands reduce their activities and there is no further cooling from evaporation. All of this is regulated by a part of the autonomic nervous system which acts as a thermostat. This system is discussed more fully in Chapter 29.

Protection from infection. On the surface of your skin at this moment there are bacteria which could cause serious infection and even death if they were able to grow inside your body. Your skin, however, provides an impervious barrier which bars their way. When this barrier is removed by injuries which break the surface of the skin, these bacteria may enter the body.

Protection from light. This is an important function of the skin. Ultraviolet rays are destructive to exposed protoplasm; they kill bacteria and other microorganisms within a few minutes. Sunlight is high in ultraviolet radiation and we could stand little exposure to it were it not for a filter of pigment in the lower part of the epidermis. This pigment known as **melanin,** varies greatly with individuals within a race and with the different races. In general, the races which have developed in tropical regions, where the sun's rays are more intense, have a heavier deposit of melanin than their fellow human beings of colder climates. This is racial adaptation. A fair-skinned person usually will develop a coat of tan after a period of judicious exposure to sunlight. Many of us have learned, through sad experience, just how damaging the ultraviolet rays can be to human protoplasm by undue exposure of the skin to the sun's rays when the protective filter of melanin is not well developed. A few people do not develop melanin and they cannot remain in bright sunlight more than a few minutes at a time without developing serious burns.

Protection through pain. The skin is liberally supplied with sense organs which can give rise to the sensation which we know as pain. Some might feel that this is not an advantage. Pain, however, is nature's warning of danger, and without it we would be liable to serious injury without knowing it. A tack in a shoe could badly lacerate the foot with great damage to skin, muscles, and even to the bone if we did not have the warning of pain at the first prick of the skin.

MUSCLES

The power to contract seems to have been one of the original properties of animal cells. It is possessed by all the cells in the simple animals. As animal bodies became more complex, the cells became specialized for specific functions, and this power of contractility is greatly reduced in many of the cells. Other cells have retained and ex-

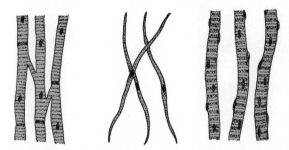

FIG. 28.11 Diagram to show the basic differences between the three kinds of muscle tissue. Left, striated; center, smooth; right, cardiac.

tended their power of contraction and take care of the body movements of higher animals. These are the muscle cells, which collectively make up the muscular system—a rather extensive system; about 40 percent of the weight of the human body is muscle, and muscle tissues are found in all regions of the body. There are three kinds of muscles: **striated, smooth,** and **cardiac.** These differ in their histology and physiology, so they will be studied separately.

Striated or Skeletal Muscle

Muscles of this type are called striated because when they are viewed under the microscope they are seen to be composed of many long, cylindrical fibers which bear small cross striations. They are also called skeletal muscles because they are closely related to the skeleton, and the majority of them are attached to some part of the skeleton. The movements that take place at the movable joints of the body are controlled by such muscles, which pass freely over the joints but are attached to the bones on each side. The muscle may be attached directly to the bone or, more commonly, by a connective tissue extension known as a **tendon.** When the muscle contracts, the bones at either end may move, but one end will probably move much more than the other. It is customary to call the point that moves the most the **insertion** and the point that moves the least the **origin** of the muscle. In the case of limb muscles, the origin is usually the end nearest the trunk, and the distal end of the muscle is the insertion.

For the muscle to exert proper control over movements of the joints, not only must it be able to initiate movements by shortening itself, but it must also be able to slow down movements by exerting a resisting force as the muscle is stretched by other forces. Without this property of muscle we would not be able to walk downhill, and all our movements would be jerky

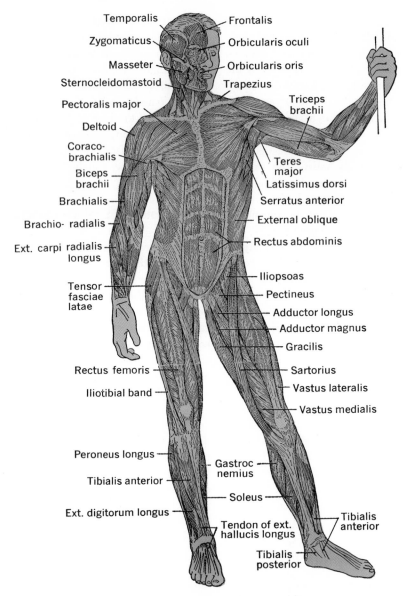

Temporalis
Zygomaticus
Masseter
Sternocleidomastoid
Pectoralis major
Deltoid
Coraco-brachialis
Biceps brachii
Brachialis
Brachio- radialis
Ext. carpi radialis longus
Tensor fasciae latae
Rectus femoris
Iliotibial band
Peroneus longus
Tibialis anterior
Ext. digitorum longus

Frontalis
Orbicularis oculi
Orbicularis oris
Trapezius
Triceps brachii
Teres major
Latissimus dorsi
Serratus anterior
External oblique
Rectus abdominis
Iliopsoas
Pectineus
Adductor longus
Adductor magnus
Gracilis
Sartorius
Vastus lateralis
Vastus medialis
Gastrocnemius
Soleus
Tendon of ext. hallucis longus
Tibialis anterior
Tibialis posterior

FIG. 28.12 Front and back view of the skeletal muscles of the human body. (Redrawn from Millard and King, *Human Physiology and Anatomy*, W. B. Saunders, 1962.)

as muscles contracted without any resistance from opposing muscles. This resisting force enables us to fix a joint in position, as is done when we maintain a standing position.

We might expect to find an extremely complicated mechanism to do all these things. Actually, the arrangement is quite simple. Each muscle consists of a large number of muscle fibers grouped together in **motor units.** The nerve which innervates the muscle is composed of many nerve fibers, and one nerve fiber runs to each motor unit. When only one of these nerve fibers stimulates the muscle, its particular group of muscle fibers is activated and only those fibers contract. Thus, by stimulating varying numbers of nerve fibers, the brain can regulate the number of motor units pulling at one time and consequently the force which the muscle as a whole exerts.

The way in which this system works may be illustrated by holding a book in your hand. If your brain stimulates enough motor units, you will be able to exert enough of a pull to lift the book against gravity. With slightly fewer motor units in action, you balance the pull of the muscle against the force of gravity and are able to hold the book in place. If still fewer motor units are stimulated, gravity will pull the book down, but you will be able to lay it down gently on a table thanks to the resisting action of the muscle fibers that are being stretched while they are being stimulated to contract. Finally, if no motor untis are being stimulated, your hand and the book will drop in the absence of any resisting force.

The movement of the body that results from such contractions is called the **action** of the muscle. Several muscle actions might be mentioned to illustrate this principle. Limb muscles that bend a part of the limb are **flexors.** Opposing muscles that straighten the limb out are **extensors.** If your hand is open and you draw it up into a fist, you have flexed your fingers. The muscles that accomplish this action are, therefore, flexors. If you open your hand, you extend your fingers. The muscles involved in this action are extensors. If you are standing and draw the lower part of your leg up toward your hips, you can feel the powerful flexor muscle on the back of the thigh contracting. To straighten your leg out again, the extensor muscles on the front of the thigh come into play.

Muscles that pull a limb away from the median body line are **abductors,** and those that pull it back toward the median line are **adductors.** If your arm is at your side and you raise it up to the level of the shoulders and hold it there, you use the abductor muscles of the shoulder and upper part of the arm. The adductor muscles, around the pit of the arm, are the ones that pull the arm down.

Muscles that rotate a body part are called **rotators;** some of them move the structure clockwise and others move it counterclockwise. Turning your head from side to side is an example of this action.

There are always such opposing sets of muscles that do just opposite things, and purposeful movements of the body are possible only through muscle coordination, with just the right degree of resistance and relaxation being exhibited by one set of muscles when its opposing set is contracting to move the body part. The great athlete is not necessarily the one with greatest muscular development but the one that is able to achieve the highest degree of coordination between the parts of a well-balanced set of skeletal muscles.

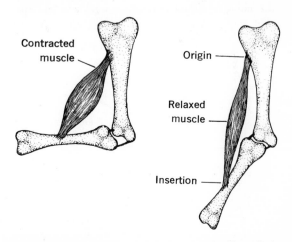

FIG. 28.13 The action of a flexor muscle. Note that the primary movement of the bone is at the point of insertion.

The Physiology of Muscle

Muscle action may be studied through isolated muscles and their connecting nerves which have been removed from an animal body. This is possible because the muscle tissue remains alive for some time after the body is technically dead. When a human body is being embalmed there may be a considerable movement of the muscles as the still living cells are stimulated by the embalming fluid. This is likely to be quite frightening to one who witnesses it for the first time, but it does not mean that the person is not really dead. The vital processes of the body have stopped, but the muscle cells retain their stored up ATP which can release energy for a time. For experimental purposes it is better to use the muscles of cold-blooded animals since these tissues stay alive longer at normal room temperatures. The calf muscle of the frog leg has long been a favorite of physiologists, and much that we have learned about muscle action has come from experiments with this muscle.

An experiment on fatigue will illustrate how the frog leg is used. The calf muscle together with its connecting nerve is dissected out. When the nerve is stimulated with a weak electric current the muscle will contract just as if it had

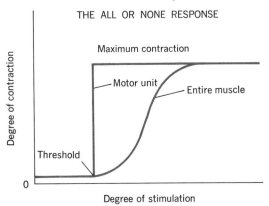

THE ALL OR NONE RESPONSE

Maximum contraction

Motor unit

Entire muscle

Degree of contraction

Threshold

0

Degree of stimulation

FIG. 28.14 The all-or-none response of a motor unit compared with the gradually increasing response of a whole muscle.

received a normal nerve stimulation in the living frog. But if the stimulations are continued in rapid sequence, the response will diminish and finally cease altogether. It might be assumed that this cessation is due to muscle fatigue, but if the electric current is then applied directly to the muscle, it will contract with its original vigor. We can conclude that the muscle fibers themselves were not fatigued; rather it is the nerve connections which become fatigued and fail to transmit the current (see Chapter 29).

Another experiment on the relationship of the force of contraction to the force of stimulation brings out an interesting aspect of muscle physiology. A very weak electric current causes a very weak contraction of the muscle. Increasing the strength of the current causes increasingly strong contractions up to the maximum degree of contraction. Then stronger currents cause no greater degree of contraction. Do the motor units respond with greater force with increasing stimulation up to a certain maximum stimulation? Experiments with individual muscle fibers show that this is not true. It is possible to dissect out a single fiber and stimulate it through its nerve connection. There is no variation in degree of contraction. A very weak current will not elicit any response at all, but when the current crosses a certain "threshold" the fiber will contract to its fullest extent. Increasing the force of the current brings no increase in the strength of the contraction. This is what is known as the **all-or-none** response, a principle which applies to many physiological reactions. It might be compared to the shooting of a rifle—the force delivered by the bullet does not come from the force used in pulling the trigger. The potential force is in the powder within the cartridge. The trigger releases this force, but to do this it must be pulled hard enough to trip the hammer which explodes the cartridge. Less force on the trigger elicits no response at all; more force does not expel the bullet with any greater power; it is all or nothing at all.

With this information on the response of individual motor units we can see that it is not the variation in response of the fibers that gives the variation in the vigor of response of a muscle. Rather it is the fact that a stronger impulse stimulates a greater number of units, and thus the muscle as a whole responds more vigorously. After the units are all stimulated, there can be no increase in vigor even though the stimulation is made much stronger.

It is possible to make records of the contractions of isolated muscle preparations through the use of an apparatus as shown in Fig. 28.15. The muscle is attached to a point on a lever in such a way that the point makes a mark on a revolving drum (kymograph) when the muscle contracts. A tuning fork which vibrates at the rate of 100 times per second may also be attached to a marking tip on the smoked drum. This permits an exact timing of the events involved in muscle contraction. For instance, a single muscle twitch can be recorded and studied. When the stimulation is applied, there will be a lag of about 0.01 second before the muscle begins to respond. Then, about 0.04 second is required before the muscle reaches its maximum degree of contraction. Finally, about 0.05 second elapses while the muscle is relaxing.

The drum can be revolved at much slower speeds to study other reactions. Fig. 28.16 shows a series of **single twitches,** but the drum is going so slowly that each twitch appears as a single vertical line; the marker goes up and then back down at about the same spot. This experiment shows the effect of **warm-up** on muscle contraction. Athletes undertake a period of warm-up before starting a game or athletic event of any kind because they have learned that their efficiency is greatly improved after this preliminary activity. It can be seen from this graph just how great the increased efficiency can be. A series of stimulations of equal strength were applied, but it can be seen that the response became more vigorous with each contraction up to a

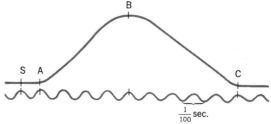

FIG. 28.15 Record of a single contraction of the calf muscle of a frog. The electric stimulus was applied at *S*. The curve *A-B* shows the contraction phase and *B-C* shows the relaxation phase. Time in 1/100 second is shown at the bottom.

certain point. This is known as the **staircase phenomenon.** Temperature is not involved, since the temperature of the muscle remains about the same. During contraction, however, muscles generate lactic acid in their metabolism and this acid seems to increase their sensitivity to stimulation.

Another interesting muscle reaction is illustrated in Fig. 28.17. Electric shocks of the same intensity are given with a gradually increasing frequency. At first there is time for complete relaxation between contractions, but as the frequency increases the muscle does not relax completely before the next shock comes. This causes a great degree of contraction, a principle known as **summation.** Finally, the shocks come so frequently that there is no time for relaxation in between, so the muscle is maintained in a state of contraction, a condition known as **tetanus.** We use a summation method to maintain contrac-

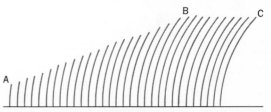

FIG. 28.16 Muscle warm-up (the staircase phenomenon). The degree of contraction of a muscle increases with repeated stimulation up to a certain point (B). The accumulation of lactic acid from the first contraction (A) makes the muscle more sensitive to stimulation.

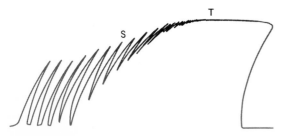

FIG. 28.17 The principles of summation and tetanus are illustrated here. When the stimulations are given in increasing frequency, a point is reached at which the muscle cannot completely relax before the next contraction. The degree of contraction increases at this point. Finally there is no relaxation at all between stimulations and the muscle remains in a state of constant contraction or tetanus.

tions of our muscles. If you wish to hold a book straight out from your body, you keep a steady stream of rapid stimulations going to the mus-

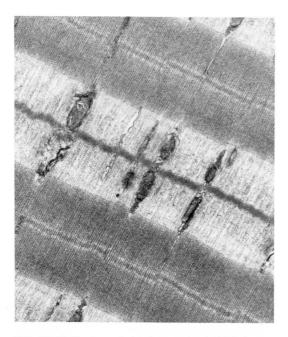

FIG. 28.18 Electron photomicrograph of striated muscle fibers The myofibrils can be seen extending longitudinally and it is their varying thickness which causes the banding. (BioInformation Associates.)

cles of the arm to fix it in this position. This cannot be continued indefinitely, however, for there will be fatigue in the nerve connections. The stimulations cannot be continued in such rapid succession and your arm will begin to waver and you must put the book down.

The Mechanism of Muscle Contraction

Much has been learned in recent years about the nature and function of striated muscles. By means of the electron microscope we have been able to obtain greatly enlarged views of the muscle fibers. These show that the fibers are composed of still smaller fibers, the **myofibrils,** which are the true contractile elements of the muscles. The myofibrils are made up of alternate light and dark bands. These bands lie side by side in the different myofibrils and this gives the entire muscle fiber its striated appearance. Chemical studies show that the myofibrils contain two kinds of proteins, **actin** and **myosin.** The dark bands appear where both of these are present and the light bands where there is only one or the other. The fine structure of these is shown in Fig. 28.18. In addition, the muscle fiber contains water, mineral ions, ATP, and **creatine phosphate (CP).**

A commonly accepted theory states that muscular contraction results when the structural units of actin and myosin slide past one another as shown in Fig. 28.19. The energy which causes this action is supplied by ATP, which is the universal storehouse of energy for cells. This can be demonstrated by experiments in which muscle fibers are treated chemically to remove all the parts except the actin and myosin. When these fibers are suspended in water containing certain critical mineral ions, such as magnesium and potassium, they contract vigorously when ATP is added to the water.

In the living muscle the carbohydrate (glycogen) which is formed in the muscle is broken down to give the energy to convert ADP into ATP. The glycogen is first broken down into glucose and then, if sufficient oxygen is present, the energy is extracted through aerobic metabolism. When muscles are used vigorously, however, the energy demands are too great to be satisfied by this means alone. The cells then obtain the extra energy needed through anaerobic metabolism. Lactic acid is an end product of such metabolism and accumulates in the muscles. We have learned that a small amount of this acid actually stimulates the muscle response and improves its performance. Some of the lactic acid diffuses into the blood and is taken up in the liver, and this prevents the lactic acid concentration in the muscle from becoming too great. When you stop to rest after violent exercise, your deep breathing continues for a time, and the extra oxygen makes it possible for some of the lactic acid to go through the Krebs cycle and replenish the store of ATP in the muscles. The rest of the lactic acid in both muscle and liver is reconverted back to glucose and glycogen. We say that you build up an **oxygen debt** during very active exercise and the continued deep breathing during rest repays this oxygen debt.

Creatine phosphate (CP) plays an important role in muscle physiology. The breakdown of glucose with the formation of ATP is not sufficiently rapid to supply all the energy needed for rapid contraction of muscles either by aerobic or anaerobic metabolism. Here is where CP comes in; during periods of inactivity the breakdown of glucose continues and supplies more energy than is needed. When ATP is formed to its full capacity, some of its energy is then transferred to CP, where the high-energy phosphate bonds are stored as an energy reserve. These bonds are quickly transferred back to ADP to make ATP in times of rapid contraction of the muscle.

It has been commonly accepted that the energy used by muscle work is needed for the contraction of the muscle. But recent delicate experiments indicate that the ATP bonds are not split

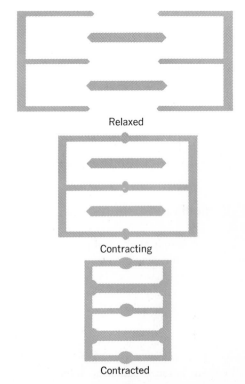

FIG. 28.19 Diagram to illustrate the postulated method of muscle contraction. The muscle fiber contains thin and thick fibrils. During contraction the thin fibrils come together and begin to crumple. As contraction continues, the thin fibrils crumple still more and the thick fibrils begin to crumple at the ends.

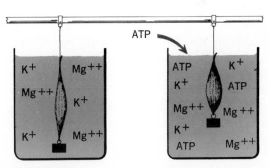

FIG. 28.20 Demonstration that ATP is necessary for muscle contraction. As soon as ATP is added to a muscle suspended in water containing potassium and magnesium salts, the muscle contracts.

during the contraction phase of a muscle twitch but rather during the relaxation phase. How could energy be needed for relaxation and not for contraction? The hypothesis has been proposed that the energy of ATP is needed to spread apart the bands of actin and myosin and that once spread, these bands contain the potential energy used in the spreading. Upon stimulation of the muscle fiber, this potential energy is then released in the contraction of the bands.

Smooth or Visceral Muscle

A microscopic study of a portion of the intestine will show many long, slender, spindle-shaped cells, each with a single nucleus. These are the smooth muscle cells, so named because they do not have striations across them. They are also called visceral muscles because they are found in the visceral organs of the body. Smooth muscles, as a whole, control the movements of the internal organs—stomach, intestine, blood vessels—while striated muscles are largely concerned with the responses of the body to the external environment. The smooth muscles contract in response to stimulations of the autonomic nervous system, which is not under our voluntary control. You cannot empty the contents of your stomach into your small intestine at any particular time you wish. Yet the contents will be emptied through contractions of the smooth muscles in the stomach wall in response to various internal reactions which affect the autonomic nervous system. Hence, we say that smooth muscles are involuntary muscles in contrast to the striated muscles which, in general, are voluntary muscles that can be made to respond in accordance with our wishes.

If we consider the power of contraction to be the measure of the efficiency of a muscle, the smooth muscles are not as efficient as striated muscles. Smooth muscles can shorten themselves to about one-sixth of their fully extended length.

Smooth muscles also respond more slowly and are more sluggish in their actions. Smooth muscles can be stretched considerably—for which we may be thankful, or else we could not eat that extra piece of apple pie after a big meal. In fact, all our meals would have to be much smaller if our stomachs did not have this power of distention. The urinary bladder is another example of distensibility. Smooth muscles mostly have dual innervation; that is, two nerves connect with the muscle; one stimulates it to contract and one inhibits its action. In the next chapter we shall learn how this arrangement regulates muscle activity.

Cardiac Muscle

Muscle of this type, which is found only in the heart, has several unique features. The fibers are long and striated as are the skeletal muscles, but there are frequent cross connections between the fibers. The arrangement of actin and myosin filaments is about the same as that found in the skeletal muscles, but this muscle is involuntary like the smooth muscle. The cardiac muscle also has a distinctive characteristic of its own; it will contract without any stimulation from nerves. The heart can be cut from an animal and it will continue to contract even though all nerve connections have been severed. There is a certain region of the heart, the pacemaker, which initiates the rhythmic contractions, and through the cross connections between fibers they spread over the entire heart in a definite pattern.

The heart muscle, like the smooth muscles, does have a dual innervation, but the nerves only slow the heart beat or speed it up; they do not initiate the stimulus for beating.

Muscle Tone

Muscles must be used if their strength is to be retained. If a muscle is completely immobilized and not used over a long period of time, the muscle fibers will actually disappear and be replaced with connective tissue that is incapable of contraction. Muscles used only slightly will remain in existence, but will tend to be soft and flabby. Muscles which are properly exercised will become firm and ready to perform when needed. When muscles are used properly, they tend to remain in a state of partial contraction, even when not being stimulated; this makes them firm. Muscles with this good tone are able to contract more quickly and forcefully than muscles with a poor tone. Our posture and general physical appearance result from the tone of the muscles of our bodies. We can recognize a person who gives proper exercise to the muscles of his body; there is a grace of bearing and carriage that is unmistakable.

Muscle tone can be maintained by massage and artificial stimulation. Some persons with a disinclination to expend the effort required for exercise are able to maintain their muscles in a good state of tone by having regular massage treatments. In the same way it is often possible to maintain the tone of dormant muscles following polio infection during the time the nerves are repairing themselves.

The smooth muscles also vary in their tone. Such variation does not seem to result from the degree of use so much as from the control of the nerves that regulate the contraction of smooth muscles. This will be studied in the next chapter. However, smooth muscles degenerate when not used normally. After a prolonged period of insufficient and improper food the muscles of the stomach and intestines may degenerate to a point where they cannot handle a normal meal.

REVIEW QUESTIONS AND PROBLEMS

1. From fossil remains of the skeletons of prehistoric animals, biologists have been able to reconstruct the appearance of the entire animals. Explain how this is possible.
2. The human skull has only one movable bone. Which one is this and why must it be movable?
3. As a person grows older there is frequently a progressive ossification of some of the cartilages. How does this affect people?
4. What changes take place in the pelvic bones of a woman during the latter part of pregnancy? Why are such changes necessary?
5. How does the growth of bone differ from the growth of cartilage?
6. How do membrane bones differ from cartilage bones in their origin and formation?
7. Name the different kinds of teeth in the human mouth and tell how many there are of each.
8. Why is the fontanel necessarily associated with the high state of mental development of mankind?
9. When astronauts return to earth after many days in space, a careful analysis of the calcium and phosphorus content of their bones is made. Why is this done?
10. What part of the skin of animals is used for leather?
11. As a person grows older, what happens to the skin which causes it to wrinkle?

12. When a person becomes cold or frightened, he develops goose pimples. Give the possible explanation for this automatic reflex.

13. Describe two ways in which the human skin shows a homeostatic response to changes in temperature.

14. In New Jersey a child who lacked sense organs of pain in his skin was born recently. What problems will this child face throughout his life?

15. What is the difference between **tendons** and **ligaments**?

16. Each motor unit of a muscle contracts to its maximum extent when it receives sufficient stimulation, yet the entire muscle can show various degrees of pull. Explain how this is possible.

17. Why is the efficiency of a muscle increased after a "warm-up" period?

18. How do smooth and striated muscles differ according to innervation?

19. The heart will continue beating even when cut completely from the body, but it has nerve connections. What is the purpose of these connections?

20. What evidence indicates that the energy expenditure comes during the relaxation phase rather than the contraction phase of a muscle?

21. What is the role of **creatin phosphate** in muscle physiology?

22. What is meant by an oxygen debt and how is it repaid?

23. How can it be demonstrated that it is the nerve connections rather than the muscle fibers which become fatigued after repeated rapid stimulations?

FURTHER READING

Hoar, W. S. 1966. *General and Comparative Physiology*. Englewood Cliffs, N. J.: Prentice-Hall.

Pantelouris, E. M. 1966. *Introduction to Animal Physiology and Physiological Genetics*. New York: Pergamon Press.

Romer, A. S. 1970. *Vertebrate Body*. 4th ed. Philadelphia: W. B. Saunders.

Wilkie, D. R., and Carlson, F. D. 1974. *Muscle Physiology*. Englewood Cliffs, N. J.: Prentice-Hall.

29

The Nervous System and Drug Abuse

The human body is a complex organization of billions of cells formed into many complicated organs, yet these are all united and function as a coordinated whole through the influence of the nervous system. A few of the simplest metazoans exist without this coordinating system, but as animal bodies become more complex the nervous system assumes an increasing importance. In the human body it is so vital that even a slight impairment of its function greatly impedes body activities, and more extensive damage results in death.

For convenience of study the nervous system is divided as follows: the **central nervous system,** consisting of the brain and spinal cord; the **peripheral nervous system,** consisting of nerves from the brain and spinal cord; the **autonomic nervous system,** which is a special part of the peripheral system consisting of nerves that control the involuntary body reflexes; and the **sense organs,** such as the ears and eyes. We shall survey these one at a time.

THE CENTRAL NERVOUS SYSTEM

The Brain

Man is superior to other forms of animal life in only one respect—his more highly developed brain. Physically he is a weakling. Take away his guns, knives, and other weapons which his intellect has devised and he is no match for

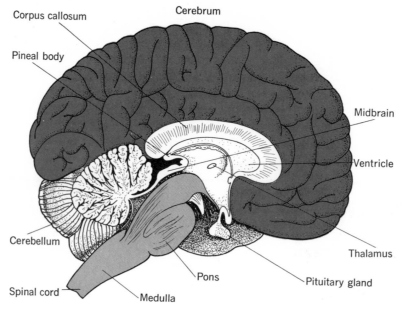

Corpus callosum

Cerebrum

Pineal body

Midbrain

Ventricle

Cerebellum

Thalamus

Spinal cord

Pons

Pituitary gland

Medulla

FIG. 29.1 Midsaggital section of the human brain.

most other animals of equal weight. A tiger or chimpanzee half his size or timber wolf could tear him to pieces in short order. His senses compare very poorly to other animals'; sight, hearing, and the sense of smell are much more acute in most contemporary mammals. There is one feature, however, in which he stands supreme—his mental capacity—and this has enabled him to assume the dominant position he now occupies on the earth.

This superiority is reflected in the anatomy of the brain. The human brain weighs about 3 pounds, which is far more in proportion to size of the body than is found in any other animals. A 600-pound gorilla, although one of the most intelligent of the subhuman mammals, has a brain weight of only about 1¼ pounds. This great difference is due primarily to the more highly developed cerebrum in man; most of the human brain is composed of the two large hemispheres of this portion of the brain, which is the center of intellect. Not only is there a difference in size,

but man's superiority is also reflected in the more extensive convolutions of the cerebrum. Later we shall learn the advantage of this construction.

People have suspected that the brain was an organ of thought since early history, but many great thinkers have opposed the idea. Even the great Aristotle argued that the function of the brain was to cool the blood. Perhaps he and others believing likewise had observed that an exposed brain in an experimental animal is somewhat insensitive to pain and concluded that it could have nothing to do with the sensations of the body. As the techniques of experimental biology developed, however, the true function of the brain became apparent. When a portion of the right cerebellum of a dog is removed, the dog becomes deaf in its left ear. This is explained by the fact that most of the body nerves cross sides before they reach the cerebral hemispheres of the brain. Destruction of the left side of the brain results in paralysis and loss of sensation of the right side of the body. By destroying small portions of the brain and studying the effect on the body, it has become possible to localize the areas of control in the brain. The brain of an animal can also be exposed and certain areas stimulated with a wire carrying a weak electric current. This elicits response in various muscles and locates motor areas. The most recent work on localization has been done with the brain waves of human subjects. It has been found that whenever a sensory impulse reaches the brain there is a change in the electrical state at the particular point where the impulse terminates. By studying the variations in the patterns of brain waves following various stimuli, it has been possible to locate many sensory areas of the brain.

Embryologically the central nervous system originates as an invagination of the ectoderm in the early embryo. This invagination develops into a hollow tube, and the anterior portion of this tube enlarges and forms three lobes—the forebrain, midbrain, and hindbrain. The adult brain is formed from these as shown in the following list.

Forebrain
1. Cerebral hemispheres
2. Diencephalon
   a. Thalamus (paired)
   b. Hypothalamus (paired)

Midbrain
1. Cerebral peduncles
2. Corpora quadrigemina

Hindbrain
1. Cerebellum
2. Pons
3. Medulla

The hollow portion of the embryonic brain persists as four cavities, **ventricles.** The first and second ventricles are located in the right and left hemispheres respectively. The third is surrounded mostly by the diencephalon and the fourth is within the medulla. These ventricles are connected with one another and with the cavity of the spinal cord by means of a narrow tube. There is a fluid, the **cerebrospinal fluid,** within these cavities, and it is under considerable pressure. This helps hold the soft brain tissue in place against the wall of the cranium. Sometimes during embryonic development, the narrow tube connecting the ventricles becomes clogged and the cerebrospinal fluid develops an abnormally high pressure in the ventricles of the cerebrum. This causes a swelling of the cerebrum as well as an enlargement of the skull, for the skull is soft during this period. This condition produces an abnormally large forehead and is often associated with mental deficiency caused by the misshapen brain. The condition is referred to as **hydrocephalus,** or "water on the brain."

The **cerebrum** is the seat of intelligence, imagination, memory, reasoning—all the higher functions of the nervous system. This portion makes up about two-thirds of the human brain. It consists of an outer **gray matter** and an inner **white matter.** The gray matter contains many cell bodies and their processes which make the proper connections for impulses coming to the brain. The gray matter folds inward with the convolutions of the brain, thus greatly increasing the amount of this vital nerve tissue in comparison with a smooth brain. Hence, the amount of convolution is as much an indication of potential intelligence as actual size of the cerebrum. The white matter consists of nerve fibers and masses of nerve tissue called **basal ganglia.** These are of great importance in coordinating muscular movements, and injury to this portion of the brain results in such uncoordinated movements as are seen in cerebral palsy.

The cerebrum is covered with a hard outer membrane, a delicate layer containing many blood vessels and a thin inner membrane. When a person has abnormally high blood pressure, there is danger that one of the small blood vessels in this outer covering may break. This may result in the accumulation of blood between the membranes and an undue pressure on the brain at that point. This is the cause of **apoplexy,** or **stroke.**

The **diencephalon** is of great importance in sensory and automatic adjustments of the body. One part, the **thalamus,** is a major relay station for sensory impulses going to the cerebrum and is also concerned with motor coordination. The **hypothalamus** integrates the various automatic body reactions. For instance, when a dog becomes overheated, it begins to pant and its tongue hangs from its mouth. There is also an increased saliva flow which helps to cool the blood flowing through the tongue. If the hypothalamus is injured, however, such reactions do not occur. The dog may become so overheated that it suffers heat prostration, yet its body never undergoes any of these cooling reactions. The hypothalamus also produces releasing factors which stimulate the release of hormones from the pituitary gland which lies just below it.

The **cerebral peduncles** are bundles of fibers joining the cerebral cortex to other portions of the brain and spinal cord. The **corpora quadrigemina** are four hemispherical bodies which are reflex centers. The turning of the head and the movements of the eyes which constantly are taking place during our waking hours are regulated by these bodies.

The **cerebellum** is concerned with the smooth coordination of muscular activities. Damage to this region results in jerky muscular movements and poor balance. The unpleasant motion sickness, which makes a sea voyage a nightmare of discomfort for many people, is believed to be caused by a disorganization of the impulses from the cerebellum which regulate the body adjustments involved in maintaining equilibrium.

The **pons** comprises a large number of fibers which pass through it and make connections between other parts of the brain. It also seems to be a center concerned with the regulation of respiration.

The **medulla** is actually an enlargement of the spinal cord. It is a major center of mediation of reflex actions and sends out impulses which regulate the heart beat and respiration, but it is quite sensitive to modification from higher centers. The pons, for instance, may influence impulses affecting respiration and the hypothalamus may influence the impulses affecting the heart beat.

The brain of the frog. When we dissect out the brain of the frog and find only a forward extension of the spinal cord with a few bulges, it may appear that there is little relation between the frog brain and the human brain. However, the parts are almost identical and differ only in the degree of development. The cerebral hemispheres are greatly reduced in the frog and are not convoluted. A conspicuous pair of optic lobes is found on the dorsal part of the midbrain rather than the corpora quadrigemina. The cerebellum is only a small disk behind the optic lobes. There is another reason that accounts for the difference in appearance. The frog brain is all in one plane, but the human brain shows three distinct bends. Embryologically, the human brain is all in one plane when it is first formed, but it later bends forward, then backward, and finally forward again to give it a shape somewhat like a Z, when viewed from the side. This makes the brain more compact and not so long as the frog brain in proportion to its width.

The Spinal Cord

The spinal cord connects with the medulla through the foramen magnum of the skull and extends posteriorly down the back, terminating in a point near the anterior end of the second lumbar vertebra. It is entirely contained within the neural canal of the vertebral column. The distribution of matter in the spinal cord is just the reverse of that in the brain—the gray matter occupies the central portion of the cord, whereas the white matter is found on the outside. There is a small central canal in the center of the gray matter—a remnant of the cavity in the embryonic tube. There are three membranes surrounding the spinal cord—the same as were found around the brain. These are called the **meninges.** (Inflammation of the meninges is meningitis.) However, in the membranes of the spinal cord there is a space, the subarachnoid space, which is filled with cerebrospinal fluid under pressure. The meninges and the fluid within continue on down the neural canal in the vertebrae posterior to the termination of the spinal cord. It is possible to make a spinal puncture safely in the caudal region and remove some fluid to study it or to inject medicines or anesthetics. The punctures are usually made between the third and fourth or between the fourth and fifth lumbar vertebrae.

THE PERIPHERAL NERVOUS SYSTEM

This part of the nervous system consists of the nerves that connect the different body parts with the central nervous system.

Structure of a Peripheral Nerve Cell

The nerves of the body are composed of cells, as are the other structures that go to make up the human body, although the cellular nature of a nerve may not be readily apparent. A nerve fiber in a leg may be over three feet long and so thin that it cannot be seen without a microscope, but it will only be a part of a single cell. The nerves of the body are like cables containing large numbers of these fibers bound together. The nerves branch and rebranch until the individual fibers connect with the motor units of the muscles or with parts of sense organs.

A typical nerve cell, **neuron,** will have a **cell body** containing cytoplasm and a nucleus, but there will be a number of processes extending out from the cell body. Most of these are **den-**

drites, but there is one, the **axon,** that will be much longer than the others. This forms the impulse-carrying part of a nerve fiber. The axon will be surrounded by a **medullary sheath** which consists of a number of cells. There will be an indentation, the **node of Ranvier,** where one cell of the medullary sheath ends and another begins. Finally, the medullary sheath is surrounded by a protective **neurilemma.**

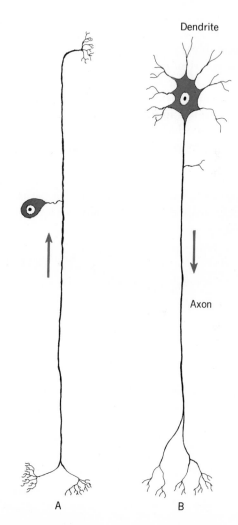

Dendrite

Axon

A   B

FIG. 29.2 Diagram of typical neurons. Afferent or sensory neuron (A). Efferent or motor neuron (B).

Neurons do not connect with one another, but impulses may pass from one to the other where their processes come close together. Such an approximation of processes is called a **synapse.** The axon terminates in a series of branches called an **end brush.** Impulses may pass from the end brush of the axon to the dendrites of another neuron, but not in the reverse direction. Thus, impulses can pass in one direction only. The neurons that carry the impulses to the central nervous system are **afferent neurons;** these originate in sense organs and may also be called sensory fibers. Those that lead from the central nervous system are **efferent neurons;** they are primarily motor neurons terminating in muscles of the body. There is a third type of neuron, the **associational neuron,** found in the brain and spinal cord, which carries impulses from one part of the central nervous system to another.

There is no neurilemma surrounding the associational neurons. This fact has a practical importance. Whenever a portion of a peripheral axon is destroyed, mechanically or by disease,

regeneration may occur if the tough, outer neurilemma remains intact. Regeneration is impossible in the central nervous system. Syphilis organisms may invade the brain and destroy some axons. This condition is called **paresis** and results in a type of insanity. Unfortunately, the damage is permanent, for the axons have no neurilemma to guide the regenerating ends together.

The Nature of a Nerve Impulse

During the early days of the study of nerve impulses, it was assumed that the impulse was electrical in nature, very much like an electric current traveling along a copper wire. When investigations were conducted to determine the speed of a nerve impulse, however, it was found to be much slower than an electric current. The nerve impulse travels at speeds varying from 25 to about 300 feet per second in those nerves which have been investigated. This may seem to be rapid, but it is far slower than the speed of

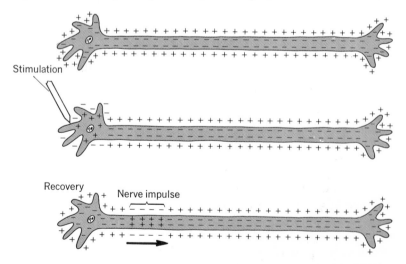

FIG. 29.3 The nature of a nerve impulse. Stimulation of a sensory nerve ending induces a brief increase in the permeability of the axon membrane which permits positive ions to enter the axon and reverse the charge on the membrane. This change stimulates an increase in permeability of adjacent parts of the axon membrane and the nerve impulse travels along the axon like a wave. After the wave passes, recovery is quick.

electricity, which travels at about 100,000 miles per second. It is apparent, therefore, that the nerve impulse must be of a different nature. Further investigation has shown that the nerve impulse is a series of chemical and electrical changes traveling as a wave along the axon of the neuron. These are metabolic reactions and use energy. Oxygen is consumed and carbon dioxide is released. The energy of the impulse comes from within the axon itself. The stimulation at one end of the axon sets in motion a chain reaction which sends the impulse the length of the axon. It is an **all-or-none** response. A certain threshold of stimulation is required to initiate the impulse, but once this threshold is reached, the strength of the impulse remains the same no matter how much stronger the stimulation.

Much of our information about the nature of a nerve impulse has come from experiments using the axons of the squid, crayfish, and certain worms. These are chosen because the axons are very large. Those of the squid, for instance, are almost 1 mm in diameter. This is large enough to permit insertion of small electrodes to measure electrical charge and delicate pipettes to determine the chemical nature of the axon at various points. The nonconducting axon was found to have an overall positive charge on the outside of its membrane and an overall negative charge on the inside of the membrane. We say that the membrane is **polarized** electrically. The charge across the membrane has been measured at about 0.075 volt. The charge is maintained by the sodium and potassium pumps, as discussed in Chapter 7. The sodium ions are maintained at a level of concentration about ten times as great on the outside as on the inside, while the potassium ions are about thirty times as concentrated within the axon as on the outside of the membrane. Both sodium and potassium ions are positive, but the sodium ions are in contact with the membrane outside, while certain negative ions are in contact with the membrane inside. The potassium ions lie deeper within the cytoplasm inside the axon. The polarity of the membrane can be maintained only so long as the energy of ATP is available to maintain the active transport of the sodium ions. When the food supply is exhausted and ATP can no longer be generated, the axon loses its ability to transmit impulses because it loses its polarity.

A variety of stimuli can cause an increase in the permeability of the membrane. When this happens, the sodium ions enter the membrane and bring about a temporary depolarization. When the depolarization causes a drop in voltage across the membrane to about 0.060 volt, the threshold for stimulation has been reached. At this point the sodium ions flow in rapidly and eliminate the voltage entirely; in fact, there is a momentary overshoot which causes the interior of the membrane to become positively charged, and the outside of the membrane is left negatively charged because of its sudden loss of positive ions. This reverse of polarization at one section of the axon acts as a stimulus to the increase of permeability of adjacent portions of the axon. Thus, a wave of depolarization proceeds along the axon. This is the nerve impulse.

The axon recovers its polarity very quickly after losing it. It is thought that this is brought about by the rapid movement of positive potassium ions from within the axon to the outside of the membrane in response to the attractive force of the negative ions on the outside. The axon is then ready to transmit another impulse. The time between the depolarization and the recovery of polarity to the normal 0.075 volt is very short, less than one-thousandth of a second in warm-blooded animals, so it is possible for as many as a thousand impulses to move along an axon in one second. A strong stimulation will cause a considerable number of rapid-sequence impulses to travel along the axon, while a weaker stimulation will result in fewer impulses. A strong stimulation can bring about a more intense response, therefore, even though each impulse behaves according to the all-or-none principle.

Only when the stimulation has ceased and the axon is at rest do the sodium and potassium pumps get back to work and restore the concentration of sodium and potassium ions to their original state inside and outside the membrane. It is estimated that as many as several hundred thousand impulses could be conducted along the giant axon of the squid before the axon had to have a rest period to restore the balance of sodium and potassium ions. In the smaller axons of other animals this figure would probably be much smaller, but even so, many impulses can be conducted in rapid sequence.

Transfer of the Nerve Impulse at the Synapse

Since there is no physical connection between neurons at synapses, there must be some way that the nerve impulse can cross the gap and continue along adjacent neurons. The transmission is accomplished by chemical means. When an impulse reaches the end brush of a neuron, it stimulates this end brush to release a chemical known as **acetylcholine.** This chemical contacts the dendrites of adjoining neurons and starts depolarization, triggering a nerve impulse along the axon.

This method of transfer of the impulse creates a problem. Why does the chemical not continue stimulating the dendrites so that there is a continuing series of impulses? If this should happen, there would be uncontrolled impulses stimulating muscles long after the original impulse got the reaction started. Such a catastrophe is forestalled by the release of a second chemical at the synapse. This is **cholinesterase,** an enzyme that splits the acetylcholine into two parts, acetyl and choline fractions, which are inactive. It seems that the acetylcholine remains intact just long enough to stimulate one impulse; another impulse must come down the axon and release more

acetylcholine before another impulse is generated in the adjacent dendrites.

We now have a basis for the understanding of nerve fatigue. As we learned in Chapter 28, repeated stimulation of a nerve within a short period results in weaker and weaker muscle response until finally there is no response at all. The muscle itself may still have sufficient ATP for full contraction, but it does not receive the impulses to stimulate the contraction. The breakdown of transmission is at the synapses between the neurons. With repeated rapid stimulation the synapse presents an increasingly steep barrier to the passage of impulses. This is thought to be caused by the temporary exhaustion of the reserves of acetylcholine in the end brushes. Thus, while a sense organ may receive stimulation and an impulse may be generated, this impulse will not be able to pass the first synapse and no reaction will be elicited. After a short period of rest the end brushes seem to regain their power to produce acetylcholine and can again transmit impulses. In man and many other animals, however, there is a gradual diminution of this power during the waking activities, and there must be the relaxation of sleep before they can be restored.

If a person has been awake for as long as forty-eight hours, the fatigue at the synapses has become so great that he can be kept awake only with the strongest stimulation. The fatigue can be postponed through the use of certain stimulants, such as caffeine, theobromine, adrenaline, and even stronger stimulants such as the so-called pep pills. These stimulants seem to decrease the permeability of the plasma membrane to sodium ions, thus lowering the threshold of stimulation required to initiate and transmit an impulse. Other drugs, such as the barbiturates, seem to bring about a decrease in the permeability of the membrane and thus raise the threshold required for initiation and transmission. These drugs are used as sleeping pills because they depress the movement of impulses from one neuron to another.

After continued use of these chemical stimulants and depressants, the nerves become adjusted to them and may actually require them for normal operation. This is one explanation for drug addiction, and a very unpleasant period of abstinence from the drugs is required before the nerves readjust to function without them.

Some chemicals have been found which inhibit, or break down, the cholinesterase at the synapses. These result in a serious upset of the mechanism of impulse transmission because there is no quenching of the acetylcholine at the synapses. Hence, an impulse generated at a sense organ will continue passage through the synapses long after the stimulation has passed. This results in uncontrolled stimulation of the muscles. Certain "nerve gases" have been developed from such chemicals which could kill many people if inhaled in very small quantities.

The Nerve Arc

The human body has forty-three nerves coming from either side of the central nervous system. From the brain twelve **cranial nerves** originate and pass out of the skull through various openings. These nerves lead to sense organs and muscles of the head and the upper part of the body. From the spinal cord thirty-one **spinal nerves** originate and make their exit from the spinal column at the points of articulation of the vertebrae. (The frog has only ten cranial and ten spinal nerves.) Each of the spinal nerves has a **dorsal root,** coming out from the dorsal portion of the spinal cord, and a **ventral root,** coming out from the ventral portion of the cord. The two roots unite as they leave the spinal cord and then separate again as they lead to the muscles or sense organs. Experiments show that the dorsal root carries only **sensory fibers,** whereas the ventral root carries only **motor fibers.** This can be demonstrated by experimental destruction of the roots. When the dorsal root of a nerve leading to

the leg is cut, the animal will feel no pain in that leg, yet it can move the leg normally as in walking. With a severed ventral root, however, the animal will feel stimulations of the leg and will move other body parts in response, but it cannot move the leg itself.

Reflex Action. The neurons of the ventral and dorsal roots terminate in the gray matter of the spinal cord. The synapses which transfer impulses to other neurons leading to the brain or other parts of the spinal cord are located here. The cycle of movement of impulses from sense organs to sensory neurons, to associational neurons, to motor neurons is known as the **nerve arc.** This arc can lead all the way to the brain and include a conscious knowledge of the reflex, or the connections can take place at the spinal cord level and the muscle reflex can occur without conscious control—a type of nerve arc known as **reflex action.**

Reflex action can be demonstrated easily in the frog. You can cut the entire top of the head from a frog, completely removing the brain; yet if you stick a pin in this frog, it can make a perfect leap. You can even strip all of the flesh from the body, leaving only the vertebral column and the legs; then if you touch a drop of acid to one

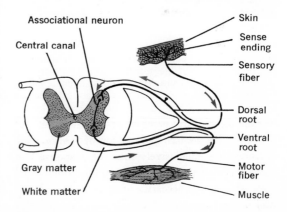

FIG. 29.4 Cross section through the spinal cord and connecting nerves, showing the pathway of impulses during reflex action.

FIG. 29.5 A complex reflex action in the frog. The entire top of the head has been cut off, including all of the brain except the medulla. The frog can still maintain normal posture. When it is stimulated with a needle, it first raises itself and then executes a coordinated leap. Since the cerebrum has been removed, this is a spinal reflex.

leg, the muscles will contract in the other leg and it will scratch the irritated portion of the leg. This demonstrates conclusively that the response can come from connections made at the spinal cord level without the necessity of a nerve arc all the way to the brain.

Reflex action is protective in nature. If you touched your finger to a hot stove, you would not have time to think, "My finger is on a hot stove; it is burning; it hurts; I must remove my finger from the stove." With all this conscious thinking and reasoning, you would not remove your finger until it was badly burned. Instead, you touch and jerk your finger away before you even realize what has happened, thus minimizing the burning. You make thousands of such adjustments every day without thinking about them.

Some reflex actions are very complex and

involve many neurons, both sensory and motor. Many of these involve parts of the brain, although not the parts at the conscious level. Some instincts may be classified as very elaborate reflex reactions.

Conditioned reflexes. It is possible to change the reflexes which involve parts of the brain. Such conditioned reflexes were first studied extensively by the Russian physiologist **Ivan Pavlov.** Dogs normally produce saliva copiously when they see a piece of fresh meat. Pavlov showed, however, that if they had never eaten meat this reaction did not occur. The reaction was actually a conditioned reflex; the nerves stimulating the flow of saliva were stimulated by

the sight of meat after the dog had learned that meat is good to eat. Pavlov also found that if he rang a bell just before giving a dog meat to eat, the dog soon became conditioned to salivate at the sound of the bell alone, even if no meat was near. The associational neurons had been "trained" to make the connections from the incoming impulse of the sound of the bell to the outgoing stimulatory impulses leading to the salivary glands.

You respond to such conditioned reflexes many times each day. If you drive an automobile, you have become conditioned to make certain responses without thinking of what you are doing. Even when riding in the back seat, you are very

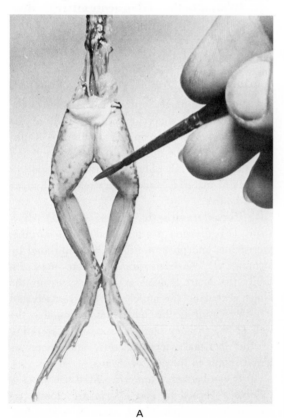

A

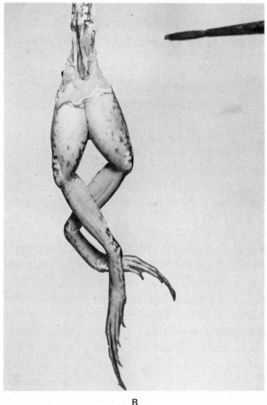

B

FIG. 29.6 Flexion reflex in the frog leg. Even though only the hind legs and a portion of the spinal cord remain, the leg will withdraw when it is touched with a brush containing acid.

likely to press down with your right foot when an emergency arises. If you use a typewriter, you soon learn to press the right keys and bars to produce the words and spaces on paper in response to the visual image of words you wish to copy or the thought of words in your mind. You do not need to think of the location of each letter on the keyboard. If you change to a typewriter on which one of the controls is in a different position, you will have difficulty for a time, because your fingers continue to move to the position of the controls on the old typewriter. Many of our everyday activities are on this reflex level, which is fortunate because the conscious level of the brain is thus freed to contemplate some of the more important creative activities of life. While driving a car you can be thinking about the proper answers to questions which you may be asked in a biology class.

THE AUTONOMIC NERVOUS SYSTEM

The autonomic nervous system is a special set of peripheral nerves that presides over the great number of body functions that are not under the control of the will. The accommodation of the pupil of the eye, the movements of the stomach and intestine, the rate of heart beat, the secretions of the sweat glands—these are but a few of the many functions of this system. The system consists of two antagonistic divisions: the **sympathetic system,** which consists of two chains of ganglia and connecting nerves in the thoracic and lumbar regions, and the **parasympathetic system,** which consists of nerves arising from the cranial and sacral regions of the central nervous system. We say these two are antagonistic because the nerves of one usually have an effect on the body just the opposite of the nerves of the other. For instance, the sympathetic system has nerves that dilate the pupils of the eye, while the parasympathetic system has

nerves that constrict the pupils of the eye. It is through the combined action of these two that our pupils open and close down according to the intensity of the light.

There are drugs which can stimulate or inhibit one of the autonomic systems without affecting the other. Atropine inhibits the parasympathetic division. This causes dilation of the pupils of the eyes, even in bright light, for the sympathetic nerves are not restrained. Fig. 29.7 shows the effects of the two systems on the involuntary body reactions. For reference purposes we will list the two divisions of the system, the effect of certain drugs on the divisions, and some of the effects of the divisions on the body organs.

Sympathetic (thoracolumbar) division. This division is stimulated by adrenaline and ephedrine; it is inhibited by ergotoxin. It dilates pupils of the eyes, constricts the blood vessels of skin and viscera, stimulates the sweat glands, accelerates the heart, dilates the air passages in the lungs, inhibits the muscles of the stomach and intestine, inhibits the wall of the urinary bladder, stimulates the sphincters (muscles which close the openings to the stomach and intestine and relax only temporarily to permit passage of material), and causes release of sugar from the liver.

Parasympathetic (craniosacral) division. This division is stimulated by pilocarpine, muscarine, and physostigmine; it is inhibited by atropine. It constricts pupils of the eyes, inhibits the heart, constricts air passages in the lungs, stimulates the muscles of the stomach and intestine, inhibits the sphincters, stimulates the wall of the urinary bladder, stimulates secretion of the lacrimal (tear) glands, and increases blood supply to the genital organs.

It can be seen from this partial list that the reflexes governed by this system are closely related to the emotions. Grief, anger, fear, and sexual stimulation all result in profound changes in the functioning of the organs controlled by the autonomic nervous system. These reactions are

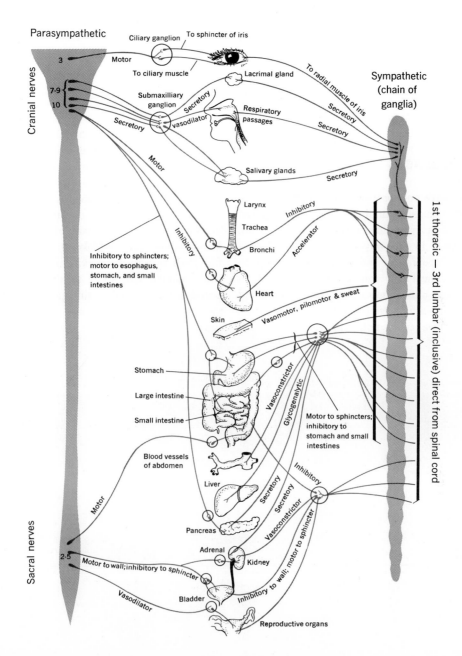

FIG. 29.7 The autonomic nervous system and its connections to the body. Note the antagonistic action of certain sympathetic and parasympathetic nerves. (Redrawn from Greisheimer, *Physiology and Anatomy,* Lippincott, 1959.)

emergency reactions, and the more normal functioning of the system is interfered with during emotional stress. It is better not to eat when emotionally upset; the stomach cannot deal with a meal properly under such circumstances. A person suffering from indigestion, constipation, and even stomach ulcers may need no more than emotional relaxation to relieve these conditions.

THE SENSES

We have only one way of maintaining contact with our surroundings—through impulses from the sensory endings of nerves in our sense organs. Only in this way can we know what is happening around us and make adjustments to our environment. We can live satisfactory lives without some of our senses. A person who is blind or deaf can learn to use other senses more extensively, but we could not exist if none of our senses functioned. Extreme old age is often accompanied by a progressive dulling of the senses until death is only a culmination of a series of gradual losses of contact with the outside world.

Sight

Man is more dependent than most other mammals upon the sense of sight for a great many of his sensory impressions. A dog, for instance, may use odor more than sight to recognize his friends and foes. A cold nose unexpectedly poked against your skin often reminds you how important the sense of smell is to a dog. Mice use vibrissae (whiskers) on their snouts to find their way in the dark. Deer and many other wild animals use scent to detect approaching enemies. Man has come to depend upon sight for most of his impression and makes less use of these other senses.

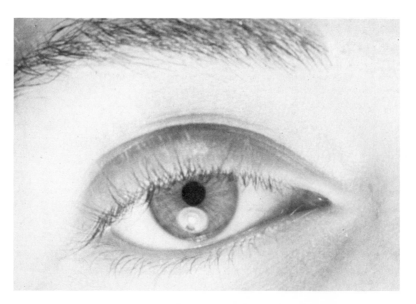

FIG. 29.8 Human eye. The outer white sclerotic coat merges into a transparent cornea at the front, revealing the colored iris which can expand or contract to regulate the size of the pupil to adapt the eye to changes in light intensity.

The human eye has three basic layers. The outer **sclerotic coat** is easily seen as the white of the eyes. This bulges out and becomes transparent in the front to form the **cornea.** The second coat is a black layer known as the **choroid coat.** It prevents light rays from bouncing around inside the eye. At the front of the eye this forms the **iris,** which can be seen through the **cornea.** The other surface of the iris varies in color from pale blue to black. In the center of the iris is an opening, the **pupil,** which admits light into the eye. The size of the pupil varies as the iris opens or closes the pupil in an autonomic response to the intensity of light entering the eye. The **lens** lies just behind the pupil. The lens focuses the image on the **retina.** The retina is the third coat which covers the rear portion of the eye. It bears two types of photo receptors: rods and cones.

Light brings about depolarization of the photoreceptor cells on the retina by altering their membrane permeability and, as a result, nerve impulses are generated. These impulses cause the liberation of acetylcholine by the receptors at synapses with adjacent neurons. The adjacent neurons are stimulated and transmit the impulses to the brain via the optic nerve. The **rods** are photoreceptors which are very easily activated and which respond to very dim light. They contain a substance, **visual purple,** which responds to light energy in such a way as to bring about a depolarization of the sensory ending. When light becomes very bright, however, the visual purple breaks down into two inactive parts and the sensitivity of the rods is decreased. The less sensitive **cones** are stimulated by more intense light. When illumination drops, the visual purple in the rods is reformed, although it takes about fifteen minutes before it is present in sufficient quantities to permit the rods to reach their maximum sensitivity. Vitamin A is necessary for the formation of visual purple and a person deficient in this vitamin may suffer from night blindness. The **cones** function normally and the person can see well in bright light, but as the light intensity

drops, his vision becomes very poor because he cannot produce visual purple in sufficient quantities for the best functioning of the rods.

Color vision is possible because the cones are sensitive to different wavelengths of light. Experimental evidence indicates that there are three kinds of cones, sensitive to red, green, and blue respectively. Through the stimulation of one, or various combinations of the three, we can see the various colors of the visible spectrum. Color vision is best in bright light and becomes progressively less acute as the light becomes less intense. In very dim light we are hardly aware of color distinctions because the cones are not being activated and the rods are not color-sensitive.

The distribution of rods and cones varies widely in the retina. They are most concentrated in the central region and become less concentrated at the peripheral regions. Hence, vision becomes less acute when the objects being viewed are seen with light entering the eye at an angle. The most acute area is directly behind the pupil, a region known as the **fovea.** This is the area we use the most. When you wish to see an object clearly you turn your head and or eyes so that the image reaches the foveal region. Here there are only cones, and they are present at a density of about 147,000

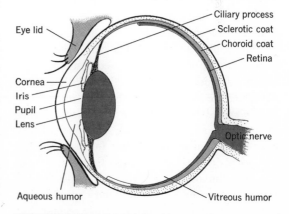

FIG. 29.9 Section through the human eye.

per square millimeter. In other parts of the retina the rods outnumber the cones, and in the peripheral regions of the retina there are no cones. Hence, you cannot detect colors in your peripheral vision. There are no rods or cones at all at the point where the optic nerve joins the retina and, as a result, this area is a **blind spot.** Fortunately, this spot lies somewhat off center on the retina and people generally are not even aware that the blind spot exists, although certain demonstrations will bring it out.

Glaucoma is a serious eye defect which develops when the circulation of the fluids within the eye is prevented and a pressure develops in the vitreous humor which is in contact with the retina. This pressure may cause destruction of the rods and cones and a permanent impairment of vision which may reach the point of total blindness. Early symptoms of this affliction include a decrease in acuity of peripheral vision since the rods in the periphery of the retina tend to be destroyed first. Eye operations can be performed which restore the normal circulation of the fluids in the eye and prevent further damage to the retina. The entire retina may pull away from the sclerotic coat, causing what is known as a **detached retina.** The retina can be reattached by eye surgery.

The human eye bears many similarities to a camera. The retina may be compared to color film; both receive and are activated by light of different wavelengths. The eye does not have a shutter like a camera, but an exposure of about a tenth of a second is required to establish a good image on the retina. With shorter exposures we detect images, but one image cannot be erased before another is formed. Moving pictures are shown at a rate of about 24 per per second, and this is too fast for us to see the individual images. As a result, the images fade one into the other and we get the illusion of movement on the screen. Ordinary mazda light bulbs flash off and on 60 times per second, but our retinas cannot detect the periods of darkness in between the flashes. The retina retains the image seen by one flash through the dark period between flashes. Telvision has a scanning beam which moves across the picture tube so rapidly that we retain an afterimage and see a whole picture on the tube.

In a camera the image is usually focused on the film by mechanical moving of the lens closer to or farther from the film. The **lens** of the human eye is focused by action of the **ciliary processes** which change the shape of the lens. When the eye muscles are relaxed, the ciliary processes exert a pull on the lens so that it is focused on infinity; all objects from about 20 feet to those as far as the eye can see are sharp. To see objcts which are closer, we contract muscles that exert a pull on the ciliary processes and cause them to release the tension on the lens. As a result, the lens becomes more convex and nearby objects come into focus. The curvature of the cornea bends the light rays to some extent, and the lens supplements this initial bending.

A camera usually has an iris diphragm which can be opened or closed to admit more or less light according to the brightness of the image to be photographed. For best results many people use an exposure meter which contains a light-sensitive photoelectrc cell to measure the light and thus know just which setting of the diaphragm on the camera to use. In some modern cameras the exposure meter is a part of the camera and automatically adjusts the diaphragm to the right aperture. The eye also has an iris and it can be opened or closed. It is regulated automatically by responses of certain autonomic nerves to the intensity of light entering the eye. The combination of iris control and the variable sensitivity of the rods and cones give the eye an enormous range of sensitivity—all the way from a fraction of a foot candle on a dark night to several hundred foot candles on a beach or on snow during a sunny day.

Some people are **farsighted;** that is, in their eyes the combined curvature of the cornea and the lens is not sufficient to bring nearby objects into focus. They need glasses with a convex curvature for reading. Middle-aged people typically need this type of accessory lens because their own lenses are not able to change shape as much as in youth. On the other hand, **nearsighted** persons have lenses which are too convex, and they require concave lenses in their glasses in order to see clearly at a distance. **Astigmatism** is another common eye trouble that afflicts most people to some degree. This is usually due to an unequal curvature of the cornea that puts objects in one plane in sharper focus than the linear objects in the opposite plane. Extreme astigmatism may require correction with lenses which have an unequal curvature in the opposite direction.

On certain occasions the lens may gradually develop an opacity which prevents the normal passage of light and leads to the loss of clear vision. This condition, known as **cataract,** may be corrected by the removal of the lens and the use of glasses with sufficient curvature to correct for the missing lens. Cataract is one of the delayed effects of heavy radiation on the human body. Many of the survivors of the bomb at Hiroshima developed cataracts of their eyes in later years.

Man has **binocular vision;** that is, the images seen by his his two eyes are so superimposed as to give one image with depth. Sometimes, such as after overindulgence in alcoholic beverages, the nerves lose their coordination and a person will see double. Only a few of the other mammals share binocular vision with man.

Lacrimal glands secrete a fluid which lubricates the eyes and keeps the delicate outer membranes moist. Blinking is an automatic reaction which helps spread this fluid over the front of the eyes. When there is any irritation, such as is caused by a foreign object in the eyes, the lacrimal glands increase their secretions and

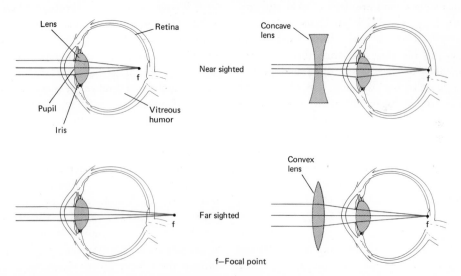

FIG. 29.10 A nearsighted person has a lens which focuses the parallel light rays at a point in front of the retina. By placing a convex lens in front of the eye, the focus point is made to fall on the retina and vision is sharp. Farsighted persons have the opposite problem, which is corrected by placing a concave lens in front of the eye.

help wash the object out. Since these glands are regulated by the autonomic nerves, they are affected by strong emotions. Grief brings an overflow in the form of tears.

The frog eye is constructed much on the order of the human eye, but each eye picks up a different image and there is no binocular vision. The frog's eyelids cannot be closed over the eyes, but there is a thin membrane, the **nictitating membrane,** that moves up from the lower eyelids to cover the eye. This membrane is present in the human body as a vestigial structure—the small membrane lying between the upper and lower eyelids next to the nose. The frog also has the interesting ability to retract his eyes. Normally they protrude well out from the head and give better vision in this position. However, if you touch one of them, it will be retracted within the head.

Hearing

The external ears are flaps of skin and cartilage which serve to pick up sound waves. In some animals these are quite important; a dog may prick up his ears and turn them in various directions to best pick up sound waves. Man has muscles like the dog, but the best he can do is wiggle his ears slightly, since these muscles are vestigial. The sound waves enter the skull through the **auditory canal** and strike the **tympanic membrane** (eardrum), which begins vibrating in response to the vibrating air. These vibrations are transmitted to three small **ear bones** in the middle ear and thence to the **cochlea** of the inner ear. The cochlea is an organ somewhat resembling a coiled snail shell. It contains a fluid which receives the vibrations from the ear bones. Within the cochlea are many nerve endings, each apparently sensitive to vibrations of different rates of speed. These sensory endings connect with nerve fibers to form the **auditory nerve,** which leads to the

brain, and we become conscious not only of the vibrations but of the pitch of the vibrations. Thus, if we strike the key of middle F on the piano, we cause a string in the piano to begin vibrating at the rate of 345 times per second. This sets the air to vibrating at the same rate. The air, upon reaching our tympanic membranes starts it vibrating at the same rate. Then the vibrations pass through the ear bones and to the fluid in the cochlea. In this fluid there are certain hairlike projections tuned to just this pitch; they are stimulated, and an impulse is generated and transmitted to particular neurons which are a part of the great cable of neurons, the auditory nerve. When the impulses reach a particular area of the lateral portion of the cerebrum, we become conscious of a sound of this particular pitch. We seldom are exposed to pure tones; most sounds are combinations of many vibrations of different rates. Hence, many individual neurons are carrying impulses to the brain and we hear the combined effect.

We cannot pick up vibrations of all rates of speed. Our sensory endings can receive only those from about 16 to 16,000 vibrations per second. Vibrations above 16,000 are said to be supersonic. But dogs have ears capable of picking up sounds of higher frequencies. A whistle giving off vibrations at about 18,000 per second cannot be heard by humans, but a dog hears it and can be trained to come when this silent whistle is blown.

The ear bones are located in the **middle ear,** an air-filled cavity. This cavity has a connection to the nasal cavity by means of the **Eustachian tube.** This tube is a necessity, for the air outside the tympanic membrane changes in pressure and the tympanic membrane would burst if there were no provision for producng a corresponding change in the middle ear. When we drive up a mountain, go up in an airplane, or even go up in an elevator in a tall building, we can feel the pressure on our tympanic membranes

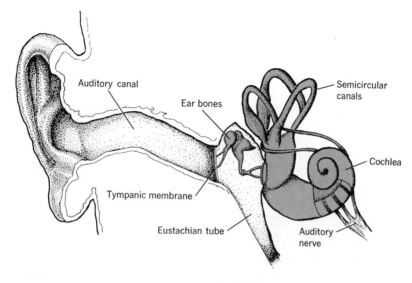

FIG. 29.11 The parts of the human ear.

as the outside pressure falls. Then as air escapes down the Eustachian tube and the pressure is equalized on either side of the tympanic membranes, we can hear normally again. On our way down, the reverse happens. This arrangement has one drawback—whenever we have an infection of the mucous membranes of the nasal cavity, it is possible to transmit the infection to the middle ear. Suppose you have a bad cold and blow your nose very hard while holding one nostril stopped. This pressure may send some of the mucus up the Eustachian tube into the middle ear. There it may set up an infection which results in the clogging of the tube. In extreme cases the infection may spread to the mastoid bone behind the external ear and surgery may be necessary to relieve the condition.

Equilibrium

We have discussed hearing first because hearing is usually associated with the word "ears." But there is a much more important function of the ears; we can live without hearing, but we cannot exist at all normally without a sense of equilibrium. Two sensory mechanisms are involved, both located in the inner ear. The first of these consists of three **semicircular canals** filled with lymph. These are concerned with the **motion sense.** At one end of each canal there are small hairs connected with sensory receptors which detect movements of the lymph and relay this information to the brain. When the head is moved, the fluid lags behind temporarily and the hairs are moved, thus stimulating the receptors. There is one canal for each of the three planes of the body, so we become aware of movements in any plane. If you spin around for a time the lymph within the canals begins to move with the body. Then if you suddenly stop, the fluid continues moving and you feel a sensation of dizziness as the moving lymph continues to stimulate the hairs. You feel as if you are moving when actually you are not. When the body is subjected to rapid and continuous changes in position, as on a ship in a rough sea, the sense related to the semicircular canals seems to become upset and certain autonomic nerves are stimulated, with a resulting nausea known as motion sickness.

Equilibrium also involves a **position sense,** so a person knows his position when he is not in motion. This is possible because of small **ear stones** suspended in a gelatinous fluid in a small sac below the semicircular canals. The sac is lined with receptors sensitive to pressure. Since gravity exerts a pull on the ear stones, you know if you are right side up or upside down or in any position in between. The crayfish has sand grains which perform a similar function (see Chapter 25).

Taste

The human sense of taste is localized in **taste buds** on the tongue. Experiments indicate that there are four kinds of taste buds, sensitive to **sweet, sour, salt** and **bitter** respectively. Each taste bud opens to the surface by means of small pores which allow a little of the substance in solution to enter the bud and stimulate the sensory endings within. If the tongue is wiped dry and dry sugar is placed on it there will be no sensation of sweetness; the sugar must be in solution to stimulate the receptors. The taste buds are unevenly distributed. Salt and sweet are more numerous on the tip of the tongue, sour along the side, and bitter at the back. A bitter substance may not seem so bad in the mouth, but as you swallow if you get the full sensation of bitterness, just as it passes into the throat. A child licking a sucker with the tip of his tongue is getting the utmost in sweet taste from the sucker.

You may feel that you have a greater variety of tastes than four, but this is because you seldom taste a food without also smelling it and you associate the two. Perhaps you have noticed how tasteless food seems when you have a cold which impairs the sense of smell. You still taste but you fail to receive the associated aroma which you normally detect when eating.

As is true of sensory receptors as a whole, those in the taste buds can be stimulated with weak electric currents. When a current is applied to an area of the tongue rich in the taste buds sensitive to salt, you can detect a sensation of a salty taste.

Smell

We have already mentioned the fact that the human sense of smell is poor in comparison to this sense in some other animals; yet even in man it is very sensitive. We detect odor when molecules of a substance in the air enter the nose and contact the olfactory receptors which are localized in an area of epithelial tissue about 10 sq cm in size in the nasal cavity. The mechanism of odor detection is not as well understood as the mechanism of taste detection, but experimental evidence has given us some plausible theories. It appears that there are about seven kinds of receptors, each of which is sensitive to molecules of a particular class. In most cases the shape of a molecule determines its class and thus to which receptor it will become temporarily attached. Each class of molecules elicits a primary odor—pungent, musky, and so on. The great complexity of odors which we can detect can arise when molecules have a shape which permits attachment to more than one receptor and when molecules of more than one class are being given off from a substance. It is known, for instance, that over one hundred different kinds of molecules are given off from geranium leaves and it is this combination which gives the distinctive geranium odor.

Senses of the Skin

The skin bears a variety of sensory receptors which detect various types of stimuli. Those related to the sense of **touch** are located near the surface and are stimulated by a very light pressure. Some are found with the hairs on the skin, and so contact with a hair brings about stimulation even though there is no contact with the skin itself. The touch receptors vary considerably in their distribution. You do not rub your shoulder against an object to determine what it feels like; you use your fingertips. Your finger tips are more sensitive because the touch receptors are much more numerous there. The receptors are most numerous of all on the skin of the lips. Perhaps this is why a baby tends to put everything he picks up to his mouth, and it could also explain a popular social contact between adults.

The distribution of receptors can be demonstrated by touching two needles held at varying distances to the skin. On the lips you can feel two distinct points even though the needles are practically together. On the fingertips you will likely feel only one point if the two needles are closer than about one-eighth inch. This shows that only one receptor is being stimulated by the two needles. On your back the needles may be as far apart as two inches and you still will feel only one point.

Lying deeper in the skin are the **pressure receptors.** These are stimulated when they are pushed out of shape. This occurs only when considerable pressure is applied to the skin.

The skin also contains **heat receptors** and **cold receptors.** If a warm object is moved over the skin, you will note that the sensation of warmth is distinct only at certain areas. Other areas show a sensitivity to cold. These actually respond to changes in temperature, one to changes in the direction of warmth and the other to changes in the direction of cold. Your

skin feels cold when you first go out on a cold day, but once it is chilled, you are no longer aware of the cold on the skin. An easy way to demonstrate this is to prepare three basins of water, one hot, one lukewarm, and one cold. Place the left hand in the hot water and the right in the cold water. Within a few minutes you will no longer be conscious of the difference in the temperature. Then place both hands in the lukewarm water: the water will seem cold to the left hand and warm to the right. This explains why a swimming pool seems to be cold on a warm day and warm on a cold day.

Finally the skin contains **pain receptors** which have a threshold of stimulation much higher than the other skin receptors. Only massive stimulations which might damage the skin bring about the sensation of pain. This is good because it would not do for us to feel pain from light contacts of the skin; we would be in constant pain. As it is, pain serves as a warning of damaging contact. Evidence indicates that damaged cells release something, probably histamine, which stimulates pain receptors. This would account for the persistence of pain after an injury. Extreme temperature changes also stimulate the pain receptors. A blindfolded person touched with a piece of dry ice may feel that he has been burned; the pain receptors are the same for all types of stimulations in the extreme.

Internal Sensory Receptors

Many sensory receptors are located within the internal organs. These function primarily to stimulate the autonomic nerves and are a part of the complex system of homeostatic regulation of the internal body functions, although some of them also transfer their stimuli to the conscious centers of the brain.

Within the skeletal muscles there are **proprioceptors** which are stimulated by the stretching and contracting of these muscles. These sense organs give us a **muscle sense** which helps to maintain equilibrium. The importance of this sense is apparent when your leg goes to sleep and you have no feeling in its muscles. Complex muscular movements would be impossible without these proprioceptors.

Within the hypothalamus of the brain there are **thirst receptors** which are sensitive to very slight changes in the osmotic concentration of the blood. If the body loses water or there is an extra intake of salt, we feel the sensation of thirst. Other areas of the hypothalamus contain **hunger centers** and **satiety centers** which are sensitive to the glucose level of the blood. Within certain blood vessels are receptors sensitive to the carbon dioxide level of the blood and others sensitive to the oxygen level. These are related to the rate and depth of breathing. Still other receptors are related to the release of enzymes and hormones within the body. These internal receptors will be studied more extensively in connection with the various organ systems which are involved.

DRUGS AFFECTING THE NERVOUS SYSTEM AND THE PROBLEM OF DRUG ABUSE

Many people do not realize that many substances which we ingest and inhale are composed in part of chemicals which affect the central and autonomic nervous systems. Any substance which affects body functions can be classified as a **drug,** which means that these chemicals are drugs. Coffee, tea, and cola drinks contain **caffeine,** which belongs to the class of drugs called **stimulants.** Wine, beer, and distilled liquors contain **ethanol,** which belongs to the class of drugs called **sedatives.** Cigarette smoke contains **nicotine,** a powerful nervous system stimulant.

Many drugs which affect the nervous system are used for medical purposes. Pain-killers, sleeping pills, tranquilizers, and diet pills all have definite medical uses. Unfortunately, these drugs are often used without medical supervision because they can produce pleasant feelings and because they can be used to avoid thinking about unpleasant problems.

Some drugs which affect the nervous system by altering perception have been widely used, primarily by young people, in recent years. Some of these drugs are used in religious ceremonies in certain cultures, while others have occasionally been used for medical purposes, but for the most part these drugs are used solely for the alteration of consciousness which they produce. In most parts of the United States, the use of these drugs is, at present, illegal.

In this section we will discuss the effects and uses of several classes of drugs which affect the nervous system. We do not intend to discuss the political and legal issues which surround the use of drugs such as marijuana, LSD, and other hallucinogens. Instead, we want to present the results of recent biological research on the effects of all these classes of drugs on the body so that you will be in a position to make up your own mind about the issues involved.

Stimulants

The most commonly used stimulant is **caffeine.** It is found naturally in coffee and tea, and it is added to cola drinks. It tends to increase alertness and counteract fatigue, which is one reason why the morning cup of coffee and coffee breaks during the day have become so commonplace. The Englishman has his "spot of tea" in the morning and during the day. In New Zealand tea is served before one gets out of bed in the morning. Overuse of caffeine causes nervousness, increased heart rate, and elevated blood pressure and can cause dependence. Caffeine stimulates gastric secretions which can corrode the stomach lining when they flow into an empty stomach. Those with stomach ulcers are usually advised to give up coffee for this reason. Caffeine is included in many headache remedies. Headaches may occur because of excess blood pressure caused by a dilation of the capillaries in regions of the brain. Caffeine causes constriction of these capillaries and can thus bring relief. Caffeine is also used in cold remedies and in tablets designed to prevent sleep. Caffeine is a purine and purines are found in gene DNA. When caffeine is applied in large quantities to smaller experimental organisms, it can be **mutagenic** by acting as a purine substitute. Evidence for a mutagenic effect in higher animals has not been found.

The **amphetamines, "pep pills"** and **"speed,"** are well known among those who take pills for "kicks." They are known as **"uppers"** because of their stimulating effect. They affect the autonomic nerves in a way similar to the hormone adrenaline (epinephrine). The amphetamines include **benzedrine, methedrine,** and **dexedrine.** Benzedrine was once used in inhalers to relieve nasal congestion, but the abuse of this drug became so great that it is no longer sold in over-the-counter preparations. Dexedrine has been widely used as an appetite depressant for those trying to lose weight. It also has a stimulating effect and was issued to combat troops going into action in past wars. In emergency it would relieve hunger and give the soldiers energy to carry on for long periods of time without fatigue. **Cocaine** has a stimulating effect similar to the amphetamines. It has been found that amphetamines will sometimes quiet overactive children. When the nervous system is overactive, these drugs apparently act as a depressant, although the way in which the drugs have this effect is not known.

Nicotine is another widely used drug which may be classified as a stimulant because it increases heart rate and blood pressure and constricts blood vessels. For this reason heart patients are better off if they do not smoke. Nicotine is an addicting drug, as any who have tried to stop smoking will testify. Nicotine mimics the synaptic transmitter substance acetylcholine, which transmits nerve impulses. To the smoker in need of a cigarette, nicotine seems to act as a tranquilizer because it relieves the nervousness which develops from nicotine deprivation after the body has become accustomed to it. Nicotine is a very powerful and deadly drug. A pill of pure nicotine no larger than an aspirin tablet would affect the nervous system enough to cause death. The amount absorbed from smoking or chewing tobacco is only a small fraction of the fatal amount.

Sedatives

Some drugs depress the nervous system. The **barbiturates,** often called **"downers,"** are in this group. They are often used as sleeping pills and in the reduction of high blood pressure. They are popular among drug abusers, but they can be dangerous and many barbiturate overdose deaths occur each year. Barbiturates are often used in combination with other drugs or alcohol. The combination of two substances can sometimes have a **synergistic** effect, which means that the combined effect is much greater than the effect of either substance alone. The combination of alcohol and sleeping pills (barbiturates) can be lethal because of a synergistic effect. Many people die from this particular combination each year.

Alcohol is a sedative which has been in widespread use since the dawn of history. Most of that time it was never stronger than fermented grape juice, apple juice, or grain extracts, and overuse was not a problem. Then man learned that he could make alcoholic drinks much stronger by distilling grain spirits, and now overuse is a problem. Some estimates indicate that more than 5 percent of the people in the United States have an alcohol problem and that many are uncontrollable alcoholics. Thousands of deaths each year can be traced to an overuse of alcohol, including deaths caused by drunken driving. Alcohol contains calories which provide energy, but those who depend upon alcohol for calories may neglect vital foods and suffer from vitamin, mineral, and amino acid deficiencies. As one alcoholic put it, "If I am hungry I can spend two dollars for a meal, or I can spend half as much for a bottle of wine and forget all about being hungry and feel much better also."

Psychedelic Drugs

Some substances can alter brain wave patterns and produce greatly altered perceptions. They are popularly called **psychedelic** (mind-expanding) **drugs.** They are also called **psychotomimetic** (psychosis-mimicking) drugs because of the similarity between reactions produced by

the drugs and severe personality disturbances. Or they may be called **hallucinogenic** (illusion-causing) drugs. For centuries the Aztec Indians used the mescal buttons of the peyote cactus for their euphoric properties in religious rites. Today these are sometimes ground into a powder to make mescaline. Certain types of mushrooms and many other plants contain substances which have psychedelic effects. They are generally poisonous when taken in excess quantities and are therefore dangerous to use.

LSD. The best known and most widely used of the psychedelic drugs is a chemical, **lysergic acid diethylamide,** better known as **LSD,** or "acid." It can be easily synthesized chemically and is very potent in small quantities. Only 1 g (about $\frac{1}{28}$ ounce) can be divided into over 10,000 high-potency doses. Cases have been reported in which a person handled packages containing a small amount of the powder on the outside and became quite high by putting his finger in his mouth. The initial reaction to the intake of LSD may be dizziness and sometimes nausea. This is followed by perceptual alterations. Bright,

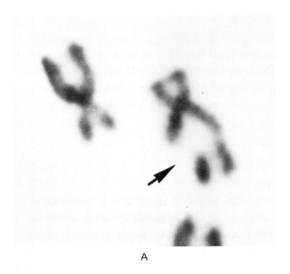

A

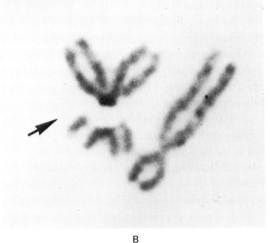

B

FIG. 29.12 Chromosome aberrations found in the blood cells of a long-time user of LSD by E. J. Egozcue. Both aberrations involve breakage of the chromatids.

It is thought that the impurities in the LSD, rather than the drug itself, causes such aberrations. (Courtesy of Dr. J. Egozcue.)

flashing colors, altered perception of sounds, distortions of images, and dissociation of self from reality are some of the perceptual alterations which users report. These alterations of perception can have tragic consequences. Recently the daughter of a well-known television personality jumped out a window in her apartment. In the euphoric "high" induced by LSD she felt that she could soar through the air like a bird. Sometimes a user has a "bummer" or "bad trip." During a bad trip a person may become panic-stricken because of delusions of being attacked by monsters or of being trapped without means of escape. There is no good way to predict when "bad trips" will occur. They are more likely to follow an overdose, but since the purity of the available LSD varies greatly, it is hard to identify the quantity in advance.

Some people who use LSD experience a "flashback" phenomenon. Hallucinations which originally appeared during a trip may appear suddenly and without warning weeks or months after the drug has been used. Apparently an environmental stimulus can trigger hallucinations which were originally experienced during an LSD trip. Prolonged psychotic reactions, in which the symptoms are similar to those of schizophrenia, are sometimes caused by the use of LSD.

Another effect of LSD should be mentioned. Some studies have detected an increased incidence of chromosome aberrations in body cells of habitual users. Dr. J. Egozcue, at the University of Washington, found that as many as 20 percent of the cells taken from the bloodstream of a long-time user had various types of chromosome aberrations. Nonusers showed a chromosome aberration of about 2 percent in control studies. Other studies, however, have had variable results, some positive and some negative. Such variations now seem to have been caused by differences in the product itself. Careful experiments on pure LSD do not seem to cause chromosome damage, but the illicit LSD bought off the street contains various amounts of contaminants which can induce the aberrations. These contaminants are also teratogenic in their effects; carefully controlled tabulations in New York City showed a marked rise in the number of birth defects among babies born to women who had used illicit LSD during early pregnancy.

Marijuana. Marijuana is another drug which has psychedelic effects. The dried leaves and stems of the plant *Cannabis sativa* are used to make marijuana cigarettes for smoking, and there are recipes for candy, tea, or other edible products which call for its inclusion. Some common names are pot, grass, tea, hay, and mary jane. The active ingredient in this plant is **tetrahydrocannabinol.** The active ingredient can be obtained in a much more concentrated form

FIG. 29.13 Marijuana derivatives. At the right the stem and leaves of an intact plant are shown. At the left are marijuana products, including pressed pollen, hashish, manicured marijuana, seeds, and stems and leaves. (Courtesy Carolina Biological Supply Company.)

by scratching the flowers, stems, and leaves of the plant and collecting the resin which oozes out. The concentrated resin is known as **hashish.** Hashish is also smoked, usually in a pipe, and it is more potent than a marijuana cigarette.

Many people claim that marijuana is not addictive and that it does not have any lasting harmful effects; some of them want to legalize its use and sale. Other people contend that, while marijuana may not be addicting in the physiological sense as heroin is, its continued use can result in psychological addiction; they claim that there is an increasing indication that long-term use may produce some harmful effects. The young person today, living in a culture where marijuana use is widespread, may have difficulty deciding whether to participate or not. An objective evaluation of current research is given to help in such decisions.

One study shows that marijuana may cause chromosome damage which could affect the health of the user and be passed on to the user's children. Morton Stenchever, at the University of Utah School of Medicine, examined leukocytes from forty-nine persons who had used marijuana for an average of three years and found an average of 3.4 percent of the cells had chromosome aberrations. Cells from control subjects which had not used the drug showed an average of 1.2 percent. While this increase in chromosome aberrations is not tremendous, it is something that should be considered. Douglas Gilmour, at New York University School of Medicine, found a comparable increase in chromosome aberrations in cells from those who had used marijuana more than twice a month but found no increase in cells from persons who had used it less frequently. Warren Nichold, at the Institute for Medical Research, gave the drug in controlled quantities to volunteer subjects for twelve days and found no increase in aberrations of their cells at the end of this time. Since chromosome damage in a fetus can result in birth defects, the studies mean that marijuana may have a teratogenic effect. More

research is needed to settle this now controversial question.

Marijuana may cause disruption of DNA synthesis and interfere with the functioning of the immune system. Gabriel Nahas, of the Columbia University College of Physicians, found that the lymphocytes of habitual marijuana users failed to respond to mitotic stimulants as well as lymphocytes from nonusers. The rate of DNA synthesis was 40 percent lower in the cells of users. Several other workers have found similar results. Since lymphocytes are involved in the immune reactions, the results imply that users may have a lowered resistance to disease.

Louis Harris, at the Medical College of Virginia, found that high doses of tetrahydrocannabinol given to mice enables them to receive grafts of skin from other mice. Mice that did not receive tetrahydrocannabinol rejected attempted grafts. This effect is comparable to the effect induced by immunosuppressive drugs given to people who receive organ transplants. People who receive these drugs are highly susceptible to infection and must take antibiotics regularly to protect themselves.

One possible beneficial application of this discovery may lie in the treatment of cancer. When tetrahydrocannabinol was given to mice with malignant tumors, their life span was increased by as much as 36 percent compared to those which did not receive this treatment. Cancer growth is dependent upon rapid DNA synthesis, and anything which reduces this synthesis will slow the growth of the cancer. Both the decreased lymphocytes and the cancer treatment are the results of the action of tetrahydrocannabinol on the immune system of the body.

Marijuana may disrupt hormone activity. Tetrahydrocannabinol may interfere with the output of certain hormones, or it may substitute for some of them. Robert Kolodny, at the Reproductive Research Foundation in St. Louis, found that the blood levels of testosterone in men who were habitual users of marijuana averaged about 44 percent lower than those in nonusers. Blood levels of testosterone were lowest in those who had used the drug most frequently. Of twenty heavy users, six had sperm counts diminished to the level of sterility and two had greatly reduced sexual potency. When these two refrained from using the drug for several weeks, potency returned. Tests showed that the testes had the power to produce testosterone but lacked the stimulus. This research leads to the hypothesis that the drug may inhibit the output of gonadotropic hormones from the pituitary gland. These are the hormones which stimulate the output of sperm and male hormone from the testes. Laboratory studies showed that the testosterone output decreased as much as 36 percent within three hours after smoking one marijuana cigarette.

At Cambridge Hospital in Massachusetts, fourteen cases of **gynecomastia** (male breast development) were reported; all these patients had been long-time users of marijuana. The examining doctors concluded that the similarity in chemical structure between tetrahydrocannabinol and the female hormone estradiol might be the reason for this response. The depression of testosterone levels was also thought to be a factor.

Marijuana causes deleterious effects on throat and lungs. This is one result of marijuana smoking about which there is no controversy. Many physicians have observed the sore throats and bronchial irritation that follow marijuana smoking. The smoke from marijuana contains about 50 percent more tar than high-tar tobacco smoke, and all the potential carcinogenic hazards of cigarette smoking apply to smoking marijuana. W. D. Paton, at Oxford University, has reported an increasing incidence of emphysema among those who have smoked marijuana for several years.

Marijuana may cause damage to the nervous system. This is a highly controversial point and there are no conclusive results on which all can agree. Harris Rosenkrantz, at the Mason Research Institute, has shown that if high doses of tetrahydrocannibol (about 150 times the amount

absorbed from smoking one cigarette) were given to rats each day for twenty-eight days, a marked reduction in RNA and protein synthesis in the brain occurred. This could mean that irreversible brain damage had been done. Doses equivalent to that obtained from an occasional cigarette did not cause measurable reductions in RNA and protein synthesis but did cause an increase in the production of ATP and an increase in the oxidation of glucose in the brain.

Alteration of personality patterns have been observed in some long-time users of marijuana. Several investigators have reported what they call an "antimotivational syndrome." Harold Kolansky and William Moore, at the University of Pennsylvania, treated thirteen patients who had smoked marijuana three to ten times a week for at least sixteen months. All appeared to be apathetic and sluggish in both mental and physical responses. These patients seemed to lack a goal in life and had lost interest in their personal appearance. They also had a reduced time sense and some difficulty with recent memory. The degree of these symptoms was related to the duration of use of the drug. When its use was stopped, the symptoms gradually disappeared, but they continued intermittently for as long as twenty-four months in very heavy users. Quite a number of other investigations agree with these findings, but negative findings have also been reported. Some of the latter show that the body develops a tolerance to marijuana in time and this results in a corresponding partial return to more normal behavior patterns.

One other interesting reaction might be mentioned. Those who smoke marijuana for the first time often report little change in their mood or feelings, but the effect is enhanced by later use. Analysis of the concentration of tetrahydrocannabinol in various parts of the body of experimental animals shows that, with the first use, most of it is concentrated in fatty tissues of the body and little goes to the brain. With later use, however, more goes to the brain. It seems as if the fatty tissue becomes saturated so that more is available to reach the brain, where it has its effect.

All these studies indicate that tetrahydrocannabinol is a powerful drug which may have long-term effects which last beyond the euphoric feelings which follow its intake.

Narcotics

Narcotics are a class of drugs which are derived from **opium.** Opium is found in the resin

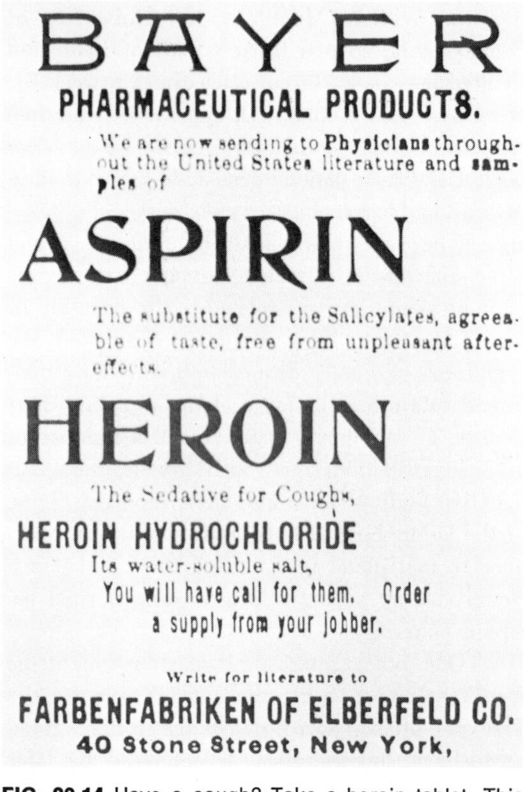

FIG. 29.14 Have a cough? Take a heroin tablet. This might have been the advice of your pharmacist in 1900. Pharmacy journals of the time ran advertisements like this one equating heroin with aspirin. In a short time, however, advertisements for drugs to treat the heroin habit began to appear. (National Audiovisual Center.)

which comes out of the opium poppy when it is scratched. This sticky resin can be taken by mouth or smoked to obtain a euphoric feeling. Opium may be further refined into **morphine** or **heroin,** which are much more potent than the original opium. **Codeine** and **paragoric** are milder opium derivatives. All these drugs have valuable medical uses. When a person is in a serious accident, an injection of morphine relieves the shock, eases the pain, acts as a sedative, and relieves extreme anxiety and tension. Codeine is often used to relieve coughs and colds or menstrual disturbances. Paragoric has long been an effective diarrhea medicine. Shortly after its discovery, heroin was recommended as a sedative for coughs and was sold over the counter for this purpose. It was included in the Sears Roebuck catalogues in the early 1900's. Heroin was also hailed as a cure for morphine addiction. It was soon discovered, however, that heroin was even more potent and addictive than morphine.

Common street names for heroin are horse, H, junk, smack, and stuff. Those who use heroin to get high often start to use the drug by inhaling heroin powder. As dependence on the drug grows, a user may begin to inject heroin. Two methods of injection are used. "Skin popping," or injecting under the skin, does not produce as strong or as rapid an effect as "mainlining," or injecting directly into a vein. When a person becomes addicted to heroin, his body develops a tolerance for it and he must increase the amount he takes to maintain the same feelings of euphoria. As an addict's habit increases, the amount of money which he needs to maintain it also increases. Many burglaries, robberies, and other crimes are thought to be committed by addicts who are trying to get money to supply themselves with drugs.

Heroin addiction is generally debilitating. Addicts lose interest in food. Malnutrition and lowered resistance to disease are often side effects of heroin addiction. Addicts sometimes share the same drug paraphernalia and many contract hepatitis from infected needles. The heroin sold on the street varies in purity and strength and many addicts overdose themselves unintentionally. It is difficult for an addict to stop using heroin because the withdrawal symptoms are severe and sometimes fatal. Heroin withdrawal causes severe cramps, vomiting, diarrhea, and convulsions. It also seems that many addicts develop a psychological as well as physical dependence on heroin.

There are many different methods of treating heroin addiction. Some methods emphasize psychological counseling and total abstinence from the drug. Other approaches use a heroin substitute, **methadone,** which is thought by some to be the most hopeful treatment currently available for heroin addiction. Methadone belongs to the same class of drugs as heroin and substitutes for heroin in the body, preventing heroin withdrawal symptoms from occurring. When given in therapeutic doses, it usually does not produce euphoria. Methadone is a potent addictive drug, however. When a large dose of methadone is taken or when it is injected rather than taken orally, it does produce a "high" which is similar to, although less intense than, a heroin "high." Patients sometimes take the methadone dispensed by methadone clinics and sell it on the street. Last year in New York City there were more drug-related deaths caused by methadone than by heroin.

Cocaine is also classified as a narcotic. It produces euphoria and is used medicinally to relieve pain. Cocaine is derived from the coca shrub of South America. Some Indian tribes in this region suck on the leaves to help them endure hunger. It is not used very much medically because of its tendency to induce paranoic tendencies, headaches, dizziness, and depression and the destruction of tissue caused by its application. It is still common among illicit drug users, however, and goes by such names as coke, snow, she, or girl. It is usually sniffed into the nostrils and many users suffer from the destruction of nasal tissues as a result. The development of tolerance and withdrawal symptoms are much less marked than with the opium derivatives.

REVIEW QUESTIONS AND PROBLEMS

1. Damage to a motor area of the left side of the brain will usually cause paralysis of a part of the right side of the body. Explain this phenomenon.
2. What important function is served by the cerebrospinal fluid in the brain?
3. The cerebrum of the mammals is much more highly developed than that of the birds, but the cerebellum of the birds is more highly developed. What do these facts indicate about the two?
4. Why are the convolutions of the brain considered to be important as a measure of the potential intelligence?
5. Damage to the hypothalamus can have serious effects on an animal's body. Explain what these effects may be and tell why.
6. Distinguish between a nerve, a neuron, an axon, and a neurilemma.
7. Describe the three kinds of neurons according to function.
8. Describe the nature of a nerve impulse.
9. An axon cannot transmit an impulse when it is deprived of oxygen. Explain why.
10. Sodium ions are somewhat larger than potassium ions. What relationship might this have to the events connected with a nerve impulse?

11. Explain how a nerve impulse is transmitted across a synapse.

12. Certain nerve gases are known to cause a breakdown of cholinesterase. How would these gases affect the human body and why?

13. Some chemicals act as stimulants, while others act as depressants of the nervous system. Explain how these function.

14. Suppose there were no such thing as reflex action. Tell how your daily activities would be altered.

15. What is a conditioned reflex and how does it differ from other types of reflexes?

16. How does the focusing of the human eye differ from the focusing of a camera?

17. When you go from bright sunlight into a darkened room your vision is very poor at first, but gradually improves until you can see quite clearly. Describe all of the events taking place in your eyes as they accommodate to the lowered light intensity.

18. A person who eats a diet deficient in vitamin A will have very poor vision when the light is dim. Explain why.

19. Color vision in very dim light is much less acute than in a bright light. Explain why.

20. What is wrong with the eyes of a nearsighted person? How can the vision of such persons be made normal?

21. When you strike the key of middle F on the piano, the piano string vibrates 345 times per second. Explain just what happens in your ears which makes you aware of a note of this pitch.

22. If you have a bad cold and blow your nose very hard, you may feel a stuffiness in your head and will not hear very well for a time. Explain what happens.

23. If you spin around rapidly and suddenly stop you will feel a dizziness and will find difficulty keeping your equilibrium. Explain why you have this sensation.

24. Certain connoisseurs can recognize hundreds of varieties of wine by taking a small sample. Since there are only four types of taste receptors, how is this possible?

25. How are odors detected in man?

26. How does a touch receptor of the skin differ from a pressure receptor and a pain receptor?

27. When you first dive into a swimming pool the water may feel quite cold, yet after a few minutes you do not notice the cold. Explain.

28. Recent research has indicated that marijuana may affect the immune system. How might this knowledge be used medically to treat people who need organ transplants?

29. What is a synergistic effect? What happens when alcohol and barbiturates are taken together?

FURTHER READING

Day, R. H. 1966. *Perception*. Dubuque, Iowa: W. C. Brown.

Droscher, V. B. 1971. *Magic of the Senses: New Discoveries in Animal Perception.*
New York: Harper and Row.

Katz, B. 1966. *Nerve, Muscle and Synapse*. New York: McGraw-Hill.

Pribham, K. H. 1969. *Brain and Behavior*. Baltimore: Penguin Books.

Wilentz. J. S. 1971. *Senses of Man*. Clifton, N. J.: Apollo Editions.

30

The Circulatory System

In a large complex animal, such as a human being, **transportation** of materials within the body is extremely important. Oxygen must reach the cells and carbon dioxide must be carried away; food must be distributed and excretory wastes removed; hormones must be carried from the glands where they are produced to the body parts which they influence. Even a brief interruption in circulation to the brain results in unconsciousness; everyone has experienced that temporary blackout that comes when he stands suddenly and the blood drains from his brain. Longer interruptions result in death to the cells affected. A tight cord around an arm which prevents circulation to the arm will result in death of this body part. Cells in a small area of the heart frequently die when a blood clot cuts off their blood supply. This is the common heart ailment known as coronary thrombosis.

The circulatory system also **fights infectious agents** which may invade the body. Special cells in the blood are adapted to engulf and destroy these invaders, and the plasma contains antibodies which react with infectious agents to render them ineffective.

Finally, the blood **equalizes the temperature** in various parts of the body and keeps it constant. Circulating blood serves a function similar to the circulating water in the radiator of an automobile. It flows around the body organs where metabolism is generating a considerable amount of heat and then flows out into the capillaries of the skin where the heat is dissipated into the surrounding air. The cool blood then flows back to the original organs and prevents them from becoming overheated. If the body is generating unusually large quantities of heat, or if the day is unusually warm, the sweat glands aid in this reaction by secreting extra quantities of sweat on the skin. The evaporation of this moisture cools the skin and thus aids in cooling the blood flowing through it.

BLOOD

The blood is the transporting part of the circulatory system. It constitutes about 9 percent of the body weight. There will be about six quarts of this precious fluid circulating in the blood vessels of a man weighing about 150 pounds. A loss of one quart of this total at one time is serious and a loss of as little as three pints is almost certain to result in death. On the other hand, one pint may be removed without serious consequences. This is often done when blood is donated for transfusions. Blood is a body tissue, a liquid tissue, with cells making up about 45 percent of the total volume in men and about 40 percent in women. The remainder is a viscid liquid known as plasma.

Plasma

Human blood plasma is a clear, straw-colored fluid which transports many products. It is composed of 90 to 92 percent water, 7 to 9 percent proteins, about 0.9 percent dissolved minerals, about 0.1 percent glucose and varying amounts of digested food, hormones, waste products, and other substances being transported. It is slightly alkaline, with a normal pH of 7.3.

The plasma proteins are found as colloidal dispersions within the plasma. The most abundant protein is **albumin,** which is similar to the albumin which makes up much of the white of

an egg. This protein gives blood a viscid consistency which must be maintained. When a person suffers a serious injury or a traumatic emotional experience, he may go into **shock.** In this condition, increased permeability of the capillary walls permits much of the water in the blood to escape, but the large albumin molecules cannot pass out of the blood vessels and so blood becomes more highly concentrated. The blood is more viscid and it does not flow well through the capillaries. Blood pressure decreases greatly and unconsciousness and even death may follow. Administration of blood plasma or water balanced with salts helps to relieve shock.

Fibrinogen is a blood protein which furnishes the fibrous material of a blood clot. It will be discussed later in this chapter.

The **globulins** are the third kind of blood proteins. Globulins are often designated by Greek letters such as alpha, beta, and gamma. The **alpha globulins** are usually called **haptoglobins;** they bind free hemoglobin molecules which escape into the plasma when red blood cells burst. Free hemoglobin molecules are toxic, but when they are bound by the haptoglobins, they are harmless. When there is a massive bursting of red blood cells and there is not enough haptoglobin to absorb all the released hemoglobin, a person may suffer toxic effects. **Ahaptoglobinemia** is a rare genetic disease characterized by an inability to produce haptoglobin. People with this defect suffer constantly from the toxic effects of free hemoglobin.

The **beta globulins** are **transferrins,** which pick up and transport iron ions. These ions would also be toxic if they were free in the blood, but as they diffuse into the blood from the intestine, they are bound by the transferrins and transported to the cells where they are needed. Higher vertebrate animals could never have developed without transferrins. A great amount of oxygen is required to maintain the warm-blooded state, and this means that vertebrates require a high concentration of hemoglobin. Since hemoglobin has iron as its central core, the body needs transferrins to bring in iron to produce a sufficient amount of hemoglobin.

The **gamma globulins** contain antibodies which will be discussed later in this chapter. A few people inherit a defect called **agammaglobulinemia.** These people cannot produce gamma globulins. They usually die early in life from a minor infection because they cannot build immunity to the infection. In Chapter 35 we shall learn how this defect can be overcome.

Erythrocytes, Red Blood Cells

About 99 percent of the blood cells are erythrocytes. They are composed of a thin outer plasma membrane. The cell is filled primarily with the protein **hemoglobin,** which has a great affinity for oxygen. In the adult, red cells are produced by the red marrow of the bones, particularly by the breastbone (**sternum**) and by the ends of the shaft bones. Red cells are needed in the embryo as early as three weeks after conception. At this time they are produced in the yolk sac. Later in fetal development they are produced in the liver and spleen.

When the red blood cells are being produced they are rather large and have nuclei like most other body cells. There is little hemoglobin in immature red cells. As they mature, however, the nuclei are lost and the amount of hemoglobin is greatly increased. Then they are released into the blood stream. In certain types of anemia the cells are released prematurely while they are still nucleated and while they are poor oxygen carriers. The absence of a nucleus seems to doom the red blood cells to a short existence. Studies using radioactive iron to mark newly released cells show that they have an average life of about 120 days. As they age they become more and more fragile until they finally burst. The remains of the cell are removed by amoeboid cells of the spleen and liver, and the hemoglobin is picked

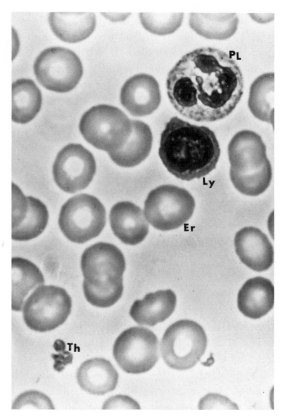

FIG. 30.1 A stained human blood smear which shows the different kinds of blood cells. PL—Polymorphonuclear leukocyte (granulocyte). Ly—Lymphocyte (agranulocyte). Er—Erythrocyte. Th—Thrombocytes (blood platelets).

up by the haptoglobins. There is a considerable turnover each day. An average human body will have about 20 trillion red blood cells. If each has a life of 120 days, the erythrocytes are being produced and destroyed at the rate of about 115 million per minute.

The erythrocytes are small, double-concave disks. In cross section they look like dumbbells. They have a diameter of only about 7.2 μ. This small size provides for an exposed surface that is large in relation to volume and makes them efficient in absorbing oxygen. It also enables them to travel through the minute capillaries of the body. Some capillaries are so fine that the red cells, small as they are, can travel through only in single file. A man will have the almost unbelievable number of about 5,000,000 red blood cells per cubic millimeter of blood. This is only

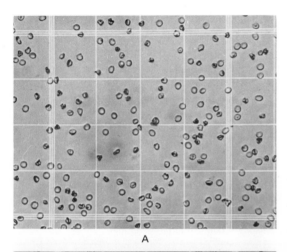

A

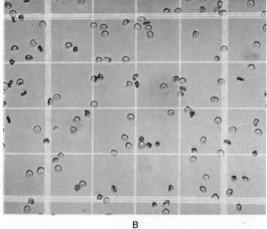

B

FIG. 30.2 Method of determining the number of red blood cells. Diluted blood is placed on a slide which is marked off in squares. Each square includes a measured volume of blood. By counting the number of cells in the squares, an accurate blood count can be made. Blood in (A) has a normal red blood cell count. (B) shows blood taken from a person with severe anemia, a disease in which the number of red blood cells is greatly reduced.

about $\frac{1}{25}$ drop. In women about 4,500,000 is considered the normal number.

If the number of red cells or the hemoglobin content of the red cells becomes deficient, a person has **anemia.** In such a condition the blood cannot supply oxygen to the body tissues in normal quantities and the person is handicapped by a lack of sufficient energy to carry on his activities. There are many causes of anemia—insufficient iron in the diet, insufficient vitamin B_{12} in the diet, excessive bleeding, inherited abnormal hemoglobin, and abnormal secretions from the stomach wall. The latter is one of the most common and better understood of the anemias. Vitamin B_{12} is necessary for the normal production and release of red blood cells. In some persons, however, there is not a proper absorption of this vitamin from the food and they have a condition known as **pernicious anemia.** During the last century it was found that such persons could benefit by eating large quantities of liver. Today we know that vitamin B_{12} is stored in the liver. The real trouble in pernicious anemia, however, lies with the secretions of the stomach; the stomach gastric juice does not contain an **intrinsic factor** which is necessary for the absorption of vitamin B_{12}. When the vitamin is injected into the body, there is a very marked improvement, or the person can take liver extract, which can be absorbed since it has already combined with the intrinsic factor from the stomach of the animal from which the liver was taken. Extract from the stomachs of slaughterhouse animals is also helpful since it will furnish the missing factor.

Leukocytes, White Blood Cells

White blood cells are usually larger than the erythrocytes but much less numerous. Their number is subject to considerable variation, but usually falls between 5,000 and 10,000 per cubic millimeter of blood, the typical number being about 7,000. They are nucleated cells and have a

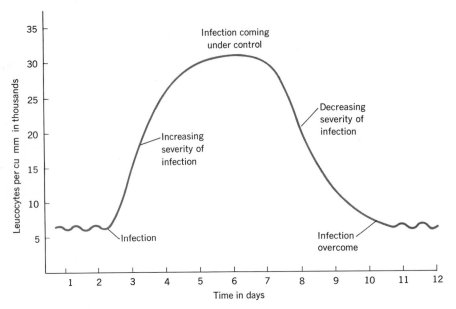

FIG. 30.3 The body responds to infection by increasing the number of leukocytes in the blood stream. When the infection is brought under control, the leukocyte count returns to normal.

much longer life expectancy than the erythrocytes. Most of the leukocytes are amoeboid in nature, throwing out pseudopodia like an amoeba and squeezing through small openings in the capillary walls. Thus, they are found in the intercellular spaces as well as in the blood. There are two major types of these cells.

1. **Polymorphonuclear leukocytes (granulocytes).** These cells have nuclei divided into distinct lobes. Typically there will be two or three lobes connected with each other by means of fine threads. These cells also have distinct granules in the cytoplasm. They are produced in the red marrow of the bones along with the red blood cells. They are important in fighting disease agents which may gain entrance to the body. Since they are amoeboid, they can squeeze themselves out of the blood vessels and mobilize at the site of an infection, where they destroy the invaders by engulfing them. The pus that accumulates at such an infection contains large numbers of these cells. The body has the power to increase the number of "polys" during an infection

and thus fight the invaders more efficiently. Whereas a normal count may be about 7,000, an infected appendix may so stimulate the mobilization of these cells that the count may rise to as high as 30,000 per cubic millimeter or more.

The polymorphonuclear leukocytes may be further subdivided into three kinds based upon the reaction of the granules to certain dyes. **Neutrophils,** the most numerous of the leukocytes, make up about 63 percent of the total white blood cell count. Their granules stain a violet color, a neutral reaction to combined acid and basic cyes. **Eosinophils** stain red with the acid dye known as eosin. They make up about 1.6 percent of the total white blood cell count. **Basophils** have granules which stain blue with basic dyes. They make up only about 0.4 percent of the total white blood cell count.

The neutrophils are the leukocytes primarily involved in fighting infectious agents. The function of the other two types of white cells is not clearly understood, but we have observed that the number of eosinophils will go up greatly in certain allergic reactions such as asthma and hay fever. It will also go up when there is an infestation of the intestine with certain parasitic worms.

2. **Agranulocytes.** These cells have a single large nucleus that fills most of the interior of the cell. There are no granules in the cytoplasm. There are two types of these white blood cells, and they are quite different in function. Both are formed in lymphatic tissue, such as the lymph nodes and the lymph tissue in the spleen. **Lymphocytes** are the smallest of the leukocytes, little larger than red blood cells. They produce some of the antibodies, but more important, they can be transformed into other cells which are of great value in fighting infection. When invading organisms enter the body, the lymphocytes aggregate at the site of infection and change into large macrophages. These large cells can engulf as many as one hundred bacteria before the engorgement brings about death. In comparison, a neutrophil can engulf only from about five to twenty-five bacteria before it dies. The importance of the lymphocytes is emphasized by their abundance; about 30 percent of the leukocyte count is lymphocytes. **Monocytes** are much larger than the lymphocytes and make up about 5 percent of the white blood cell count. They are amoeboid in nature and they are found in the body tissues. They engulf foreign bodies.

When the leukocyte count goes up during an infection, there is not an even rise in the numbers of all kinds of leukocytes. The neutrophils are all primary ones that increase, so the proportion of polymorphonuclear cells to agranulocytes will become greater. In normal blood the percentage of granulocytes may be 65, but it can easily reach 90 percent in acute appendicitis or other infections. Certain chronic infections may stimulate the production of extra numbers of lymphocytes so that the number of this type of leukocyte will become much greater than 30 percent of the total. We have already mentioned how the eosinophils increase in certain allergic and other conditions. A diagnostician often makes a differential blood count on the leukocytes as well

as a total white blood cell count, since valuable information can be obtained by studying the different proportions of the leukocytes.

In some tragic cases there is an uncontrolled production of white blood cells in the bone marrow or the lymph nodes. This condition is known as **leukemia,** or cancer of the blood. In severe infections the white blood cell count may rise as high as 50,000 per cubic millimeter, but in leukemia it can rise as high as 800,000 without any infection.

Thrombocytes (Blood Platelets)

When a smear of a drop of blood is made on a microscope slide, stained, and examined under the microscope, only two types of cells will be seen—the red and white blood cells. There is a third type, however, which is not revealed by this technique. There are about 250,000 thrombocytes per cubic millimeter of blood and each is about half the size of an erythrocyte. They burst shortly after they leave the body when they encounter foreign material. By special techniques this bursting can be prevented and the cells can be seen.

The thrombocytes contain a chemical which is related to the chain of reactions involved in blood clotting. The more we study this vital reaction of the blood, the more involved we realize it

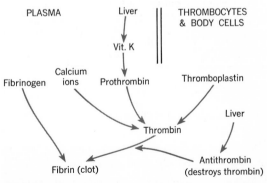

FIG. 30.4 Many factors are involved in the regulation of blood clotting. See text for details.

is. Clotting serves to seal off the breaks in blood vessels. Otherwise even a small injury would be fatal, for there would be nothing to stop the blood from flowing from the body. We know that the blood contains the protein **fibrinogen,** which can be converted into solid fibrin which forms the clot. The conversion normally does not take place within the blood vessels, but when there is an internal injury it occurs. What causes the conversion? Several things are necessary. **Calcium ions** are needed and we can prevent clotting by mixing sodium oxalate or sodium citrate with freshly drawn blood. The oxalate or citrate unites with the calcium ions and there is no clotting. We use these chemicals to prevent clotting of blood which is donated for transfusion purposes. Second, the protein **prothrombin** is needed. This is produced in the liver and vitamin K is necessary for its production. These three essentials for blood clotting are always present in the blood plasma, but a fourth ingredient is necessary before the reaction will take place. This is **thromboplastin,** which is within the thrombocytes as well as within the cells of the body tissues in general. When the body is injured as in a cut, some of the cells will be injured and release thromboplastin. As the blood flows out, more thromboplastin is released as the thrombocytes burst. The combination of thromboplastin, calcium, and prothrombin forms **thrombin.** This then unites with the fibrinogen and the **fibrin** is formed. It can be summarized as follows:

$$\text{thromboplastin} + \text{calcium} + \text{prothrombin}$$
$$= \text{thrombin}$$
$$\text{thrombin} + \text{fibrinogen} = \text{fibrin}$$

Bleeding Diseases

Other plasma components besides platelets are also required in small quantities for normal clotting. A deficiency of any one of the components causes prolonged bleeding after an injury.

Deficiencies may be caused by a dietary deficiency which can easily be corrected by adding the needed substance to the diet. When women give birth, they are often given injections of vitamin K and calcium to be sure they have no deficiency of these necessary substances.

There are at least four hereditary bleeding diseases. The most common is known as **hemophilia.** Hemophilia is caused by a recessive, sex-linked gene. (Details on the inheritance of this trait are discussed in Chapter 36.) The recessive gene fails to direct the production of one of the plasma components needed for thrombin formation. Another gene fails to direct the production of another of the plasma components needed for thrombin formation. This defect is called **Christmas disease.** A third gene directs the production of thrombocytes which lack normal adhesiveness. These defective thrombocytes do not adhere well to rough surfaces and do not burst open readily when there is an injury, so the clotting reaction is not properly initiated. This condition is called **pseudohemophilia.** A fourth gene causes **afibrinogenemia,** a defect characterized by a lack of fibrinogen. In all these defects the bleeding will eventually stop if the blood vessels are held together until they form a seal by growing together. Today the vital substances have been isolated from blood plasma and may be given by injection to stop bleeding.

Heart Attack and Stroke

Heart attack and stroke are the leading causes of death in the United States today. Both are caused by small internal clots which clog blood vessels which supply important body organs. Ruptures of thrombocytes can occur within the blood vessels, and these ruptures might cause internal clots frequently were it not for a safety mechanism which usually prevents their formation. The liver produces **antithrombin,** which can combine with and neutralize small quantities of thromboplastin which may be released internally. Sometimes the thromboplastin may be released in too great a quantity and small clots do form. When clots lodge in small vessels supplying the heart, the result is a heart attack, or a **coronary thrombosis.** The part of the heart muscle which is supplied by the blocked vessel will die. The chance of a heart attack is increased when the vessels are narrowed by cholesterol deposits on their walls. If a clot lodges in a small vessel supplying the brain, the result is a **stroke.** A stroke can cause partial paralysis or death. Those who are in danger of having a stroke or heart attack may take extra quantities of antithrombin to prevent the formation of internal clots. People taking antithrombin bleed a little longer when they are injured, but this is better than suffering a possibly fatal heart attack or stroke.

Antibody Production

Vertebrates can produce antibodies to neutralize foreign proteins which enter their bodies. Antibodies are not produced before birth or shortly after birth, but the foundation for their production has been laid. **Thymic cells** are produced by the thymus gland, which is located in the upper part of the chest in human beings. The thymus is quite large in the fetus and newborn baby, but it gradually disintegrates until in a mature person it is only a small mass of connective tissue. The thymic cells migrate to the lymph nodes, spleen, liver, and bone marrow. After birth thymic cells become plasma cells which react with foreign antigens and produce antibodies.

One of the great mysteries of the process has been the way in which the plasma cells "recognize" the proteins of other cells of the same organism and do not produce antibodies against them. One theory states that there are many different kinds of original thymic cells. Each type of cell is coded to react with a particular kind of antigen. When cells coded to produce antibodies

against one of the antigens of the organism's own body contact that antigen, they are destroyed. As a result, there are no plasma cells which can be sensitized to the body's own proteins. Occasionally a few thymic cells may fail to make contact with a body antigen and there may be some antibody production against that antigen. For instance, one affliction, *Myasthenia gravis*, is characterized by self-destruction of one's own muscles by antibodies which attack some muscle proteins.

Experiments in other vertebrates lend support to this theory. When the thymus gland from one newborn animal is injected into another, the recipient can later receive transplants from the donor without rejection. This has led to speculation that in the future a group of newborn babies may exchange parts of their thymus glands with one another. The cells of each baby would thus

THE BLOOD TYPES

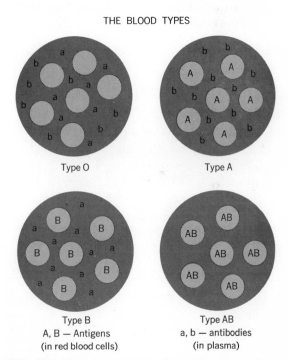

Type O   Type A

Type B   Type AB
A, B — Antigens   a, b — antibodies
(in red blood cells)   (in plasma)

FIG. 30.5 Antigens and antibodies of the four major blood types. Antigens are found on the membranes of the red blood cells. Antibodies are contained in the plasma.

"learn to recognize" the cells of the other babies as their own. Then if there should ever be a need for an organ transplant, the organ could be taken from anyone in the group without giving the recipient immunosuppressive treatment to depress antibody formation. Research is progressing rapidly in this field and we will have many new discoveries in the future.

The Blood Types

Blood transfusion is one of the great medical discoveries of the century. Many lives have been saved by blood transfusion. Early attempts at blood transfusion sometimes ended in tragedy when the blood cells being introduced would clump and cause the death of the recipient. The discovery of **blood types** has made transfusion a safe procedure.

The four major blood types are **O, A, B,** and **AB.** The percentage of each of these types varies with different population groups. Among the Caucasian population of the United States, type O is the most common type. About 47 percent of the Caucasian population possesses this type. About 40 percent have type A. About 10 percent have type B and only about 3 percent have type AB. The black population in the United States has a higher proportion of type A and a lower percentage of type O, and those of Asian origin have a higher proportion of type B blood. Each racial group has distinctive proportions of blood types.

The letters refer to antigens found on the membranes of the red blood cells. Type A people have the A antigen; type B people have the B antigen; type AB people have both of the antigens; type O people have neither antigen. Blood plasma may contain anti-A antibodies or anti-B

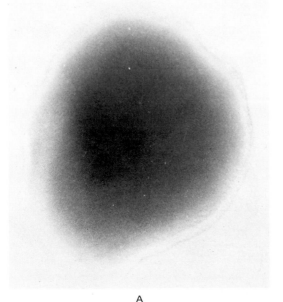

A

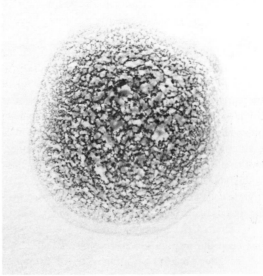

B

FIG. 30.6 Blood typing. Blood is mixed with serum containing anti-B antibodies and does not agglutinate (A). Blood from the same sample is mixed with serum containing anti-A antibodies and there is agglutination (B). This set of reactions shows that the blood being tested is type A.

antibodies, but no one has antibodies which would cause his own red blood cells to clump. Each person has antibodies for the antigens he does not have. A type A person has anti-B; a type B person has anti-A; a type O person has both types of antibodies; a type AB person has neither. Blood can be typed by mixing it with sera from type A and type B blood. Serum is defibrinated plasma which still contains the gamma globulins. If you have the A antigen, your red cells will be agglutinated (clumped) by the anti-A antibodies in serum from type B blood, but not by the anti-B antibodies from type A blood. You can easily figure out the reactions of the four types of blood to these two sera.

It is not clear why antibodies for A or B antigens are present in those who do not have these antigens in their red blood cells. In other cases of antibody formation, it seems that contact with an antigen is necessary before there is any antibody production, but anti-A and anti-B appear apparently without any such contact. One theory states that the young baby contacts many antigens during its early life and that some of these may be similar to A and B. If the baby lacks the A antigen, he will develop anti-B antibodies which will react with the B antigen. If the baby lacks the B antigen, he will develop anti-A antibodies.

In blood transfusions it is important that no cells containing an antigen matched by an antibody in the recipient's blood go into his body. If these cells are clumped in the blood vessels there will be a clogging of the capillaries and

death is almost certain to follow. It is always best to give blood of the recipient's own type, although in emergencies it is possible to use some other types in some cases (see Table 30.1). It is possible to use plasma of the kind which will affect cells within the recipient's body. A certain concentration of antibodies is required to cause clumping and the antibodies being put in will be quickly diluted by the plasma already in the person's body. This reduces the concentration so that no great harm is done. For this reason, it is possible to give blood plasma from any type blood to anyone.

The antigens and the antibodies of the blood types are determined by heredity. Other blood antigens have also been discovered, but since the plasma does not normally contain antibodies for these antigens they seldom have significance in blood transfusions. There is one exception to this, the Rh factor, which is discussed below. The pattern of inheritance of these antigens is so clearly known that the blood antigens can be used in legal cases involving disputed parentage. By a study of the antigens in the mother's blood and in the child's blood, it is possible to determine which antigens must have been in the father's blood. This study has been so highly refined that the exact combination of antigens found in any one person's body are almost as distinctive as his fingerprints as a means of identification.

The Rh Factor

This antigen is named from the rhesus monkey, in whose blood the factor was first demonstrated. About 85 percent of the people in the United States carry this factor and are Rh-positive; 15 percent do not carry it and are Rh-negative. Normally the blood of an Rh-negative person carries no antibodies that react with this factor, but antibodies may be produced if an Rh-negative person is exposed to the factor. For in-

TABLE 30.1. Compatibility of Blood Transfusions

Blood Type	Can Give Transfusion To	Can Receive Transfusion From
O	O, A, B, AB	O
A	A, AB	O, A
B	B, AB	O, B
AB	AB	O, A, B, AB

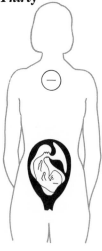 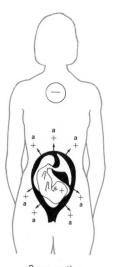 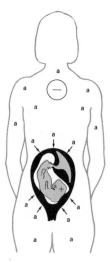

Negative mother
Negative child
Baby normal

Same mother
Positive child
Baby normal
Mother sensitized

Same mother
Second positive child
Baby born with
erythroblastosis

FIG. 30.7 An Rh-negative woman can become sensitized to the Rh factor if she bears an Rh-positive child. Future Rh-positive babies may be born with erythroblastosis because antibodies from the mother have destroyed red blood cells in the fetus. In this illustration the woman's first child was Rh-negative, but her second child was Rh-positive. The woman was sensitized to the Rh factor by her second child. Her third child was also Rh-positive and was born with erythroblastosis.

---

stance, if an Rh-negative person receives a transfusion from an Rh-positive person, there will be no reaction. This contact with the Rh-positive factor may sensitize the negative person so that a second transfusion from an Rh-positive person may cause serious consequences. This is similar to the development of an allergy. A person is not allergic to a food when he first eats it, but if he is sensitized by this first contact he will show an allergic reaction upon eating the food for a second time. An Rh-negative woman may also become sensitive if she carries an Rh-positive child in her uterus. The first Rh-positive child will be normal, but if it sensitizes the mother, the second Rh-positive child that she bears may be born with a serious blood defect, **erythroblastosis.** The antibodies of the mother's blood react with the Rh factor in the child's blood to cause this condition. Most children with this condition used to die shortly after birth, but today when trouble is expected, the doctors will be ready to give blood transfusions to the newborn child and can save most of them by replacing the defective blood with normal blood. The Rh factor is hereditary, so an Rh-negative woman and an Rh-negative man need not worry—all their children will be Rh-negative. The potential danger lies in children of Rh-negative women and Rh-positive men. These children may be either Rh-positive or Rh-negative. If the children are Rh-negative there is no danger; if they are Rh-positive there is no danger to the first child, but there may be to a second Rh-positive child. The first Rh-positive child may be abnormal if the mother has been sensitized by an Rh-positive blood transfusion sometime before the pregnancy.

Rh-induced erythroblastosis can now be prevented. A serum is available which contains Rh antibodies. A negative woman is sensitized when positive cells from the embryo's blood enter her blood during birth. If she receives the serum, known as **rhogam,** shortly after the birth, the foreign cells are destroyed before she can react to them and begin producing Rh antibodies.

THE MECHANICS OF CIRCULATION

When we dissect a laboratory animal and see the arteries and veins and their connections to the heart, it seems quite obvious that the blood must make a circuit through the body. Before the seventeenth century, anatomists had made extensive dissections of the human body as well as of lower animals, but they had numerous false conceptions of circulation. Some thought that the blood flowed out from the heart and was then returned by the same vessels—an ebb and flow; others believed blood was produced in the heart and flowed out never to return; none conceived of the idea of a connection between the arteries and veins. It remained for William Harvey, in 1633, to demonstrate the fact that the blood made a complete circuit of the body, traveling away from the heart in the arteries and returning through the veins. Harvey never saw the connecting capillaries, but he postulated that they must be there. They were discovered four years after his death. Harvey's experiments were simple. He found that a severed artery bled only from the cut end toward the heart, while a severed vein bled only from the end away from the heart. If the thoracic cavity was opened and arteries connecting to the heart were tied with a string, the arteries would swell greatly as the blood was forced out into the occluded arteries. On the other hand, when veins were tied up, they would collapse and the atria of the heart would also collapse where the veins joined it. Harvey also calculated the volume of blood passing through a major artery and found it to be far too great to be manufactured in the heart or to be absorbed by the tissues receiving it. Therefore, he concluded that it must pass into the veins and return to the heart for redistribution.

Blood Circuits

The heart is the center of the circulatory system. It consists of four chambers—two atria which receive blood from the veins and two ventricles which pump the blood into the arteries. The **right atrium** receives the blood from all parts of the body except the lungs. The **superior vena cava** carries the blood to the heart from the parts of the body above the heart and the **inferior vena cava** carries the blood to the heart from the parts of the body below the heart. As the atrium expands, the blood flows in; as it contracts, the blood is forced down into the **right ventricle** through the **tricuspid valve.** The right ventricle beats and forces the blood out through the **semilunar valve** into the **pulmonary artery** and on to the lungs. In the lungs the arteries branch out into microscopic capillaries. The erythrocytes absorb the oxygen from the air in the alveoli and the plasma gives up most of its carbon dioxide into the same air.

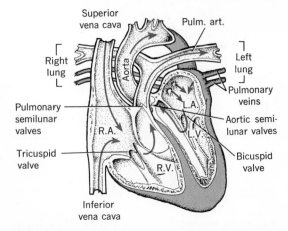

FIG. 30.8 Human heart and connecting blood vessels. (Millard and King, *Human Anatomy and Physiology,* W. B. Saunders, 1962.)

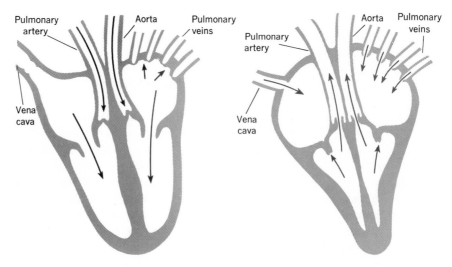

FIG. 30.9 The circuit of blood in the heart shown in the two phases of heart beat. At left, the atria contract and the ventricles dilate. At right, the reverse occurs.

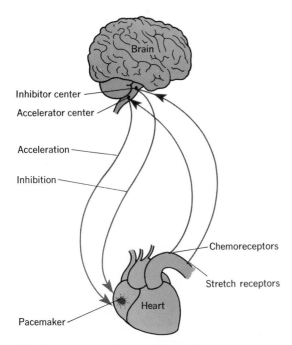

FIG. 30.10 Homeostatic regulation of the rate of heartbeat. The heart contains a pacemaker which stimulates the heart to beat, but the rate may be regulated by nerves.

The color of the blood changes from a dull, dark red to the brilliant scarlet of purified blood. The capillaries reunite to form the **pulmonary veins** (four of them) which lead back to the heart and enter the **left atrium.** As the left atrium contracts, the blood passes through the **mitral valve** to the left ventricle. This is the largest and most powerful chamber of the heart because it must pump the blood to all parts of the body. As it contracts, the blood is forced past the **semilunar valves** into the **aorta.** The aorta comes out from the anterior end of the heart, and makes a loop to the left, and runs posteriorly down the dorsal body wall just ventral to the spinal column. All along the aorta, branches come out to carry blood to all parts of the body except the lungs. Eventually the blood reaches the tiny capillaries of some part of the body where the oxygen is given up and the blood again assumes the dark red color of hemoglobin without oxygen. At the same time the food and other products pass into the cells and the absorption of carbon dioxide and excretory wastes takes place. Finally the capillaries reunite to form veins which return the blood to the right atrium. This circuit may seem rather extensive, yet a corpuscle can make a complete trip through the circuit in less than one minute.

During the **heart beat,** both atria beat in unison and then, as the ventricles fill with the blood from the atria, the ventricles beat. This is followed by a short rest and then the beat is repeated. The stimulus for the heart beat lies in the heart itself. The organ continues to beat even when removed from the body. The stimulus originates in the upper right-hand side of the right atrium and spreads out to the other parts of the heart. If this part of the heart is removed, the heart will no longer beat, but removal of other parts does not interfere with the heart beat. There are also nerve connections to the heart that further regulate the rate of beating. There are two types—accelerator and inhibitor—which speed up or slow down the rate of heart beat. They might be compared to the accelerator and the brakes of an automobile; by proper use of the two we can achieve a control over the speed of the car and regulate it to suit our needs. The nerves are autonomic and automatically make adjustments. If we sever only the inhibitor nerves, the heart beats wildly—far faster than it should beat. If we sever only the accelerator nerves, the heart beats slowly—too slowly for the needs of the body. If we sever both types of nerves, the heart continues to beat at a steady uninterrupted pace, making no adjustments to the needs of the body.

It is possible to study the heart beat by making an **electrocardiograph.** Each time a muscle contracts, it generates a weak electrical impulse which can be detected by delicate instruments. If an electrode is attached to each of the two arms, the changes in electrical potential during a heart beat can be detected and tracings of the changes transferred to paper.

The **arteries** of the body are relatively thick-walled and elastic, for they must absorb the full force of the powerful beat of the ventricles

without bursting. When the blood comes coursing out from the heart, the aorta swells with the force of the blood as it passes through. This swelling continues along all the arteries and it can be felt as a **pulse** in several parts of the body. The inner, thumb side of the wrist is the most popular spot for taking the pulse. Since each pulsation of the artery in the wrist comes from a beat of the left ventricle, the rate of heart beat can be determined by taking the pulse rate. Pulse rate is quite variable. The normal man at rest has a pulse rate of about 68 per minute, and a woman has a pulse rate of about 75. The embryonic heart beats about 140 times per minute,

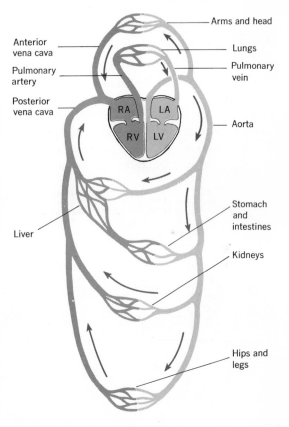

FIG. 30.11 A schematic representation of circulation in the human body. The oxygenated blood is shown in the lighter colored vessels to the right.

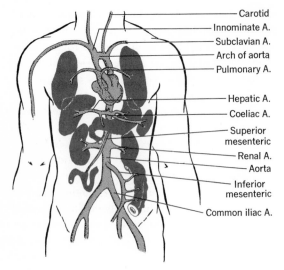

FIG. 30.12 The larger arteries of the human body.

the infant's about 120, the child's about 82. The rate increases during strong emotions—we have all noticed the pounding of the heart that follows a sudden fright. It increases during exercise—it may reach 180 beats per minute during very vigorous exercise. However, if a person is

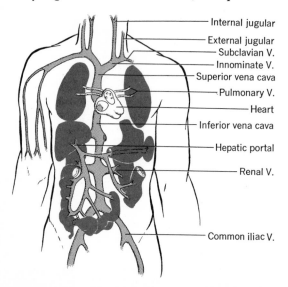

FIG. 30.13 The larger veins of the human body.

accustomed to vigorous exercise the rate will not rise so high and will return to normal more quickly. The heart supposedly develops tone with exercise, just as the skeletal muscles, and increases its capacity with the increased demands of vigorous exercise.

The **capillaries** connect the arteries to the veins. Individually they are very tiny in diameter, but collectively they are very large—the combined diameter of the capillaries leading from an artery is ten times greater than the diameter of the artery. This causes the pressure of the blood to drop greatly and the blood flows more slowly. The capillary walls dilate and contract under the stimulation of the autonomic nervous system, adrenaline, and certain chemicals. Dilated vessels tend to lower the blood pressure by allowing the blood from the arteries to flow into the capillaries more freely, while constriction raises the pressure by restricting this flow.

The **vein** receiving the blood from the capillaries is about twice as large as the artery that brought the blood. The veins have thinner walls than the arteries and the blood flows along at a much lower pressure than in the arteries. The veins possess valves that prevent a backward flow of the blood. These seem to be necessary because of the lower pressure in the veins.

Blood Pressure

There is probably not a reader of these lines who has not had his blood pressure taken. It is done painlessly today, but two hundred years ago the technique was not so simple. An English clergyman, Stephen Hales, was the first person to take blood pressure, so far as we know. He used an old mare that was about to be killed as unfit for service. The mare was tied down and a needle connected to a long glass tube was inserted in the large carotid artery of the neck. The blood rose to a height of $9\frac{1}{2}$ feet in the tube. In other words the blood pressure was sufficient to support the weight of a column of blood or water at this height. However, the level of this column did not remain constant —it would ascend to this height as the force of the heart beat came into the artery, but would drop in between the beats. This shows that the pressure is not as great between the pulsations of the artery as it is during the pulsations that arise from the heart beat.

Today we reduce the length of the tube by using mercury rather than blood or water, since mercury is more than thirteen times as heavy as water. It is unnecessary to break the skin; we balance the pressure of the blood with air pressure and read the height of a mercury column connected with the air. The blood pressure varies in different parts of the body. It is much greater in the aorta than it is in an artery in the foot. It is easy to take in the upper arm, and the pressure at this point is used as a relative indication of the pressure in the entire body. A rubber cuff is wrapped around the upper arm, and air, connected to a mercury column, is pumped into the cuff until the circulation is entirely cut off from the lower part of the arm. Let us assume that the column has now risen to a height of 150 mm. We place a stethoscope above the large artery in the crook of the elbow and begin to reduce the pressure. When the column of mercury has dropped to 120 mm, we begin to hear a regular series of sounds somewhat like pistol shots at a great distance. These sounds are caused by the spurts of blood that are forced through the collapsed artery to strike the stationary column of blood below. We now know that the blood pressure during these spurts is slightly greater than the air pressure around the arm, so we can say that the individual has a blood pressure of 120. This is called the **systolic blood pressure** because it originates from the beat of the heart which is the systole. As we continue to lower the air pressure in the cuff, the sounds become louder because the spurts of blood pass through the collapsed artery

with a greater force. When the mercury has dropped to 70 mm, however, we cease to hear the sounds altogether. This indicates that the pressure between the blood spurts is equal to the outside pressure on the arm and the artery does not collapse between the spurts. This is called the **diastolic blood pressure** because it is the pressure in the arteries during the diastole (the resting period) of the heart. From what sources does this pressure arise? We remember that as the heart forces the blood out into the arteries, they swell, absorbing much of this great force. Between the beats, the elasticity of the arteries causes them to contract to their original size and keeps the blood flowing at this reduced pressure. If the arteries were as hard and unyielding as glass tubes, the blood pressure would be much higher during the beat of the heart but would drop to almost zero in between the beats.

In normal adults adults at rest the systolic blood pressure may range from 100 to 140 and the diastolic from about 55 to 80. It rises rapidly with rigorous exercise, strong emotions, and general nervous tension.

Disorders of the Mechanism of Circulation

Hypertension, high blood pressure, is a common accompaniment of modern civilization. It is most common in elderly people, but is often found in young adults and even children. Obesity and nervous strain are predisposing conditions that encourage its development. There are several other factors that also enter the picture. Kidney trouble tends to cause a rise in the blood pressure. The kidneys, unable to remove the wastes from the blood properly at normal pressure, secrete a substance that stimulates the heart to increased effort. The extra pressure may aid in the removal of the wastes. A diet low in nitrogenous wastes, as found in meat, and a limitation of salt intake will help relieve this type of high blood pressure. High blood pressure can also be caused by deposits of a fatty material, **cholesterol,** along the artery walls. This narrows the openings and the heart must pump harder to get the blood to the body. In time calcium deposits may harden this lining and a person is said to have a hardening of the arteries. These arteries cannot swell with each heart beat as they would if they had normal elasticity. Some evidence indicates that a diet rich in saturated fats and oils, primarily animal fats, increases the chance of such deposits, but this is disputed by others. Another type results when the capillaries tend to become somewhat permanently constricted, interfering with the flow of blood into them and increasing the pressure in the arteries. It seems that when the capillaries have reacted so frequently to worry and other emotional strain, they just cannot return to their normal state. We cannot blame high blood pressure on work, for work doesn't cause it if one is happy and contented with his work.

High blood pressure is a leading cause of death in the United States. The increased load on the heart may result in heart failure or the rupture of a small blood vessel in the brain causing stroke. Varicose veins in the legs and other body parts may be caused by the high pressure dilating the walls of the veins.

Hypotension, low blood pressure, is much less common and is certainly the lesser of the two evils. It may be found in endocrine and nervous disorders and in advanced cases of tuberculosis and cancer. It is characterized by general fatigue and lack of physical endurance, because the blood cannot carry the needed oxygen and remove the wastes at the necessary rate.

Syphilis is a serious blood disease that may cause havoc in the organs of circulation. It may infect the valves of the heart so as to permanently weaken them and cause "leakage of the heart." The heart loses some of its effectiveness as some of the blood leaks back through the valves during the beat, and the heart may increase the power of its beat as a result. If the aorta is attacked, it may be weakened so that it swells like a balloon during the systole and it may burst if the pressure is suddenly elevated by emotional crisis.

Coronary thrombosis is a condition involving a small clot which lodges in a branch of the coronary artery which supplies the heart.

TISSUE FLUID AND LYMPH

The pressure of the blood in the capillaries causes water and dissolved substances to be squeezed out of the capillaries by filtration. The capillaries have very thin walls—only one cell in thickness—so water and dissolved substances are easily passed out, but the plasma proteins, which are large molecules, cannot pass out easily, and the tissue fluid has a low concentration of protein. The erythrocytes are too large to pass out of an unruptured capillary, but the amoeboid leukocytes can squeeze through the intercellular spaces of the capillary walls and are found in the tissue fluid.

Tissue fluid is important in distributing food, hormones, and other products to the cells and in removing wastes. As the capillary network nears the veins, the blood pressure is reduced and the blood reabsorbs some water and dissolved materials from the tissue fluid by osmosis and diffusion, but most of the tissue fluid enters the **lymph vessels,** where it is known as **lymph.** The lymph moves through lymph vessels and passes through lymph nodes. **Lymph nodes** are glandular masses made of small tubes lined with leukocytes. The leukocytes engulf bacteria and other foreign material which have entered the body. Lymph nodes may become swollen and tender to the touch when they are working to full capacity during an invasion of the body by bacteria. An infection on the hand may cause swollen lymph nodes in the armpit. The lymph nodes are thought to be the place where the lymphocytes are produced.

Lymph flows through progressively larger vessels which eventually unite in the chest region to form a duct which empties into a large vein in the shoulder region, where it then becomes a part of the blood plasma again. In its circuit, lymph not only distributes and picks up dissolved products but also removes excess proteins, cell fragments, and other debris from the intercellular spaces. Movement of lymph is aided by contractions of the muscles surrounding the lymph vessels. This is one reason why physical activity is necessary for a normal healthy body. The swollen ankles and feet which occur after a long bus ride are caused by the accumulation of lymph in the lower extremities. Lymph is produced continuously, but without muscle contractions it is not carried along. Gravity causes it to settle and remain in the extremities.

REVIEW QUESTIONS AND PROBLEMS

1. Why would we still need a blood system if we had other systems which made blood unnecessary for transportation?
2. The normal liberation of hemoglobin from hemolysis of erythrocytes causes no difficulty, but massive bursting of these cells can produce a toxic effect. Explain.
3. The appearance of **transferrins** in the blood is correlated with the development of the warm-blooded state. Why do the two go together?
4. A person with **agammaglobulinemia** would have no trouble receiving a transplant, nor would he develop allergies, but he would suffer greatly from infections. Explain.
5. The frog has large nucleated red blood cells, but a person could not survive with this kind of cell. Explain why.
6. Why do you think the human male erythrocyte count averages about 10 percent higher than that of the female?
7. Many television commercials warn us about iron-deficiency anemia, yet many people who take iron do not benefit. What other factors are not mentioned in these commercials which must be taken into consideration?
8. Why is a differential leukocyte count needed, as well as a whole leukocyte count, in order to diagnose some diseases correctly?
9. Suppose you are searching for a way to control leukemia. What line of investigation would you pursue and why?
10. When blood is withdrawn into a plastic bag for use in transfusion, the bag is usually first moistened with a solution of a calcium compound. Why is this done?
11. Those who have suffered from heart attacks may take antithrombin regularly for the rest of their lives. Why?
12. A boy with hemophilia recently underwent open heart surgery and recovered. This would have been impossible a few years ago. Explain why.
13. Suggest a reason why it is better for a fetus not to produce antibodies while it is still in the mother's body.
14. How is it possible for plasma cells to "recognize" native proteins and not produce antibodies against them?

15. Mr. Jones gave a blood transfusion to his friend Mr. Smith, but now Mr. Jones needs a transfusion and the physician will not use blood from Mr. Smith. Give a possible reason for this decision.

16. When an Rh-negative woman gives birth to an Rh-positive child, the woman is immediately given an injection of anti-Rh (rhogam). Why is this done?

17. Some people have too rapid a heart beat. A surgeon may sever one of the nerves leading to the heart to relieve the condition. What nerve would he sever and why?

18. What happens to blood pressure when artery walls become narrowed by cholesterol and calcium deposits are formed? Explain.

19. What changes take place in the body to cause an increase in blood pressure during emotional tension?

20. A specialist in allergic diseases reports that those who tend to develop allergies easily are usually more resistant to infectious diseases. What is the possible connection between these two conditions?

21. A swelling of the lymph nodes in the arm pits or the groin of the leg is a cause for concern. Explain why.

22. What functions do tissue fluid and lymph accomplish that could not be accomplished by the whole blood?

FURTHER READING

Graubard, M. 1965. *Circulation and Respiration.* New York: Harcourt, Brace, and World.

Guyton, A. C. 1973. *Circulatory Physiology.* 2nd ed. Philadelphia: W. B. Saunders.

Race, R. R., and Sanger, R. 1968. *Blood Groups in Man.* 5th ed. Oxford, England: Blackwell.

Vander, A. J., et al. 1970. *Human Physiology.* New York: McGraw-Hill.

31

Nutrition

Food of some sort is a requirement for all forms of life. Photosynthetic and chemosynthetic organisms manufacture their own food, but those without autotropic abilities must depend upon an external food source. Most animals spend the greater part of their waking hours in food gathering. Only man has been able to achieve a high degree of control over food production which allows him to use time for other pursuits. We would never have had the great achievements in science, music, literature, and the arts without our control over food production. Our soaring population, however, threatens such pursuits. For many people in the world food gathering is still the primary concern in life. In the United States there is a great abundance of food, yet many people in this country suffer from nutritional deficiencies. Many nutritional deficiencies are caused by ignorance of the principles of nutrition rather than by a lack of money to buy food. Many modern "convenience" foods are rather low in nutritional value. Instead of squeezing oranges to make fresh orange juice, we may drink a concoction of sugar, citric acid, flavoring, and a little orange juice because it is more convenient to prepare. Many foods at the supermarket have been so refined and treated, they have lost much of their original nutritional value. It is important that everyone have some knowledge of nutrition so that they can make intelligent decisions about the food they eat.

We should first learn something about the nature of the digestive tract which processes the assortment of materials which we ingest and delivers usable nutrients to our cells.

THE DIGESTIVE TRACT

The digestive reactions start even before food enters the body. When the time for a meal nears, you may hear the pleasing rattle of dishes in the kitchen, and the aroma of the cooking food reaches your nostrils. The autonomic nervous system becomes active and your mouth waters as the flow of saliva is increased. Also the flow of gastric juice in the stomach begins and the stomach is prepared for food. All this creates a great desire to eat. Only through exertion of will power can you postpone your desires in deference to social customs that prescribe a polite approach to the beginning of a meal.

Mouth

The process of digestion begins in the mouth. Food is masticated and broken down into smaller particles which are more easily acted upon by the digestive enzymes. It is also mixed with saliva, which serves to moisten and lubricate it. It is not possible to swallow dry food; any attempt to do so causes a person to gag and perhaps vomit. When mixed with saliva, however, food may be swallowed easily, because the saliva not only is moist (about 99 percent water) but contains **mucus,** a viscid substance which gives the saliva its lubricating quality. Saliva also contains a starch-splitting enzyme known as **salivary amylase** and **maltase** which breaks maltose down to glucose. Food is usually kept in the mouth for such a short time that there is little chance for much digestion here, but it can continue within the small masses of food after they have reached the stomach. Saliva also contains an antibacterial substance, **lysozyme,** which

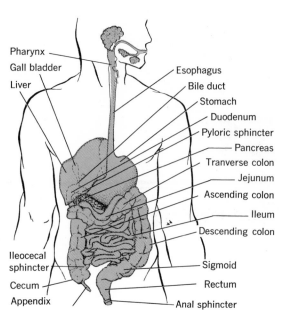

FIG. 31.1 The human digestive system. A part of the liver has been removed to show the stomach lying beneath it.

helps keep down the bacterial population of the mouth. Saliva is secreted by three pairs of **salivary glands** located around the mouth. They are constantly active, so the mouth is always moist, but they are stimulated to extra activity by the odor or taste of food.

When we feel that the food has been sufficiently masticated, we push it to the back of the mouth with the tongue. Here it enters the **pharynx** (throat) and powerful peristaltic contractions take over and move it to the **esophagus,** where similar contractions take it on to the stomach.

Stomach

The stomach is closed at its anterior end by the **cardiac sphincter.** A sphincter is a constriction of a tube which remains closed most of the time, opening only upon the proper stimulation to allow material to pass through. As the peristaltic contractions of the esophagus bring the food close to the stomach this sphincter will open, allow the food to pass, and then close again. This prevents the food from passing back into the esophagus and out of the mouth when the stomach walls contract or when a person is in a position where the stomach is higher than the mouth.

The stomach is shaped somewhat like a thick J. The upper portion which receives the food is known as the **fundus** and the hook of the J is the **pylorus.** The fundus is capable of great distension and serves mainly for storage. We can be thankful for this; otherwise we would have to greatly restrict the size of our meals. The pylorus is a very active portion of the stomach and undergoes churning actions which mix the food with the gastric secretions.

Gastric juice is secreted by thousands of tiny glands embedded in the walls of the stomach. This juice contains the following four substances.

Pepsinogen is an inactive form of pepsin. When mixed with hydrochloric acid it becomes the active **pepsin,** a protein-splitting enzyme. It is probably secreted in an inactive form because otherwise it would digest the protein of the glands which produce it.

Hydrochloric acid (HCl), a powerful acid in full strength, will dissolve iron. A small drop on the skin will cause a painful burn, but it is produced in surprisingly high concentrations by cells of the stomach wall. Pure gastric juice will have a concentration of about 0.5 percent of this acid and a pH of about 2.0. As already indicated HCl is needed to activate the pepsinogen. The stomach wall itself is exposed to this powerful combination and yet resists digestion because the mucous membrane on the surface of the stomach produces ammonia which neutralizes the acid. If there is a small injury or an impairment of the blood supply to a part of the membrane, the cells may cease to produce ammonia and the stomach lining itself may be digested in this region. This causes an **ulcer** to develop which is continually irritated by the action of the hydrochloric acid and pepsin. Persons with stomach ulcers often take alkaline powders or other chemicals which absorb acid to avoid further injury to the ulcers. This, however, will interfere with protein digestion if there is insufficient acid to activate the pepsinogen.

Rennin, another component of gastric juice, is especially abundant in the stomachs of infants, but it is present in only small quantities in adults. It causes milk to curdle—the milk protein is coagulated and milk forms a solid. This helps digestion because milk remains in the stomach longer in such a solid state. Rennin is extracted from the stomachs of calves and has commercial uses. Most cheese is curdled by rennin before aging and further treatment. Rennin is also the active ingredient of a dessert powder which forms a solid when mixed with milk.

Lipase. There is a small amount of this fat-splitting enzyme in gastric juice, but it plays a very small part in total fat digestion.

The secretion of gastric juice, like that of saliva, is stimulated by the autonomic nerves when the odor, sight, or thought of food comes before a person who has not eaten recently. The juice will already be in the stomach when the food reaches it. It is necessary, however, for the secretions to continue after the meal is completed. This comes about from stimulation within the stomach. The contact of food with the stomach wall stimulates certain cells to release a hormone, **gastrin.** This is absorbed and circulated by the blood. As it passes over certain nerve centers of the brain, a stimulus to continue secretion of gastric juice is relayed to the stomach. These are autonomic nerves, and since the autonomic nerves are closely related to the emotions, the flow of gastric juice may be affected by strong emotions. A person under any sort of emotional tension does not have any appetite. It is usually better not to eat under such circumstances, since there cannot be proper digestion of the food ingested. In violent emotional states, the stomach may actually eject its contents through **vomiting.** The autonomic nerves permit a relaxation of the cardiac sphincter and involuntary contractions of the diaphragm and lower chest muscles press against the stomach and force the contents back out through the mouth. On the other hand, relaxation is conducive to proper digestion. Soft music, candlelight, and flowers on the table at dinnertime are more than just a romantic idea; they are positive adjuncts to a good appetite and proper digestion.

Food in the stomach is thoroughly mixed with gastric juice through waves of powerful contractions in the pylorus. If there is any air in the stomach, this may cause embarrassing, gurgling sounds. The pyloric sphincter remains closed for a time, however, and the food remains in the stomach for from one to several hours, depending upon the size of the meal, the type of food eaten, and the emotional state. In time the pyloric sphinc-

ter will open and allow a small portion of the stomach contents to escape. Then it closes again; but it will reopen to repeat the process a short time later. This continues until the stomach is emptied—four or five hours after a typical meal. Such a method ensures that the food will be well mixed with the juices in the small intestine.

Many think that an empty stomach is the stimulus which induces hunger, but this has little to do with it. Actually, there is a **hunger center** in the hypothalamus of the brain. When this is stimulated by tiny electrodes in experimental rats, the animals eat ravenously even though they have just been fed. In normal circumstances this center is stimulated when the blood glucose concentration drops below a certain level. This will come when food is no longer being digested and absorbed from the intestine and is accentuated by heavy use of glucose in the muscles during exercise. You tend to have a ravenous appetite after strenuous exercise. After eating, however, this glucose level will rise and affect the **satiety center** in the hypothalamus and we feel satisfied. This is the reason we often take a sweet dessert to get that satisfied feeling, and it is also why between-meal sweets tend to dampen the normal appetite. One drug, dexedrine, is much used in reducing diets to keep down the great hunger that may accompany a drastic restriction in the intake of food. Dexedrine may stimulate the satiety center as if there were a high concentration of glucose in the blood and produce a false sense of appetite satisfaction. Its use must be supervised by a physician because of its other effects on the autonomic nervous system.

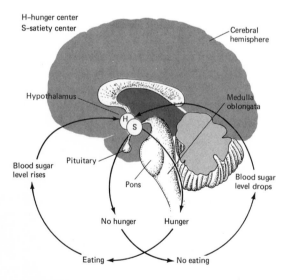

FIG. 31.2 Diagram to show how blood sugar level regulates hunger.

Small Intestine

The small intestine of an adult averages about twenty-five feet in length. For convenience of study it is divided into three parts. The first loop, which is about one foot long, is known as the **duodenum.** This is followed by a second loop, the **jejunum,** which is about five feet long. The remaining portion of the small intestine is known as the **ileum.** Food from the stomach reaches the duodenum first. As a result, this region of the small intestine is subject to ulceration because the highly acid contents of the stomach encounter its walls. Within the duodenum food is mixed with secretions from two important glands. As it passes down the jejunum, it is also mixed with secretions from the wall of the small intestine itself. We shall discuss these three secretions.

Pancreatic juice is secreted by the pancreas and passes through the pancreatic duct into the duodenum. This juice is alkaline and helps neutralize the hydrochloric acid which comes in from the stomach. Pancreatic juice is rich in the fat-splitting enzyme **lipase,** and most fat digestion is done by this secretion. It also contains two protein-splitting enzymes, **trypsin** and **chymotrypsin.** These act upon any proteins which were not digested in the stomach. As was true of pepsin, these enzymes are secreted in inactive forms, trypsinogen and chymotrypsinogen. The trypsinogen is activated by a secretion of the intestine, **enterokinase,** and the chymotrypsinogen is activated by the trypsin itself. There is also a fourth enzyme, **amylase,** which acts upon any starch remaining in the food and changes it to disaccharides. In addition, the pancreatic juice contains sodium bicarbonate ($NaHCO_3$) which neutralizes the food coming from the stomach and raises the pH to about 8.

As was true of gastric juice, pancreatic juice is secreted in response to hormonal action. As food reaches the duodenum, special cells in this part of the intestine produce the hormone **secretin,** which goes into the blood and stimulates the secretion of pancreatic juice.

Bile is a digestive juice which reaches the duodenum by means of the bile duct leading down from the gallbladder of the liver. While this secretion does not contain digestive enzymes, it plays an important part in the digestive process. It is alkaline and aids in neutralizing the food coming from the stomach. It also plays an important role in the emulsification of fats. When emulsified (broken down into small parts), the fats are much more easily digested by the lipase. Bile also plays a part in the absorption of digested fats. The liver produces bile at a constant rate, but there is a sphincter which closes the bile duct. This causes the bile to back up into the gallbladder, where it is stored and concentrated. The presence of fat in the duodenum stimulates certain cells to produce the hormone **cholecystokinin,** which is absorbed by the blood and stimulates the relaxation of the sphincter together with contractions of the gallbladder. These two events cause bile to flow into the duodenum.

Bile contains pigments derived from the hemoglobin of red blood cells. Each day billions of these cells die and are replaced by fresh cells. The worn-out blood cells are seized by cells of the spleen, liver, and other organs; the iron is removed and will be used for the construction of more hemoglobin. The old hemoglobin, deprived of its iron, is then taken up by the liver and is excreted in the form of **bilirubin,** which has a golden-red color. Some of this bilirubin may be oxidized into **biliverdin,** which is yellow-green. Most of the pigment in human bile is bilirubin, but as it is acted upon by bacteria in the intestine it is oxidized into the biliverdin. This undergoes still further chemical changes until it is a deep yellow or brown by the time it passes from the body. This gives the characteristic color to the feces; a sure sign of bile deficiency is a pale, almost colorless, condition of the feces. In extreme diarrhea the feces may show the characteristic yellow-green color of biliverdin because they do not stay in the body long enough for the further changes to take place.

Should the bile duct be clogged and bile not pass through properly, the broken-down hemoglobin will accumulate in the blood and a yellow or even greenish color develops in the skin. This condition is known as **jaundice** and can come about from any condition that tends to block the bile duct. Gallstones can do this, or various infections which settle in this region may cause the duct to swell and be clogged. In recent years there seems to be an increasing number of cases of **infectious hepatitis,** a virus infection which results in jaundice.

Succus entericus is the name given to the digestive juices secreted by the glands in the walls of the small intestine. Much of this secretion is mucus which aids the movement of the digesting food through the tract, but there are also enzymes which complete several digestive processes begun by other enzymes. For instance, there are **peptidases** which carry the polypeptides of partial protein digestion on to the absorbable amino acids. There are also enzymes which break down the disaccharides into the simple monosaccharides, and there is a small amount of **lipase** which can act upon fats. Secretion of the succus entericus is stimulated primarily by the mechanical stimulation of the intestinal wall by food passing through.

The small intestine is the primary area of food absorption. There is an enormous exposed surface, not only because of the length of this organ, but because of the many folds in its lining and the millions of small fingerlike processes, **villi,** which project into the interior. Each

TABLE 31.1   Digestive Juices and their Functions

Digestive Juice	Enzymes	Action of Enzymes	Other Products and Functions
Saliva	Amylase	Starch to maltose	Mucus for lubrication
	Maltase	Maltose to glucose	Lysozyme, antibacterial agent
Gastric Juice	Pepsinogen[a]	Proteins to proteoses and peptones	Hydrochloric acid; changes pepsinogen to active pepsin
	Rennin	Curdles milk	
	Lipase (weak)	Simple fats to fatty acids and glycerol	
Pancreatic Juice	Lipase	Fats to fatty acids and glycerol	Sodium bicarbonate; neutralizes acid, makes food basic
	Trypsinogen[a]	Proteins to proteoses and peptones	
	Chymotrypsinogen[a]	Proteins to proteoses and peptones	
	Amylase	Starch to maltose	
	Carboxypeptidase	Acts to split polypeptides into amino acids	
Bile	—	—	Emulsifies fats
Succus Entericus	Peptidases	Peptones to amino acids	Enterokinase, converts trypsinogen to trypsin
	Lipase	Fats to fatty acids and glycerol	Mucus, for lubrication
	Maltase	Maltose to glucose	Secretin, stimulates flow of pancreatic juice
	Sucrase	Sucrose to fructose and glucose	
	Lacrase	Lactose to galactose and glucose	

[a]Must be activated before it can function as an enzyme.

villus is highly specialized for absorption; each contains a loop of fine capillaries, and in the center of each there is a small lymph vessel, a **lacteal.** As the peristaltic waves move the digesting food down the small intestine, the villi are bathed with the food in solution. Simple diffusion plays some part in absorption, but other factors are certainly involved. Glucose, for instance, will be absorbed until there is none left in the intestine. This would indicate that there is some active transport when the concentration within the intestine becomes less than that in the blood. Further, the bile forms some sort of temporary union with fatty acids and acts to conduct them across the cell plasma membranes. Simple sugars and amino acids are absorbed primarily by the blood in the small capillaries, while the fatty acids and glycerol from fat digestion are absorbed primarily by the lacteals.

Large Intestine (Colon)

The small intestine terminates where it joins the large intestine. There is an **ileocaecal sphincter** at this junction which prevents the contents of the large intestine from re-entering the small intestine. The large intestine is concerned primarily with storing, processing, and eliminating the residue which remains after digestion and absorption have taken place. Much of the food which we eat cannot be digested by the enzymes in our bodies. The cellulose from the cell walls of most plant foods is a good example. Many of the tough fibers of meat also resist the digesting enzymes. This waste arrives at the large intestine in a rather fluid state, since it is mixed with a considerable amount of water. One function of the large intestine is to absorb much of this water; without such absorption, our water requirements would be multiplied many times. We realize how important this is when we have diarrhea. The contents of the large intestine pass through too rapidly to allow for water absorption and the feces are quite liquid. This can lead to a serious dehydration of the body if it continues very long.

The large intestine is swarming with literally billions of bacteria which attack the food residue and bring about certain changes in it. They have enzymes which man does not have and can derive nourishment after man has obtained all that he can from it. Some of these bacteria are purely commensal and derive nourishment without contributing anything or harming their host. In some cases, however, there is mutualism—some of the bacteria in their breakdown of the waste release certain vitamins (vitamin K and some of the vitamin B complex) and these can be absorbed through the wall of the large intestine. Excessive and unnecessary use of certain antibiotics can be harmful to man because many of these useful bacteria may be destroyed.

The small intestine opens into the large intestine from the side, leaving a blind pouch, the **caecum,** at the beginning of the large intestine. There is a small, wormlike projection, the **appendix,** extending from the posterior part of the caecum. It may become clogged with decaying food and infection may develop. This is known as appendicitis and is commonly corrected by surgical removal of the appendix. If an infected appendix becomes so swollen with pus that it bursts, many bacteria are liberated into the coelum, causing a general infection of the peritoneum which lines this cavity. Such an infection, known as **peritonitis** is very serious. Before the days of modern antibiotics, deaths from such infections were frequent and many appendixes were removed just on the chance that they might be infected. Today peritonitis can usually be brought under control by antibiotics and the danger is much less. The appendix is probably a rudiment of a much larger caecum in man's ancient ancestry when perhaps his diet was different. In some modern animals, such as the horse and the rabbit, this enlarged caecum is present and acts as a fermentation vat where there can be bacterial digestion of plant cellulose.

For convenience of study the large intestine is divided into the **ascending colon, transverse colon, descending colon,** and **rectum.** The rectum is the terminal portion and is closed by the **anal sphincter.** As the feces descend into the rectum there is a desire for defecation and the anal sphincter can be opened to allow the waste to pass from the body.

The Functions of the Liver

The liver is one of the largest organs of the human body, and judging by its many functions, it is one of the most important. We shall list some of these functions.

Bile secretion. We have already stressed the importance of this function and its relation to the removal of worn-out hemoglobin.

Carbohydrate storage. The blood from the intestine with its absorbed food goes through the liver before going to the heart and over the body. The carbohydrates, after digestion, are in the form of monosaccharides (such as glucose). These are taken from the blood by the liver and transformed into glycogen, which is a complex polysaccharide often spoken of as animal starch.

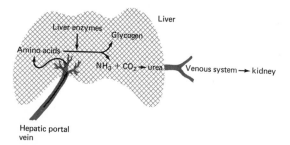

FIG. 31.3 Deamination of amino acids takes place in the liver. When amino acids in excess of those needed for protein synthesis are present in the blood, liver enzymes split off the amine group and convert the remainder to glycogen, which is stored in the liver until it is needed.

In this form it is stored, and from this store the liver can produce glucose which is released into the blood stream.

Blood glucose regulation. The blood always contains a certain concentration of glucose within narrow limits of tolerance. If there is a marked rise in blood glucose, a person goes into a coma and death follows shortly. A marked decline results in shock and unconsciousness, also followed by death. The liver plays an important part in maintaining the glucose within the critical limits required for life. After a meal rich in carbohydrates, the blood coming to the liver contains a high concentration of glucose. If this were allowed to go directly to the body cells, death would result in a few hours. Instead, the liver absorbs most of this glucose and converts it into glycogen. On the other hand, when we have not eaten for some time the glycogen is converted into glucose and released into the blood. Thus, the glucose concentration of the blood is held within the rather close limits of tolerance in spite of great variation in the amount being supplied by absorption from the intestine at different times. Later we shall see how the hormone insulin is related to this homeostatic regulation.

Fat conversion and storage. Of course, the amount of glycogen which the liver can store is limited and in a well-fed person there comes a time when it is saturated, so to speak. Then what happens to the monosaccharides from the intestine? They are converted into fats and stored in the liver. In times of need these fats can be converted back to simple sugar and released into the blood stream. The liver is limited in the amount of fat it can store, and it will release into the blood some of the fat in the form of fatty acids and glycerol. These are carried to various other parts of the body where fats are stored—under the skin and in fat pads on the abdomen and hips. The liver can also bring about a recombination of fatty acids and glycerol into fats when these products of digestion tend to become excessive in the blood.

Deamination of amino acids. A healthy adult, with access to a plentiful food supply, will generally eat more protein food than is required to supply the animo acids needed for the repair of tissue proteins which are destroyed in normal metabolism. Amino acids cannot be stored like fats and carbohydrates, and they cannot be allowed to accumulate in the blood. The liver plays a part here to prevent amino acid excess. It deaminates these amino acids; that is, it removes the amino groups (NH_2) and combines them with carbon dioxide to form a chemical known as urea. This is carried by the blood to the kidneys, where it is removed and excreted as a major waste in the urine. The remainder of the amino acid molecules are then converted into glycogen or fatty acids. Protein foods can be fattening if they are eaten in excess.

Storage of vitamins and other vital substances. Fresh fruits and vegetables are usually rich in provitamin A and vitamin D. In these days of frozen food techniques, canning, and rapid transportation from all corners of the globe, we have the opportunity to partake of such foods the year around. In the comparatively recent past these were seasonal foods and during long periods they were quite scarce. The liver, however, has the power to absorb and store these valuable vitamins and release them when needed. The liver oils of certain fish are so rich in the stored vitamins that they are used as a vitamin source to supplement human nutrition. The liver also stores the antianemic factor which is produced by the stomach wall.

Poison removal and detoxification. If a person dies under conditions which might lead to a suspicion of poisoning, a small piece of the liver may be removed and analyzed. This is done because the liver will show the greatest concentration of the poison of any organ of the body. The liver removes and can hold out of the circulation a large amount of poison. We would be much more easily poisoned were this not true. If the dose is not too great, the absorbed poison will gradually be altered chemically and released into the blood where it can be removed by the kidneys.

Production of embryonic red blood cells. Blood is needed early in embryonic life; but the bone marrow which is the source of blood cells in the adult does not appear until later. The liver produces these cells for the embryo. They contain a special kind of hemoglobin which is normally not found in adult blood. There are special cases of anemia, however, in which the liver seems to be reactivated or to remain activated and continues to produce blood cells with this fetal type of hemoglobin.

Prothrombin production. The blood plasma protein, prothrombin, is produced by the liver. It is one of the necessary ingredients for the clotting of blood. Vitamin K is necessary for prothrombin synthesis.

NUTRITIONAL NEEDS

We can obtain energy from three main classes of foods, **proteins, fats,** and **carbohydrates.** There was a time in the recent past when it was believed that a person needed only to eat a balanced quantity of these three classes of foods in order to be properly nourished. We have found, however, that a person will die more quickly on a diet which contains only pure proteins, fats, and carbohydrates than if he eats nothing at all. Certain vitamins and minerals are essential, and a deficiency of even one of these can result in serious defects and even death. Let us consider the various requirements.

Calories

Calories are units of energy and must certainly be included in any diet. Most Americans, however, are concerned with consuming too many

Calories. We may obtain Calories from proteins, fats, or carbohydrates, but those which we consume in excess of our energy requirements are converted into fat.

The source of the Calories is important. Americans usually get far too many from fats by eating foods like french fries, fried chicken, sour cream, and rich desserts. There is a tendency to consume too many "empty Calories" by eating food which contains litle nutritive value. For example, sugar is the primary ingredient in many ready-to-eat breakfast cereals, and refined sugar is little more than Calories. Highly refined grains which have lost much of their nutritional value make up the rest of these cereal products. A few vitamins and minerals may be added, but the total product is far from a truly nutritious food. Unfortunately, while there are many television commercials for these refined foods, there are few commercials for the highly nutritious fresh fruits, vegetables, whole grains, and meats which are not sold under a brand name.

We learned in Chapter 9 that genes code the production of polypeptide chains from twenty different kinds of amino acids. These chains form the proteins which are the basis for all growth and cellular activity. If one of the twenty amino acids is not present, complete chains cannot be synthesized. We can synthesize ten of these amino acids with the help of enzymes, **transaminases,** which alter other amino acids. These ten amino acids are known as the **nonessential amino acids.** The ten **essential amino acids,** however, cannot be synthesized and must be included in our diet. Animal proteins generally have all twenty amino acids, so they are called **complete proteins.** Complete protein foods include, meat, fish, eggs, milk, and cheese. Vegetable foods also contain proteins, but these proteins are usually not complete. One or more of the essential amino acids may be missing. Corn, for instance, is deficient in lysine, so a diet of corn alone would not allow proper growth and development. Beans are deficient in methionine, so beans alone would

not provide for proper growth. Corn and beans supplement each other if they are eaten at the same meal and provide all the necessary amino acids.

Protein deficiency is a great problem in many of the underdeveloped nations. Many people eat a diet consisting primarily of vegetable products which may be deficient in certain essential amino acids. Protein deficiency especially affects growing children. You have seen pictures of children with the distended abdomens and the thin, underdeveloped limbs which are characteristic of protein deficiency. In some areas this disease is known as **kwashiorkor,** which means "red boy," because the skin turns red as a result of dermatitis. The children will show a remarkable recovery if given a protein which contains the deficient amino acid, even though it is another vegetable protein. A team at Purdue University, directed by E. T. Mertz, has developed a strain of corn which is high in lysine. Pigs which were fed this corn grew 300 percent faster than those fed other varieties of corn. It may be possible to grow this corn in areas of the world where protein deficiency is a problem. Another research team has developed an all-vegetable product which contains all the essential amino acids. This product, called INCAP Vegetable Mixture #8, is used to treat kwashiorkor in some parts of South America.

The amount of protein needed per day is not great. If it is a complete protein, only 25 to 37 g (around 1 ounce) is sufficient for a man and somewhat less for a woman. This would be only one good mouthful. As a safety measure, about four times this quantity is recommended. Excess amino acids cannot be stored for future use. They are converted into glucose in the liver. It is better to use carbohydrates for energy rather than excess proteins because proteins are expensive and the excretion of the waste products of protein digestion places an extra load on the kidneys. The nitrogenous part of proteins, which is split off in the liver, must be excreted. About

90 percent of the amino acids consumed are converted into glucose.

There are many who suffer from amino acid deficiency. Young children, especially, must have balanced amino acids if there is to be proper growth. Without sufficient amino acids, there may be permanent mental retardation. The genetic inheritance of the individual provides the potential for brain growth and development, but this potential cannot be realized without the amino acids which are coded into the polypeptide chains and proteins by the genes. Pregnant women who have amino acid deficiencies may give birth to children who have already lost part of their potential.

The Vitamins

Vitamins contain no Calories, but they are needed for normal nutrition. Many of them serve as coenzymes and perform other vital functions in cellular metabolism. A deficiency of one of the essential vitamins can result in serious defects. It is true that recommendations to take vitamins in pill form have been overdone, but it is well to be familiar with the vitamins and what they do in order to be able to recognize signs of deficiency.

Vitamin A. This vitamin is synthesized in the liver from **carotene,** a yellow material found abundantly in carrots, spinach, and most other fresh vegetables and fruits. Vitamin A is the only one of the vitamins that is not found in plants— it can only be produced in the livers of animals, but they must obtain the carotene from plants to produce it. We can obtain vitamin A, however, from certain animal products. Butter, eggs, cheese, and the fish-liver oils are quite rich in it. Both the carotene and the converted vitamin are fat-soluble. This is of significance in the use of mineral oil as a laxative. Mineral oil will absorb carotene if it is mixed with digesting food and will deprive the body of potential vitamin A.

For this reason, mineral oil should not be taken along with food; it may cause some loss of vitamin A from the body. Vitamin A is stored in the oil of the liver.

Vitamin A is necessary for normal **growth.** Deficiencies in early childhood result in stunted, underdeveloped bodies. It is also necessary for the health of the mucous membranes. Infections of the eyes and membranes of the nasal cavity result when vitamin A is deficient because of the lowered resistance of these membranes. It is necessary for normal **vision,** as was brought out in Chapter 29. Deficiencies first result in night blindness—the inability to see well in poor light. If there is serious deficiency over a prolonged period, total blindness will result. This is often seen in cattle that have not received a sufficient amount of this vitamin. Vitamin A is also related to **reproduction.** In the male there will be degeneration of the germinal epithelium of the testes when this vitamin is deficient. The effect also extends to the cells producing the male hormone, and the quantity of this hormone diminishes as this tissue degenerates. A prolonged deficiency results in complete sterility. In the female a deficiency of vitamin A causes delayed or completely suppressed ovulation (the release of eggs from the ovaries). This results in low fertility or sterility.

Vitamin-B complex. Vitamin B was one of the first vitamins to be discovered. In 1897 Christian Eijkman, a Dutch physician who was working in Java, noticed that chickens fed on polished rice often developed a nervous affliction, polyneuritis, which was similar to the very common Oriental disease known as beriberi. Chickens fed on rice from which the outer hull had not been removed had no such difficulty, and afflicted chickens quickly recovered when fed the whole rice or the hulls that had been removed from polished rice. Eijkman accordingly concluded that the outer covering of the rice grain contained a substance necessary for normal nutrition of the chickens. This same substance was also found to be effective in treating beriberi in humans. It was later called vitamin B. As biochemists turned their attention to the analysis of this vitamin, they found that it was made of many different components. To date, at least twelve have been identified, with the possibility that the number will increase as the studies continue. We now refer to the entire group as the vitamin-B complex and can conveniently study them as a group, for they usually occur together in natural foods. Whole grain cereals, brewer's yeast, and egg yolk are rich food sources, but the complex is also found in fresh fruits and vegetables, milk, liver, and lean meat. We will briefly survey some of the better known factors of the B complex, together with some of the effects of deficiency. It is often easier to speak of the vitamins in terms of their effect when deficient than their

A

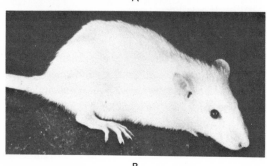

B

FIG. 31.4 The rat at the top has beriberi caused by a deficiency of vitamin B₁, thiamine. The rat in the bottom photograph is the same rat which has received 0.008 mg of thiamine daily for ten days.

positive value. When one has a sufficient quantity of vitamins, he is just normal.

Vitamin B_1 (thiamine) deficiency results in digestive upsets, loss of appetite, and nervous disorders which eventually terminate in the paralytic disease **beriberi** if the deficiency is great and long extended. This factor is also necessary for plant growth, which is the reason why plants store such an abundant quantity of it in their seed. In its absence a germinating seed cannot form a root system. In a few plants the addition of thiamine to the soil will result in increased vigor and growth of the plants, but most plants synthesize their own vitamins and naturally possess all the thiamine necessary for normal growth. Thiamine forms a coenzyme needed for the breakdown of pyruvic acid in the cell.

Vitamin B_2 (riboflavin) deficiency causes an itching and burning of the eyes and an inflammation of the tongue. In rats the deficiency causes cataract of the eyes, but deficiency in human beings has not been proved to have the same result. Riboflavin is used in the formation of FAD, which is needed in cellular respiration (see Chapter 11).

Niacin deficiency causes nervous and mental disorders accompanied by typical changes in the skin. The skin thickens and becomes spotted with brown pigmentation. These spots are subject to easy infection, so the skin is usually covered with many small sores and eruptions. In its extreme form this condition is known as **pellagra**. Niacin is an essential ingredient in the formation of NADP, which is a vital part of the cellular respiratory cycle (see Chapter 11).

Vitamin B_6 (pyridoxine) deficiency causes a scaliness of the skin and a loss of hair in rats. It is assumed to be necessary in the human diet, but this has not definitely been established as a fact.

Pantothenic acid is needed for the formation of coenzyme A, which is another essential in the breakdown of food within the cell.

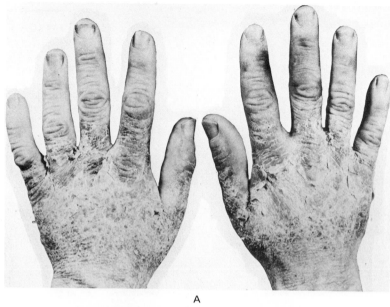

A

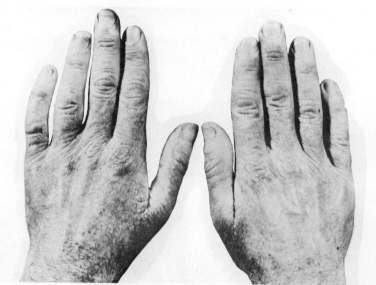

B

FIG. 31.5 Hands of a person with pellagra caused by a deficiency of niacin, one of the B-complex vitamins (A). After three weeks on a diet enriched with extra niacin, the hands are back to normal and nervous disorders have also disappeared (B).

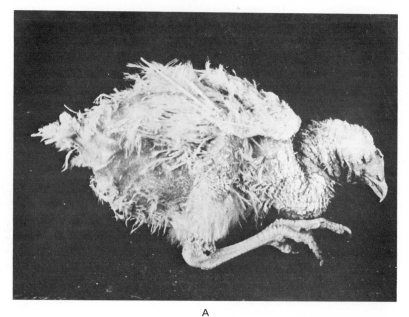

A

B

FIG. 31.6 The chicken above is suffering from advanced scurvy caused by deprivation of vitamin C. The same chicken is shown below. It has recovered after three weeks on foods with a normal vitamin C content (B). (E. R. Squibb.)

Vitamin B$_{12}$ includes the metal cobalt in its molecules. It is necessary for the maturation of red blood cells and a deficiency causes anemia. Remarkable recoveries from certain types of anemia have been the result of treatment with this vitamin. It is stored in the liver. Even before we knew of the existence of this vitamin, liver was much used for the treatment of certain types of anemia.

Folic acid is another of the B group which is necessary for the enzymatic synthesis of the purine and pyrimidine bases which form the vital rungs of the DNA ladder. The dietary requirements for man have not definitely been established because this vitamin seems to be synthesized to some extent by intestinal bacteria.

Biotin is necessary for enzymatic reactions involved in fatty acid synthesis. As a rule, we obtain sufficient quantities of this vitamin from its synthesis by intestinal bacteria, but when these bacteria are inhibited by antibiotics or other antibacterial agents, we may suffer from a deficiency of this vitamin.

Unlike vitamin A, the B complex has no great storage depot in the body and therefore requires frequent replenishment. Some of the vitamins in the group are heat labile—that is, they are destroyed by prolonged cooking. These vitamins are also soluble in water, and much of the B vitamin content of foods goes down the drain when vegetables are cooked in water and the water is then discarded.

Vitamin C (ascorbic acid). This vitamin is most abundant in citrus fruits and tomatoes, but is also moderately abundant in most fresh fruits and vegetables It is heat labile and will be destroyed by prolonged cooking or even storage for long periods at warm temperatures. Canned tomatoes, citrus fruits, and other such foods, however, retain some of this vitamin because they are canned in the absence of oxygen; it seems that the combination of heat and oxygen is most destructive. Deficiency of this vitamin results in scurvy, a disease formerly prevalent among sailors who took long voyages without provision for fresh food. A great part of Great Britain's sea power during the last century can be traced to her discovery that sailors did not get scurvy if they were given daily rations of "lime" juice (it was really lemon juice). We still call British sailors "limeys" because of this practice. This vitamin is not stored in the body and needs replenishment almost daily. Babies often suffer from vitamin C deficiency because pasteurization destroys the natural vitamin C of milk. Many doctors advise the addition of orange juice to the diet of the baby as early as three weeks after birth.

Vitamin D (calciferol). This is called the sunshine vitamin because it is produced when ultraviolet light (either from sunlight or artificial sources) strikes certain precursors of the vitamin. Practically all plant foods contain **ergosterol,** which becomes vitamin D when exposed to sunlight. It is also generated in the skin of animals exposed to sunlight. Milk is a rich source of vitamin D when the cows have been raised in sunlight, but milk may be very low in vitamin D when they are kept indoors. Canned milk and some fresh milk is irradiated with ultraviolet light to increase the vitamin D content. This is very important in the diet of children, for when the vitamin is deficient, children will develop **rickets,** a disease characterized by serious malformations of the bones and teeth. Vitamin D is necessary for calcium utilization by the body, and symptoms similar to calcium deficiency develop when it is deficient. We have the ergosterol in our body cells, and vitamin D is produced when the sunlight strikes our exposed skin. Of course, there are times of the year when the weather prevents much manufacture in such a manner. Then we must depend on the vitamin in our food, although the liver stores this vitamin in the liver oil and thus helps take care of this seasonal fluctuation, just as with vitamin A. The fish-liver oils are valuable sources of vitamin D as well as of vitamin A.

Vitamin E (tocopherol). This vitamin is most abundant in the oil found in whole wheat and in lettuce, but it is also found in a wide variety of unprocessed foods. There is little chance that a normal varied diet will be deficient in it. Experiments with animals indicate that an insufficient quantity causes sterility in the male, through degeneration of the testes, and sterility in the female, through the premature death and abortion of the fetus.

Vitamin K. This vitamin is essential for normal blood clotting, since the liver uses it in the synthesis of prothrombin. It is a widely distributed vitamin, found in a great many vegetable foods and especially abundant in spinach, kale, cabbage, and tomatoes. We also obtain it in liver.

There is a closely related compound—**dicumarol,** found in sweet clover—which is so closely similar to vitamin K that the enzyme systems of the liver utilize it in place of vitamin K. The important prothrombin is not produced, however. This is a good example of a substance known as an **antimetabolite** which can take the place of an essential ingredient in body synthesis but which results in a product that is not the normal product. In this case, the antimetabolite can have medical value. Dicumarol can be administered to persons who are subject to the formation of blood clots, and it will inhibit these clot formations just as antithrombin does.

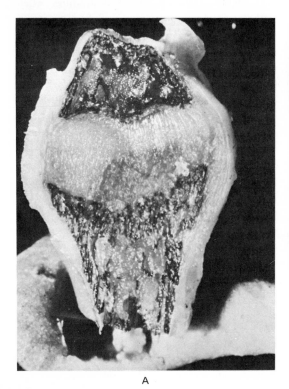

A

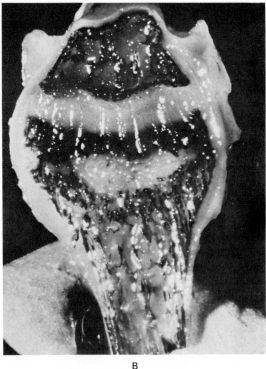

B

FIG. 31.7 Effect of vitamin D on calcium deposit in the rat femur. The picture at left shows the end of the femur taken from a rat which has rickets caused by vitamin D deficiency. This bone is deficient in calcium deposits, as shown by the dark areas. At right is a femur taken from a rat of the same age which received 0.00025 mg, one USP unit, of vitamin D daily. The calcium deposits are normal for a rat of this age.

The Minerals

Certain inorganic minerals must also be included in the diet. These also have no caloric value but are necessary for building certain body tissues and fluids. We shall list some of those which are better known.

Calcium. This is the most abundant mineral of the body. Calcium salts make up the hard parts of the bones and teeth. Calcium is especially important in the diets of children and expectant mothers, but it is needed by others as well because the bones are in constant turnover. Old bone is constantly being replaced by newly mineralized tissue and there must always be an adequate supply of calcium for this purpose. Calcium is also necessary for normal blood clotting and for muscle contraction.

Phosphorus. This is the second most abundant mineral in the body. It is an essential part of bones and is also found widely distributed in the softer body tissues. The genes themselves (DNA) contain phosphorus as an essential part of their molecules. We also learned that the energy currency of the cells (ATP) must have phosphorus before it can be formed or utilized in cell metabolism. The proportions of calcium and phosphorus are very important and are regulated by one of the endocrine glands.

Iron. This mineral occupies a central position in the hemoglobin molecule. Anemia will result if there is iron deficiency. Iron also forms a part of the cytochromes which act as acceptors in both photosynthesis and respiration.

Iodine. This mineral is required in very minute amounts. Only about 0.0003 g per day is needed by an adult, but serious consequences result when this small amount is not obtained. There are regions of the United States in which the deficiency was once common. These regions were far removed from the seacoasts, and there was little iodine in the soil. Since their food came primarily from this soil, the inhabitants did not obtain the minimum iodine needed and they developed **goiter,** sometimes characterized by a swollen thyroid gland. Today, with our modern system of food transportation and preservation, a person in South Dakota can feast on iodine-rich sea foods whenever he wishes. Table salt is also available today that contains a small amount of an iodine salt. Iodine is a necessary element for the hormone thyroxin, made by the thyroid gland.

Many other minerals are needed. Sodium and chlorine are two which are usually obtained in considerable excess, since these make up our common table salt. The kidneys remove the excess, but there are times when this load may be too heavy. Persons having heart ailments, for instance, are usually advised to greatly reduce intake of table salt. Some minerals are needed in such minute quantities that we call them trace elements. These include copper, zinc, cobalt, manganese and fluorine.

REVIEW QUESTIONS AND PROBLEMS

1. There are some persons alive today whose stomachs have been entirely removed. In such cases the esophagus is connected directly onto the duodenum. They require no special diet. Consider the digestive functions of the stomach and describe the alternate means of digestion for each of the three classes of food in such cases.

2. What limitation in eating habits might be required by persons who have had their stomachs removed?

3. The greatest accomplishments in the arts, sciences, and literature have been in those countries with the most highly developed agricultural systems. Explain why.

4. The enzymes for protein digestion in both the stomach and the small intestine are secreted in an inactive form. Why is this necessary?

5. What conditions lead to the development of ulcers in the stomach and why?

6. Why do between-meal candy bars tend to reduce the appetite at mealtime?

7. Name and locate the sphincters of the digestive tract.

8. What structural features of the small intestine add to its absorptive capacity?

9. Pancreatic juice is normally secreted into the intestine only when food is entering the duodenum from the stomach. Explain how this is controlled.

10. Bile has a somewhat laxative action on the movement of materials through the intestine. When a person has diarrhea he is usually advised to reduce his intake of fats. Explain why.

11. Why does a person's skin turn yellow when he has a clogged bile duct?

12. Bile is not an enzyme, yet digestion is not normal without bile. Explain.

13. In what respect may the bacteria in the colon be of value?

14. How is the glucose content of the blood kept within rather close levels of concentration even though the intake of carbohydrates may vary greatly?

15. Persons with kidney trouble may at times be advised to restrict the protein content of the food they eat. What relation between protein content of food and kidney function would make this advisable?

16. Which vitamins are manufactured in the body from precursors in foods? Explain the process in each case.

17. A piece of the liver is often removed for analysis when a person has died under mysterious circumstances. Why is this done?

18. Which vitamins are likely to be lost from foods fried in oil? Explain.

19. What is an antimetabolite? Illustrate with a case relating to vitamin K.

20. Sometimes vitamin K is administered by a physician and sometimes its antimetabolite, dicumarol, is administered. Under what conditions would each be of value?

21. Why is phosphorus such a vital mineral?

FURTHER READING

Davis, A. 1970. *Let's Eat Right to Keep Fit*. New York: Harcourt Brace Jovanovich.

Gordon, M. S. 1972. *Animal Physiology*. 2nd ed. New York: Macmillan.

Guyton, A. C. 1969. *Function of the Human Body*. 3rd ed. Philadelphia: W. B. Saunders.

Jennings, J. B. 1965. *Feeding, Digestion and Assimilation in Animals*. New York: Pergamon Press.

Lappe, F. M. 1971. *Diet for A Small Planet*. New York: Ballantine Books.

Wood, D. W. 1970. *Principles of Animal Physiology*. New York: American Elsevier.

32

Respiration and Excretion

Before food materials can be used they must be combined with oxygen in cellular respiration, which was described in Chapter 11. The method of obtaining oxygen in animals is known as external respiration. Just as vital to life is the elimination of the wastes generated by cellular respiration. Carbon dioxide is one of these wastes; others include soluble solids, which are known as excretory wastes. Very small animals can obtain oxygen and eliminate wastes by simple diffusion and active transport, but as animals become larger and more complex, special organs are required. We have described some of these in our survey of the animal kingdom. This chapter reviews how these essential processes occur in man.

ORGANS OF RESPIRATION

The Upper Respiratory Tract

Air normally enters through the **nostrils.** Coarse hairs line these openings and remove the larger foreign particles that might be in the air. Air then passes into the **nasal cavity,** which is lined with a mucous membrane that is moist and sticky with mucus. It also circulates through the **sinuses,** where similar conditions exist. In these cavities most of the remaining foreign particles are removed as they adhere to the mucus. The air is also warmed and moistened so that it does not arrive in the lungs dry and cold; this would interfere with the exchange of gases. The sinuses have another important function; they act as a resonance chamber for the voice. Musical instruments differ in tone quality largely because of the structure of the resonance chambers; in the same manner people have distinctive voices due to the structure of their sinuses. A singer with a pleasing voice may owe much to thorough training, but he must also thank his parents for providing the genes that produce sinuses with a pleasing resonance. When a person has a bad cold, the sinuses are clogged and the voice sounds different because of a lack of resonance. There is a mass of **adenoid tissue** at the back of the nasal cavity that may show abnormal growth and interfere with breathing through the nostrils. The excess growth is often removed to eliminate mouth breathing in children. The two small openings to the **Eustachian tubes** are also found in this region; these tubes run to the middle ear. The air passes from the nasal cavity

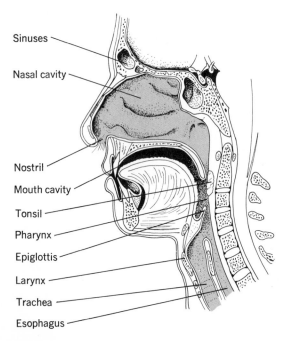

Sinuses

Nasal cavity

Nostril

Mouth cavity

Tonsil

Pharynx

Epiglottis

Larynx

Trachea

Esophagus

FIG. 32.1 Section through the human head showing the upper respiratory tract and related body parts.

into the **pharynx** (throat). There are two **tonsils** in the throat, one located on either side where the mouth cavity joins the throat. These are lymph nodes and, as such, have the important function of removing infectious agents. Sometimes, however, they become badly infected and may spread poisons over the body. Children may even have arthritis from such poison. Dramatic recovery may occur when the tonsils are removed, but routine removal of tonsils of all children is not advisable.

The Lower Respiratory Tract

The air next passes into the **larynx** (voice box or Adam's apple) through an opening, the **glottis.** A serious problem must be overcome at this juncture. Food as well as air passes through the pharynx, and each of the two must be sent into its proper tube. The glottis is normally open, allowing the air to pass freely into the larynx. When we swallow, however, the larynx moves up so that the glottis is closed by the **epiglottis** and the food or liquid passes by and into the esophagus. The separation is not always accomplished. If a person has food in the mouth and suddenly laughs or gasps, he may inhale through the mouth and carry some of the food down into the lower respiratory tract. This produces most unpleasant symptoms as he coughs violently in attempts to remove the foreign material from the tract. Children sometimes inhale objects such as rocks, marbles, coins, and safety pins which require surgical removal. Air often goes into the stomach; some people seem to enjoy their soup better when they can suck it in with large quantities of air; babies need to be "burped" when they are feeding because they swallow air along with their milk.

The larynx contains the **vocal cords,** a pair of tough membranes, one on either side, that partially close the air passage. These are muscular and can be stretched or relaxed at will. When stretched they vibrate as the air passes

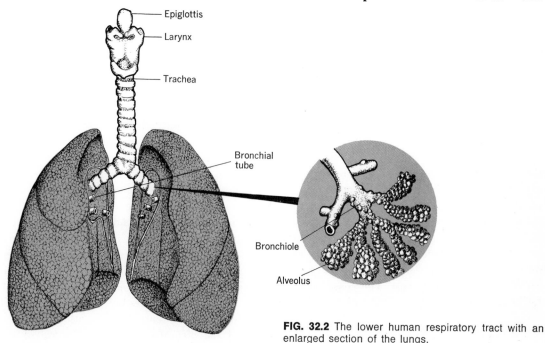

Epiglottis

Larynx

Trachea

Bronchial tube

Bronchiole

Alveolus

FIG. 32.2 The lower human respiratory tract with an enlarged section of the lungs.

over them producing the sounds from which we form speech. As the sounds pass out of our mouths, we flip them with our tongues, hiss them through our teeth, and manipulate them with our lips to make words. We can speak without using the vocal cords—whisper—by using the sound of air being expelled through the respiratory tract. We can alter the tone of the voice by the degree of stretching of the vocal cords. Stretched tightly they produce a high tone, and stretched less tightly they produce a low tone. The range of tone is limited by the stretching ability, but may be greatly extended by voice training which utilizes the upper and lower tones of the range more than they are used in speaking. The male sex hormones cause an enlargement of the larynx and a thickening of the vocal cords which results in a distinctive difference in the voices of the two sexes after puberty.

The air goes from the larynx to the **trachea** (windpipe), which forks to form two **bronchial tubes,** one leading to each side of the chest.

The trachea and bronchial tubes are composed of rings of cartilage embedded in softer tissue. Since breathing works on the principle of a vacuum suction, the cartilage is necessary to prevent collapse of the tube. The bronchial tubes branch into many smaller tubes, the **bronchioles,** which eventually terminate in tiny sacs, the **alveoli,** that make up the main part of the **lungs.** This alveolar structure of the lungs gives

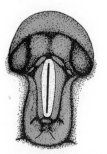
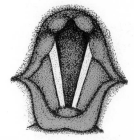

FIG. 32.3 The vocal cords as they are seen by looking down through the mouth. They are shown relaxed on the left and contracted to produce sound on the right.

much more exposed surface in the exchange of gases than if the lungs were only hollow inside. It is estimated that there are about 100 square yards of exposed surface in these small sacs in the lungs of an adult man.

MECHANISM OF BREATHING

The lungs lie within the two **pleural cavities,** one on each side of the chest. These cavities are lined with moist mucous membranes, the **pleura,** that lubricate the movements of the lungs within. When a person has pleurisy, this lining is inflamed and roughened, and there is pain when the delicate lung tissue slides over the inflamed portion. The lungs fill the pleural cavities of the normal chest. The space between the lungs and the pleura is filled with the mucous

Diaphragm breathing

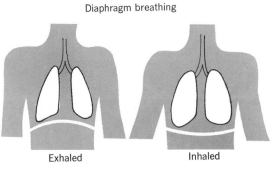

Exhaled   Inhaled

Chest breathing

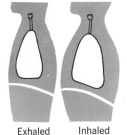

Exhaled   Inhaled

FIG. 32.4 Method of inhalation of air. When the chest is expanded or the diaphragm is lowered, the volume of the chest cavity is increased and air comes in. Most people use a combination of these two methods in normal respiration.

secretion of the pleura. Inspiration and expiration of air are accomplished through changes in the size of the pleural cavities and the consequent compression and expansion of the lungs. We accomplish this in two ways. **Chest expansion** is the most noticeable method of inspiration. This is accomplished by contracting muscles which bring about an elevation of the sternum along with the ventral portion of the ribs attached to it. As a result, there is an expansion of the chest cavity, as is illustrated in Fig. 32.4 and this tends to create a vacuum in the pleural cavities. The lungs expand in response to the vacuum, and air is drawn in through the nostrils and fills the extra space in the lungs as they expand. When the muscles are relaxed, the natural elasticity of the ribs and the lungs draws the ribs back down to a resting position and air is forced out of the lungs. We can also change the size of the pleural cavities through use of the **diaphragm,** a muscular organ which separates the chest from the abdominal cavity. When the diaphragm is lowered, the pleural cavities are made large and air is inhaled. When the muscles of the diaphragm are relaxed, it tends to return to its original position and air is exhaled. For maximum exhalation we can bring muscles into play which depress the rib cage more than its relaxed state and we can bring the diaphragm higher by contracting muscles of the abdomen which force the diaphragm higher into the pleural cavities.

Breathing is generally an automatic reaction controlled by the autonomic nerves, but the muscles involved are also under conscious control. You can hold your breath and breathe more rapidly or more deeply than normal at will.

Normal breathing at rest is accomplished by a combination of chest and diaphragm movements. Each breath takes in about 500 ml (approximately 1 pint) of air and exhales an equal amount. There are about sixteen to eighteen respiratory cycles per minute. The trachea and the bronchial tubes leading to the lungs hold about 150 ml, so only about 350 ml reaches the lungs at each breath. The total capacity of the lungs averages about 4,000 to 5,000 ml. You can see that only a small amount of air is changed with each breath. The maximum amount of air which can be expired after a maximum inspiration is known as the **vital capacity.** This varies according to sex, height, body weight, and age to some extent, but should be about 3000 ml in a young woman who is 5 feet 7 inches tall and weighs 130 pounds. A young man of the same height and weight, on the other hand, should have a vital capacity of about 3,750 ml. A 10 percent deviation below normal is of no great consequence, but lesser vital capacities can bring about difficulties in breathing.

A respiratory disease which seems to be becoming more prevalent today is **emphysema.** This condition occurs when the walls of the alveoli become irritated and break down. As a result, the surface area available for gas exchange is reduced. The person involved develops a shortness of breath and even small exertions will bring on very heavy breathing. Some emphysema is likely to develop in most elderly persons, but statistical evidence indicates that early incidence and severity is related to heavy smoking. There is also some evidence that the smog and smoke of some cities may be a contributing factor.

Regardless of vital capacity, no person can empty the lungs completely during expiration. Even after the most extreme expiration, there will still be about 1000 ml of **residual air** in the lungs. Since breathing represents only a partial

emptying and refilling of the lungs with each breath, the expired air does not vary greatly from the inspired air. Table 32.1 shows the comparison as averaged from many samples.

The actual amount of nitrogen is not increased, but the percentage is slightly increased because of the absorption of oxygen which is not completely replaced by carbon dioxide released into the alveoli. The argon and other rare gases have no physiological significance. These measurements were made when the inspired air was saturated with water vapor. Actually, this is rarely the case; the air varies considerably in its humidity when it is inhaled, yet will be practically saturated when exhaled because of the contact with the moist membranes of the respiratory system. This is readily apparent when we exhale on a cold day or when we breathe on a cold object; the vapor in the exhaled air condenses into water droplets.

Regulation of Breathing

With the first cry of a newborn baby, a breathing rhythm begins which must continue until the final gasp at death. At no time during this period can this process be suspended for more than a few minutes. We need not worry that we will forget to breathe; an inborn mechanism regulates this vital reaction without any conscious effort, even though voluntary muscles are involved. We can interrupt the process temporarily when we wish, as when swimming under water, but it is impossible to stop it altogether. No one can commit suicide by holding his breath. The stimulations to breathe are so strong that they will eventually overcome any voluntary effort to stop them. What is the nature of these regulatory mechanisms?

We have already learned that the control centers are located in the medulla of the brain. On either side there is an **inspiratory center** and an **expiratory center.** The inspiratory cen-

TABLE 32.1.

Component	Inspired Air	Expired Air	Alveolar Air
Nitrogen	78.08%	78.26%	78.36%
Oxygen	20.94	16.30	14.20
Carbon dioxide	0.04	4.50	6.50
Argon and other rare gases	0.94	0.94	0.94

ter sends impulses to the muscles of the chest and diaphragm and this causes the contractions which enlarge the chest cavity. As the lungs expand, however, there are **stretch receptors** in the lungs which are stimulated by the stretching of the lung tissue. These nerve endings relay impulses to the expiratory centers of the medulla, which in turn send impulses to the inspiratory centers and inhibit them. Thus inhibited, these centers cease sending impulses to the respiratory muscles, and the muscles relax. This causes the chest cavity to become smaller as it returns to a resting state and the lungs are thus passively deflated. As they deflate, the stretch receptors are no longer stimulated, the inhibition is released, and a new stimulation to contract comes to the muscles. Thus we have a homeostatic respiratory cycle, which can be interrupted temporarily by impulses from the voluntary centers of the cerebrum.

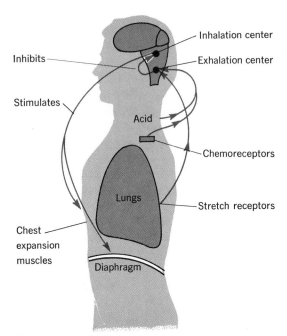

FIG. 32.5 The homeostatic adjustments involved in the regulation of breathing.

This accounts for the rhythm of breathing, but it does not account for the variations in rate and depth. The average rate for an adult is about sixteen to eighteen respiration cycles per minute at rest. During sleep, however, it is much lower. It reaches its height during and immediately following vigorous exercise. This is correlated with the rate of metabolism of the cells. How can the cell metabolism, which creates a need for oxygen, signal the respiratory centers to speed up the supply by faster and deeper breathing? The primary stimulus seems to be carbon dioxide which is released from the cells. We have learned that this gas, which is generated in cell respiration, passes into the blood, where a part of it is carried as carbonic acid. The concentration of this acid in the blood, therefore, will be increased as a result of increased cell respiration. Such an increase stimulates the inspiratory centers of the medulla as the blood passes through them, and they step up the rate of stimulations to the muscles of inspiration. The increased rate of breathing increases the rate of removal of carbon dioxide from the blood, and as the carbonic acid decreases, there will be a corresponding reduction in the rate of breathing. We have learned that lactic acid is also generated during intense muscle activity. Some of this is absorbed by the blood and acts as an additional stimulus to the inspiratory centers of the medulla.

There are **chemoreceptors** in the large blood vessel, the aortic arch, coming from the heart which are sensitive to an extreme lowering of the oxygen level of the blood. In such cases these receptors will send messages to the inspiratory centers of the medulla and stimulate breathing. Normally, however, the much more sensitive reactions to the increase in acid in the blood prevents the blood oxygen level from reaching a very low level. The oxygen content of the air must be very low for conditions to be such that there would be a drastic lowering of the oxygen content in the blood without a corresponding rise in the carbonic acid level. In such cases this safety mechanism would take over and increase the rate and depth of breathing and thus prevent acute oxygen deficiency.

Air Pressure and Respiration

If you have ever driven to the top of Pikes Peak you surely noticed that your automobile lacked its usual power as you neared the top. It was suffering from oxygen deficiency. The carburetor was adjusted to the air at lower elevations, and while the same volume of air was entering your engine as at base level, it contained less oxygen because the air was "thinner." Air has weight and the air at lower altitudes is compressed by the volume of air above it so that there are more molecules per pint. Although the air on top of Pikes Peak is still about 21 percent oxygen, there are just about half as many molecules of this gas in any given volume as in the air at sea level. A normal inhalation of air will include only about half as much oxygen as would be present if you were at sea level. If you walk around a bit at this altitude, about three miles above sea level, you will notice that you tire very easily and begin breathing heavily with the slightest exertion. The chemoreceptors in the aortic arch probably are functioning.

The engines of automobiles can be adjusted to function more efficiently at high altitudes by increasing the air intake and thus compensating for the reduced quantity of oxygen in the air. Our bodies can also make adjustments. A healthy man will have about five million red blood corpuscles per cubic millimeter of blood. These red cells will contain sufficient hemoglobin to absorb and distribute oxygen for the body's needs at a lower altitude. If a person stays at a high altitude for a time, however, the number of red blood cells will begin to rise. The number will not stay high, however, if the person remains in the region; instead there will be a gradual increase in the hemoglobin content of each red blood corpuscle.

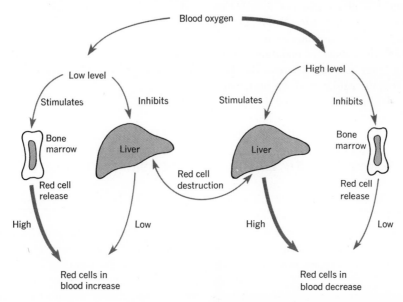

FIG. 32.6 How the concentration of oxygen in the blood homeostatically regulates the number of red blood cells released from the red bone marrow.

This will increase to such an extent that the total number will return to about five million, yet the blood will retain its increased oxygen-carrying capacity. Mountain climbers do much better if they give themselves a period of adjustment to high altitudes before scaling high peaks. One group actually made it to the top of Mount Everest in India, which towers to a height of 29,000 feet, without oxygen masks. This was possible because they made gradual ascents and remained for a time at each station in order that their bodies might become adjusted to the rarefied air found at the peak.

In the early days of high-altitude flying there were great problems concerning respiration. At an altitude of about 37,000 feet (seven miles) the air has a density only one-fifth of that found at sea level. The air does not contain a sufficient concentration of oxygen to sustain human life. Unconsciousness comes within a few seconds and death within a few minutes if a person is breathing this air. With the help of oxygen masks, however, the amount of oxygen per breath can be raised enough to sustain life. Today practically all high-altitude flying is done in pressurized cabins where the air density is maintained at a level found at a lower altitude. There is the danger, however, that the cabins may be perforated and lose their pressure. Commercial jet planes that fly around 30,000 feet and higher have emergency oxygen masks for this contingency. Such accidents are not expected, of course, but it is comforting to know that a safety measure is provided.

A deep-sea diver or a tunnel worker who is active where the air is kept under high pressure has just the opposite problem. He may get too much oxygen with each breath, since the number of molecules of oxygen in each volume is greater than at normal pressure. At five times sea level pressure, which is what is required by a diver working 200 feet below the surface, oxygen in air is found at the same concentration as pure oxygen at sea level. This may cause a man to get oxygen drunk and greatly lower his efficiency. Artificially mixed air with a smaller than normal percentage of oxygen enables him to absorb a more normal amount of oxygen.

Another problem faces such workers. Although nitrogen is an inert gas with respect to body tissues, some of it diffuses into plasma. At a high pressure the amount of absorption increases. Hence, the pressure must be released slowly so that the nitrogen can diffuse back into the lungs, or some of it may form gas bubbles in the body tissues. This causes severe pains in the joints and muscles, or perhaps more severe symptoms such as paralysis, and may even result in death if the person is not promptly put into a decompression chamber where the pressure is again raised and then lowered very gradually. To prevent these occurrences, known as the **bends,** such wokers must undergo a gradual decompression to allow a gradual release of the nitrogen. In recent times the U.S. Navy has had men spend several days in a deep underwater apparatus at very high pressure. Helium was substituted for nitrogen in the artificially mixed air. Helium, being much lighter than nitrogen, is released from the blood much more rapidly, thus minimizing the danger of the bends.

ABSORPTION AND TRANSPORTATION OF GASES IN THE BODY

As the blood passes through the alveoli, some of the oxygen within the alveoli diffuses into the plasma. The oxygen requirements of the human body are so great that they could not be supplied if plasma were the sole means of transportation. The **hemoglobin** of the red cells functions as the primary means of oxygen transportation. Hemoglobin is a very large protein molecule in which iron occupies a central position. It is made of four component chains with an atom of iron at the end of each. It has a great affinity for oxygen. Soon after the plasma has absorbed the oxygen, the oxygen molecule attaches itself to the iron atom of each chain. This forms **oxyhemoglobin,** which is lighter in color and a more brilliant red than the dark red of hemoglobin alone. As the hemoglobin combines with the oxygen, the concentration of oxygen in the plasma is lowered, thus causing more oxygen from the lungs to diffuse in until almost all the hemoglobin has been converted into oxyhemoglobin. The hemoglobin increases the oxygen-carrying capacity of the blood about sixtyfold. In other words, we would need about sixty times as much blood as we now have if we had no hemoglobin.

When blood reaches the small capillaries of the other body tissues, a reversal of the process of oxygen absorption takes place. The cells around the capillaries are continually using oxygen, and they have a concentration of this gas which is much lower than in the blood. As a result, there is a diffusion from the plasma into the surrounding cells. This lowers the oxygen level in the plasma, and oxygen begins to break away from the hemoglobin to replenish it. This oxygen in turn diffuses into the cells. The process continues until most of the oxygen has passed from the hemoglobin. Oxygen combines with hemoglobin when the oxygen is abundant in its surroundings, and is released from hemoglobin when it is deficient in the surroundings. When man is at a high altitude, the concentration of oxygen in the air may be so low that the hemoglobin does not readily unite with it in the lungs. The hemoglobin will be low in oxygen when it reaches the cells. Such reduced oxygen stimulates a homeostatic mechanism which causes a release of extra red blood cells into the blood stream. This increases the oxygen-carrying capacity of the blood, but several days are required before it is brought up significantly.

Two other factors are involved in the affinity of hemoglobin for oxygen. Hemoglobin gives up its oxygen more readily when the carbon dioxide level of the plasma increases and also when the temperature rises. Cells actively undergoing metabolism, such as muscle cells at work, give off large amounts of carbon dioxide and also release heat. Both of these factors cause the hemoglobin to release its oxygen more readily, and this oxygen is available to the cells which need it most. At the lungs the carbon dioxide level of the plasma drops drastically and the hemoglobin can combine with oxygen readily.

We can represent the hemoglobin-oxygen union as a reversible equation using Hb as a symbol for hemoglobin:

$$Hb + O_2 \rightleftharpoons HbO_2$$

Hemoglobin also has an affinity for carbon monoxide and readily unites with it:

$$Hb + CO \rightleftharpoons HbCo$$

Hemoglobin does not readily give up carbon monoxide, however, and thus loses its ability to transport oxygen. This accounts for the highly poisonous effects of carbon monoxide.

Carbon dioxide, which is generated during the catabolic phase of metabolism, becomes more concentrated in the cells than in the plasma. Hence, it diffuses from the body tissues into the plasma. The cells of an average man's body at rest generate about 200 ml of carbon dioxide per minute. Plasma can absorb and carry only about 4.3 ml of carbon dioxide per liter of blood. This means that the blood would have to circulate through the body at the very high rate of 47 liters per minute if plasma alone carried the carbon dioxide. This rate is about ten times faster than the blood is known to circulate. The plasma actually carries only a small percentage of the carbon dioxide in solution. Hemoglobin unites with about 10 percent of the carbon dioxide and carries it as $HbCO_2$. The major part of the carbon dioxide, however, reacts with water chemically and forms hydrogen and bicarbonate ions with carbonic acid as an intermediate product, as shown below:

$$CO_2 + H_2O \rightleftharpoons H_2CO_3 \rightleftharpoons H^+ + HCO_3^-$$

This is a reversible reaction; at the lungs, where the carbon dioxide concentration is low, the reaction goes to the left of the equation, and carbon dioxide diffuses out into the alveoli of the lungs.

The formation of carbonic acid from carbon dioxide and water takes place mainly in the red blood cells because these cells contain an enzyme, **carbonic anhydrase,** which greatly speeds this reaction. Therefore, much of the carbon dioxide passes into the red blood cells and is united with water to form carbonic acid, which then is ionized into hydrogen and bicarbonate ions. The bicarbonate ions then pass out into the plasma and are carried to the lungs. At the lungs, as the carbon dioxide from the hemoglobin passes out of the red blood cells, the bicarbonate ions re-enter these cells, and the carbonic anhydrase now converts them back to carbon dioxide and water. This carbon dioxide is now free to diffuse out of the cells and into the air spaces of the alveoli.

The story of carbon dioxide transfer is not complete; two important problems remain to be solved. We have noted that, for every bicarbonate ion produced, a hydrogen ion is also produced. Since hydrogen ions cause a lowering of the pH, we might expect the blood plasma to become highly acid after absorbing the carbon dioxide at the body tissues. We know this does not happen; the blood pH remains constant at about 7.3; death results if the blood pH ever drops below 7.0, which is neutrality. This maintenance of pH is possible because the blood contains certain inorganic ions and proteins which act as **buffers;** that is, they take up most of the free hydrogen ions and thus prevent acidification. When the elimination of carbon dioxide from the body through the lungs is below par, the buffering capacity of the blood may be overtaxed and the blood pH will drop. This can happen when the disease of pneumonia causes a clogging of many of the air spaces of the lungs or when a person is in an atmosphere with a high carbon dioxide content.

A second problem to be solved concerns the electrostatic charge of the red blood corpuscles. When bicarbonate ions leave the cells, the hydrogen ions remain, and even though buffered, they would tend to cause the cells to lose their electrical neutrality and become positively charged. This is prevented by what is known as the **chlo-**

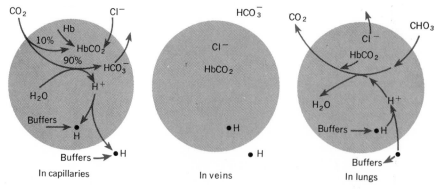

FIG. 32.7 The chloride shift is a homeostatic mechanism which regulates the transportation of carbon dioxide in the blood.

ride shift. The blood plasma always contains some free ions of chlorine, which are negative in charge. Each time a bicarbonate ion leaves a corpuscle, a chloride ion enters the corpuscle and maintains the electrostatic neutrality of the corpuscle. At the lungs when the bicarbonate ion re-enters the corpuscle, the chloride ion shifts back out into the plasma.

THE BASAL METABOLIC RATE

When a person first awakes in the morning his metabolism is probably at its lowest ebb. If we assume that he has had a good night's rest, does not have any infectious disease, is comfortably warm, has not been out on a spree the night before, and is not awakened by a discordant shock to the nervous system such as the jangling of an alarm clock, his body will be utilizing only sufficient food and oxygen to maintain the living state. This is known as his basal metabolic rate. The basic metabolic rate varies with age, height, weight, and sex, but it can also be influenced by certain abnormalities such as thyroid disease. It is often desirable to know if basal metabolism is being carried on at a normal rate. Since heat is a byproduct of metabolism, the rate can be determined if the person is placed in a closed chamber and the heat given off from his body is carefully measured. A much less cumbersome method involves measuring the amount of oxygen utilized in a given period, since oxygen is needed for cell respiration. Many a person has had his BMR taken by breathing through his mouth from a tube supplying pure oxygen while his nose was closed by a clamp resembling a clothespin. In recent years there has been a tendency to use a method which is simpler and more accurate in many ways. A small volume of blood is analyzed for the amount of protein-bound iodine. Iodine is a basic part of the thyroxin hormone and there is no other protein-bound form of iodine in the blood. Hence, the amount of this form of iodine

is an accurate indication of the amount of thyroxin in the blood, and since this is the hormone which stimulates cell metabolism, its quantity is an indication of the rate of metabolism.

It is customary to express a person's BMR in terms of percentage above or below normal. For instance, a BMR of +10 means that it is 10 percent higher than would be expected for a person of this particular height, weight, sex, and age. Variations of less than 5 percent in either direction are considered to fall within the range of normality. When they become greater than 10 percent, they are considered significant enough to warrant treatment. It is easy to raise a person's BMR by administration of thyroid pills which contain the dried substance of thyroid glands of animals such as cattle. These contain thyroxin which can be absorbed in the intestines and raise the amount of this hormone in the blood. When the BMR is too high, however, the treatment is more difficult. It depends on drugs which suppress the thyroid, on surgical removal of part of the gland, or on administration of enough radioactive iodine to destroy some of the cells of the gland.

DISORDERS OF THE RESPIRATORY TRACT

The respiratory tract serves as the portal of entry for the majority of the infectious diseases that afflict mankind. Air is continually inhaled and it often contains the agents of disease. Many diseases infect the respiratory tract itself. No part of the tract is exempt; there is **rhinitis** of the nasal cavity, **sinusitis** of the sinuses, **pharyngitis** of the throat, **laryngitis** of the larynx, **bronchitis** of the bronchial tubes, and **pneumonia** of the lungs. There is a wide range of organisms that may cause these infections. We often start with a simple bad cold. This is caused by a virus in times of low body resistance. The mucous membrane of the nasal cavity and sinuses becomes inflamed and swollen and secretes a copious watery fluid. Bacteria often come in at this point and infect the damaged tissue. The sinuses may become clogged and we develop serious pain in the infected region. The infection may spread downward and we develop a sore throat with swollen tonsils. Next comes the larynx; we become hoarse and speak with difficulty. If we still cannot get the infection under control, it spreads to the bronchial tubes and we cough up quantities of viscid mucus produced in the inflamed tubes. Finally the lungs themselves may develop pneumonia. This is quite dangerous, because the mucus secretions may so fill the alveoli that is impossible to get sufficient air into the lungs to support life. **Tuberculosis** is a common lung in-

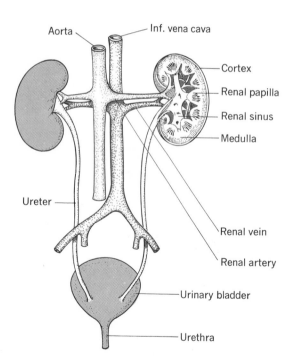

FIG. 32.8 The human excretory system. The kidney on the right has been opened to show the interior structure.

fection which is gradually being conquered by modern medical science.

Serious allergic manifestations may occur in the respiratory tract because of the presence of foreign matter in the air. Some plants produce large quantities of pollen which is carried by air currents. A person sensitive to a specific pollen will react to it when the pollen is inhaled. The reaction often takes place in the nasal cavity and sinuses, and the person has **hay fever.** The mucous membranes become swollen and secrete large amounts of watery fluid in a rhinitis similar to infectious rhinitis. This may become so serious that the person cannot breathe through the nostrils without special medication to reduce the swollen membranes.

If some of the pollen passes by the natural filters of the nostrils, it may set up inflammation of the bronchioles, a condition called **allergic asthma.** In certain occupations workers inhale fine powdery dust that penetrates to the lungs and causes trouble. Glass workers, granite workers, pottery workers, and others who grind or work with similar materials are susceptible. The small dust particles are sharp and may so injure the lungs that the worker develops a type of pneumonia called **silicosis.**

Emphysema, the breakdown of the alveoli with a reduction in the absorptive area of the lungs, is a common lung disorder which can have serious consequences. It has been discussed earlier in the chapter.

EXCRETION

In addition to the wastes of metabolism which must be excreted, there are many substances in the blood which are needed in certain quantities but which can be harmful when present in excess. These include vitamins, hormones, glucose, amino acids, and the inorganic minerals.

Water is also a vital compound, but the amount of water must be kept within narrow limits regardless of the variation in intake. The excretory system of man is a master homeostatic mechanism which maintains at optimum level the amount of water and the concentration of many soluble substances in the blood.

The kidneys are the primary excretory organs of man. The sweat glands carry off some dissolved wastes, but they lack the selective capacity of the kidneys. The large intestine also serves as an accessory excretory organ; some minerals may be actively transported from the blood into the large intestine and thus expelled along with the feces. Some of the excess water of the body is given off as vapor in the breath along with carbon dioxide. Everything considered, however, the kidneys must do the primary job of maintaining the proper balance of water and dissolved minerals in the body.

Human Organs of Excretion

The human **kidneys** are each about 4½ inches long and lie against the dorsal body wall in the upper part of the abdominal cavity. They are embedded in a protective layer of fat and connective tissue which holds them in place. A "floating kidney" may result from a very hard blow which jars the kidney loose from its attachments. A tube, the **ureter,** extends from the median surface of each kidney and leads down to the **urinary bladder** in the lower part of the abdominal cavity. This bladder is closed by a sphincter muscle, and thus the urine coming down from the ureters backs up into the elastic bladder. As the bladder becomes distended, nerves which arouse the desire to empty the bladder are stimulated. The sphincter can be relaxed by voluntary control and the urine is then carried through a tube, the **urethra,** to the out-

side. Each kidney receives a renal artery from the aorta. This brings blood into the kidney, and a renal vein extends from each kidney to the inferior vena cava, which carries the blood back to the heart. The entire blood supply of the body passes through the kidneys in about 45 minutes. Since an average person has about 5 to 6 quarts of blood, this means that about 150 quarts of blood are filtered through the small capillaries in the kidneys each 24 hours. From this great volume of blood about 1½ quarts of urine will be extracted. If the kidney is cut in a lateral, longitudinal section, you can see that it is made of two types of tissue, an outer **cortex** of dark tissue and an inner **medulla** of lighter tissue. A cavity, the **renal sinus,** is in the center, and several extensions of the medulla, the **renal papillae,** extend out into the renal sinus. The urine drips into the renal sinus through many tiny pores on the renal papillae, and the sinus, in turn, is drained by the ureter. A more detailed study of the kidney tissue shows it to be made up of about one million **nephrons** together with collecting tubules. A detailed study of one nephron will show how the kidney accomplishes its excretory function.

A nephron begins with a hollow **Bowman's capsule** which lies in the cortex of the kidney. This capsule receives a small arteriole, which branches into a tuft of capillaries, the **glomerulus,** within the capsule. The capillaries then reunite and form another small arteriole (rather than a vein) which leaves the capsule. This arteriole is smaller than the one entering, and as a result, a pressure builds up within the capillaries. The pressure causes a filtration of the water and dissolved substances out of the capillaries and into the capsule. Delicate microdissection needles have been inserted into a capsule and the fluid withdrawn has been found to be of the same concentration as blood. In fact, it is about the same as blood with the exception of the cells and blood proteins. Hence, there has been no selective re-

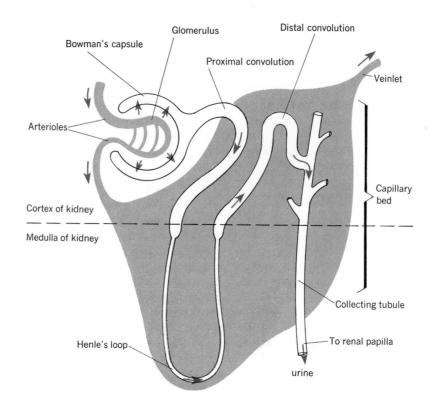

FIG. 32.9 Diagram of a nephron and its relationship to the blood vessels, collecting tubule, and parts of the kidney.

moval of wastes in the capsule. Urine is formed by a complex combination of osmosis, diffusion, and active transport as the filtrate passes along the tubule which drains the capsule.

The tubule from the capsule can be divided into several parts. First there is a **proximal convolution** which remains in the cortex. Next, there is a hairpin-shaped **loop of Henle** which extends down into the medulla as a **descending portion** and comes up into the cortex as an **ascending portion.** Finally, there is a **distal convolution** which then joins a **collecting tubule** in the cortex. The collecting tubule picks up the

products of many nephrons and extends down into the medulla, finally draining into the renal sinus. The arteriole from the glomerulus in the capsule again branches into many capillaries which surround the various tubules leading from the capsule. Blood in these capillaries is not under pressure, so it can reabsorb water and dissolved substances. There is also tissue fluid around the nephrons and collecting tubules.

The exact method of changing the capsular filtrate into concentrated urine has been subject to much investigation. The **countercurrent theory,** originated by the Swiss biologist Werner

Kuhn, has much evidence to support it. According to this theory, the capsular filtrate is concentrated as it goes down the descending portion of the loop of Henle, then diluted in the ascending portion, and finally concentrated again in the collecting tubules. High-threshold substances like glucose are taken into the blood by active transport, and low-threshold substances like urea are not so reabsorbed. Frozen sections of rat kidneys have proved this to be true. Fluids that are more concentrated thaw at lower temperatures than more dilute fluids and the concentration at different points along the tubules can be determined by the temperature at which the fluids thaw.

Testing the fluids by micropipette removal also shows this to be true.

As the capsular filtrate flows into the proximal tubule and then into the descending loop, water passes out of the tubule by osmosis because the tissue fluid in the medulla is rather highly concentrated, especially in sodium chloride. The ascending loop, on the other hand, is relatively impermeable to water, but some of the solute passes out by active transport. This is especially true for sodium chloride. The positive sodium ions are actively transported outside the tubule, and the chloride ions, because they are negative, go along because of the electrostatic attraction.

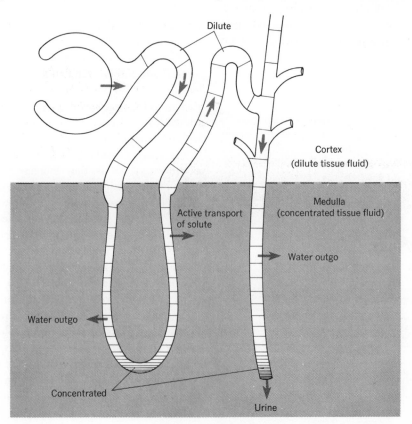

FIG. 32.10 The method of formation of urine by the countercurrent mechanism.

Because of this active transport of solute, the medullary tissue fluid is kept at a high concentration. When the liquid in the loop reaches the distal convolution in the cortex, it is again isotonic with the surrounding blood and tissue fluid. This happens because so much of the solute has been lost by active transport in the ascending loop. Finally, as the liquid passes into the water-permeable collecting tubule and back down into the medulla, with its hypertonic surroundings, water is reabsorbed from the tubule all the way down to the papilla. On the average, about 99 percent of the water is reabsorbed from the collecting tubules. Without such reabsorption we would have to drink about 100 times the amount of water we now drink and urinate 100 times more frequently. It would be inconvenient, to say the least.

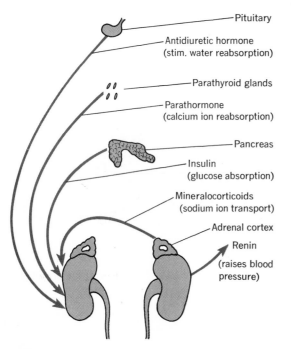

FIG. 32.11 Diagram to illustrate the variety of hormones which are involved in kidney function.

The tubule liquid is now urine as it drips into the renal sinus. The concentration of this urine is influenced by the **antidiuretic hormone** from the pituitary gland. This hormone makes the epithelium of the collecting tubules freely permeable to water, and this is necessary if there is to be the proper reabsorption of water. When a person has a deficiency of this hormone, water is not freely passed out of the tubules in spite of the hypertonic surroundings. As a result, large quantities of very dilute urine are excreted. Such persons have a constant thirst and drink large quantities of water. They are said to have **diabetes insipidus.**

At the opposite extreme, too much water removal can result in too great a concentration of urine. The urine will also be more concentrated when a person's diet is very rich in proteins and when there is little water intake plus great water loss through sweating. Sometimes dissolved substances in highly concentrated urine form crystals which grow with accretion to form kidney stones within the renal sinus. Such stones can also form in the urinary bladder. Caffeine and alcohol act as diuretics; that is, they increase the volume of urine by decreasing the water permeability of the collecting tubules. Among beer drinkers, the high intake of liquid and the diuretic effect of the alcohol lead to a desire for frequent urination.

Occasionally there may be a partial breakdown of the capillaries of the glomeruli and the blood proteins may escape into the capsular filtrate. Since this protein is in the form of a colloid, it cannot be reabsorbed and passes from the body in the urine. This constant loss of blood proteins is very serious if they are lost in any great quantity, and plasma transfusions may be necessary to replenish the proteins. The condition is known as **albuminuria** since albumins are included in the proteins.

A homeostatic mechanism tends to maintain the blood pressure within the glomeruli at the optimum level. If the blood pressure goes up, there is a constriction of the arteries leading into the capsule, thus lowering the pressure to the proper level. If the pressure to the kidneys is reduced because of some obstruction, the kidneys have cells which release **renin** into the blood. (Do not confuse with rennin produced by the stomach.) Renin causes compounds to be released into the blood which stimulate arteries all over the body to contract. Such contraction gives the blood narrower channels through which to pass, and as a result, the blood pressure rises and the pressure within the glomeruli is greater. Renin is also released when the diet is rich in nitrogenous foods or minerals that must be excreted. Kidney troubles may be accompanied by high blood pressure because of this homeostatic adjustment. Those with high blood pressure from this or other causes usually restrict the intake of salt and of foods that leave quantities of wastes that must be excreted.

Still other hormones are known to be involved in the complex homeostatic regulation of kidney function. A hormone from the parathyroid gland regulates the absorption of calcium ions in all parts of the body, including the kidneys. Another hormone from the adrenal glands functions in the reabsorption of sodium ions. Since the concentration of these ions in the tissues of the medulla of the kidneys is an important phase of the regulation of kidney function, both of these hormones are involved in the total picture of excretion. These hormones will be considered in Chapter 33.

REGULATORY FUNCTIONS OF THE KIDNEYS

The kidneys are among the most important body organs in maintaining the concentration of dissolved substances at their proper level within the blood and other body fluids. A slight impairment in kidney function can result in serious upsets in this balance, and death occurs if the impairment of kidney function is severe. A few people have been saved from death by kidney transplants. The person's antibody-producing mechanism must be depressed before such a transplant can be successful, however, and this creates problems related to infections. Some patients are kept alive by artificial kidneys. Blood from the person's arm is passed through cellophane membranes immersed in water containing various concentrations of dissolved substances. Wastes diffuse from the blood into the surrounding water. Selectivity in waste absorption can be obtained by varying the concentrations of solutes in the water. The treatment is done about once a week and is very expensive.

In order to better understand the regulatory functions of the kidneys, let us tabulate their important activities.

Removal of Waste Products of Nitrogen Metabolism

Protein foods all contain amino acids, and amino acids all contain nitrogen. In digestion proteins are broken down into their constituent amino acids, which serve as the building materials for the construction of the proteins within the cells. Some amino acids are left over, however, and they would have a toxic effect if they were allowed to accumulate in the blood. Normally enzymes within the liver break down the amino acids and prevent their accumulation. A few people lack the enzyme needed to break down the amino acid phenylalanine and it accumulates in the blood in such quantities as to act as a poison on the developing brain. Babies with this inherited deficiency will be mentally retarded unless this amino acid is removed from their diets. Some of the whole amino acid will be excreted by the kidneys, but not enough to keep the phenylalanine level of the blood down to normal. Such persons are said to have **phenylketonuria, PKU.** (See Chapter 9 for more details of this condition and how it may be corrected.)

In persons with normal liver enzymes, the amino acids are deaminated (the amino group NH_2 is removed). The rest of the amino acid can then be converted into the energy-yielding glucose, glycogen, or fatty acids. The amino group is then converted into ammonia, NH_3. But ammonia is highly toxic; a rabbit, for instance, will die if the concentration of ammonia in its blood reaches the level of 1 part in 20,000. Still, ammonia is the primary nitrogenous waste excreted by water animals, such as fish. In their environment there is plenty of water available, so the ammonia can be held in a very dilute form and excreted along with large quantities of water. Land animals do not have such enormous quantities of water available for excretion. They have developed liver enzymes which can convert the ammonia into less toxic products. In man the product is **urea,** $CO(NH_2)_2$, formed through a combination of ammonia with carbon dioxide. A small amount of nitrogenous waste may also be excreted as **uric acid,** $C_5H_4N_4O_3$.

Uric acid is the primary nitrogenous waste of insects, reptiles, and birds. Because it is the least toxic of the three nitrogenous wastes, uric acid can be highly concentrated before excretion without harm to the organism. It does not require the great dilution that highly toxic wastes do, and therefore insects, reptiles, and birds do not need to take in large quantities of water. Uric acid is not highly soluble and it easily crystallizes as it becomes concentrated. Most of the excretory wastes of insects, reptiles, and birds is in the form of uric acid crystals, which are contained in the whitish material passed from the body along with the feces.

Some uric acid is produced by all people, but in excess it circulates over the body and may become sufficiently concentrated to form small needlelike crystals. These tend to accumulate in joints, usually of the lower extremities because of gravity, and cause great pain when the joint is moved. An afflicted person is said to have the **gout.** He must be careful of his intake of protein foods, seeing that there is not a great excess of amino acids to yield the uric acid. The joints of the big toe seem to be a common settling place for the crystals, and a person with the gout may often be seen with the feet propped high in the air in the hope that the crystals will become dislodged and dissolve in the blood so that the uric acid can be removed by the kidneys.

An average sample of human urine contains about 50 g of solids in solution per liter. About half is urea, but there are a few other organic wastes. **Creatinine,** a byproduct of muscle metabolism, is the next most prevalent, but it represents only about 1.5 g. **Uric acid** averages about 0.5 g, and there are smaller quantities of other organic compounds such as **urochrome,** which gives the amber color to urine. Urochrome comes from the breakdown of hemoglobin.

Removal of Excess Inorganic Salts

Sodium chloride (table salt) and potassium chloride are the major inorganic salts in the human diet and are also the major salts excreted in the urine. A certain quantity of the ions of these salts is necessary for normal body physiology, but these salts are usually ingested in such quantities as to result in an excess of the ions in the blood. Progressively smaller quantities of the salts of sulfate, phosphate, ammonia, calcium, and magnesium are also present. These inorganic salts are excreted from the sweat in abundance, but elimination is not selective, so the body can be depleted of needed salts through excessive perspiration. Persons working under conditions where they perspire heavily may be advised to take salt tablets to replenish the lost salt and thus avoid the muscle cramps and heat stroke which can result from salt depletion.

Regulation of Blood pH

Urine is usually acid, but the pH may vary considerably and even be on the basic side, depending on the diet. Some foods leave an excess of hydrogen ions, while others leave an excess of hydroxide ions upon their metabolism in the body. Since the excess is usually greater for the hydrogen ions, urine is usually acid. The buffering action of blood proteins tends to prevent fluctuations in the pH of blood, but the buffers would become saturated were it not for removal of the ions at the kidneys.

Regulation of Water Balance of Blood

The osmotic pressure of blood and the total blood volume are maintained to a large extent by the selective removal of water by the kidneys. On one day a person may drink great quantities of water and on the next day very little, yet his blood volume and the concentration of dissolved substances will remain about the same on both days. On the first day urine will be excreted in very large quantities and will be very dilute, while it will be small in quantity and very concentrated on the second day. There is a limit to the possible concentration of urine, however, and a certain minimum quantity of water is required. A very slight increase in the salt content of the blood stimulates the thirst center of the brain and creates the desire to drink. If you eat salty foods, you become thirsty. Extra water is needed to carry out the excess salt. Those who enjoy drinking beer often eat salty foods along with the beer to increase their thirst for more beer.

A person will die of thirst if he drinks much sea water. One pint of sea water has so much salt that a pint and a half of water is needed to carry it off from the kidneys. Hence, drinking sea water depletes the water already in the blood and increases rather than decreases thirst. Fish and other marine animals do not have this problem because their blood is isotonic to sea water. Some mammals living in the desert have excretory systems which can achieve a higher concentration of urine, and therefore they need less water for excretion. The kangaroo rat which lives in the southwestern desert regions of the United States has a kidney which can concentrate its urine four times as much as is possible for the human kidney.

Removal of Other Excess Dissolved Substances

Urine frequently contains other substances which are vital to the body, including vitamins, hormones, amino acids, and glucose. Some people have the idea that if a certain vitamin is good for you, then a great amount of it is even better. The body can use only a certain quantity, however, and in the case of many vitamins, any excess is excreted in the urine. An analysis of the urine reveals whether a person is getting as much thiamin as he can use, for instance. If thiamin is present in the urine, the person is getting as much as he can use, and the addition of more will merely increase the amount removed by the kidneys. This is not true of the fat-soluble vitamins, A and D. They do not go into true solution and so are not removed by the kidneys when present in excess. They can be stored in fats within the liver, but can bring about harmful reactions when the intake is far in excess of needs.

Normally, almost all glucose in the glomerulan filtrate is reabsorbed, so the urine is entirely free of this sugar. A child (or adult, for that matter) who gorges himself on sweets on a single occasion will have glucose in the urine temporarily. Persons with sugar diabetes are deficient in a hormone (insulin) which stimulates the absorption of glucose by certain body tissues. As a result, the concentration of the glucose becomes abnormally high in the blood and some is excreted in the urine (see Chapter 33).

The kidneys also help maintain the proper level of hormones in the blood by removing those present in excess. The pregnancy test in which a woman's urine is injected into a frog to detect the presence of a hormone is based on this principle.

REVIEW QUESTIONS AND PROBLEMS

1. Both food and air pass through the pharynx. How are they channeled into the proper tubes for transportation to the stomach and lungs respectively?
2. Explain the homeostatic reactions which cause you to breathe more heavily after vigorous exercise.
3. A person wishes to stay under water and breathe through a long tube. The tube contains 1,000 ml of air. What difficulties would he have and why?
4. Deep sea divers are likely to get the bends if they return to the surface too quickly. Explain why.
5. Sometimes a person's ribs fuse so that he cannot expand his chest, but he still continues to breathe. Explain how.
6. When a person has a reduced vital capacity, it may mean that he has emphysema. Explain.
7. Under what circumstances do the chemoreceptors of the aortic arch function to increase the rate and depth of respiration?
8. Why is carbon monoxide a dangerous poisonous gas?
9. What role do the red blood cells play in the transporation of carbon dioxide?
10. Describe three ways in which your basal metabolic rate can be determined.
11. What happens at the glomerulus in the Bowman's capsule?
12. Describe what happens in the descending and the ascending tubules of the Henle's loop.
13. When a person has serious albuminuria, he may be given plasma transfusions. How would this benefit him?
14. When the kidneys do not remove wastes properly, the blood pressure may go up. Explain.
15. Uric acid is a major excretory waste of some animals living in a very dry environment. Why is this advantageous?
16. The salt concentration of sea water is so great that a person will die if he drinks only sea water, yet some animals live in the sea. How is this possible?
17. Vitamins are essential to human physiology, yet some of the B vitamins may be found in the urine at times. Why do the kidneys remove such substances which are vital to existence?
18. Kidney stones can cause pain and suffering. How do stones get in the kidneys?

FURTHER READING

Gordon, M. S. 1972. *Animal Physiology. 2nd ed.* New York: Macmillan.

Pitts, R. F. 1968. *Physiology of the Kidney and Body Fluids.* Chicago: Yearbook Medical Publishers.

Potts, W. T., and Parry, G. 1963. *Osmotic and Ionic Regulation in Animals.* New York: Pergamon Press.

Wood, D. W. 1970. *Principles of Animal Physiology.* New York: American Elsevier.

33

Regulation Through Hormones

The activities of the human body are remarkably complex. All these activities must be regulated so that they take place at the proper rate in spite of the great variations in conditions under which the body is living. Cellular respiration must not vary above or below certain prescribed limits, glucose absorption and concentration in the blood must be maintained at the right level, the balance of each of many ions in the blood must be maintained, and sexual development must be held in abeyance during the early periods of growth and then allowed to come into full expression with adulthood. When we consider all the many adjustments that must be made, we may wonder how any person can be normal. A breakdown of any of the homeostatic mechanisms involved will result in serious deviations from normal.

Many of these conducted regulatory activities are by **hormones**—secretions which originate in a specific part of the body and are carried by the blood to all parts of the body. Hormones may influence body functions far removed from their source. Many of the hormones are secreted by special glands, **endocrine glands,** which are also called ductless glands because they have no ducts to carry their products.

THYROID HORMONES

The **thyroid gland** is located in the neck, one lobe on each side of the trachea, connected across the front by a narrow isthmus of thyroid tissue. Surgeons had seen the gland for centuries without knowing its function, although various theories were proposed. One held that it served to cool the blood flowing to the head, another held that it produced a lubricating fluid for the vocal cords, and still another held that it was necessary merely to fill out the neck region. Its true function became apparent in 1883 when a Swiss surgeon, Emil Kocher, described the symptoms which developed in sixteen of his patients after removal of the thyroid glands because of tumors on them or diseases which were causing a breakdown of the tissue. Soon after the removal of the glands, his patients showed great loss of energy, the skin became swollen and puffy, and they became mentally dull. These symptoms are similar to those found in **cretinism,** a condition in which a person is physically dwarfed, mentally retarded, and immature sexually. Autopsies performed upon cretins after death showed underdeveloped thyroids.

In 1891 the English physician G. R. Murray found that persons suffering from thyroid hormone deficiency would greatly improve if fed the ground up thyroid glands taken fresh from other persons who had recently died. Soon it was found that the thyroid glands of other animals were just as effective as those from human beings, and by 1900 many persons were eating bits of thyroid tissue from sheep as a treatment of thyroid hormone deficiency. A baby showing the early symptoms of cretinism would develop into a normal adult if the thyroid treatment was given. Today dried and powdered thyroid tissue from slaughterhouse animals is formed into pills of carefully measured strength and administered to many people who do not have thyroid secretions in sufficient quantities to permit them to maintain maximum body efficiency.

Chemical analysis of the secretions of the thyroid gland shows that it produces several closely related hormones, all of which function to regulate the rate of cell respiration. **Thy-**

roxine and **triiodothyronine** are the best known of these hormones. Persons who have an oversecretion, **hyperthyroidism,** have a high rate of cell respiration, an increased rate of heart beat, and a high heat output. They are also likely to be highly nervous because of the increased sensitivity of the nerve receptors. They show a typical bulging of the eyes because the eyelids remain open very widely. Persons with an undersecretion, **hypothyroidism,** are just the opposite; they have a slow heart beat and low blood pressure, lack energy, and tend to be somewhat cold all the time. Some persons with hypothyroidism are greatly overweight because most of

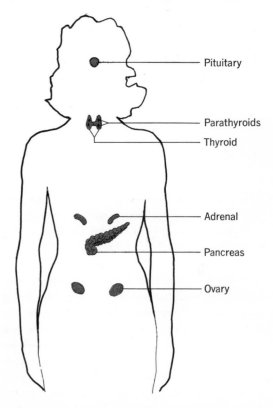

FIG. 33.1 The location of the major endocrine glands. These are shown for a woman; a man is the same except that there are testes instead of ovaries. The pineal gland, lying near the center of the head, is also sometimes included.

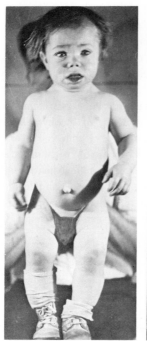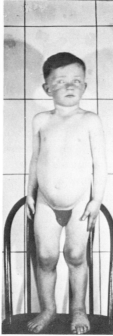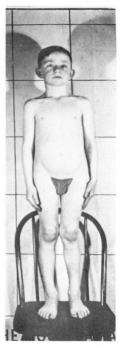

FIG. 33.2 Cretinism, which is caused by defective secretions of the thyroid gland, is shown at the left. The child has inherited this recessive trait. He develops normally when he is given thyroxine regularly. He is shown at the ages of four and nine years. (Good Samaritan Clinic, Atlanta.)

their food intake is stored as fat rather than consumed for energy production within the cells.

Iodine occupies a central position in the molecules of the thyroid hormones. When the diet is deficient in iodine, a swelling of the thyroid gland, a condition known as **goiter,** may occur. This condition was once widespread among people living in regions of the United States far from the seacoasts. Iodine is abundant in sea water, and all seafoods are, as a result, rich in iodine. Land near the oceans is also usually rich in iodine, which is incorporated into food crops grown on it. Hence, people living near the seacoasts get plenty of iodine. Today goiter is very rare in the United States because of the availability of seafoods in all sections and because iodine salts are added to much commercial table salt.

We learned in Chapter 32 that a **basal metabolism test,** by measuring the amount of oxygen consumed, can determine rate of cell respiration, which will show whether the thyroid hormones are being secreted in proper quantities. The **protein-bound iodine test** is also coming into more frequent use because it directly measures the thyroid hormone concentration of the blood. The thyroid hormones are produced as amino acid derivatives (from tyrosine), but before being released they are bound to a protein, **thyroglobulin,** which serves as a carrier for the hormones. No other blood protein is associated with iodine, so the amount of protein-bound iodine is a reflection of the amount of thyroid hormones in the blood.

In the frog the thyroid hormones, also control metamorphosis. If the thyroid gland is removed from a frog tadpole, the tadpole will never undergo metamorphosis but will grow into a giant-sized tadpole. If extra thyroid hormones are administered to very young tadpoles, they will undergo metamorpohis and form tiny frogs long before the time for normal metamorphosis. Sometimes the simple addition of a small amount of iodine compounds to the water in which the tadpoles live will bring about early metamorphosis, indicating that they are not getting sufficient iodine for maximum secretion of their thyroid glands.

PARATHYROID HORMONES

For many years it was thought that the thyroid glands produced hormones which controlled muscle contractions in addition to regulating the rate of cell respiration. This concept prevailed because when the thyroid glands were removed from experimental animals, within a day or two there developed involuntary muscular twitchings, followed by muscular convulsions and death. Similar symptoms developed in human patients when a part of a diseased thyroid gland was removed. A careful study of the anatomy of the gland, however, revealed the presence of four pea-sized bodies embedded in the thyroid gland. If care was taken to leave these intact, the thyroid gland could be removed without the muscle abnormalities developing. These four bodies were named the **parathyroid glands.**

The secretion of the parathyroids, **parathormone,** seems to be a polypeptide chain composed of about seventy-six amino acids. It is known to regulate the absorption and excretion of calcium and phosphate ions. The hormone enhances the absorption of calcium ions at the intestine and decreases the excretion of these ions at the kidney. Hence, the calcium ion concentration in the blood increases. At the same time the excretion of phosphate ions is increased by the hormone, so the concentration of phosphate ions in the blood is decreased.

An injection of parathormone into an experimental animal will result in a rise in the concentration of calcium ions in the blood but at the expense of this element in the bones and teeth. If a person has a tumor of these glands which produces an overgrowth and oversecretion of the hormone, the bones and teeth will soften because of a reduced calcium content. A chronic deficiency has just the opposite effect. Extra calcium is deposited in the bones and they become thick and brittle. Calcium ions also seem to be necessary to maintain the normal permeability of the axon membranes to sodium and potassium ions. The ability of the axon to transmit impulses depends upon its permeability to sodium and potassium ions. A lowering of the calcium ions in the blood increases the permeability of these membranes and increases nerve sensitivity. The body soon loses control over the stimuli going to the muscles, and this causes muscular twitchings and convulsions which are characteristic of parathormone deficiency. An experimental animal may be in extreme convulsions because of the removal of the parathyroids, but an injection of calcium salts into the blood stream relieves the convulsions almost immediately. Within a few minutes the animal will be up and behaving in a normal manner.

The blood level of the calcium ions acts as a homeostatic control for the release of parathormone. A low level stimulates the hormone release, and a high level inhibits the release. It is only when the homeostatic mechanism breaks down that abnormal concentrations of calcium ions are present in the blood. Deficiencies of calcium in the diet can bring on symptoms similar to parathormone hyposecretion for obvious reasons.

PANCREATIC HORMONES

The **pancreas** is a gland which produces digestive enzymes which are emptied into the duodenum of the small intestine through a duct, but it is also an important source of hormones picked up by the blood circulating through this organ. Histological studies of the pancreas reveal two distinct types of tissue. Many small secretory cells line the ducts; these secrete the pancreatic juice. In addition, there are small islands of cells having no connection with the ducts. These are the **isles of Langerhans,** which produce the pancreatic hormones. Two kinds of cells are found in the isles of Langerhans; each kind produces a different hormone. One kind, the **alpha cells,** produce glucagon, and the other kind, **beta cells,** produce insulin. Since insulin is the better known, we shall discuss that first.

Insulin is an important hormone in the homeostatic regulation of the glucose level of the blood. It is a protein made of two polypep-itde chains of twenty-one and thirty amino acids respectively. Being a protein, it is quickly broken down by digestive enzymes, and so it must be injected rather than given by mouth to those suffering from a deficiency. Insulin functions in two ways. First, it promotes the absorption of glucose by the cells, possibly by making the plasma membranes more permeable to glucose. At least we know that when insulin is deficient, the glucose level of the blood will rise, but it will actually be deficient within the cells. Second, insulin promotes the absorption of glucose and its conversion into glycogen in the liver. Both of these effects reduce the glucose level of the blood.

A hyposecretion of insulin results in the disease of **diabetes mellitus,** sugar diabetes. The glucose level in the blood rises, and the urine begins to react positively to a sugar test as the kidneys remove some of the excess glucose from the blood. If the insulin deficiency is great, the concentration of glucose in the blood will reach such a high level that the kidneys cannot remove the excess. In this case the afflicted person may go into a **diabetic coma** and may die unless insulin is injected promptly.

Diabetics can lead a normal life if they take regular injections of insulin, but the amount injected must be carefully measured in terms of the expected carbohydrate intake. Some persons with only a mild insulin deficiency can get along without insulin injections if they greatly restrict the carbohydrates in their diet. Many special foods are available for such persons.

A person taking insulin must be careful not to inject too much. An overdose can lower the glucose level of the blood so that a person can go into **insulin shock** and die unless glucose is promptly available. Diabetics must also learn to control their emotions, because strong emotions cause extra glucose to be released from the liver and this can cause diabetic coma even though insulin has been injected in a quantity which would be normal except for the emotional stimulation. Some diabetics carry candy in one pocket and insulin in another so they are prepared for fluctuations in either direction.

Diabetics must also be careful about their fat intake. The shortage of glucose within the cells causes a shift from carbohydrate to fat breakdown for ATP formation. The oxidation of fatty acids is not complete, however, and this causes an accumulation of ketone bodies. One of these ketones is **acetone,** which can be detected on the breath and in the urine of a person with diabetes if he has not taken insulin. These ketone bodies cause a shift in the ion balance of the blood, which produces the coma.

Glucagon, the hormone produced by the alpha cells, is a very small protein molecule made of a single polypeptide chain of only twenty-nine amino acids. It has an effect just the opposite of insulin. It brings about an increase of blood glucose by stimulating the conversion of glycogen to glucose in the liver. The glucose then escapes into the blood stream. Other hormones, such as adrenalin, also serve to stimulate glycogen breakdown in the liver, so removal of the pancreas results in the symptoms of insulin deficiency only.

The maintenance of the proper concentration of glucose is much more complicated than our review of the pancreatic hormones alone would indicate. A number of other hormones, yet

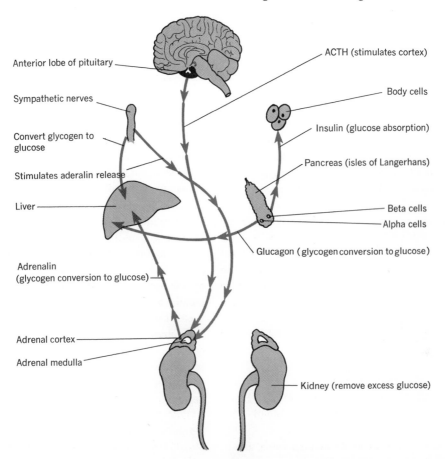

Anterior lobe of pituitary

ACTH (stimulates cortex)

Sympathetic nerves

Body cells

Convert glycogen to glucose

Insulin (glucose absorption)

Stimulates aderalin release

Pancreas (isles of Langerhans)

Liver

Beta cells

Alpha cells

Glucagon (glycogen conversion to glucose)

Adrenalin (glycogen conversion to glucose)

Adrenal cortex

Adrenal medulla

Kidney (remove excess glucose)

FIG. 33.3 A number of hormones are involved in the homeostatic regulation of the glucose level of the blood.

to be studied, are involved. All the hormones together make up one of the most elaborate homeostatic mechanisms of the entire body. Figure 33.3 shows the interrelationship of these various hormones.

THE ADRENAL HORMONES

Every person has two **adrenal glands,** one adhering closely to the anterior surface of each kidney. Like the kidney, each adrenal gland consists of two portions—an outer **cortex** and an inner **medulla.** The hormones of the medulla have been longest known, so we shall consider them first.

There are two types of cells in the adrenal medulla, each secreting a separate hormone. The two hormones are best known as **adrenaline** and **noradrenaline.** They, like the thyroid hormones, are derived from the amino acid tyrosine. In man 70 to 90 percent of the hormone secretion from the medulla is adrenaline. The two hormones have related yet somewhat different functions. The medulla of the adrenal glands is the only endocrine gland with a nerve supply and under control of nerve impulses. There is also some evidence that the pituitary gland may play a role in the control of the secretion of the medullary hormones.

Injections of adrenaline into the human body has profound effects. The capillaries in the skin and viscera contract, but they become dilated in the muscles. The heart beats faster and stronger, which elevates blood pressure. Blood clotting time is reduced because of the contraction of capillaries in the skin. The glucose level of the blood rises as the liver is stimulated to convert glycogen into glucose and release it into the blood. All of this tends to make a person more alert and ready for action. Thus, adrenaline is an emergency hormone. Small quantities are continually released, but in times of emotional stress, a message from the pituitary stimulates extra release. Continued emotional stress, however, will tend to deplete the liver of its reserve glycogen and to maintain the body at a state of high blood pressure and uneven distribution of blood. This can cause many disorders.

Adrenaline is of great medical value. It can be injected directly into the heart to restore the heart beat when the heart stops. It can relieve attacks of asthma by shrinking the capillaries of the bronchial tubes. It can relieve the stuffy nose of hay fever victims. It can stop bleeding of the gums after dentists' operations on the teeth. It can be injected into a person who is in insulin shock to stimulate glucose release from the liver.

Noradrenaline lacks some of the stimulating properties of adrenaline. It does not affect the heart beat or blood pressure, and it causes constriction rather than dilation of certain internal blood vessels. Noradrenaline is also released by the end brushes of the neurons of the sympathetic nerves. The adjacent dendrites are stimulated by noradrenaline. It is easy to understand how emotional stress which affects the sympathetic nerves

has an effect similar to the effect of medullary hormones.

The **cortex** of the adrenal glands produces a large number of hormones, but these can be organized into three primary groups.

1. **Sex Hormones.** Male and female sex hormones are produced by the adrenal cortex. The effects of these hormones on the body is the same as those produced by the gonads. Both male and female hormones are produced in both sexes. Occasionally there will be a tumorlike overgrowth of the cortex which causes an excessive production of sex hormones. If this occurs in young children, it can cause a premature sexual maturity. Some years ago a five-year-old girl in Peru gave birth to a son. She had matured at this age because of overactivity of the adrenal gland. Similar precocious sexual maturity has been observed in boys. If the overgrowth occurs in a mature woman, the androgenic hormones cause masculinization, or the appearance of male secondary sex characters such as beard growth, deepening of the voice, and the development of heavier musculature.

2. **Mineralocorticoids.** These hormones regulate the mineral ion concentration in the blood. One, for instance, controls sodium excretion. When the hormone is deficient there will be excessive excretion of this mineral, and an ion imbalance in the body results. High intake of table salt can help restore the normal balance by adding the sodium.

3. **Glucocorticoids.** These hormones play an important role in carbohydrate metabolism. They promote the conversion of glucose to glycogen in the liver and slow down the utilization of glucose by the cells. They also promote the conversion of body proteins to sugar which occurs in starvation. They reduce inflammation and irritation caused by allergies and certain types of arthritis. **Cortisone** is one of the best known glucocorticoids because it has been helpful in relieving the pain associated with certain types of arthritis.

The hormones from the cortex are known as **steroid hormones.** They are derived from cholesterol, the fatty substance which often causes circulatory disorders when it accumulates on the lining of the blood vessels. The steroid hormones always have four carbon rings and vary only in the attachments to these rings.

Hypofunction of the adrenal cortex leads to a combination of symptoms known as **Addison's disease.** This disease is characterized by salt wastage and a drop in the volume of extracellular fluid; the kidneys do not function properly and excretory wastes accumulate in the body. There is a reduced glucose content of the blood. Heavy deposits of melanin occur in the skin. The symptoms are complex because of the many functions of the glucocorticoids. Addison's disease was usually fatal in the past, but it can now be controlled through injections of cortex hormones and a high salt intake.

GONADAL HORMONES

The gonads of both sexes are dual purpose organs. They produce gametes, or sex cells, and they also produce hormones which influence sexual function and body growth and development. These hormones, like those from the adrenal gland cortex, are steroid hormones.

Hormones from the Testes

A microscopic study of a human testis shows that the testis contains many small tubules the **seminiferous tubules,** which produce and release sperm. Located between the seminiferous tubules are the **interstitial cells** which produce male sex hormones. These hormones are known as **androgens,** the most common of which is **testosterone.** We often use the latter term to refer to all male sex hormones. These hormones stimulate the development of the male

secondary sexual characteristics. If a boy is castrated before puberty, his voice will remain high pitched, his beard will not grow, his muscles will not develop into adult form, and he will not grow pubic or axillary hair. He will not be able to maintain penile erection or to ejaculate in response to sexual stimulation. A boy with a low output of the androgenic hormones will not show full development of the male secondary sexual characteristics described above. He may find it hard to adjust to a normal social life because of his appearance is different from that of his peers.

Androgen production begins in the early embryo and is responsible for the differentiation of the male sex organs. When the embryo is seven weeks of age, the external genitalia have not differentiated and they appear the same in both sexes. If androgens are produced at this time, which usually happens in male embryos, the genital tubercle will enlarge to form a penis and the urogenital opening will close. In female embryos, this opening persists and becomes the opening to the vagina. The early production of testosterone is also thought to condition certain target cells in the hypothalamus so that they will be sensitive to androgen stimulation in the future. If a pregnant woman takes certain drugs or treatments which depress androgen production by the male embryo, the development of the male sex organs may be suppressed and the androgen sensitivity of certain cells in the hypothalamus, which is necessary for normal male sexual response, may also fail to develop. We shall learn more about the role of the hypothalamus in sexual response later in this chapter.

During childhood the level of testosterone output by the testes is low and a boy is not greatly different from a girl in body form and musculature. During puberty, boys experience a sudden increase in androgen output and many changes associated with sexual maturity occur. The sex organs enlarge and become functional. Marked changes in the skeletal musculature occur. The proportion of protein to fat increases in muscle tissue. This makes the muscles heavier, because protein is denser than fat. Also, since protein forms the muscle fibers which are the contractile elements of muscle, the strength of the muscles increases in proportion to their size. Girls also undergo a similar transformation of muscle tissue at puberty because of the effects of the female sex hormones, but the changes are not as pronounced. Because of the difference in specific gravity caused by the different protein content of male and female muscles, women float more easily in water. The increased androgen output during puberty also causes different patterns of hair growth. Pubic hair develops in the groin; hair appears in the armpits and on the chest; the hairline on the forehead changes.

Testosterone also stimulates bone growth. An adolescent boy seems to shoot up almost overnight because, under androgen stimulation, the long bones of the arms and legs lengthen rapidly. At the same time, androgen stimulation causes the ossification of the growth area, or **epiphyses,** of these bones, and within several years the epiphyses close and growth stops. Eunuchs are often portrayed as being tall and heavy. Their height is the result of their early castration which delayed the closure of the epiphyses and prolonged the period of bone growth.

Anabolic Steroids

Chemists have been able to split the testosterone molecule into two fractions. One fraction is an **anabolic steroid,** which produces the anabolic, protein-synthesizing effects of testosterone. The other fraction is responsible for the stimulation of male sexual characteristics.

The anabolic steroids are used as a medical treatment to restore muscle strength in those who have been debilitated by disease or malnutrition. Anabolic steroids are also used by athletes who wish to develop their muscles beyond their natural capacities for growth. One source estimates that about 75 percent of professional football

players have used anabolic steroids at some time. Because of reports that some members of Olympic teams use anabolic steroids, tests are now being developed to detect their presence in the body so that their use may be effectively barred in the Olympics.

Anabolic steroids, like any steroid hormone, should never be used without close medical supervision and without good medical reason. Frivolous use of powerful hormones can cause lasting physiological damage. Some athletes have taken as much as twenty times the recommended dosage of anabolic steroids and have suffered liver damage. If boys who have not attained their full growth use these steroids, a premature closing of the epiphyses of the long bones can occur and this may stunt their growth. Excessive use of anabolic steroids will cause the pituitary to stop producing the gonadotrophins which stimulate the production of the sex hormones. When gonadotrophin production ceases, the testes regress and decrease their output of testosterone. This can adversely affect the male sexual characteristics which are maintained by testosterone.

Hormones from the Ovaries

Two types of hormones are produced by the ovaries. One type, the **estrogens,** is found primarily in two forms, **estradiol-17β** and **estrone.** All the estrogens are often called **estrogen** for convenience. The second type of hormones is the **progesterones.**

Before puberty the ovaries of a female child do not produce a great amount of the female sex hormones. As puberty approaches, the ovarian hormone output increases until, by maturity, it will have increased about twentyfold. The ovarian hormones cause the maturation of the female sex organs and they also stimulate the growth of the long bones. The sex hormone surge occurs about a year earlier in girls than in boys, so girls of junior high school age tend to be taller than their male classmates. The estrogens cause the closure of the epiphyses of the long bones, so girls stop growing earlier than boys too. Girls are about five inches shorter than boys at maturity, on the average. Estrogens stimulate increased growth of the shoulder and hip girdles and cause the laying down of fat deposits in these areas. Estrogens also promote an increase in the protein content of the skeletal muscles which leads to an increase in muscular strength, but this effect is not as pronounced as the effect of testosterone on the musculature.

At puberty the gonadotrophins produced by the pituitary stimulate the ova contained in the ovaries to undergo the final meiotic division and become mature. Each month one ovum matures in this way. The ovarian hormones also begin to prepare the uterus to accept and nurture a fertilized ovum. These changes cause the **menarche,** or the beginning of the menstrual cycle.

The concentration of estrogens in the blood of a mature woman fluctuates in a regular way during each menstrual cycle. The estrogens are produced in greatest quantity in the follicle which surrounds the maturing ovum before ovulation. The estrogens cause the walls of the uterus to thicken and develop new blood supplies in preparation for the reception of a fertilized ovum. At ovulation the ovum is released from the follicle. The follicle then becomes a **corpus luteum** which secretes progesterone. This hormone brings about the final changes which prepare the uterus for the implantation of a fertilized ovum. When a fertilized ovum implants itself in the uterine wall, the corpus luteum continues to secrete progesterone. If the mature ovum is not fertilized and does not implant in the uterine wall, the corpus luteum disintegrates and ceases to secrete progesterone. When this happens, the inner layer, the **endometrium,** of the uterine wall degenerates and the dead tissue and the extra blood supply is passed from the body during **menstruation.**

Progesterone may be injected into a woman who is threatened with abortion since it maintains the uterus in a condition favorable to the implantation and growth of the embryo. It may also help to relieve the nausea of early pregnancy. **Diethylstilbesterol (DES),** a synthetic hormone, was found to have properties similar to those of progesterone. DES was used as a substitute for progesterone in cases of threatened abortion in the 1940's and 1950's. It is now found that the daughters of women who were given DES during pregnancy have an increased susceptibility to vaginal cancer in childhood and early adulthood. DES was also used extensively to fatten cattle for market since it has a potent protein-anabolic effect. Cattle grew faster and had more tender meat when they were given this synthetic hormone. After its carcinogenic properties were discovered, DES was removed from the market, but it is now being used again in small quantities in the **"morning-after" pill.**

Birth control pills contain a mixture of synthetic progesterones and estrogens. This hormonal combination simulates the hormonal conditions of pregnancy in nonpregnant women. This condition inhibits ovulation and prevents conception.

Estrogen is taken by many women during and after menopause when the ovaries stop producing female sex hormones. Estrogen alleviates some of the unpleasant effects of menopause such as "hot flashes" and depression. Estrogen also prevents some of the degenerative changes, such as increased brittleness of the bones, wrinkling of the skin, and increased susceptibility to heart disease, which often occur after menopause.

PITUITARY HORMONES

The pituitary gland is attached to the lower surface of the brain. It extends down from the hypothalamus—a part of the brain which controls the release of many pituitary hormones. Embryologically, the pituitary gland has a dual origin; the anterior lobe arises as a pouch which pushes up from the roof of the primitive mouth, and part of it, the posterior lobe, arises as a

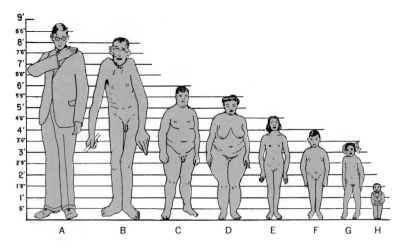

FIG. 33.4 Growth abnormalities caused by endocrine disturbances of the pituitary gland. Gigantism due to excessive somatotropic hormone in youth (A). Gigantism and acromegaly due to excessive somatotropic hormone in youth and in later life (B). Male and female cases of adiposogenital dystrophy, retarded sexual development and excessive fat deposits caused by gonadotropic and thyrotropic hormone deficiency (C and D). Female and male cases of infantilism from gonadotropic and somatotropic deficiency; the girl is twenty-one and the boy is twenty-seven years of age (E and F). Cretin with dwarfed body and low intellect due to thyrotropic deficiency and resulting hypothyroidism (G). Midget, resulting from extreme somatotropic deficiency (H). (Redrawn from Patten, *Human Embryology,* Blakiston, 1958.)

downward growth of the brain itself. In the adult the lobes are united into a small gland weighing only about one-half gram and attached to the base of the brain by a narrow stalk extending up from the posterior lobe. This small size, however, belies the tremendous importance of this gland. It is sometimes called the master gland of the body because so many of the other glands depend upon its secretions for their homeostatic regulation. Since the lobes have a different embryonic origin, it is to be expected that they have different histological structure and that they produce different hormones. This is the case, and because they are so different, we will study them separately.

The Posterior Lobe

Two hormones are released from the posterior lobe of the pituitary gland. We say "released from" rather than "secreted by" because it has been shown that the hormones actually originate within the hypothalamus of the brain and migrate through the nerve fibers to the posterior lobe. This would lead us to suspect that the secretion and release of these hormones is closely related to the nervous system, especially the autonomic nerves; and this is true. Both hormones are formed of eight amino acid residues with two amino acid differences between them.

One of the hormones, **oxytocin,** stimulates the contraction of the smooth muscles of the uterus and to a lesser extent the smooth muscles of other parts of the body, such as the intestines and the urinary bladder. The desire for urination and perhaps defecation under emotional stress may be due to the release of extra quantities of this hormone into the blood stream. This would also help explain the frequency of premature births and miscarriages among pregnant women placed under conditions of great emotional crisis. The hormone is frequently administered by injection to initiate labor at a time convenient for a woman and her physician. The hormone may also be administered during birth when the contractions of the uterine muscles are sluggish, and after birth to stimulate the expulsion of the placenta and to help bring the uterus down to its normal size.

The ejection of milk during nursing also appears to be controlled by oxytocin. The mechanical stimulation of the sucking baby sends nerve impulses to the hypothalamus, which in turn stimulates oxytocin release, and this stimulates the smooth muscle contraction of the milk ducts and causes the milk ejection. This should not be confused with milk secretion, which is stimulated by another hormone from the anterior lobe of the pituitary.

Vasopressin is the second hormone from the posterior lobe of the pituitary. It is sometimes called the **antidiuretic hormone** because it increases the reabsorption of water in the kidney tubules and thereby prevents excessive loss of water in the urine. A very slight increase in osmotic concentration of the blood seems to stimulate the release of this hormone and thus to increase water reabsorption in the kidneys. When there is a deficiency, large quantities of a very dilute urine are excreted. This causes a great thirst difficult to quench. The condition is known as **diabetes insipidus** (no relation to diabetes mellitus). Injection of vasopressin into persons suffering from this condition brings relief.

Vasopressin also stimulates constriction of the walls of the arterioles and capillaries, bringing about a rise in blood pressure. Hence, it should be used with caution by persons suffering from high blood pressure or coronary thrombosis.

The Intermediate Lobe

The anterior portion of the posterior lobe of the pituitary is distinct histologically from the rest and is sometimes classed as the intermediate lobe. It secretes two hormones, known as the **melanocyte-stimulating hormones (MSH),** polypeptides made from fifteen to eighteen amino acids. These hormones cause darkening of the skin in fishes, amphibians, and reptiles. A frog injected with an extract of the intermediate lobe will become distinctly darker within a few minutes as the melanin-containing melanocytes in the epidermal cells of the skin spread out wide. If this lobe is removed, the skin becomes very light in color as the melanocytes contract into small spots around the nuclei of the cells. The release of these hormones is regulated by nerves which in turn respond to the environment. In a dark environment the hormones are released in great amounts, and the animal changes and blends with the dark environment. In a light environment the stimulation is withheld, and the animal develops a light skin. It is suspected, but not known, that the pigmentation of the human skin is related to these hormones, although other hormones are known to have an effect as well.

The Anterior Lobe

The pituitary is sometimes called the master gland of the body because of its regulatory effects on many other hormone secretions. This is particularly the role of the hormones of the anterior lobe. At least eight separate hormones are produced, and these all play important roles in the homeostatic regulation of body functions, usually in conjunction with other hormones, through **negative feedback** loops. Such a relationship can be illustrated with the **thyrotropic hormone** produced by the anterior lobe of the pituitary. This hormone, a glycoprotein, stimulates the secretion of hormones by the thyroid gland, but the thyroid hormones inhibit the secretion of the thyrotropic hormone from the pituitary. Thus, a balance is maintained; when the concentration of thyroid hormones in the blood drops, thyrotropic hormone is released. This hormone stimulates the thyroid gland to release more thyroid hormones. But as the level of thyroid hormones rises, the thyrotropic hormone release is shut off. This is negative feedback. Injection

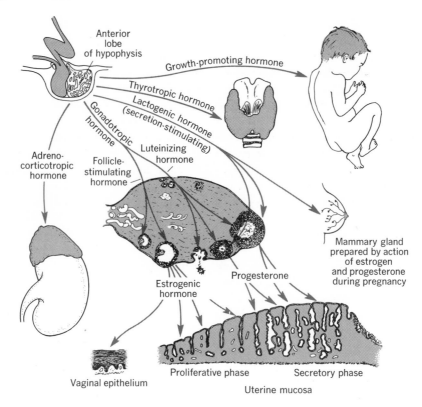

Anterior lobe of hypophysis

Growth-promoting hormone

Thyrotropic hormone

Lactogenic hormone (secretion-stimulating)

Gonadotropic hormone

Luteinizing hormone

Follicle-stimulating hormone

Adreno-corticotropic hormone

Estrogenic hormone

Progesterone

Mammary gland prepared by action of estrogen and progesterone during pregnancy

Vaginal epithelium

Proliferative phase

Secretory phase

Uterine mucosa

FIG. 33.5 Diagram to show the effects of some hormones from the anterior lobe of the pituitary gland on the body of a woman.

of thyrotropic hormone creates all the symptoms of hyperthyroidism in an experimental animal, and continued injection brings about a great increase in size of the thyroid gland.

The **somatropic (growth) hormone,** a protein, was one of the first to be discovered from the pituitary gland. When the gland was removed, there was a striking retardation of growth. Injections of extracts of the gland, on the other hand, produced accelerated growth. The skeletal and muscular tissues are most responsive to the somatotropic hormone. **Pituitary dwarfism** is well known to result from an underdeveloped gland in man, while the opposite, **gigantism,** results from overgrowth of the gland. When the gland devel-

ops a tumorous growth in an adult, there will be extra bone growth in certain places where growth potentials still exist. These areas are in the face, hands, and feet. Deformities in these areas develop, often accompanied by pain. This disease is called **acromegaly.** We have not yet discovered the homeostatic mechanism which regulates the secretion of the somatotropic hormone.

The **adrenocorticotropic hormone, ACTH,** is a protein which stimulates the secretions of the cortex of the adrenal glands, primarily the portion which produces the glucocorticoids. It is a negative feedback mechanism, as was illustrated with the thyroid hormones. ACTH is often given instead of cortex hormones to alleviate symptoms of arthritis, since ACTH stimulates a person's own glands to release more of the cortisone and related hormones.

The **lipotropic hormone** is a newly discovered secretion of the anterior lobe of the pituitary. It stimulates the liberation of stored body fats. Some cases of extreme obesity and extreme thinness may be due to hyposecretion and hypersecretion of this hormone. It is a protein hormone.

Three of the hormones produced by the anterior lobe are classified as **gonadotropic hormones**—that is, they stimulate the gonads. When fresh, whole pituitary glands taken from frogs or other vertebrates are transplanted into newly hatched male baby chicks, the chicks will show precocious sexual development just as if they had received injections of testosterone. Internal examination reveals testes which are several times normal size. The glands from one species do not actually grow within an animal of another species, but the glands contain the hormones, and hormones tend to have similar structures in all vertebrate animals. There are some small differences, but generally the hormone from one species will affect another species. This is true not only of pituitary hormones but of all hormones in the vertebrate animals.

One of the gonadotropic hormones is known as the **follicle-stimulating hormone, FSH.** It is a glycoprotein which stimulates the production of estrogens in the follicles of the ovaries in the female. In the male this hormone stimulates spermatogenesis in the germinal epithelial cells of the testes. FSH increases greatly in concentration in the blood at puberty.

The second gonadotropic hormone is the **luteinizing hormone, LH,** and is also a glycoprotein. It stimulates the release of the egg from the follicle of the ovary and the development of the corpus luteum at the site of ovulation. Progesterone is then released from the corpus luteum.

A   B

FIG. 33.6 Precocious development of male sexual characteristics shown by a five-week-old chick (A). This chick received an injection of testosterone when two weeks old. An uninjected control male is also shown (B).

In the male, LH stimulates the production and release of androgens from the interstitial cells of the testes. For this reason, LH is sometimes called the **interstitial-cell-stimulating hormone, ICSH,** when its function in the male is discussed. A negative feedback loop maintains the androgen level at the proper concentration in the blood. Nerves in the hypothalamus secrete releasing factors which enter the pituitary through the hypophyseal portal veins and stimulate the release of the gonadotropic hormones. One of these releasing factors is ICSH. It stimulates the interstitial cells of the testes to step up production of testosterone. As the level of testosterone in the blood increases, it reaches target cells in the hypothalamus and inhibits them from producing releasing

factors which affect ICSH to the pituitary. ICSH secretion is inhibited in the pituitary and this in turn inhibits the production of testosterone. This system is called a negative feedback loop.

If the testes are removed, the inhibiting action of the androgens is no longer present and the hypothalamus continues to produce the gonadotrophin-releasing factors and the pituitary secretes extra quantities of gonadotrophins. The gonadotrophins are then found in high concentrations in the blood and urine. Blood or urine taken from a castrated animal can be administered to other animals—even those of different species—and can stimulate the sexual characteristics of these animals. Extracts of blood and urine from castrated animals are sometimes given to

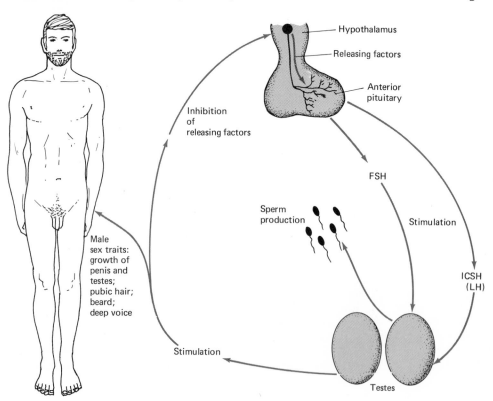

FIG. 33.7 The homeostatic relationship between the gonadotropic hormones from the pituitary and the output of sperm and androgens from the testes.

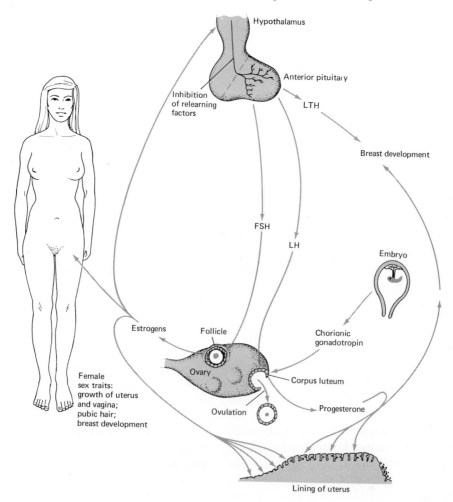

Hypothalamus

Anterior pituitary

LTH

Inhibition
of relearning
factors

Breast development

FSH

LH

Embryo

Estrogens

Follicle

Chorionic
gonadotropin

Ovary

Corpus luteum

Ovulation

Progesterone

Female
sex traits:
growth of uterus
and vagina;
pubic hair;
breast development

Lining of uterus

FIG. 33.8 Homeostatic relationships between the hormones of a sexually mature woman.

men whose testes do not produce sufficient androgens. This treatment works just as well for women who do not secrete enough estrogen, since the same gonadotrophins stimulate the production of ovarian hormones. The blood and urine of pregnant women are also rich in a gonadal-stimulating hormone, **chorionic gonadotrophin,** which is produced by the placenta. Pregnancy tests are based upon the detection of this hormone in a woman's urine.

The reproductive cycle of a woman involves both negative and positive feedback loops. FSH stimulates estrogen production in the ovarian follicles, and as blood estrogen increases, it inhibits the production of the gonadotrophin-releasing factor in the hypothalamus, which in turn inhibits the secretion of FSH by the pituitary. When the level of estrogen is high, however, it seems to stimulate the release of LH by the pituitary. LH stimulates ovulation and the development of the corpus luteum, which secretes progesterone. Progesterone then inhibits the production of the releasing factor which stimulates LH production.

The delicate balance of pituitary and ovarian hormones during the menstrual cycle is thought to prevent multiple ovulation in women. LH is sometimes called the "fertility hormone" because it is given to women who are infertile because of irregular or absent ovulation. Often in this treatment too much LH is administered and multiple ovulations and births occur.

When pregnancy occurs, chorionic gonadotrophin from the placenta stimulates the corpus luteum to continue to produce progesterone, which prevents further ovulation and maintains the uterus in a condition favorable for the development of the embryo.

A third pituitary hormone is involved in pregnancy. This is **leuteotropic hormone,** abbreviated **LTH,** which stimulates the corpus luteum to continue to secrete progesterone during the early months of pregnancy. LTH also stimulates the release of milk from the breasts after birth. When its milk-releasing function is discussed, it is sometimes called the **lactogenic hormone,** or **prolactin.** A woman's breasts are fully prepared for milk secretion long before she gives birth, but the placental hormones appear to inhibit LTH. When the placenta is expelled after birth, the inhibition is removed, secretion of LTH increases, and within a few days milk begins to flow.

THE PINEAL BODY

The human pineal body is a small rounded structure attached to the lower part of the brain near the pituitary gland. In a few reptiles it extends all the way to the outside of the skull and forms what is called the "third eye" because it is sensitive to light. The pineal body produces a hormone, **melatonin,** which causes color changes in the skin of some animals. Some researchers believe that the pineal body may play a role in regulating reproductive cycles in some vertebrates. We have much yet to learn about the functions of this gland, which for so long has not been understood.

THE PROSTAGLANDINS

In 1930 two New York gynecologists, Raphael Kurzrok and C. C. Lieb, found that human semen would stimulate contraction in strips of tissue taken from the uterus. The stimulating substance came from the secretions of the prostrate gland, so these substances were called **prostaglandins.** Later it was found that cells in all parts of the body produce similar substances. The prostaglandins produced by these cells have been found to vary slightly in chemical structure, but all are fatty acids. They are still called prostaglandins, however, Great interest has developed in prostaglandins recently because it is thought they may provide means of birth control, correction of male sterility, and abortion. All these uses are possibilities because the uterus is primarily composed of smooth muscle and prostaglandins stimulate smooth muscle contraction. One "once-a-month" contraceptive pill contains prostaglandins. This pill is taken at the time menstruation is expected and it stimulates the uterus to contract and shed its lining even though implantation of a fertilized ovum may have taken place. A woman would probably need to take this pill only about three or four times a year since unprotected intercourse will result in fertilization only about that often, on the average. The pill could be taken only when menstruation is delayed and it appears as if conception has taken place.

With the liberalization of abortion laws, research has been conducted to find a simple method of inducing abortion with a minimum of trauma to the woman. The use of prostaglandins is one such method. When prostaglandins are injected directly into the amniotic fluid, they spread out into the placenta and stimulate uterine contractions which expel the embryo. They can also be used to stimulate labor during childbirth when the contractions are sluggish. It should be mentioned that there is a difference of opinion on the advisability of using prostaglandins for birth control and abortion-inducing purposes.

It was found that about 60 percent of sterile men have small amounts of prostaglandins in their semen. It appears that prostaglandins stimulate the uterus to contract after intercourse and these contractions move the semen up through the cervix and into the uterus and Fallopian tubes. Some cases of male sterility have been corrected by mixing the semen of sterile men with prostaglandins from the semen of other men. The mixture was then used in artificial insemination.

The main drawback to the use of prostglandins has been their ability to affect other parts of the body. Prostaglandins alter blood pressure, reduce blood clotting time, cause nausea, and alter kidney function. Fourteen different kinds of prostaglandins have been identified, and it is hoped that it will be possible to separate those with undesirable effects from those which have desirable effects. They can now be synthesized, which makes experimentation simpler.

OTHER HORMONES

A number of other hormones are produced by the stomach and duodenum as part of the homeostatic regulation of the secretion of digestive enzymes. These hormones have already been discussed in the chapter on nutrition. The kidneys have also been found to produce **renin,** which raises blood pressure. In addition, the kidneys produce **erythropoietin,** which stimulates the release of red blood cells from the bone marrow. A deficiency of oxygen in the blood stimulates the release of erythropoietin. It has recently

been found that both the thyroid and parathyroid glands produce **calcitonin,** which depresses the concentration of calcium ions in the blood and stimulates their deposit in the bones. This effect is the opposite of that of the parathyroid hormone. The balance between the two hormones keeps calcium concentration of the blood at the equilibrium level.

METHOD OF HORMONE ACTION

Some of the hormones are large protein molecules, and we know that such molecules ordinarily do not pass readily into cells. How do these hormones exert their effect on the metabolism of cells? Why do hormones have powerful effects on some cells and no effects on other cells? There are several theories about the method of hormone action on cells.

One theory hypothesizes that target cells for a hormone have acceptor sites on their outer membrane which serve as attachment points for the hormone. The enzyme **adenine cyclase** has been found bound to the plasma membrane of cells. Once the hormone is attached, this enzyme is released. Adenine stimulates the production of **cyclic AMP, adenine monophosphate,** a monophosphate relative of ADP and ATP, the phosphates which are so important in energy pathways. Cyclic AMP is thought to trigger the metabolic changes in the cell. The hormone does not need to enter the cell; it needs only to attach itself to a suitable acceptor site on the membrane. Certain cells in the liver, for instance, have acceptor sites for the hormone adrenaline. When this hormone attaches itself to the plasma membrane, cyclic AMP is produced and stimulates the conversion of glycogen to glucose. Certain cells of the adrenal cortex have acceptor sites for ACTH, which is secreted by the pituitary. The presence of this hormone stimulates these cells to produce their steroids.

Several other theories are mentioned by

C. D. Turner and J. T. Bagnara in their book *General Endocrinology*. One theory proposes that hormones may alter the permeability of the plasma membrane or the membranes of intracellular organelles. In this way the hormones could influence the movement of materials into the cells or between intracellular structures. The varying permeability of the membrane could influence the rate of biochemical reactions occurring within the cell. Some studies have shown that insulin influences the transfer of glucose into the cells of certain tissues. Certain cells become more permeable to glucose in the presence of insulin. The growth hormone secreted by the pituitary gland seems to make certain cells more permeable to amino acids. Facilitating the entry of amino acids into a cell makes possible more rapid

synthesis of protein and more rapid growth. A third theory proposes that hormones activate or suppress particular genes. Certain hormones in insects are known to stimulate chromosomes to synthesize RNA, which then leads to increased production of specific proteins in the cell. When estrogen is placed on the uterus of castrated rats, there is a great increase in the output of RNA, protein synthesis, and other changes. The speed of the reaction indicates that estrogen may directly affect the genes.

All these possibilities may be correct, in part at least. Some hormones might work in one way and some in other ways. A combination of all the theories might account for hormonal effects.

REVIEW QUESTIONS AND PROBLEMS

1. How does a hormone differ from an enzyme in its effect on the body?
2. Goiter has always been more prevalent in regions away from seacoasts. Why is this true and why do we have fewer cases of goiter today than in the past?
3. A diet deficient in calcium and vitamin D produces the same sort of defects as a parathormone deficiency. Explain.
4. A man who has diabetes mellitus has a sudden fit of anger over a minor traffic accident. He goes into a coma and is soon unconscious. Explain the reason for the coma and what should be done to relieve it.
5. We say that glucagon is antagonistic to insulin. Explain what is meant by this statement.
6. Some people take adrenaline for the relief of allergies, but if they have high blood pressure, they are advised not to do this. Explain why.
7. Some hormones can be given in pill form by mouth, but others must be injected into the body. Name some hormones in each group and explain why they differ in this respect.
8. A boy who is only six years old develops a deep voice, pubic hair, and enlarged sex organs. What hormonal explanation might be given for this precocious sexual development?
9. At a certain age girls tend to be taller and larger than boys, but the boys then catch up and overtake the girls in size. Explain why.
10. The birth control pill is rich in progesterone. Why does this hormone prevent conception?

11. A person excretes large quantities of a vary dilute urine and is constantly thirsty. What hormone might be injected to relieve this condition and why?

12. An excess of the somatotropic hormone in the adult has an effect quite different from those observed in a child. Explain these effects and why they are different.

13. Through hormone injections a male cat has been stimulated to produce milk and nurse kittens. What hormones do you think were given and why?

14. Both cortisone and ACTH are effective in treating certain types of arthritis. Explain why in each case.

15. Some evidence indicates that certain strong drugs taken in early pregnancy may affect male embryos so that they never develop fully sexually. Explain why.

16. Injections of urine from a pregnant woman into a male frog will cause him to release sperm. This urine will also cause virgin female rabbits to develop corpora lutea in their ovaries. Explain these two effects.

17. Prostaglandins can be used both to correct sterility and for birth control. Explain these two functions.

18. Sexually undeveloped men are sometimes given an extract of the blood of a castrated horse. How can blood from a castrated animal stimulate sexual development?

19. How can hormones whose molecules are too large to pass into cells cause metabolic reactions in the cells?

FURTHER READING

Frye, B. E. 1967. *Hormonal Control in Vertebrates*. New York: Macmillan.

Green, R. 1970. *Human Hormones*. New York: McGraw-Hill.

Hamilton, T. H. 1971. *The Vertebrate Endocrine System and its Regulation*. New York: Harper and Row.

Turner, C. D., and Bagnara, J. T. 1971. *General Endocrinology*. 5th ed. Philadelphia: W. B. Saunders.

Whalen, R. E. 1967. *Hormones and Behavior*. New York: Van Nostrand Reinhold.

34

Reproduction and Development

All organisms must have some way of leaving descendants to continue their species because no organisms can live forever. Reproduction is therefore an essential life process. In this chapter we shall consider human reproduction and the method of the development of a new life from a fertilized egg.

THE MALE REPRODUCTIVE SYSTEM

The **testes** are the male gonads. They are suspended from the inguinal region of the body by means of a loose fold of skin, the **scrotum.** It may seem unusual that such delicate and vital organs are in a position where they may be subject to external injury. Sperm cannot develop in the high temperature that is found within the body cavity and this method of suspending the testes outside the body cavity allows the sperm to develop at a lower temperature. The temperature is regulated more exactly by a thermostatic action of the scrotum. If the weather is warm and the body is radiating considerable heat, the skin of the scrotum loosens so that the testes hang down away from the body. In cool weather the skin contracts and draws the testes in close to the body, preventing excessive lowering of the temperature. Before birth the testes lie within the body cavity in a position about the same as is occupied by the ovaries of the female. Shortly before birth or perhaps shortly afterward, the testes normally descend into the scrotum. Sometimes they do not descend, because of obstructions of the inguinal canal or a deficiency of the male hormone. This is **cryptorchidism,** which usually results in sterility unless the victim is treated to produce descent of the testes.

Microscopic study of the testes reveals two types of tissue. There are many highly coiled tubules, the **seminiferous tubules,** which generate sperm from the germinal epithelium lining the tubules. Between the seminiferous tubules are the **interstitial cells** that produce the male sex hormones. Sperm pass through the tiny tubules in the testes to the **epididymis.** This structure, seen as an enlargement that partially encircles the testis, is composed of a small, coiled tube about twenty feet long. This connects with the **vas deferens,** which passes into the abdominal cavity through the inguinal canal. The two vasa deferentia, one from each testis, join the **urethra** just below its connection with the urinary bladder. Shortly before they join the urethra, they enlarge to form the **ejaculatory ducts,** capable of contraction which expels their contents. There is a glandular **seminal vesicle** that joins each ejaculatory duct. These glands secrete a clear viscid fluid which passes into the ejaculatory duct and serves as a medium for the transportation of sperm. The mixture of this fluid with the sperm produces a milky spermatic fluid, **semen.** The seminal vesicle secretion contains some factor that activates the sperm. In the vas deferens the sperm are sluggish, but after mixture with this secretion they become quite active.

The urethra is surrounded by the **prostate gland** at the point where the ejaculatory ducts join it. The prostate gland secretes a clear viscid fluid of an alkaline nature. There is another pair of glands, the small **Cowper's glands,** attached to the urethra about two inches below the prostate gland. They also secrete a viscid fluid of an alkaline nature. The urethra serves a dual function in man; it carries urine from the urinary bladder and it also carries semen. However, urine is usually rather acid in nature and sperm cannot live

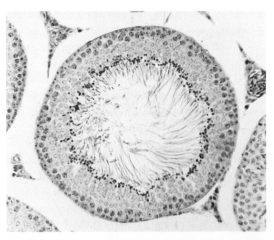

FIG. 34.1 Cross section through a seminiferous tubule from the testes of a man. The cells undergo meiosis as they move from the outside of the tubule toward the inside, and finally the fully formed sperm are pushed into the center of the tubule. The heads of the sperm appear as dark dots at the interior lining of the tubule and the tails extend inward.

in the presence of acid. The secretions of the prostate and Cowper's glands alkalize the urethra and lubricate it for the possibile passage of semen. The urethra continues through the **penis** to terminate at its tip. The penis consists of three col-

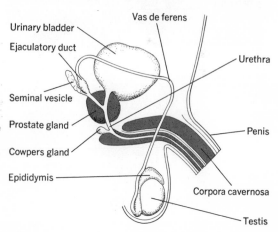

FIG. 34.2 Organs of the human male reproductive system.

umns of spongy tissues, the **corpora cavernosa,** surrounding the urethra, and a layer of skin on the outside. The tip of the penis enlarges slightly to form the **glans,** which is covered by a very thin, sensitive layer of skin. However, a thicker layer of skin normally folds over the glans to form the **prepuce.** Most of the prepuce is cut off in circumcision, an operation that is now routinely performed on nearly all male babies born in the United States.

The functioning of the reproductive system is largely governed by the autonomic nervous system and therefore is not under voluntary control. The sperm are produced constantly during the fertile period of a man's life. They make their way up the long tubes to the ejaculatory duct, where they are stored until mixed with the fluid from the seminal vesicles and released through the urethra. Even very mild sexual stimulation will cause the prostate and Cowper's glands to begin their secretions. Stronger stimulation causes a constriction of the blood vessels leaving the penis. This causes an enlargement of the organ, as the

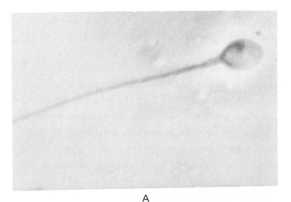

A

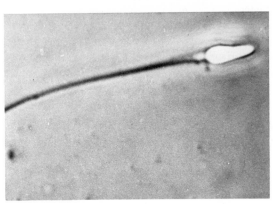

B

FIG. 34.4 Living human sperm greatly magnified. In the top photograph the head of the sperm is lying flat. In the bottom picture the head is turned on its side.

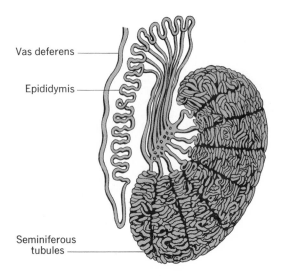

Vas deferens

Epididymis

Seminiferous tubules

FIG. 34.3 Human testis and connecting tubules. The outer covering of the testis has been removed to reveal the intricate coiled seminiferous tubules within.

spongy corpora cavernosa become engorged with blood, and makes it possible to transfer the semen in the copulatory relationship. Extreme stimulation causes a rhythmic contraction of the ejaculatory ducts which expels the semen. Normally about 4 or 5 ml of semen is ejaculated at one time. This contains the enormous number of approximately 400 million sperm, enough to fertilize every woman in this hemisphere. Any excess of semen that accumulates in the ejaculatory ducts is expelled along with the urine or through nocturnal emissions. In this way a fresh supply of sperm is always in the ducts.

THE FEMALE REPRODUCTIVE SYSTEM

The human **ovaries** are found on a broad ligament in the abdominal cavity of a woman, one on each side of the body just below the level of the navel. Each ovary is only about as large as a shelled almond. The germinal epithelium is found on the outside of the ovaries. The cells destined to become eggs are formed here and are pushed inward toward the center of the ovary. Here each cell is surrounded by a mass of cells that forms a vesicle, the **Graafian follicle.** A fluid, the follicular fluid, which contains the female sex hormone, forms in this follicle. The follicle gradually enlarges and migrates to the outside of the ovary, and the egg and the surrounding follicular fluid are released as the follicle ruptures through the outer ovarian wall. Normally only one egg is released during any reproductive cycle. This release, **ovulation,** occurs approximately fourteen days before the beginning of menstruation.

The eggs are technically shed in the body cavity, but the **ostium** of the oviduct or **Fallopian tube,** almost surrounds the ovary, so there is little chance that the egg will not follow its normal course. The egg passes down the Fallopian tube to the **uterus,** the organ in which the embryo develops. The neck, **cervix,** of the uterus projects slightly into the **vagina,** which opens to the outside through the **vaginal orifice.** There are **bulbourethral glands** near the opening of the vagina that secrete a fluid similar to that of the glands connected to the urethra of the male. The urethra of the female opens just anterior to the vaginal orifice and there is a small mass of tissue, the **clitoris,** just anterior to the opening of the urethra. The clitoris is homologous to the penis of the male; it has a glans and prepuce and is capable of engorgement like the penis. All the external structures are surrounded by two pairs of lips, the inner **labia minora** and the outer **labia majora.**

The reactions of the female reproductive organs are largely governed by the autonomic nervous system, as are those of the male. Sexual stimulation causes secretion by the bulbourethral glands that alkalizes and lubricates the vaginal tract. There is also a distention of the vaginal walls. The semen is received in the vagina near the cervix of the uterus. The female reproductive tract is lined with ciliated epithelium which keeps a thin layer of mucus constantly flowing outward. Sperm have a tropism which causes them to swim against the flow of this mucus, so they swim up the female reproductive tract. In the upper region of the Fallopian tube they may meet an egg, if ovulation has recently occurred. Some studies indicate that the period of fertility lasts only a day or two, at most, after ovulation. If the egg is not fertilized during this time, it undergoes degeneration. There is a slight elevation of the basal body temperature following ovulation, and it is possible to determine when ovulation takes place by taking the temperature every morning upon first awakening. There is a gradual decline of this temperature, beginning with menstruation and lasting until ovulation, when there will be a sudden rise.

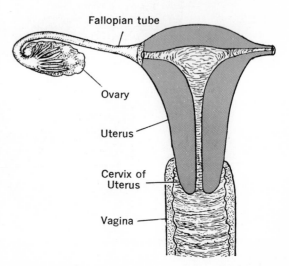

FIG. 34.5 Organs of the female reproductive system.

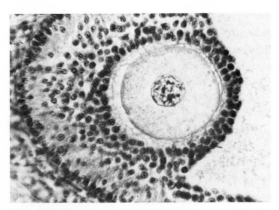

FIG. 34.6 Section through the human ovary showing an egg with its corona of follicle cells within a Graafian follicle.

The maximum chance of fertilization occurs during the rise of temperature.

Just preceding and following ovulation the uterine wall undergoes changes in preparation for the reception of the fertilized egg and the development of the embryo. The inner lining of the uterus becomes thickened and highly vascularized. If there is no fertilization, the entire inner lining is shed. This is accompanied by a bleeding of the vascular layer that is being destroyed. This blood helps to wash the shed lining from the uterus. This is **menstruation,** which occurs about fourteen days after ovulation. After several days the uterus begins to repair itself and the entire cycle is repeated. A twenty-eight-day cycle seems to be most common, but there is normal variation from twenty-one to thirty-five days, depending on the hormone balance of the individual woman. Menstruation ceases at about forty-eight years of age through a readjustment of the hormonal secretions. This is known as the **menopause** and is often accompanied by emotional disturbances as the nervous system adjusts itself to the changing hormone concentration.

If the egg is fertilized, it begins development in the Fallopian tube and, upon reaching the uterus, implants itself firmly in the uterine wall. This **implantation** is accomplished by an

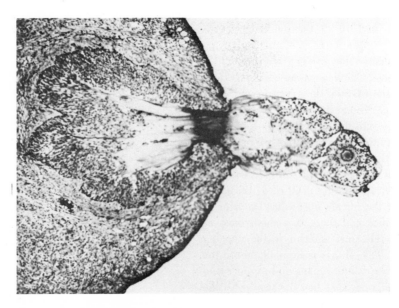

FIG. 34.7 Ovulation. The surface of the ovary ruptures and the egg, with its surrounding fluid is expelled. The egg can be seen near the end of the erupting mass.

(Courtesy of R. J. Blandau, University of Washington School of Medicine.)

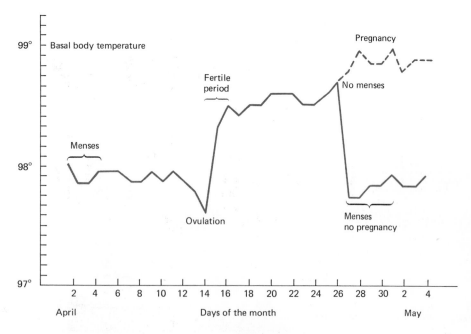

FIG. 34.8 Relationship between basal body temperature and ovulation.

enzyme secreted by the fertilized egg which dissolves the surrounding cells and forms a hole for itself in the wall. Such an implantation produces **pregnancy.** Menstruation ceases and the uterus continues its development to take care of the developing embryo. About 266 days after fertilization the embryo will have completed its prenatal development and will be born. However, the duration of pregnancy is usually calculated from the time of the last menstruation, which is about 280 days or about nine calendar months.

The **breasts** are an important part of the female reproductive system. They are generally sensitive to sexual stimulation, and they are capable of providing a baby with its sole source of nourishment. The breasts are modified sweat glands. The breast tissue is dormant for about twelve years until it is activated by the surge of sex hormones at puberty. In boys there is usually some enlargement of the breasts at puberty. In a girl they continue to enlarge and the area around the nipple, the **areola,** becomes larger and more deeply pigmented.

Breasts vary in size and shape, but this variation does not affect their efficiency as milk producers. Each breast has about seventeen milk-producing units. Each unit is like a highly branched tree with a milk-producing alveolus at its terminal tip. The breasts also contain strands of elastic tissue which connect to the chest wall and support the glandular tissue.

The breasts are remarkable chemical factories. They take elements from the blood and convert them into a milk which supplies all the nutritional needs of a young infant. Glucose (blood sugar) is converted into **lactose** (milk sugar). The amino acids are synthesized into **casein,** the milk protein. Fatty acids and glycerol are put together to form the fat found in the milk. Vitamins are removed from the blood and incorporated in the milk in the proportions needed by an infant. A wide assortment of min-

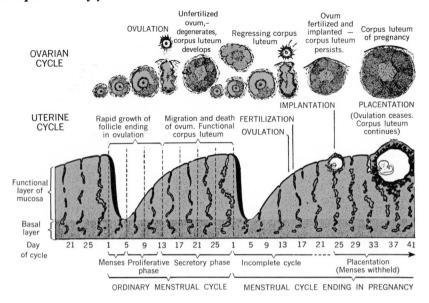

OVARIAN CYCLE

OVULATION · Unfertilized ovum,- degenerates, corpus luteum develops · Regressing corpus luteum · Ovum fertilized and implanted — corpus luteum persists. · Corpus luteum of pregnancy

UTERINE CYCLE

Rapid growth of follicle ending in ovulation · Migration and death of ovum. Functional corpus luteum · FERTILIZATION OVULATION · IMPLANTATION · PLACENTATION (Ovulation ceases. Corpus luteum continues)

Functional layer of mucosa

Basal layer

Day of cycle   21 25 1 5 9 13 17 21 25 1 5 9 13 17 21 25 29 33 37 41

Menses Proliferative phase Secretory phase Incomplete cycle Placentation (Menses withheld)

ORDINARY MENSTRUAL CYCLE   MENSTRUAL CYCLE ENDING IN PREGNANCY

FIG. 34.9 Changes in the tissue of the ovary and in the endometrium of the uterus during a menstrual cycle and pregnancy. (Redrawn from Patten, *Human Embryology,* Blakiston, 1958.)

erals are incorporated into milk in the correct proportions.

It is interesting to note that the milk

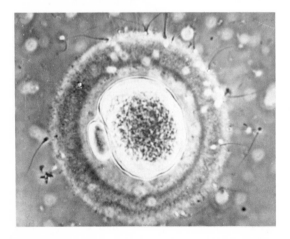

FIG. 34.10 Human fertilization. This photograph shows living human sperm surrounding a living ·egg. The sperm are attacking the corona of the egg and some have penetrated part way through to the egg. Note the polar body to the left of the egg. (Courtesy of Landrum Shettles, Columbia University.)

changes in composition as the baby grows older and its requirements change. During the first few days after birth the breasts produce a yellowish fluid, **colustrum.** Colustrum contains little nourishment and the baby usually loses weight during these first days, but it helps to clear the infant's digestive tract of mucus and to prepare it for the milk which is produced after about three days. The stimulation of the breasts by the sucking infant helps to stimulate the hormones which control milk secretion.

The breasts are a common site for cancer development in women, and all women should examine themselves for lumps or other abnormal growths regularly and should have regular medical examinations.

DISORDERS OF THE REPRODUCTIVE SYSTEM

Many things can go wrong with the reproductive system and many of the disorders of middle and old age can be traced to defects in this system. There are also some disorders of the reproductive system which are of primary interest to young people.

Sterility

Records of the census bureau show that more than 15 percent of all couples who have been married more than ten years do not have children of their own. Today some people choose not to have any children because of their feelings about population growth. Many people, however, want children but cannot have them. In about 60 percent of these cases, the inability to have children is caused by some defect in the woman's reproductive system. In 40 percent of the cases, the cause is a defect in the man's reproductive system. About one-third of those with sterile marriages can have children with the help of modern medical techniques.

A woman's Fallopian tubes may be obstructed so that, even though she ovulates regularly, the eggs cannot travel from the Fallopian tubes to the uterus and the sperm cannot ascend the Fallopian tubes to meet the eggs. Sometimes it is possible to open the tubes surgically. Some women fail to ovulate regularly and are completely or partially sterile as a result. Fertility hormones, as discussed in Chapter 33, can stimulate ovulation and permit pregnancy, but they may also induce multiple births. In other women a thick mucus in the opening of the cervix may block the entry of the sperm. Injection of semen directly into the uterus may induce pregnancy. In other cases, fertilization may occur frequently, but the woman may not have the proper hormone balance for implantation of the embryo in the uterus or for maintaining the embryo after implantation. Hormone therapy can often remedy the imbalance.

A man may be sterile because of a low sperm count in his semen. Normal semen has as many as 400 million sperm in one ejaculation of 4 or 5 ml. This may seem to be far too many, but sterility or low fertility will result if the count drops much below about 50 million. Poor nutrition, chronic alcoholism, disease, or malformed organs can be responsible for a low sperm count. Some men have a normal sperm count, but the sperm are abnormal in shape or are not motile. Collection of the semen and injection of the semen directly into the uterus of the wife can help many such men to become fathers. Many men are sterilized by an untreated gonorrheal infection of the vas deferens or of the epididymis. After the infection the tubes may grow together and prevent the passage of sperm to the ejaculatory duct. In a few cases a clogged vas deferens has been opened surgically. Artificial insemination is often used when the husband is sterile. Semen from another man is collected and injected into the uterus of the woman so that she may become pregnant and bear a child.

The Venereal Diseases

The sex act makes it possible to transfer infectious agents from the sex organs of one person to those of another. Infections resulting from such contacts are called **venereal diseases.** **Gonorrhea** and **syphilis** are the most common venereal diseases and both can cause great damage if not properly and promptly treated.

Before the discovery of antibiotics, venereal diseases were very difficult to treat successfully and often led to blindness, insanity, heart defects, paralysis, and death. When penicillin and other antibiotics became available, it was believed that these diseases could be eliminated, and in the United States they almost were. By the late 1950's the incidence of venereal infection had dropped to such low levels that public health officials thought that the diseases were under control. In the 1960's however, the incidence of venereal diseases began to increase again for several reasons.

Public health officials relaxed their vigilance in finding and treating contacts of infected persons because they felt that it was only a matter of time until venereal diseases would be eliminated.

Some of the venereal disease organisms became resistant to the antibiotic treatments. In time the resistant strains become widely distributed.

The birth control pill became widely available. Many people stopped using condoms to prevent conception. The condom is one of the best preventives against venereal disease.

A shift in public attitude toward premarital sexual intercourse took place. Premarital sex was no longer so widely condemned. This change in attitude coincided with the greater mobility of young people and their greater opportunity for sexual contacts.

This four-way punch caused a resurgence of venereal diseases, and as of this writing, the incidence of venereal diseases has reached epidemic proportions. In 1974 over 2.5 million cases of gonorrhea were reported in the United States, and certainly there were many more which went unreported. This means that, in a group of people of all ages, about one in every one hundred will have, has had, or now has a reported case of gonorrhea. The incidence of cases among young people is at least five times this high, or one in twenty. Syphilis now occurs at approximately one-seventh this frequency, but this still ranks it high among infectious diseases. Unfortunately, both men and women carry the organisms which cause both diseases for some time without knowing they have the infection and they can spread it to others during this time.

Dysmenorrhea

Some women suffer discomfort immediately preceding and during menstruation. Both physio-logical and psychological factors contribute to this discomfort. Progesterone, one of the female sex hormones, is secreted in great quantities in the week preceding the onset of menstruation. This hormone can cause water retention which leads to the "bloated" feeling described by some women. Increased water retention may also contribute to the irritability, increased nervous tension, and depression which sometimes precede menstruation. This set of symptoms is called the premenstrual tension syndrome. Diuretics or lowered salt intake is sometimes prescribed for these symptoms. Occasionally menstrual difficulties may be caused by a misplaced uterus or a completely intact hymen.

A woman's attitude toward menstruation is also thought to influence the amount of discomfort she may experience. Margaret Mead found that the women in the South Seas island cultures which she studied rarely seemed to have painful menstruation. Dr. Mead thought the acceptance of menstruation as a natural process had much to do with the lack of pain experienced by these women. Although sex education is given more frequently today in this country than it was fifty years ago, many woman still refer to menstruation as "the curse."

Prostate Trouble

In older men, the prostrate gland sometimes becomes enlarged and interferes with the passage of urine down the urethra. This may cause painful urination and a need for frequent urination because the bladder is not emptied each time. The prostate gland is sometimes removed to correct this condition. The prostrate is also a frequent site of cancer. Prostate cancer is not a rapidly growing cancer and some physicians choose to leave it undisturbed, although prostate removal and treatment with radioactive cobalt are the usual treatments recommended.

THE BEGINNING OF A NEW LIFE

The reproductive systems and all the many social and psychological reactions associated with sex have one biological end, the bringing together of a sperm and an egg which will start a new life. In Chapter 8 we learned how sperm and eggs are produced. Now let us see how they are brought together.

Fertilization

Semen is released in the vagina, near the uterine cervix. At first the semen is quite viscid; and this helps it to stick to the walls of the vagina. Soon, however, the semen becomes thin and watery so the sperm can swim about. We have already mentioned that the semen contains prostaglandins which stimulate the uterine mus-cles. These contractions may help suck the semen up into the uterus and then into the Fallopian tubes. Studies on experimental animals show that sperm is found in the outer part of the Fallopian tubes a few minutes after intercourse is completed. Sperm could not possibly swim this distance unaided in such a short time. They do have a tendency to swim against the flow of mucus which comes down the Fallopian tubes, and this directs them toward the distal ends of the Fallopian tubes where a newly ovulated egg may be found.

The egg is surrounded by a **corona** of hundreds of small follicle cells, which block the way of the sperm. Within the **acrosome,** or head, of the sperm there is an enzyme called **hyaluronidase** which liquifies the cement which holds the cells of the corona together. No one sperm has enough of this enzyme to break through the barrier, but the combined action of

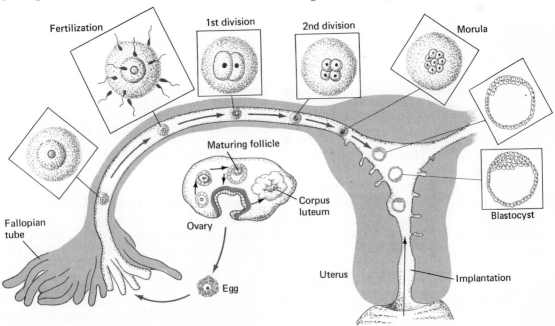

FIG. 34.11 Fertilization, early development, and implantation of the embryo are shown in this diagram. The end of the Fallopian tube has been pulled away from the ovary in order to show ovulation. There is a time lapse of about seven days between fertilization and implantation. (Redrawn from Winchester, *Human Sexuality*, Charles Merrill, 1973.)

the enzyme from many sperm cells can make openings in the corona large enough for some sperm to squeeze through. This may be the reason why a man is sterile when his sperm count is low. Only one sperm is needed for fertilization, but many are needed to break this barrier. Once a sperm head contacts the surface of the egg, it sticks to the membrane and the egg engulfs the head and middle piece, leaving the tail behind.

When the sperm head enters the egg, it brings in a small amount of an enzyme that triggers a chain reaction that releases enzymes already in the egg. These enzymes, which are contained in the lysosomes, bring about several important changes. They alter the surface potential of the eggs so that more sperm will not adhere to the surface and any which may already be stuck will be released. The change in surface potential prevents fertilization of the egg by more than one sperm in all but rare cases in which two sperm enter the egg at about the same time (resulting in an abnormal pattern of mitosis and death of the zygote). The rate of oxygen consumption goes up to about five times its previous rate, indicating an increased metabolic rate. The outer membrane of the egg thickens and becomes more permeable. The increased permeability assists the passage of nutrients into the egg and the elimination of wastes from the egg. The sperm head absorbs water and soon becomes about as large as the nucleus of the egg. The nuclei of the sperm and the egg fuse to make one nucleus with the diploid chromosome number. About thirty hours after fertilization the first mitosis takes places in the zygote.

Parthenogenesis

The egg has a complete haploid set of genes along with the food needed to nourish the embryo. The sperm stimulates the reactions needed to start development of an embryo in most species. In some animals, sperm entry is not needed to start embryo development. Male honeybees are produced by parthenogenetic development. Aphids and plant lice produce several generations of females this way during the summer months. Some eggs of a breed of white turkeys will hatch without fertilization. Parthenogenetic development can also be induced in some lower animals. The egg of the sea urchin can be "fooled into thinking" that it has been fertilized by pricking it with a needle. Exposing these eggs to ice water or strong salt solution has a similar effect.

Can parthenogenesis occur in human beings? Women not wanting to admit sexual relations with men have claimed that it can. Cases which have been investigated, however, have shown that the child possessed genes which could not have come from the mother, and therefore the child much have been conceived in the usual way. As a final test, attempts to make successful skin grafts from the child to the mother can be made. If parthenogenesis did occur, the mother should be able to accept a skin graft from the child because she would have all the genes possessed by the child. So far, in all cases attempts at skin grafts have failed.

Tubular Development

The cells divide synchronously as the young embryo moves along the Fallopian tube toward the uterus. These divisions produce two, four, eight, sixteen, and thirty-two cells in five mitotic divisions. After about three days the embryo has thirty-two cells clustered together in a shape resembling a mulberry. This stage of development is known as the **morula.** Then the cells begin to form a hollow ball. Within the cavity of the ball there is an inner cell mass which will produce the body of the embryo. The mass of cells is now known as a **blastocyst.**

If the embryo becomes stuck in the Fallopian tube, it will continue to develop there and it may even form a placental attachment to the wall of the tube. This is known as a **tubular pregnancy.** In rare cases the egg will drop free in the body cavity instead of being caught up in the Fallopian tube during ovulation. It can be fertilized there and can develop for a short time by attachment to various body organs. All these abnormal placements of the fertilized egg are called **ectopic pregnancies.** These pregnancies must be terminated surgically.

Implantation

At about the seventh day after fertilization, the blastocyst implants itself in the uterine wall. The seven-day embryo develops **trophoblastic cells** on the outer surface of the blastocyst. These cells produce enzymes which digest the surface of the uterus so that the embryo can settle into a little cavity which it makes for itself. At about the twelfth day of life the trophoblastic cells send out small fingerlike processes which penetrate the wall of the uterus. These **chorionic villi,** as they are called, anchor the embryo and begin absorbing food from the uterus. They also secrete chorionic gonadotropin, a hormone which is necessary to maintain the uterus in proper condition for pregnancy. The villi continue to grow and they form the placenta of the embryo.

The Embryonic Membranes

As the embryo continues growth, a thin membrane, the **amnion,** forms around it. This membrane is only two layers of cells thick, but it is very tough. It contains the amniotic fluid in which the embryo floats. Human beings are really water animals during the first part of their existence. Each human embryo has its own "private lake" in which it is immersed until the time of birth. The **amniotic fluid** is balanced with salts

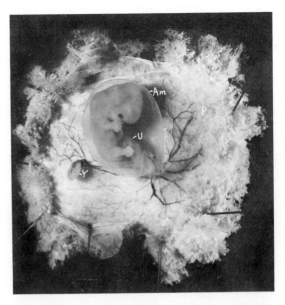

FIG. 34.12 A human embryo, forty-two days of age, with its connecting membranes. Note the placenta which has been cut open to expose the embryo. The amnion contains the amniotic fluid in which the embryo floats. The umbilical cord attaches the embryo to the placenta. The yolk sac is rudimentary. (Courtesy of C. F. Reather, Carnegie Institution of Washington.)

FIG. 34.13 The amnion is peeled from the placenta so that it can be used for identification of chromosomal sex. This is a routine procedure for all births in many large hospitals.

and is somewhat like dilute sea water in its composition. The embryo is very soft and would be easily injured were it not for this supporting fluid. Some of this fluid is removed for analysis in tests for suspected genetic defects. **Amniocentesis** is a technique which involves inserting a needle through the mother's abdomen into the amniotic cavity. Tests of this fluid and the cells which it contains can diagnose many genetic abnormalities. The **chorion** is a membrane formed on the outside of the amnion. Some of its cells play a part in the establishment of the chorionic villi.

A third membrane extends from the abdominal region of the embryo and forms the **yolk sac.** Birds and reptiles have a very large yolk sac which contains most of the nourishment within the egg. In placental mammals no large amount of stored food is needed, but the yolk sac has persisted and has an important function. Blood must be formed early in embryonic life. It must contain red blood cells which can pick up oxygen from the placenta. The organs which produce red blood cells in later life are not present in the early embryo. It is the yolk sac which produces the embryonic hemoglobin and red blood cells.

The **placenta** forms a life-sustaining con-

nection with the uterus of the mother. It is well formed by the fifth week. It sends processes deep into the uterine wall and blood circulating within the placenta comes in close contact with blood within the maternal circulation. Oxygen, food, and other products are absorbed and wastes are passed back into the mother's blood.

Uterine Development

After four weeks of development, the embryo is about one-fifth inch (5 mm) long. Its heart beats and pumps blood out to the developing placenta and back to the embryo through the umbilical cord. Cell differentiation has produced the beginning of eyes, brain, spinal cord, lungs, and digestive organs.

At six weeks the embryo is about one-half inch (12 mm) long and limb buds have developed. Fingers and toes are beginning to differentiate. The head makes up about one-third of the total body length to accommodate the developing brain. Many organs are in the most critical stage of development. Anything which slows their development at this time will cause lasting deformity. If the mother takes some kinds of drugs, is exposed to low oxygen pressure or heavy radiation, or contracts certain viral diseases, the embryo may be born blind, deaf, mentally retarded, with heart defects, or with other deformities. The embryo can now swallow some of the amniotic fluid which is cycled through the intestinal tract. Fingers and toes are formed, but they are webbed. The duct systems which will develop into the reproductive organs are formed, but they are not sexually differentiated.

At eight weeks physicians begin to call the embryo a **fetus.** It is about an inch long. Calcium deposits begin in the cartilage, and this is the beginning of bone. The eyes are sealed shut as protection against moving fingers when the

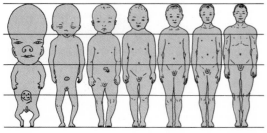

2 mo. 4th mo. Newborn 2 yr. 6 yr. 12 yr 25 yr.
fetal   fetal

FIG. 34.14 Two fetal and five postnatal stages, drawn to the same height, to show the changes in body proportions which come with growth. (Patten, *Human Embryology,* Blakiston, 1958.)

fetus develops the power of movement. The differentiation of the sex organs begins.

At twelve weeks the fetus is about three inches long and weighs about one ounce. The skeleton becomes more rigid as ossification continues. The muscles are sufficiently developed to permit some movements, although the mother may not yet feel them. The initial development of the organs is complete and the fetus becomes less susceptible to damage through maternal exposure to noxious agents. From this point on, until birth, there is mainly an increase in size rather than differentiation.

At twenty-eight weeks the fetus reaches an important stage of development. It is about ten inches long and weighs about two and one-half pounds and now has a chance to survive if it is separated from its mother's body. It would have to be kept in an incubator because its own metabolism is not sufficiently mature to maintain the proper temperature. It might also have to have respiratory stimulation because its breathing muscles are not yet fully devloped.

At about thirty-eight weeks or 266 days from conception, the time for normal birth arrives. The fetus, which has had its head pointing somewhat upward, turns so that the head points downward. Sometimes it fails to make the turn (a breech presentation), or it turns part way (a shoulder presentation), but birth is usually normal in either case. The amnion breaks, releasing the amniotic fluid. The cervix dilates and labor contractions begin, forcing the head down through the dilated cervix. The baby is born, and the umbilical cord is cut near the body. The uterus continues its contractions and expels the placenta.

Upon birth the baby makes several important adjustments. It suddenly changes from a liquid to an atmospheric environment. The oxygen supply from the mother is cut off and expands its own lungs promptly and begins to breathe. The heart must also change the route of blood circulation. It has been pumping blood into the placenta, and now a valve which has connected the two ventricles closes so that blood can go to the lungs. An imperfect closure of this valve produces a "blue baby." If this valve fails to close, unoxygenated blood is pumped through the body as well as oxygenated blood and gives the skin a blue color. The kidneys begin to remove waste products from the blood and to adjust the body's water balance. If any one of these vital changes does not take place, the life which has been developing for so long will be terminated.

MULTIPLE BIRTHS

About one birth in every hundred is a twin birth. This means that about one person in fifty in the United States has a twin brother or sister.

Kinds of Twins

Monozygotic twins, also called **identical twins,** arise from a single fertilized egg, or zygote. These twins have identical genes and they look very much alike. Any differences between them are probably the result of environmental effects. They are always of the same sex. The splitting of the zygote may occur during the tubular development stage and two separate blastocysts arrive at the uterus for implantation. Each blastocyst develops a separate membrane. Occasionally, one twin may arise as a bud from the other twin after implantation and they both share some of the same embryonic membranes. These are called **dichorionic** and **monochorionic** twins respectively.

Dizygotic or **fraternal twins** arise from separate eggs fertilized by separate sperm. They are no more alike than brothers or sisters born at different times. They may be of opposite sexes. They always have separate implantations and are therefore dichorionic. When there is any doubt about whether a pair of twins are identical or fraternal, skin grafting will give the answer. Monozygotic twins can accept skin grafts from each other, but fraternal twins reject grafts because of protein differences caused by genetic differences.

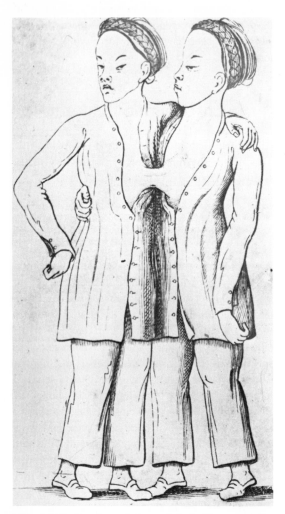

FIG. 34.15 Drawing of the original Siamese twins which was used on an advertising poster by the P. T. Barnum Shows.

Inheritance of Twinning

Does the chance of having twins run in families? One woman in Russia had twenty-seven pregnancies without one single birth. She had sixteen twins, seven triplets, and four quadruplets, for a total of sixty-nine. None were identical twins. This and other cases indicate that genes do influence the chance of having fraternal twins. Identical twins seem to occur by accident. They occur in about one pregnancy in 300 in all races. The Nigerians have fraternal twins about once in 25 pregnancies, the Japanese have such twins once in 400 pregnancies, while in the United States, including all races, there is about one twin birth in 130.

Hormones and Twinning

Hormone balance determines the chance for multiple ovulation and heredity influences the output of the hormones involved. This balance also tends to favor multiple ovulation with increasing age and number of children. A woman who has had six children and is at least thirty-five years of age has over six times the chance of bearing twins as women in their twenties who are having their first pregnancy. The fertility pill, mentioned in Chapter 33, can alter the hormone balance to increase the chance of ovulation and it often causes multiple ovulation. A woman in Australia gave birth to nine babies after taking such pills, but all were too premature to live more than a few days.

Conjoined Twins

About one pair of twins in 500 are born with some junction between their bodies. This may vary from a small area covering only a few inches to large junctions in which many body organs are shared in common. As long as each baby has a complete set of body organs, they can usually be successfully separated, but those with many body parts in common cannot and they usually die shortly after birth.

TERATOGENIC AGENTS

The birth of a child with a physical or a mental defect is often tragic. Sometimes birth defects are genetically caused or the result of embryonic accidents over which the parents have no control. Many birth defects are the result of exposure to some agent which interferes with normal development. These are known as **teratogenic agents** (from the Greek words *terato,* meaning "monster," and *genic,* meaning "origin of"). All persons should be familiar with these agents and all prospective mothers should avoid contact with anything which is suspected of being teratogenic.

A number of drugs are teratogenic. In Chapter 2 we discussed the tragedy of the hundreds of children born in Europe without arms and/or legs because their mothers had taken the drug **thalidomide** during early pregnancy. Some of the **antibiotics,** especially those in the streptomyces group, as well as some of the widely used pain killers are teratogenic when they are given to experimental animals in high concentrations. These drugs are not recommended during pregnancy. **Cortisone,** the adrenal cortex hormone, which is used to treat arthritis and allergies, can increase the incidence of hairlip and cleft palate when given to pregnant rats and its use in early human pregnancy is forbidden. The use of **LSD** in early pregnancy produces a higher incidence of birth defects.

Some herbicides and insecticides are teratogenic. The herbicide used to defoliate the jungles in Vietnam caused many birth defects in babies born to women who had been in the region. Dioxan, the carrier of the herbicidal chemical, not the herbicide itself, was the teratogenic agent. Indian women in Arizona and New Mexico who ate animals which had grazed near streams which had been sprayed with herbicides to kill the brush had an abnormal number of babies with birth defects.

Exposure to a **low oxygen level** can be teratogenic. A pregnant woman who takes a trip to a high mountain may feel faint and short of breath, but she will recover as soon as she comes down. An embryo in the early stages of development, however, may experience real difficulty and its growth may be retarded.

High-energy radiation is teratogenic. Many women in Hiroshima who were exposed to radiation from the atom bomb had defective babies. Most physicians do not use diagnostic X rays on pregnant women. Some physicians restrict the use of extensive radiation on mature women to the first ten days after menstruation because this is the only time when they can be sure that the women are not pregnant.

Some **viral infections** can be teratogenic. The best known is **rubella,** or German measles, which produces only mild symptoms in the mother. The virus penetrates the placental barrier and severely affects the developing organs of the fetus. It inhibits mitosis so that organs in a critcal stage of development fail to develop properly. When the infection is contracted during the second to fifth week, the embryo is likely to have heart defects because this is the time when the heart is developing. Eye defects appear when the infection occurs during the third and fourth week. Infection during the fifth week causes deafness. In the sixth to eighth week rubella can cause mental retardation and motor defects. One study showed that about 80 percent of the women who had rubella during the first twelve weeks gave birth to babies with some abnormality. Other viruses which may be teratogenic are those which cause hepatitis, mononucleosis, shingles, and even influenza. Many doctors recommend therapeutic abortion when a woman contracts one of these diseases during early pregnancy.

The **spirochete of syphilis** can cross the placenta and cause congenital syphilis in the fetus. A baby with congenital syphilis may be born with serious defects of the heart and blood vessels, mental retardation, or other defects.

Genes can also be teratogenic agents. If both parents carry harmful recessive genes, these may be expressed in their child. Hemophilia, alcaptonuria, phenylketonuria, and color blindness are genetically determined birth defects. We shall consider these defects more fully in Chapter 36.

EXPERIMENTAL EMBRYOLOGY

Many facts about embryonic development have been uncovered through the use of delicate microdissection instruments which can be used to remove and transfer parts of embryos. One set of these experiments has helped us to understand why some genes function at one time and place and not at other times and places.

Embryonic Induction

Early work on this subject was conducted with the embryos of frogs and salamanders. It was found that the **dorsal lip** of the **blastopore**—the opening which will form the mouth—was important as an organizer of the development of body parts. When a portion of this lip was transplanted to the ventral region of another embryo, a separate head and upper body developed there. The result was that the salamander had two anterior body parts, both joined to a single posterior body part. This experiment demonstrated that cells from one part of the body of the embryo can develop into cells characteristic of other body parts if they are properly stimulated. They can be stimulated by an organizer

which is a chemical secreted by cells of the dorsal lip. Even when the dorsal lip cells are killed by boiling, the organizer is still effective. If the dead cells are placed on a young embryo, an anterior body part will develop there.

Hans Spemann, the German embryologist, did pioneering work on organizers and received the Nobel Prize for his discoveries. He also found that the organizers which continue the differentiation, organ after organ. For instance, the brain forms an optic vesicle which will become the retina of the eye. This vesicle produces organizers which stimulate the adjacent ectoderm to form the lens of the eye. If an optic vesicle is removed from an embryo and inserted under the skin on the side of the body of this embryo (or another embryo), it will cause the ectoderm to form a lens.

The ectoderm along the dorsal region folds inward and forms the **neural tube** which becomes the brain and spinal cord. When this layer of ectoderm is cut out and placed in a nutrient solution, it continues to grow, but it does not form a brain or spinal cord. The mesoderm, underneath the ectoderm, produces the organizers which cause the ectoderm to differentiate into the brain and spinal cord. When some of the mesoderm from the dorsal region is placed beneath the ectoderm on the ventral region, this ectoderm forms a brain and spinal cord. Ectoderm from any part of the body will form a brain and spinal cord in tissue culture if some mesoderm from the dorsal region is placed with it. This differentiation will occur even if ectoderm is placed in a solution in which the mesoderm has been growing but is now removed. This experiment shows that some chemical agent has been produced by the mesoderm which acts as an organizer which stimulates the ectoderm to develop into a brain and spinal cord.

When the embryo reaches a more advanced stage of development, the tissues continue to develop in the original way even when they are transplanted. Once certain cells from the head have received the stimulus to produce an eye, they will produce an eye even if they are transplanted to other body regions. A frog with an eye on the side of his body is shown as an example in Figure 34.16. Cells which form a leg can be transplanted to any part of the body after they have received the inducing stimulus and they will form a leg. Accidental induction can result in body parts growing at the wrong place. Some people have a third leg growing from the center of the back.

Regeneration

Many of the simpler invertebrates have an extensive power of regeneration. Small bits of tissue from almost any part of the body will develop into an entire new animal. With the higher invertebrates, such as arthropods, the regenerative ability is decreased, but arthropods can regenerate a new leg if one is lost. In the vertebrates the power is limited to regeneration of smaller parts, as a rule, and the process is reversible. If a lizard's leg is cut off, no new leg will appear, but if the tail is cut off, a new tail will develop. If the nerve which normally goes to the tail is diverted into the cut stump of a leg, however, a new leg grows out. If a leg of a salamander is cut off, a new leg will grow back. If the nerve to the leg region is cut, no new leg will grow. It appears that the nerves produce some inducing substance which stimulates the genes which are responsible for leg production to function.

The ultimate end to discoveries on organizers and inducers would be the ability to stimulate a single human cell into the development of an entire embryo. We have already found ways to stimulate leukocytes to undergo mitosis and some degree of differentiation. When extracts of kidneys are added to tissue cultures of such cells, kidney nephrons are produced. If we could find a way to control the stimulation of development, a cell could be stimulated to develop a blastocyst and this could eventually lead to the development of an entire organism.

Nuclear Transplants

One method of human asexual reproduction may develop from the following experimental discoveries. Experiments on frogs show that a nucleus can be taken from a body cell and transplanted into an egg in which the nucleus has been destroyed. The frog that is produced has all the characteristics of the one which furnished the nucleus. The nucleus can come from a

FIG. 34.16 A three-eyed frog. The extra eye on the right side of the body was induced by transplantation of tissue during early embryonic life. (Courtesy of C. L. Markert, Yale University.)

different species of frog. This method of reproduction is called cloning. If it were possible to do this with people, the child produced would have only one true parent and two grandparents. This would be one way to prevent genetic defects. Clones of persons with particularly desirable traits could be produced. If such a practice were to become possible, you can imagine some professional football team managers placing orders for seven duplicates of an outstanding 260-pound lineman.

Embryologists have also produced embryos with four parents. This is done by combining two embryos to make one. Young mouse embryos in the two to thirty-two cell stage lie within the fertilization membrane, and this membrane can be digested by proteolytic enzymes. When two embryos are placed in contact, they fuse and form one double-sized embryo and another membrane develops around it. This embryo can then be transplanted into females which have been mated to sterile males to induce the proper hormonal conditions for implantation. The embryos become normal in size and are born in a normal gestation time. The offspring may show mosaic color coat patterns which correspond to the location of the cells from each parent. It is even possible to combine embryos of different species. A combination of mouse and rat embryos has been produced, as has a combination of a mouse and human embryo. No attempt was made to implant the mouse-human embryo since the organisms are so different that it is very unlikely it would develop.

REVIEW QUESTIONS AND PROBLEMS

1. Occasionally a boy is born with **cryptorchidism** (undescended tests). If this is not corrected, the boy will probably be sterile. Why?
2. A man mowing his lawn on a hot day probably has a relaxed scrotum and his testes hang well down from the body. When he goes in and takes a cold shower, his scrotum contracts and brings the testes up close to the body. Explain these automatic reactions.
3. Sperm taken directly from the testis or epididymis are very sluggish, but when ejaculated in semen, they are very active. What accounts for the difference?
4. When the prostate gland is removed, a man cannot father any more children, even though his ability to have intercourse in unimpaired. Explain.
5. Sterility often follows untreated cases of gonorrhea. Explain why.
6. Sometimes a woman becomes pregnant and finds that the embryo is developing in her body cavity rather than in the uterus. Explain how this can happen.
7. At what day, counting from the first day of menstruation, should a couple have intercourse if they want to induce pregnancy? Explain.
8. Only one sperm is needed for fertilization, yet sperm are released by the millions. Why are so many necessary?
9. What advantage does breast feeding have over bottle feeding for babies. Give reasons.
10. If you were a physician and a woman came to you who had been childless for five years in spite of repeated unprotected intercourse, what factors would you check as possible causes for her sterility?

11. Suppose a woman cannot maintain an embryo in her uterus and she hires a surrogate mother to bear a child from her egg and the sperm from her husband. What difficulties might arise legally and emotionally?

12. How is the widespread use of the birth control pill related to the increase in venereal disease in the United States?

13. Suppose we discovered a way to induce parthenogenesis in human beings. What advantages and what disadvantages might such a discovery have?

14. Many women conceive frequently but never know about it. How is this possible?

15. How can amniocentesis be used to detect abnormalities in the fetus long before time for birth?

16. Why is the heart one of the first organs to develop and function in the embryo?

17. Why is twenty-eight weeks an important point in the development of a human embryo?

18. Explain the adjustments which must be made by a newborn baby.

19. Suppose you are a newly married young woman and wonder about the chance of having twins. Two of your aunts had twins, your grandmother had twins, and there were twins born to some of your great grandparents. What investigations would you make to determine if you might have inherited the tendency to twinning?

20. Why are geneticists particularly interested in identical twins?

21. Why are teratogenic agents more dangerous during early pregnancy than during later pregnancy?

22. A widespread program to vaccinate school children for rubella has been undertaken. Since rubella causes only very mild symptoms in children, why are they carrying out such a program?

23. Suppose we learn how to induce latent genes to function. What very practical application of this knowledge might be made on those with damaged body parts?

24. If human cloning becomes possible, how might it be used to ensure the birth of a genetically normal child when both parents carry defective recessive genes?

FURTHER READING

Deschin, C. S. 1969. *The Teenager and VD*. New York: Richards Rosen Press.

Lloyd, C. W. 1964. *Human Reproduction and Sexual Behavior*. Philadelphia: Lea and Febiger.

Morris, D. 1968. *The Naked Ape: A Zoologist's Study of the Human Animal*. New York: McGraw-Hill.

Winchester, A. M. 1973. *The Nature of Human Sexuality. Columbus*, Ohio: Charles E. Merrill Books.

35

Human Sexuality

Biological, psychological, and cultural factors shape human sexuality and its expression by individuals. It is difficult to discuss these aspects of human sexuality and it is also difficult to provide even a brief introduction to this subject in one short chapter. In this chapter, therefore, we will discuss only a few topics which pertain to this subject. The topics we will discuss are all related to biological development or physiological functioning. We will discuss the determination of sexual identity, the physiological aspects of sexual arousal, the variety of sexual expression or outlet, and birth control methods. These topics are by no means inclusive, but they do introduce you to parts of the subject of human sexuality which have been studied by biologists.

HISTORICAL BACKGROUND OF SEX RESEARCH

Before we begin to discuss recent biological research into aspects of human sexuality, it is interesting to look briefly at the history of sex research in this country. In this country and in many parts of the Western world, sex and sexual activities were not considered fit topics either for polite discussion or for scientific research during the nineteenth century. All expressions of sexuality except for heterosexual coitus in marriage were frowned upon. Masturbation was thought to cause blindness, nervous disorders, and general debility. Women were considered incapable of orgasm, and children were thought of as nonsexual beings. During this period there were some scientists, mostly physicians or psychologists, who did conduct investigations into both the physiological and psychological aspects of sexuality. The two whose contributions led most directly to modern research into human sexuality were Havelock Ellis and Sigmund Freud.

Havelock Ellis was an English physician and literary editor who devoted much of his life to the study of sex. In 1896 he published a series of essays on sexuality called *Studies in the Psychology of Sex*. The final revised edition was published in 1936. Ellis's work was based on the collection of written sexual histories, his observations of his own behavior, and the writings of other biologists, psychologists, and anthropologists. His work anticipated the findings of later researchers even though his research techniques were not as thorough as theirs. He believed that sexual feelings and responses manifested themselves long before puberty, that masturbation was common in both sexes and at all ages, that homosexuality and heterosexuality were not exclusive states of mind or behavior, and that orgasm occurred in women as well as in men.

Sigmund Freud, the father of psychoanalysis, contributed many insights into the nature of human sexuality. Like Ellis, Freud believed that sexuality appeared in young children long before puberty. He also believed that children of both sexes were born with the capacity for both male and female sexual behavior and that events occurring early in childhood shaped the form of sexuality which the individual would exhibit later in life.

Other, less well-known physicians and scientists also studied aspects of sexuality. In 1855 Felix Roubaud contributed the earliest published description of the physiological reactions which take place during sexual arousal. This description is quite complete in technical detail and was one source of inspiration for studies conducted by Dr. Kinsey and his associates almost a century later. Dr. Robert Dickinson, an American gynecologist, anticipated much of the work done by the Kinsey group on the sexual response of women, and in fact, he shared some of his data with Kinsey. Dickinson was particularly interested in the physiological changes which occur during masturbation in women.

These sex researchers and a number of others set the stage for the work of the well-known sex researchers of recent years: Dr. Alfred C. Kinsey and his associates, Dr. William H. Masters and Mrs. Virginia E. Johnson, and Dr. John Money and his associates.

The "Kinsey Reports" are, of course, famous and notorious. They are famous because they supplied information about the sexual activities of people of a detail and completeness never before achieved. They were notorious because they answered questions about human sexuality which most people, at that time, thought improper to ask. The first two volumes published by Kinsey and his associates, Wendell Pomeroy, Charles Martin and Paul Gebhardt, were *Sexual Behavior in the Human Male*, published in 1948, and *Sexual Behavior in the Human Female*, published in 1953. These reports have often been criticized because they do not explore the relationship of sexual activity to emotional relationships. This kind of criticism is spurious because Kinsey never intended to answer this sort of question. The goal of the Kinsey group was to find out as much information as possible about the sexual activities of men and women of all ages in order to provide a factual basis for sound programs of sex education and marriage counseling. Kinsey had no prejudice or preconception about what was "normal" or "perverse" in sexual behavior. He and his associates interviewed 18,000 people and asked them what they did, and when and how often they did it. The information they collected formed the basis for a revision of information in medical textbooks, sex education courses, and marriage counseling procedures. You must read these reports in their original form to appreciate the amount of work that went into writing them and the amount of information contained in them.

Kinsey also made some direct observations of sexual activity and he made films of some of these activities in the laboratory. This project was the beginning of the work conducted by Dr. William Masters and Mrs. Virginia Johnson. These two researchers studied human sexual response in the laboratory. They recorded and measured the physiological reactions which take place during sexual arousal by observing people who volunteered to perform various sexual acts, including masturbation and heterosexual coitus, in the laboratory. From these observations and recordings, Masters and Johnson gained an impressive knowledge of the physiological responses involved in sexual arousal. This knowledge has enabled them to help many people who have suffered from sexual disfunctions.

Another modern sex researcher is Dr. John Money. Money and his associates at The John Hopkins Hospital and Medical School have studied the biological, psychological, and cultural factors which contribute to the determination of an individual's sexual identity. Money and his associate Dr. Anke Ehrhardt have been able to assess the importance of the components which contribute to sexual identity by studying people who, because of genetic defect, hormonal or metabolic abnormality, or surgical accident, have some features of both sexes. Money and his associates have studied these people before and after correction and treatment of their disorders, and in this way they have been able to determine the influence of genetic sex, hormonal sex, morphological sex, and gender identity on the formation of sexual identity.

THE DEVELOPMENT OF SEXUAL IDENTITY

John Money and his associates in the psychohormonal research unit of The John Hopkins Hospital and Medical School have contributed much information to our understanding of the processes through which a child achieves his or her sexual identity—how a child comes to know that he or she is a boy or girl. Their findings come from studies of **hermaphrodites**—people whose biological sex is ambiguous because of a chromosomal, genetic, or metabolic defect. These studies have made it clear that a person's sexual identity is not determined by one factor alone but by a combination of four components: genetic sex, hormonal sex, morphological (anatomical) sex, and gender identity.

Components of Sexual Identity

Sexual identity begins to develop at the moment of conception when the **genetic sex** of the child is determined. Genetic sex is determined by whether the child receives two X chromosomes or one X and one Y chromosome. Genetic sex determines whether the gonads which develop will be male (testes) or female (ovaries). If testes develop, they will secrete androgens for a brief time during fetal development. If androgens are secreted, the male internal sex organs—seminal vesicles, the prostate gland, ejaculatory ducts, and Cowper's glands—and a penis and scrotum will develop. If ovaries develop, no sex hormones are secreted during fetal development, and female internal sex organs—Fallopian tubes and uterus—and a vagina and vulva develop. The hormones secreted by the gonads determine the child's **hormonal sex.** The type of internal sex organs and external genitalia determine the child's **morphological sex.** The androgens secreted during fetal development by a male fetus are also thought to affect the development of particular parts of the hypothalamus which become active at puberty. At puberty the hypothalamus stimulates the pituitary to secrete hormones which in turn stimulate the gonads to produce sex hormones. If androgens are present during fetal development, the pituitary secretes hormones continuously. If androgens are absent, the pituitary secrete hormones cyclicly. The cyclic secretion of hormones is the basis for the estrous cycle seen in most female mammals and the menstrual cycle seen in female primates, including human females.

In some mammalian species the presence of fetal androgens is also thought to influence the kind of sexual behavior which the individual will exhibit during adulthood, but current experimental evidence seems to indicate that this is not true in humans.

When development is normal during fetal life, a child is born with the same genetic, hormonal, and morphological sex. In other words, if the child is male, he has one X and one Y chromosome, he has testes which secreted androgens during fetal development and he has male accessory sexual organs and external genitalia. If the child is female, she has two X chromosomes, ovaries which did not secrete androgens during fetal development, a uterus and Fallopian tubes, and a vagina and vulva.

The appearance of a child at birth usually determines the sex which will be assigned to it. When the parents learn that their baby is a boy or girl, they immediately begin to react to the baby in special ways which are related to its sex. Their reactions help the child to learn whether it is a boy or girl. This knowledge, called **gender identity,** is usually well established by the time the child is three years old.

Importance of Gender Identity

Money and his associates have studied and counseled many children whose genetic, hormonal, and morphological sexes were not the same and who therefore had ambiguous sexual identities at birth. When children with this problem are born, they must be assigned to one sex or the other and treated medically and surgically to bring their hormonal sex and morphological sex into line with their assigned sex. If the condition which causes this mixture of genetic, hormonal,

and morphological sex is identified at birth, treatment can often be started very early.

One hormonal disorder which leads to this kind of problem is called the **adrenogenital syndrome.** In this disorder, the adrenals secrete an excess of androgens (male sex hormones). The overproduction of androgens begins in fetal life, so a child who is genetically female may be born with masculinized external genitalia. This disorder does not affect the child's internal sex organs, so with the proper hormonal treatment and surgical correction of the genitalia, these girls can grow up to be perfectly normal, fertile women. Sometimes the adrenogenital syndrome is not recognized at birth, especially when the masculinization of the genitalia is not pronounced or when the child is born in a hospital where the hospital personnel are not familiar with this syndrome. When the disorder is not treated, the child continues to be affected by excessive amounts of androgen and the genitalia become increasingly masculinized. If the child is raised as a girl and establishes a female gender identity, she will often come to medical attention at the time puberty should occur because she does not develop properly. Her breasts do not enlarge, she does not develop the female type of fat and muscle deposits, and she does not begin to menstruate. Instead she develops male secondary sex characteristics such as a deepened voice, facial hair, and the male type of musculature. This happens because the excessive amount of androgens produced by the adrenals mask the effect of the estrogens (female sex hormones) produced by the ovaries. Children with this syndrome who have been raised as girls and who have developed a female gender identity do not want to become boys, even though they develop male secondary sexual characteristics. They do not develop a male gender identity; instead they want to receive treatment which will bring their body image into line with their gender identity.

Cases like this seem to indicate that in humans hormonal sex is not as important as gender identity in determining the way in which a child develops sexual identity. Morphological sex is also not the determining factor since a child can have genitalia which appear to belong to one sex and think of itself as belonging to the other sex —the sex of the gender identity.

Hormonal Sex: Effects on Development and Behavior

Sex hormones are important in determining sexual identity at two times during life: during fetal development and during and after puberty During fetal development the presence or absence of androgens determines whether the internal sex organs will differentiate into the male or female form and whether the external genitalia will become male or female. If development is normal, a fetus which is genetically male will develop testes which secrete androgens. The testes also secrete a substance which suppresses the development of the female internal sex organs and genitalia.

The effect of androgens on fetal development of the sex organs in humans has been studied by observing development in male children who have a hormonal disorder called the **androgen-insensitivity syndrome.** Genetically male children who have this syndrome develop testes during fetal life and these testes secrete androgens, but the cells in these children are not sensitive to androgen. As a result, the children develop female, rather than male, external genitalia. They do not develop the internal female sex organs because they are affected by the female-organ-suppressing substance which their testes produce. The testes usually atrophy and remain in the body cavity rather than descending through the inguinal canal. At birth these children usually appear to be female, although the vagina may be very shallow. They are usually raised as girls, and they develop normal female gender identities. Their condition is often not detected until puberty, when they fail to mature. The disorder is usually treated by estrogen therapy and corrective surgery on the genitalia. With the proper hormone therapy, these children grow up as normal-looking, although sterile, females, and they usually maintain a perfectly normal female gender identity.

The development of children with the androgen-insensitivity syndrome provides evidence for a well-known concept in mammalian embryology: to make a male you must add or substract something from a female. The female embryo is the basic form in mammalian development and the male sex hormones are needed to change this form into the male form.

At puberty the gonads, which have been quiescent since birth, begin to secrete sex hormones. The testes secrete androgens, which stimulate the growth and development of the male internal sex organs and external genitalia, the growth of facial, axillary, and pubic hair, the development of muscle deposition, and the growth of the long bones. The ovaries secrete estrogens, which stimulate the development of the female internal sex organs and external genitalia, the development of the breasts, and the development of muscle and fat deposits which are characteristic of the adult female.

In a person whose sexual development follows the normal course, the sex hormones which are appropriate to the person's genetic and morphological sex are secreted at puberty. This leads to the development of an adult body type which is consistent with the person's juvenile gender identity and which reinforces it. Secretion of sex hormones of the opposite sex during puberty can lead to disturbances of sexual identity. Most people to whom this happens, such as the adrenogenital syndrome girls discussed previously, do not adopt a new gender identity to conform with the way their bodies are developing. Instead they wish to receive hormonal therapy which will bring their body type into line with their gender identity.

In humans, therefore, the male sex hormones are important in determining the sex to which a person will be assigned at birth, since the presence or absence of male hormones determines the way in which the external genitalia will develop. The influence of male sex hormones at puberty, however, is greatly modified by the influence of the person's gender identity.

Of course, the sex hormones do have some influence on sexual identity at puberty. This is especially true in men because the presence of androgens is necessary to maintain sexual responsiveness. In the absence of androgens, a fully developed male may be able to achieve and maintain penile erections, but he will not be able to ejaculate, since androgens influence the production of sperm and semen. Some men who are castrated in adulthood undergo severe depressions which can be alleviated by hormonal therapy with androgens.

In some mammalian species early experience with the male sex hormones greatly influences later sexual behavior. Female rats, for example, who are given androgens during their prenatal development or soon after birth will show a higher incidence of male sexual responses, such as mounting, during their adult lives. Early hormonal experience in humans does not seem to program rigid sexual responses which then appear invariably in adult life. Money and his associates have noted, however, that androgens do appear to influence the kind of erotic stimuli by which a person is aroused. This observation was made in the study of a girl with the adrenogenital syndrome who was not treated until after the time at which puberty should have occurred. While this girl was under the influence of androgens from her adrenal glands, she found visual stimuli, such as the pictures of "cute" boys in the Sears Roebuck catalog, erotically stimulating. Visual stimuli are usually more exciting to men than women. It is interesting to note that the stimuli were appropriate to her gender identity and not her hormonal sex. After she was treated and began to undergo puberty under the influence of estrogens, she found romantic imagery,

rather than visual stimuli, erotically arousing. This kind of imagery is the type which most girls find erotic.

Influence of Genetic Sex on Development

The basic embryological form in mammals is female. In order to create a male, something must be added to the female embryo. In mammals this is first the Y chromosome. The presence of the Y chromosome causes the gonads to differentiate as testes. The testes then secrete androgens during fetal development and the fetus differentiates male internal sex organs and male genitalia.

If a fetus receives an abnormal number or assortment of sex chromosomes it will differentiate as male or female on the basis of the presence or absence of the Y chromosome. For example, a fetus which receives only one X chromosome and no other sex chromosomes (**Turner's Syndrome**) will differentiate as a female. A fetus which receives two X chromosomes and one Y chromosome (**Kleinfelter's Syndrome**) will develop as a male, although his sexual organs may be malformed and his androgen output is usually low. A fetus which receives one X chromosome and two Y chromosomes will also develop as a male. A person with this chromosomal combination may be sterile and may show various kinds of behavior disorders. None of these chromosomal abnormalities appear to lead directly to deviations in sexual behavior, although mental retardation and other psychopathologies to which these individuals appear to be predisposed sometimes include abnormal sexual behavior.

Genetic sex seems to influence final sexual identity by its effects on the developing fetus. Genetic sex determines whether the fetus will develop testes or ovaries. When everything goes right during fetal development, the type of gonads which develop determine whether androgens will be secreted, and androgens affect the morphological development of the fetus. The effect of genetic sex, therefore, is expressed through hormonal action. When the hormonal activity which occurs during fetal development is not normal and is not consistent with genetic sex, then genetic sex has very little influence on sexual development and identity.

Pathways Leading to the Establishment of Sexual Identity

The information which we have presented in the previous sections shows that sexual identity in humans is not determined by genetic sex, hormonal sex, morphological sex, or gender identity alone, but by a combination of all four components. Money has summed up the contribution of each of the four components in the following way.

Sexual identity begins to develop with the determination of genetic sex at the moment of conception. Genetic sex determines the fetal gonadal sex, which in turn determines the fetal hormonal sex. The fetal hormonal sex influences genital and sex organ development and brain development. Fetal hormonal sex thus sets the stage for adult hormonal sex by determining whether the adult will have a cyclic or tonic pattern of sex hormone secretion. Genital development determines sex assignment at birth and sex assignment determines the parents' behavior toward the child. The parents' behavior and the child's body image help the child to develop a gender identity.

At puberty the hormonal sex determines the type of body changes which will occur. These changes lead to the adult body image which influences adult gender identity. Hormonal influences at puberty also influence the type of erotic stimuli which are sexually arousing. The child's gender identity, the pubertal body changes, and

pubertal erotic stimulation lead to the development of the final, adult sexual identity.

THE PHYSIOLOGY OF SEXUAL AROUSAL

Masters and Johnson are the latest in a long line of sex researchers to document the physiological changes which occur during sexual arousal in adult men and women. Because their work provides the most complete description of the physiology of sexual arousal and because their research has enabled them to treat many people suffering from sexual disfunctions, we present a brief summary of their most basic findings in this section.

Phases in Sexual Arousal

Masters and Johnson have divided the physiological reactions which occur during sexual arousal into four stages: the excitement phase, the plateau phase, the orgasmic phase, and the resolution phase. These phases are seen in both men and women, although there are variations in male and female responses in the orgasmic phase and the resolution phase.

The **excitement phase** begins after an individual has been stimulated by sexually effective stimuli. The length of time it takes an individual to enter this phase depends on the intensity of the stimulation and the receptivity of the individual to stimulation. Stimulation may be visual, olfactory, or tactile. During this phase, the nipples of the breasts may erect in both men and women, although this is seen more commonly in women than in men. A **sex flush** may appear on the abdomen, neck, and breasts in women. A sex flush sometimes appears in men during this phase, but it usually does not appear until the plateau phase. The vagina enlarges in length and width and the vaginal walls begin to "sweat" a clear substance which provides lubrication. The clitoris becomes tumescent (erect) and increases in size and diameter. The penis becomes erect and increases in size and diameter. Erection may come and go several times during this phase. **Myotonia** (increased muscle tension) is seen in both men and women during this phase. Heart rate and blood pressure increase. The length of the excitement phase is variable, ranging from a few minutes to hours.

The **plateau phase** which follows the excitement phase is short and intense. The physiological reactions which began in the excitement phase continue to develop in this phase. The nipples of the breasts remain erect or become erect for the first time. The sex flush deepens and spreads over a wider area of the body in women and appears in some men for the first time. Myotonia becomes more marked. The outer third of the vagina dilates and the clitoris withdraws under its foreskin or hood. The uterus elevates itself in the abdomen and irregular contractions, which began in the excitement phase, increase in intensity. The urethral bulb of the penis enlarges, as a prelude to ejaculation, at the end of this period. The testes are elevated in the scrotum and pulled in closer to the body in preparation for ejaculation. Blood pressure and heart rate increase and respiration rate becomes elevated.

The **orgasmic phase** which follows the plateau phase is very brief, lasting three to ten seconds. It culminates in ejaculation in men and in orgasm in women. During this phase the myotonic reactions seen in the earlier phases increase greatly in intensity and extent. Muscular contractions occur in voluntary and involuntary muscles throughout the body. The sex flush, if present, remains intense. The vagina contracts rhythmically and uterine contractions similar to those seen in the first stage of labor are seen. The vaginal and uterine contractions appear after the subjective experience of orgasm in women. Clitoral changes cannot be seen during this phase since the clitoris remains withdrawn, beneath its

foreskin. The intensity of the vaginal and uterine contractions varies from woman to woman and from one experience to the next. Contractions appear to be more intense after orgasms achieved through masturbation rather than coitus. The subjective experience of orgasm also varies from woman to woman. Some woman report localization of sensation to the clitoral region, others report sensation in the vagina, still others report a general feeling of "warmth" extending over the entire body. The distention of the urethral bulb marks the onset of ejaculation and orgasm in men. The urethra and muscles at the base of the penis contract rhythmically to produce the ejaculatory response. Sensation is usually reported to be localized in the penis, but intensity of orgasm varies from man to man and from time to time. Blood pressure, heart rate, and respiratory rate reach maximum levels during this phase.

The **resolution phase** begins after orgasm. During the resolution phase all physiological systems return to their nonstimulated state. Myotonia disappears, as does the sex flush. The nipples lose their erection. The vagina decreases in size and the uterus returns to its normal position. The penis becomes detumescent and loses its erection. The testes move down in the scrotum. Men usually enter a refractory period during this phase during which they cannot be sexually aroused. This period may last from twenty minutes to a number of hours. Women do not usually enter a refractory period and they can be aroused to orgasm again, if the proper stimulation is available.

Types of Sexual Expression or Outlet

The type of sexual expression or outlet which is considered "normal" or "proper" has always been a topic of great interest in human societies. From the physiological viewpoint, all forms of sexual expression are basically the same, since the same physiological reactions occur in all sexual activities. The forms of sexual expression which are considered acceptable vary from culture to culture.

In our society it is often said that heterosexual coitus is the only "natural" form of sexual expression and that the desire to express sexual feelings in this way is inborn. It is said that since it is necessary for the species to reproduce and since reproduction is accomplished through heterosexual coitus, this form of sexual expression is the only biologically sound one. This kind of argument stems from a misunderstanding of certain biological principles. First it should be understood that for a species to reproduce itself, it is not necessary, or even desirable, for every member of that species to reproduce. Sexual activity, particularly sexual activity leading to reproduction, is not necessary for the survival of the individual. It is just as necessary to control population density as it is to reproduce to ensure the survival of a species. In all species, therefore, a balance must be achieved between the number of individuals which can successfully reproduce themselves and the number which must not reproduce. When this balance is upset and an area becomes overpopulated with individuals of a certain species, sexual and maternal behavior in that species often undergoes changes so that the number of offspring are reduced. Heterosexual coitus, therefore can be regarded as one biologically acceptable sexual outlet which individuals engage in under some, but not all conditions.

Another facet of the belief that heterosexual coitus is the only acceptable form of sexual expression is the belief that certain elements of sexual behavior are exclusively male or exclusively female. Studies of mammalian sexual behavior in many species including man indicate that male and female elements of sexual behavior are exhibited by both males and females in most species. The degree to which male behavior is found in females and female behavior is found in males depends on the species. In some species, such as the rat, in which hormonal influences during fetal life determine adult sexual behavior, the inter-

change of male and female elements of sexual behavior is rather small. This is particularly true with respect to males in whom female elements of behavior appear to be suppressed by the presence of androgens. Female rats exhibit a fair number of male elements of sexual behavior.

As we pointed out earlier, however, human sexual identity and sexual behavior are influenced as much, or more, by gender identity as they are by hormonal influences. This means that we would expect to find human sexual behavior to be more flexible than that of rats. Studies of human sexual behavior, such as the Kinsey reports, tend to support the idea that human sexual behavior is very flexible with respect to the kind of outlets used and the interchange of "male" and "female" behavioral elements.

There are a number of ways in which humans achieve orgasm. Kinsey distinguished six methods which account for most sexual outlets. These include masturbation, nocturnal orgasm, heterosexual petting, homosexual relations, sexual contacts with animals, and heterosexual relations.

Kinsey found that about 95 percent of the male population interviewed had masturbated at some time. Studies of female masturbation showed that, depending on the study and the sample, 50 to 80 percent of all women masturbate at some time.

Nocturnal orgasm is experienced by boys and men as "wet dreams." Ejaculation occurs during sleep, usually accompanied by erotic dreams. Women do not experience ejaculation, of course, but they can experience orgasm during sleep accompanied by erotic dreams.

Heterosexual petting is usually defined to mean sexual contact between people of the opposite sex which does not lead to coitus. Heterosexual petting is often used to mean premarital petting, although petting continues, of course, in marriage. Kinsey found that the number of men who participated in this kind of sexual outlet was influenced by their level of education. Men with lower educational levels petted less (and began heterosexual coitus earlier) than men with higher educational levels. Eighty-nine percent of all men sampled by Kinsey had petted at some time during their lives. One hundred percent of all married women had petting experience prior to marriage, according to Kinsey's findings, and 90 percent had petted at one time, regardless of marital status.

Homosexual relations constitute another major sexual outlet. Kinsey's study of sexual habits showed that more than one-third of the male population and one-fifth of the female population had had at least one homosexual experience. Later studies tend to confirm these figures. It should be noted, however, that Kinsey found that very few people choose homosexual relations as an exclusive form of sexual outlet.

It should also be pointed out that there are a number of societies (the Batak people of Sumatra and the Marin-Amin tribe of New Guinea, for example) in which male homosexuality during adolescence and sometimes at other times of life is an institutionalized and accepted practice. This kind of attitude toward homosexuality is often found with societies which prize female virginity at marriage. In these societies, boys make a smooth transition from homosexual relations to heterosexual relations when they marry.

Sexual contacts with animals are not a common sexual outlet in our urban society. Kinsey found that 40 to 50 percent of all boys reared on farms had some sexual contact with animals. Boys raised in the city had occasional contacts with household pets. Very few girls had sexual contacts with animals.

Heterosexual coitus is a form of sexual outlet experienced by almost everyone in our society. Kinsey found that 67 to 98 percent of all men (depending on educational level) had experience with heterosexual coitus before marriage. Studies done in the 1960's and 1970's replicate these results. Approximately 50 percent of all women in Kinsey's sample had experienced heterosexual coitus before marriage.

After marriage almost all men and women participate in coitus. A small percentage do not because of medical or psychological difficulties or because of ignorance of how to proceed. Kinsey found that marriage did not preclude the utilization of sexual outlets other than coitus. For example, by age fifty-five marital coitus supplied only 68 percent of all sexual outlets of married men. The remaining 38 percent was supplied by relations with other men and women, by masturbation, or by nocturnal orgasms.

BIRTH CONTROL

One area in which biological research has greatly affected the expression of human sexuality has been in the development of new methods of birth control and the refinement of old methods. Effective methods of birth control separate heterosexual coitus from its former consequence, conception. Today the rapidly increasing human population has caused many people to predict that we will soon outstrip our resources for feeding, clothing, and supplying energy for all the people in the world. Population control is a problem considered to be one of the most serious, or the most serious, problem facing the world today. Birth control techniques are therefore of great interest to us all.

Contraceptive Techniques

Contraception means the prevention of conception. Contraceptive techniques are methods for preventing the formation and implantation of a fertilized egg which can follow heterosexual coitus. There are many contraceptive techniques available for use today. Some are methods which have been in use for centuries; others are the product of modern scientific research. We will describe briefly these techniques and their relative effectiveness.

Abstinence means self-denial. In some cultures abstinence from heterosexual relations is prescribed at certain times of life—before marriage or for a certain period after the birth of a child. This method of contraception is not usually effective over lifelong periods.

Coitus interruptus (withdrawal) is one of the oldest contraceptive techniques known to man. It is mentioned in the Bible. *Coitus interruptus* involves the withdrawal of the penis from the vagina before ejaculation occurs. This technique has the advantage of requiring no mechanical devices or prior planning. Its disadvantages are many, however. First, it requires that the male maintain complete conscious control of his actions throughout coitus. Since coitus involves many involuntary physiological reactions, complete control is very difficult. Second, the first few drops of the ejaculate contain the greatest concentration of sperm. This means that even a slight hesitation in withdrawal at the proper moment is likely to drastically reduce the effectiveness of this method of contraception. Third, some sperm cells are present in the secretions of Cowper's glands which moisten the tip of the penis before ejaculation occurs. Thus, even if withdrawal is accom-

FIG. 35.1 Various kinds of contraceptive foams, jellies, and creams are readily available without prescription today.

plished prior to ejaculation, sperm may be released into the vagina. This method is not considered very effective.

Douching flushes the vagina with some fluid (vinegar in water, for example). The purpose of douching after coitus is to remove the sperm from the vagina before they can travel up to the uterus and Fallopian tubes. Sperm in the ejaculate move very rapidly, however, and unless the woman douches immediately after coitus, the chance of removing any sperm is very small. Douching is not recommended as an effective means of contraception.

A number of different **spermacidal jellies, creams, and foams** are available for use as contraceptive agents. These jellies, creams, and foams are inserted into the vagina five to fifteen minutes before ejaculation is to take place. The advantages of using these spermacidal agents are that they are readily available in most drug stores, where they can be obtained without a doctor's prescription. Their effects are localized to the vagina. The disadvantages of this method of contraception are that it requires prior planning and it is messy. Used alone, these spermacidal agents are moderately effective. When they are used with a condom or with a diaphragm, their effectiveness is good.

The **condom** is a sheath of rubber or animal skin which fits over the erect penis. It is placed on the penis before it is introduced into the vagina. The condom provides a physical barrier which prevents the sperm from reaching the vagina, uterus, and Fallopian tubes. When the condom is checked before use to make sure that it is free from defect and when it is removed from the vagina immediately after ejaculation before the penis loses erection and can slip out, it is an effective means of contraception. Advantages of the condom are that it is readily available without a doctor's prescription, it is effective, and it does not interfere with any physiological functions. Disadvantages are the feeling of some men that the condom reduces sensation, the need for interruption of lovemaking to put it on, and the chance of mechanical failure.

The **diaphragm** is a rubber cup stretched over a collapsible metal spring coil. It is designed to fit over the cervix (the mouth of the uterus) and to mechanically prevent the entrance of sperm into the uterus. It must be used with a spermacidal jelly or cream which is applied to the inside and rim of the diaphragm before it is inserted in the vagina. It must be inserted in the vagina at least twenty minutes before ejaculation takes place. If coitus is repeated after the initial time, more jelly or cream must be applied with a special inserter. The diaphragm must be left in place for six hours after coitus. It is considered a very effective method of birth control. The advantages of the diaphragm are that, when used correctly and consistently, it is very effective and it does not interfere with the sensation of either partner. The disadvantages of the diaphragm are that it must be obtained with a doctor's prescription and it must be fitted by a doctor. Its use requires planning and reduces spontaneity and its insertion and removal can be messy. The necessity for keeping it in place for six hours after coitus is inconvenient. Of the mechanical means of contraception, it is about equal in effectiveness to the condom.

Oral contraceptives, which are popularly known as "the pill," are combinations of synthetic female sex hormones (estrogens and progesterone). When oral contraceptives are taken in the prescribed regime, they prevent ovulation and cause certain changes in the cervical mucus. These physiological reactions prevent conception. There are several types of oral contraceptives. The combination type has a combination of synthetic estrogen and progesterone in each pill. The sequential type has some pills which contain only estrogen and other pills which contain only progesterone. The combination pill is thought to be more effective.

The advantages of oral contraceptives are that they are the most highly effective contracep-

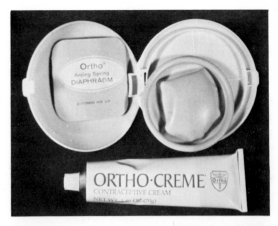

FIG. 35.2 The use of a diaphragm, which fits over the cervix of the uterus, with a spermicidal cream or jelly is one of the most reliable methods of contraception.

tive now available, they are not tied to coitus, so spontaneity is preserved, and they alleviate difficult menstrual problems in some women. The disadvantages stem mostly from the fact that these pills are very powerful hormones. Like any other potent drug, they not only produce the desired effect, but they also produce unwanted side ef-

FIG. 35.3 Oral contraceptives are a highly effective method of contraception. The package on the left has the days of the week marked as a reminder to avoid skipping a day or taking two pills in one day. Oral contraceptives can produce dangerous side-effects and should not be taken without a doctor's supervision.

fects. The most dangerous side effect of oral contraceptives is the possibility that the pills may induce or may predispose a woman to have a thromboembolism, or blood clot. Other side effects include nausea, weight gain, water retention, and changes in skin pigmentation. The long-range effects of the use of oral contraceptives over periods of years are not yet known. The benefits of the use of oral contraceptives, like any drug, must be weighed against the danger involved in their use. Oral contraceptives should only be used under close medical supervision.

The **intrauterine contraceptive device (IUD)** is a small metal or plastic device which

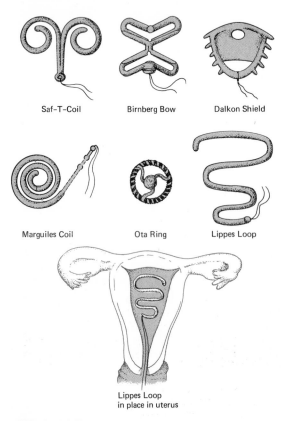

Saf-T-Coil   Birnberg Bow   Dalkon Shield

Marguiles Coil   Ota Ring   Lippes Loop

Lippes Loop
in place in uterus

FIG. 35.4 IUDs used today. The Dalkon shield has recently been removed from the market because it has been associated with cases of septic abortion in women who became pregnant while using it.

is designed to fit inside the uterus. It is not known exactly how an IUD prevents conception, but the current theory is that it acts as an irritant to the uterine wall and prevents the development of the conditions in the uterus which are necessary for the implantation of a fertilized egg to take place. An IUD must be fitted by a doctor and it must be removed by a doctor.

Advantages of the IUD include its effectiveness (a little less effective than the pill), its non-interference with the body's hormonal mechanisms, and its permanence. Disadvantages include the lack of knowledge about how it works, the inability of most women who have not had a child to retain it, the pain which follows insertion, and the increased severity of cramping and menstrual blood flow which are often experienced by women using an IUD. Some types of IUDs have been removed from use because a few women who have become pregnant while wearing them have suffered septic (infected) abortions.

None of the contraceptive techniques currently available are fail-safe or entirely without disadvantage. Each person who wishes to use some kind of contraception should be well informed about the methods available so that he or she may make an intelligent choice.

Abortion

When contraceptive techniques fail or are not used and an unwanted or medically undesirable conception takes place, abortion may be used as a means of birth control. Abortion of pregnancies up to twelve weeks duration (as measured from the last menstrual period) is accomplished by one of two methods: vacuum suction or dilation and curettage.

The **vacuum suction method** is the least traumatic and least expensive method of performing early abortion. In this method the cervix is dilated and a sterile tube is passed into the uterus. The contents of the uterus are then sucked out through the tube. This procedure requires only local anesthesia and can be performed in an outpatient clinic.

The **dilation and curettage method (D and C)** often requires overnight hospitalization because it is usually performed under general anesthesia. In this method the cervix is dilated and the contents of the uterus are removed by scraping with an instrument called a curette. This technique is also used to diagnose conditions causing menstrual problems.

After twelve weeks of pregnancy, abortion can be induced by injecting saline into the uterus to induce early labor or by a **hysterotomy**—a procedure in which the fetus is surgically removed from the uterus. Both of these procedures require hospitalization. These procedures are expensive and unpleasant. This means that if a woman is considering abortion, she should act as early in the pregnancy as possible.

Sterilization

When a couple has had all the children they want or when an individual decides that he or she does not wish to have children, sterilization may become a desirable alternative to contraception as a birth control technique. Both men and women may be sterilized without losing their ability to function sexually.

In men the sterilization procedure is called a **vasectomy.** In this procedure the *vasa deferentia,* the tubes which lead from the testes to the ejaculatory ducts, are cut so that the sperm produced in the testes cannot reach the ejaculatory ducts to enter the ejaculate. A man who has a vasectomy will not notice any difference in his sexual response or in the volume of his ejaculate since the semen which forms the bulk of the ejaculate is not manufactured by the testes but by the seminal vesicles, the prostate gland, and Cowper's glands. This procedure can be carried out in a doctor's office or in an outpatient clinic under local anesthesia.

The sterilization procedure in women involves a procedure called a **tubal ligation.** In this procedure the Fallopian tubes, which transport the egg from the ovaries to the uterus, are cut and tied off. The ovaries are not affected by this procedure and they continue to produce female sex hormones and the woman continues to ovulate and menstruate, but since the eggs can no longer reach the uterus or Fallopian tubes, she can no longer become pregnant. This procedure involves abdominal surgery and requires hospitalization.

Sterilization is an excellent form of birth control for people who are sure that they do not want to have any more children, but these procedures are irrevocable ones and the decision to undergo sterilization should not be made without a great deal of thought. It is now possible for men to store sperm in sperm banks for limited periods of time, and so a man who undergoes a vasectomy can still have the option of having a child by artificial insemination, but the procedures for preserving sperm are not yet perfected and this back-up system cannot be guaranteed.

Birth Control Techniques of the Future

A number of contraceptive techniques which are improvements on those available today are currently being studied. One method which is now being tested is the use of the "mini-pill." The "mini-pill" differs from other oral contraceptives in that it contains very small amounts of only one hormone—progesterone. This pill works by changing the consistency of the cervical mucus. Since estrogen is the component of conventional oral contraceptives which causes the side effects, the "mini-pill," if proven effective, should be a better kind of oral contraceptive.

Another experimental method of birth control also uses synthetic progesterone. Long capsules of progesterone are implanted under a woman's skin. These capsules release very small amounts of progesterone over a long period of time. These progesterone implants can last for a year and possibly longer. If this method proves to be safe and effective it would provide a very convenient long-term technique of birth control.

Prostaglandins, hormonelike substances produced by many body tissues, may also provide a new contraceptive technique. Prostaglandins can be used to induce abortion by inserting prostaglandin tablets into the vagina of a pregnant woman. Prostaglandins are thought to have some role in causing the uterine contractions of labor, and these contractions may be artificially induced by prostaglandin vaginal tablets. The tablets could be used not only to induce abortion but also to induce a late menstrual period. Theoretically, a woman who found her menstrual period was delayed would simply insert the prostaglandin tablets until her period began. At this point she would not know if she were pregnant or if her menstrual period had been delayed for some other reason. If a woman had a normal menstrual cycle and could predict fairly accurately when her menstrual period should begin, she would have to use prostaglandins only once or twice a year, since this is the average number of pregnancies which would occur by chance during unprotected coitus during a year.

Another line of research has been directed at producing a vaccine which would immunize a woman against her sexual partner's sperm. This research is based on the observation that some women are naturally immune to the sperm of their partners and therefore cannot conceive.

An oral contraceptive for men has been developed, but because of undesirable side effects, such as acute nausea following ingestion of alcohol and a reduction of sexual drive, it has never been marketed. Research into this type of contraceptive still continues.

Another line of research on male contraceptive techniques involves the development of a reversible vasectomy. A slicone plug is being worked on which could be implanted in the vas deferens to block the passage of sperm and then removed when fertility is desired.

REVIEW QUESTIONS AND PROBLEMS

1. How does genetic sex determine hormonal sex?
2. How do androgens affect the developing fetus?
3. Do female sex hormones (estrogens) have any effect on fetal development?
4. Which is the most important component of human sexual identity? Give evidence to support your answer.
5. What information can we obtain by studying children with the adrenogenital syndrome and the androgen-insensitivity syndrome?
6. What sort of physiological changes occur during the excitation phase of sexual arousal?
7. How do men and women differ in their physiological reactions during the resolution phase of sexual arousal?
8. What kinds of sexual outlets are there? Do physiological reactions to sexual arousal differ with the different kinds of outlets?
9. What determines choice of sexual outlet?

10. What kinds of birth control techniques are currently available? Name five and list their advantages and disadvantages.
11. What are the advantages and disadvantages of sterilization as a birth control technique?
12. Why should oral contraceptives be used only under medical supervision?

FURTHER READING

Boston's Women's Health Book Collective. 1973. *Our Bodies, Ourselves.* New York: Simon and Schuster.

Brecher, E. M. 1971. *The Sex Researchers.* New York: New American Library.

Ellis, H. 1938. *Psychology of Sex.* New York: New American Library.

Freud, S. 1962. *Three Essays on the Theory of Sexuality.* New York: Avon Books.

Kinsey, A. C., Pomeroy, W. B., and Martin, C. E. 1948. *Sexual Behavior in the Human Male.* Philadelphia: W. B. Saunders.

—,—,—, and Gebhart, P. H. 1953. *Sexual Behavior in the Human Female.* Philadelphia: W. B. Saunders

Masters W, H., and Johnson, V. E. 1966. *Human Sexual Response.* Boston: Little, Brown.

Money, J., and Ehrhardt, A. A. 1972. *Man & Woman, Boy & Girl.* Baltimore: The Johns Hopkins University Press.

36

The Principles of Heredity

Mankind has realized the importance of heredity for a long time. It is easy to see that children resemble their parents and that certain family traits appear in different generations. Racial groups have distinctive features which are passed on to all members of their group. Each person also has his own particular characteristics which are different from all others. Plant and animal breeders have found that they can maintain desirable traits by controlled breeding and that they can establish new and desirable strains by selecting animals with desired characteristics. The great variety of cultivated plants and domestic animals which we use today are the result of this practice. Although man has long applied some of the principles of heredity, an understanding of the mechanism involved was not reached until about the beginning of the present century. Investigations since then have been extensive, especially since midcentury, and our knowledge has expanded rapidly. Many speculations about heredity were made before we had any verifiable knowledge. Let us see how some of these have led to our modern concepts.

EARLY SPECULATIONS ON HEREDITY

One ancient concept about heredity was that blood was the force of heredity and that a new life was created by a mingling of blood from the two parents. We still use such terms as blood relative, blue blood, bad blood, and royal blood as a carry-over from this ancient idea. Aristotle incorporated this idea in a hypothesis around 350 B.C. He suggested that human semen was highly purified blood which mingled with the less purified blood, the menstrual fluid, of a woman during intercourse. These mingled "bloods" then would coagulate and form an embryo. Others accepted this idea and suggested that the blood came from various parts of the body and carried with it information for construction of similar body parts in the embryo. Thus, if a man had a pug nose, blood from his nose would move to his reproductive organs and be purified into semen which would create a pug nose in his offspring.

When sperm and eggs were discovered, it became apparent that these were carriers of heredity material. Some of the first scientists to use microscopes even imagined that they could see tiny human embryos within the sperm head and suggested that these needed only to enter a woman's uterus and develop. This idea, known as **preformation,** had to be discarded as microscopes were improved and observations failed to confirm it.

Inheritance of Acquired Characteristics

During the early part of the nineteenth century **Jean Baptiste Lamarck,** a Frenchman, proposed a theory which was widely accepted then and which still believed by some people today. He thought that reproductive cells were influenced by the activities of the carriers of these cells and that traits developed by the parents could be transmitted to the offspring. A giraffe was supposed to have a long neck because it continually stretched it in an effort to reach leaves on high trees. Each generation's stretching was passed on to its descendants, with the neck becoming longer and longer each generation.

The concept of **use and disuse** came to be accepted. This concept states that body parts which are used extensively enlarge and become more functional, while those which are not used

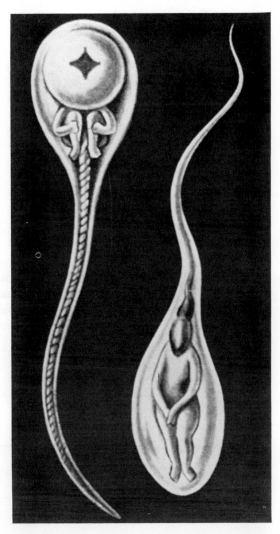

FIG. 36.1 Preformation was a theory of evolution which was widely accepted during the late seventeenth century. When sperm were viewed under the primitive microscopes of the time, some imagined that they could see tiny human embryos. These are drawings after Hartsoeker, 1694 (A), and after Dalempatius, 1699 (B). (From Winchester, *Genetics,* Houghton Mifflin, 1972.)

tend to regress. These changes are transmitted to the offspring. Lamarck got into difficulty because he did not distinguish between individual adaptation and species adaptation, a topic which was discussed in Chapter 1. Certainly individual organisms show adaptative changes, but we have found no way in which these changes can be incorporated into the genes.

Charles Darwin, whose extensive studies led him to propose the theory of natural selection, also believed in the inheritance of acquired characteristics as a method of producing species adaptations. He proposed the theory of **pangenesis,** which states that tiny pangenes are generated in the various parts of the body and migrate down to the reproductive cells. The pangenes carry directions which lead to the production of similar parts in offspring. If a man developed his arms through active use, for example, pangenes from these arms would be transmitted to his child, who, as a result, would have larger arms.

The Germ Plasm Theory

In the latter part of the nineteenth century a German, **August Weismann,** began an experiment to test the concept of inheritance of acquired characteristics. He cut the tails off newborn mice for twenty-one generations. Then, in the twenty-second generation, he allowed the tails to grow and they were just as long as the tails of the first generation of mice. This convinced him that the reproductive cells were separated from the rest of the body and were not influenced by it. His **germ plasm theory** holds that certain parts of the embryo are set aside for reproduction and grow somewhat like a parasite on the rest of the body. These parts are known as the germ plasm, while the rest of the body is called the **somatoplasm.** The somatoplasm can be modified in response to environment, but the germ plasm remains the same throughout life.

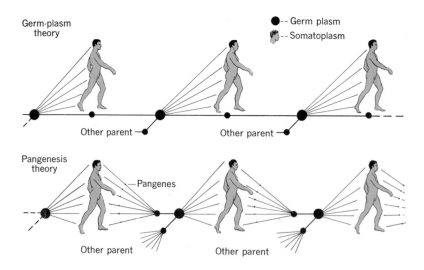

FIG. 36.2 Two theories of inheritance held during the nineteenth century. Darwin suggested the pangenesis theory which held that germ plasm (reproductive cells) was formed from the somatoplasm (general body tissue). Weismann felt that the germ plasm remained distinct from the somatoplasm and that the germ plasm produced the somatoplasm but was not influenced by it. (A. M. Winchester, *Genetics,* Houghton Mifflin, 1972.)

The Mutation Theory

The concept of an unchanging germ plasm had to be modified when **Hugo DeVries, a** Dutchman, proposed the mutation theory. He noticed the appearance of a few variant types of evening primroses among the many which grow so abundantly in Holland in spring. He brought some of these into his garden and found that they could be propagated from season to season from seed. He proposed the mutation theory, which says that the germ plasm can undergo sudden, random changes in rare instances.

We now accept the germ plasm concept as amended by the mutation theory, although we have found that the rate of mutation can be increased by various agents. We know that individual adaptations are not passed to descendants through genes.

Gregor Mendel's Discoveries

While other scientists were concerned with the merits of the proposals of Lamarck, Darwin, and Weismann, a modest monk in Brno, Czeckoslovakia, was conducing experiments with garden peas which were to demonstrate the basic principles of heredity which are accepted today. **Gregor Mendel** reported his findings and pub-

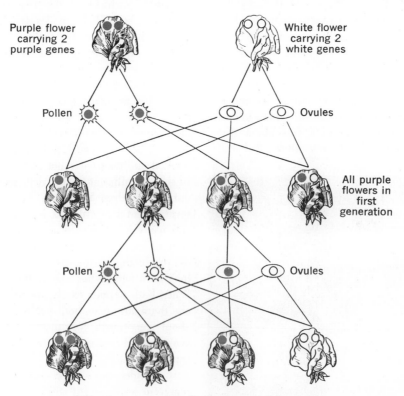

Purple flower carrying 2 purple genes

White flower carrying 2 white genes

Pollen

Ovules

All purple flowers in first generation

Pollen

Ovules

3 Purple to one white flower in second generation

FIG. 36.3 One of Mendel's garden pea crosses. Results from these experiments led him to develop his Laws of Inheritance.

lished them in 1865, but their great significance was not realized at the time, and not until 1900 did these results become the basis for modern genetics.

MENDEL'S LAWS

Mendel used garden peas because he could obtain them in a number of pure-breeding varieties. Some had purple flowers and some had white; some had green seed and others had yellow seed; some had flowers only at the ends of the stems while others had them at each axil. The garden pea is different from many organisms in that it can fertilize itself, since both male and female organs are enclosed in the flower. Peas can be cross-pollinated by transferring pollen from one flower to another. One of Mendel's crosses was between purple and white flowered plants. The seed produced by the cross grew into plants with purple flowers, but when two plants from this second generation were crossed, the third generation was about three-fourths purple and one-fouth white—3:1 ratio.

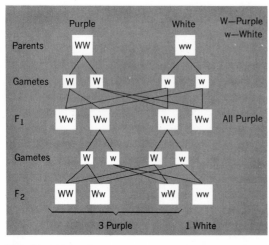

FIG. 36.4 Using letters to represent factors or genes, Mendel diagrammed a monohybrid cross as shown here.

Law of Dominance and Recessiveness

Crosses involving other traits produced similar results, and Mendel reasoned that each plant must carry two factors (now called genes) for each trait and that the pollen (male gametes) and the ovules (female gametes) each carry one of these factors. A pure-breeding strain with purple flowers would have both genes for purple and a pure-breeding white strain would have both genes for white. A cross of these two strains would produce offspring with one gene for purple and one for white. The plants will have purple flowers, however, because the gene for purple is dominant over the gene for white. Chance determines which of the two genes go into a gamete, so half the ovules and half the pollen grains will have the gene for purple and the other half will have the gene for white. Figure 36.3 shows how these combine to give the 3:1 ratio in the next generation.

This principle was found to hold true for many traits in many different species, including the human species. It explains how traits can be carried in one generation without being expressed and then come out in future generations. In some families normal parents may have several normal children and then have a child with severe mental retardation because both the parents carry a recessive gene which causes a deficiency of some vital cellular enzyme. The gene for the production of the enzyme is dominant and as long as a person has one of them he is normal. When both parents carry one recessive gene and one dominant gene, both parents are normal, but about one-fourth of their children, on the average, will receive both recessive genes and will express the defect.

Mendel devised a system of using letters to represent genes and it still is used today. A capital letter represents the dominant gene and a lowercase form of the same letter represents the recessive gene. The letter is the first letter of the trait which deviates from normal. For the color of pea flowers the letter is *w* since white is a recessive trait and less common than the dominant purple, which is designated *W*. Albinism in man is due to the gene *a*, and the dominant normal gene or alternate is *A*. Blaze, a white forelock of hair, is determined by a dominant gene, but it is less common than nonblaze, so the letter *B* stands for blaze and *b* for the more common nonblaze.

Modern Terminology

Certain genetic terms need to be understood before we go further in this discussion. A cross between two organisms which differ primarily in a trait produced by a single pair of genes is called a **monohybrid cross.** Let us illustrate some of the terms by using, as an example, a cross between a pure-breeding black guinea pig and a pure-breeding white guinea pig. Both guinea pigs are **homozygous** in their genes for coat color; that is, they carry two genes of the same kind. The offspring of this cross will have one gene for black and one for white; they will be **heterozygous** for these genes. The offspring will all be black, because the gene for black is dominant over the gene for white. During meiosis in these heterozygous guinea pigs, the gene number is reduced to one-half, and one-half the sperm and one-half the eggs will have the gene for black and the other half will have the gene for white. When this first generation is mated among itself, three-fourths of the offspring will have at least one

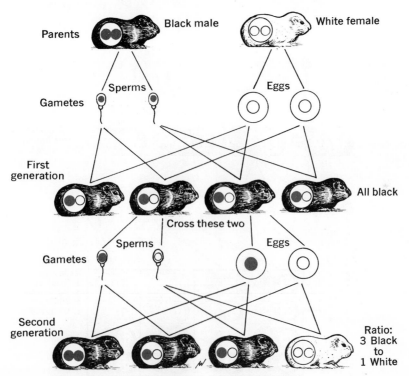

FIG. 36.5 This monohybrid cross of guinea pigs shows that Mendel's theories apply to animals as well as to plants. The genes for black coat color are shown as colored circles; those for white coat color are shown as clear circles.

gene for black and will be black. The other one-fourth will receive two genes for white and will be white.

The original parents are designated as the P₁, the first generation is called the F₁, and the second generation is called the F₂. The genes for black and white are called **alleles,** a term which refers to alternate forms of a gene which occupy the same locus on the chromosome and determine the same trait. Remember that chromosomes are paired and an allele of a particular gene located at a certain spot on one chromosome matches an allele at the same spot on its partner chromosome. (An exception to this is the male sex chromosome, which we will study later.) The

genotype of an organism is composed of all the particular alleles it possesses, while the **phenotype** is the expression of these alleles. Black coat color is the phenotype of guinea pigs carrying either one or two alleles for black coat color, but only those homozygous for the white allele have the white coat color phenotype. Two or more genotypes may be expressed as the same phenotype.

Genetic Ratios

Genetic ratios which have been worked out statistically do not represent the actual propor-

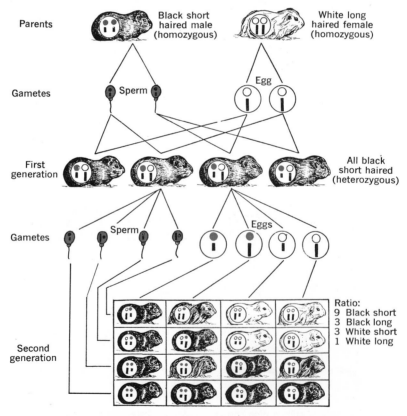

FIG. 36.6 Mendel's second law, independent segregation, is demonstrated by this dihybrid cross of guinea pigs. The genes for coat color and length of hair are recombined in the second generation.

FIG. 36.7 Genetic ratios are illustrated by the kernels on an ear of corn. Top: a second generation ear from a cross of pure, starchy, full kernel corn with pure, sweet, shrunken kernel corn. The 3:1 ratio is evident. Middle: the second generation results of a cross of pure, purple (dominant) corn with pure, white (recessive) corn. Bottom: a dihybrid combination of the two traits.

tions of genotype and phenotype which will occur experimentally. Especially when there are small numbers of offspring, considerable variation from the expected ratios may occur. Human albinism is the result of possessing a homozygous genotype for a certain recessive gene which fails to produce an enzyme needed for melanin formation. One couple with normal melanin pigmentation had a child who was an albino. When the rules of heredity were explained, the mother said, "I am so glad to know that now that we have had the one albino child, we can have three more children who will be normal." Two alleles for albinism are just as likely to come together at a second pregnancy, however, as they were at the first. Some such families have had three albino children and none with normal pigmentation. If many families in which the parents both have this allele are surveyed, however, the ratio will come out almost exactly 3:1.

Law of Independent Assortment

After Mendel had verified his hypothesis on dominance and recessiveness through many monohybrid crosses, he began to wonder if each gene was assorted independently of other genes. Chromosomes had just been seen and no one had yet associated them with heredity. Mendel made some crosses to answer this question. He crossed peas which were yellow and round (spherical) in

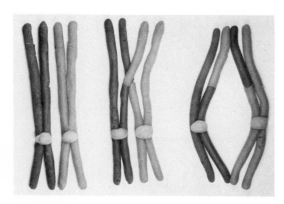

FIG. 36.8 Models to demonstrate crossing-over between chromatids of paired chromosomes during the prophase of the first meiosis. The two chromatids of each chromosome are held together by the centromeres. Crossing-over allows recombination of genes on the same chromosome.

shape with some which were green and wrinkled. All the offspring were yellow and round, since these two are dominant traits. In the F_2, however, he obtained a ratio of about 9:3:3:1. The way this works out is shown in Figure 36.6. These results showed that genes segregate independently in the reproductive cells. The recombinations of factors were just as likely to be found as the original combinations in the P_1.

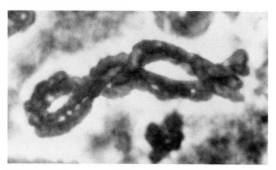

FIG. 36.9 Crossing-over in a cell taken from a grasshopper testis. This highly magnified view of a single pair of chromosomes shows exchanges taking place at four points.

Linked Genes and Crossing Over

Mendel's results were rediscovered in 1900 at about the time that chromosomes were being studied and the principle of meiosis had been discovered. The behavior of chromosomes was found to parallel Mendel's postulated behavior of the factors of heredity, and it was soon found that genes are located on chromosomes. Genes are much more numerous than chromosomes, and each chromosome is composed of many genes. Any two genes on the same chromosome, therefore, will not show independent segregation but are linked together. Mendel did not happen to make any dihybrid crosses involving genes of the same choromosome, but this kind of linkage was demonstrated in the sweet pea by two English geneticists, William Bateson and R. C. Punnet, in 1906. They found that a pair of alleles that influences flower color is linked with a pair that influences the shape of pollen grains. A cross between a variety of sweet peats with purple flowers and cylindrical pollen grains and one with red flowers and spherical pollen produced F_1 offspring with purple flowers and cylindrical pollen, indicating that these two traits are dominant. The F_1 generation was crossed with the double recessive (red and spherical). Bateson and Punnet expected a 1:1:1:1 ratio from this cross, but instead they obtained a 7:1:1:7 ratio. There were seven of each parental type and one of each recombination type.

This ratio led the investigators to suspect that the two genes were linked by being on the same chromosome. If this were true, they still had to explain how they obtained some recombinations. It appears that segments of homologous chromosomes can be exchanged. During the first meiotic division, the paired chromosomes are already duplicated, so they form tetrads of four chromatids each. Cytological studies show that the chromatid tetrads can exchange pieces of membranes at this time. See Figures 36.8 and 36.9 for an illustration of this exchange.

This exchange is known as **crossing over** and it has evolutionary value. It permits selection for individual genes rather than entire chromosomes. Without crossing over, selection would favor or eliminate chromosomes according to their overall effect. One chromosome might have some harmful genes and be eliminated even though it might also have some beneficial genes. Crossing over permits the most beneficial combinations of genes to be selected.

OTHER PATTERNS OF HEREDITY

Intermediate Inheritance

As genetic investigations continued, it became apparent that Mendel's laws did not explain the inheritance of all traits. Some crosses produced offspring which were intermediate between the two parents. When red and white short-horned cattle were crossed, the offspring were roan, an intermediate shade. This has been called **blending inheritance.** It should be noted that the genes for red and white remain distinct and can be expressed as red and white in future generations. This is illustrated in Figure 36.10. We cannot use capital and lowercase letters as gene symbols, since there is no dominance, so we use superscripts. The gene for red can be presented as C^R, C for color and R for red, while the gene for white would be C^W.

Codominant Inheritance

In some cases heterozygous individuals express both alleles fully and are said to exhibit **codominant inheritance.** Human blood types illustrate this principle. Three alleles account for the four basic blood types. The allele for type O is recessive to the alleles for types A and B, but the alleles for types A and B are codominant. We use the symbols, a, A, and A^B for the three genes.

The six possible genotypes and the four phenotypes are shown below.

Genotypes	Phenotypes
a/a	O
A/A or A/a	A
A^B/A^B or A^B/a	B
A/A^B	AB

Heterozygous Expression of Recessive Genes

We have shown the recessive genes are not expressed when a dominant allele is present, but delicate tests can sometimes show the presence of recessive genes in heterozygous individuals. The fruit fly, *Drosophila*, usually has red eyes, but a certain mutation leads to a failure to produce one of the pigments needed for red and the eyes of flies homozygous for this trait are sepia (brownish). Flies heterozygous for sepia appear to have eyes which are equally as red as those who are homozygous for the dominant allele. If the eye pigments are separated by chromotography, however, a difference can be ceen. The heterozygous fly contains less of a certain pigment.

A genetic defect called **Tay-Sachs disease** appears in children who are homozygous for an allele which fails to direct the production of an enzyme (hexoaminidase A) which is involved in the metabolism of fatty material. These children are normal at birth, but fatty deposits accumulate around nerve cells. This leads to a gradual loss of mental and motor abilities and death occurs within a few years. Carriers of this allele can be identified because their cells produce less of this enzyme than do the cells of homozygous normal persons. This allele is widespread among Jewish people whose ancestors came from eastern Europe. Recently a widespread testing program was undertaken in the United States to uncover the carriers of this allele. If the carriers do not have

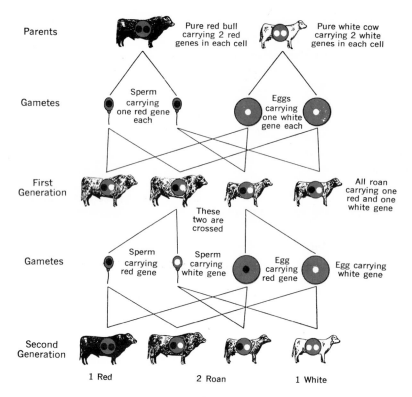

FIG. 36.10 Intermediate inheritance of hair color in short-horned cattle. Cattle which are heterozygous for red and white coat color are roan.

children, then in one generation this tragic disease could be wiped out. About one Jewish person in every thirty in the United States is a carrier, and all of them would have to forego parenthood.

A similar testing program was instituted for **sickle cell anemia** among blacks in the United States. About one in every ten blacks is a carrier of the allele for this type of anemia. The red blood cells become sickle-shaped and do not carry oxygen well. Death is common in early childhood, although some people can survive with continued treatment. This defect is caused by the alteration of one amino acid in the hemoglobin molecule (see Chapter 9). Those who are heterozygous for the trait have both normal and aberrant hemoglobin. They may not have anemia, but separa-

tion of their hemoglobin molecules can detect the two types of hemoglobin. Carriers of this trait can, of course, pass it on to their children.

Carriers of other harmful recessive genes can be detected by techniques which detect differences at the cellular level even though there are no differences in the total organism. These tests are valuable for genetic counselors who advise couples on their chances of having abnormal children. A serious hereditary disease such as **cystic fibrosis** may be carried in the families of both husband and wife. They can be tested to see if either one of them is a carrier for this trait. If both are carriers, they are advised that the chance of their having a child with cystic fibrosis is 1:4 for each child they have.

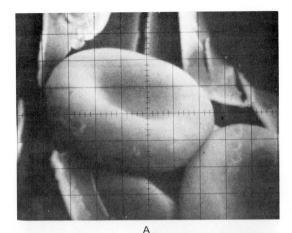

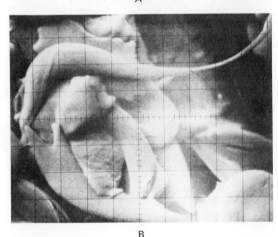

FIG. 36.11 Normal red blood cells (A) and blood cells from a person with sickle-cell anemia (B). Sickle-cell anemia is a hereditary disease which causes hemoglobin to form rigid chains inside the red cells. This produces distortion of the cells. Sickle-cells cannot carry oxygen well and the afflicted person suffers from extreme anemia. The squares are 1 micron each way.

Quantitative Inheritance

Not all characteristics are inherited as dominant-recessive, intermediate, or codominant traits. Some traits show a continuous variation from one extreme to the other. Human stature is an ex-

ample. We know that environment plays an important part in growth, but there is no doubt that heredity also influences growth. Like many other body traits, body stature is influenced by a number of genes located at different points on the chromosomes. These genes may be called **polygenes.** Most of them are expressed equally, so if a person receives an approximately equal number of genes for tallness and shortness, he will be medium in height, assuming he receives proper nourishment. Two parents of medium stature, however, can have a very tall child if the genes for tallness segregate as shown in Figure 36.12.

The amount of melanin in the skin is another example of the great quantitative variation in a trait which can result from polygenic inheritance. Some races inherit genes for heavy deposits of melanin, while others inherit genes for light deposits. Crosses between members of these races will produce children with intermediate skin

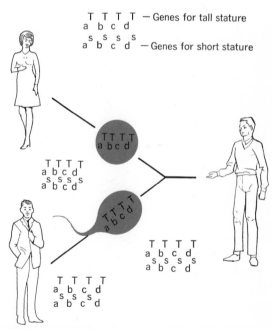

FIG. 36.12 Polygenic inheritance illustrated by human stature. Chance segregation of genes for tallness allows medium-sized parents to have a very tall child.

shades. Segregation of the genes during meiosis in these children can produce a wide range of pigmentation in their children.

SEX DETERMINATION

There are many differences between males and females. The sex organs are different, and other differences are found in many parts of the body. The skin, the skeleton, the blood, facial hair growth, and mammary gland development show characteristic sex differences. We cannot explain all the differences on the basis of a single gene or polygene; a mechanism which induces the development of sex differences must be involved. All people have the genes necessary to express characteristics of both sexes. Sex determination, therefore, involves some sort of trigger

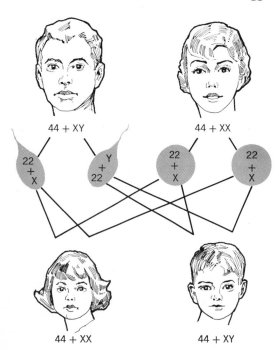

FIG. 36.13 Relationship between chromosomes and sex determination.

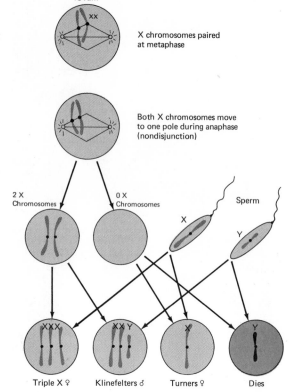

FIG. 36.14 Sexual abnormalities are produced by the nondisjunction of the X-chromosomes during oogenesis.

which stimulates one set of genes to operate while inhibiting the expression of another set of genes.

Sex Chromosomes

A clue to the mechanism of sex determination came early in this century when it was found that in some insects there was a difference in the chromosomes of males and females. The female chromosome could be matched up in perfect pairs, but the males had one pair which was unequal in size and shape. This unequal pair became known as the **X chromosome** and the **Y chromosome.** The female had two X's and the male had an X and a Y. This chromosome difference is found in many other species and in quite a number of plants which have separate sexes. We can use human beings to illustrate this.

The X and Y are known as **sex chromosomes,** while the other chromosomes are known as **autosomes.** A woman has a pair of X chromosomes and twenty-two pairs of autosomes. During oogenesis, the number of chromosomes is halved and each egg receives one X chromosome and twenty-two autosomes. A man has an X and a Y chromosome and twenty-two pairs of autosomes in his somatic cells. During spermatogenesis the X and Y pair segregate so that one-half the sperm receive an X and the other one-half

receive a Y. Sex determination depends upon which of the two types of sperm fertilize an egg.

We have learned much about the function of the sex chromosomes from cases in which they have not been distributed normally during meiosis. In rare cases the paired X chromosomes in a woman fail to disjoin after their pairing and an egg containing two X chromosomes and twenty-two autosomes is produced. An equal number of eggs are produced which contain no X chromosomes and twenty-two autosomes. When these abnormal eggs are fertilized with normal sperm, four abnormal sex chromosome distributions occur, as shown in Figure 36.14. The XO zygote becomes a female with **Turner's syndrome.** A

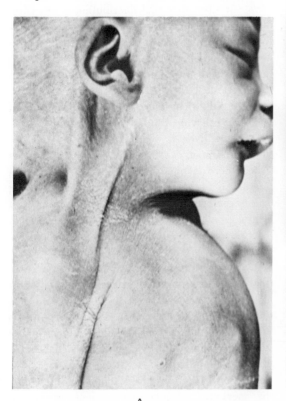

A

B

FIG. 36.15 Turner's syndrome. The baby shows the fold of skin on the neck at birth which is characteristic of this syndrome (A). A mature woman with Turner's syndrome is shown in (B). She takes female sex hormones to produce female secondary sexual characteristics.

person with this syndrome may appear normal at birth, but her sex organs do not mature properly at puberty. The condition can be recognized at birth by the presence of a webbed neck. Hormone treatment can help these girls develop normal secondary sexual characteristics, but they can never be fertile because their ovaries are not functional.

A zygote receiving only a Y chromosome dies, since the X chromosome contains many vital genes and no one can live without an X chromosome. The egg with three X chromosomes is called a **tri-X female,** or sometimes a **super**

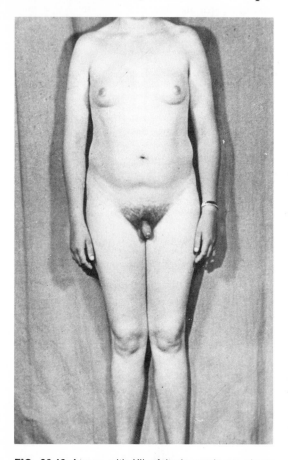

FIG. 36.16 A man with Klinefelter's syndrome shows breast development, feminine musculature sex organs and undersized. (Courtesy of Polv Riis, M.D., Gentofte University Hospital, Hellerup, Denmark.)

female. These woman are often normal in appearance, but some have menstrual disturbances and may be mentally defective.

The XXY combination produces a male with **Klinefelter's syndrome.** The sex organs are male, but they are only about half normal size. Although the testes do not produce sperm or adequate amounts of male hormones, males with Klinefelter's syndrome are capable of sexual intercourse. The musculature is somewhat feminine and some breast development may occur. Mental retardation may also be part of this syndrome.

Nondisjunction can also occur during spermatogenesis, but this is less common than in oogenesis.

Another abnormal distribution of sex chromosomes is the genotype XYY. Some men in penal institutions who have commited sex-oriented crimes have this distribution. These men are all taller than average and somewhat mentally retarded. This has led to speculation that the extra Y might cause a tendency for this type of criminal behavior. Routine genotype of men leading normal lives showed there are many who also have this genotype. The XYY combination arises when there is nondisjunction of the two chromatids of the duplicated Y chromosome during the second meiotic division. This produces a sperm containing two Y chromosomes which, combined with a normal egg, produces the XYY karyotype.

These abnormalities indicate that a Y chromosome is necessary for maleness, and that extra X chromosomes can upset the balance and produce female characteristics even when a Y is present. The more X chromosomes there are over the normal number, the greater the chance for mental retardation. A few cases of XXXX females and XXXY males have been found, and they may be mentally defective. Occasionally a person who appears to be a woman but who has a genotype of XY is found. This happens when a genetic male inherits a gene for androgen insensitivity and does not respond to its own output of androgen. Testes develop and they secrete androgens, but this gene makes the target cells unresponsive to the androgens. A person with the androgen insensitivity syndrome will appear to be female until puberty, when the secondary sex characteristics fail to develop. Development of the secondary sex characteristics can be controlled by administering hormones, but this individual is sterile because she possesses no internal female reproductive organs.

The XY method of sex determination is by far the most common in animals with separate sexes, but there are some deviations. In birds, butterflies, moths, and some fish, the two unlike sex chromosomes belong to the female and sex is determined by the type of sex chromosome in the egg. We say that the females are WZ and the males are ZZ. In a few insects such as the grasshopper, there is no Y chromosome and the male has only a single X. Half the sperm carry the X and the other half do not. In the honeybee and a few other related insects, there are no sex chromosomes; males hatch from haploid, unfertilized eggs and females from diploid, fertilized eggs.

Sex Hormones

Sex hormones have much to do with the development of sex characteristics, as we learned in Chapter 33. How are the hormones correlated with the sex chromosomes? Human sex is determined at the moment of conception, but the developing embryo will show no indication of its sex until there is a differentiation of the sex organs in the eighth week of development. It seems that the sex chromosomes trigger the primitive glands to secrete androgen or not to secrete them. Once this starts, the rest of sexual differentiation is determined by hormones.

In invertebrate animals there are no sex hormones and sex determination happens on a cell by cell basis. In each cell the sex chromosomes stimulate the genes involved to express one sex and the genes for the opposite sex are suppressed.

HEREDITY INFLUENCED BY SEX

X-Linked Inheritance

Genes on the X chromosome show different phenotypic ratios than those on the autosomes. The allele for the common form of **color blindness** is located on the human X chromosome. This is a recessive allele, but since a male has only one X chromosome, this allele will be ex-

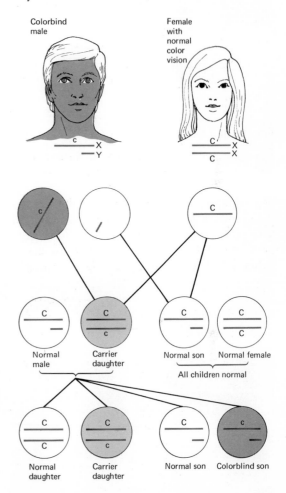

FIG. 36.17 Color blindness is caused by an X-linked gene. A color-blind man can pass the gene to his daughters, who inherit his X-chromosome, but not to his sons. About one-half of his daughters' sons, however, will inherit the trait.

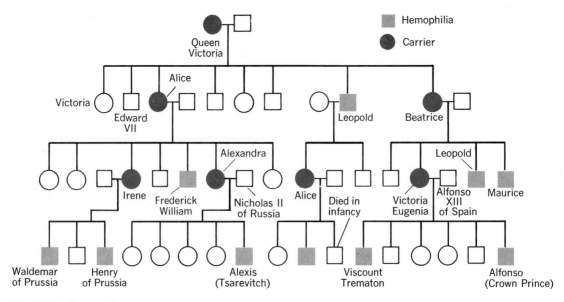

FIG. 36.18 Hemophilia, the bleeder's disease, appeared frequently in the royal families of Europe after first appearing in Queen Victoria of England as a mutation. The afflicted males are shown as the dark squares and the carrier women are shown as circles of a lighter shade.

pressed. This allele causes a defect in the cones of the eye which perceive green, and green and red perception are confused. The Y chromosome does not carry alleles of the genes found on the X chromosome. A female who has two X chromosomes, however, must receive the defective allele from both parents if she is to be color blind. Sex-linked genes follow a sort of skip-generation method of inheritance, as shown in Figure 36.17.

Genes on the X chromosome are known as **X-linked** or **sex-linked** genes. **Hemophilia,** excessive bleeding, is caused by a recessive X-linked gene. This gene fails to produce one of the plasma fractions necessary for clotting, as described in Chapter 30. Most of the genes on the X have no alleles on the Y, but there must be a few which are allelic or there would be no pairing of these chromosomes in meiosis.

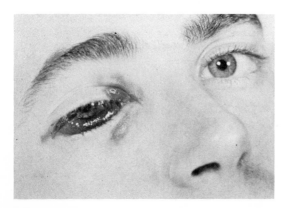

FIG. 36.19 This man has hemophilia and the rupture of a small blood vessel in his lower eyelid has caused extensive bleeding beneath the skin. (J. V. Neel, Heredity Clinic, University of Michigan.)

Sex-Limited Inheritance

Some genes are normally expressed only in one sex. The male beard and the female breast development are examples of traits which are normally found only in one sex. The type of beard is influenced just as much by genes from the mother as from the father, and the type of breast development is just as much influenced by genes from the father as from the mother. Cattle breeders have learned that the genes from the bull are just as important as those from the cow in determining the milk yield of female offspring. A rooster determines the egg-laying ability of his offspring just as fully as the hen.

Sex-limited genes are typically autosomal genes which are inhibited or stimulated by sex hormones in those animals which have sex hormones. In the insects and other invertebrates there are still sex-limited genes and the expression is on a cellular basis.

CHROMOSOME ABERRATIONS

Aberrations Involving Entire Chromosomes

Nondisjunction can occur in the autosomes as well as in the sex chromosomes. This produces a gamete with one extra autosome and one with too few autosomes. Union with normal gametes produces some zygotes which are **trisomic**—possess three of a chromosome—and some which are **monosomics**—possess only one of a chromosome. Even though all genes are normal, the balance between the quantity of genes is upset and serious abnormalities occur. Plants seem to be better able to survive such an upset of gene balance, and many plants are monosomic or trisomic for specific chromosomes. Most abnormalities in chromosome number are lethal in animals, although a few survive.

The most frequent of the nonlethal trisomies in man is **Down's syndrome,** or **mongolism.** Those with this syndrome have a downward fold of the upper eyelid which resembles the lid fold of the Mongolian race. They also have broad hands, stubby fingers, a wide rounded face, a large tongue, and greatly retarded mental development. About one in 600 babies born in the

United States has this syndrome. It is caused by **trisomy-21;** there are three chromosomes designated as 21, which is one of the shortest of the twenty-two autosomes. The chance of a woman having a baby with this syndrome increases greatly with the age of the mother. A woman in her twenties has about one chance in 1,500 of having a mongoloid baby, but a woman past forty has about one chance in 35. Women over thirty-five produce half the mongoloid babies born in this country, even though they have fewer babies, as a group, than women under thirty-five. The long contact of the chromatids during the second meiotic division in the ovary seems to increase the chance that they will not disjoin properly. See Chapter 8 for a further discussion of meiosis.

Trisomies of chromosomes 18 and 13 have also been found in live-born babies, but the effects are so extreme that they usually do not live more than a few months. Many other trisomies and even monosomies occur, however. A study of the chromosomes of spontaneously aborted fetuses show that at least 40 percent have chromosome aberrations of some kind. Spontaneous abortion is a natural way of purging defective embryos. The effects of most chromosome aberrations are so extreme that the fetus cannot survive to full term.

Aberrations Involving Pieces of Chromosomes

Sometimes chromosomes break and pieces without a centromere will be lost. This results in **deletions** of blocks of genes and again this upsets the gene balance in the cell. Persons with small deletions can live, although they may be abnormal. **Chronic granulocytic leukemia** is caused by the loss of a part of chromosome 21. Small deletions of other chromosomes cause various other abnormalities. Sometimes a broken piece will become attached to another chromosome and

A

B

FIG. 36.20 One chromosome too many. This little girl has Down's syndrome (mongolism) which is caused by the presence of three chromosomes 21. This imbalance in chromosome number causes mental retardation and other symptoms.

this can also produce genetic defects in future generations. This is known as a **translocation.**

The new techniques for analyzing chromosomes, which were discussed in Chapter 8, have made it possible to recognize many small chromosome abnormalities which were formerly overlooked. This has shown us that many human abnormalities for which we had no satisfactory explanation previously can be traced to small chromosome alterations. High-energy radiation and other agents can increase the number of chromosome breaks and thus the chance for aberrations.

PRENATAL DETECTION OF GENETIC DEFECTS

With the great improvement in prenatal care and diet for expectant mothers, the number of birth defects from prenatal deficiencies and injuries has declined. This means that more birth defects today are of genetic origin. We can recognize the cause of these defects by studying the chromosomes of the newborn baby. We now have a method of identifying these defects early in fetal development. Recent changes in abortion laws mean that the law now permits therapeutic abortion when it is found that the fetus is abnormal.

The technique of **amniocentesis** can be used to detect fetal abnormalities. A bit of the amniotic fluid is removed from around the fetus by a slender needle inserted through the woman's abdomen. Chemical tests of the fluid will show the presence of any abnormal metabolic products which indicate that certain genes are not functioning properly. Cells found in this fluid come from the epithelial cells of the fetus. These cells can be cultured and tested to see if they can produce vital cellular enzymes. Karyotypes can be made of the cell chromosomes to detect abnormalities. If abnormalities are detected by this technique, the parents may elect to have a therapeutic abortion. If no abnormalities are detected, then the parents may have the baby and peace of mind as well.

Suppose a woman forty-two years of age becomes pregnant. She knows that there is a high risk of mongolism and other defects in babies born to women in her age group and she is tempted to have an abortion rather than risk bearing a defective child. She learns of amniocentesis, however, and has it performed. The tests show that the cells and chromosomes are normal, so she continues the pregnancy and has a normal baby. Another couple find that they are both carriers for Tay-Sachs disease. They know that the chance that this affliction will appear in their children is one in four and have decided not to have children. When they learn about amniocentesis, they decide that they will begin a pregnancy and rely on amniocentesis to indicate whether it should be continued to term. When the fetus has generated sufficient amniotic fluid, amniocentesis is performed. The test shows that the fetal cells are producing the enzyme necessary for fat metabolism which is lacking in Tay-Sachs disease. They have a normal, healthy child.

Over forty different genetic defects can be recognized by amniocentesis and this technique is being used more and more. We also now have

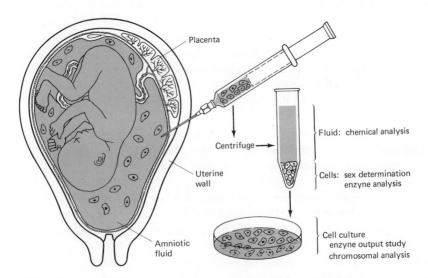

FIG. 36.21 Amniocentesis, the removal of some amniotic fluid, makes it possible to recognize genetic abnormalities during fetal developments. Information about a serious abnormality can be used to decide if a pregnancy should be terminated. (Adapted from T. Friedman, *Scientific American*, 1971.)

a **fetoscope,** which permits a visual examination of the fetus through a tube inserted through the cervix. This tool allows the detection of gross abnormalities of body structure. Use of these techniques can reduce the number of babies born with defects.

REVIEW QUESTIONS AND PROBLEMS

1. Some salamanders have been found in caves in Kentucky where there is no light. Their eyes are rudimentary and nonfunctional. How would Lamarck have explained the way in which they became blind? How would you explain it in the light of recent discoveries?

2. Mendel crossed tall pea plants with dwarf plants and found that tall was dominant. He then crossed the F_1 heterozygous plants with dwarf plants. Using the correct letter symbols, give a diagram of this cross, showing both genotype and phenotype of the offspring obtained.

3. A couple with normal pigmentation have a child who is an albino. Diagram this pedigree, showing the genotypes of both parents and the child.

4. A dog breeder wants solid-colored pups but gets a few which are spotted because some of the solid-colored dogs are heterozygous for the recessive gene for spotted color. How, through genetic crosses, could he determine which dogs are heterozygous for the recessive gene?

5. A breed of cats, the Manx, have short tails. When two Manx cats are crossed, however, only half the offspring have short tails. About a fourth have long tails, and a fourth have no tails. Give a diagram which explains this situation.

6. Which of the following human traits do you think would be due to variation in a single pair of alleles and which would be due to multiple alleles? **Anirida,** absence of the iris of the eye; blaze, a white forelock of hair; the shape of the nose; intellectual potential; cystic fibrosis, production of a very sticky mucus.

7. Black and short hair are both dominant traits in guinea pigs, while white and long hair are recessive alternates. Use a diagram to show the expected proportion of each kind of offspring from a cross of a heterozygous black, short-haired guinea pig and a white, long-haired one.

8. What biological disadvantage would be encountered if a species did not have chromosome crossing over during meiosis?

9. How does intermediate inheritance differ from codominant inheritance?

10. A woman who has blood type A has a child who is type O. She sues a man for support, claiming that he is the father. He has blood type B. Could he have been the father? Use a diagram to explain.

11. Of what value is the ability to recognize human carriers of recessive genes?

12. A newly married black couple find that the husband carries the gene for sickle-cell anemia but the wife does not. Should they be concerned about the chance for sickle-cell anemia in their children? What about their grandchildren?

13. Some blonde German women had babies fathered by black U.S. soldiers during

the occupation following World War II. These children are now grown and some have married fair-skinned Germans. Could any of the children of these marriages have skin pigmentation darker than that of the parents? Explain.

14. Suppose nondisjunction of the sex chromosomes occurs in the first meiosis of spermatogenesis. Show the kinds of sperm which would be produced and the abnormalities of sex which might occur if these sperm fertilize normal eggs.

15. How does sex determination in birds differ from that in people?

16. A dominant gene for androgen insensitivity causes human XY zygotes to fail to develop into males. The gene has no effect on XX zygotes. A woman is heterozygous for this gene. Use a diagram to show the sex ratio expected in her children.

17. Show by diagram the types of children expected from a color-blind woman married to a man with normal color vision.

18. A man and his wife have normal blood clotting time, but the wife's father had hemophilia. What are the chances that this couple might have children with hemophilia?

19. Doctors will not perform amniocentesis on all pregnant women. Under what circumstances do you think it would be justified?

FURTHER READING

Bergsma, D. 1973. *Birth Defects Atlas and Compendium*. Baltimore: Williams and Wilkins.

McKusick, V. 1971. *Mendelian Inheritance in Man*. 3rd ed. Baltimore: John Hopkins Press.

Stern, C. 1973. *Principles of Human Genetics*. 3rd ed. San Francisco: W. H. Freeman.

Winchester, A. M. 1972. *Genetics*. Boston: Houghton Mifflin.

———— 1975. *Human Genetics*. Columbus, Ohio: Charles E. Merrill Books.

———— *Heredity*. New York: 1975. Barnes and Noble College Outline Series.

37

Patterns of Behavior

Biologists in the past have tended to leave the study of behavior to psychologists, but recent research in this field has shown that behavior is an integral part of biology. Its study fits logically into any study of biology as a whole. In this chapter we shall explore some interesting results of research on animal behavior with some reference to plant behavior where pertinent information exists.

HEREDITY AND BEHAVIOR

Are patterns of behavior learned or inherited, or do they represent some blend of both these factors? An understanding of the biological basis of behavior depends upon an answer to this question, so we shall first consider the possible role of heredity in behavior.

Examples in Man

If you clasp your hands together repeatedly you will find that you always put the same thumb on top. Reversing the position of the thumbs seems awkward and does not feel natural. Have you learned to clasp your hands in this manner? Did chance positioning of the thumbs when you first began to clasp your hands cause you to learn this? Or have you inherited some predisposition which determines the position? Studies of the hand-clasping trait in many fami-

624

lies indicate that a single gene determines which thumb you are inclined to place on top.

Many other body mannerisms and responses are influenced by your heredity. To say this is not to belittle the importance of learning in many human responses. The complex pattern of reflexes which mesh together in driving an automobile is certainly not inherited. Yet you do inherit a nervous system which can be conditioned by constant repetition of specific responses to certain situations. As a result, your learned responses become more or less automatic. Some people learn to play the piano easily and with a minimum amount of training; others may practice for many years with only very limited success. It is apparent that while people do not inherit such specific skills, they may inherit an aptitude for learning specific skills with ease.

Examples in Other Mammals

Since human beings are poor subjects for experimentation, we often turn to other mammals for objective studies of behavior. Some horses are known as pacers; they tend to move both right legs forward in unison, and then both left legs go forward while the right legs are going backward. Other horses have the trotting gait; they move the right front leg and the left hind leg forward at the same time, and then the left front and the right back leg come forward together. These horses need to be trained before they are ready for a race which allows only one type of gait, but a single gene determines which type of gait they can most readily be trained to follow.

Hunting dogs are well known for their inherited behavior patterns. Some dogs, the setters, respond to the sight of a game bird by sitting down and gazing intently at the bird. Other dogs, the pointers, stand rigidly with one foot off the ground and the nose pointing toward the game. A single gene determines the particular pattern which these dogs can best be trained to follow. Many

centuries of selection by man have established a gene for a particular behavior pattern in a particular breed of dogs. Some dogs have been selected for genetic endowments which cause them to bark when they are trailing animals. The baying of the fox hound on a hunt is a phenomenon familiar to all, even though most of us have come no closer to a fox hunt than to see it in movies or television programs. Other dogs inherit a disposition to be silent while trailing. Through breeding, man can select the type of behavior pattern he desires. Retrieving game is another behavior pattern which is so strong in some breeds of dogs that they retrieve balls, sticks, and other objects over and over again. As is true of horses,

FIG. 37.1 Many behavior patterns, such as the method of clasping the hands, seem to be influenced by heredity. Which thumb do you place on top? You will find that this response varies with different people. This response tends to follow a Mendelian pattern of inheritance.

training is necessary to bring about the expression of these traits as man desires them, but the basic pattern of response is inherited.

Examples in Invertebrate Animals

Among the invertebrates, which lack the complex parts of the brain associated with learning found in the vertebrates, heredity plays a major role in establishing behavioral patterns. The silkworm typically spins a single cocoon of silk when it is ready to go into the pupa stage; a dominant mutation of a gene causes it to spin a double cocoon, one inside the other, with an air space between. The larvae of the giant American silkworm, the Cecropia moth, always spin the double cocoon, so it appears that selection has chosen the dominant gene and established this pattern of behavior. The double cocoon has survival value to the pupa; it is made of two layers with a dead air space between, insulation which protects the larva from great fluctuations in temperatures. The Cecropia moth inhabits regions which generally have more severe winters than those found in the habitats of the silkworm. It appears probable that the pattern of spinning a double cocoon might have been established through natural selection because it has survival value to the Cecropia moth. On the basis of this and other findings we can assume that at least some behavioral patterns have become established through natural selection, just as other characteristics have been established.

Studies in sexual behavior in the fruit fly, *Drosophila*, show that heredity plays an important role in sexual vigor and response of the male. Single males placed in vials with numerous virgin females show great variability in their mating reactions. Some accomplish only a single mating over a twelve-hour period, while a few mate with as many as a dozen females over a period of a few hours. Through selection it has been possible to establish some strains in which the males show strong sexual vigor and other strains in which males exhibit low sexual vigor. Certainly this behavioral characteristic is inherited, and in a natural environment it is assumed that the sexually vigorous males would accomplish the greater number of matings. Selection would favor this group, provided that there is not some survival advantage to the less vigorous sexual characteristic.

Internal Stimulation of Responses

These examples indicate that heredity influences the responses of an organism to a particular stimulus. We generally think of such a stimulus as having an external origin, but many responses come about as a result of internal stimulations. The silkworm larva must spin its cocoon at a particular time in its life, just at the end of the larval stage, if the cocoon is to serve its proper function. The stimulus for the spinning behavior comes from within the larva, from its endocrine system. In the head region of the larva there is a pair of small glands, the corpora allata, which produce a hormone that prevents the spinning reaction. If these glands are removed early in the life of the larva, it will spin a tiny cocoon around itself when it is not ready to pupate and will die. It is also possible to localize the particular area of the brain which contains the information for spinning. Large sections of the brain can be removed without affecting the spinning reaction, but when one small region of the brain is destroyed there will be no spinning of silk regardless of stimuli.

The stimulus to drink or eat in mammals is conditioned by internal stimulation of salt and glucose concentrations in the blood as it flows over sensory receptors in the brain (Chapter 31). In one experiment a very small plastic tube was implanted in a goat's brain so that drops of strong salt solution could be introduced into the area of the brain which is sensitive to salt concentration of the blood. The goat might be grazing in a field, but when a drop of salt water was sent down the tube the animal would immediately go to a watering trough and drink. The process could be repeated over and over again with similar results even though the goat became so full of water that it could hardly swallow any more. These two examples show that we must include internal stimuli whenever we attempt to explain animal behavior.

Alterations of Behavioral Patterns

Can inherited patterns of response be altered through training? The answer is yes, to a certain extent, but certain difficulties may arise in their alteration. In the past it was common practice in the schools to train left-handed children to use the right hand for writing. We live in a civilization which is geared to the right-handed person, and it was thought best to make all conform to this pattern. Such alteration of natural patterns of behavior, however, may bring about emotional conflicts which sometimes are manifest in such nervous symptoms as stuttering. Today's schools allow left-handers to follow their natural inclinations.

Drugs can bring about altered patterns of response. Experiments on spiders show that the web-spinning reaction is affected in a specific manner by specific drugs. Some drugs cause the spiders to spin a disorderly web, and each drug results in a particular type of disordered web. One drug actually resulted in a more orderly web. This drug is **lysergic-acid diethylamide, LSD,** which has become a problem with respect to its human use. A spider given this drug will spin a web with such precision that the result looks as if it had been laid out with engineer's instruments. One of the difficulties arising from the use of the drug by man lies in the altered patterns of behavior which it may cause, sometimes with tragic results.

THE ROLE OF DNA AND RNA IN BEHAVIOR

The examples of patterns of behavior we have considered show that responses to stimuli may be inherited in many cases and do not require learning. Learning may sometimes alter inherited patterns of behavior. In other cases heredity endows an organism with the ability to become conditioned to certain stimuli so that a specific behavior pattern follows. Since genes are made of DNA, it follows that DNA contains the information about inherited patterns. RNA, of course, is involved in the transmission of the information from DNA to the brain cells where they stimulate the motor nerves to elicit a particular response.

Recent research indicates that RNA plays an important role in the learned responses too. One of the simplest animals which seems to have the ability to learn is the flatworm, the planarian. Planarians are small enough and simple enough in structure that simple experiments can elucidate their behavior. In one experiment planarians were placed in a T-maze where they had a choice of turning right or left. If they turned left they would receive an electric shock, but a right turn led them to a dark moist chamber which resembled their natural habitat. After many trials, the planarians learned to turn right most of the time. They had learned the direction of the shock and turned in the direction which would lead away from the shock and toward the moist chamber. After training, some planarians were cut in half and allowed to regenerate. Those regenerating from the posterior halves "remembered" the correct direction to turn in the maze, as did those regenerating from the anterior halves. Thus, it appears that memory in the planarian is not localized exclusively in the so-called "brain" of the head region.

The experiment was repeated, but the cut halves of the planarians were allowed to regenerate in a solution of ribonuclease, an enzyme which destroys RNA. This time the worms from the anterior halves retained some memory of the right direction to turn, but those from the posterior halves did not. These results indicate that the RNA must play some role in the learning reaction and that RNA can be distributed over the body. This concept was supported by the reports of one experimenter who trained some planarians and then cut them up in pieces and fed them to untrained planarians. The untrained, cannibal planarians seemed to acquire some degree of proficiency in turning in the right direction in the maze. If the cut-up pieces of planarians were treated with ribonuclease before the feeding, however, there would be no transmission of learning.

These results are reminiscent of the widespread belief among some human cannibals that it is possible to acquire the cunning and wisdom of their enemies by eating their brains and hearts. We should point out that there is no evidence that such transfer is possible in higher animals. It does seem, however, that RNA plays a definite role in learning, and this has led to some wild speculations, especially by newspaper reporters anxious for a sensational story. Some have even predicted that the day will come when a person will not have to go through the laborious process of learning. If he wants to know how to speak the German language, he may simply take an RNA pill which has been programmed to convey the information and he will immediately be able to speak and write the language. Needless to say, it would not be wise for the college students of today to wait for this product to become available. For the foreseeable future, learning must continue to come only from the same laborious application of mental endeavor as it has in the past.

INSTINCT

A female bird hatched in an incubator and kept in captivity will, when mature, build a nest of the same materials and in the same shape and size as are characteristic of other females of her species. Such behavior can be explained only on the basis of a genetic endowment and is classed under the heading of **instinct.** The bird will build the nest only when certain stimuli are present; in this case it is the hormonal changes taking place within her body, which in turn are influenced by changes of the seasons and possibly by the courtship of the male. To sum up, we can say that instinctive behavior is an elaborate response genetically determined which is brought about by both internal and external stimulation. In the vertebrates, with their more advanced centers of learning in the brain, learning may play a part in modifying instinctive behavior. In the invertebrates, with their comparatively low level of learning capacity, behavior tends to be more extensively governed by strictly inherited forces. A dog can be trained to overcome its instinctive drive to chase chickens, but a moth continues to flutter near a flame even after its wings are singed and until it finally is consumed by the flame.

MIGRATION

Seasonal migration is one of the most interesting and complex of instinctive animal behavior patterns. Such widely diverse animals as turtles, eels, salmon, seals, birds, and butterflies are migratory. Migration is practiced by groups rather than by individuals, as a rule. As the time for migration comes, the various members of the society or even entire populations begin a mass movement which may take them great distances.

A population of green sea turtles lives most of the year near the coast of Brazil, but each year the entire population begins to swim east across the Atlantic Ocean and ends up in early April on Ascencion Island, a small dot on the map 1,400 miles from the Brazilian coast. The island is only seven miles long, but these turtles hit it exactly, without benefit of compasses or any navigational aid which man would require. After laying their eggs on the sandy beaches of the island, the turtles return to their Brazilian home. The young turtles which hatch out later have the inherited instinctive navigational ability to swim to the coast of Brazil, where they join their ancestors which they have never seen before.

Great herds of reindeer in Canada travel as much as 800 miles north in the spring to reach their summer grazing grounds in the tundra. Seals from islands off the coast of Alaska migrate to islands off the coast of California during the fall. Eels in the rivers of the interior of the United States migrate down the rivers to a particular spot in the Atlantic Ocean where they reproduce. The young eels, when hatched, reverse the pattern and migrate into the rivers ending up far from the coastal waters. As we learned in Chapter 26, salmon migrate from the ocean up into the rivers and finally spawn in the small freshwater streams where they were hatched years previously.

Because of their ability to fly, birds are the greatest migrants. The long-distance record for migration seems to be held by the Arctic **tern,** which we discussed in Chapter 27.

Even the tiny broad-tailed hummingbird, which has a body no larger than a large moth, makes a migration of considerable distance. They spend the summer months in the Rocky Mountain region of the United States and southern Canada, but in late August they take off for their winter home in southern Mexico, flying at speeds of about 50 miles an hour en route.

These are only a few of the many examples of migration which could be described, but they suffice to show how widespread migration is. The great question to be explained in all cases is: How do these animals find their directions in their migrations? No one answer will explain all cases, but a few examples will serve to illustrate.

Some birds, such as the geese, which fly in formation with a leader at the head of a V-shaped pattern, seem to have learned landmarks as a guide. An older bird which has flown the

pathway before acts as a guide, and the younger birds learn the landmarks so that they may someday serve as leaders. These birds are known to fly low when the weather is overcast and to stop altogether if the visibility gets too poor.

This method could not explain the success of birds who hit a tiny island after flights of thousands of miles over open water. There is evidence that some migrating animals use the position of the sun and stars in the sky as a guide in their migrations, but since the position of these celestial bodies varies with the time of day or night, the animals also need an accurate sense of time to use such navigational aids. Here we get into the topic of biological clocks, and we shall

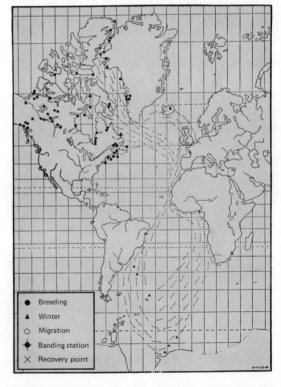

FIG. 37.2 One of the very complex behavior patterns which is influenced by heredity is migration in animals. This diagram shows the migration routes of the Arctic tern, a yearly round trip of about 25,000 miles, much of it over open water. (United States Fish and Wildlife Service.)

defer further discussion of navigation until later in this chapter.

COMMUNICATION

Two great developments have made possible the high degree of complexity of life found on the earth. Both involve the transfer of information. First, there was the development of the genetic code in DNA which made possible the storage and transfer of information from cell to cell and from generation to generation. Once an efficient set of characteristics was established through natural selection, the genes made it possible to maintain and continue those characteristics because the information in the genetic code can be transferred from one organism to another. Second, the development of means of communication between animals has made possible the transfer of learned information. Information can be passed along and accumulated through the generations. Suppose man had no means of communication of ideas from one person to another through either the written or spoken word. If we could survive at all, we would live on a level which would make the caveman's existence seem superior by comparison.

When we speak of communication, we commonly think in terms of human speech and writing, but there are many other ways that a message can be transferred. Facial expressions, body position, and hand gestures are important signals in human communications. We can understand the "language" of other animals to a certain extent. A housecat has a certain type of meow which means "I am hungry," another which means "I want to go outside," another which denotes anger, and still another which denotes pain. Wolves inform others of their moods by body positions. One position indicates that attack is imminent, another is a friendly signal, still another indicates fear, and so on. It is a surprising sight to many big game

hunters in Africa to see antelopes and lions drinking from the same water hole, paying no attention to one another. Let a hungry lioness start out for food, however, and the antelopes give her a wide berth. She gives warning of her intent by her body positions and the antelopes react accordingly. The pronghorn antelope has a patch of white fur just under the tail which can serve as a silent warning of danger. When one in a group senses danger, he raises the tail and exposes the patch of white. As others see this sudden exposure of white, they get the message, and soon all bound off toward safety.

The Language of the Bees

Communication between the honeybees has been investigated extensively by **Karl von Frisch** of the University of Munich in Germany. He has found that bees have an elaborate means of communication. A worker bee scouting far from the hive may come upon a cherry tree with its blossoms laden with nectar. She flies back to the hive and soon hundreds of workers will be collecting nectar from the tree. How did the scout let the others know where the tree was located? She did not lead them there. She remained in the hive and, communicated her find to others. To try to find the answer, Dr. von Frisch placed sugar water at various distances and various directions from a hive which had glass windows to allow observation of what went on inside the hive. When one scout discovered the sugar water, she was dabbed with a colored dye for identification. Then her actions back at the hive were observed. Communication seemed to be accompanied by a special dance. The scout would go in a circle and then cut across the circle wagging her abdomen. The speed of wagging indicated the distance of the food from the hive. Forty wags per minute indicate that food is located slightly more than 75 yards away, and eight wags per minute show the other workers in the hive that they must travel about 3.7 miles to find the food. Other distances are in proportion. But distance information above is not sufficient. The bees must know the direction they must fly to find the food. This information is communicated by the direction of the path across the center of the circle. If the bee goes straight up, the food is toward the sun; straight down means it is directly opposite the sun. A path 60 degrees to the left of vertical coming up indicates a food source 60 degrees to the left of the direction of the sun. This would seem to work nicely in sunny weather, but what about overcast days when the sun is not visible in the sky? Do the bees then have to give up this means of communication? Bees have eyes which are polarized; they can tell the direction of the sun even on overcast days by the polarization of the light. You can do the same with the aid of a polarizing filter, but not with your unaided eye.

Courtship

One of the most important uses of communication between animals is known as **courtship**. A male must know if a female is receptive or the female must know that conditions are right for mating. Each species of bisexual animal has developed particular patterns of courtship which accomplish these purposes. The absence of mating between members of different species is frequently due to different patterns of courtship. In the fruit fly, *Drosophila*, the males have patterns of courship which vary from species to species. When males of one species are placed with females of another species, the males carry out their particular courtship ritual, but this is a strange behavior to the females and they fail to become aroused sexually. As a result, the females do not give the correct response, and mating does not take place.

Some of the most elaborate patterns of courtship are to be found in birds. In many

species, days or even weeks of courtship precede the actual mating. In prairie chickens, for instance, the males select a territory during the breeding season and defend it against all other males. They attract the females to their territory by a booming sound which they produce by inflating and deflating air sacs on either side of the neck. They also do a ritual dance in which they rush forward, stop suddenly, and stamp their feet. This may continue for as long as six weeks before females will enter the male territory and permit mating. Many male birds spread and display their plumage before the females, like the peacock spreading its brilliantly colored tail feathers during the courtship period.

For some birds courtship is necessary if they are to produce eggs. The male lyrebird of Australia, famous for his song and the lyre-shaped tail, spends several months on his mound displaying his feathers and singing. A female

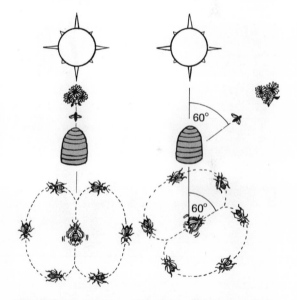

FIG. 37.3 The language of the bees. When a bee finds a source of nectar, it returns to the hive and communicates information about the distance and direction of the food to the other bees through what is known as the wag-tail dance. This behavior pattern is inherited.

may be watching from the surrounding trees, and if she is not stimulated by the display, she will lay no eggs. Hormonal changes take place in the female as a result of the courtship. A female pigeon reared in isolation will not produce eggs, but if she is placed where she can see a male strutting, she will lay eggs whether fertilization takes place or not. She will also lay eggs without a male being near if her neck is tickled in a way which simulates the tickling done by the male with his beak during courtship.

In some birds the two sexes look so much alike that courtship is necessary for them to distinguish one from the other. Males of the

FIG. 37.4 Courtship of the emperor penguins in the Antarctic. The two sexes look so much alike that they distinguish one another only through their courtship patterns. (Courtesy of Paul Richard, University of Northern Colorado.)

common tern will catch a fish and lay it at the feet of some likely recipient. If the fish is proffered to a female, she will accept it and the male will bow to her and make a nest in the sand, and they will mate. Should the fish be proffered to another male, it will be rejected with an aggressive peck.

Many animals which have a more highly developed sense of smell than birds use odor as a means of recognition of sex. Even though the two sexes of dogs may look very much alike to human beings, the dogs have no difficulty distinguishing the difference when they are near each other. Dogs have a highly developed sense of smell and they can distinguish sex by smelling trails where other dogs have been. Odor also plays an important role in insects, and males of certain species have highly elaborate antennae which pick up the scent of females. This is especially true of the night-flying insects which cannot depend upon vision. Experiments with one type of moth showed that the females have special glands which they protrude from their bodies during mating season. The scent can be detected by males up to several miles away; they follow the scent and find the female.

The flashing of fireflies on a summer night is a courtship pattern which aids them in finding each other. The males fly around while the females remain on the ground. When a male flashes in the air, the female on the ground gives a flash in response if she is in a receptive mood on that particular night. A definite time interval elapses between one flash and the other, so the male can detect a flash which is in direct response to his signal. After several exchanges of flashes, the male descends and mates with the female.

What is the evolution of these and the countless other courtship patterns? There is one species of balloon fly in which the male spins a rather large balloon of silk and presents it to the female. She accepts this and holds on to it tightly while the male mates with her. Why should the male go to such trouble? Why not just approach the female and mate? In some species of this fly the balloon is not made, and frequently the predaceous female will turn on her suitor and eat him. Perhaps in this mishap we have a clue to the evolution of some courtship rites. In other species of the fly, the male will first capture an insect and present it to the female. While she is busily engaged in devouring the insect, the male mates with her and escapes in safety. In still other species, the male may capture a smaller insect, but he wraps this insect with silk and presents this to the female— a present in a wrapping so to speak. The ultimate end to this line of development is the case in which the male makes the large balloon of silk and presents this wrapping with no present inside to the female. The female's instinctive response is to accept this and give it her full attention during the mating. The male achieves protection from his ferocious mate without having to capture an insect.

BIOLOGICAL RHYTHMS

Many behavioral patterns of animals are subject to rhythmic variation in response to various rhythmic variations in the environment. When a person flies from New York to London, he finds that his "biological clock" is not in agreement with the man-made clocks of the area. Suppose he leaves New York at 8 P.M. on a fast jet, has a leisurely dinner on the plane, watches a movie, and then, just as he is settling down for a night's sleep, a voice over the loudspeaker announces the imminent landing in London. He groggily gets off the plane, sees the bright daylight, the people scurrying around in their early morning activities, and smells breakfast cooking. The clocks show a time of 7 A.M., but his body "clock" indicates that it is only 2 A.M. His body metabolism is at the same low level he would experience in New York at 2 A.M. Later, when everyone is ready to go to bed that

evening, his body metabolism is just reaching its height of activity. Executives have found that it is not wise to ask people newly arrived from long east or west flights to make important decisions. In fact, it is difficult even to add a simple column of numbers in the confused conflict on man-made clocks and body clocks. Such observations demonstrate the principle of biological rhythms which are a part of the homeostatic mechanisms of all forms of living organisms and which aid in adjustment to rhythmical variations in the environment.

In addition to the rhythms of day and night —the **solar day** which is based upon the time of revolution of the earth—there are also rhythms based upon other periods to which organisms may become adjusted. There is a **lunar month** of about twenty-nine solar days which is based upon the time required for the moon to orbit the earth. Such a rhythm affects the phase of the moon in the sky. Likewise, there is the **lunar day** of about 24 hours and 50 minutes, which is based upon the time of the rising of the moon. The lunar cycles are related to various physical phenomena on the earth, such as the time and height of the tides, and many animals have lunar cycle timers. There are also **yearly rhythms** associated with the variations in the length of day and night and the seasons. Most living things have built-in biological clocks which correlate activities with some or all of these rhythms.

The Rhythms of the Fiddler Crab

The little fiddler crab lives on the beaches of many eastern states. The males have one large claw and one small claw, roughly resembling a fiddle and a bow. These crabs undergo a daily cycle of color changes. During the daylight hours they are dark brown, but at night the exoskeleton turns pale brown. The darker daytime color evidently protects them against the sun's rays, while the paler night color may make them blend more with the sand on the beach. Upon first observing this phenomenon, one might surmise that this is an automatic adjustment to light and dark. Such an assumption can easily be tested by bringing some of the crabs into a room which is kept dark at all times and at a constant temperature. When such crabs are examined under a very weak light, no brighter than moonlight on the beach at night, it can be seen that they still turn dark at dawn and become pale again at sunset. Such continuing responses indicate that the crabs must have some internal timing mechanism, a biological clock which accurately measures the time of day and night.

The clock can be reset by providing artificial periods of light and darkness. Suppose we turn on a bright light in the room at 12 noon each day and turn it off at midnight. After several days of this treatment the crabs will began to turn dark at noon each day and become pale at midnight. Once this new rhythm is established, it will continue even though we return the crabs to darkness for the entire 24-hour period again. We can also reset the "clock" while the crabs are in darkness by placing them in ice water. Such treatment seems to stop the clock, and when they are removed from the ice water, the clock starts

again but it has lost time. It they are in the ice water for six hours, the light and dark color phases will be six hours later than night and day. These experiments lead us to believe that the clocks are controlled by metabolic rhythms, since the cold temperature brings metabolism almost to a halt and also stops the biological clocks.

The crabs have another rhythm associated with the tides. They are very active and run around on the beaches at low tide, but at high tide they burrow in the sand and are inactive. Thus, they are not swept out to sea as the waves break over the beaches. In a dark room they keep up this rhythm of activity and inactivity correlated with the lunar day. Since there are two low tides each 24 hours and 50 minutes, there are two periods of activity correlated with these two tides. The amazing thing about this reaction is that it can be adjusted to the tides in any region to which the crabs are taken. If they are shipped to Chicago, for instance, they gradually reset their activity rhythms to the times that low tides would occur in Chicago if Chicago were on an ocean beach. Hence, we must conclude that these crabs have some sensory mechanism for detecting the gravitational pull of the moon which brings about the tides, and they respond to this even

FIG. 37.5 Fiddler crabs. These crustaceans have a delicately adjusted biological clock which responds to the cycle of days, tides, and seasons.

though they are in a dark room far removed from any beaches.

In addition, the crabs have a yearly timer related to the seasons; they mate at about the same time each year. Thus, with only one animal as an illustration, we can see how organisms respond to a series of variables in their environment through some built-in mechanism that is sensitive to factors in the environment by which the ordinary senses are generally considered to be unaffected.

Plant Rhythms

Plants as well as animals respond to rhythms of the environment. Frank Brown of Northwestern University, who has done more than any other person in investigations of biological rhythms, reported an interesting example with bush beans. Growing bean seedlings show peaks of respiratory activity at specific hours of the day, namely 7 A.M., 12 noon, and 6 P.M. Brown kept some bean seeds in a dark room at a constant temperature for three years and then germinated them while still in the dark. Amazingly, the seedlings growing from these seeds exhibited the three daily periods of maximum respiratory activity so accurately that one could set a watch by these activities. How was this possible? Did the beans have a timing mechanism which kept track of time to the minute over all those years in the dark? The bean seedlings were probably responding to certain changes which accompany the light and dark of day and night. A radio set can detect these changes; if you were closed in a dark room with a radio you could tell when dawn and sunset came by the stations you would pick up. There is a change in the layers of ionization in the atmosphere correlated with day and night, and this affects the distance of reception of radio signals. The bean seedlings evidently have a mechanism sensitive to such changes which accompany day and night, and they can detect these changes in a dark room.

Like the crabs, bean plants can reset their clocks to new times when light conditions vary. Young bean plants exhibit a rhythm known as "sleep movements." The leaves droop down at night and extend out fully during the day. When the plants are taken into a room with very dim light, about the intensity of moonlight, they still undergo the regular cycle of drooping during the night hours. If they are given a bright flash of light during the night they begin to extend the leaves, but then they droop again. They have been influenced by this flash, however, and the next night they will also begin to extend their leaves at exactly the time corresponding to the time the flash was given the night before. This rhythm continues for many days.

Location of Biological Clocks

Where is the timing mechanism located in a plant or animal body? For most forms of life this remains a mystery, but in a few animals specific glands are involved. In **cockroaches,**

FIG. 37.6 Plants also have biological clocks which respond to varying conditions. Flowers appear on very young seedlings of the weed lamb's quarters (*Chaenopodium rubrum*). These seedlings are only about an inch tall. The plant normally grows to several feet before flowering. Manipulation of light cycles and temperature caused the early flowering.

for instance, running activity is correlated with the dark, and they become inactive in the daytime. Such correlation is maintained even in a room kept constantly at a very low light intensity. Like other timing mechanisms, this one can be reset by keeping the cockroaches in the light at night and in the dark during the day. Gradually the period of activity will be shifted to the daytime hours. The clocks can also be reset for any desired timing by refrigerating the cockroaches The timing mechanism in the cockroach has been found to be a gland in the head which releases a hormone which stimulates activity. If this gland is removed from a roach which has been timed for daylight activity and transplanted into a roach timed for the normal nightime activity will be only in the daytime.

Monthly and Annual Rhythms

Many organisms have cycles related to the lunar month. On the Pacific coast of the United States there is a little fish, the **grunion,** which has a monthly reproductive rhythm correlated with the time when the moon and the sun are in the same direction, the time of the new moon. The combined gravitational pull of both these bodies causes the ocean tides to be unusually high at such times. Each month when these very high tides come, the grunion ride the waves up onto the beach and quickly drill holes in the sand where they deposit their eggs and sperm before the next high wave comes in and carries them back into the ocean. California newspapers carry reports of the exact time of the "grunion runs" and many people go out onto the beaches to gather these edible fish. The eggs hatch and the young develop in the moist sand until the next extra high tide when they are carried into the ocean.

The reproductive rhythms of many land animals are also correlated to some extent with the lunar months. The cycle of a woman is roughly about one lunar month, although it is not in phase with any particular position of the moon. In guinea pigs, sheep, and domestic swine there are about two reproductive cycles each lunar month.

Annual rhythms seem to be responsive to the relative length of day and night, the photoperiod, as has been discussed for plants in Chapter 22 and for birds in Chapter 27. There is evidence that such rhythms can persist even under constant conditions of light and temperature. The eyes of **potatoes** have a metabolic rate which is nearly twice as high in April as in October and November. This rhythm continues even if the potatoes are sealed in containers where conditions of light, temperature, moisture, and pressure are unvarying throughout the year. Potatoes are normally planted in the spring and their calendar has been set for maximum activity at this season. Likewise, most seeds planted at various times of the year under exactly the same conditions in the laboratory show a great difference in the speed of germination and early growth. They are by far most active in the spring.

Just as daily biological clocks can be reset by varying the time of light and dark, annual biological calendars can be reset by varying the photoperiod. The common weed known as **lamb's quarters,** *Chaenopodium rubrum,* normally flowers in the summer, and it maintains this pattern of flowering if it is grown in a room with constant light. But if it is subjected to a photoperiod of twelve hours of daylight alternating with twelve hours of darkness in the very early spring, it will flower then. In fact, small seedlings no more than a quarter of an inch high will produce flowers, whereas under normal conditions the plant will be two or three feet tall when it flowers.

We could continue with examples of biological rhythms almost indefinitely, but all indicate that both plants and animals have mechanisms for adjustment to the various physical rhythms on the earth and that these mechanisms can be altered to adjust for a changed position on the earth's surface. The experiments on these phenomena also indicate that living things can retain their rhythms even when kept under constant conditions in a laboratory. Such rhyth- mic activity may be due to responses to various forces such as magnetic fields, gravitational pull, and ionic changes in the atmosphere, forces to which man has no conscious response, or to endogenous causes.

REVIEW QUESTIONS AND PROBLEMS

1. List one of your behavior patterns which you think you acquired through inheritance and give reasons for your answer. Do not use a trait discussed in this chapter.
2. List one of your behavior patterns which you think you have acquired through training. Does this mean that heredity could not have a part in the expression of this behavior pattern? Explain your answer.
3. Heredity seems to play a much more important role than learning in the behavior patterns of invertebrates in contrast to vertebrates. Tell why this is true.
4. Explain how survival value may have played a part in the establishment of a particular behavior pattern, using an example to illustrate.
5. How might natural selection be of importance in establishing sexual patterns of behavior?
6. How has the role of internal body secretions in behavior been demonstrated in silkworms?
7. Can instinctive behavior be altered through training? Illustrate with an example to show that it can or cannot be altered.
8. How have experiments on the planarians indicated that RNA might have a role in learned behavior patterns?
9. Explain the possible role of natural selection in the establishment of the instinct to migrate.
10. Explain why it might be said that societies of animals would be impossible without communication.
11. How does a worker bee communicate the distance and direction of food from a hive of honeybees?
12. Explain how patterns of courtship may establish a barrier to hybridization between members of similar species.
13. In some species of penguins a male suitor may present a pebble before a female and an acceptance of the pebble indicates that the female is sexually receptive. In light of the reports of studies on courtship in this chapter, formulate a hypothesis about the origin of this courtship pattern.
14. Describe some courtship pattern which you have observed in some species of animals not discussed in this chapter and see if you can find any biological advantage to this pattern.
15. How is it possible to reset the biological clock of a fiddler crab?

16. How is it possible for a fiddler crab to adjust its activity to the tides even when it is kept in an enclosed laboratory far from the ocean?

17. Bean seedlings react to day and night even though they are kept in the dark at constant temperatures. How might this be explained?

18. Roaches are normally active only at night. How can you produce a roach which will show activity twenty-four hours each day?

19. Give two examples of daily rhythms and two examples of annual rhythms which have not been mentioned in this chapter.

FURTHER READING

Allen, T. B. 1972. *The Marvels of Animal Behavior*. Washington, D. C.: National Geographic Society.

Cloudsley-Thompson, J. L. 1961. *Rhythmic Activity in Animal Physiology and Behavior*. New York: Academic Press.

Davis, D. E. 1966. *Integral Animal Behavior*. New York: Macmillan.

Jolly, A. 1972. *The Evolution of Primate Behavior*. New York: Macmillan.

Tavolga, W. N. 1969. *Principles of Animal Behavior*. New York: Harper and Row.

Wendt, H. 1965. *Sex Life of the Animals*. New York: Simon and Schuster.

38

The Organism and Its Environment

Ecology is the study of organisms in relation to their environment. It has come to be a well-known word as we have realized the importance of this relationship. The deterioration of the earth's environment has become a matter of major concern. It is not possible for any organism to live alone. The earth is populated so densely with so many different forms of life than any one organism is certain to be associated with many other living things no matter where it lives. Such associations have a profound effect upon the type of existence it leads. The physical environment is also very important, and much of the organism's energy is expended in adapting to physical conditions in its environment.

POPULATIONS

The term **population** is often used to denote human groups, but in a biological sense the word applies to groups of any given species living together in a relatively permanent association. The pine trees in a forest, dandelions on a lawn, bullfrogs in a pond, grasshoppers in a field, typhoid bacteria in a human body, or people in a city are all examples of populations.

Population Boundaries

The number of organisms within a population may show variation. Some new ones may be added through reproduction or migration from other populations; others may be lost through death or migration away from the group. Still the population continues as an entity. The boundaries which limit a population vary considerably. A small pond may limit the boundary of a population of yellow perch, while a population of moose may occupy hundreds of square miles of a forested area of western Canada. The density of the population also varies greatly. The pond may be swarming with yellow perch, but in the Canadian forest a hunter may search for days without sighting a single moose.

The unifying link that distinguishes a population is the free interbreeding of its members with one another. The limits of a population are determined by various barriers. Water plants and animals may be limited by land barriers, while land plants and animals are limited by water barriers. Other factors, such as climate, mountains, and other barriers also are important. The great deserts of the world are as effective as oceans in preventing the mingling of members of different populations.

The Balance of Numbers Within a Population

The number of organisms within a population tends to reach an equilibrium and to remain relatively stable as long as environmental conditions remain constant. Since the environment may change, however, the population fluctuates in numbers in response to fluctuations in favorable or unfavorable environmental conditions. Such adjustments are known as **ecological homeostasis.** When food is plentiful, enemies scarce, and the climate favorable, the number of individuals in a population will increase. When any or all of these conditions become less favorable, the populaton number declines.

Usually a population includes as many individual members as can be supported and can survive within the limits of the boundary. We can illustrate this principle by a classic survey of the number of rats in a square city block of Baltimore. The public health department there made a careful study of the rat populations of typical blocks in the older residential areas. Such blocks limit the rat population, because the wide city streets with heavy automobile traffic discourage migration from one block to another. Suppose a study shows a certain block has a population of 87 rats. This number tends to remain nearly constant from year to year as long as conditions in the block remain the same. The birth rate among the rats is very high, but the death rate is also very high, and the great majority of the rats never live long enough to reproduce. Some are killed by dogs, some by traps, some by fights among themselves, and some by starvation or disease. When food becomes scarce, the females tend to neglect their young and the young may be eaten by their mothers or by other rats. The struggle for existence is so keen that constant vigilance is necessary to stay alive.

Small variations in the environment can alter the number of rats that can live in this block. If a new family moves in with an extra garbage can, this may become a 96-rat block. A new dog may cut it to an 83-rat block. Sporadic poisoning campaigns alter the population only temporarily. Most of the rats may be poisoned, but some will survive; some will be too wary to eat the poisoned food, or as other rats die, there is so much food that they need not turn to the poisoned food. Let us assume that only nine survive a poisoning campaign. For a time these rats will live very well; they will be fat and sleek with plenty of food and little competition for living space. Soon, however, the laws of natural selection will begin to operate. Each surviving female will bear from twelve to fourteen young at each gestation, and a gestation may occur as often as every six weeks. Within a few months the population number will be back to its original figure. A permanent reduction is possible only by making the environment hostile. Destroying nesting places, using tight-fitting covers on garbage cans, and maintaining a constant campaign of destruction are ways in which this can be done.

Plagues

Some populations show considerable variation between two extremes because of great cyclic variations in climate. In most tropical areas of the earth, where climate is relatively stable and does not show great year-to-year fluctuations, the population number of any particular species tends to remain constant. In regions with great fluctuations of climate, however, the numbers may vary greatly. The midwestern plains of the United States and Canada are examples of such regions. The temperature may drop far below zero in winter blizzards and rise to well above 100 de-

grees in the summer. The amount of rainfall also fluctuates. Both plant and animal populations show great variations from season to season and from year to year. In certain years conditions may be almost perfect for grasshoppers and there will be a plague of grasshoppers. They will destroy almost all green plant tissue over hundreds of square miles. Crops will be a total loss in these areas. The next year, however, there will be a normal population. The reproductive potential of grasshoppers is great; each female may lay hundreds of eggs. However, great attrition occurs in most years. In those rare years when the winters are mild, when there is little disease to kill the young, and other factors are favorable, most of the young survive and the plague results. In the Bible we read of plagues of grasshoppers (locusts) which destroyed crops and led to great famines.

The arctic **lemmings** are small rodents about 6 inches long which live in the arctic tundra of North America, Siberia, Greenland, and Lap-

FIG. 38.1 Arctic lemmings. At times these small mammals multiply very rapidly and become a plague. (© Walt Disney Productions.)

A

B

FIG. 38.2 Mounted remains of a passenger pigeon. Flocks of this large, dovelike bird used to darken the sky in huge clouds, as shown in this old drawing which appeared in a Chicago paper in 1885. Hunters killed them by the thousands and destroyed their nesting places, so that by 1914 they were extinct.

land. They are the main source of food for many of the arctic predators. Snowy owls, wolves, foxes, and to some extent polar bears feed on lemmings. In some years there is a plague of lemmings and their food supply becomes scarce as the result of competition among themselves. This may cause a mass migration. In Lapland they have been observed moving westward across northern Sweden and Norway as a great horde, devouring almost everything of an organic nature. Nothing seems to stop them; when they come to a lake they do not attempt to go around, they plunge right in and swim across. When they come to a haystack, they do not go around but eat their way through. When they reach a steep cliff, they plunge right over it without hesitation. The first to fall are killed on the rocks below, but their bodies serve as a cushion for those still falling and many survive the fall and continue the migration. Foxes, wolves, and bears all follow this migration and have great feasts. When the surviving lemmings reach the ocean they plunge in and swim westward until they become exhausted and drown. There are still sufficient "stay-at-homes," however, to repopulate the area, and there is no permanent reduction in lemming population.

Although there may be plagues of almost any form of life, the word is popularly associated with plagues of **disease organisms.** There were plagues in Europe during the fourteenth century which killed more than one-fourth of the human population. The Roman Empire lost about one-half of its people to a plague in the sixth century. Sometimes entire populations of plants or animals are wiped out by plagues of disease organisms. The population of beautiful **chestnut trees** in the forests of eastern United States is now gone because of a plague known as the chestnut blight.

Extinction

When the environment of a population is altered too greatly in an unfavorable manner, we have the opposite of plague—**extinction.** Most species of plants and animals which have existed on the earth have become extinct because they could not adapt to a changing environment. Only 150 years ago **passenger pigeons** were the most abundant game birds in North America. They were present in such great numbers that droves of them looked like a huge cloud in the sky. They nested in rookeries; one rookery near Shelbyville, Kentucky, was several miles wide and over forty miles long. Market hunters often went to these rookeries at night and killed thousands of birds, at the same time destroying the nesting places. By 1898 the last nesting place had been destroyed and only a few scattered flocks remained. The last known individual died in a Cincinnati zoo in 1914. This population could not adapt to the changes in environment caused by

FIG. 38.3 A whooping crane in a protected area near Port Aransas, Texas. This large bird was once near to extinction, but it has made a comeback in recent years when it was declared a protected species.

species of plants and animals which became extinct because of extreme alterations in population environments.

Introduction into a New Environment

Whenever a species reaches a new environment in which it has never lived before and finds favorable conditions, a sudden, explosive increase in numbers may occur until a stable population is formed.

Man has had some unfortunate experiences in making such introductions of new species. A man living in Florida visited the Columbian Exposition in Chicago in 1890 and saw a display of **water hyacinths** from Japan. They were so beautiful that he thought it would be nice to grow

them in the lily ponds of his Florida estate. He brought some back, but they did not grow well in his ponds, so he threw the remains into the St. Johns River, which flowed alongside his estate. He lived to see the river so clogged with water hyacinths that he could not move his yacht from its anchorage. They have spread to waters all through the southern states and millions of dollars are spent each year spraying with 2,4D in an effort to keep them under control.

Some Englishmen in Australia once thought it would be great sport to hunt **rabbits** in the great plains there, but there were no native rabbits, so they imported some in 1788. In Australia there are no native carnivorous animals which in other continents keep rabbits under control, and conditions were favorable for rabbit growth. Soon rabbits were so numerous that they were

A

B

FIG. 38.4 Water hyacinths have beautiful flowers (A), but they can clog waterways when they multiply in great numbers (B). These hyacinths were photographed on the St. Johns River in Florida.

man. Fossil records reveal the remains of many destroying crops and grazing plants. Today the Australians are experimenting with imported rabbit diseases and other means of rabbit population control.

With man moving about so much and so swiftly today, we have a major problem of accidental importation of plant and animal pests. The **Mediterranean fruit fly,** which was brought to Florida in some citrus fruit from the Mediterranean region in 1927, threatened to wipe out the citrus industry of Florida. The fly was eliminated by a vigorous program of quarantine on fruit shipments and the destruction of many infected orange groves. The **Russian thistle** was started in our western states from seeds clinging to goods shipped from Russia; it is now a great nuisance of the western plains.

The Population Growth Curve

When a species reaches a new environment favorable for its growth and reproduction, its numbers will increase following a curve which starts slowly, gains height rapidly, gradually levels off and then declines to a plateau of relative stability. This is known as the sigmoid curve because of its shape when plotted on a graph. We can illustrate population growth with the growth of **bacteria** in milk. Milk is normally sterile within the mammary glands of the cow, but a few bacteria are introduced as the milk flows from the external openings of the udders. Let us assume that milk just removed from the cow has 10 bacteria per ml (about 25 drops). If the milk is not refrigerated, the bacteria have ideal conditions for growth; plenty of food is available and there is

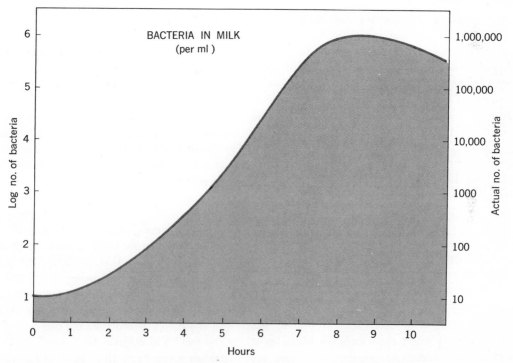

FIG. 38.5 Growth curve for bacteria in warm milk. This is a logarithmic graph with each vertical division representing a tenfold increase in numbers. A curve of this nature is characteristic of the expansion of species when they are introduced into a new and favorable environment.

little competition for this food. Under such conditions the bacteria can grow and duplicate themselves in about thirty minutes. Figure 38.5 shows how this growth can be plotted on a graph. Note that the growth rate curve gets ever steeper as the numbers increase. In only six hours the numbers will be about 45,000 and would reach about 3 million in nine hours if this geometrical increase continued. The population is not 3 million at nine hours, however, because the growth rate begins to slow down at about eight hours, when it reaches about 1 million. At this point crowding occurs and food is not so plentiful. There is also an accumulation of the waste products of metabolism, which are not favorable to continuing growth. The growth rate declines until it reaches a plateau when the bacteria begin to die at the same rate at which new ones are produced. Then a decline begins.

This type of curve describes the growth of species in a new environment of all forms of life; rabbits in Australia, water hyacinths in Florida, disease bacteria in a person's body, English sparrows in the United States, or the white man in America. Changes in the environment can produce fluctuations in the curve and alter the final plateau. If the snakes of a region are destroyed, the rodents of the area will increase to a new plateau. If the nesting places of a population of birds are destroyed when man clears the land for farming, the birds will decrease to a new and lower plateau.

SOCIETIES

Within many populations there are associations of individuals into smaller groups known as societies. Schools of fish, packs of wolves, and herds of deer are typical societies which exist within large vertebrate populations. These are groups associated for their mutual benefit. A herd of deer has an advantage over solitary deer. Many eyes and many nostrils can detect approaching enemies better than one pair of each. A closely huddled herd in winter stays warmer than a single animal. A herd can better defend itself and its young against attack if they are huddled than if they are widely dispersed.

Families Within Societies

There may be **family** organizations within the societies. Within a flock of geese the female, her mate, and her offspring tend to stay together in a family group until the young are mature. A herd of **seals** on an island is divided into families. A bull is surrounded by his harem of females and their young, while the maturing males live in another part of the island.

The Honeybee Society

The greatest complexity of organization of societies is found among certain insects. Ants, termites, and bees have large societies with a division of labor among the various members. The honeybee society, known as a hive, has been most widely investigated. Each hive contains one reproductive female, the queen; several hundred males, the drones; and many thousands of immature females, the workers. The workers are divided into groups which perform different tasks. Some construct new comb, others take care of the developing larvae, others are "fanners" which ventilate the hive by fanning their wings, others are guards, and still others are field workers which gather nectar and pollen. When the number of bees in a hive becomes too great, a new queen is produced and the old queen, together with about half the workers and drones, will form a "swarm" and migrate to a new location and establish a new hive. Thus, there is reproduction of the societies as well as reproduction of the individuals within a society. Such a society has been compared to a complex organism with each member of the society equivalent to a single cell of the organism. Both the cells of an organism and the individuals of a society perform functions which benefit the whole. Both organism and society can continue even though some smaller units within it are destroyed.

Hierarchies Within Societies

In vertebrate societies we frequently observe a pattern of dominance known as social hierarchies. If you observe a cage of monkeys at a zoo this is quite evident. One monkey seems to be the leader and all others give this one a wide berth. He can attack any of the others and they usually run away without a fight. One who may run away from the leader, however, may attack all the others without much opposition. This pattern of social hierarchy was first studied in detail by W. C. Allee at the University of Chicago. In a flock of white leghorn chickens he noted what he called a **"peck order."** One dominant chicken, A, can peck all the other chickens, which do not resist. The second in dominance, B, can peck all except A, and so on down the line to the poor chicken at the bottom of the hierarchy, which can be pecked by all but cannot peck any

FIG. 38.6 Social hierarchy among chipmunks. One helps himself to some food while another, lower in the social hierarchy, looks on but does not share the food.

in return. If a new chicken is introduced into the society, there are many fights until the new chicken learns his niche in the society. Sometimes there is a change in the order, especially as older chickens become weak with old age. Challengers of the right to peck cause fights which settle the new order. Not all societies of birds have such a rigid peck order. In pigeons, for instance, the peck order is more flexible. More fights occur and different pigeons are winners on different occasions. In invertebrate societies, such as hives of bees, social behavior does not involve fights to establish dominance.

COMMUNITIES

The group of populations living together in a given area is known as a community. This is a biological unit which is self-sustaining as long as energy continues to come into it from the sun. The community, like the species which populate it, has been established through natural selection. The numbers of species within the populations tend to remain stable in an established community, each species fitting into a niche which may be related to many other species in the community. Much of this relationship is in the form of **food chains** which are intricately interwoven into **food webs.**

The Food Chain

A food chain generally has its beginning in the autotrophic organisms, most commonly the green plants of the area. These capture the energy of the sun and make it available to all members of the community. We call these plants the **producers.** Next in the chain are the plant eaters, **herbivores,** which get their food from the producers. These are also called **first-order consumers.** Next we find animals, and even a few plants, which are **carnivores** and feed upon the herbivores. These are the **second-order consumers.** Still other carnivores prey upon these and become **third-order consumers.** The chain may go higher to fourth- or fifth-order consumers. Some are **omnivores** and feed upon both producers and consumers. Man is a good example.

Since each level of life must utilize some of the energy for its various life processes, it is obvious that there are more producers, in total volume of protoplasm, than there are herbivores. Each order of consumers is smaller in volume than organisms which furnish its food. This tends to form a food pyramid (energy pyramid). Let us illustrate with a short food chain on the plains of Central Africa. The nonliving parts of the area form the greatest total volume. The air, soil, and water are needed to support life. Next we find the producers—tall grass, some small trees, and bushes that flourish in this region. Many herbivores, such as zebras and antelopes, live on this abundant food material and are present in great numbers. Next come the lions, large carnivores which prey upon the herbivores. These predators,

however, are never as numerous as the herbivores in the region.

The Food Web

This rather simple food chain becomes a complex **food web** when we explore all the different organisms deriving nourishment through it. When a lion kills a zebra, he does not eat all of it. Even though several lions share the feast, some of the zebra carcass remains. When the lions leave, hyenas come out to feed. Hyenas have the strongest jaws of any animal and chew up the bones and other tough body parts which the lions leave. Vultures also see the dead animal and circle down to feed on it. Hyenas and vultures are known as **scavengers.** Bacteria and molds attack any bits of zebra flesh which are not eaten by the larger forms of life. These are known as **decomposers.** They extract any remaining energy and return the mineral matter back to the soil where it can be reused by grow-

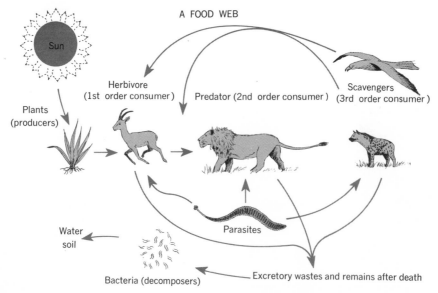

FIG. 38.7 A simplified food web on the grasslands of Africa. This diagram shows some of the complicated interrelationships between various species and their sources of food.

ing plants. The decomposers also attack the fecal matter and excretory wastes of the animals and obtain the energy remaining in these waste products. This also restores to the soil minerals which were removed by the producers.

The food web is still not complete. Some plants and animals are **parasites.** They get their food from other living things. There are blood-sucking flies in Africa which get food from zebras, lions, and hyenas. The lions may have various intestinal worms which absorb food which the lions have eaten. Bacterial parasites may invade any of the animals and derive nourishment while harming their hosts. Thus, we see how many organisms in a community are joined together in a food web and how each has a role. A change in any one organism in the food web may greatly alter the others.

The Role of the Predator

In Yellowstone Park in the northwestern part of Wyoming there was once a stable community which included elk, deer, and antelope along with mountain lions, which were **predators** on these herbivores. As this area was developed for a national park, it was thought best to get rid of the mountain lions. This was done through a planned program of trapping and shooting. The elimination of the mountain lions caused a great problem. Elk, for instance, tend to reproduce at a rate which allows many of their numbers to be killed by these predators, while enough are left to keep the population going. Without this elimination of some of the offspring, the elk population increased greatly. A food shortage resulted, and many of the elk turned to eating the bark of trees in the winter. Many trees were threatened with destruction as a result, so the park rangers had to turn to human selection to compensate for natural selection. Each winter about 1,500 elk are rounded up and shipped off to keep the population at a level which the coun-

try can support. The rounded-up elk are sent to stock game preserves in Montana and other parts of the nation. The cost is high, about $2.50 per elk, but this thinning of the herd is necessary.

The predator has a definite role in the balance of nature. In addition to keeping population numbers under control, the predators keep them healthy. Rabbits sometimes have a disease known as **tularemia** which can be spread to man. It has been noted that the disease is found only in those regions where there are no rabbit predators. A predator such as a wolf, fox, or lynx chasing rab-

FIG. 38.8 Elk damage to the bark of trees. When there is insufficient grazing food, the elk turn to tree bark. When the bark is heavily grazed, the trees can be damaged.

bits is much more likely to catch the sick one that cannot run fast. This prevents the disease from spreading among the entire rabbit population. In the absence of predators, many infected rabbits may be found.

SYMBIOSIS

Within a community two different species often form a close association which benefits at least one of them. Such associations are known as **symbiosis.** We can easily see how they might have evolved from a casual association which proved to be of advantage and developed into a more permanent relationship. Some of these associations are **facultative,** and either of the two species can live without the other. Others are **obligatory** associations, in which at least one of the species must have the other in order to survive. We recognize three major types of symbiosis.

Mutualism

An association that benefits both species is called **mutualism.** It can be a loose association

FIG. 38.9 Mutualism between a rhinoceros and a tick bird. Both forms of life benefit from this association.

such as exists between the **tick bird** and the **rhinoceros** in Africa. Wherever a rhinoceros is found, tick birds are also found. The birds feed on the ticks and other skin parasites of the rhinoceros and benefit by getting food. The rhinoceros in turn benefits from the sharp eyes of the birds, which can see enemies approaching at a great distance. The rhinoceros is very nearsighted, but when the birds jump up and down excitedly and emit warning cries, the rhinoceros runs away or turns to fight.

A closer association is found between the **dairy ants** and the plant lice, **aphids.** The aphids are protected by the ants and may actually be carried into the ant nest at night. The ants benefit because the aphids secrete a sweet liquid, honeydew, which serves as food for the ants.

The **termites** and the **protozoans** which live in the termite intestine exhibit **obligatory mutualism.** Termites eat wood, but they have no enzymes which can digest the cellulose of the wood. Within their intestines live several species of protozoans which do have enzymes that can break down the complex carbohydrate structure of cellulose. Neither the termites nor the protozoans can live alone. The protozoans must have the protection of the termite intestine and they must have the pulverized food as it is eaten by the termite. Both share in the use of the digested cellulose. By special techniques it is possible to kill the protozoans in the termites. This causes the termites to starve even though they eat wood.

Mutualism may exist between two plants or between a plant and an animal. **Lichens,** as we learned in Chapter 18, are a combination of algae and fungi. The algae manufacture food for both, while the fungi give protection. The **Smyrna fig** tree is a plant which must have a small wasp of a certain species to accomplish pollination or it cannot reproduce. When this plant was imported to California from the Near East, it failed to produce figs. Only when the wasps were brought in did the trees begin to bear fruit.

Commensalism

When only one of the two organisms living in close association benefits and the other is not harmed, we say it is an example of **commensalism.** Wherever **sharks** are found, there are often smaller fish known as **shark suckers.** These have a special suction cup on top of the head which enables them to attach themselves to the ventral surface of sharks. They obtain a free ride with the sharks, and when the sharks feed, the suckers may share in the food. The shark is neither harmed nor benefited by the association. Many **bacteria** which live in the human mouth are commensals. Some green plants live on other larger plants, but do not derive any food from the host. These commensals are known as **epiphytes.** The beautiful **orchid** is an example of an epiphyte.

Parasitism

In this association the parasite species receives benefit to the harm of the other, the host. Most of the higher animals of the earth have worm parasites which may live in various parts of the body. The **tapeworm** is a long flat worm which may inhabit man's intestine. It absorbs food which the host eats, and it releases waste products that are absorbed by the host and cause a loss of vitality. Almost any wild animal will have from one to several worm parasites in its body. These are **endoparasites. Ectoparasites** live on the outside of the body. Examples are fleas, lice, ticks, and the fungi which cause mange, ringworm, and athlete's foot. Plants also have their share of parasites. The shelf fungus destroys trees; parasitic insects bore into the bark and wood and harm the tree.

All parasites have one characteristic in common; they are highly specialized for the type of existence they lead, and there are very few parasites that are able to live without their specific hosts. They have so adapted to a parasitic existence that they have lost the characteristics which enable them to exist as free-living animals or plants. Consider *Ascaris.* It lives in the intestine and cannot obtain free atmospheric oxygen; it has anaerobic respiration. It has lost any digestive glands that its ancestors may have had, for it needs no enzymes while it is living in the presence of digesting food. It may seem strange that *Ascaris* is not digested by the enzymes which are digesting the food around it. Studies of this worm show that if one dies and is lodged in the intestine and not expelled, it will be digested. This shows that there is something about the physiology of the living worm that resists digestion. It produces enzyme-neutralizing substances which protect it from the surrounding enzymes. These substances are somewhat like the antitoxins produced by our own bodies which neutralize the toxins of disease organisms and other foreign bodies that may get into our bodies. This is one reason why internal parasites are usually specific about their hosts. The intestinal secretions of different animals may vary. A parasite produces the neutralizing agents for one animal; an entirely different set might be necessary for another animal. We, no doubt, eat many eggs and larvae of parasites which inhabit other animals, but we are not affected because the young parasites are digested by our enzymes. The gametes of the human malarial parasites can withstand the digestive juices of the *Anopheles* mosquito, but not those of the *Culex* mosquito. The *Ascaris* found in man shows no distinguishable morphological difference from that found in pigs, but the human *Ascaris* will not infect the pig and vice versa.

ECOSYSTEMS OF THE LAND

Communities of living things are found in almost all parts of the earth, but they are quite variable. Temperature, moisture, available mineral matter, light, other forms of life, and other limiting factors determine which organisms shall live in specific localities. These factors organize the earth into ecosystems.

Tropical Rain Forests

On the land areas which lie near the equator where there is little variation in temperature throughout the year and where the rainfall is 90 inches or more each year there are tropical rain forests. In this great humid belt most of the plants are treelike in size; even the ferns grow to the size of trees. There may be several hundred different species of trees in such a community, and they rise to a height of 300 feet or more from the forest floor. Their leaves form a dense canopy which blocks out the sunlight, leaving the lower areas in a sort of twilight even on a bright day. Many smaller ground plants grow on the dense humus of decaying plants on the forest floor. This abundant plant life supports a rich assortment of animal life. Snakes, lizards, amphibians, birds, and mammals are found in great variety. The tropical rain forest represents the greatest concentration of living matter to be found on the earth.

The Grasslands

North and south of the tropical rain forests there are great land areas known as grasslands. Here the average annual rainfall lies between about 10 and 30 inches. This area supports many herbivorous animals which serve as food for many other animals including man. Many large predators are to be found in the grasslands, as we learned earlier in the chapter with the example of an African grassland as a typical community.

The Deserts

Where the rainfall on land areas is less than 10 inches each year, deserts are to be found. In

FIG. 38.10 A desert biome in Arizona. Deserts are of great biological interest because of the adaptations found in plants and animals which live in the dry environment.

a few desert regions the rainfall may be greater than this, but is not evenly spaced, so the climate is very dry most of the time. Desert plants and animals are of great interest to biologists because they show the extremes of adaptation which are possible. Most desert plants have spines or thorns for protection and special ways of storing water. Others sprout from seed, grow, bloom, and produce seeds on the water from one good shower. The deserts bloom out in great beauty at such times, but the plants wither and die as the moisture is exhausted and only the seeds remain alive. The desert animals find no drinking water over most of the area. They have developed special ways of conserving water. They usually burrow underground during the day and come out at night when the air is cooler and more humid. The **kangaroo rat** of our southwestern desert lives on dry seeds and obtains water only from his own body metabolism. The kidneys of this rat are remarkably efficient in concentrating the urine and little water is lost in excretion. Other desert animals, such as rattlesnakes, obtain water from the blood of these rats when they are eaten.

Some deserts, such as the Sahara in northern Africa, have so little rainfall that they consist mainly of bare rock and shifting sands. Only in oases, where underground water comes to the surface, is there to be found much plant or animal life. Some parts of the southwestern desert of the United States are of this nature.

Most deserts are very hot, but some lie in cold regions. The large Gobi Desert of Asia has extremely cold winter temperatures.

Deciduous Forests

In the temperate zones where the rainfall is greater than 30 inches annually we find deciduous forests. During the cold winters of this region, the trees lose their leaves and remain dormant, but in the warm summers, the growth can be very luxuriant. In contrast to the great variety of species of trees in the rain forest, there will be only about ten to twenty species of trees in a deciduous forest. Almost all deciduous forests are to be found in the Northern Hemisphere because there are few land masses south of the equator in the temperate zone which have the rainfall needed for this ecosystem. The floor of a deciduous forest is usually covered with a thick layer of decaying leaves. This organic matter supports many bacteria, slime molds, mushrooms, and other fungi. Many small spring-flowering plants grow, and mammals such as deer, squirrels, racoons, foxes, and opossums make their homes here. Birds are especially abundant during the spring and summer, but most of them migrate to southern regions during the winter. Many insects are also present in the summer.

Taiga

North of the deciduous forests and northern grasslands and deserts, there is a broad band 400 to 800 miles wide which is known as the taiga or northern coniferous forest. It is found in Canada, northern Europe, Siberia, and parts of the northern United States. Here the summers

FIG. 38.11 A kangaroo rat jumps from the threatened strike of the sidewinder rattlesnake. The kangaroo rat is an important link in the food chain of the southwestern desert of the United States.

can be warm and the summer days are very long. This encourages plant growth, but winter comes early and the ground is frozen and covered with snow during the winter months. The trees are evergreens such as spruce, hemlock, larch, pine, fir, and Douglas fir. Any one forest, however, is likely to consist of only one species of tree. A spruce forest may extend for hundreds of miles with hardly any other species of trees in the area. Mammals include some large species—moose, elk, antelope, and bears—along with smaller species like rabbits, marmots, chipmunks, wolves, and foxes. Insects are abundant in the summer. Most of the comparable area of the Southern Hemisphere is covered by water, so we find no southern taiga.

Tundra

On the northern boundaries of the taiga the trees become smaller, being dwarfed by the severe climate. The taiga merges into a vast barren area known as the tundra. The ground is frozen to a depth of hundreds of feet and only the top few inches thaw out during the brief summers. The plant life includes lichens, mosses, and a few seed plants. There are twenty-four hours of daylight in summer, but this changes to twenty-four hours of darkness in the winter, with the lowest temperatures on earth. In the Siberian tundra a temperature of 93 degrees below zero has been recorded. Many animals make this inhospitable region their home. The musk ox is similar to our domestic cattle, but has a long shaggy coat of hair. Musk oxen feed by pawing through the snow to reach the lichens underneath during the winter.

MOUNTAIN LIFE ZONES

The life zones of the earth or **biomes,** are not as sharply defined as the description up to this point might seem to imply. Overlappings and mergings sometimes make it difficult to give a clear-cut classification of one particular region. The zones do not follow lines of latitude exactly,

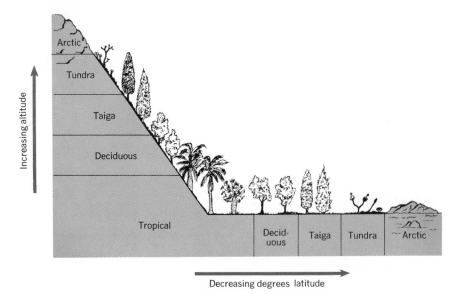

FIG. 38.12 Correspondence between the life zones at different latitudes and altitudes. All the life zones, from tropical to arctic, can be found on a high mountain located in the tropics.

since there are great variations in temperature and rainfall in the same latitude due to ocean currents, nearness to great bodies of water, and altitude. Altitude especially can make a crazy-quilt patchwork of zones within a limited area. In our own Appalachian or Rocky Mountain regions we find several of the great life zones within an area of a few square miles.

To illustrate, let us start at the base of a tall mountain in southern Mexico where we find ourselves in a typical tropical rain forest. As we ascend the mountain side, we find this blends into a typical deciduous forest where the climate is similar to that found in our own eastern states and which would be exactly similar but for the absence of the seasons typical of the middle latitudes. Then we reach an area of coniferous forests with climate conditions similar to those found in southern Canada. As we continue upward, we find dwarfed trees. Then we pass the timberline and enter the tundra biome. At the top of the mountain is a zone of perpetual snow and ice such as is found in the polar regions. As we descend the other side of the mountain, we see the tundra blending into grasslands and into a desert biome. The Sierra Nevada range of California exemplifies this difference brought about by rainfall. The western slopes show the zones requiring heavy rainfall, but the eastern slopes are mainly desertlike.

Many plants of the mountain biomes are the same as those of a similar biome of a more northern latitude. The animal life, however, is very likely to be considerably different. Birds act as agents which spread the seeds of many plants, but animals in general do not have such an efficient method of air transport.

ECOSYSTEMS OF THE WATER

About 70 percent of the earth's surface is covered by oceans, which support more life than all the land areas of the earth. Ocean ecosystems are determined more by depth and distance from the shore than by latitude because ocean water does not show the great temperature variations which are found on land masses. Ocean basins are generally shaped somewhat like inverted hats. Extending from the continents there is a gradual slope, the **continental shelf,** equivalent to the brim of the hat. Then there is a rather abrupt increase in the steepness of descent, the **continental slope.** This levels off into a deep **abyssal plain,** scarred with deep rifts which may extend down as far as seven miles under the surface in regions of the Pacific Ocean off the Philippine and Japanese coasts. Also, great mountains rise from the abyssal plain. They sometimes extend above the surface and form oceanic islands.

The sea which lies over the continental shelf is known as the **littoral zone.** Most of this area is not more than 600 feet deep. The limit of sunlight penetration varies from 250 to about 600 feet, depending upon how much suspended material there is in the water; therefore, most of this zone can support photosynthetic plants. Most of the plants are algae, and they manufacture an enormous amount of food, which forms the base of a food pyramid supporting a great variety of animal life. The free-floating forms of

FIG. 38.13 Timberline in the Rocky Mountains of Colorado. Although this picture was made in June, much of the area is still covered with snow. At about 12,000 feet the weather conditions are like those of the arctic tundra.

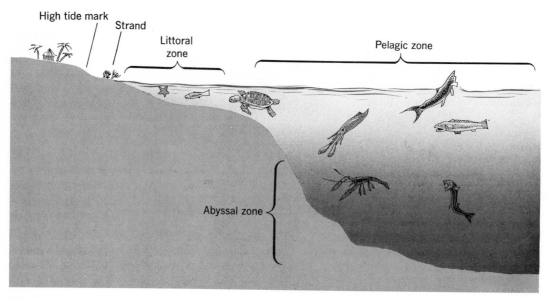

High tide mark

Strand

Littoral zone

Pelagic zone

Abyssal zone

FIG. 38.14 The ocean zones.

life are known as **plankton.** Then there are the actively swimming forms, the **nekton,** and the bottom crawlers of the **benthos.** Man obtains a great volume of food from the littoral zone. One part of the zone, the intertidal region, is known as the **strand.** Life in this region must be able to endure the drying sun and pounding surf during part of each day. Many animals, such as clams and marine worms, burrow into the sand and remain there during low tide, to emerge when the tide comes in. Plants are usually attached to rocks in this region.

Out beyond the continental shelf we find the **pelagic zone,** which extends to the depth of light penetration. All life here must either swim or float, since there is no place for anchorage. The larger nekton forms are the whales, porpoises, and sharks, but there are also great schools of bony fish feeding upon the abundant plankton.

The **abyssal zone** occupies the ocean depths beyond the point of light penetration. Here there is eternal darkness and all life is dependent upon a steady shower of the remains of organisms from the upper pelagic zone. No photosynthetic plants

are found here; bacteria and a few molds are the only representatives of the plant kingdom. Life is very competitive; it is usually a case of eat or be eaten. Many animals have delicate sense organs which detect the presence of nearby animals by slight changes in the pressure brought about by a moving body. Many also have light-producing organs which may act as lures to attract food and help the animals find one another during the reproductive season.

There seems to be no limit to the depths at which life can exist. The Scripps Institute of Oceanography recently sent a specially designed vessel into the deepest known ocean waters, the Mindanao Sea near the Philippine Islands. Here, at a depth of seven miles and under a pressure which would crush a man, crabs, shrimp, and other animals were seen and photographed.

The **freshwater** rivers, lakes, streams, and ponds support somewhat different forms of life. The salt content of the water is much less and the plants and animals must have ways of preventing the absorption of too much water by osmosis. Most saltwater forms of life die when moved to fresh water, although a few have the capability of surviving in both environments. Life in many fresh waters may also be exposed to swift moving currents, which means that plankton cannot exist. Living things must either have an anchorage or be strong swimmers. Finally, fresh water is subject to great variation according to the surrounding climate. In winter the water may become very cold and even freeze over. In summer it can become rather warm, and many freshwater habitats may dry up altogether. The smaller forms must produce spores or cysts to withstand this drying. There are even some fish which can burrow down into the mud when the dry period approaches and remain alive even though the mud becomes dry and hard as a brick. When the rains come and water softens the mud, the fish wriggle free and resume an active existence.

THE DISPERSAL OF LIFE ON THE EARTH

Not all species of organisms grow in all regions of the earth where they might thrive. Often this is simply a matter of dispersal. Rabbits did not live in Australia before the eighteenth century because they had never reached Australia until that time. In general, the older a species is in geologic time, the more widely it is distributed, because it has had a longer time to reach the various parts of the earth. Another important factor is the means of dispersal available. Migrating birds help the spread of the seeds of plants and encysted algae and protozoans which may be carried in the mud on the feet of waterfowl. Many animals, however, are restricted in their habitats by various barriers.

Geographical Barriers to Dispersal

Mountain ranges, deep valleys, oceans, and land may serve as barriers. These barriers are transient as measured by great geological eras, but they can be formidable for considerable periods of time. The Rocky Mountains are high enough to block the movement of many animals and, as a result, the animals on the eastern slopes of these mountains are different from those on the western slopes. The Allegheny Mountains, on the other hand, are usually less than a mile above sea level at their highest peaks and there are many gaps through which animals can migrate. As a result, life on the two sides of these mountains is very similar. They are high enough, however, to furnish an avenue for the southern migration of both plants and animals from Canada, which extend their ranges southward as far as northern Georgia. On the tops of the Great Smoky Mountains there is a fir-spruce zone in which birds that are typical of this zone in Canada nest.

Land is a big barrier to both marine and freshwater fish. Before the Panama Canal was built, the fish on the west coast of Panama were different from those on the east coast even though they lived only a few miles apart. During the last forty years, however, a few have managed to move from one ocean to the other through the canal and are now established in a new ocean. Freshwater fish are often confined to a particular drainage system and cannot cross the bridge of land into another river which may be only a few miles away. Waterfalls often bar the progress of animals. Niagara Falls formerly kept the marine lampreys out of the Great Lakes, but when shiping canals were built from the St. Lawrence River, the lampreys entered the Great Lakes and became a serious menace to the fish.

Climatic Barriers to Dispersal

Tropical and warm-temperate plants and animals are restrained from northward movements by the long, cold, and dark arctic winters. All the monkeys are tropical or warm-temperate. None inhabit the United States and only one is found in Europe—in southern Spain where a number of subtropical species are found. On the other hand, the tropical heat and related factors form an impassable barrier for arctic and cold-temperate animals. The arctic fox, the polar bear, the snowshoe rabbit, and the reindeer are highly adapted for snow and ice and could not long survive in tropical jungles. The amount of rainfall is also important. Deserts like the Sahara, which extends nearly all the way across northern Africa, form uncrossable barriers for most animals.

Ecological Barriers to Dispersal

An animal is limited in its range by the kind of food that it eats, and plants by the soil condition. The American buffalo, or bison, was a plains animal and rarely entered the eastern forests. Squirrels and woodpeckers, on the other hand, are confined to forested areas by the nature of their feeding and nesting habist. Any parasitic animal is limited to the regions which are

inhabited by its host or by agents upon which it is dependent for its spread. The parasite of malaria has recently had its range greatly reduced by a world-wide program of control of the *Anopheles* mosquito. It is often difficult to determine the ecological factors which control range. A species may be common in one locality and rare or absent in another that appears to be equally favorable. There are some lungless salamanders which are confined to a very small range in the cold, damp slopes of the Great Smoky Mountains where it rains almost daily. Here their need for a very moist habitat is clearly the limiting factor that prevents their spread into the surrounding lowlands. This hardly explains why they are not found on other slopes in the same mountains where there is an equally heavy rainfall. The parula warbler in Canada and northeastern United States nests almost exclusively in *Usnea* lichen. In the southeastern United States it nests in Spanish moss, which is a seed plant in the pineapple family, but resembles *Usnea* lichen in appearance.

FIG. 38.15 A lamprey eel attached to a large trout taken from one of the Great Lakes. Lamprey eels found their way into the Great Lakes from the Atlantic Ocean when the St. Lawrence Seaway was built. They have caused havoc among the fish life. A new predator can upset the ecosystem. (George Skadding, LIFE Magazine.)

Aids to Dispersal

Animals capable of rapid locomotion on land, in water, or in the air will spread rapidly, but more sluggish animals must depend upon other means of dispersal, as a rule. Mud taken from the feet of migrating birds has been found to contain eggs and cysts of many small crustaceans and protozoans. Lice, ticks, and fleas hitch rides on their host animals. Clams move about very little as adults, but produce larvae which attach themselves to the gills of fish and are carried great distances by these more active hosts. Many animals make their way downstream floating on logs and debris caught by the crest of floods. Even in the ocean there are currents which carry sea animals for great distances. Winds play their part also. Hurricane winds in Florida bring salt-marsh mosquitoes and other insects far inland. The beaches of the North Atlantic states may be covered with thousands of Portuguese man-of-wars after a hurricane in the Atlantic has blown them in from the Gulf Stream. There are forty-four species of European birds that have been found in North America, apparently blown across the ocean, since many of them were found after a storm. The animals which are dispersed in these ways do not usually become established in their new environment, but there is no doubt that such accidents do occasionally expand the range of animals.

Man has, been one of the greatest aids to dispersal of animals, since he has developed means of rapid locomotion to all parts of the earth. Early explorers let some of their horses escape and these established herds of wild horses in the southwestern region of the United States. The Norway rat has been carried as a stowaway in the holds of ships to every port in the world. The European periwinkle clings to the hulls of ships and drops off in our own coastal waters. The Colorado potato beetle formerly lived on wild plants belonging to the potato family in the foothills of the Rocky Mountains. When man planted a bridge of potato plants all the way across the country, this beetle spread to the Atlantic Coast.

Land bridges, both past and present, which connect some of the continents are used by animals as migration routes. The isthmus of Panama allows tropical forms from South America to move into Central America and Mexico; many monkeys have taken this route. It is known from geological studies that Alaska was connected with Siberia rather recently, during a period when the northern climate was much warmer than it is today. At that time there was an extensive migration of both animals and plants between the Old World and the New World. As a result, the species and genera are very similar in eastern Asia and parts of North America.

BIOLOGICAL SUCCESSION

Life in a mature biological community remains very nearly the same as long as it is not disturbed by some natural or human force which greatly alters the balance of life. When a biological community is completely destroyed, however, one type of vegetation after another will appear in a regular series of steps which may require hundreds or thousands of years. Eventually, however, the plant and animal life reach a **climax** in development and the dominant species of trees and animals for that biome take over and remain until another great disturbance takes place. For example, after the continental glaciers melted in the northeastern states some 10,000 years ago, they left the landscape completely bare of soil and denuded of all life, both plant and animal. Many **seral stages,** one after the other, occurred in the region before the climax forest was restored. As the glaciers receded, caribou, arctic hares, arctic foxes, and many birds migrated across the region carrying the seeds and spores of plants which started growing, and soon some plant life was re-established. The first plants

BARE BOTTOM (Pioneer) STAGE

SUBMERGED VEGETATION

EMERGING VEGETATION

TEMPORARY POND and PRAIRIE

BEECH and MAPLE FOREST

(Climax) STAGE

FIG. 38.16 Biological succession in a pond as it gradually fills with soil. (Redrawn from Buchsbaum, *Readings in Ecology,* University of Chicago Press.)

were species of lichens and mosses which require little soil, and the first animals were insects, the larvae of which feed on such low plants. As soon as a little soil had accumulated in low patches, fast-growing annuals became established. These were followed by perennial herbs and pioneer shrubs, such as blackberries, dewberries, raspberries, blueberries, and huckleberries. Ground-nesting birds, such as pipits, longspurs, and horned larks, now found a suitable habitat along with some predatory mammals and birds. As the soil deepened, larger shrubs became common, and tough pioneer trees became established. Rabbits now found suitable cover, deer began to browse on the shrubs and small trees, a varied fauna of birds which nest in shrubbery became established, and the larger carnivores, such as foxes and wolves, followed. Next a growth of pioneer trees—swamp maples, sassafras, persimmon, cedars, and pines—began to crowd out the low vegetation, and squirrels, tree-nesting birds, wood-boring insects, and many other forms of animal life found a suitable home. Birds and mammals of the open country were then crowded out or reduced to a few colonies in open spots. Finally, such slow-growing but dominant trees as sugar maple and beech, or white spruce farther north, gained a foothold and raised their crowns above the forest, gradually crowding out many of the smaller, short-lived trees. Rabbits no longer found their favorite brier patches, deer had to search elsewhere for tender twigs on which to browse, and birds of the shrubby habitats became scarce. However, as old trees developed dead branches and hollow trunks, more woodpeckers found food and homes, and more dens became available for raccoons, squirrels, and other den-loving species. After these seral stages, the forest finally became relatively static again and remained that way until our ancestors destroyed it once more with the ax and saw.

Another interesting succession occurs along the shores of a shallow lake as it gradually fills up with sediments and aquatic vegetation. This is illustrated by Reelfoot Lake in western Tennessee, a lake that was formed when an earthquake in 1811 caused the land to settle. In the century and a half since then, many parts of the shallow lake have filled up with aquatic plants and debris. A large patch of open water at the southern end of the lake contains many fish; diving birds, such as ducks, grebes, and coots, swim on the

surface, and a varied plankton furnishes an abundance of food. Pondweeds and water lilies float on the surface of the shallower portions of the lake, many of them with their roots in the mud bottom. Here catfish and turtles feed, herbivorous ducks dive for roots or tip-up for them, and aquatic insects are common. Closer to the shore or in very shallow places, plants send their roots into the muddy bottoms and their leaves and flowering stalks into the air. These include saw grass, cattails, and sweet flag. Here bitterns and redwing blackbirds nest a few inches above the water, the kingrail piles up its moundlike nest on the bottom, grebes construct their floating rafts, and muskrats build their dens. Wading birds, such as herons and egrets, feed on tadpoles, frogs, and water insects. Water snakes are most common in this zone. Cypress trees grow to large size in one area and play host to a great colony of nesting egrets and herons. A peregrine falcon also nests in a similar location. On the shores of the lake, where they may be occasionally flooded in wet seasons, grow many woody plants that need semiaquatic conditions, including swamp maples, black willows, and

FIG. 38.17 A beaver dam of logs and mud and the lake formed behind it. These lakes are an important part of the ecosystem in many regions, but their creation has been stopped by the extermination of the beavers.

button bush. Here may be found raccoon and mink, and such nesting birds as the yellowthroat, water thrush and prothonotary warbler, the latter particularly abundant in dead snags often standing in the water. Finally, on the dry land beyond the influence of the lake, the typical trees and shrubs of the region grow and are inhabited by upland birds and mammals.

REVIEW QUESTIONS AND PROBLEMS

1. What is meant by the terms **population, society,** and **community** in the biological sense?

2. Some populations are limited to a small area, while others extend over a very wide range. What are the factors which determine the extent of a population?

3. Why are sporadic poisoning campaigns not very satisfactory as a means of controlling a rat population in a city area?

4. As a rule, the populations in tropical regions are much more stable than those of the temperate or arctic regions. Explain.

5. In 1907 a great plague of mice in Nebraska destroyed about three-fourths of the alfalfa crop. In 1906 and 1908 there were only the normal numbers of mice. Explain what might have caused the plague in 1907.

6. The trappers of the arctic regions are always happy to see a plague of lemmings. Since they do not trap lemmings, why should this make them happy?

7. Weeds are plants which grow in greatest abundance in regions where man has destroyed the natural growth of plants; they are very scarce in an undisturbed habitat. Explain the ecological principle involved.

8. The population of people in Ceylon remained rather stable for many centuries. Recently a campaign was instituted to rid the island of malaria. In the following ten years the population doubled. Explain.

9. What advantages come to animals living in societies which are not available to animals living an individual existence?

10. Define the terms **herbivore, carnivore, scavenger, decomposer.** Show how each of these is related to a food web.

11. A drought on the central plains of Africa is always followed by a decrease in the lion population, even though there may still be plenty of water for them to drink in the water holes. Explain.

12. Show by an example the important role of a predator in maintaining the balance of nature.

13. How does facultative symbiotic association differ from obligatory symbiotic association? Illustrate with an example of each.

14. Distinguish between mutualism and commensalism through the use of examples of each.

15. Deserts, grasslands, deciduous forests, and coniferous forests are all found at the latitude of 30 degrees north. How can so many ecosystems exist at the same latitude?

16. The large carnivorous animals are usually most abundant in the grasslands, but they are not grass eaters. Explain.
17. Describe some of the methods of survival of desert animals.
18. Much life is found in the abyssal zone of the ocean, where no food can be manufactured. Describe the food chain involved.
19. Describe the barriers to dispersal which may keep plants and animals from spreading to other regions. Give examples.
20. What is the role played by birds in the distribution of plants and animals, on the earth?
21. What is meant by succession and climax as related to biological communities?

FURTHER READING

Emmel, T. C. 1973. *An Introduction to Ecology.* New York: W. W. Norton.

Hardy, A. 1971. *Open Sea: Its Natural History.* Boston: Houghton Mifflin.

Kormandy, E. 1966. *Organisms, Populations, and Ecosystems.* Dubuque, Iowa: W. C. Brown.

Odum, E. P. 1971. *Fundamentals of Ecology.* 3rd ed. Philadelphia: W. B. Saunders.

Read, C. P. 1970. *Parasitism and Symbiology: An Introductory Text.* New York: Ronald Press.

Smith, R. L. 1966. *Ecology and Field Biology.* New York: Harper and Row.

39

Our Ecological Crisis

The deterioration of our habitat poses a great threat to mankind. The human population of the world is expanding at an explosive rate. There is an ever increasing demand for the earth's resources, and we are approaching a point at which further deterioration of our ecological surroundings presents a threat to the continued existence of the earth. The earth contains a limited amount of resources and can support a limited number of people. How it can support them depends on the standard of living they require. If we all reduce our standard of living to the bare subsistence level, there would be room for a few billion more people. This would mean, however, that most other species which do not directly contribute to man's welfare would be crowded out. If each person is to have more than a subsistence standard of living, the present population represents the outer limit. We have been wasteful in the use of our precious resources in the past and we must face the consequences.

POPULATION GROWTH

Many of our problems are caused by too many people competing for the limited available resources. In 1798 **Robert Malthus,** an Englishman, warned that the human population was multiplying in a geometrical ratio, while the food supply was increasing in an arithmetical ratio. He predicted that there would have to be regular wars, famines, and other castastrophes to reduce

FIG. 39.1 A small planet in the vastness of space. This view of the earth, taken from the moon, shows the limited nature of our habitat and the need to conserve our resources if the quality of human life is not to deteriorate drastically. (Courtesy of NASA.)

the population. The Industrial Revolution had just begun at this time, and the higher standard of living which many people were enjoying made his warning seem invalid. Malthus's reasoning was correct, but the effects his theory predicted were postponed by the Industrial Revolution. In the United States, especially, people were unable to imagine any shortage of food when the government was paying farmers not to raise crops to avoid market surpluses. Even in countries with marginal food supplies, no attempt at population control was made. These countries received food from the surplus grown by the United States. This food and improved medical care allowed their population growth to continue. Now we are beginning to see that the human population cannot expand indefinitely; it will be checked by starvation or disease if man does not have the sense to reduce it before this point is reached. This point has already been reached in many parts of the world. We cannot practice death control without corresponding birth control.

Ecological Position of Prehistoric Man

Man has not always held the dominant position on the earth that he now holds. During 99 percent of the time in which he has existed, he was just one of many species competing for food and living space. There were probably times when man was in danger of extinction. Primarily because of man's intelligence, however, individuals were able to stay alive long enough to reproduce and to preserve the species.

Until approximately 12,000 years ago human societies were limited to small roving groups which moved from place to place as they exhausted the food supplies in each area. This type of food gathering severely limited the number of people which the land could support. Then something happened which greatly altered man's lifestyle and made it possible for the land to support larger human populations. Man learned about agriculture. He found that he could plant seeds and grow plant crops which he could use for food and fibers rather than moving from place to place to gather wild plants. Man also discovered that wild animals could be tamed and used for food and labor. Because of these discoveries, the land could support more people. Groups of people could remain in one place and build permanent places of habitation.

FIG. 39.2 Until recent times, plagues regularly reduced the human population. This painting by Pierre Mignard, made in the mid-seventeenth century, shows a plague at Epirus. A physician in the foreground collapses with the disease as he treats a victim. On the temple steps at right priests offer sacrifices to appease the divine wrath symbolized by the angels in the sky pouring miasmas from urns. These miasmas were thought to be the cause of the disease. (Armed Forces Medical Library.)

The Population Curve

Before the development of agriculture, the growth of the world's population averaged about 0.1 percent per century, a growth rate barely above stability. At this rate the population increased from about 50,000 in 300,000 B.C. to about 12 million in 12,000 B.C. The stability of agricultural societies produced a rate of population growth of about 5.3 percent per century. This seems to be a slow rate by today's standards, but by A.D. 1700 the world's population had risen to about 600 million, which is the population of India today. Then a number of important discoveries greatly increased man's chance for survival.

First, the discovery of fossil fuels gave man greatly superior energy resources. Coal and then oil were taken from the energy-rich deposits composed of ancient forms of life which had existed many millions of years ago. The invention of machines powered by this abundant energy source eliminated many tasks previously done by human labor. Improved transportation made it possible to move food for great distances. The people congregated in large cities to supply labor for industry. Greatly improved medical care, surgery, public health, and sanitation began to prevent and cure many diseases which had previously accounted for most human deaths. Engineers controlled flood waters and irrigated many areas where the rainfall had been inadequate for raising crops. This increased the food supply.

The Population Explosion

Because of these changes, the death rate dropped dramatically but the rate of reproduction remained high. The resulting population increase can only be characterized as an explosion. By 1800 the world population was almost a billion and by 1900 it was 1.5 billion. At present it is about 4 billion and is predicted to double in about thirty-five years. This rate of growth far outstrips the rate of growth of our food supplies and energy resources.

At last many people are realizing the seriousness of the situation. India, which has one of the highest rates of population increase in the world, now has an active program to promote birth control. The Indian government has started a vast educational program to teach people the dangers of overpopulation. Public health workers instruct women on birth control methods. Mobile units perform vasectomies upon request and even offer a transistor radio to men as an incentive. It has been difficult, however, to alter reproductive patterns which have existed for centuries, even though the people can see the results of over-

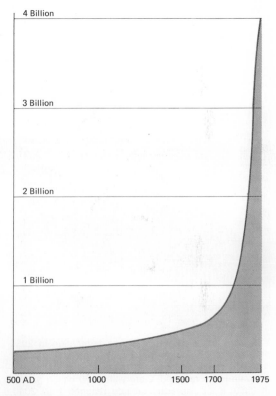

FIG. 39.3 World population curve. It is obvious that such an explosive growth rate cannot continue indefinitely without exhausting the world's resources.

crowding all around them. Many of the communist countries, which formerly encouraged large families, have instituted programs to emphasize birth control. The area which has the greatest population growth today is Latin America. Mexico is finding that efforts to improve the country's standard of living are being buried under a flood of new babies. Mexico has increased its agricultural output by 80 percent in the past twenty years, but at the same time the population has increased by 80 percent, so many people are still not able to get enough food. Many countries are losing ground in their fight for improved living conditions. The population grows faster than the improvements.

Changing Proportions of the World's Populations

The rate of population increase is not uniform in all parts of the world. Although the United States experienced a baby boom after World War II, the population increase in this country has declined in recent years until it is now at the replacement level. The total number of people will continue to increase for a time because most of the population is in the age group which can reproduce, but the population should level off in thirty to forty years if the present birth rate continues. The same is true for Canada and Europe. In other areas, however, there has not been a corresponding decline. There will be proportionately more Asians, Latin Americans, and Africans than North Americans and Europeans in the future. Figure 39.5 shows projections for the year 2020 according to present trends. Southern Asia and Latin America will show the greatest proportional gains. Of course, great famines, plagues, nuclear wars, and other factors could alter this.

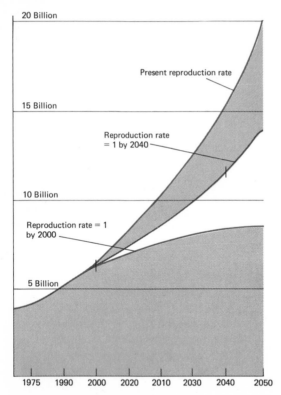

FIG. 39.4 The size of the human population in 2050 depends on how soon the reproduction rate (ratio of births to deaths) reaches 1. If the present reproduction rate is not reduced, there will be 20 billion people in 2050. If the reproduction rate reaches 1 by 2040, the world population in 2050 will be 14 billion. If the reproduction rate reaches 1 by 2000, there will be only 8 billion people by the year 2050.

ECOLOGICAL RESULTS OF OVERPOPULATION

An area is said to be overpopulated when the organisms use resources faster than they can be replaced. We have denuded the forests and it will be hundreds of years before they will grow back even if they are not disturbed. One of our most valuable natural resources is the few inches of topsoil found on many areas of the land surface. The topsoil contains humus, bacteria, decay-

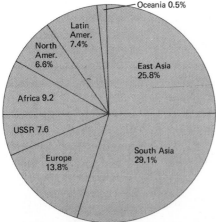

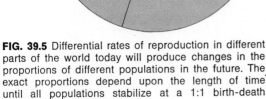

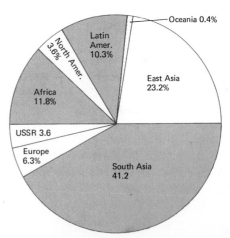

FIG. 39.5 Differential rates of reproduction in different parts of the world today will produce changes in the proportions of different populations in the future. The exact proportions depend upon the length of time until all populations stabilize at a 1:1 birth-death ratio. This figure shows the proportions of different populations if we assume that the present rate of population growth will continue until the year 2020 and then reach the 1:1 ratio.

ing organic matter, and minerals which are necessary for plant growth. Overcultivation, overgrazing, and the removal of the natural plant cover which retains topsoil are causing a loss of this asset faster than it can be replaced.

Erosion

Water which flows over unprotected topsoil carries it away. The Mississippi River in flood stage carries enough topsoil into the Gulf of Mexico every minute to cover forty-five acres of land seven inches deep. New Orleans was once on the Gulf of Mexico, but it is now many miles inland because silt carried by the Mississippi has built up an extensive land area below the city. Plants minimize erosion because their roots twine around the soil particles, making the soil loose and spongy and retentive of moisture. Destruction of the plants on the watersheds leaves the soil open to erosion and more water goes into the rivers, increasing the flood danger.

There is no way to prevent all erosion, especially if land is cultivated. Land can regenerate topsoil at a certain rate, so if the rate of erosion does not exceed the rate of regeneration, the land's productivity will not be reduced. When steep hillsides are plowed up and cleared of brush and trees, erosion of the topsoil by running water soon makes the land unable to support crops. Terracing and contour farming are two soil-conserving techniques that produce cultivatable areas with gentle slopes. Some areas should always be left with natural plant cover.

Carthage was once a great city in the Middle East. It lay in the center of great forests and fertile fields. Today the area is mostly desert and wasteland. Overcultivation and the overgrazing in areas with marginal rainfall can set up a chain of events which can produce these changes. When the sun's heat falls on unprotected areas, it is reflected back into the air, causing the air temperature to increase and the relative humidity to fall. Rainfall decreases because rain evaporates before it reaches the ground. These changing climatic conditions lead to the creation of a desert.

FIG. 39.6 In the dust bowl of Oklahoma during the 1930's huge dust storms destroyed much of the topsoil and left the land barren. This shows the piles of sand drifted against a fence. (United States Department of Agriculture.)

Soil erosion caused by wind can be just as serious as erosion caused by water. Some of the midwestern plains in the United States were in danger of becoming deserts during the 1930's. For a number of years there was abundant rainfall in this area. Farmers were encouraged to plow up large areas to plant wheat. Then, as would be expected with the law of averages, there were several dry years. The crops withered in the ground, leaving the soil with no plant cover, and the strong winds typical of this region whipped up the soil into great dust storms. Sometimes the dust was so thick that visibility was reduced to a few feet. The dust was carried for hundreds of miles, leaving behind only coarse sand which could support neither crops nor grass for grazing. The area became known as the Dust Bowl. There has been some concern this condition may return to the Midwest with the recurrence of several years of very dry weather.

FIG. 39.7 Nuclear generating plant near Platteville, Colorado. Our increasing population and our better standard of living has created great demands for energy. To satisfy these energy demands we have turned to nuclear power plants. The disposal of radioactive wastes and the potential pollution of the air, soil, and water are very serious problems, however, and nuclear power plants are not the best solution to our energy needs problem.

Soil Pollution

Some fertile land has been rendered unsuitable for crop growth by pollution from toxic by-products of industrial complexes. The Copper Hill region of Tennessee contains a large area where no plants grow. A copper smelting factory in this region gave off small amounts of copper into the air from its smoke stacks. Copper is highly toxic, and in time it accumulated in the soil until the copper concentration was lethal to plant growth.

In an area just northwest of Denver, the Rocky Flats Complex of the Atomic Energy Commission produces triggers made of plutonium for nuclear weapons. It was recently discovered that small amounts of plutonium were escaping into the air during the manufacturing process. Soil samples taken from an area extending for several miles showed traces of plutonium. While the concentration of plutonium was small, it was such that safety standards rule out any human consumption of plants or animals raised in this area.

Strip Mining

When deposits of coal or other minerals lie near the surface, it is more economical to strip off the surface and scoop out the product than it is to dig conventional mines. Near Cypress Gardens, Florida, there is a huge area which was once covered with orange groves but which is now a gaping hole. This area was strip-mined for phosphate deposits. With the present energy crisis increasing the demand for coal, there is great pressure for strip mining in Wyoming. Pennsylvania and West Virginia, however, have been extensively strip-mined the land was left in an unproductive condition for many years after the mining was completed.

FIG. 39.8 Strip mining for coal near Fall Brook, Pennsylvania. This type of mining destroys the valuable topsoil and leaves an unproductive gap on the earth's surface. New strip-mining contracts require that the land be restored to its former state before the mining starts. (Wide World Photo.)

Superhighways and Sprawling Suburbs

As the number of people and automobiles continued to increase, there were demands for more superhighways. Many of these superhighways became inadequate before they were completed and more were planned to try to keep up with the demand. They have cut a wide swath through some of the richest farmland of the country. Millions of square miles have been taken over for highways and we are now feeling the effects of having inadequate amounts of arable land. With increasing deterioration and crime in the central cities, suburbs have sprung up like mushrooms in the surrounding countryside, taking over flat farmland and thus further reducing the land available for food production. Many square miles have also been taken over each year by the so-called "land developers" who divide the land up into tracts and sell them. These tracts may remain empty for many years, taking the land out of productivity.

Garbage Disposal

Getting rid of the wastes of our society is a major problem. It is estimated that each person throws away five pounds of garbage per day. Every item we purchase comes in an elaborate wrapping of plastic and paper. Many items come in cans and throwaway bottles. Our mail boxes are filled with circulars and "junk mail." We are sometimes hard put to get rid of this waste. Landfill is the most common method of garbage disposal. Big holes are scooped out of the land, filled with garbage, covered, and abandoned. New York City hauls garbage about twenty-five miles out into the Atlantic Ocean and dumps it.

FIG. 39.9 Garbage disposal has become a major problem in this country. Americans put out 135 million tons of garbage each year. This is a small portion of the daily garbage output from New York City being loaded on a barge to be hauled out into the Atlantic Ocean for dumping. (Wide World Photo.)

Ocean currents are bringing this garbage back toward the shores of Long Island at the rate of about one mile a year. Some states have passed laws prohibiting throwaway bottles and the use of cans for beer and soft drinks. Others have set up recycling plants to reuse glass and metal. Plastics are a particular problem because they do not decompose. Plastics may be used less in the future, however, because they are made from petroleum products, which are in short supply. Much garbage is flammable and some cities have begun burning it as a fuel to generate electricity or to create methane gas which can be used instead of natural gas. Experiments have been conducted on the feasibility of compressing garbage into bricks which can be used for building purposes.

Use of Pesticides

The insects and other arthropods are man's chief competitors for the world's food supply. Chemical poisons are widely used to kill many of these pests. If their use were suddenly to stop, the world food supply would drop drastically. When **DDT** was discovered, it was hailed as the ideal pesticide. Small amounts killed insects and did not seem to be toxic to human beings. Dusting clothing and spraying of houses with DDT in war-torn areas drastically reduced typhus fever, which is spread by lice. Malaria was brought under control in many areas by spraying with DDT to kill the mosquitoes which carry it. The greatest use of DDT, however, has been on crops. We are now beginning to find that DDT can be harmful. For one thing, it does not degrade. It tends to remain in the soil or to be washed into rivers and oceans, where it is absorbed by organisms. Even in the Antarctic, DDT has been found in the fatty tissues of penguins. Ocean currents and food chains have spread this pesticide all over the world. DDT has a harmful effect on birds which has recently been

discovered. DDT affects the shells of their eggs so that they become thin and tend to break easily.

As a result, many uses of DDT have been banned and alternate pesticides have been substituted. Other methods of pest control are being tried. Still no matter how nontoxic a pesticide may be for vertebrate life, any pesticide which is spread widely will kill many useful insects, as well as many harmful ones. Bees, for instance, are needed for the cross-pollination of many flowers. Pest control methods which are highly specific are badly needed. The application of insect hormones can be used to control specific insects. When maturation hormones for a particular insect are applied to a crop, the larvae of that insect will mature too early, turning into miniature adults which are incapable of reproduction. No other species of insects is affected.

FIG. 39.10 Even in the Antarctic, pesticides have found their way into the penguin population. Through food chains and ocean currents, DDT has reached these birds and can be detected in the fat deposits of their bodies. (Paul Richard, University of Northern Colorado.)

Biological warfare against the insects is also possible. All insects are susceptible to disease and if we can spread infective agents for these diseases, many will be killed. The Japanese beetle which destroys many vegetables and fruit crops, is killed by a bacillus which infects the larval grub of this beetle. This bacillus causes the blood of the grub to become milky white and the disease is fatal to the grub. This bacillus is not harmful to other insects or to people, so it can be used for pest control purposes.

An interesting genetic control method may also be used in some cases. In Burma millions of mosquitos of a species found in California have been released. These Californian mosquitos mate with native mosquitos to produce a sterile hybrid (see Chapter 12). In a region near Rangoon, the native mosquitos were wiped out by this program and mosquito-borne infections were brought under control. Experiments are under way to find other insect species which will yield sterile hybrids when they are crossed with major insect pests. Other genetic control methods involve introducing insects which carry genes for male sterility or female lethality into an area where they will mate with the native insects. Radiation sterilized males have been released in some areas. The screw worn fly was practically eliminated from Florida by this method. (This procedure was described in Chapter 4.)

WATER POLLUTION

An abundant source of clean, pure water is taken for granted in America today, but supplying water is becoming an increasing problem. The average family of four, living in a city, uses about 600 gallons of water each day. Industry requires about five gallons of water to produce one gallon of gasoline, ten gallons to produce a can of vegetables, and about 30 gallons to produce the paper for the Sunday newspaper. We usually think that the amount of food which can

be produced determines the density of the population, but it may well be the water supply which is the critical limiting factor. Pollution is threatening much of our water supply.

Industrial Wastes

Most manufacturing plants use water. These plants usually return the water, rich in waste products, to a nearby river or lake. With increasing industrialization, water pollution has become a major problem. The Hudson River, once described by Henry Hudson as beautifully clear and sweet to the taste, now contains so many pollutants that it is murky and unfit for human use and the abundant fish life of the river has disappeared. Other life forms are also destroyed by pollution. Murky water does not allow light to penetrate and, without light, green plants and algae cannot live. These organisms lie at the base of the food chain which leads to larger forms of animal life, including edible fish. These photosynthetic organisms are also the source of oxygen which is needed by animal life in the water as well as on the land. Some people are concerned that the pollution of ocean waters will kill off photosynthetic organisms, such as diatoms, which are a major source of the world's oxygen. Continued reduction in the numbers of small photosynthetic organisms would lower the concentration of oxygen in the earth's atmosphere. When the photosynthetic organisms die, animal life is also reduced and saprophytic bacteria and molds which live on dead organic matter thrive. Many saprophytic organisms are anaerobic and give off noxious products which cause more animal deaths.

Modern detergents also contribute to water pollution. For many years we used some detergents which were not easily degradable. These tended to build up to levels which adversely affected organisms living in the water. Now detergents are degradable, but they present another problem. The whiteners and brighteners added to many detergents contain phosphates which overstimulate the growth of certain algae when the detergents reach the water. This stimulation causes an overgrowth of algae which use up all available nutrition. These algae die and add huge masses of decaying matter to the water. The overgrowth of algae also crowds out other species who live in the water.

Extensive use of fertilizers containing weed killers and insecticides can also contribute to water pollution. These fertilizers are washed into the waterways where they overstimulate the growth of algae. The weed killers and insecticides cause the death of many small forms of animal life. In Chapter 16 we described the red tide which appears off the west coast of Florida. It is caused by the overgrowth of a small organism, *Gymnodinium*. Some studies indicate that the red tide is caused by large amounts of fertilizer washing into the Gulf of Mexico from the surrounding area.

Thermal pollution is another aspect of the water pollution problem which is not often considered. Many manufacturing plants use water from rivers and lakes to cool machines and then return the heated water to its source. Eight thermonuclear reactors have been built on the shores of Lake Michigan at the present time. Each reactor has an eight foot pipe which takes in the lake waters to cool the nuclear generators. The water is returned to the lake 20 degrees warmer. Many forms of water life can live only within a restricted temperature range and elevating the temperature a few degrees kills them. Warm water does not hold oxygen as well as cool water and organisms living in the lake are deprived of oxygen as a result of the increased water temperature.

We cannot overlook the pollution by **radioactive wastes** from nuclear generating plants. While each nuclear plant releases only a small amount of radioactive waste, if many plants along the same waterway each contribute a share, it can builds up to a sizable amount. The Hanford, Washington center, maintained by the Atomic Energy Commission, has buried its dangerous radioactive wastes in containers for many years. Some of these containers have sprung leaks and there is the danger that some of the wastes will reach the Columbia river and make it too radioactive for fish life or for human use.

Water pollution from **sewage disposal** has become a major problem. Many cities have inadequate numbers of sewage disposal plants and expanding populations, and much raw sewage is dumped into rivers, lakes, and oceans. Flowing water can purify itself, given enough time and distance. Bacteria break down many organic wastes and photosynthetic organisms remove carbon dioxide and phosphates and oxygenate the water, but, when the level of pollution becomes too great, this method breaks down. In many regions of the Orient, it has been customary to use human waste for fertilizer. Fertilizer has been too scarce to waste this source of minerals. Today many American cities have begun to dry and pulverize the sludge from sewage and sell it as fertilizer. One man in Miami was startled to see tomato plants and other vegetables growing in his lawn. He had used sewage sludge fertilizer. The treatment of the sewage had apparently not killed the tomato seeds which pass through the human intestinal tract without injury. Decomposing sewage gives off a flammable gas, methane, and some cities are beginning to use it commercially.

The Great Lakes are a good object lesson in the destructive effects of excessive water pollution. There was once a large fishing industry on these lakes, but, today, fewer fish are caught each year. In some regions the fish which are caught are not fit for human consumption. Lake Erie has suffered the most. It is the smallest lake, and many industrial plants surround it. A thick sludge covers the bottom of the lake and extends up onto its formerly sandy beaches. Lake Sinclair, a smaller branch of Lake Erie near Detroit, is dangerously polluted by mercury-containing wastes. The USDA has prohibited human consumption of fish from this lake because they contain a dangerously high concentration of mercury compounds. When mercury is consumed by man, it accumulates in the body and eventually its accumulation causes permanent mental damage and death. Many rivers in Alabama and other southern states have also been placed off limits to fishermen for the same reason.

Oil slicks have also become a major pollution. Several huge oil tankers have sprung leaks and discharged millions of gallons of crude oil which has washed up on the beaches. Offshore drilling rigs have sprung leaks which have had devastating effects on the beaches in the areas of California and the Gulf coast states. The water fowl have been the greatest losers from these oil

FIG. 39.11 This duck is a casualty of an oil slick on a beach at Tampa Bay, Florida. Its feathers are coated with oil and it cannot fly. It has already ingested some oil in an effort to clean its feathers, so that it cannot be saved even though it is cleaned with detergents. (Wide World Photos.)

slicks. Once their feathers are soaked with oil these birds cannot fly and, if they consume too much oil, they can die. People have saved many birds by washing their feathers with detergents, but thousands of birds have lost their lives in each oil spill.

Irrigation

Irrigation has done wonders in increasing the amount of arable land in the world, but it can also contribute to water pollution. Extensive use of irrigation in hot regions can make the land unproductive because salts accumulate in the soil. When water is spread in a thin layer over land in hot climates, much of it evaporates, thus concentrating its mineral content. Since there is no heavy rainfall in regions which need irrigation, rain does not wash these minerals out of the soil and they build up to high levels. Eventually the land becomes toxic to plant life. Some areas of the Imperial Valley in California suffer from this kind of pollution problem.

Dams are usually built to hold water for irrigation. Much water evaporates from the surface of the damned water and the evaporation of surface water concentrates the salt content of the remaining water. The Colorado River originates in the mountains of Colorado and is an important source of water for Arizona, California, and Mexico. Arizona takes a lot of water from the Colorado River to supply its expanding population. This leaves less water from the Colorado for California, and what is left by the time the Colorado reaches Mexico is so high in mineral content that it is useless for irrigation purposes.

The Aswan High Dam in Egypt has provided water for irrigation to a large area of desert land which is now productive. It remains to be seen whether mineral buildup will be a problem in the continued use of irrigation on this land. Before the dam was constructed the Nile River carried mineral matter into the Medi-

FIG. 39.12 The Aswan High Dam in Egypt. This dam generates electricity for millions of people and has made possible the cultivation of millions of acres of land, but it has created ecological problems. (Wide World Photos.)

terranean Sea and stimulated the growth of microorganisms which are a food source for sardines. Now, with most of the water being used for irrigation and less minerals reaching the Mediterranean, there are fewer sardines.

Pollution of water was originally considered to mean contamination with human excrement. Many diseases and deaths were caused by the use of drinking water containing pathogenic organisms. Some diseases which can be spread though contaminated water are typhoid fever, amoebic dysentery, bacillary dysentary, and cholera. It was only about a hundred years ago that the medical profession began to accept the idea that microorganisms cause disease. Cities began to treat their water to rid it of pathogenic organisms and there was a dramatic reduction in the number of cases of typhoid fever and other water-borne infections. Philadelphia saw its cases of typhoid fever grow with its population, but when the city began to filter the water, and later to chlorinate it, the number of typhoid cases dropped greatly. There has been a decline from 10,000 cases of typhoid in 1905 to less than 100 cases per year in recent times, in spite of great increases in population numbers. Supplying purified water is becoming a problem, however. Chlorine, which is by far the best way to kill harmful bacteria in a water supply, is another scarce resource.

AIR POLLUTION

Most of the earth's atmosphere is found within three miles above sea level. There is some air above this, but it is so dispersed that it will not support animal life. Like water, however, the air is a convenient place for dumping gaseous wastes and it is becoming polluted. It is estimated that about 150 million tons of gaseous garbage is given off to the atmosphere by industrial plants each year.

Internal Combustion Engines

The primary source of air pollution is the automobile. Automobiles have internal combustion engines which release harmful by-products of gasoline combustion into the air. One of these by-products is **carbon monoxide**. Most people are familar with the danger of remaining in a closed building while running a car engine. The buildup of carbon monoxide in a closed area causes carbon monoxide poisoning. Carbon monoxide can also build up to dangerous concentrations in heavy traffic. Carbon monoxide poisoning, in its early stages, causes drowsiness and delayed reaction time which can lead to traffic fatalities.

Jets are a great convenience, but they also emit large quantities of polluting gases on take-off and landing and they contribute considerably to the air pollution of cities. In addition they scatter their exhausts high in the air with consequences which we cannot now assess completely. The supersonic jets which were to be the mode of travel of the future have mostly been scrapped as impractical. This is probably good because they were to fly much higher than subsonic jets and they would have given off exhaust which might have stayed in the higher layers of the atmosphere for many years, unmoved by air streams.

The Greenhouse Effect

We have always had some air pollution from volcanic eruptions, forest fires, and other causes, but the industrialization process has greatly added to this "natural" pollution. Some

people are concerned about the buildup of these wastes because they tend to shield the earth from sunlight. This shielding might cause what is called the **greenhouse effect.** Just as we use glass or plastic to cover plants in a greenhouse to maintain a higher temperature during cold months, the layer of wastes in the atmosphere may retain the warmth of the earth and prevent it from rising upward. An overall warming of the earth of only a few degrees might cause the melting of much of the frozen water at the poles and this could cause a rise in the level of ocean water. Many large coastal cities of the world would be inundated by such a melting. Another theory is that the shielding prevents the penetration of the sun's rays to a certain extent and this protection might cause a cooling of the earth's temperature. An overall cooling of a few degrees would cause the return of the ice age during which glaciers covered much of the United States. We really do not know which, if either, of these alternatives might occur, but it would not be a good idea to try to find out by allowing ever-increasing atmospheric pollution.

Another possible danger of atmospheric pollution is this. We are shielded from the high-energy radiation from outer space by a layer of **ozone**, O_3, high in our outer atmosphere. Some authorities feel that the continued outpouring of exhausts, especially from supersonic jets, may combine with and lower the ozone concentration. The decrease in ozone concentration would allow the penetration of more high-energy radiation to the earth's surface. As we learned in Chapter 4, this could have harmful genetic and somatic consequences.

City Smog

About 40 years ago the officials of Los Angeles invited people to come to the "beautiful city of the angels." Its clear, clean air was a major inducement to many people to move to this region. Smog is now a way of life in Los Angeles. The exhaust from millions of automobiles reacts with the air, in the presence of sunlight, and generates a murky gas known as smog. In some cities, smog has become so heavy that it is a health hazard. Many deaths from heart conditions and chronic respiratory ailments occurred recently in London when there were several windless days and the smog was not removed from the atmosphere. Many cities now have smog alerts and ban automobile traffic and the atmospheric disposal of wastes at critical times. An example of the effects of smog was felt in Donora, Pennsylvania recently. The air was very still for several days, while the concentration of smog built up to dangerous levels. People began to cough and sneeze and their eyes burned. Many people were hospitalized and 20 died from the effects of smog. Those who died were mainly older people who already suffered from cardiovascular problems, but many younger people were also affected.

Noise

We should not close a discussion of air pollution without reference to noise, since noise is composed of air-borne sound waves. Men working in noisy factories have been found to have twice as many hearing difficulties as those of comparable ages working in relatively quiet offices. Hearing loss among city dwellers is significantly greater than that found among those who live in rural regions. In one region of Africa where the sound level is very low, it was found that even elderly people had very acute hearing. Continued exposure to high levels of sound seems to cause a gradual destruction of the nerves and ear structures involved in hearing. Very loud music, such as the music played at some rock concerts, can also cause damage which can lead to a hearing loss later in life.

Experiments also show that people exposed to high noise levels first feel irritability and nervousness, then discomfort, and, finally, pain. The body responds physiologically to these sounds. A narrowing of the capillaries which causes an increase in blood pressure occurs. This is not good for people who have high blood pressure or heart disease.

This concludes our partial listing of some environmental conditions which are deteriorating. More public attention to these conditions is needed, with more public pressure being brought to bear on our elected representatives so that they will pass meaningful legislation aimed at checking this deterioration. Checking environmental deterioration will not be easy and it will mean an acceptance of some decrease in our very high living standards, but we really have no choice.

FIG. 39.13 Smog over New York City. Air pollution has become a critical ecological problem in many large cities of the world. (Wide World Photos.)

REVIEW QUESTIONS AND PROBLEMS

1. The human population is now increasing at an explosive rate. What factor do you think will cause an eventual leveling off of the numbers? Explain why you chose this factor.

2. Prehistoric man did not practice birth control and he did not have problems with a deteriorating environment as a result of overpopulation. Explain why.

3. If some worldwide catastrophe disrupted most of man's transportation and communication networks, the people in the more highly developed countries would suffer more than those in less developed countries. Explain why.

4. Some idealistic people advocate the abandonment of technological achievements and a return to the simple life where each family lived by their own wits, hunting, fishing and raising crops for their individual needs. What problems would arise if we attempted to do this?

5. Why did the discovery of fossil fuels start a rapid increase in the world's population? What do you think will happen when the supply of these fuels runs out, if we do not find a satisfactory substitute?

6. Soil erosion tends to be worst in areas where the population is most dense. Explain why this is true.

7. Some have suggested that we solve two problems at once by allowing strip mining and then filling the holes with our mountains of garbage. What problems would be involved?

8. A farmer in Iowa complains about a law banning the use of DDT to control insects in his fields, saying that his yield will go down considerably without it. How would you evaluate his arguments and what alternatives would you suggest?

9. What local restrictions against air pollution have gone into effect in the area where you live?

10. When a federal cabinet member heard that an oil slick in California had killed many birds, he is reported to have said, "What are a few birds when we are trying to get the oil to supply our industrial needs." What sort of evaluation would you make of this statement?

11. Some of the farmers in Mexico accused the Americans of adding salt to the Colorado river because the water is too salty to be used for irrigation. How would you explain the salt content of the water to the Mexican farmers?

12. Many citizens of the United States develop diarrhea when they visit many of the lesser developed areas of the world. Why do you think this happens so frequently, when people who live in these regions do not seem to be bothered?

13. The efforts of ecologists caused a cancellation of the supersonic transport program in the United States. What were the ecologists worried about?

FURTHER READING

Bernarde, M. 1973. *Our Precarious Habitat.* rev ed. New York: W. W. Norton.

Borgstrom, G. 1973. *World Food Resources.* New York: Intext.

Burns, W. 1969. *Noise and Man.* Philadelphia: J. B. Lippincott.

Ehrlich, P. R. 1970. *Population, Resources, Environment.* San Francisco: W. H. Freeman.

Johnson, and Steere. 1974. *The Environmental Challenge.* New York: Holt, Rinehart and Winston.

Winchester, A. M. 1973. *The Nature of Human Sexuality.* Columbus, Ohio: Charles E. Merrill Books.

40

Biological Evolution

A Greek philosopher **Heraclitus** wrote in 575 B.C., "All is change and only change is changeless." Today we express the same thought by the phrase, "The only variable law of nature is variation." Nothing remains the same when it is considered over long periods of time. Astronomy shows that the planets, stars, and nebulae are constantly changing in their relative positions, levels of illumination, and in other characteristics. Geological studies show that the earth has undergone great changes in the past and is undergoing changes now which will make our current maps of oceans, continents, rivers and mountains obsolete in the course of time. Fossils of deep-sea life which have been found in Kansas show that the midpoint of the United States was once under deep ocean water. Fossil remains and geological deposits show that large climatic changes have occurred in the United States. At some times this country was a lush tropical jungle, while at other times a large ice sheet covered most of it.

Species of organisms must change to accommodate to environmental changes, if they are to survive. A few species have remained relatively unchanged through millions of years, but most species have either undergone extensive adaptations or have become extinct. The species found on the earth today are only a small portion of the species which have existed in the history of the earth. In Chapter 5 we learned something about the possible method of development of life on the earth. In this chapter we shall consider some of changes which have taken place since that time.

CONCEPTS OF THE METHOD OF EVOLUTION

Let us turn now to some concepts about the ways in which these species adaptations have taken place.

Older Accounts

One of the oldest records we have of the development of life on the earth is found in the first book of the Bible, which was written in about 4000 B.C. The Genesis account reads:

". . . and He said; let the earth bring forth the green herb, such as may see, and the fruit tree yielding fruit after its kind. And God said: let the waters bring forth the creeping creatures having life, and the fowl that may fly over the earth. And God created the great whales and every living and moving creature, which the waters brought forth. And God said: Let the earth bring forth the living creature in its kind, cattle, and creeping things, and beasts of the earth, according to their kind."

This account has been given various interpretations, but it does list the appearance of phyla and species, somewhat in the order established by studies of fossil forms.

Empedocles, a Greek philosopher who lived around 500 B.C., thought of the early earth as a giant cauldron which contained many body parts. These parts came together in various combinations but only those which were efficient survived. This theory was a recognition of the process of natural selection. According to this theory, the body of a man might be joined to the legs of a horse and the neck of a swan. This would not be a good combination and this organism would not survive.

Many early leaders of the Christian church believed in evolution. Around A.D. 500 St. Augustine and St. Thomas Aquinas suggested that living matter came from the earth in a simple form which underwent changes in the course of time until various species were created. They believed in divine creation, but they also believed that living organisms, once created, could change.

Charles Darwin

No account of evolution would be complete without mentioning **Charles Darwin** whose name is associated with the concept. Darwin was born in Manchester, England in 1809, the son of a physician who wanted his son to follow in his footsteps. Darwin was sent to medical school, but he disliked it and he spent most of his time collecting insects, observing birds and going hunting. He dropped out of medical school, and then enrolled in a theological seminary. Darwin was not interested in theology either, and although he received a university degree in theology, he spent more time studying natural history. During this time he met Dr. Henslow, a botanist, who recommended that he join the crew of the HMS Beagle which set out to circumnavigate the globe in 1831. Darwin was appointed the naturalist of the ship for this voyage which lasted five years. Four of these years were spent off the coast of South America where Darwin made observations on many species which lived there and on the fossil remains of some extinct species. Darwin began the voyage believing in the generally accepted idea of the time that all species were created by a divine power in the exact form in which they existed at the present. The observations which he made on this voyage led him to doubt the validity of this idea, however.

For example, he found the fossilized remains of a giant armadillo which had been over 13 feet in length. The living species of armadillos in this area never reached a length of more than three feet. He found that many insects in Brazil had unusual markings and shapes which protected them from predators existing only in this area. In the Galapagos Islands, he found that different species of finches lived on different islands and that each species was adapted to conditions on its particular island. He observed that snails living on these islands showed gradual changes in form as he moved from island to island.

When Darwin returned to England, he visited animal breeders, especially pigeon breeders, and he observed that they could breed animals with a great variety of different characteristics by selecting the animals which were to be bred to each other. Darwin reasoned that selective breeding could also occur in nature by the elimination from the breeding population of those individuals which were less fit to survive.

Darwin read the essay by Thomas Malthus

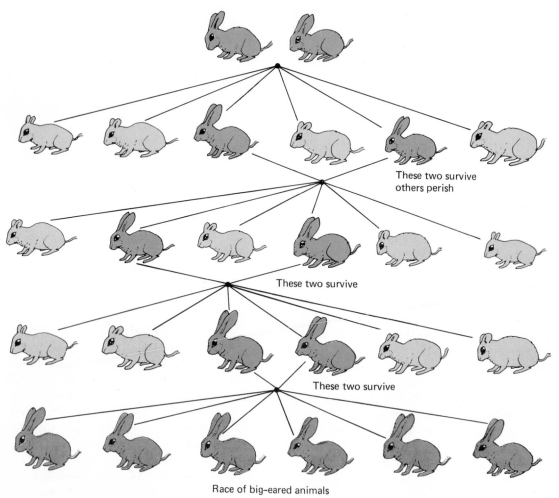

These two survive
others perish

These two survive

These two survive

Race of big-eared animals

FIG. 40.1 The principle of natural selection is based upon overproduction of young, variety among the young, and a selective elimination of the less fit. This diagram assumes that each female produces an average of six offspring, a very low figure.

on the "Principles of Population." This essay pointed out that all species produce more offspring than the number which survive to maturity. This essay reinforced Darwin's belief that the offspring which die before maturity must be those which have inherited characteristics which are not adaptive to their environment.

Darwin combined all his data, and from this body of evidence, he proposed a theory of evolution called The Theory of Natural Selection. This theory has four basic postulates.

1. Each species produces more offspring than survive to maturity.
2. Variation in characteristics exist among these offspring.
3. There is a competition for resources among the offspring.
4. Those offspring which possess the most favorable characteristics survive to maturity and reproduce. This idea is called the "survival of the fittest."

Darwin hesitated to publish his data and his theory, knowing that any theory which stated that all species were not created by a devine being would cause a great controversy. In 1858, his friend Alfred Wallace sent him an essay called "On the Tendency of Varieties to Depart Indefinitely from the Original Type." This essay presented the same ideas that Darwin had developed from his observations and its publication stimulated Darwin to publish his own findings in a book called *The Origin of Species by Means of Natural Selection of the Preservation of Favored Races in the Struggle for Life.* The publishers of the book printed 1250 copies, thinking that a scientific book of this nature would sell only a few copies. Instead all copies were sold the first day the book was available, and additional copies were sold as fast as they were printed. The book would probably not have generated such interest if Darwin had not included man among the species which changed through evolution.

Many people felt that Darwin's theory challenged religious ideas about divine creation and a great controversy arose over the acceptance of the idea of evolution. This controversy reached its peak in the United States during the famous Scopes "monkey trial" in Tennessee in which a schoolteacher named John Scopes was prosecuted for teaching Darwin's theory of evolution in a Tennessee high school. Scopes was found guilty in the original trial and was forbidden to teach the theory of evolution in a public school. His conviction was later reversed on appeal and, in time, Darwin's theory was accepted as the scientifically correct one.

EVOLUTION TODAY

Those who are unfamiliar with the principles of theories of evolution often ask why evolution took place in the past but does not occur today. The answer to this question is that it does occur today. Evolution is a continuous process. Most people do not realize that evolutionary changes occur over very long periods of time—hundreds and thousands of years. Large scale evolutionary changes occur only over very long periods of time, although some very small organisms with rapid generation times can evolve more quickly.

Natural Selection

Three factors influence the rapidity with which natural selection occurs. First, the length of the life cycle of a species is important. Fruit flies, which have an average generation length of 15 days, are able to adapt to a changing environment more quickly than elephants, which have an average generation length of over 30 years. Selection for desirable characteristics takes place with each generation and it stands to reason that if more generations are produced in a particular time span, there will be more rapid adaptation.

Second, the survival rate of a population is

important. Under conditions in which only one percent of a population survives, selection is much more effective than under conditions in which 50 percent of the offspring survive.

Third, the number of offspring which are produced during the lifetime of the female of a species is a factor. If a female of one species produces an average of 100 offspring during her lifetime and a female of another produces an average of eight, the selective elimination of the less fit can occur more rigorously in the first species. On the average, only two percent of the offspring of the first species need to survive to reproductive age to maintain the species. Because so many offspring are born, there is an intense competition for resources, and all but the two percent of the offspring, who must survive if the species is to survive, are eliminated. In the second species, 25 percent need to survive to reproductive age to maintain the species. In a species like this, the moderate number of offspring reduces the competition for resources and more survive to reproductive age. The greater the percentage of offspring surviving to reproductive age, the less effective the selection which occurs in any one generation.

FIG. 40.2 One of 37,500. This bullfrog represents the sole survivor of 37,500 frog eggs. All the others have perished before maturity. This frog must be a very fit specimen to have survived.

All organisms tend to produce offspring in numbers which are correlated with their chance for survival. Plants whose seedlings have a poor chance for survival produce an enormous number of seeds. Some parasitic worms produce millions of eggs because the chance that any one egg will infect the proper host is very small.

Those animals which care for their young produce fewer offspring than those which provide no care. The bullfrog, *Rana catesbiena*, is an example of an animal which provides no care to its offspring. A mature female bullfrog lays about 15,000 eggs each year during the five years of her reproductive life. If all these eggs hatched and grew into mature frogs, the earth would be covered with bullfrogs in a few generations. Most of the bullfrog eggs hatch, but tadpoles and young frogs are food sources for many predators such as fish, snakes, some birds, small carnivorous mammals and other frogs. If a female bullfrog lays her eggs in a favorable, balanced habitat, such as an isolated Louisiana swamp, about two eggs out of the 15,000 will hatch and survive to a reproductive age. These two survivors are the result of intense selection pressures. Only the tadpoles and frogs which are the fastest swimmers, the most protectively colored, the quickest to detect and catch food, and the most resistant to disease will survive.

Most of the characteristics which enable bullfrogs to survive to adulthood are hereditary traits and these traits are passed on by the survivors to their offspring. Therefore bullfrogs tend to become better adapted for survival in their habitat with each generation. You might think that eventually all bullfrogs would become so well adapted to their habitat that they would all survive, but you must remember that habitat conditions never remain constant. Continual adaptation and continual evolution occur in all species.

The elephant is a slow-breeding species which produces a much smaller number of offspring per female than the frog, but the elephant cares for its offspring over a long period of time.

FIG. 40.3 A bull seal, on the left, and his harem of females. Through fierce battles this bull has established a territory in which the females collect.

The gestation period is about two years. The mother nurses the calf for two years and protects it against predators until the young elephant is large enough and fast enough to protect itself. The periods between gestations in elephant species is approximately ten years. This seems like a small number of offspring when it is compared to the number produced by the bullfrog, but because of their long life span, this small number of offspring is quite adequate to maintain a stable elephant population.

Sometimes we can see the results of natural selection within a few years, as a result of severe selection pressures caused by rapidly changing environmental conditions. For example, there is a species of moth in England called the pepper moth which usually has a light-colored speckled body which blends well with the bark of the trees upon which it rests. In a heavily industrialized area near Manchester, the bark of these trees has become darker because of the accumulation of factory soot. The pepper moths which live near Manchester have a darker body color which blends with the darker bark color of the trees in this area. Those pepper moths with the lighter body color are more visible to predators in this area and are eaten before they can reproduce, while the pepper moths whose body color is darker survive to reproduce. In a few generations, only darker colored pepper moths were found in this area. Thus sudden changes in a species' habitat can increase the severity of selection pressures and cause quick evolution.

Inheritance of Traits Involving Reproduction

It stands to reason that those individuals in any species which are the most successful in reproducing will produce a larger number of offspring than individuals who are less successful.

A

B

FIG. 40.4 Two varieties of the peppered moth. At left, both varieties are on a soot-covered tree trunk near Birmingham, England. At right, they are both on a lichen-covered tree trunk in an unpolluted area. Arrows point to the head of the camouflaged form in both photographs. The black variety developed during the industrial revolution as a result of natural selection in a changed environment. (H. B. D. Kettlewell, University of Oxford, England.)

Success in reproduction involves the inheritance of traits which make reproduction more likely. In order for an individual member of any species to reproduce successfully, it must first survive to reproductive age. It must then be able to reproduce itself. In species which reproduce asexually, successful reproduction involves only the ability of the individual to undergo fission or budding. However in species which reproduce sexually, male and female gametes must come together to produce a zygote from which a new individual develops. There are many aspects of sexual reproduction which are genetically determined and which are, therefore, susceptible to natural selection.

In some species of plants selection has affected the speed with which pollen grains grow pollen tubes which penetrate the ovule of the female flower. The sperm nuclei of the pollen grain move down this tube into the ovule and fertilize the egg cell which is contained in the ovule. Some plants produce pollen grains which grow pollen tubes faster than others. Those individuals which produce pollen grains with faster growing pollen tubes stand a better chance of fertilizing the eggs of female plants than those individuals which produce pollen grains with slow growing pollen tubes. It is likely, therefore, that over time, those plants which produce pollen grains with rapidly growing pollen tubes will form the main body of the plant population. In other cases selection affects the shape of the pollen grain with those pollen grains whose shape best suits them for transportation by wind or animals being favored.

Some animal species are **sexually dimorphic.** That is, in these species, the males and females look different. These differences can become part of the courtship rituals which often precede copulation in birds and mammals. When this happens, those animals which posses the most distinctive appearance become more sexually attractive and are consequently more successful in reproducing. It should be pointed out that traits which enhance the reproductive potential of an individual are not necessarily tied to other characteristics which enhance survival, although this is often the case. In many sexually dimorphic birds, for example, the female, who sits on the nest most of the time, is dull colored, while the male, who does not protect the eggs as often, is bright colored. The female chooses the male for his bright coloration, but this bright coloration is not allowed to interfere with the protection of her eggs and thus the survival of the species. Sometimes the distinctive dimorphic characteristics become so elaborate that they do not enhance the survival of the animal in question. The fantastic tail of the male peacock is an example.

Artificial Selection

Man has established varieties and breeds of organisms which are specialized to meet human requirements. These breeds are established by **artificial selection.** Selecting characteristics which make the organism useful to people does not mean that the selected characteristics are useful to the species in question, and many domestic species cannot survive without human care.

FIG. 40.5 A peacock with spread tail. In birds, as a rule, the male has brightly colored plumage which is used as a display to attract females during courtship. Natural selection may be a factor in establishing the plumage patterns.

The sweet juice of one of the artificially-selected varieties of oranges is very different from the sour juice of the fruit of the wild orange tree. In selecting for fruit with sweet juice, however, we have produced a plant which cannot survive without human help. In order to obtain the desirable variety of fruit, we must graft the fruit

FIG. 40.6 Artificial selection has produced many plant and animal varieties which are desirable to people. The sweet, juicy, thin-skinned, almost seedless orange (A) was produced from the sour, heavily seeded wild orange from which it was derived (B).

stock onto the root system of a wild orange tree so that the hardy root system of the wild orange can support the fruit of the artificially bred juice orange.

The longhorn cattle which are native to the plains of Texas have been replaced by fat, heavy beef cattle which have been bred for the quality of their meat. These beef cattle need human care to stay alive. It is often necessary to help the cows to give birth. The native longhorns are well adapted to their native plains and they do not require special care. Some cattle ranchers are now returning to longhorns because they require no special care and because their leaner meat is now popular because of medical advice against excessive intake of saturated fats and cholesterol.

Mammals usually produce milk only during the period during which they suckle their young and they produce only enough milk to supply the needs of their offspring. Man has developed breeds of cows and goats by artificial selection

FIG. 40.8 This is not a Shetland pony but a full-grown horse. It is the result of artificial selection for small size. For thirteen generations, only the smallest offspring from a group of normal-sized horses were bred. (Jungleland, Thousand Oaks, California.)

which produce milk in large quantities throughout their lives. Domestic fowl with high egg yields are also products of selective breeding.

The existence of the many dog breeds shows the extreme variation which can be produced by artificial selection. A Great Dane and a cocker spaniel are very different in body size, build, length of hair, disposition, and many other characteristics, yet both have the same ancestor, a wild dog which was domesticated by some of our ancestors.

We have been able to obtain a variety of vegetables from one common ancestor by artificial selection. Today we have kale, kohlrabi,

FIG. 40.7 The size of the eggs in (A) are distinctly larger than those in (B). The large eggs produced by hens at the Mt. Hope Farm, Williamstown, Massachusetts, after only three generations of selection. The largest eggs from flock were allowed to hatch so that hens which laid large eggs could be selected.

Brussel sprouts, cabbage, and cauliflower—all descendants of a small, wild European cabbage which was only a weed.

Artificial selection can produce evolutionary changes at a rate somewhat faster than natural selection because we exercise absolute control over the breeding of the organism but the principle is the same and only the selective force is different.

Genetic Drift

When a population of a single species is separated into two groups which cannot breed with one another because of geographic separation, these groups will tend to diverge until after long periods of time, they become quite different. Many of these differences are caused by natural selection taking place under different environmental conditions, but some of it may be accounted for by **genetic drift,** which is a random chance establishment of genes. Some individuals with certain types of genes may have more offspring by pure chance than those with other genes. This is especially possible when a small group from one population migrates to another area. The migrating group may not have been a representative sample of the gene pool of the population from which they came. Chance elimination of those with certain genes and reproductive fluctuations can produce a gene pool in a small group which deviates considerably from the original gene pool of the population.

Some studies on human populations illustrate this process. In 1719 a group of 28 people migrated from the Rhine river region of Germany to an area in Pennsylvania. These people were Dunkers, members of a religious sect which required that marriage take place only between members of their own group. A study of the blood types of these people today show that about 60 percent have type A blood, while only about 45 percent of the Rhineland Germans have type A blood. The Pennsylvanians around the Dunkers have only about 40 percent type A blood. It seems that the migrating group, by chance, included a higher percentage of those with type A blood than did the general population from which they came and differential reproduction after their separation from the main population resulted in a higher percentage of type A blood in the descendants of this small group.

The Role of Mutations in Evolution

Selection would be greatly limited were it not possible for genes to mutate. Variety lies at the base of selection, and without mutation variety would be limited to the genes already in existence. With mutation, however, there is practically no limit to the amount of selection which may occur. Many highly desirable varieties of

FIG. 40.9 A gene mutation which produces the very small wings in the fruit fly, *Drosophila*. The gene for vestigial wings, like most mutations, is harmful. There are a few mutations which are beneficial, and these few are of evolutionary significance.

cultivated plants and domestic animals are the result of single mutations. The naval orange, which has no seed in its fruit, is the result of a single gene mutation which caused the rudimentary seed to be concentrated in a growth in the "naval" of the orange. The Concord grape is a mutation of a very small tart wild grape. The ancon sheep is the result of a mutation of a single gene in a flock of sheep in New England.

Since mutations are random gene changes, and since all forms of life which have survived are highly adapted already, most mutations cause harmful changes. A few, however, will be of advantage and selection can establish these and eliminate those which are undesirable.

EVIDENCES OF ORGANIC EVOLUTION

Since man is a relative newcomer to the face of the earth, he was not around to observe most evolutionary changes which took place, but we can learn much about the extent and direction of evolution from fossil remains and the geographical distribution of species. We learned in Chapter 12 how new species can arise through mutation and chromosome changes. Let us now review some of the other evidence.

Evidence from Geographical Distribution

The geographical distribution of species strongly impressed Charles Darwin on his famous voyage. The plants and animals on the Pacific islands near the coast of South America were similar to those in South America, but they showed some differences. These differences became progressively greater in the islands further from the coast. It appeared evident that these species originally came from South America, and after long periods of isolation, had undergone evolutionary changes to adapt to their new location.

Camel species are good illustrations of geographical distribution. The one-humped Arabian camel lives in the desert and is well adapted for desert life. It has broad pads on its feet which give it traction in the sand. It can exist for long periods without water and extra food is stored in its hump. The two-humped Batrician camel lives in colder regions in Central Asia and has developed long shaggy hair which protects it from cold weather. Its feet are hard and resist injury from the rocky terrain. The South American llama is another member of the camel family. Actually, camels seem to have first developed in America and to have then spread to Africa across a land bridge which once connected the two continents. The separation of these two continents was so long ago, however, that extensive changes in both groups of camels have taken place. Another South American camel species, the alpaca, migrated up into the mountains. It has developed a heavy, shaggy coat which protects it from the cold. Still another species, the vicuna, has occupied another region and is smaller than the alpaca.

We can observe this geographical distribution in many different groups of plant and animal species. Those found closest together geographically show the greatest similarities and as distances between groups increase gradual increases in differences occur.

Evidence from Paleontology

The fossil records of ancient forms of life give us the most convincing evidence of many evolutionary changes which have taken place. One of the most complete fossil records is that of the horse. The earliest horses which left fossil ramains were about the size of a house cat. They had four functional toes on the front legs and three on the hind legs. In more recently deposited strata, fossil remains of ancestors of the modern horse have also been found. These are about the size of a small dog and have the same toe development as the earlier species. Deposits on top of these reveal a horse species which is about the size of a sheep, but which has only three functional toes on all four feet. In still later deposits the ancestors of the horse are about the size of a Shetland pony. These horse ancestors have three toes on each foot, but the center toe is enlarged and the other two do not touch the ground. The most recent horse fossils are about the size of a modern horse. These fossils have only one toe which is greatly enlarged. Similar progressive

FIG. 40.10 Fossil remains enable us to trace the evolutionary development of many species. A monosaur which lived 80 million years ago, found in Kansas (A). A ground sloth which lived about 1 million years ago, found in Bolivia (B). Fish which lived about 45 million years ago in the area of Green River, Wyoming (C).

FIG. 40.11 Mesohippus, ancestor of the modern horse, which lived about 30 million years ago. This is a picture of a reconstruction made using fossilized remains as a guide. (Courtesy of the Field Museum of Natural History, Chicago.)

changes can also be observed in the length of the skull and the development of the teeth in successive fossil remains.

The development of the elephant has been

FIG. 40.12 Fossilized leaf of *Fagolopsis* which fell in the mud about 25 million years ago near Florissant, Colorado. This genus is now extinct, but plants which have descended from it are alive today.

studied in a similar way, but we also have more concrete evidence in the form of preserved specimens. In the frozen ground of Siberia, the bodies of about fifty prehistoric elephant ancestors—**mammoths**—have been found. The mammoths have been preserved in this natural deep freeze. One is on display at the American Museum of Natural History in New York City. It is kept behind glass in its frozen state.

Plant fossils are more extensively found than animal fossils and they enable us to trace the development of the seed plants through a series of coniferous relatives. No person who takes the time to study fossil remains can doubt that extensive revolutionary changes have taken place in all species.

Evidence from Animal Embryology

Even though adult animals undergo rather extensive changes in the course of evolution, the embryo tends to undergo development in the same way that it always did. Hence, the embryo is a valuable source of evidence for the study of past forms of animals. The lamprey eel, as an adult, is very different from the small *Amphioxus*. If we studied only the adults we might conclude that they were not closely related to *Amphioxus* at all. But, as we learned in Chapter 26, the larvae of the lamprey and *Amphioxus* are so much alike that the lamprey was placed in the same subphylum as *Amphioxus* until it was noted that the larval lamprey changed into a very different adult form.

The development of gill slits among all the vertebrates, which was discussed in Chapter 26, seems to indicate a similar evolutionary background. Why should the embryo of a snake develop gill slits which are similar to those of a fish unless there is some relationship between the two? The vertebrate heart is composed of two chambers in fish, three chambers in the amphibians, three chambers with a partial partition in the ventricle in the reptiles, and four chambers in birds and mammals. The embryos of the entire group develop first a two-chambered heart which later divides again if the adult form has more than two chambers. They all appear very much alike in the embryonic state and the changes which distinguish the various species appear as development progresses.

Evidence from Comparative Anatomy and Physiology

A comparison of anatomy among different species of plants and animals show structural features which can hardly be explained by anything except evolutionary relationships. The huge python snakes of Africa have vestigial legs projecting from their bodies on either side of the cloaca. Other snakes have vestigial legs as well. The vestigial legs indicate that snakes are descended from species which had legs. The asparagus plant has rudimentary leaves which form scales. These scales are functionless. The stem of the asparagus plant is enlarged and possesses chlorophyll so that photosynthesis occurs in the stem. The scales are vestiges of leaves which were once functional. The horse stands on the tip of one toe, but rudiments of other toes can be seen higher up on the leg. Birds have a third eyelid which moves across the eye from the anterior corner. Mammals have this eyelid in a vestigial form which is seen at the inner corner of the eyelids.

The comparative study of body parts and their functions also reveals evolutionary relationships. For example, all vertebrates have some sort

FIG. 40.13 A fossil of a trilobite, a primitive crustacean which was abundant during the early Paleozoic era about 500 million years ago. Although this species is extinct today, many modern crustaceans are descended from it.

of forelimb, although it may be vestigial in some species. The bones in the forelimb and their relationship to each other are similar in all vertebrate species, even though the forelimb may be specialized to perform different functions in different species. The wing of a bird, the front leg of a horse, and the arm of a monkey all contain similar bones although the forelimb performs very different functions in each of these species. This sort of relationship is evidence of evolutionary kinship.

Physiological similarities between different species are also used to document evolutionary relationships. Hormones taken from one vertebrate species can be used to stimulate hormonal responses in another vertebrate species. Testosterone can be extracted from the testes of a bull and injected into a male baby chicken which will then then undergo precocious sexual development. Many people whose thyroid glands produce insufficient thyroxine, take a crude thyroid extract which is prepared from the thyroid glands of cattle to supplement their own insufficient hormone output.

Chemical analyses of hormones taken from different vertebrate species show that differences exist among the structures of these hormones. There are small differences in structure in closely related species, while greater differences are found between hormones which come from more distantly related species.

Disease organisms are specifically adapted to the physiological environment found in their host species. In some cases, however, disease organisms may also successfully invade tissues of species closely related to their usual host. The polio virus which grows in certain human tissues, can also grow in tissues coming from monkeys and apes, but it cannot grow in tissues coming from other vertebrate species. A fungus which attacks wheat also infects closely related cereal plants, such as barley, rye, and oats, but it cannot infect corn or grass. Worm parasites are highly specific with respect to their host organism, but they can sometimes infect closely related species as well.

These physiological relationships can be used to demonstrate the closeness of the evolutionary relationship between two species.

Evidence from Serology

Relationships among animals can best be indicated by similarities and differences in the composition of their proteins. Differences in protein composition can be detected by serology. This technique was described in Chapter 12.

LIFE IN THE GEOLOGICAL ERAS

Geological time is divided into **eras.** We shall survey the various forms of life which records indicate existed during each geological era.

Archeozoic Era (2,000,000,000 to 1,200,-000,000 years ago)

Fossils are not found in geological deposits from this era, although there is evidence of organic material which indicates that some form of life existed on the earth as far back as the beginning of this era. Life of this time, however, was very primitive. The organisms of this period had soft bodies without any of the hard structures that lend themselves readily to the process of fossilization. The rocks from this period were subjected to high temperatures and pressures as they sank below the surface of the earth. This tended to melt the rocks and obliterate any impressions that might have been made by these primitive organisms. Therefore, we can only surmise the exact form of life in this dim, distant past.

Proterozoic Era (1,200,000,000 to 550,-000,000 years ago)

During this period, we find records of primitive blue-green algae which are bacterialike in their structure, but contain chlorophyll. Definite fossils of protozoa (Radiolaria) have also been found in deposits of this era. These developed a siliceous shell that could be preserved easily. Fossil sponges have also been found in these deposits. There was a great amount of volcanic activity during this era that no doubt destroyed many possible fossils.

Paleozoic Era (550,000,000 to 190,000,-000 years ago)

This was the first era in which conditions for fossil formation were favorable, and fossils are found in deposits of this time in great abundance. After the poor and incomplete record of life in the first two eras, the Paleozoic with its well-preserved fossils stands in sharp contrast.

Most of the phyla of the animal kingdom were represented in the fossils of the early part of this era, although they were entirely aquatic in their habitat. The trilobites—small crustaceans which groveled on the bottom of the ancient seas —made up the majority of the animals of the time. The algae were abundant in these ancient seas and furnished the food necessary for the maintenance of the abundant animal life.

All of the life forms were restricted to the water. There the environment was more constant and the struggle against the forces of nature less rigorous than on the land. The rocks, mountains, hills, and ravines stood bleak and desolate above the waters teeming with life. However, as the number of living things increased, the struggle

FIG. 40.14 A forest scene, as it would have appeared during the Carboniferous period of the latter Paleozoic era. Giant tree ferns, horsetails, and club mosses were the dominant plants at this time. Their remains formed the raw materials for coal deposits which are a valuable energy source today. (Courtesy of the American Museum of Natural History.)

for existence became more acute; there was pressure for living room. The land had plenty of room, bleak though it was, and, at about the middle of this era, life became established on the land. Primitive fernlike plants are found in the stratifications of rock that were formed at this time. The animals remained in the seas for a time, but the fishes became dominant over the crustaceans.

As we examine the fossils of the later pe-

FIG. 40.15 Huge reptiles were the dominant animal species during the Mesozoic era. These paintings by Charles R. Knight show the probable appearance, determined from their fossil remains, of two species. *Tyranosaurus*, above, was a fierce carnivore. *Brontosaurus*, below, was a herbivore but grew to be the largest of the land animals. (Courtesy of the Field Museum of Natural History, Chicago.)

riods of the Paleozoic era we find that primitive amphibians had followed the plants to land. Huge sea scorpions, up to nine feet long, were found in the seas. The trilobites were greatly diminished. There were great forests of giant cycads and gymnosperms growing in the peat bogs that were abundant on the continents of Europe and America. The warm humid climate of the time was excellent for their growth, and their bodies packed tightly in the bogs as they died and fell to earth. As these areas became covered, they formed the foundation of our coal beds. If you look through a microscope, the cellular structure of these plants can be seen in a piece of coal that has been ground and polished. Thus, coal deposits are fossils, and when burned in our stoves, the coal releases the energy captured by these plants from the sun shining on the earth some 250,000,000 years ago. These luxuriant forests provided habitation for the increasing numbers of land animals. Toward the end of the era the amphibians had to share their land habitat with huge dragonflies having a wing spread of more than two feet, cockroaches, spiders, snails, and even a few primitive reptiles.

Mesozoic Era (190,000,000 to 55,000,000 years ago).

As the Mesozoic era opens we find a beginning of the decline of the amphibians and a rise of the reptiles. This era is called the Age of Reptiles because of the extensive radiation of this group to every type of habitat. Huge dinosaurs, as well as many smaller reptiles, dominated the land; flying reptiles, some with a wing spread as great as 20 feet, dominated the skies; and swimming reptiles, both fishlike and lizardlike, dominated the seas. Hundreds of species of reptiles have been identified and classified from the rich fossils of early Mesozoic times. At the same time, primitive mammals no larger than rats scurried around trying to avoid destruction by the domi-

nant reptiles. Most of the cycads had died out, and new and more advanced groups of gymnosperms had appeared and were spreading. Many of the smaller ferns, somewhat as we know them today, were abundant.

The middle period of the Mesozoic era was characterized by a spread of the primitive mammals and the appearance of birds. These birds were not like modern birds—they had teeth, long bony tails, claws on their wings, and were probably not capable of long sustained flight. Yet they definitely had feathers and were true birds.

The last period of the era was characterized by a cooler climate that created conditions unfavorable to the majority of the plants and animals that had evolved through adaptations to the warm, humid climate. The huge reptiles, made stiff and inactive by the cold, found themselves unable to cope with the increasing numbers of mammals, which were stimulated to greater activity by the cool weather. One by one entire species of reptiles became extinct until the great majority of the vast assemblage of reptiles—once so prevalent—disappeared. Birds similar to those of today appeared and replaced the toothed birds. The landscape began to assume a modern appearance as the flowering plants, the angiosperms, multiplied and crowded out many of the gymnosperms. Figs, elms, oaks, magnolias, willows, and maples are well-known trees found at this time.

Cenozoic Era (55,000,000 years ago).

At the beginning of this era the mammals multiplied and spread, replacing the now extinct reptiles. Only a few reptiles remained as representatives of their class. The grasses spread and supported the great group of grass-eating ungulates. Deciduous trees became abundant. Toward the middle of the era the mammals became recognizable as ancestors of modern forms. Elephantlike mastodons and mammoths, rhinoceroses, giant ground sloths, and tiny prehistoric horses roamed the continents. The anthropoid apes made their appearance. It was upon such a scene that the eyes of the first primitive human beings gazed as they developed and spread over the face of the earth, eventually to dominate their surroundings.

The climate became rather variable toward the close of this era. Periodic glaciations in which a large part of the earth was covered with great ice caps were followed by more temperate times when the ice melted away at all except the polar regions. Many of the large mammals became extinct during these glaciations, their few remaining descendants retreating to the tropical regions of the earth for their continued existence. The arthropods continued to multiply and extend themselves to every part of the globe. As time draws close to the present, we find man rising to a point of dominance over almost the entire earth. We will discuss the development of mankind in the next section of this chapter.

Our extensive knowledge of life in the geological eras of time, presented in this brief summary, is a testimonial to the great contribution of paleontology to an understanding of the life of today. The record is by no means complete—there are great gaps covering millions of years in which absolutely no records have been found. It is somewhat as if we are permitted to view isolated individual frames of a gigantic motion picture of the caravan of life through the ages. Yet, from these views we have been able to piece together the sequence of events as surveyed in this chapter. New, hitherto unknown fossils are being discovered each year to help fill the gaps and extend our knowledge of life as it existed in the past.

HUMAN LIFE OF THE PAST

History gives us a comparatively full account of man and his activities from the time when he learned to make records that could be handed down through the generations. What of man before this? What was man like as he existed before the dawn of history? To answer this question we must turn to paleontology—to the fossil records. Unfortunately, the prehistoric records of human life are not abundant and there are many gaps that make the story incomplete.

There are several reasons for the comparative scarcity of human fossils. Man has not existed on the earth very long in comparison with many other animals, and hence there has been a shorter time for fossil formation. Then too, man was not an abundant species. It is remarkable that we have uncovered enough human and subhuman fossils to give us the somewhat sketchy picture that we have of man in prehistoric times. Perhaps with future discoveries our picture will be more complete. For the present we must be content to give careful study to the human remains that have been discovered and draw what conclusions we can.

How long has man lived on earth? The answer depends somewhat on the interpretation of the word "man." In 1972 Richard Leakey found a skull and upper jaw in Africa of a species which was sufficiently like modern man to be included in the same genus, but dissimilar enough to be given a different species name. This species was named *Homo erectus*. *Homo erectus* is estimated to be about 2.5 million years old. Evidence appears to indicate that hominids sufficiently like modern man to be classified in the same species, *Homo sapiens*, first appeared about 300,000 years ago. Hence, in terms of the length of the life forms have existed on the earth, man is a newcomer.

Australiopithecus

In 1924, Raymond Dart, a professor at a university in South Africa, found the remains of several skeletons of upright creatures with a tooth structure somewhat like that of modern man. He gave these specimens the genus and species name, *Australopithecus africanus*. These early hominids had a brain case capacity of about 600 ml, which is about the same as that of a gorilla. Modern man has a brain capacity of about 1500 ml. In spite of his limited brain size, there is some evidence that *Australopithecus* learned to make and use tools. Other examples of this genus were found by Dr. and Mrs. Louis Leakey (parents of Richard Leakey) in the Olduvai Gorge in Tanganyika which lies in east, central Africa. Potassium-argon dating of the lava rock in which some of the bones were found showed them to be about 1,750,000 years old. Before the recent discoveries, it was presumed that this genus was an ancestor of *Homo erectus,* but now we must consider it a contemporary species which became extinct.

Homo erectus

Before Richard Leakey's find, several other fossilized remains of *Homo erectus* had been found in such widely scattered areas as Java, China, Central Africa, and South Africa. These did not seem to be more than 500,000 years of age. They had a brain case capacity of about 1300 ml, which is not quite as much as that of modern man.

Homo sapiens neanderthalensis

The exact time at which the first members of *Homo sapiens* appeared is not known but the earliest fossil remains in Europe are estimated to date from 350,000 to 250,000 years ago.

The remains of **Neanderthal man** were first discovered in the Neander Valley in Germany. At that time, the distinctive appearance of the skull found there led archeologists to believe that Neanderthal man was a separate species in the

Hominid family. Other Neanderthal remains have been discovered at many sites throughout Europe, Asia, and Africa. Archeologists now believe that Neanderthal man was a subspecies of *Homo sapiens* and they have named this subspecies *Homo sapiens neanderthalensis.*

Reconstructions of the first Neanderthal skulls and bones led scientists to believe that Neanderthal man did not walk fully erect, but later evidence contradicts this idea. It is now thought that Neanderthal man walked fully erect. Members of this subspecies were short and stocky in appearance with large heads and brain cases and curved bones. Neanderthal man used fire and hunted large animals with weapons made of bone and stone. These early humans lived in caves or rock shelters which they apparently occupied for long periods of time. There is some evidence that they constructed parts of their shelters, since excavations and rock pilings have been found inside caves and shelters. The Neanderthals buried their dead and sometimes included ritual arrangements of animal remains with the dead person.

Remains of Neanderthal man disappeared from the fossil record rather abruptly sometime around the beginning of the upper Pleistocene age. The cause or causes of the extinction of this subspecies are not known.

Homo sapiens (Cro-Magnon Man)

One of the more modern groups of *Homo sapiens* is **Cro-Magnon** man, a race of *Homo sapiens* which appeared in Europe around 25,000 years ago. Cro-Magnon remains indicate that these people were a rather tall, upright race who had long skulls with a large brain case and no pronounced brow ridges. In appearance they were indistinguishable from some races of modern man.

Cro-Magnon man lived in caves and rock shelters and there is some evidence that these people may have built huts as well. They hunted

FIG. 40.16 Cro-Magnon men painting on the walls of a cavern at Font de Gaume, France, according to a painting by Charles R. Knight. These paintings are the earliest examples of this art form which have been preserved. (Courtesy of the Field Museum of Natural History, Chicago.)

large animals. They are best known for the beauful cave paintings of animals and people which they left behind them. They also decorated their tools and carved human statuettes.

Cro-Magnon man undoubtedly contributed genes to the present day European gene pool. In a few places, relatively pure Cro Magnon stock is still maintained even today. The Dals of Dalnaria, Sweden and the Guandies of the Canary Islands are examples of people with direct Cro-Magnon ancestry.

Fossil human remains in other parts of the world show that Cro-Magnon man had contemporaries of other races; but other regions of the world have not received the thorough investigation that Europe has, and our knowledge of these races is somewhat limited. America has been an almost sterile ground for ancient human fossils, leading to the conclusion that this continent was not inhabited by human beings until more recent times. In all probability man migrated across a land bridge from Asia to Alaska about 130,000 years ago and then spread down over the North and South American continents. The physical similarities of the American Indian and the East Asian Mongoloids and recent blood studies bear out this conception.

About 12,000 years ago man emerged from caves, built huts of animal skins out in the open, and began a sedentary, communal life in fixed dwellings. He produced advanced tools of stone; he domesticated the wild dog. This carnivore developed a liking for the habitat and food of man, and man soon learned that he was a valuable aid in hunting and a protection against intruders. Later, man learned to plant crops and store food so as not to be at the mercies of the vagaries of nature. Other animals were domesticated. Man learned to forge his tools and weapons of bronze and iron. Communal groups increased in size, bringing widespread infectious diseases due to increased close contact among people. Nutritional diseases made their appearance because of unbalanced diets. At this point we find our survey blends into the history that brings the story of man down to the present day.

REVIEW QUESTIONS AND PROBLEMS

1. Suppose a certain species is highly successful in its environment. Still, we say that it must constantly undergo evolutionary changes or it will become extinct. Explain.
2. Suppose no one had ever suggested the idea of evolution. What observations have you made personally which might lead you to conclude that evolution has taken place?
3. How did geographical distribution of species indicate to Charles Darwin that evolution had taken place?
4. What one factor do many people fail to take into consideration when they do not understand why evolution is so extensive?
5. One species of animal produces an average of one thousand offspring per female, while another species produces only ten. Would there be a difference between these two with respect to their ability to adapt to changing environment? Explain.
6. One species has an average generation cycle of six weeks, while another takes two years to complete its average cycle. Which species would adapt most quickly to changing environment and why?
7. Suppose, because of some great catastrophe, everyone moved away from Florida. What would happen to the groves of cultivated oranges as compared to groves of wild oranges? Explain why this would take place.
8. Suppose you wanted to produce a breed of Great Danes, half the size of the present average size. Tell how you would proceed. How would your procedure compare with natural selection?
9. We know that most mutations are harmful, yet we say that mutations are necessary for natural selection. Explain this statement.
10. Some populations which have lived apart for a long time from the original population show differences which seem to have no adaptive value at all. Explain how these differences might have arisen.
11. Why do biologists often study embryos to try to determine evolutionary relationships?
12. What is a vestigial structure and how are such structures significant in studies of evolution?
13. Why is the evolutionary history of man harder to trace than that of some other animal species?

FURTHER READING

Briggs, D. and Walters, M. 1969. *Plant Variation and Evolution*. New York: McGraw-Hill.

Dodson, E. O. 1960. *Evolution, Process and Product.* Princeton: Van Nostrand.

Hamilton, T. H. 1967. *Process and Pattern in Evoluion.* New York: Macmillan.

Lasker, G. W. 1961. *The Evolution of Man.* New York: Holt, Rinehart and Winston.

Romer, A. S. 1972. *The Procession of Life.* New York: Doubleday.

Stebbins, G. L. 1971. *Processes of Organic Evolution.* Englewood Cliffs, N. J.: Prentice-Hall.

Van Lawick-Goodall, J. 1971. *In the Shadow of Man.* Boston: Houghton Mifflin.

Volpe, E. P. 1967. *Understanding Evolution.* Dubuque, Iowa: William C. Brown.

Index